Perovskite Solar Cells

Perovskite Solar Cells

Materials, Processes, and Devices

Edited by Shahzada Ahmad, Samrana Kazim, and Michael Grätzel

WILEY-VCH

Editors

Prof. Shahzada Ahmad
BCMaterials, Basque Center for Materials, Applications & Nanostructures
EHU/UPV Science Park
48940 Leioa
Spain

Ikerbasque, Basque Foundation for Science, Bilbao
Spain

Dr. Samrana Kazim
BCMaterials, Basque Center for Materials, Applications & Nanostructures
EHU/UPV Science Park
48940 Leioa
Spain

Ikerbasque, Basque Foundation for Science, Bilbao
Spain

Prof. Michael Grätzel
École polytechnique fédérale de Lausanne (EPFL)
Institut des sciences et ingénierie chimiques
Station 6
1015 Lausanne
Switzerland

Cover Design: FORMGEBER, Mannheim, Germany
Cover Image: based on "Angew. Chem. Int. Ed. 2016, 55, 14522 – 1454" (@ Laura Calio)

Library of Congress Card No.: applied for

British Library Cataloguing-in-Publication Data
A catalogue record for this book is available from the British Library.

Bibliographic information published by the Deutsche Nationalbibliothek
The Deutsche Nationalbibliothek lists this publication in the Deutsche Nationalbibliografie; detailed bibliographic data are available on the Internet at <http://dnb.d-nb.de>.

Print ISBN: 978-3-527-34715-5
ePDF ISBN: 978-3-527-82578-3
ePub ISBN: 978-3-527-82580-6
oBook ISBN: 978-3-527-82579-0

Typesetting Straive, Chennai, India
Printing and Binding CPI Group (UK) Ltd, Croydon CR0 4YY
Printed on acid-free paper

C9783527347155_220724

Contents

Foreword

The recent report released from the Intergovernmental Panel on Climate Change (IPCC) suggests greenhouse gas emissions from human activities are responsible for approximately 1.1 °C of warming since 1850–1900. Over the next two decades, global temperatures are estimated to reach or exceed 1.5 °C due to climate warming. Harvesting energy from renewable sources is of paramount importance to achieve sustainable development. The sun is the best gift to humanity and harvesting solar energy is a cost-effective solution to substitute fossil fuels and address demanding energy crises and environmental problems. Photovoltaics is an ideal approach to convert sun rays into electricity, and currently, it is cheaper to produce electricity in most of the world than generating energy from traditional sources. However, wide-scale deployment and grid parity demand new approaches in solar cell industrialization. For this, we need revolutionary technology. It requires state-of-the-art research and development to make next-generation photovoltaics cost effective and sustainable by using earth-abundant materials. Perovskite solar cells (PSCs), facilitated by the knowledge gain from dye-sensitized solar cells, have gained a surge of interest in R&D worldwide owing to their spectacular performance in thin-film technology. The interest in large-scale R&D on PSCs is monumental and aroused interest not only from photovoltaics companies but also from process- and materials-based enterprises that are also betting to tap into the opportunities. Currently, the NREL certified power conversion efficiency has reached 25.5% in 2020, and supercedes the efficiency value of other thin-film photovoltaic technologies. Furthermore, it is also equipped with unique and advantageous properties compared to conventional silicon or thin-film photovoltaic technology, with interest ranging from integrated photovoltaics in buildings and automobiles to various device architectures. Perovskite itself is a semiconducting mineral that has multiple applications. The PSC efficiencies continue to progress rapidly, while the scientific community pursues to enhance the performance of PSCs through the engineering of perovskites, innovative processes for thin-film deposition, reduction of electrical losses by designing a better interface, and use of advanced charge selective layers through molecular engineering. This unlocks new horizons to optimize device fabrication steps and accelerate the process of lab-to-fab. This book covers almost all aspects of perovskite solar cells from materials, device architecture, process, and device kinetics to designing charge selective layers, which will provide an improved and holistic understanding of developing perovskite materials for solar cell applications. Design, fabrication, passivation approaches and stability evaluation are also discussed, which are crucial for basic and technological research of the new-generation solar cells. Moreover, the perovskite materials and processes discussed in this book can be treated as a valuable pathway towards developing light-emitting diodes, lasing, radiation scintillators, quantum, and neuromorphic devices. The authors bring diversity and inclusion in science, in addition to up-to-date knowledge and experiences over various aspects of PSC technologies. I would like to thank the editors and chapter authors profusely for their efforts and significant contributions to the book that sets an outstanding landmark in this fast-evolving research field, which will guide and motivate the next generation of researchers. With the interdisciplinary nature of the work involved in perovskite solar cells technology, this book will be of interest to the readers from a diverse field of physical sciences and engineering including materials science, physics, chemistry, nanotechnology, and electronic and process engineering, among others.

Professor Subodh Mhaisalkar
Tan Chin Tuan Centennial Professor, School of Materials Science and Engineering
Executive Director, Energy Research Institute @ NTU (ERI@N)
Associate Vice President (Strategy and Partnerships)
Nanyang Technological University, Singapore.

1

Chemical Processing of Mixed-Cation Hybrid Perovskites: Stabilizing Effects of Configurational Entropy

*Feray Ünlü, Eunhwan Jung, Senol Öz, Heechae Choi, Thomas Fischer, and Sanjay Mathur**

University of Cologne, Institute of Inorganic Chemistry, Chemistry Department, Greinstr. 6, Cologne, 50939, Germany

1.1 Introduction

The worldwide deployment of photovoltaic technology to combat the growing energy demand often encounters the "golden triangle" of solar technology defined by an optimal balance among cost, efficiency, and stability parameters [1]. During recent years, perovskite-based photoabsorber materials have emerged as a promising alternative for cost-effective and solution-processable photovoltaic devices. The diversity of options related to processing, composition, and device setups makes this research an interesting and rapidly growing domain of energy materials. In particular, methylammonium lead iodide ($CH_3NH_3PbI_3$, $MAPbI_3$), which is the prototype among organic–inorganic hybrid perovskites, represents a semiconductor with a direct band gap of 1.55 eV corresponding to an absorption onset of 800 nm, making it an efficient absorber of the visible light. The photon-generated excitons in $MAPbI_3$ have a weak binding energy of less than 50 meV, which means that most of them can dissociate very rapidly into free carriers at room temperature [2–5]. Compared with organic photoabsorbers, the electrons and holes produced in this material exhibit small effective masses comparable with inorganic Si, resulting in high carrier mobility [6]. The recombination of charge carriers occurs on a timescale of hundreds of nanoseconds, resulting in long carrier-diffusion lengths ranging between 100 and 1000 nm in thin films with thickness less than a micrometer [7]. However, $MAPbI_3$ solar cells lack thermal stability due to the structural transition from tetragonal to cubic phase at 55 °C, which is within the operation window of perovskite solar cells (PSCs). In addition, $MAPbI_3$ compositions tend to degrade under the attack of moisture and oxygen to lose their light harvesting properties. Moreover, fundamental problems such as light-induced trap-state formation are less understood, concerning mainly the reproducibility of devices and posing a barrier on the commercialization of organic–inorganic hybrid perovskites.

*Corresponding author.

Perovskite Solar Cells: Materials, Processes, and Devices, First Edition.
Edited by Shahzada Ahmad, Samrana Kazim, and Michael Grätzel.

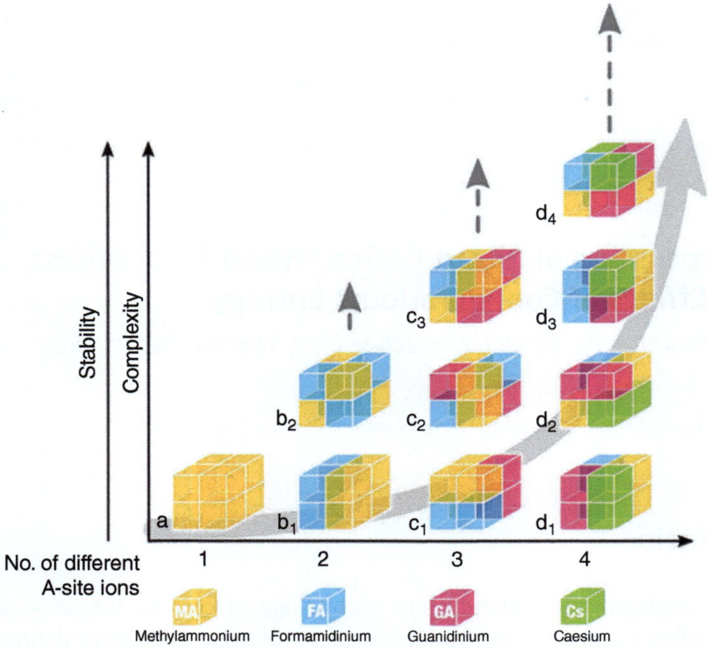

Figure 1.1 The structural complexity of organic–inorganic hybrid perovskites and the related raise of efficiency and stability. Sources: Based on Saliba et al. [8], McMeekin et al. [9], Gong et al. [10].

In this chapter, chemical processing of mixed-cation perovskites based on multiple substitutions on the A-site of the perovskite is discussed in terms of their entropic stability and resulting improvement in the fundamental properties. During the last years, the landscape of synthesis and structure of organic–inorganic perovskites has become more complex due to attempted iterations in compositions to achieve high(er) efficiency and long-term stability of perovskites in photovoltaic (PV) devices. Figure 1.1 schematically illustrates the configurational complexity possible in multiple-cation perovskites that has led to the increased efficiency [8–10]. Nevertheless, the more complex the perovskite, the higher the versatility of functional properties, depending on configurational structure, grain morphology, and optoelectronic properties. Apart from that, an important concept concerning the crystal structure, namely, the Goldschmidt tolerance factor, which allows the prediction of stable crystal phases for certain elements and molecules and their stability, is explained.

1.1.1 Stability Issues of Organic–Inorganic Hybrid Perovskites

Ideal solar modules, which in most cases are placed on the rooftop of the buildings, must be robust against moisture, heat, and light for a long period of time. Despite the high efficiency of PSCs, their long-term stability is still behind the commercial requirements. Although, the organic–inorganic hybrid lead halide perovskites, $MAPbI_3$ and $FAPbI_3$, are excellent semiconductors for PV applications having demonstrated optimized photovoltaic efficiencies of 21.6% [11] and 25.2% [12], respectively, they are chemically vulnerable against water and superoxide molecules, as well as high intensity light. Especially, moisture is known to initiate the degradation process [13], and various degradation mechanisms have been postulated. However, detailed experimental evidences for a unifying mechanistic model are missing.

Niu et al. [14] reported that the perovskite decomposes back into methylammonium iodide (MAI) and PbI_2 with water (3), HI decomposes from methylamine (4), and after oxidation (5) or under radiation (6), molecular iodine is produced. In other studies, hydrate complexes, or even metallic Pb clusters, are found in degraded perovskite [13], which will not be discussed further here:

$$(3)\ CH_3NH_3PbI_3 \rightleftarrows CH_3NH_3I + PbI_2$$

$$(4)\ CH_3NH_3I \rightleftarrows CH_3NH_2 + HI$$

$$(5)\ HI + O_2 \rightleftarrows 2I_2 + 2H_2O$$

$$(6)\ 2HI \rightleftarrows H_2 + I_2$$

In the case of degradation under light and oxygen, Aristidou et al. [15] claimed that the free electron in the excited $CH_3NH_3PbI_3$ (7) can transfer to oxygen to form superoxide molecule (8), which can deprotonate the MA moiety to release methylamine and iodine (9):

$$(7)\ CH_3NH_3PbI_3 \xrightarrow{h\upsilon} CH_3NH_3PbI_3^*$$

$$(8)\ O_2 \xrightarrow{CH_3NH_3PbI_3^*} O_2^{*-}$$

$$(9)\ CH_3NH_3PbI_3 + O_2^{*-} \xrightarrow{\text{Deprotonation}} CH_3NH_2 + PbI_2 + 1/2I_2 + H_2O$$

Wang et al. [16] investigated the degradation behavior of perovskite polycrystalline films depending on the grain morphology and found that the perovskite single crystals were stable in ambient air for more than two years, which highlights the intrinsic stability of the hybrid perovskites and the effect of grain size and grain boundaries on the stability. Therefore, increased grain sizes were found to show better stability. For instance, thin films with an average grain size of 678 nm showed 10% degradation under humid atmosphere (relative humidity [RH] 85%), whereas 85% degradation occurred for smaller average grain size (297 nm).

Besides the chemical reactions and their implications, the limited thermal instability of perovskites is strongly connected to temperature-driven phase transitions. Pristine $MAPbI_3$ forms the tetragonal phase at room temperature, while a phase transition to the cubic phase occurs at 55 °C, which is within the operating temperature range of a solar cell device and below the surface temperatures (up to 85 °C) generally achieved in the field tests (Figure 1.2). The phase transition not only causes changes in the crystal lattice but also can affect optoelectronic behavior and film morphology. Goldschmidt developed a concept [17] that gives fundamental guidelines to predict stable perovskite structures. Since the phase transitions are strongly related to thermodynamic parameters, the entropy of mixing or configurational entropy plays a critical role in understanding the

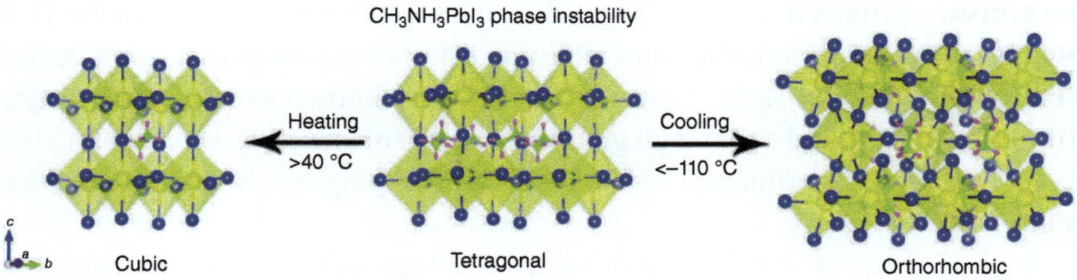

Figure 1.2 Phase transition in $MAPbI_3$ perovskite at different temperatures leads to the phase instability within the solar cell operation frame. Source: Arabpour Roghabadi et al. [13].

structure–property–stability relationship and in designing new perovskites. Section 1.2 briefly reviews the perovskite structure and Goldschmidt's concept revealing the role of A-site cation and its thermodynamic behavior in the lattice. Following this, the chemical processing of multiple-cation perovskites is extensively discussed considering the state of the art of organic–inorganic hybrid perovskite and recent literature reports.

1.2 Crystal Structure of Perovskites

Discovered in 1839 by Gustav Rose and named after Russian mineralogist *Lew Alexejewitsch Perovski*, perovskites represent a family of crystalline compounds that adopt a similar crystal structure as the parent mineral calcium titanate $CaTiO_3$. This class of materials can be generally described by the ABX_3. In the ideal ABX_3 perovskite structure, the large monovalent cation A occupies 12-fold coordination site in the middle of the cube to form an AX_{12} cuboctahedron, while the smaller divalent cation B forms corner sharing BX_6 octahedra in a 3D perovskite framework (Figure 1.3). The ideal cubic perovskite structure is not very common due to the sizes of the cations A and B, and the anion X and the mineral perovskite itself are slightly distorted. Distorted perovskites have reduced symmetry, affecting the

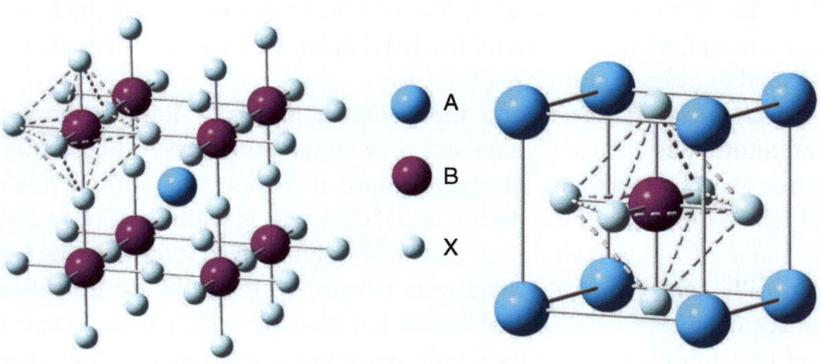

Figure 1.3 The ideal cubic perovskite structure ABX_3 that has a three-dimensional network of corner sharing $[BX_6]$ octahedra surrounding a larger A-site cation.

crystal structure transition, which is one of the key aspects of the structural stability issue for perovskite materials. Such structural distortions, when reversible, can be used for mechanical energy harvesting purposes as demonstrated for the perovskite barium titanate ($BaTiO_3$) [18, 19].

1.2.1 Goldschmidt Tolerance Factor for 3D Structure

In the ideal cubic perovskite structure case, the lattice constant, a, is described in Eqs. (1.1, 1.2):

$$a = \sqrt{2}(r_A + r_X) = 2(r_B + r_X) \tag{1.1}$$

where r_i is the ionic radii of constituting ions (A, B, and X). The ratio of the equations for the lattice constant is given by the Goldschmidt tolerance factor, t, which has been used extensively to predict the stability of the perovskite structure based on chemical formula (ABX_3) and to estimate the degree of distortion [17] that is defined as

$$t = \frac{(r_A + r_X)}{\sqrt{2}(r_B + r_X)} \tag{1.2}$$

It is well known that $t = 1$ stands for a perfect cubic lattice, with the boundary conditions for stable cubic perovskite structures being $0.9 \leq t \leq 1$ and for a distorted perovskite structure being $0.8 \leq t \leq 0.89$, leading to a tetragonal or orthorhombic crystal structure. When the A-site cation is smaller than the ideal value, then t becomes smaller than 1, and as a result the BX_6 octahedra will be distorted to fill space. However, when the A-site cation is too large to form a perovskite structure ($t > 1$), non-perovskite structures such as a hexagonal structure are formed. Similarly, different structures such as ilmenite-type $FeTiO_3$ are formed for $t < 0.8$, when the A-site cation is too small to meet the boundary conditions and form a perovskite structure (Figure 1.4).

The concept of the Goldschmidt tolerance factor works reasonably well with the organic–inorganic halide perovskite materials, which have been used to determine whether a combination of several ions can form a stable perovskite structure [17, 21, 22]. However, there are too many exceptions that cannot be entirely explained by Goldschmidt tolerance factor. Therefore, in such cases, another numerical parameter is necessary to anticipate the structure of a new perovskite configuration especially the substituted ones [23].

1.2.2 Octahedral Factor

The octahedral factor is also known as constraint factor for perovskite structure that is defined as the ratio of the ionic radii of the B-site cation and the halide anion X in the general composition ABX_3 (Eq. (1.3)):

$$\mu = \frac{r_B}{r_X} \tag{1.3}$$

This factor represents the structural stability of the BX_6 octahedron. When it is in the range of $0.44 \leq \mu \leq 0.9$, it generally leads to a stable perovskite structure. Along

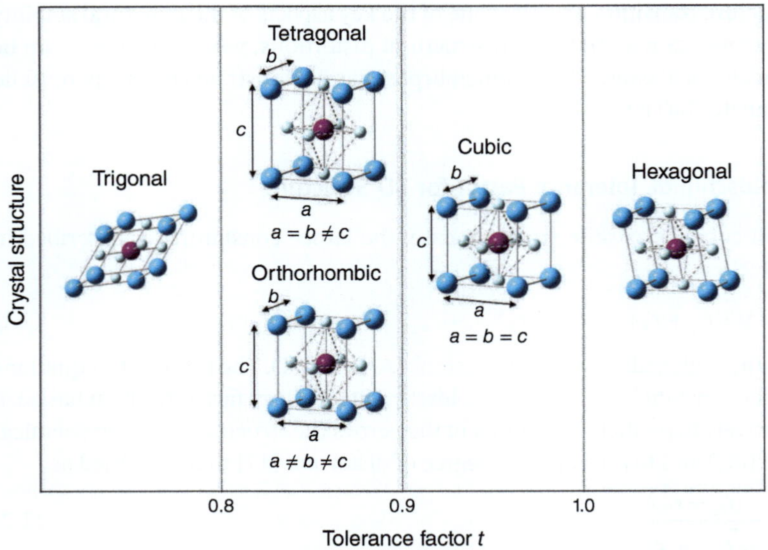

Figure 1.4 The tolerance factor t and the crystal structure of perovskite. Source: Adapted from Yu [20].

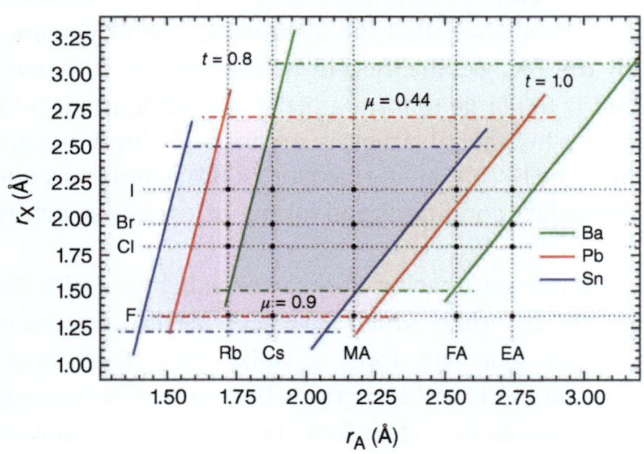

Figure 1.5 Tolerance limit map of parametric space for potentially stable perovskites based on B-site cations Pb, Sn, and Ba. Ionic radii of A-site cations and X-site anions are marked. Source: Yu [20]. Licensed under CC BY 3.0.

with the Goldschmidt tolerance factor, octahedral factor can provide an insight in selecting ions and combinations thereof capable of forming stable halide perovskites among many theoretically possible configurations as indicated in Figure 1.5.

It should be noted that in the case of hybrid perovskites, defining ionic radii for non-symmetric A-site organic molecular cations is nontrivial, and also the reduced electronegativity of heavy halide anions needs to be considered in terms of their polarizability (size/charge) [20, 23]. To address these ambiguities, some modifications have been proposed in elucidating the structural parameters, such as the consideration of effective ionic radii [23].

1.2.3 Role of A-Site Cation

The optoelectronic properties of the hybrid organic–inorganic halide perovskite compounds are strongly governed by the metal halide framework. The valence band is built by the s-lone pair of Pb^{2+} cation and p orbitals of the halide anion, while the conduction band consists of unfilled 6p and 3p orbitals of lead and halogen ions (Figure 1.6). Although the A-site cation does not directly play an important role for the optoelectronic properties [25, 26], the shape, size, and charge distribution of A cations are crucial for forming the perovskite structure. For instance, it was shown that incorporation of a small amount of methylammonium (A = $CH_3NH_3^+$, MA) can stabilize the black phase of perovskite based on formamidinium ($CH_3(NH_2)_2^+$, FA), leading to photoconversion efficiencies >20% [27]. The device performance could be improved by addition of an inorganic (Cs^+) or organic (—$C(NH_2)_3^+$, GA) cation that reportedly stabilized the power output of the device as well [8, 10, 28–31].

Highly symmetric cubic structure is the ideal symmetry for enhanced optoelectronic properties of perovskite in which B cation assembled around X anions forms the BX_6 octahedron and A cation occupies the interstitial site (Figure 1.7) [22, 32, 33]. Previous reports showed that there is a dependence between the Goldschmidt tolerance factor and band gap of the perovskite. The higher the tolerance factor ranging $t = 0.8–1.0$, the lower the band gap. However, in the case of GA cation, the compound is not photoactive anymore because it is too large to adopt the perovskite structure.

A-site cations do not directly influence the electronic structure, but they rather play a structure-directing role. The inorganic Pb–I framework is deformed by A-site cation through steric and Coulomb interactions causing an octahedral tilting, which is responsible for the changes in the electronic structure near the band edges. The octahedral tilting and the deviation from the ideal perovskite structure can be

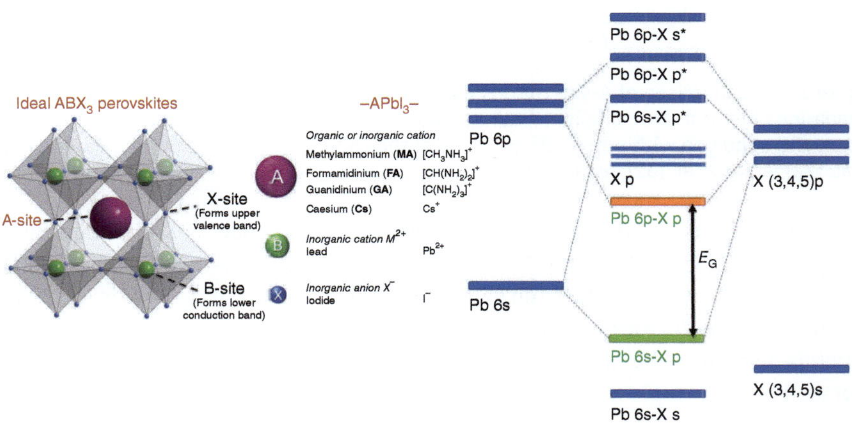

Figure 1.6 The electronic structure of the ideal $APbX_3$ perovskite: valence band (green) consists of s (Pb) and p (halide) orbitals and conduction band (orange) consists of Pb and halide p orbitals, leading to a direct band gap (E_G). Source: Based on Ünlü et al. [24]

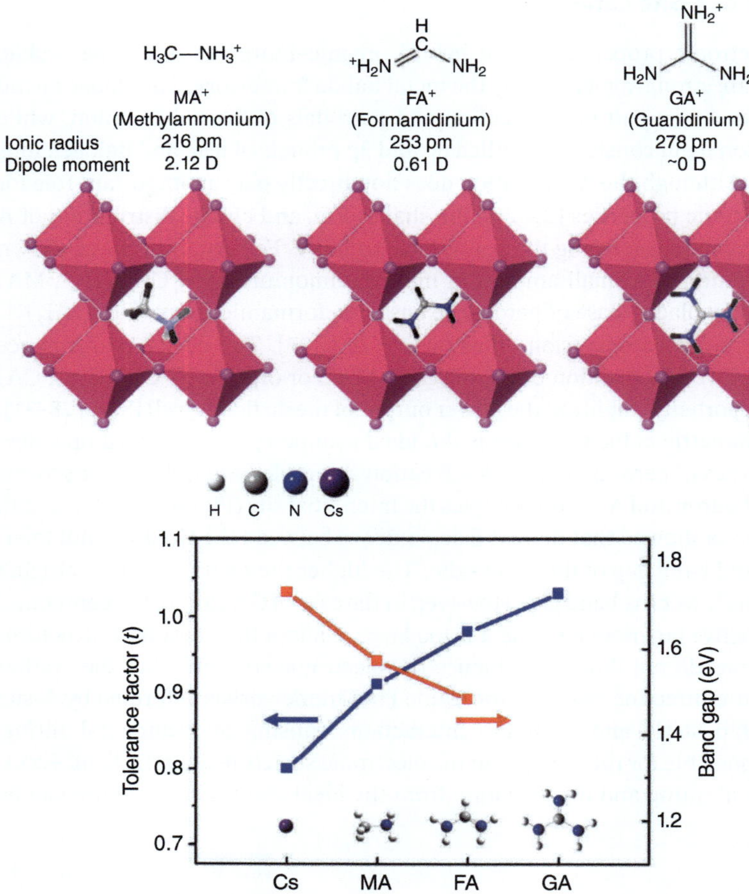

Figure 1.7 Comparison of A-site cations for organic–inorganic hybrid perovskite comparing chemical structure, ionic radius, dipole moment, Goldschmidt tolerance factor, and band gap with schematic structure of the cubic perovskites. Sources: Based on Li et al. [22], Amat et al. [32], Jena et al. [33].

assessed by the octahedra connection bridge Pb–X–Pb angle Φ, which decreases from α-$FAPbI_3$ ($\Phi = 179.9°$) to β-$MAPbI_3$ ($\Phi = 163.6°$) and to γ-$CsPbI_3$ ($\Phi = 153.2°$), resulting in an increase of the band gap of 1.48 eV in α-$FAPbI_3$ to 1.67 eV in γ-$CsPbI_3$ (Figure 1.8). The degree of deviation can be quantified as a function of the angle ψ, which is defined as the average Pb–X–Pb-angle in Pb–X framework. The larger the average Pb–X–Pb angle ψ, the smaller the band gap [34, 35].

1.2.4 Theoretical Calculations: Molecular Dynamics of A-Site Cation

Since the first emergence of the halide PSC [36], a tremendous amount of experimental and theoretical research efforts have been devoted to the engineering and fundamental understanding of PSC [7, 37–43]. On the theoretical side, many of

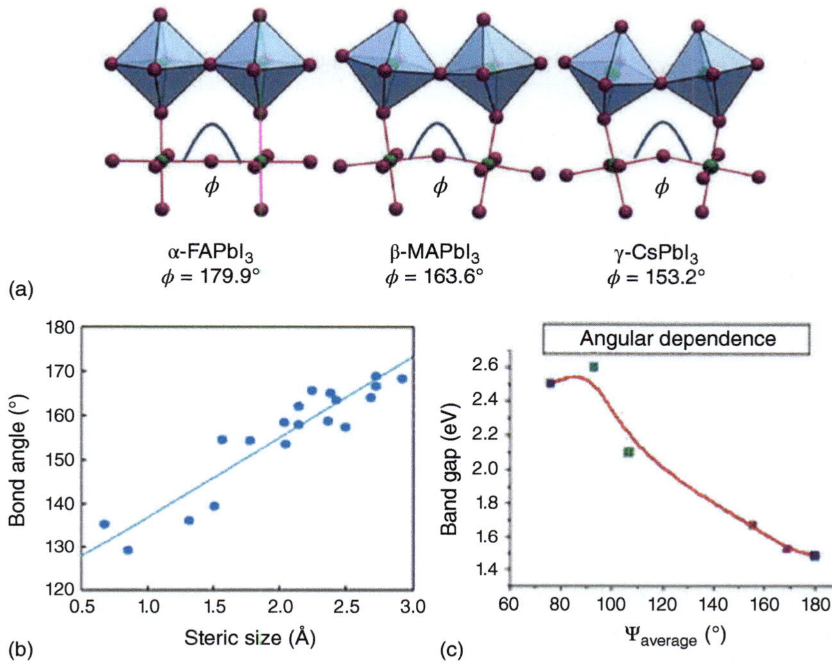

Figure 1.8 Octahedral tilting in the inorganic perovskite lattice because of A-site cation. (a) X–Pb–X angles in organic–inorganic halide perovskites. (b) Bond angle as a function of A-site cation size. Source: Manser et al. [34]. (c) Band-gap dependency from average X–Pb–X bond angle. Source: Stoumpos and Kanatzidis [35].

the intriguing physics hidden in PSC have been explored to discuss the hysteresis, ferroelectricity [40], role of native defects in optical properties [41], Rashba [42], and Dresselhaus [43] effects both in bulk and two-dimensional geometry, prolonged carrier lifetime by rotating methylammonium molecule ions [44], strain-engineered band gap by its unique molecular orbital configuration [45], and so on. These theoretical research work could be mostly done using density functional theory (DFT) [46] calculations, since these are built on quantum mechanical formula and are very appropriate to directly study the orbital-level electronic structures, bond characteristics, and optical transitions.

The simple DFT calculations however often underestimate the perovskite band gaps due to the discontinuity of the exchange–correlation energy, which is a typical shortfall of such techniques [47]. The introduction of relativistic corrections, such as first-order scalar relativistic (SR-DFT) and higher-order spin-orbit coupling effects (SOC-DFT), leads to more realistic values, but these do not exactly match the experimental results. As shown in Figure 1.9, SR-DFT gives nearly similar band gap (E_g) values for $MAPbI_3$ as a result of the compensatory influence of electron self-energy/many-body effects, which tend to increase E_g, and relativistic interactions that reduce E_g. However, it fails to accurately describe the dispersion of the valence band (VB) in Pb perovskites or to calculate exact values for Sn perovskites

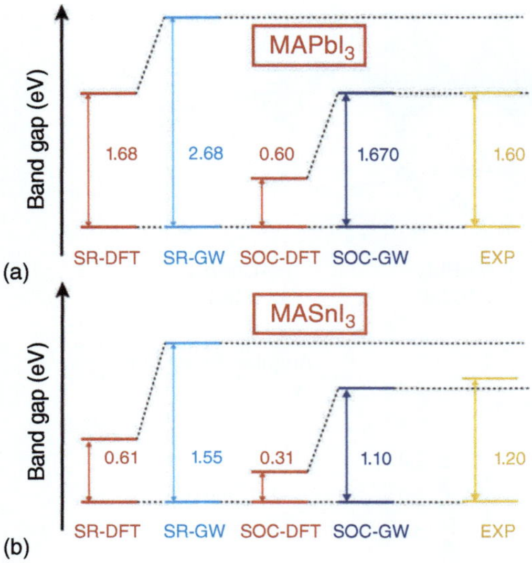

Figure 1.9 Theoretical calculations of perovskites' band gap: comparison of different theoretical methods for (a) $MAPbI_3$ and (b) $MASnI_3$. In both cases, spin-orbit coupling GW (SOC-GW) method could reach better agreement with experimental data. Source: Umari et al. [48].

(Figure 1.9b), as the relativistic effects are less in the lighter metal center. Given the complexity of this system, the state-of-the-art, self-consistent, quasi-particle GW with SOC corrections (spin-orbit coupling quasiparticle self-consistent [SOC-QSGW]) has proven to be the only approach to accurately determine the electronic structure of inorganic and hybrid metal halide perovskites. However, DFT can still provide structural and in some cases semiquantitative information to compare the electronic properties of homologous halide perovskites but should be used with caution [34, 48].

Organic A-site cations show several interesting features, which were investigated in the recent years. For instance, they have a permanent dipole moment, rotate around the C–N axis at their cuboctahedral position, and thus create dynamic and asymmetric structures, which is one of the unique characteristics of organic–inorganic hybrid perovskites compared with their all-inorganic counterparts. The A-site cation interacts with the metal halide framework through Coulomb interactions and hydrogen bonding, influencing the whole perovskite lattice depending on temperature due to the dynamic reorientation [34].

Motta et al. [49] showed through DFT calculations that this dynamic reorientation influences the band structure of the perovskite lattice. They performed DFT calculations for the cubic $MAPbI_3$ considering the fast rotation of MA cation. Therefore, the calculation was not only limited to $CH_3NH_3^+$ in the (100) or (111) direction, for which a higher symmetry is maintained, but also examined cases where the symmetry is reduced. It led to such a symmetry reduction having profound consequences for the electronic structure, namely, that the band gap changes from direct to indirect, when the orientation of the MA cation changes from (111) to (011) orientation (Figure 1.10).

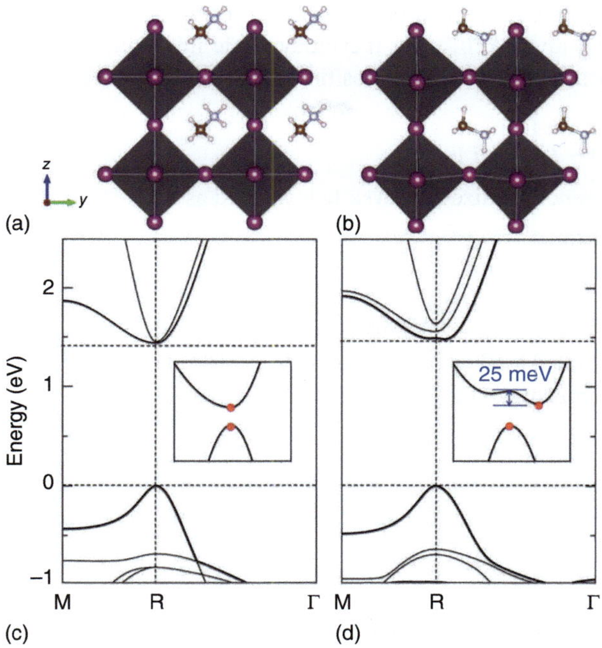

Figure 1.10 Dynamic reorientation of MA cation: cubic $MAPbI_3$ cubic phases with (a) MA in the (111) orientation and (b) in the (011) orientation. Band structures for (c) MA in (111) and (d) in (011) orientation, which shows a band-gap shift from direct to indirect. Source: Motta et al. [49]. Licensed under CC BY 4.0.

1.2.5 Entropy of Mixing: Configurational Effects in Mixed-Cation Perovskites

The configurational entropy of solid materials is related to the way how atoms or molecules are arranged in a mixture, alloy, or crystal lattice. An ideal defect-free and fully ordered material has an entropy of zero. However, this is not the case for real materials where the configurational entropy is defined as "disorder in the site occupancy on crystal sublattices, including defects" [50], and is described with the following equation (Eq. (1.4)), where n is the number of particles and x_n is the mole fraction:

$$S_{\text{conf}} = -nR(x_1 \ln x_1 + x_2 \ln x_2) \tag{1.4}$$

Theoretical calculations and simulations allow to retrieve information on the entropy and enthalpies of systems. In addition, the Goldschmidt's concept including tolerance factor forms the basis to apply DFT methods to obtain deeper insights into the thermodynamics of organic–inorganic hybrid perovskites.

In the case of mixed-cation perovskite, it is important to understand that the perovskite is stable against phase separation. To describe the disorder of mixed-cation perovskite, the configurational entropy can be introduced. In mixing two components, for instance, $MAPbI_3$ and $FAPbI_3$, the thermodynamic miscibility is given by Helmholtz free energy as function of composition x and temperature T [51] (Eq. (1.5)):

$$\Delta F(x, T) = \Delta U(x, T) - T\Delta S(x, T) \tag{1.5}$$

The internal energy difference of mixed perovskite is defined as

$$\Delta U(x, T) = \frac{\sum_k \Delta U_k(x) g_k e^{-E_k(x)/k_B T}}{\sum_k g_k e^{-E_k(x)/k_B T}} \tag{1.6}$$

$$\Delta U_k(x) = E_k(x) - [(1-x)E_{\mathrm{MAPbI3}} + xE_{\mathrm{FAPbI3}}] \tag{1.7}$$

where E_k is from DFT calculated total energy of the alloy.

The entropy of mixing according to Eq. (1.4) is given by

$$\Delta S(x) = -k_B[x \ln x + (1-x)\ln(1-x)] \tag{1.8}$$

These theoretical calculations can be used to build phase and miscibility diagrams that could eventually support multiple-cation models and explain the alloying experiments.

1.3 Multiple A-Site Cation Perovskites

$MAPbI_3$ is the most popular composition among all organic–inorganic hybrid perovskites, where numerous approaches were reported to optimize structure, morphology of crystals and fabrication of thin films, halide mixing to tune band gap, and crystallization. In contrast to its phase transition at low temperature (>40 °C), formamidinium-based perovskites ($HC(NH_2)_2PbI_3$, $FAPbI_3$) undergo phase transition above 150 °C to form the photoactive black α-phase, which has a lower band gap than $MAPbI_3$ (1.47 vs. 1.55 eV for $MAPbI_3$), and an absorption edge shifted toward the red region allowing the increase of photon absorption and power conversion efficiency (PCE). Yang et al. [52] introduced a direct intramolecular exchange procedure to fabricate $FAPbI_3$ solar cells and observed a PCE of 20.2% and a short-circuit current density (J_{sc}) of 24.7 mA cm^{-2}. However, the black phase of $FAPbI_3$ is not stable at room temperature, at which it forms the inactive yellow δ-phase [53]. Finally, all-inorganic perovskite $CsPbI_3$ is also used in solar cell application due to its higher thermal stability but has a band gap higher (1.73 eV) than the other compositions and active black phase stable at high temperatures (>300 °C). The polymorphism can be controlled by A-site cation engineering to stabilize desired photoactive phase and to tune optical properties and thermal stability. In the following sections, alloying of different A-site cations to reach the configurational stability of hybrid perovskites is presented.

1.3.1 FA^+/MA^+ Alloying for Higher Phase Stability and Photovoltaic Efficiency

Pellet et al. [54] probed the mixing of formamidinium (FA) and methylammonium (MA) cations in 3D hybrid perovskites to expand the absorption spectrum and lower

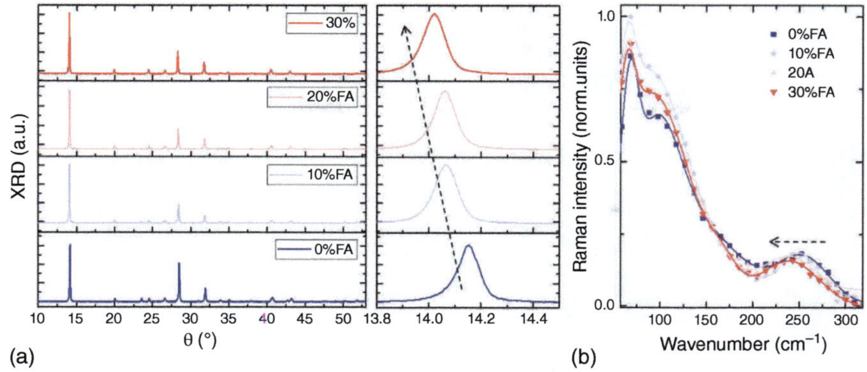

Figure 1.11 Effect of FA/MA alloying on the hybrid perovskite structure. (a) X-ray diffractograms of $FA_xMA_{(1-x)}PbI_3$ thin films showing the peaks shifting to smaller angles while increasing FA amount. (b) Raman intensity of $FA_xMA_{(1-x)}PbI_3$ thin films underlying the shifting related to FA amount. Source: Zhang et al. [55].

the band gap. The combination of A-site cation mixing and sequential deposition method led to phase-pure double-cation black perovskite. Using the double-cation $(MA)_{0.6}(FA)_{0.4}PbI_3$ system, high-performing solar cells could be obtained reaching 14.9% PCE [54]. In their study, Zhang et al. [55] present that by increasing the FA^+ amount into $MAPbI_3$ structure, the photoluminescence (PL) lifetimes increase due to better crystallinity, which leads to reduction of trap states and therefore enhancement of carrier-diffusion length. It was observed that the positions of the diffraction peaks at $2\theta = 14.1°$, 20.0°, 24.4°, 28.4° and 31.8° decreased with increasing FA^+ amount, indicating the enlargement of the crystal lattice to a cubic or quasi-cubic crystal phase (Figure 1.11), giving rise to longer PL lifetimes.

In another approach, where MA^+ was mixed into $FAPbI_3$, Binek et al. [56] have showed that 15% of MA cations are sufficient to stabilize the α-phase at low temperature (Figure 1.12a). They proposed that the intercalation of small-cation MA^+ having high dipole moment enhances the hydrogen bonding to the inorganic Pb–I cage. Moreover, due to better Coulomb interactions between high dipole moment and Pb–I octahedra, the Madelung energy increases, achieving a better stability (Figure 1.12b). In addition, other various chemical methods, e.g. cation exchange [57], mechanosynthesis [58], vapor assisted routes [59], and halide mixing [9, 60], were employed to gain FA/MA-mixed perovskite, aiming at improving grain morphologies, phase purity, and solar cell efficiency.

1.3.2 Cesium Inclusion for Thermal Stability

Apart from the instability of the black phase of $FAPbI_3$ at room temperature, the organic–inorganic hybrid perovskite, although stabilized with $CH_3NH_3^+$ cations, showed lower photostability and thermal stability due to the volatile MA organic moiety. Indeed MA^+ cations induce the crystallization of α-$FAPbI_3$ perovskite, but at a much slower rate due to only slightly smaller size of MA that makes the kinetics sluggish. This leaves a large part of the yellow phase non-transformed.

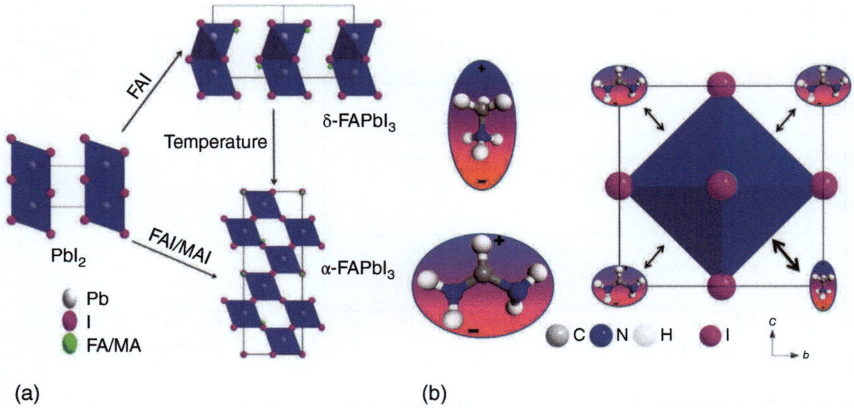

Figure 1.12 Phase stabilization effect of MA/FA alloying. (a) Intercalation of formamidinium iodide (FAI)/MAI in to PbI_2 network leads to the α-phase crystallization of the perovskite, while pure FA-based perovskite forms δ-phase. (b) Optimized structure model of MA-stabilized $FAPbI_3$ showing the different dipole moments and the impact on interaction within the perovskite lattice. Coulomb interaction is stronger in between high dipole moment MA and Pb–I cage. Source: Binek et al. [56].

Alternatively, incorporation of Cs^+ on the A-site of the $FAPbI_3$ or $MAPbI_3$ perovskites could provide better thermal stability and photostability, along with the stabilization of black $FAPbI_3$ at room temperature. The $Cs_{0.1}FA_{0.9}I_3$ turned black even at room temperature with a shrunken perovskite lattice, owing to the smaller cesium cations that led to the reduction of the cuboctahedral A-site vacancy and hence stronger interaction between A-site cation and iodide anions. This reduced A-site volume enables the enhancement in photostability and thermal stability as the FA^+ cations are more enclosed [61]. Furthermore, Cs^+ incorporation lead to larger grains, lower trap states, and longer charge carrier lifetime, improving the solar cell performance to 19% compared with pristine $FAPbI_3$ with PCE of 17.4% [61]. In the case of $Cs_xMA_{1-x}PbI_3$, Niu and coworkers obtained 18.1%, with a J_{SC} of 22.57 mA cm^{-2}, V_{OC} of 1.06 V, and FF of 0.76. In addition, the thermal stability of the unencapsulated devices for $x = 0.09$ was also improved compared with $MAPbI_3$ [62].

Formation of stable cubic perovskite is not favored because $FAPbI_3$ has a larger non-spherical A-site cation, whereas $CsPbI_3$ leads to a smaller tolerance factor (0.81). However, Li and coworkers could show that the alloying of both materials could tune the tolerance factor to a stabilized cubic perovskite structure (Figure 1.13) [22], achieving a new concept to determine effective tolerance factor (Eq. (1.10)):

$$r_{\text{effective}} = xr_A + (1 - x)r_B \tag{1.9}$$

$$t_{\text{effective}} = \frac{r_{\text{effective}} + r_X}{\sqrt{2}(r_{Pb^{2+}} + r_X)} \tag{1.10}$$

This methodology in general could be used in finding new perovskites for photostable and thermally stable solar cell applications.

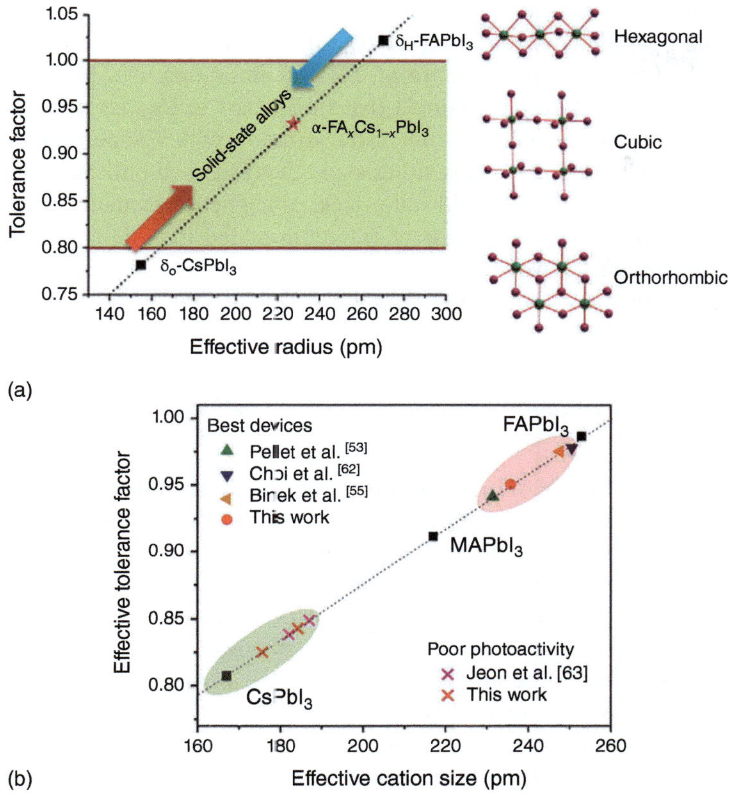

Figure 1.13 Tolerance factor-induced crystallization of perovskites. (a) Combination of small and large cations for tuning of ideal tolerance factor giving cubic structure. (b) Effective tolerance factors obtained from different double-cation perovskites in literature. Source: Li et al. [22].

1.3.3 Rb^+ Small-Cation Influence on Perovskite Structure for Thermal Stability

Apart from cesium cation alloying into $FAPbI_3$, smaller rubidium cations are also suitable for incorporation into the $FAPbI_3$ lattice to maintain the active black phase of formamidinium lead perovskite. However, $RbPbI_3$ forms a yellow δ-phase orthorhombic non-perovskite structure, which is not photoactive probably due to its smaller size that strongly affects the perovskite crystal structure of $FAPbI_3$. Moreover, compared with Cs-containing perovskites, the other advantages are the lower toxicity and better solubility in dimethylformamide, which is commonly used for perovskite solutions. In addition, through the fluorescence lifetime image microscopy (FLIM), Park et al. [65] found that $Rb_{0.05}FA_{0.95}PbI_3$ has longer PL lifetime compared with pristine $FAPbI_3$ due to the suppression of yellow phase formation. The fabricated solar cell exhibited remarkable enhancement of PCE from 13.56% ($FAPbI_3$) to 17.16%, which is related to higher short-circuit current density (J_{sc}), open-circuit voltage (V_{oc}), and fill factor (FF). The best performing solar cells showed operational stability for 1000 hours under ambient conditions at an average RH of 55% and temperature of 25 °C, maintaining 97% of initial PCE. In contrary, Kubicki et al. [66] claimed that in contrast to Cs that is incorporated in the $FAPbI_3$ lattice, Rb does not form an alloy with FA perovskite. They prepared perovskites via mechanosynthesis and characterized bulk material and thin films with ^{133}Cs and ^{87}Rb solid-state nuclear magnetic resonance spectroscopy (NMR) spectroscopy. In the case of rubidium added to $FAPbI_3$ and to multiple-cation perovskites, it was found to separate into a mixture of rubidium halides acting as passivation layer rather than Rb occupying the position of an interstitial cation.

1.3.4 Guanidinium Large-Cation Influence on Perovskite Structure for Stability

Further attempts to stabilize $MAPbI_3$ or $FAPbI_3$ were made by introducing the larger guanidinium cation ($C(NH_2)_3^+$, GA) into the $MAPbI_3$ lattice. Guanidinium lead iodide ($GAPbI_3$) itself exceeds the limit of tolerance factor ($t = 1.03$), producing low-dimensional perovskites that show temperature-dependent phase instability [67, 68]. However, it has been shown that when GA was incorporated in $MAPbI_3$, 3D structure was maintained in a specific composition range of $MA_{1-x}GA_xPbI_3$ ($0 < x < 0.25$) [69]. As long as the amount of GA was not higher than 25%, 3D structure was maintained without any phase separation of $GAPbI_3$ or $MAPbI_3$ that is thermodynamically less favored. The introduction of small amounts of GA^+ cations led to better stability due to the increased number of H—I bonds, which are shorter (Figure 1.14a). The improvement of H-bonding is also assumed to enhance the grain size and interfacing among grains and further passivation of undercoordinated iodine present within the grain boundaries. This was found to prolong charge carrier lifetimes (Figure 1.14b) and open-circuit voltages (>1.1 V), thus upgrading solar cell performances. In addition, Hou et al. [70] demonstrated the use of guanidinium chloride (GACl) as additive to ameliorate the grain quality of $MAPbI_3$ thin films and for surface passivation and reduced charge recombination.

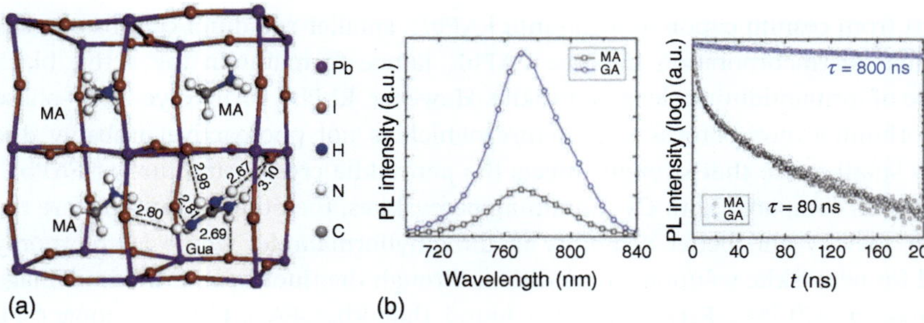

Figure 1.14 (a) Influence of the larger-cation GA on perovskite structure. Source: Jodlowski et al. [69]. Licensed under CC BY 4.0. (b) Enhanced PL lifetimes for mixed GA/MA perovskite compared with pristine $MAPbI_3$. Source: De Marco et al. [30].

They observed that GA does not intercalate into the perovskite crystal lattice, but rather it stays at the grain boundaries to enhance the grain interconnectivity via H-bonds.

1.3.5 Triple- and Quadruple-Cation Hybrid Perovskites for Stability and Optimum Performance

The application of Goldschmidt tolerance factor for 3D perovskites allows not only alloying the two cations on the A-site but also inducing more cations and even having an intermixing of the halide anions. Table 1.1 provides an overview of perovskite compositions used so far, which could achieve high efficiencies through compositional engineering. However, it is theoretically possible to upgrade the solar cell efficiency up to 33% for single-junction PSC depending on the band gap. The Shockley–Queisser limit [78] for best performing single-junction solar cell can be obtained by a semiconductor of 1.34 eV band gap [79]. Among double-cation perovskites, best efficiencies (>20%) could be reached by FA/MA mixing [12, 55, 73, 74]. Although the double-cation strategy could solve some of the problems mentioned, one approach could even outmatch this by using cesium (Cs) as the third A-site cation. Inclusion of Cs^+ in FA/MA perovskite was established by Saliba et al. [8] leading to stable black cubic phase at room temperature without any impurities of the photo inactive δ-phase of $FAPbI_3$. The smaller radius of Cs^+ (1.81 Å) can be employed to decrease the effective tolerance factor and hence gain cubic or pseudo-cubic perovskite structure. The mixed halide triple-cation perovskite $Cs_x(MA_{0.17}FA_{0.83})_{(1-x)}Pb(I_{0.83}Br_{0.17})_3$ yielded stable PCEs above 21% (Figure 1.15b) showing higher thermal stability and reproducibility under several fabrication conditions such as different preparation protocols, temperature and atmospheric changes. This outstanding stability and device reproducibility was an important step towards large-scale application. Singh and Miyasaka [28] could confirm this outstanding property of triple-cation perovskite by its preparation under ambient atmosphere with controlled RH of 25%, achieving stabilized PCE of over 20%. Different antisolvents were used to engineer the grain morphology that showed comparable results, corroborating the insensitivity of the Cs/MA/FA perovskite toward different solvents. The grain size and morphology were affected by Cs concentration under the ambient air process, giving the best performance of device with large micrometer size grains for 5% Cs incorporated in the triple-cation perovskite (Figure 1.15a).

The use of A-site cation engineering enables to increase the entropy of perovskite at the lead site to enlarge the complexity of composition in perovskite, which can stabilize unstable phases of the normal pristine perovskites ($MAPbI_3$, $FAPbI_3$, and $CsPbI_3$). Apart from the abovementioned strategies like mixing two or three cations, the inclusion of further cations can be employed to increase the already achieved stability and charge carrier transport optimizations. Saliba et al. [80] reported on quadruple-cation perovskites employing Rb as fourth cation in their Cs/FA/MA-mixed halide perovskite. Alkali metal cations gained wide interest owing to their stability against oxidation that in turn could stabilize lead- or tin-based perovskites against oxidation. $RbPbI_3$ crystallized in a yellow

Table 1.1 Overview of recently reported highly efficient organic–inorganic multiple A-site cation perovskites with their characteristics.

Perovskite	*t*	Special feature	Record PCE	Stability	References
$MAPbI_3$	0.91	Band gap 1.55 eV High absorption coefficient Low exciton binding energy	21.6%	Fabrication under ambient conditions, room temperature (RT), RH 25%	[11]
$FAPbI_3$	0.99	Band-gap black phase 1.47 eV Broad light harvesting up to 840 nm	25.2%	Certified, 1000 h, RH 20%, retained 90% of PCE	[12]
$CsPbI_3$	0.81	Band-gap black phase 1.73 eV Robust against moisture and air	19.03%	500 h under illumination, retained 90% of PCE	[72]
$FA_{0.1}MA_{0.9}PbI_3$	0.92	Quasi-cubic crystal phase PL redshift Stabilized black phase	20.2%	300 h under full illumination MPP tracking at 45 °C under N_2	[55]
$(FAPbI_3)_{0.92}(MAPbBr_3)_{0.08}$	0.98	Loss in potential of 340 mV EL EQE 8.9% 2D passivation layer on top of perovskite	23.4%	500 h under full illumination MPP tracking, 22.6% PCE (certified), retained 85% of PCE	[73]
$(FAPb_3)_{0.95}(MAPbBr_3)_{0.05}$	0.98	Double-layered solar cell with passivation layer Poly(3-)hexylthiophene HTM	22.7%	200 h, RT, RH 85%, and 1370 h (encapsulated) device under illumination, retained 95% of PCE	[74]
$FA_{0.85}Cs_{0.15}PbI_3$	0.95	Stabilized black cubic phase Band gap 1.52 eV	18.9%	1000 h in air, retained 88% of PCE	[75]
$Cs_{0.05}(MA_{0.17}FA_{0.83})_{1-x}Pb(I_{0.83}Br_{0.17})_3$	0.99	Purer, defect-free perovskite Thermally more stable Highly reproducible	21.1%	250 h under full illumination MPP tracking at RT, 18% PCE (stabilized) Long-lived component 5032 h	[8]
$GA_{0.015}Cs_{0.046}MA_{0.152}FA_{0.787}Pb(I_{0.815}Br_{0.185})_3$	0.97	Bang gap 1.62 eV Long PL lifetime 3.2 μs	20.96%	60 days, under ambient atmosphere at 80 °C, retained 70% of PCE	[76, 77]

Stability is related to the reported conditions during measurements, fabrication, or exclusive stability test. Tolerance factor *t* were calculated according to Eq. (1.3).

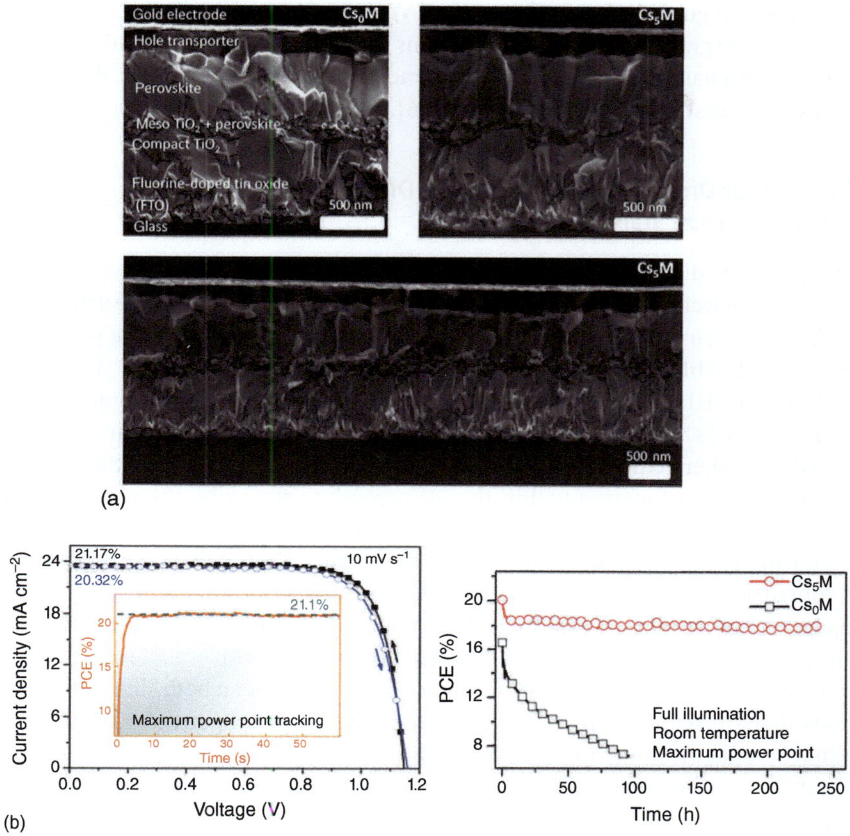

Figure 1.15 Triple- and quadruple-cation hybrid perovskites. (a) Cesium impact on grain morphology of triple-cation Cs/MA/FA perovskites inducing better grain sizes and alignment. (b) Cs/MA/FA-based perovskite solar cells attaining PCE increase and stability under full illumination up to 250 hours of maximum power point (MPP) tracking. Source: (a) Reprinted with the permission from Saliba et al. [8], (b) Saliba et al. [8].

non-perovskite structure with tolerance factor below 0.8. However, with 5% Rb in Rb/Cs/MA/FA, the group could achieve PCE values of over 20.6% with high V_{oc} of 1.24 eV. The loss in potential (difference between V_{oc} and band-gap value) was only 0.39 V, which is due to low nonradiative recombination losses [80]. Theoretically, these values lie between 0.23 V (for band gap of 1.0 eV) and 0.3 V (band gap of 2 eV) for defect-free materials having only radiative recombination [81]. Besides Rb and Cs, the other alkali metals deviate greatly from the ideal tolerance factor, so they are not suitable for intercalation or A-site intermixing. Apart from alkali metal cations, the inclusion of guanidinium ($C(NH_2)_3^+$, GA) cations in the perovskite was shown to passivate the grain boundaries of thin films to minimize the nonradiative charge carrier recombination and enhance the nonradiative recombination time due to the suppressed formation of halide vacancies [76].

1.3.6 Larger Organic Cations: Reducing Dimensionality for Improved Thermal Stability

Although three-dimensional (3D) hybrid perovskites are preferred due to their excellent optoelectronic properties, the stability issue related to moisture sensitivity remains a challenge toward commercialization. The vulnerability of the prototype compound $MAPbI_3$ against water, which can deprotonate the $CH_3NH_3^+$ moiety to form CH_3NH_2, HI, and PbI_2, is a persisting problem for the environmental stability of perovskites [82–84]. Furthermore, these small organic cations can easily be released at higher temperatures and under illumination [85, 86]. In the context of A-site cation engineering within the 3D perovskite structure, the use of bigger organic cations achieving lower-dimensional perovskite structures seems to be a promising solution against these issues [87]. Since there are no size rules for low-dimensional networked perovskites, A-site cations of different length can be adopted, enabling more flexibility in terms of configurational tuning. Mitzi et al. [88] reported on two-dimensional (2D) layered perovskites with organic ammonium cations ($R—NH_3^+$). Here, the MX_4^{2-} anions are surrounded on both side by organic cations that are stabilized by hydrogen bonds. These layered perovskite structures are analogous to the inorganic *Ruddlesden–Popper* compound K_2NiF_4 (Figure 1.16a) that can be considered as ⟨100⟩ slice of the 3D perovskite structure [88]. Moreover, the 2D layered perovskites are primarily classified into ⟨100⟩, ⟨111⟩, and ⟨110⟩ families (Figure 1.16b), from which the ⟨100⟩ Ruddlesden–Popper (RP) phase is most commonly used for optoelectronic applications. The general formula of RP phase is $(RNH_3)_2\ A_{n-1}BnX_{3n+1}$ where RNH_3 is the organic cation spacer, increasing the lattice of the perovskite due to its larger sizes.

Tsai et al. reported on 2D *Ruddlesden–Popper* perovskite $(BA)_2(MA)_3Pb_4I_{13}$ (BA: *n*-butyl ammonium) with a crystal growth orientation perpendicular to the substrate. This orientation facilitates the charge transport and could yield 12.5% efficiency [89]. Low-dimensional perovskites are generally more stable than 3D perovskites. For example, Smith et al. could fabricate layered perovskite with phenyl ammonium cation, which leads to stability under moisture due to the hydrophobicity of the benzene group [90]. Furthermore, the use of different functional organic spacers, for example, fullerene derivatives, conjugated oligomers, or organic dyes, allows tuning the optoelectronic properties and charge carrier transport abilities of the designed 2D perovskites [91–93]. The size and nature of the organic spacer and the hydrogen bonding to the inorganic frame influence the Pb–I–Pb bond angles, which impact the valence bandwidths where s-orbital (Pb) and p-orbitals have stronger overlap integrals for angles close to 180°, in turn reducing the band gap. Additionally, for 2D perovskites, the interlayer spacing between inorganic layers is the determining factor for band-gap tuning. The longer the interlayer spacing,

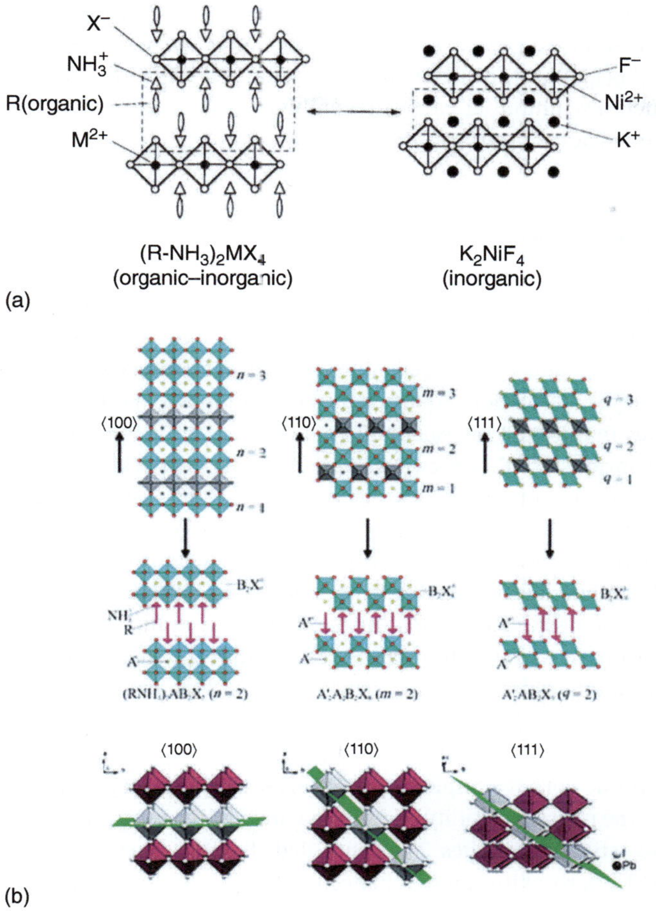

Figure 1.16 Low-dimensional perovskites. (a) 2D organic–inorganic hybrid perovskites and their similarity to Ruddlesden–Popper compound K_2NiF_4. Source: Mitzi [88]. (b)⟨100⟩, ⟨111⟩, and ⟨110⟩ phases of 2D perovskites. Source: Thrithamarassery Gangadharan and Ma [87].

the wider the band gap. That is why pure 2D layered perovskites could not achieve comparable solar cell performance in relation to the 3D AMX_3 perovskites majorly due to the wider band gaps attributed to bulkier organic spacer cations. Hence, a combination of 2D and 3D perovskites is becoming increasingly interesting in terms of unifying high efficiency and long-term stability in PSCs. In 2D/3D system, the small organic cations are partially replaced by bulkier organic spacer cations. For instance, Grancini et al. [94] reported one-year stable and HTM-free 2D/3D $(AVA)_{0.3}(MA)_{0.7}PbI_3$ (AVA: amino valeric acid, $HOOC(CH_2)_4NH_3$) PSCs, without any obvious loss in efficiency (PCE 12.9%) under ambient air with 1 sun illumination at 55 °C. The partially substitution of the A-site with the organic spacer cation widens the band gap at the interface, which functions as electron recombination barriers due to the lower lying conduction band (CB) of the 2D perovskite than the CB of the 3D phase. This combination of wide and low band gap perovskites leads to stable power conversion efficiencies of up to 19% for $BA_{0.05}(FA_{0.83}Cs_{0.17})_{0.95}Pb(I_{0.8}Br_{0.2})_3$ designed by Wang et al. [95] The 2D/3D alloying creates heterojunctions inhibiting carrier recombination at grain boundaries consisting of heterodimensional interfaces.

1.4 Conclusion and Perspectives

Organic–inorganic hybrid perovskites gained interest in the past decade, owing to their excellent properties for optoelectronic application such as high absorption coefficient, long carrier lifetimes, and very uniquely the high defect tolerance that allows facile and cheap thin-layer fabrication. Apart from these fundamental characteristics, hybrid perovskites manifested physical and chemical versatilities in terms of crystal structure with various Pb–X framework from 0D to 3D networking, achieving a wide scope of application opportunities, e.g. solar cells, light-emitting diodes, photodetectors, or ferroelectrics. Moreover, the dynamic organic A-site cations lead to a great flexibility of the perovskite lattice compared with the classical inorganic oxide perovskites. Altogether, hybrid perovskites developed as highly efficient new technology and can be the next-generation solar technology. To fulfill the requirements for commercialization of PSCs, the stability is recently the most important challenge, which researchers address.

The formation of an intrinsic stable perovskite structure can be predicted by the concept of Goldschmidt's tolerance factor, allowing the versatility of elements in the sites of the ABX_3 framework if complied with $0.9 < t < 1$ [33]. A very important approach to overcome the phase instability and thermal instability was the chemical processing of multiple-cation perovskites. The controlled alloying of various cations at the A-site of the ABX_3 perovskite could enhance the stability of the usually sensitive $MAPbI_3$ perovskite toward phase transition and moisture. The main role of this cation is to template the perovskite inorganic framework and thus influence the electronic structure indirectly. Especially organic polar A-site cations, which rotate within the inorganic frame, lead to a dynamic disorder [96–98]. This could be controlled by alloying inorganic cations such as Cs and Rb into the multiple-cation perovskite [65, 80, 99–101].

To provide external stability, great efforts have been made to improve the moisture resistance of the vulnerable perovskite layers, either by incorporating hydrophobic layers such as polymers/carbon composites [102], by incorporating non-hygroscopic intermediate layers between perovskite and hole transport material (HTM) [103, 104], or by passivating the perovskite surface by small molecules [105, 106]. Furthermore, lowering the dimensionality of the inorganic octahedral network by large organic A-site cations turned out to be a possible solution for the stability issue [87, 94, 95].

Apart from organic–inorganic hybrid perovskites, all-inorganic lead halide perovskites have emerged as an important class of strongly emissive materials with promising applications in PV, photodetectors, and light-emitting diodes (LEDs).

$CsPbX_3$ materials exhibit several advantages, including higher melting point (>500 °C), higher thermal stability, and higher stability against photobleaching, making them suitable candidates for optoelectronic applications [107]. Therefore, $CsPbX_3$ nanocrystal colloids have become the most popular perovskite studied for LED applications. For PV applications, $CsPbX_3$ perovskites have also recently made great progress in terms of efficiency, reaching over 18% [108].

Apart from PSCs, perovskite-based LEDs also gained interest in recent years, achieving a quantum efficiency of above 20% (in comparison with 30% reported for best performing organic LEDs) in about four years. Moreover, the flexibility in perovskite material composition allows to tune the exciton emissions in the desired wavelength region that decides the emissive color [109]. The major issues with perovskite LEDs at the present stage of progress are material stability/thermal stability, which is being addressed by all-inorganic perovskites, and color stability, i.e. color changing with duration of operation under the constant voltage, likely due to ion migration and operational stability, required as a standard evaluation for successful commercialization. Surface defect passivation and interface engineering will be the approaches used to enhance the performance of both PV and LED technologies [110].

Organic–inorganic hybrid perovskites and their all-inorganic $CsPbX_3$ counterparts stand out due to their excellent optoelectronic properties, which have been improved immensely by a wide variety of methods worldwide within a remarkably short time. Due to the variation in the composition, problems such as structural instability could be eliminated. Interface engineering has made it possible to produce not only highly efficient solar cells but also solar cell stability in laboratories up to 1000 hours. However, for commercialization, other important factors play an important role. Apart from structure-based intrinsic stability, which could be eliminated so far by this A-site cation engineering, other reasons caused by ion migration, defect dynamics, thermal strain between substrate and perovskite film, crystal strain, and crystal orientation should be investigated more extensively.

The commercial feasibility of emerging generation technologies such as PSCs also depends on improved costs per watt peak (Wp) of the installation compared with the already used classical PV technology on the market [111]. The outstanding advantage of PSCs is the possible use of solution-based techniques at low temperatures such as spraying, doctor blading, and roll-to-roll printing and the promise of highly efficient, lightweight, and low-cost options, leading to the reasonable energy payback times.

Now there are two main challenges in this area: fundamental understanding of the physics, chemistry, and engineering of perovskite technology and scaling up and optimization of the production toward commercialization. For this, it is not only important to optimize thin-film fabrication or achieve high efficiencies. Today, it is rather important to comply with the general restrictions, such as more sustainable large-scale production. One small step toward more sustainable fabrication could be the avoidance of toxic and environmentally harmful solvents, which are mainly used in perovskite thin-film fabrication.

Meanwhile, our group has established protic ionic liquids (PILs) in solution-based hybrid perovskite processing [27], which resulted in a substantial decrease of annealing temperature and time required for the formation of polycrystalline perovskite thin films without compromising the final film quality. Furthermore, it was possible to use green solvents such as water, alcohols, or acetonitrile. However, we further aim to focus on the development of such perovskite inks including the fundamental investigation of solution dynamics.

Acknowledgments

The financial and infrastructural support provided by the University of Cologne is thankfully acknowledged. The German Science Foundation (DFG) is gratefully acknowledged for the project funding received within the priority programme SPP 2196 "Perovskite semiconductors: From fundamental properties to devices" and in the frame of the SPP 1959 "Manipulation of matter controlled by electric and magnetic field: Towards novel synthesis and processing routes of inorganic materials."

References

1 Snaith, H. and Perovskites, J. (2013). The emergence of a new era for low-cost, high-efficiency solar cells. *J. Phys. Chem. Lett.* 4: 3623–3630.
2 Miyata, A., Mitioglu, A., Plochocka, P. et al. (2015). Direct measurement of the exciton binding energy and effective masses for charge carriers in organic-inorganic tri-halide perovskites. *Nat. Phys.* 11: 582–587.
3 Hirasawa, M., Ishihara, T., Goto, T. et al. (1994). Magnetoabsorption of the lowest exciton in perovskite-type compound $(CH_3NH_3)PbI_3$. *Phys. B Phys. Condens. Matter* 201: 427–430.
4 D'Innocenzo, V., Grancini, G., Alcocer, M.J.P. et al. (2014). Excitons versus free charges in organo-lead tri-halide perovskites. *Nat. Commun.* 5: 1–6.
5 Valverde-Chávez, D.A., Ponseca, C.S., Stoumpos, C.C. et al. (2015). Intrinsic femtosecond charge generation dynamics in single crystal $CH_3NH_3PbI_3$. *Energy Environ. Sci.* 8: 3700–3707.
6 Giorgi, G., Fujisawa, J.I., Segawa, H., and Yamashita, K. (2013). Small photocarrier effective masses featuring ambipolar transport in methylammonium lead iodide perovskite: a density functional analysis. *J. Phys. Chem. Lett.* 4: 4213–4216.
7 Cui, J., Yuan, H., Li, J. et al. (2015). Recent progress in efficient hybrid lead halide perovskite solar cells. *Sci. Technol. Adv. Mater.* 16: 036004.
8 Saliba, M., Matsui, T., Seo, J.Y. et al. (2016). Cesium-containing triple cation perovskite solar cells: improved stability, reproducibility and high efficiency. *Energy Environ. Sci.* 9: 1989–1997.

9 McMeekin, D.P., Sadoughi, G., Rehman, W. et al. (2016). A mixed-cation lead mixed-halide perovskite absorber for tandem solar cells. *Science (80-.)* 351: 151–155.

10 Gong, J., Guo, P., Benjamin, S.E. et al. (2018). Cation engineering on lead iodide perovskites for stable and high-performance photovoltaic applications. *J. Energy Chem.* 27: 1017–1039.

11 Kogo, A., Sanehira, Y., Numata, Y. et al. (2018). Amorphous metal oxide blocking layers for highly efficient low-temperature brookite TiO_2-based perovskite solar cells. *ACS Appl. Mater. Interfaces* 10: 2224–2229.

12 Jeon, J., Kim, M., Seo, J. et al. (2021). Pseudo-halide anion engineering for α-$FAPbI_3$ perovskite solar cells. *Nature* 592 (7854): 381–385.

13 Arabpour Roghabadi, F., Alidaei, M., Mousavi, S.M. et al. (2019). Stability progress of perovskite solar cells dependent on the crystalline structure: from 3D ABX3 to 2D Ruddlesden–Popper perovskite absorbers. *J. Mater. Chem. A* 7: 5898–5933.

14 Niu, G., Li, W., Meng, F. et al. (2014). Study on the stability of $CH_3NH_3PbI_3$ films and the effect of post-modification by aluminum oxide in all-solid-state hybrid solar cells. *J. Mater. Chem. A* 2: 705–710.

15 Aristidou, N., Sanchez-Molina, I., Chotchuangchutchaval, T. et al. (2015). The role of oxygen in the degradation of methylammonium lead trihalide perovskite photoactive layers. *Angew. Chem. Int. Ed.* 54: 8208–8212.

16 Wang, Q., Chen, B., Liu, Y. et al. (2017). Scaling behavior of moisture-induced grain degradation in polycrystalline hybrid perovskite thin films. *Energy Environ. Sci.* 10: 516–522.

17 Goldschmidt, V.M. (1926). Die Gesetze der Krystallochemie. *Naturwissenschaften* 14: 477–485.

18 Acosta, M., Novak, N., Rojas, V. et al. (2017). $BaTiO_3$-based piezoelectrics: fundamentals, current status, and perspectives. *Appl. Phys. Rev.* 4: 041305.

19 Burns, G. and Dacol, F.H. (1982). Polarization in the cubic phase of $BaTIO_3$. *Solid State Commun.* 42: 9. https://doi.org/10.1016/0038-1098(82)91018-3.

20 Yu, C.-J. (2019). Advances in modelling and simulation of halide perovskites for solar cell applications. *J. Phys. Energy* 1: 022001.

21 Ji, D., Feng, S.Z., Wang, L. et al. (2019). Regulatory tolerance and octahedral factors by using vacancy in $APbI_3$ perovskites. *Vacuum* 164: 186–193.

22 Li, Z., Yang, M., Park, J.S. et al. (2016). Stabilizing perovskite structures by tuning tolerance factor: formation of formamidinium and cesium lead iodide solid-state alloys. *Chem. Mater.* 28: 284–292.

23 Kieslich, G., Sun, S., and Cheetham, A.K. (2015). An extended tolerance factor approach for organic–inorganic perovskites. *Chem. Sci.* 6: 3430–3433.

24 Ünlü, F., Jung, E., Haddad, J. et al. (2020). Understanding the interplay of stability and efficiency in A-site engineered lead halide perovskites. *APL Mater.* 8 (7): 070901.

25 Chang, Y.H., Park, C.H., and Matsuishi, K. (2004). First-principles study of the structural and the electronic properties of the lead-halide-based

inorganic–organic perovskites ($(CH_3NH_3)PbX_3$ and $CsPbX_3$ (X = Cl, Br, I). *J. Korean Phys. Soc.* 44: 889–893.

26 Umebayashi, T., Asai, K., Kondo, T., and Nakao, A. (2003). Electronic structures of lead iodide based low-dimensional crystals. *Phys. Rev. B – Condens. Matter Mater. Phys.* 67: 2–7.

27 Öz, S., Burschka, J., Jung, E. et al. (2018). Protic ionic liquid assisted solution processing of lead halide perovskites with water, alcohols and acetonitrile. *Nano Energy* 51: 632–638.

28 Singh, T. and Miyasaka, T. (2018). Stabilizing the efficiency beyond 20% with a mixed cation perovskite solar cell fabricated in ambient air under controlled humidity. *Adv. Energy Mater.* 8: 1–9.

29 Giorgi, G. and Yamashita, K. (2015). Zero-dipole molecular organic cations in mixed organic-inorganic halide perovskites: possible chemical solution for the reported anomalous hysteresis in the current-voltage curve measurements. *Nanotechnology* 26: 442001.

30 De Marco, N., Zhou, H., Chen, Q. et al. (2016). Guanidinium: a route to enhanced carrier lifetime and open-circuit voltage in hybrid perovskite solar cells. *Nano Lett.* 16: 1009–1016.

31 Kubicki, D.J., Prochowicz, D., Hofstetter, A. et al. (2018). Formation of stable mixed guanidinium-methylammonium phases with exceptionally long carrier lifetimes for high-efficiency lead iodide-based perovskite photovoltaics. *J. Am. Chem. Soc.* 140: 3345–3351.

32 Amat, A., Mosconi, E., Ronca, E. et al. (2014). Cation-induced band-gap tuning in organohalide perovskites: interplay of spin-orbit coupling and octahedra tilting. *Nano Lett.* 14: 3608–3616.

33 Jena, A.K., Kulkarni, A., and Miyasaka, T. (2019). Halide perovskite photovoltaics: background, status, and future prospects. *Chem. Rev.* 119: 3036–3103.

34 Manser, J.S., Christians, J.A., and Kamat, P.V. (2016). Intriguing optoelectronic properties of metal halide perovskites. *Chem. Rev.* 116: 12956–13008.

35 Stoumpos, C.C. and Kanatzidis, M.G. (2015). The renaissance of halide perovskites and their evolution as emerging semiconductors. *Acc. Chem. Res.* 48: 2791–2802.

36 Kojima, A., Teshima, K., Shirai, Y., and Miyasaka, T. (2009). Organometal halide perovskites as visible- light sensitizers for photovoltaic cells. *J. Am. Chem. Soc.* 131: 6050–6051.

37 Hariz, A. (2016). Perspectives on organolead halide perovskite photovoltaics. *J. Photonics Energy* 6: 032001.

38 Ivanovska, T., Dionigi, C., Mosconi, E. et al. (2017). Long-lived photoinduced polarons in organohalide perovskites. *J. Phys. Chem. Lett.* 8: 3081–3086.

39 Khenkin, M.V., Anoop, K.M., Visoly-Fisher, I. et al. (2018). Reconsidering figures of merit for performance and stability of perovskite photovoltaics. *Energy Environ. Sci.* 11: 739–743.

40 Yuan, Y., Xiao, Z., Yang, B., and Huang, J. (2014). Arising applications of ferroelectric materials in photovoltaic devices. *J. Mater. Chem. A* 2: 6027–6041.

41 Kim, J., Lee, S.H., Lee, J.H., and Hong, K.H. (2014). The role of intrinsic defects in methylammonium lead iodide perovskite. *J. Phys. Chem. Lett.* 5: 1312–1317.

42 Zhai, Y., Baniya, S., Zhang, C. et al. (2017). Giant Rashba splitting in 2D organic-inorganic halide perovskites measured by transient spectroscopies. *Sci. Adv.* 3: 1–7.

43 Kepenekian, M. and Even, J. (2017). Rashba and Dresselhaus couplings in halide perovskites: accomplishments and opportunities for spintronics and spin-orbitronics. *J. Phys. Chem. Lett.* 8: 3362–3370.

44 Chen, T., Chen, W.-L., Foley, B.J. et al. (2017). Origin of long lifetime of band-edge charge carriers in organic–inorganic lead iodide perovskites. *Proc. Natl. Acad. Sci.* 114: 7519–7524.

45 Kim, J., Lee, S.H., Chung, C.H., and Hong, K.H. (2016). Systematic analysis of the unique band gap modulation of mixed halide perovskites. *Phys. Chem. Chem. Phys.* 18: 4423–4428.

46 Kohn, W. and Sham, L.J. (1965). Self-consistent equations including exchange and correlation effects. *Physiol. Rev.* 140: A1133.

47 Perdew, J.P. (1985). Density functional theory and the band gap problem. *Int. J. Quantum Chem.* 28: 497–523.

48 Umari, P., Mosconi, E., and De Angelis, F. (2014). Relativistic GW calculations on $CH_3NH_3PbI_3$ and $CH_3NH_3SnI_3$ perovskites for solar cell applications. *Sci. Rep.* 4: 4467.

49 Motta, C., El-Mellouhi, F., Kais, S. et al. (2015). Revealing the role of organic cations in hybrid halide perovskite $CH_3NH_3PbI_3$. *Nat. Commun.* 6: 7026.

50 Butler, K.T., Walsh, A., Cheetham, A.K., and Kieslich, G. (2016). Organised chaos: entropy in hybrid inorganic-organic systems and other materials. *Chem. Sci.* 7: 6316–6324.

51 Brivio, F., Caetano, C., and Walsh, A. (2016). Thermodynamic origin of photoinstability in the $CH_3NH_3Pb(I_{1-x}Br_x)_3$ hybrid halide perovskite alloy. *J. Phys. Chem. Lett.* 7: 1083–1087.

52 Yang, W.S., Noh, J.H., Jeon, N.J. et al. (2015). High-performance photovoltaic perovskite layers fabricated through intramolecular exchange. *Science* 348 (6240): 1234–1237.

53 Koh, T.M., Fu, K., Fang, Y. et al. (2014). Formamidinium-containing metal-halide: an alternative material for near-IR absorption perovskite solar cells. *J. Phys. Chem. C* 118: 16458–16462.

54 Pellet, N., Gao, P., Gregori, G. et al. (2014). Mixed-organic-cation perovskite photovoltaics for enhanced solar-light harvesting. *Angew. Chem. Int. Ed.* 53: 3151–3157.

55 Zhang, Y., Grancini, G., Feng, Y. et al. (2017). Optimization of stable quasi-cubic $FA_xMA_{1-x}PbI_3$ perovskite structure for solar cells with efficiency beyond 20%. *ACS Energy Lett.* 2: 802–806.

56 Binek, A., Hanusch, F.C., Docampo, P., and Bein, T. (2015). Stabilization of the trigonal high-temperature phase of formamidinium lead iodide. *J. Phys. Chem. Lett.* 6: 1249–1253.

57 Ji, F., Wang, L., Pang, S. et al. (2016). A balanced cation exchange reaction toward highly uniform and pure phase FA1: XMA_xPbI_3 perovskite films. *J. Mater. Chem. A* 4: 14437–14443.

58 Prochowicz, D., Yadav, P., Saliba, M. et al. (2017). Mechanosynthesis of pure phase mixed-cation MA: $XFA_{1-x}PbI_3$ hybrid perovskites: photovoltaic performance and electrochemical properties. *Sustain. Energy Fuels* 1: 689–693.

59 Chen, J., Xu, J., Xiao, L. et al. (2017). Mixed-organic-cation $(FA)_x$ $(MA)_{1-x}PbI_3$ planar perovskite solar cells with 16.48% efficiency via a low-pressure vapor-assisted solution process. *ACS Appl. Mater. Interfaces* 9: 2449–2458.

60 Isikgor, F.H., Li, B., Zhu, H. et al. (2016). High performance planar perovskite solar cells with a perovskite of mixed organic cations and mixed halides, $MA_{1-x}FA_xPbI_{3-y}Cl_y$. *J. Mater. Chem. A* 4: 12543–12553.

61 Lee, J.W., Kim, D.-H., Kim, H.-S. et al. (2015). Formamidinium and cesium hybridization for photo- and moisture-stable perovskite solar cell. *Adv. Energy Mater.* 5: 1501310.

62 Niu, G., Li, W., Li, J. et al. (2017). Enhancement of thermal stability for perovskite solar cells through cesium doping. *RSC Adv.* 7: 17473–17479.

63 Choi, H., Jeong, J., Kim, H. et al. (2014). Cesium-doped methylammonium lead iodide perovskite light absorber for hybrid solar cells. *Nano Energy* 7: 80–85.

64 Jeon, N.J., Noh, J.H., Yang, W.S. et al. (2015). Compositional engineering of perovskite materials for highperformance solar cells. *Nature* 517: 476–480.

65 Park, Y.H., Jeong, I., Bae, S. et al. (2017). Inorganic rubidium cation as an enhancer for photovoltaic performance and moisture stability of $HC(NH_2)_2PbI_3$ perovskite solar cells. *Adv. Funct. Mater.* 27: 21–27.

66 Kubicki, D.J., Prochowicz, D., Hofstetter, A. et al. (2017). Phase segregation in Cs-, Rb- and K-doped mixed-cation $(MA)_x(FA)_{1-x}PbI_3$ hybrid perovskites from solid-state NMR. *J. Am. Chem. Soc.* 139: 14173–14180.

67 Szafrański, M. (1997). Investigation of phase instabilities in guanidinium halogenoplumbates(II). *Thermochim. Acta* 307: 177–183.

68 Szafrański, M. and Katrusiak, A. (2000). Phase transitions in the layered structure of diguanidinium tetraiodoplumbate. *Phys. Rev. B: Condens. Matter Mater. Phys.* 61: 1026–1035.

69 Jodlowski, A.D., Roldán-Carmona, C., Grancini, G. et al. (2017). Large guanidinium cation mixed with methylammonium in lead iodide perovskites for 19% efficient solar cells. *Nat. Energy* 2: 972–979.

70 Hou, X., Hu, Y., Liu, H. et al. (2017). Effect of guanidinium on mesoscopic perovskite solar cells. *J. Mater. Chem. A* 5: 73–78.

71 Jeong, M., Choi, I.W., and Go, E.M. (2020). Stable perovskite solar cells with efficiency exceeding 24.8% and 0.3-V voltage loss. *Science (80-.)* 369: 1615–1620.

72 Wang, Y., Liu, X., and Zhang, T. (2019). The role of dimethylammonium iodide in $CsPbI_3$ perovskite fabrication: additive or dopant? *Angew. Chem. Int. Ed.* 58: 16691–16696.

73 Yoo, J.J., Wieghold, S., Sponseller, M.C. et al. (2019). An interface stabilized perovskite solar cell with high stabilized efficiency and low voltage loss. *Energy Environ. Sci.* 12: 2192–2199.

74 Jung, E.H., Jeon, N.J., Park, E.Y. et al. (2019). Efficient, stable and scalable perovskite solar cells using poly(3-hexylthiophene). *Nature* 567: 511–515.

75 Li, N., Liu, J., and Li, C. (2020). Zwitterion-stabilizing scalable bladed α-phase $Cs_{0.1}FA_{0.9}PbI_3$ films for efficient inverted planar perovskite solar cells. *ACS Sustainable Chem. Eng.* 8: 7020–7030.

76 Jung, E., Budzinauskas, K., Öz, S. et al. (2020). Femto-to microsecond dynamics of excited electrons in a quadruple cation perovskite. *ACS Energy Lett.*: 785–792. https://doi.org/10.1021/acsenergylett.9b02684.

77 Zhang, W., Xiong, J., Li, J., and Daoud, W.A. (2019). Guanidinium induced phase separated perovskite layer for efficient and highly stable solar cells. *J. Mater. Chem. A* 7: 9486–9496.

78 Shockley, W. and Queisser, H.J. (1961). Detailed balance limit of efficiency of p–n junction solar cells. *J. Appl. Phys.* 32: 510–519.

79 Rühle, S. (2016). Tabulated values of the Shockley–Queisser limit for single junction solar cells. *Sol. Energy* 130: 139–147.

80 Saliba, M., Matsui, T., Domanski, K. et al. (2016). Incorporation of rubidium cations into perovskite solar cells improves photovoltaic performance. *Science (80-.)* 354.

81 Green, M.A. (2012). Radiative efficiency of state-of-the-art photovoltaic cells. *Prog. Photovoltaics Res. Appl.* 20: 472–476.

82 Niu, G., Guo, X., and Wang, L. (2015). Review of recent progress in chemical stability of perovskite solar cells. *J. Mater. Chem. A* 3: 8970–8980.

83 Manser, J.S., Saidaminov, M.I., Christians, J.A. et al. (2016). Making and breaking of lead halide perovskites. *Acc. Chem. Res.* 49: 330–338.

84 Juarez-Perez, E.J., Ono, L.K., Maeda, M. et al. (2018). Photodecomposition and thermal decomposition in methylammonium halide lead perovskites and inferred design principles to increase photovoltaic device stability. *J. Mater. Chem. A* 6: 9604–9612.

85 Aristidou, N., Eames, C., Sanchez-Molina, I. et al. (2017). Fast oxygen diffusion and iodide defects mediate oxygen-induced degradation of perovskite solar cells. *Nat. Commun.* 8: 1–10.

86 Xiang, Y., Zhang, F., He, J. et al. (2018). Light-current-induced acceleration of degradation of methylammonium lead iodide perovskite solar cells. *J. Power Sources* 384: 303–311.

87 Thrithamarassery Gangadharan, D. and Ma, D. (2019). Searching for stability at lower dimensions: current trends and future prospects of layered perovskite solar cells. *Energy Environ. Sci.* 12: 2860–2889.

88 Mitzi, D. and Synthesis, B. (1999). Structure , and properties of organic-inorganic perovskites and related materials. *Prog. Inorg. Chem.* 48: 1–121.

89 Tsai, H. and Nie, W. (2016). High-efficiency two-dimensional Ruddlesden–Popper perovskite solar cells. *Nature* 536: 312–316.

90 Smith, I.C., Hoke, E.T., Solis-Ibarra, D. et al. (2014). A layered hybrid perovskite solar-cell absorber with enhanced moisture stability. *Angew. Chem. Int. Ed.* 53: 11232–11235.

91 Leveillee, J., Katan, C., Zhou, L. et al. (2018). Influence of π-conjugated cations and halogen substitution on the optoelectronic and excitonic properties of layered hybrid perovskites. *Phys. Rev. Mater.* 2: 1–12.

92 Passarelli, J.V., Fairfield, D.J., Sather, N.A. et al. (2018). Enhanced out-of-plane conductivity and photovoltaic performance in $n = 1$ layered perovskites through organic cation design. *J. Am. Chem. Soc.* 140: 7313–7323.

93 Kikuchi, K., Takeoka, Y., Rikukawa, M., and Sanui, K. (2005). Fabrication and characterization of organic-inorganic perovskite films containing fullerene derivatives. *Colloids Surfaces A Physicochem. Eng. Asp.* 257–258: 199–202.

94 Grancini, G., Roldán-Carmona, C., Zimmermann, I. et al. (2017). One-year stable perovskite solar cells by 2D/3D interface engineering. *Nat. Commun.* 8: 1–8.

95 Wang, Z., Lin, Q., Chmiel, F.P. et al. (2017). Efficient ambient-air-stable solar cells with 2D-3D heterostructured butylammonium-caesium-formamidinium lead halide perovskites. *Nat. Energy* 2: 1–10.

96 Poglitsch, A. and Weber, D. (1987). Dynamic disorder in methylammoniumtri-halogenoplumbates (II) observed by millimeter-wave spectroscopy. *J. Chem. Phys.* 87: 6373–6378.

97 Guo, Y., Yaffe, O., Paley, D.W. et al. (2017). Interplay between organic cations and inorganic framework and incommensurability in hybrid lead-halide perovskite $CH_3NH_3PbBr_3$. *Phys. Rev. Mater.* 1: 1–6.

98 Kieslich, G., Skelton, J.M., Armstrong, J. et al. (2018). Hydrogen bonding versus entropy: revealing the underlying thermodynamics of the hybrid organic-inorganic perovskite $[CH_3NH_3]PbBr_3$. *Chem. Mater.* 30: 8782–8788.

99 Pham, H.T., Duong, T., Rickard, W.D.A. et al. (2019). Understanding the chemical and structural properties of multiple-cation mixed halide perovskite. *J. Phys. Chem. C* 123: 26718–26726.

100 Hu, Y., Hutter, E.M., Rieder, P. et al. (2018). Understanding the role of cesium and rubidium additives in perovskite solar cells: trap states, charge transport, and recombination. *Adv. Energy Mater.* 8: 1–11.

101 Zhou, G. et al. (2018). Application of cesium on the restriction of precursor crystallization for highly reproducible perovskite solar cells exceeding 20% efficiency. *ACS Appl. Mater. Interfaces* 10: 9503–9513.

102 Habisreutinger, S.N. et al. (2014). Carbon nanotube/polymer composites as a highly stable hole collection layer in perovskite solar cells. *Nano Lett.* 14: 5561–5568.

103 Hou, Y., Zhou, Z.R., Wen, T.Y. et al. (2019). Enhanced moisture stability of metal halide perovskite solar cells based on sulfur-oleylamine surface modification. *Nanoscale Horizons* 4: 208–213.

104 Chaudhary, B., Kulkarni, A., Jena, A.K. et al. (2017). Poly(4-vinylpyridine)-based interfacial passivation to enhance voltage and moisture stability of lead halide perovskite solar cells. *ChemSusChem* 10: 2473–2479.

105 Cao, J., Yin, J., Yuan, S. et al. (2015). Thiols as interfacial modifiers to enhance the performance and stability of perovskite solar cells. *Nanoscale* 7: 9443–9447.

106 Abdelmageed, G., Sully, H.R., Bonabi Naghadeh, S. et al. (2018). Improved stability of organometal halide perovskite films and solar cells toward humidity via surface passivation with oleic acid. *ACS Appl. Energy Mater.* 1: 387–392.

107 Swarnkar, A., Chulliyil, R., Ravi, V.K. et al. (2015). Colloidal $CsPbBr_3$ perovskite nanocrystals: luminescence beyond traditional quantum dots. *Angew. Chem. Int. Ed.* 54: 15424–15428.

108 Wang, Y., Dar, M.I., Ono, L.K. et al. (2019). Thermodynamically stabilized β-$CsPbI_3$–based perovskite solar cells with efficiencies >18%. *Science (80-.)* 365: 591–595.

109 Hao, F., Stoumpos, C.C., Cao, D.H. et al. (2014). Lead-free solid-state organic-inorganic halide perovskite solar cells. *Nat. Photonics* 8: 489–494.

110 Li, X., Wu, Y., Zhang, S. et al. (2016). $CsPbX_3$ quantum dots for lighting and displays: roomerature synthesis, photoluminescence superiorities, underlying origins and white light-emitting diodes. *Adv. Funct. Mater.* 26: 2435–2445.

111 Boix, P.P., Nonomura, K., Mathews, N., and Mhaisalkar, S.G. (2014). Current progress and future perspectives for organic/inorganic perovskite solar cells. *Mater. Today* 17: 16–23.

2

Flash Infrared Annealing for Processing of Perovskite Solar Cells

Sandy Sánchez and Anders Hagfeldt

Institut des sciences et ingénierie chimiques (ISIC), Laboratory of Photomolecular Science (LSPM), École Polytechnique Fédérale de Lausanne (EPFL), Rue A. Argand 6, 1015 Lausanne, Switzerland

2.1 Introduction

Many research groups are putting much effort into achieve the commercialization of the perovskite solar cells (PSCs). Scaling these up for industrial production is a complex task, mainly due to the limited number of tests and knowledge about material stability. All perovskite chemical compositions are water soluble, meaning that in normal environmental conditions (more than 25% of relative humidity), these lead halide perovskites tend to degrade [1–5]. Numerous approaches have been proposed to overcome this problem, and different strategies can be divided into essentially three categories: one-chemical passivation, two-cation addition/replacement, and three-new synthesis methods.

Synthesis methods for scaling up the current PSC preparation procedures are neither practical nor environmentally friendly. Indeed, the "antisolvent method" is one of the usually adopted and most effective lab-scale methods to induce perovskite crystallization, requiring a relatively large amount of solvent (more than 50 $\mu l\,cm^{-2}$) that is typically not reclaimed [6]. This method is effective because it enables a dense and uniform coverage of perovskite films on suitable substrates, which is a key requirement to achieve highly efficient PSCs [6]. It works on small areas (several square centimeters, i.e. for lab-scale devices) but is difficult to apply on larger surfaces, causing practical issues for PSC's scale-up and commercialization.

In searching for alternatives in scaling up PSCs, this chapter introduces a new synthesis method called flash infrared annealing (FIRA) [7–9]. This method is based on thermal rapid processing, which has been employed to synthesize thin films for semiconductor applications, and allows preparing lead iodide PSCs with a record stabilized power conversion efficiency (PCE) of 20%. An irradiation time of fewer than two seconds, enables the coating of glass and plastic substrates with pinhole-free perovskite films that exhibit micrometer-sized crystalline domains. Here, the main concept is to quickly heat the material to control the crystallization kinetics without thermal degradation. A setup has been developed and upgraded,

Perovskite Solar Cells: Materials, Processes, and Devices, First Edition.
Edited by Shahzada Ahmad, Samrana Kazim, and Michael Grätzel.

containing a camera, a high-intensity IR emitter, and a cooling system. The emitter can increase the temperature up to 1000 °C in a few seconds, and the cooling system dissipates quickly the heat within the chamber. FIRA setup has been used to synthesize highly crystalline organic–inorganic perovskite films and transparent inorganic conductive layers.

In making FIRA effective, it is necessary to understand the physicochemical phenomena that take place during the crystallization of the synthesized film. From FIRA optimization parameters to experimental approaches, this chapter addresses how to quickly produce defect-tolerant perovskite thin films that enable the manufacture of highly efficient solar cells. The kinetics of the nucleation and crystal growth process is also addressed, from the fundamental studies of particular FIRA cases, disclosing the mechanism involved in driving film crystallization. These fundamental studies aim to use FIRA for manufacturing stable PSCs, increase their PCE, and enable the scale-up of this technology.

2.2 Perovskite Crystal Nucleation and Growth from Solution

2.2.1 The Antisolvent Dripping Method

The solution processability of perovskite materials is one of the most important advantages of PSCs. A one-step spin-coating method is most often used owing to its simplicity and low cost. During spin coating and annealing, solvent evaporation induces crystallization of the perovskite films [10]. The humidity plays a fundamental role in the preparation of the solution. It is well-known that perovskite $MAPbI_3$ films are sensitive to moisture, causing $MAPbI_3$ to decompose into PbI_2 [11, 12]. Therefore, the manufacture of PSCs should be carried out in controlled atmosphere conditions with low humidity values. To perform the antisolvent (AS) precipitation procedure, it is necessary to deposit a thin wet yellow film, e.g. $MAI:PbI_2$ (1 : 1) from solution in an aprotic solvent. Spin coating can be used as deposition method and then spread a so-called antisolvent, typically a nonpolar solvent such as chlorobenzene, on the newly formed film to immediately form a dark brown/red perovskite layer [13]. This implies that the antisolvent needs to be deposited at a certain point (and quite quickly) during spinning, which is difficult to reproduce. To complete the perovskite crystallization and remove the excess of the primary solution solvent, an annealing treatment is needed, typically below 100 °C.

2.2.2 Thermodynamics of Nucleation and Crystal Growth

A nucleus can be defined as the minimum amount of a new phase, which can be a cluster of a few atoms or molecules. An initially metastable phase of nuclei is called nucleation, which is a major mechanism of first-order phase transition. Then, this is followed by the growth stage that is governed by the diffusion of particles, called the growth units, to the surface of the existing nuclei and their incorporation in the structure of the crystal lattice. Almost all the nucleation processes are heterogeneous and may depend on the nucleating agent involved. Homogeneous nucleation involves only the material that crystallizes; it is therefore intrinsic to the material [14]. Its occurring conditions represent a limit for the stability of the amorphous phase [15, 16]. This can be analyzed more readily than heterogeneous nucleation that involves a foreign, often unknown, material.

Considering homogeneous nucleation for solution-processed perovskite materials, the perovskite layer is typically deposited by a solution coating method [17]. During the coating process, a so-called antisolvent is dropped on top of the layer for the perovskite crystal growth. Additional thermal annealing is necessary to remove the remaining solvent and reaches a purer perovskite phase. As the solvent evaporates, the chemical potential of the solution changes, eventually reaching supersaturation, allowing the formation of a new phase by crystal growth [18]. Once the supersaturation is reached, nucleation clusters appear, followed by the growth of these crystals. The nucleation process has Gibbs energetic barriers that need to be surpassed. These barriers are a function of the critical nucleus size [19–21].

The relationship between supersaturation and critical nucleus size inducing crystal growth can be described in terms of Gibbs free energy of the system [16]. Assume that spherical nucleus cluster is formed on a substrate surface, with r being the nucleation radius, θ the contact angle the nucleus forms with the substrate, ΔG_v the free energy per unit volume, σ_0 the nucleus surface tension, σ_1 the nucleus–substrate interfacial tension, and σ_2 the substrate surface tension. In equilibrium, Young's equation is given by

$$\sigma_0 \cos\theta = \sigma_2 - \sigma_1 \tag{2.1}$$

The change in surface and interfacial free energy can be expressed as

$$\Delta G_s = 2\pi r^2(1-\cos\theta) + \pi(r\sin\theta)^2(\sigma_1-\sigma_2) = \pi r^2\sigma_0(2-3\cos\theta+\cos^3\theta) \tag{2.2}$$

For the spherical cap nucleus, its volume V is

$$V = \frac{\pi r^3(2-3\cos\theta+\cos^3\theta)}{3} \tag{2.3}$$

Therefore, the change of the volume free energy is

$$\Delta G_V = V \cdot \Delta G_V = \frac{\pi r^3(2-3\cos\theta+\cos^3\theta)}{3}\Delta G_V \tag{2.4}$$

and the change of total free energy is

$$\Delta G = \Delta G_s + \Delta G_V = \pi\left(\frac{r^3}{3}\Delta G_V + r^2\sigma_0\right)(2-3\cos\theta+\cos^3\theta) \tag{2.5}$$

Setting

$$\frac{\partial(\Delta G)}{\partial r} = 0 \tag{2.6}$$

the critical nucleus radius is given by

$$r_0 = \frac{-2\sigma_0}{\Delta G_V}\ (\Delta G_V < 0) \tag{2.7}$$

Since ΔG_v is the energy needed from equilibrium state to supersaturation, it can be expressed as

$$\Delta G_V = -Z(T) \cdot \ln\left(\frac{\rho_s}{\rho_e}\right) \tag{2.8}$$

where *Z(T)* is a temperature and growth-type-dependent coefficient (shape of the stable clusters), ρ_s is the concentration of supersaturation, and ρ_e is the equilibrium concentration [16, 22]. The formation of stable nuclei only happens if the increase in the interfacial energy resulting from the formation of a solid–liquid interface is offset by a decrease in energy released by the formation of a (thermodynamically favored) solid. At low supersaturations, the energy required to form a solid–liquid interface is bigger than the energy released upon formation of a cluster, which is consequently unstable, resulting in a very low nucleation rate. Supersaturation can be defined by the relation of concentration of solution with respect to the equilibrium concentration at the same temperature:

$$\xi = \frac{\rho_s}{\rho_e} - 1 \tag{2.9}$$

and the free energy needed for the equilibrium state ΔG_v is given by

$$\Delta G_V = Z(T) \cdot \ln(\xi + 1) \tag{2.10}$$

With this, the critical nucleus radius can be expressed as

$$r_0 = \frac{2\sigma_0}{Z(T)\ln(\xi + 1)} \tag{2.11}$$

The antisolvent method initiates the crystallization of the perovskite solution by changing its solubility, which is one way to reach supersaturation. When this occurs, the balance of chemical potentials can surpass the nucleation barrier for certain nuclei sizes [20, 23, 24]. The free energy changes are associated with the formation of the nuclei and as a consequence with the formation of the new phase. At the surface, the solution molecules interact differently with their neighbors compared with those in bulk, and their contribution to the free energy of the new phase differs and is often greater. The difference between the free energy per molecule in bulk and at a surface gives rise to the interfacial free energy [20].

A critical nucleus size is required for a nucleus to form from a supersaturation solution, which has a number of implications, as previously discussed. The critical size of the new phase is reached as a result of fluctuations that bring together sufficient numbers of molecules. Therefore, the probability of nucleation is strongly affected by the value of the critical size. The nucleation can be controlled by modulating the critical size, which is a function of the interfacial energy [20]. The smaller the interfacial energy, the smaller the critical size, and the more likely the nucleation becomes for any given supersaturation. As a consequence, by varying either the critical nucleus size, the probability of nucleation can be manipulated [25, 26].

2.2.3 Kinetic Process for Rapid Thermal Growth

The driving force for the crystallization process depends on the supersaturation that is fundamental to the crystal growth rate [20, 24, 27]. Crystal growth kinetics is a complex phenomenon, which usually takes place through an intermediate phase. Overall, fast nucleation can improve the coverage of films, and the control of the crystal growth can contribute to form a high-crystalline material [28–31]. To grow a crystal, a cluster has to be transported to the crystal surface and may need to rearrange to fit the lattice structure. It can proceed in a successive transport through solution and surface reaction processes. Both these processes can be growth rate controlling. The cluster transport can be controlled by diffusion, which is a transport of matter as a consequence of the thermal motion of the molecules and by convection, which is mass movement due to an energetic gradient in the system. Surface-controlled growth can be determined by nucleation and dislocation growth [14, 32, 33].

In general, short-time annealing at high temperature can produce an energy process equivalent to the longer annealing at a lower temperature. However, different activation energies or driving forces are necessary to reach the supersaturated state in both cases. Therefore, they depend differently on the temperature. For rapid thermal annealing at high temperatures, the controlled growth rate depends mainly on the thermal convention. The final crystalline morphology can be molded by the temperature gradient of the thermal process. The crystalline micro-arrangement will depend on the interfacial energy and different chemical potential of the molecules (i.e. crystal growth from solution), where instabilities can provide a different directional growth, led by their anisotropy. This is the case of dendritic growth, where a metastable phase can grow in an energetically favorable direction, different from the final phase.

2.3 Rapid Thermal Annealing

2.3.1 The FIRA Method

Rapid thermal processes are commonly used in the semiconductor community, mostly for conventional silicon solar cells [34–36]. This method is used to grow pure crystal phases without losing material quality, for example, to manufacture pure silicon monocrystalline material. Large open-circuit voltage improvements by rapid thermal annealing have been reported in addition to defect removal and dopant activation [35]. Silicon solar cells with efficiencies of 17.1% have been achieved using low-cost rapid thermal processing in conjunction with surface passivation [28]. David Payne and colleagues showed that the degradation of multicrystalline silicon solar cells can be reduced by 80% by using a 10 seconds annealing treatment at 200 °C by illumination with a high intensity of 44.8 kW m^{-2} [36]. In PSCs, rapid thermal process synthesis has already been used as a cost-effective route to improve some of the structural properties in perovskite films [37–40]. Troughton et al. have proposed a short exposure to high-intensity near-infrared radiation to crystallize perovskite films and thus to potentially prepare high-efficiency and large-area PSCs [38].

The last version of the FIRA oven is composed of an array of near-infrared halogen lamps with an illumination power of 3000 kW m^{-2} [9, 41]. A hollow aluminum body

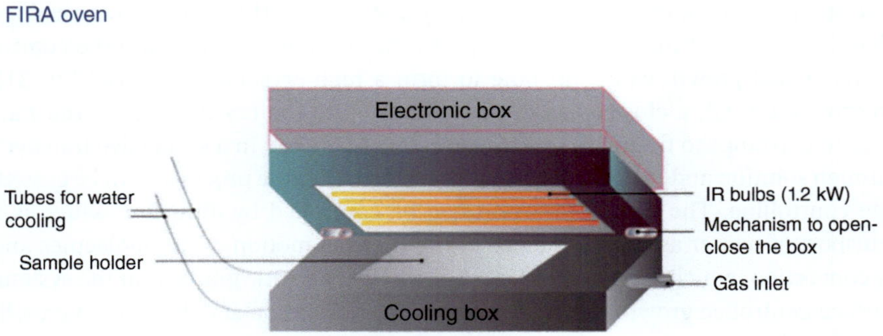

Figure 2.1 Schematic representation of the flash annealing oven used for the first application of the FIRA method (currently, FIRA oven has 1.9 kW of IR power output).

provides an effective water-cooling system, as shown in Figure 2.1 (see technical details in Section 10.3). In a typical experiment, a perovskite precursor solution is spin-cast onto an FTO substrate, which is then transferred into the FIRA oven. This FTO substrate absorbs the IR light and quickly heats up the precursor solution film [42]. The absorption spectra of FTO and ITO partially overlap with the spectral radiance of the halogen lamps [43–46].

During this process, rapid crystallization results in micrometer-sized crystal grains arranged in a dense perovskite film. Note that FIRA does not significantly heat the perovskite film and is thus compatible with low-temperature processing. As shown in Figure 2.2, the FTO substrate absorbs the IR light and quickly heats up the perovskite solution layer. For perovskite layers on a glass substrate, also shown in Figure 2.2, black perovskite phase is not formed under IR irradiation. Because none of the components absorb the IR light, the solution film remains at low temperature and therefore does not crystallize.

During FIRA, crystal nucleation and growth can be controlled by three distinct parameters, the chamber temperature, the pulse intensity, and the pulse duration. Once nucleation has occurred, it is followed by the growth of a crystalline phase.

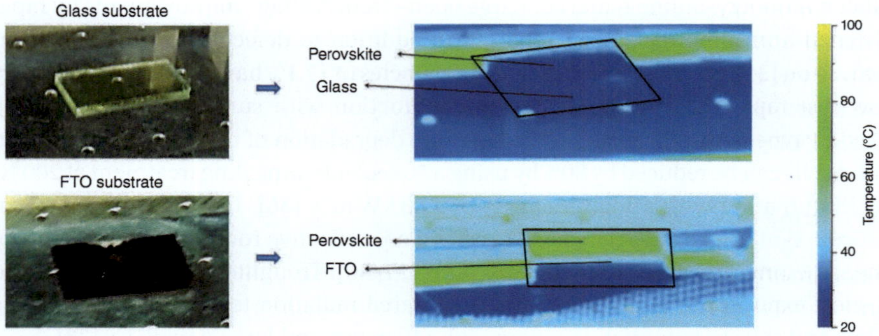

Figure 2.2 Cross-sectional temperature distributions in a perovskite film on glass and on FTO-covered glass substrates (two seconds FIRA pulse), measured by an IR camera.

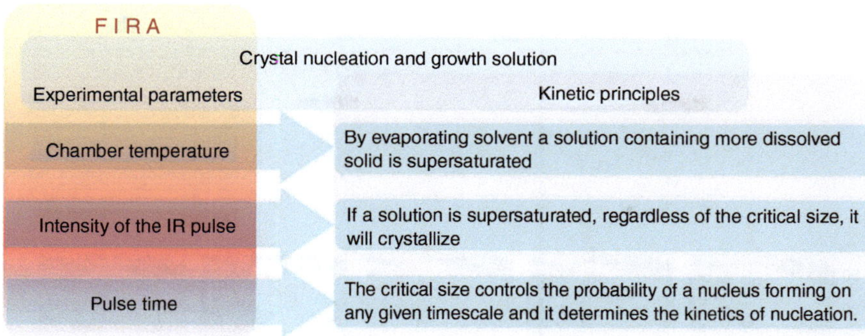

Figure 2.3 Schematic of the kinetic process in the nucleation and growth of a crystalline perovskite film from a solution and its relationship with the FIRA working parameters.

Moreover, the perovskite crystal formation can involve numerous intermediate phases [29]. The intense FIRA flash causes the rapid evaporation of the solvent and provides the driving force required for crystal nucleation and growth. Here, molecular diffusion gradients play a fundamental role on how the system evolves far from thermodynamic equilibrium through intermediate or metastable phases [32]. If sufficiently low, diffusion kinetically limits crystal growth. In conclusion, the temperature, the speed of the heating, and the heating time control how supersaturation is reached. As shown in Figure 2.3, supersaturation can be reached by controlling these main parameters.

To elucidate the perovskite crystallization dynamics, an in situ experiment was performed using $MAPbI_3$ spin-coated solutions [47]. Here, FIRA power was only 10% of its maximum intensity to slow down and thus give the opportunity to study the process in more detail at a longer time frame of 10 minutes of seconds instead of <2 seconds. Crystals start to grow from nucleation centers, going through intermediate phases along energetically favorable crystallographic directions in dendritic growth, as shown in Figure 2.4a [47–49]. The increase in crystal coverage can be seen by the change in sample color in Figure 2.4b, from yellow to brown, which is also observed in the transmittance spectra. Initially, the transmittance varies almost linearly from 450 to 650 nm, which is characteristic of the liquid phase of the film. With increasing IR pulse time, a distinct absorption band spanning most of the visible spectral range develops. Note that the film does not reach its pure black phase in this experiment, indicating a substantial presence of unreacted material within the film.

2.3.2 FIRA and Antisolvent

To examine the effects of antisolvents on the perovskite crystal size, other influential parameters were fixed: a 1.7 seconds FIRA pulse time, the number of seconds into the spin-coating program at which antisolvent was deposited, and the stoichiometric composition of the prepared perovskite solution. Using conventional antisolvent method without FIRA creates a homogenous perovskite crystal morphology, as

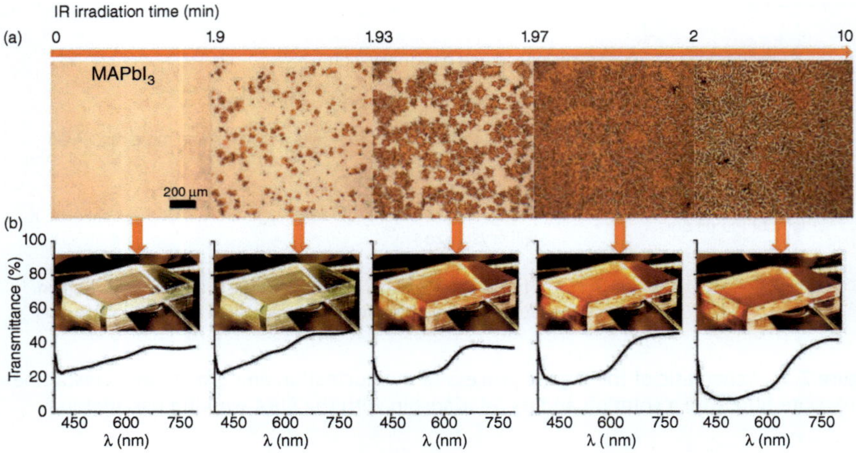

Figure 2.4 (a) Optical transmittance micrographs of in situ perovskite crystallization from solution during low-intensity FIRA annealing (10% of the maximum intensity). (b) Transmittance spectra and images of the samples. Source: Reproduced with permission from Sanchez et al. [47], © (2020) American Chemical Society.

shown in Figure 2.5a. The antisolvent induces a high nucleation density by the rapid reduction in solubility of the perovskite precursors. Followed by crystal growth, this yields relatively small crystal grain sizes due to the abundance of nucleation. Crystal nucleation from solution occurs by supersaturating the solution. FIRA achieves the same effect by irradiating a high amount of energy in a short IR light pulse, which causes the solvent to evaporate without degrading the perovskite.

Due to the high-intensity heat input, FIRA generates larger grain sizes compared with the antisolvent method, as seen in Figure 2.5a. Combining these two methods induces larger grain sizes than the conventional antisolvent method, but these are smaller than FIRA alone. The polarity of the antisolvents is essential to tune the grain size of the perovskite films. The large crystal domains of FIRA-annealed films have an internal structure, as shown in the AFM image found in Figure 2.5b. Section 2.4.4 shows specific features of the internal structure of the large crystal domains.

2.3.3 Perovskite Film Crystallization for a Single IR Pulse

Single-pulse FIRA refers to the application of a single irradiation pulse onto a sample placed inside a reflective IR chamber. This chamber is an aluminum body, which has been constructed to irradiate perovskite-coated electrodes under well-defined parameters. Using this method, perovskite film crystallization is completed within two seconds followed by eight seconds in the dark in the FIRA oven to completely remove the solvent from the film.

Crystallization from a solution typically occurs by self-seeding. Figure 2.6a schematically shows a typical concentration profile (LaMer curve), where the solution is concentrated above the solubility limit C_s [7]. Self-seeding is induced

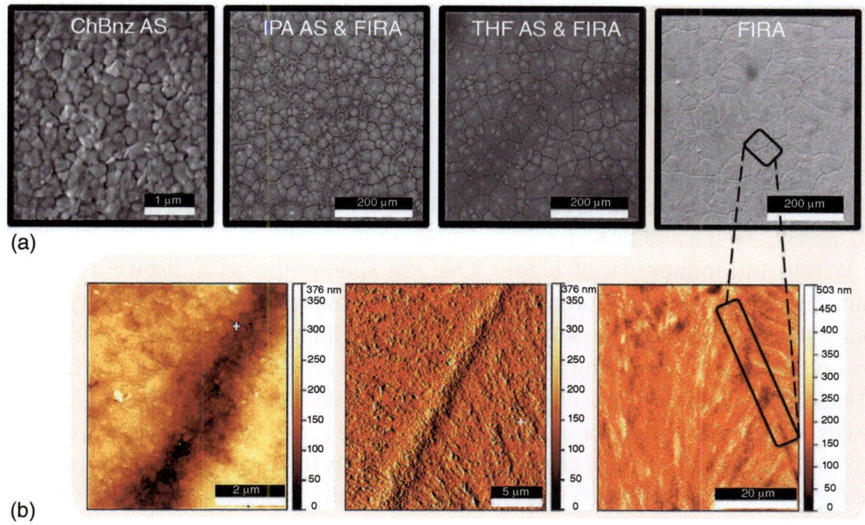

Figure 2.5 SEM images. (a) Conventional antisolvent (AS) method combined with FIRA annealing for several antisolvents such chlorobenzene, isopropyl alcohol (IPA), and tetrahydrofuran (THF). (b) Atomic force microscopy (AFM) images of the FIRA-annealed film. All films have a perovskite chemical composition of $MAPbI_3$.

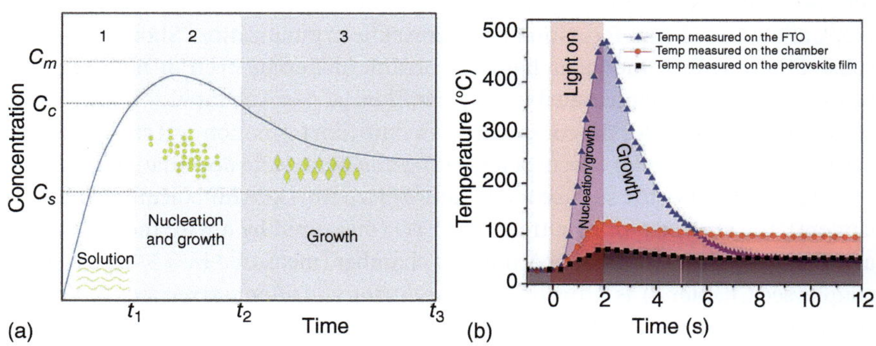

Figure 2.6 (a) Schematic representation of the concentration variation during crystal nucleation and growth. (b) Temperature measured in the chamber and on the FTO surface with thermocouples and surface temperature of the perovskite film measured by an IR camera. Source: Adapted with permission from Sanchez et al. [7], © (2018) WILEY-VCH Verlag GmbH & Co. KGaA, Weinheim.

upon surpassing a certain critical concentration value C_c, which should lie below the maximal supersaturation C_m, above which the substance (e.g. the perovskite) precipitates uncontrollably. In stage (i) delimited by time t_1, the solution concentration is increased by the removal of solvent. This is followed by stage (ii) until time t_2 where crystal nuclei form by self-seeding. Crystal nucleation and growth remove the crystallizing substance from the solution, thereby reducing the concentration. This in turn reduces the nucleation of further crystals. The interplay of further solvent removal and crystal growth often results in a steady-state concentration that lies between C_s and C_c, as indicated in stage (iii), from time t_2 to t_3 [50, 51]. The nucleation rate is predominantly influenced by the depth of the quench in stage 2), e.g. the maximum concentration reached in Figure 2.6a. High nucleation rates arise from supersaturation, which is caused by a rapid variation of the concentration [52–54]. In the antisolvent method, rather than removing the solvent as described above, high supersaturation-induced nucleation is caused through a change in C_s, C_c, and C_m by the addition of a poor solvent for the perovskite.

Within the framework of Figure 2.6a, the state-of-the-art antisolvent method induces a high nucleation density by the abrupt reduction of the solubility of the perovskite precursors in the solution, resulting in a high nucleation density followed by crystal growth, providing a laterally continuous perovskite crystal morphology [19, 55, 56]. In a more general approach, crystal nucleation from the solution proceeds by supersaturating the solution, which can be achieved by solvent evaporation [57–59]. Yi-Bing Cheng and coworkers [23] illustrate that through the manipulation of the solution concentration and degree of supersaturation, it is possible to vary the spacing between nucleation sites from a few hundred nanometers to a few hundred microns [23, 34, 42, 44, 49, 60]. They noted that the perovskite microstructure growth by heterogeneous nucleation from solution is unlikely to form a single crystal.

Concentrating the perovskite solution through solvent evaporation by heating, however, faces the challenge of maintaining the temperature within the perovskite film below its degradation temperature while preventing the dewetting of the film. FIRA allows rapid film formation and perovskite crystallization. Short heat pulses substantially reduce the degradation of organic materials even at high temperatures, as has been demonstrated in polymer films [1, 61]. The rapid solvent removal afforded by FIRA allows the control of perovskite film formation and crystallization, avoiding the detrimental effects of dewetting and perovskite degradation.

Figure 2.6b shows the surface temperature of a perovskite film (acquired by an IR camera), the surface temperature of the FTO (measured by a wire thermocouple), and the temperature of the FIRA internal chamber (measured by a K-type thermocouple; see Chapter 4) as a function of time during and after two seconds of IR flash. In particular, the perovskite surface temperature indicates a crystallization process at quite low temperatures (70 °C peak). This is advantageous and consistent with recent studies by Nayak et al. [62] and Dualeh et al. [63], showing the relationship between low-temperature perovskite crystal growth (below 100 °C) and improvement of the structural and optoelectronic properties of PSCs. Nevertheless, with the FTO layer and a high IR absorber, the temperature at its surface reaches 480 °C after two seconds of light exposure, confirming the high energy that FIRA can achieve in a short IR pulse.

2.3.4 Perovskite Crystallization with Pulse Duration

As described above, the perovskite pulse length is likely to control the nucleation density. The energy deposited in the film must be sufficiently large to allow the complete crystallization of the film (by crystal growth) without inducing parasitic effects

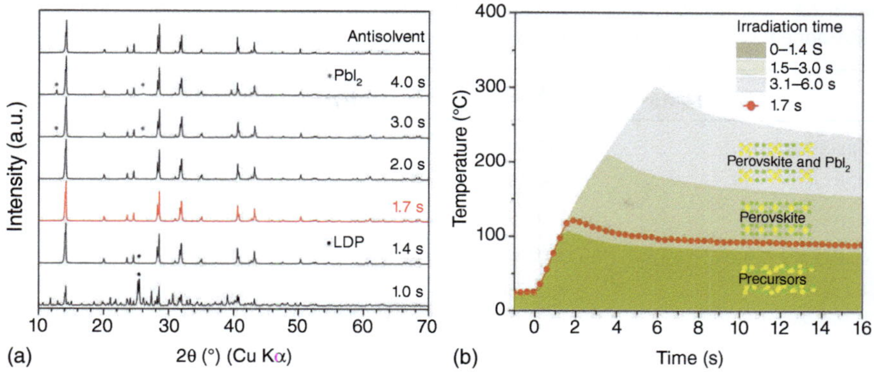

Figure 2.7 IR annealing time and its relation to the perovskite crystal formation. (a) XRD pattern of selected annealed films, where the black circles and asterisks mark the main reflections of the lower-dimensional perovskite (LDP) and PbI_2, respectively. (b) Temperature profiles for different FIRA-annealing pulse lengths. Source: Adapted with permission from Sanchez et al. [7], © (2018) WILEY-VCH Verlag GmbH & Co. KGaA, Weinheim.

(perovskite degradation) [62, 64]. Here, the role of the FIRA flash time was studied, as well as the way it controls the film morphology and crystallinity, keeping all other experimental parameters constant. The temperature in the chamber was used to measure different annealing , and the samples were kept in the FIRA chamber for eight additional seconds after the IR irradiation, as previously stated.

X-ray diffraction (XRD) shown in Figure 2.7a was employed to investigate the crystal structure in the films as function of the FIRA pulse duration. Their patterns show that all samples, apart from the 1.0 seconds sample, are dominated by the tetragonal (*I*4/*mcm*) $MAPbI_3$ perovskite phase. Although the 1.0 seconds sample also shows the presence of the $MAPbI_3$, it has a much larger contribution from a low-dimensional perovskite. This phase has a lower symmetry that probably stems from the incomplete crystallization of the precursor solution, i.e. the inclusion of solvent within the crystal morphology [65, 66]. A small amount of this phase can also be identified in the 1.4 seconds sample. The 1.7 and 2.0 seconds samples show almost identical patterns that resemble the data from the antisolvent sample. Figure 2.7b exhibits the temperature profile for different annealing time and the defined area for perovskite ($MAPbI_3$) crystal structure, with an optimal time set at 1.7 seconds. Tables 2.1, 2.2, and 2.3 list the refined crystal parameters. From 3.0 to 4.0 seconds, a growing contribution of PbI_2 can be observed, which is due to the degradation of $MAPbI_3$ caused by the long irradiation time and thus higher local temperatures (Section 2.4 discusses about the structural perovskite crystal formation under different IR pulse times).

Figure 2.8a shows top-view scanning electron microscope images of perovskite film morphologies for four FIRA irradiation times. All images show large-crystalline domains with diameters of around 40 μm. The shortest 1.4 seconds sample image reveals individual domains separated by seemingly non-crystalline regions, in good agreement with the LDP perovskite impurities of Figure 2.7a. The 1.7 and 3 seconds irradiation times result in large, laterally space-filling, pinhole-free crystalline

Table 2.1 Lattice constants from the Rietveld refinements for 1.7s FIRA-annealed and antisolvent perovskite samples.

	FIRA 1.7s	**Antisolvent**
Space group	$I4/mcm$	$I4/mcm$
a, b (Å)	8.8435(0)	8.8864(8)
c (Å)	12.6297(2)	12.6897(1)
α, β, γ (°)	90.000	90.000
V (Å^3)	987.73(8)	1002.10(0)

Table 2.2 Atomic positions from the Rietveld refinement for 1.7s FIRA-annealed sample ($R_w = 6.88\%$).

	x	**y**	**z**	**Fraction**	**Uiso (Å)**
C	0.5786(7)	−0.0786(7)	0.2715(5)	0.25	0.0100
N	0.4656(7)	0.0343(3)	0.1990(2)	0.25	0.0100
I1	0.2146(8)	0.2853(2)	0.0000	1.00	0.0856(0)
I2	0.0000	0.0000	0.2500	1.00	0.0846(6)
Pb	0.0000	0.0000	0.0000	1.00	0.0359(3)

Table 2.3 Atomic positions from the Rietveld refinement for the antisolvent sample ($R_w = 0.94\%$).

	x	**y**	**z**	**Fraction**	**Uiso (Å)**
C	0.5550	−0.0550	0.2640	0.25	0.0100
N	0.4590	0.0410	0.2020	0.25	0.0100
I1	0.2163(8)	0.2836(2)	0.0000	1.00	0.0687(5)
I2	0.0000	0.0000	0.2500	1.00	0.0796(3)
Pb	0.0000	0.0000	0.0000	1.00	0.0398(4)

domains in the films. To support this, an optical image and an SEM top-view image show similar features in Figure 2.9a,b. Similarly, top-view images from an energy-dispersive X-ray spectroscopy (EDS) analysis of 1.7 seconds FIRA-annealed perovskite films and a distribution profile of the chemical elements present in the perovskite film are shown in Figure 2.9c. For FIRA irradiation times exceeding four seconds, the change in morphology indicates a further phase change. Supported by the XRD analysis, this is attributed to the partial degradation of the $MAPbI_3$ to PbI_2 (disclosed in Section 2.4). FIRA thus enables the manufacture of highly compact, pinhole-free perovskite films with crystalline domains that are significantly larger than those produced by the antisolvent method.

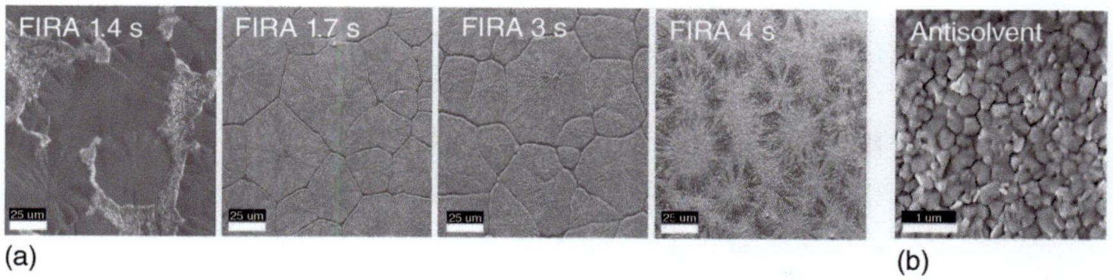

Figure 2.8 IR annealing time and its relation to the perovskite morphology. (a) Top view of FIRA-annealed perovskite films for different four annealing times, scale bar of 25 μm. (b) Top view of a reference film made by the antisolvent method followed by annealing at 100 °C for one hour on a standard hotplate, scale bar of 1 μm. Source: Adapted with permission from Sanchez et al. [7], © (2018) WILEY-VCH Verlag GmbH & Co. KGaA, Weinheim.

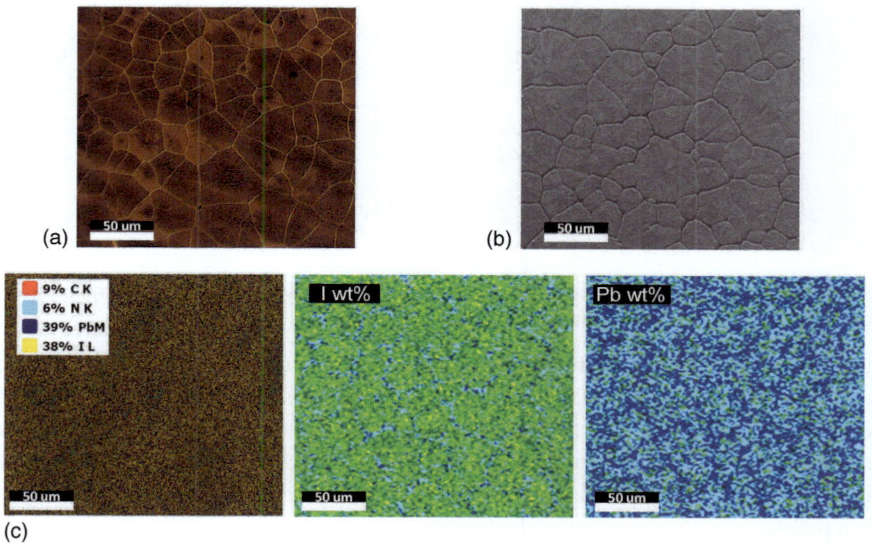

Figure 2.9 Optical and SEM images of $MAPbI_3$ films synthesized by 1.7 seconds FIRA annealing. (a) Optical transmission image, scale bar of 50 μm. (b) SEM top-view image, scale bar of 50 μm. (c) EDS images and chemical element analysis, scale bar of 50 μm.

2.3.5 Pulsed FIRA Method for Inorganic Perovskite Composition

Following the FIRA perovskite synthesis, pulsed FIRA (*p*-FIRA) allows to drive the nucleation and crystal growth through short and high intense pulses. *P*-FIRA consists of short IR pulses followed by a longer flash irradiation. Essentially, the so-called "pulses" and the following "flash" differ in length; the duration of each pulse is 20 ms and the flash is usually more than 1 second long. This allows the synthesis of compact and smooth films. In addition, the *p*-FIRA protocol can be used to synthesize inorganic perovskite films that are not stable when made in any other way.

A sequence of *p-FIRA* crystal formation is shown in Figure 2.10, where the film quality does not significantly change for more than c. 23 pulses. This suggests that

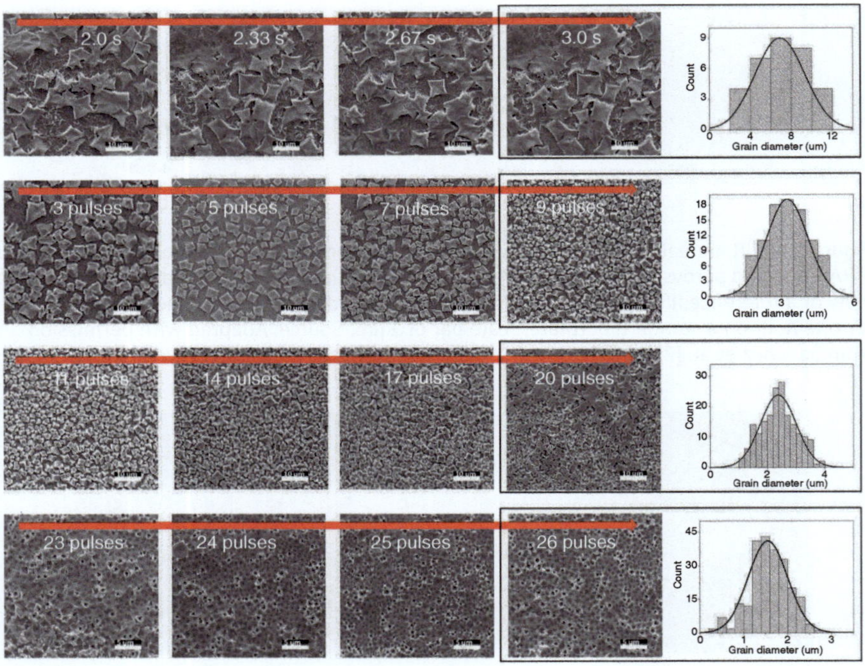

Figure 2.10 Scanning electron microscope top-view images and histogram of the crystal grain diameters for cesium lead halide films.

the flashes initiate perovskite nucleation, adding new nuclei with each flash of an extended series of flashes. The long final annealing pulse is less effective in causing crystal nucleation and instead enables already nucleated crystals to grow. Note that the overall irradiation process takes place in less than 50 seconds and stabilizes the perovskite phase without halide evaporation. The compactness of the 25-pulse film is a prerequisite for device manufacturing.

2.3.6 Warmed-Pulsed FIRA Method

Warmed-pulsed FIRA (*wp*-FIRA) similar to *p*-FIRA makes use of short IR irradiation pulses (20 ms) to drive the nucleation and crystal growth of the thin film. The main feature of *wp*-FIRA consists of the synthesis time of 15–20 seconds. *wp*-FIRA is a direct improvement of *p*-FIRA in two aspects, time and film compactness. Figure 2.11 shows the temperature as a function of the time for *wp*-FIRA; the entire synthesis of the coated solution can be described in three steps: (i) warm-up in the dark, (ii) pulses, and (iii) long pulse. Here, the electrode with the spun solution is submitted to a first warm-up. The chamber is heated by 2–4 seconds of IR, keeping the sample in a shadow. The irradiation time under shadow depends on the material composition; for hybrid perovskites, it is two seconds, while for all-inorganic perovskites, four seconds were used. The warm-up step is followed by six short pulses and an IR flash for two seconds (the sample is placed from the shadow to illuminate through a manual mechanism).

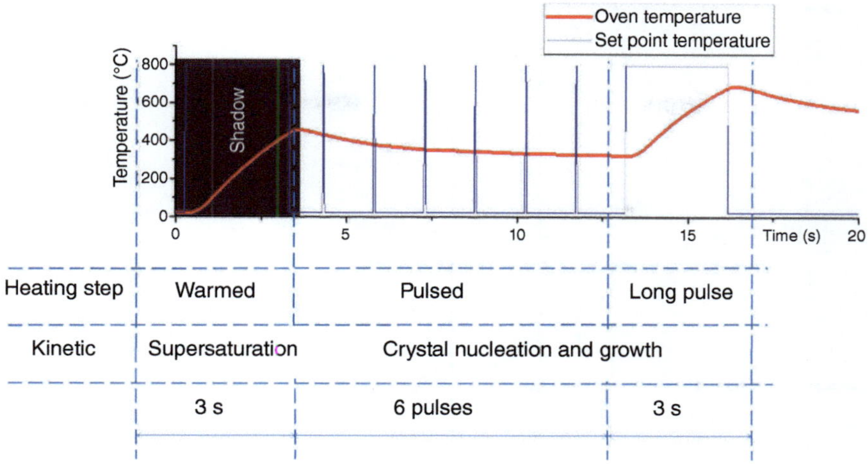

Figure 2.11 Illustration of the temperature as a function of the time for the warmed-pulsed FIRA method and its different kinetic steps. Source: Adapted with permission from Sanchez et al. [47], © (2020) American Chemical Society.

wp-FIRA achieves the synthesis of a higher film quality in a shorter processing time. Here, the introduction of the warming step or warm-up of the electrode convectively preheats the perovskite solution (not through the FTO) by generating a temperature gradient across the chamber. It initiates nucleation in a more effective fashion than applying short pulses, followed by only six pulses for additional nucleation. Using the standard $MAPbI_3$ perovskite, large needle-shaped crystallites emerge from a nucleation center, which is characteristic of faceted or crystal spiky growth [67, 68]. Mixing cesium cations and bromide anions together, the *wp*-FIRA-annealed morphology results in a hexagonal grain arrangement, which is different from single-flash FIRA annealing of the same perovskite. When entirely replacing MA by Cs, an all-inorganic perovskite composition results in a smooth layer that is even more compact than the layer synthesized by *p*-FIRA (Figure 2.12).

2.3.7 Crystallization Behavior of Mixed Perovskite Solutions

The crystallization of perovskites from precursor solution can be considered the most crucial factor for perovskite film quality, which gives rise to a variety of strategies to govern the nucleation and crystal growth of perovskite films [69]. Hybrid lead halide perovskite precursor can form colloidal aggregates in solution causing complex nucleation and growth pathways during drying [56, 70]. The diffusion of the molecules, solvent removal rate, and crystallization processes are critical in generating the kinetically trapped morphologies within the perovskite films. For thin film formation, many aspects affect the crystallization and, as a consequence, the microstructure formation, such as precursor solvents, substrates, and synthesis method. Cation replacement, solvent engineering, and molecular additives, among others, have been developed to yield the desired morphologies in perovskite films [50, 63, 64, 69, 71–77].

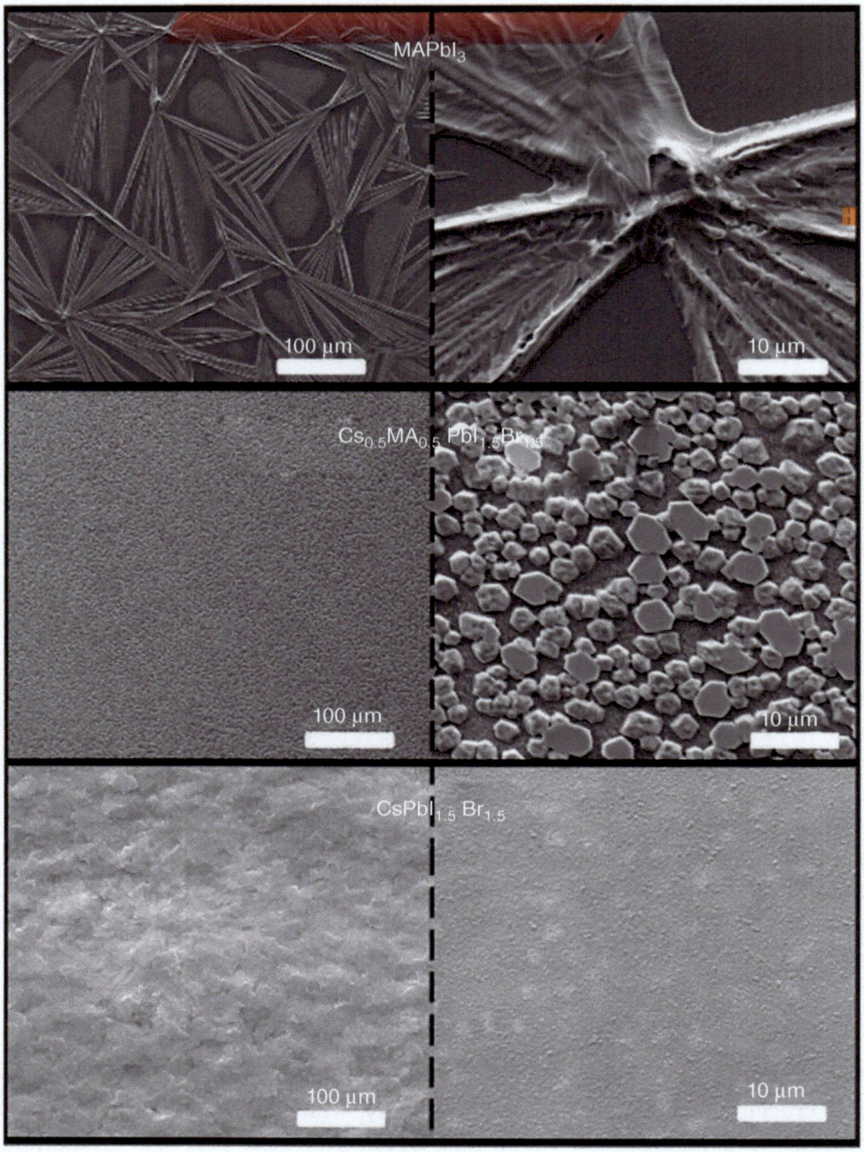

Figure 2.12 SEM top-view image of a cesium lead halide perovskite film synthesized by warmed-pulsed FIRA. Source: Adapted with permission from Sanchez et al. [47], © (2020) American Chemical Society.

Figure 2.13 shows different perovskite chemical compositions annealed by a 1.2 second FIRA pulse. All the films were cast from DMSO solution, varying the MA/Cs cation and the I/Br halide compositions. The resulting morphology differs in the films, e.g. for $MAPbI_3$, large and compact domains are observed. Mixing the bromide and iodide ($MAPbBr_{1.5}I_{1.5}$) leads to smaller domains. When the cesium

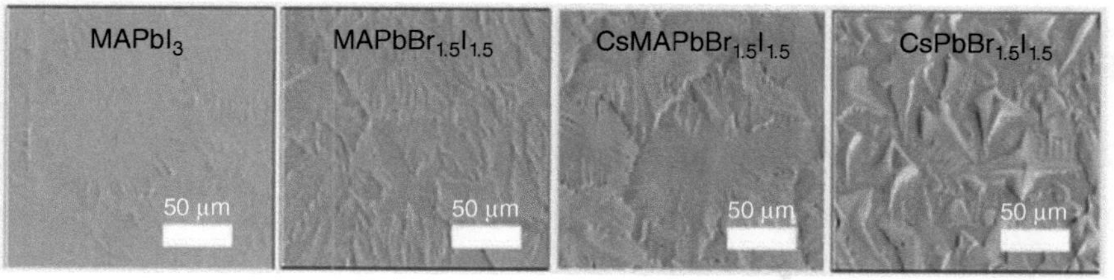

Figure 2.13 Top-view SEM of the different perovskite chemical compositions synthesized by FIRA.

is mixed with MA ($CsMAPbBr_{1.5}I_{1.5}$) or MA is replaced by Cs ($CsPbBr_{1.5}I_{1.5}$), the crystal growth changes. Increasing the cesium and bromide concentration leads to smaller crystal domains. This arises from a change in chemical potential difference driving the crystal growth, as different precursors are added [16, 21, 23, 24, 58, 78, 79]. For example, the MA organic salt is suitable for the formation of compact films by FIRA annealing.

The use of a solution process and FIRA to grow perovskite thin films allows extending the material processability. It is known that the physicochemical properties of the perovskite material can be tuned by altering the solution precursors. However, through FIRA, we can also control the morphology of the film. This necessarily implies the need for an understanding of the kinetic phenomena that are involved in thin film formation. These advantages open up a new empirical parameter space. All this knowledge is useful in the improvement of the perovskite synthesis and the quality of the thin films, which will result in higher device performances.

The synthesis method and perovskite crystallization are the keys for high-performance devices. High-quality films are compact and highly crystalline, which is indicative of a defect-free material. Table 2.4 shows the main features of the synthesis methods proposed in this review. For example, a single FIRA method requires less than two seconds, which is appropriate for the industry. Instead, pulsed and warmed-pulsed FIRA needs longer synthesis times of 1 minute and 20 seconds, respectively. However, the pulsed route can be used to prepare thin films of both hybrid and all-inorganic perovskite materials. All the methods are suitable to scale up PSC manufacture, considering not only the synthesis time but also the simplicity of the process that allows low production costs, as discussed in Section 2.5.

Table 2.4 Characteristic of the three FIRA methods used for the perovskite film synthesis.

		Perovskite film morphology	
FIRA methods	**Synthesis time**	**Hybrid**	**Inorganic**
Single flash	~2 seconds	Compact film	Single island
Pulsed	~1 minutes	Non-compact film	Compact single cuboid
Warmed pulsed	~15–20 seconds	Non-compact film	Compact film

Source: Adapted with permission from reference [47], © (2020) American Chemical Society.

2.4 Structural Analysis of FIRA-Annealed Perovskite Films with Variable Pulse Time

2.4.1 Planar and Mesoporous Substrates

To analyze the structural mechanisms in $MAPbI_3$ films, perovskite films are exposed to IR pulse from 1.4 to 4 seconds onto FTO and mesostructured TiO_2 electrodes. XRD and SAXS are the two main characterization methods used to unveil structural material changes. The pair distribution function (PDF) calculation has been used to analyze the local structure of the film, also characterizing the generated chemical products. The gained understanding of the fundamental processes that take place in reversible and irreversible degradation pathways is a step forward to enhance the stability of PSCs.

The comparison of the FIRA synthesized films for two electrode architectures and for varying FIRA pulse times [80] is showed in Figure 2.14. Figure 2.14a shows SEM images of the thin films deposited directly onto FTO substrates exhibiting similar surface structures for FIRA pulse times between 1.7 and 3.0 seconds. Based on visual observation, these samples show larger lateral crystal domains than the thin film prepared using the conventional AS method, which is consistent with what was previously reported [8]. Mesostructured TiO_2 thin layer deposited on top of FTO substrates does not lead to discernible changes in the surface morphology for AS and FIRA-2.0s samples shown in Figure 2.14b. It, however, results in noticeable differences in the lateral crystal sizes for FIRA-1.7 and -3.0 seconds samples

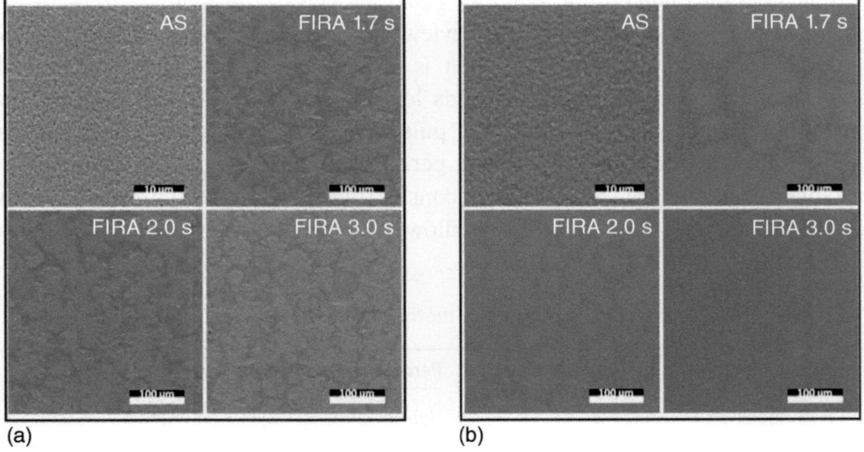

Figure 2.14 Scanning electron microscope (SEM) images of the $MAPbI_3$ films made by the AS and FIRA methods. SEM top-view images of perovskite films deposited onto (a) FTO and (b) mesostructured TiO_2/FTO electrodes by AS and FIRA methods. Source: Adapted with permission from Sanchez et al. [80], © 2019 American Chemical Society.

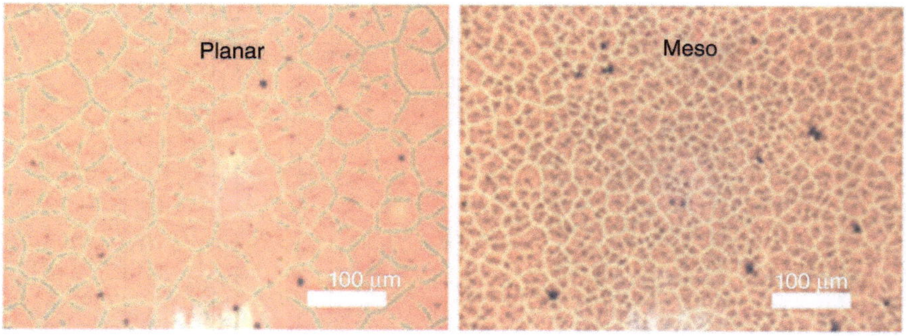

Figure 2.15 Mixed perovskite (composition) films fabricated by FIRA on top of planar and meso-architectures, optical microscope images taken in transmittance mode. Source: Adapted with permission from Sanchez et al. [9], © (2019) The Authors. Published by Elsevier Ltd.

shown in Figure 2.14b, leading to the larger apparent lateral crystal domain sizes of Figure 2.14a.

These observations suggest a different mechanism of the crystal nucleation and growth in the presence of a nanostructured mesoporous TiO_2 layer compared with the compact TiO_2 electrode [13]. The difference between the two is the interfacial energy. Because of its porosity, the mesostructured TiO_2 layer has a higher effective interfacial energy than the planar TiO_2 electrode. The optical images shown in Figure 2.15 describes how the perovskite grains grow over the different substrates, similar to the SEM images from Figure 2.14, larger for the planar case, and shorter for the mesoscopic TiO_2 layer as substrates [9].

2.4.2 Crystal Structure Analysis

Comparative observations were made from a preliminary visual examination of the XRD data. While both perovskite films prepared on planar and mesostructured electrodes shown in Figure 2.16a,b, respectively, using AS and FIRA-2.0s methods exhibit similar XRD patterns, 1.7 and 3.0 seconds FIRA-annealed thin films show pronounced differences between the two substrate types, with a broad feature centering around $2\theta = 25°$ that is visible only for the mesostructured electrodes. Since the background contribution from the sample holder has been appropriately removed, this feature thus contains information reflecting a structural deviation from a perfect periodic lattice, possibly non-crystalline components [81, 82]. Note that of the low perovskite volume probed in thin film X-ray scattering experiments, this broad feature is easier and more reliably captured using high-flux synchrotron radiation than a laboratory X-ray source because of its more improved signal-to-noise ratio [83].

Figure 2.16a shows samples with FTO electrodes, exhibiting similar diffraction patterns with some notable differences that seem to be related to PbI_2. These differences are reflected by various apparent intensities and, more importantly, a shift of the peak position for the (001) and (011) reflections to higher 2θ angles. The (010)

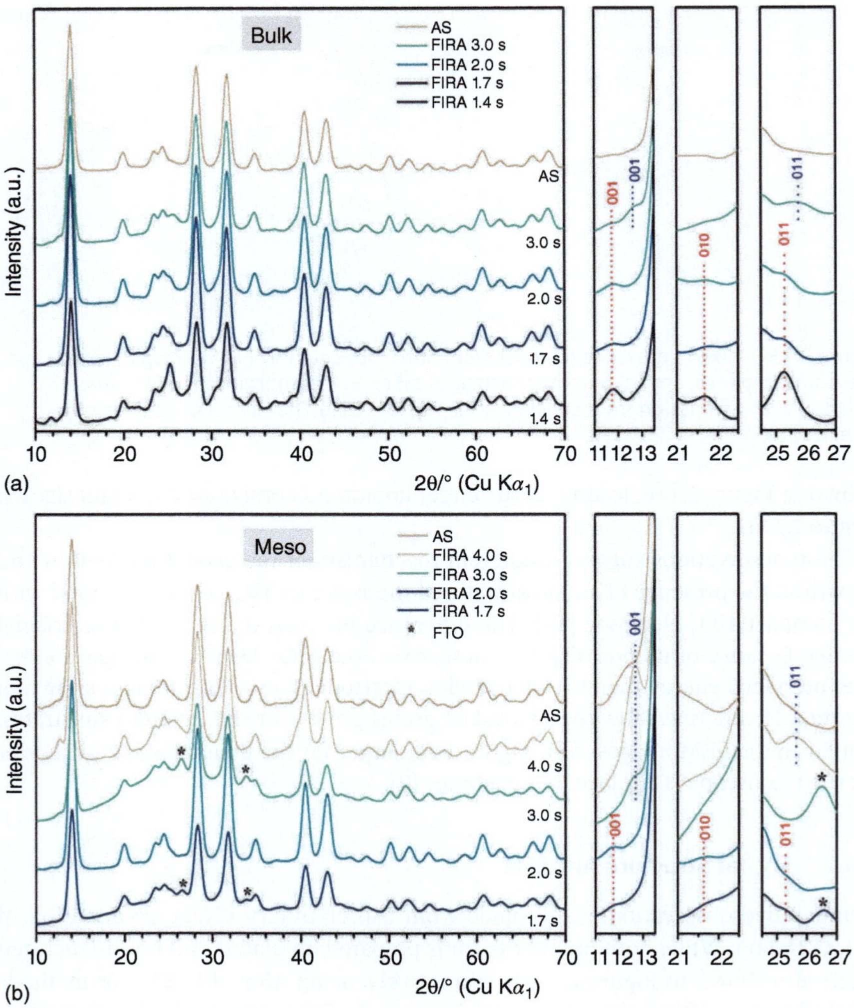

Figure 2.16 XRD patterns of $MAPbI_3$ powders (scrapped from synthesized films) (a) on FTO and (b) mesostructured electrodes obtained by X-ray scattering. Asterisks mark the main reflection peaks of SnO_2 (*P42/mnm*) arising from the FTO substrate. The highlighted low 2θ regions are magnified on the right, where Bragg peaks corresponding to reflections of PbI_2 (*P*-3m1) are labeled in red and blue. Source: Reproduced with permission from Sanchez et al. [80], © 2019 American Chemical Society.

peak position does not show a pronounced shift, indicating that the lattice change for the PbI_2 phase in the "planar FIRA 3.0 s" sample is mainly in the *c*-direction with a shortened length [15, 84]. The same observation can be made for samples on mesostructured surfaces shown in Figure 2.16b. In particular, the "FIRA 4.0 s" sample exhibits more enhanced peak intensities for the PbI_2 phases, implying a higher crystallinity. Note that this is by comparison of the right graphs, showing low-angle PbI_2-reflected phases. In addition, "meso FIRA 1.7s and 3.0 s" samples show a particularly pronounced hump feature centering around $2\theta = 25°$, indicating a considerable amount of non-crystalline components in these two samples.

2.4.3 Structure of the Intermediate Phases

Based on the evolution of the lattice parameters of the intermediate R_xPbI_y phase derived from the refinement analysis, we notice considerable crystallographic differences between R_xPbI_y structures in various FIRA samples, despite no apparent change in the structure space group [80]. These differences are particularly pronounced between 1.4 and 1.7 seconds samples, represented by a sudden increase of the *b*- and *c*-lattices with a sharp decrease of the *a*-lattice and β-constant. The increase in the *c*-lattice along with the reduction in β-constant further results in a sharp increase of the *d*-spacing between the (001) planes. As the annealing progresses (>1.7 seconds), evolutions of these structure parameters either follow a reverse trend or become less drastic. These observations indicate that the R_xPbI_y structure at 1.4 seconds is notably different from the other samples. Note that in the following discussion where specific structure differences are explored, two R_xPbI_y intermediates are denoted as R_xPbI_y (α) and (β), corresponding to R_xPbI_y phases from the FIRA 1.4 seconds and FIRA 1.7–4.0 seconds samples, respectively. Particularly, R_xPbI_y (FIRA 2.0 seconds) structure is chosen to represent the average structure of the R_xPbI_y (β) phase.

Similar to the layered PbI_2 as shown in Figure 2.17a, R_xPbI_y ($P12_1/a1$) also adopts a stacked spatial configuration, where the interlayers are in parallel with the (001)/*ab*-planes (Figure 2.17b,c). However, they are composed of corner-sharing $[PbI_6]$ octahedra instead of an edge-sharing linkage seen in PbI_2. Unlike the perfect $[PbI_6]$ octahedra in PbI_2, the ones in R_xPbI_y are distorted, each consisting of Pb (Wyckoff position: 2*a*) at the octahedral center with two I1 (4*e*) and four I2 (4*e*) atoms at the terminal and bridging positions, respectively. Comparing R_xPbI_y (α) with (β) shown in Figure 2.17b,c, noticeable changes can be observed, mainly in the tilt of the $[PbI_6]$ octahedra, leading to a reduced β-constant and an increased $d_{(001)}$. This octahedra tilt associated with the reduced β-constant signifies a topotactic transformation from the layered building block to the final 3D network in the perovskite (Figure 2.17). The progressive change of the $d_{(001)}$ in R_xPbI_y implies a modification of the interaction between the interlayer molecular components and the $[PbI_6]$ sheet.

Although experimental limitations do not allow us to determine the nature of the molecules and their atomic positions, based on earlier studies [85, 86], it is expected that these molecules (mainly MA) align in a similar way with those in R_2PbI_4, forming hydrogen bonds between the ammonium protons and I atoms [85]. Despite being weak compared with larger organic cations (formamidinium and guanidinium) [87], these interactions, along with the van der Waals forces between the $[PbI_6]$ sheets [88], result in crystal cohesion along the *c*-axis to stabilize the structure [85]. Note that it is difficult to obtain a reliable iodide occupancy in R_xPbI_y through PDF refinement because of the minor presence of this phase and the compositional and structural complexity of the composite. Hence, the occupancy

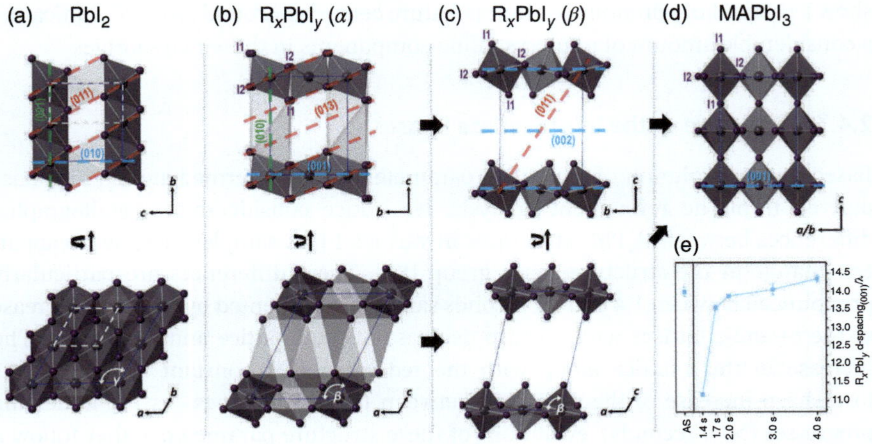

Figure 2.17 $[PbI_6]$ structures of (a) PbI_2, (b) R_xPbI_y (α), (c) R_xPbI_y (β), and (d) $MAPbI_3$ viewed from different directions. Note that the unit cell lattice vectors are different between PbI_2 and the other structures. $[PbI_6]$ units are shown as dark gray octahedra where Pb and I atoms are denoted by gray and purple spheres, respectively. R molecules (including MA) are omitted for a clearer view. Unit cells are marked with dotted blue lines. For PbI_2, the original unit cell is marked using white dashed lines, and the blue dotted lines indicate the supercell. Two different iodine atoms in R_xPbI_y and $MAPbI_3$ are labeled as I1 and I2. The important crystal planes are marked and labeled with different colors. Note that the light gray sites in (a,b) are empty sites, highlighted to show the correlation between the PbI_2 and R_xPbI_y (α) structures. (e) The evolution of the *d*-spacing between the (001) planes in R_xPbI_y. Source: Reproduced with permission from Sanchez et al. [80], © 2019 American Chemical Society.

for both I1 and I2 was initially set off-stoichiometric during the refinement to ensure a sensible (1 : 3) Pb/I ratio. Interestingly, the perovskite octahedron also contains two I1 (4a) and four I2 (8h) atoms, which align perpendicular and along the (001) planes, respectively. Despite a structure symmetry change from R_xPbI_y to perovskite, the fact that their $[PbI_6]$ units both share two I1 and four I2 atoms indicates the retention of structure coherence upon the phase transition. Based on this coherence and the multiplicity of each I site, we deduce that the fully occupied I1 (4a) and I2 (8h) in perovskites might be translated from half-filled I1 (4e) and fully occupied I2 (4e) in R_xPbI_y. This scenario may also offer a cogent explanation to the role of I1 in the transition from a terminal (in R_xPbI_y) to a bridging (in perovskite) iodide during the topotactic 2D-to-3D conversion.

While the spatial transformation from the R_xPbI_y to $MAPbI_3$ seems evident, there is no clear connection of the arrangement of their $[PbI_6]$ units to the PbI_2 precursor. To unravel a potential structural correlation, it is necessary to convert the PbI_2 crystal structure into a $2 \times 2 \times 1$ supercell (Figure 2.17a) and to highlight the significantly distorted empty octahedra sites within the interlayers of R_xPbI_y (α) (Figure 2.17b). The link between the two structures thus becomes clearer. If the octahedra in the PbI_2 (001) layer are partially occupied, this configuration seems to resemble the (010) layer in the R_xPbI_y(α) structure. Similarly, a selected occupation of the octahedra sites in the PbI_2(010) layer would resonate with the octahedra arrangement along the R_xPbI_y(001) plane. We also noticed that the initial expansion of $d_{(001)}$ in

R_xPbI_y (α), presumably induced by molecule insertion, appears as an elongation of the interlayer octahedra along the *c*-axis, which leads to an enlarged volume about twice of their original size. Despite this notable distortion, β-angle (~116°) of the R_xPbI_y (α) unit cell, however, largely retains the original γ-constant (120°) of the trigonal PbI_2 cell. These two factors concurrently result in a $d_{(013)}$ in R_xPbI_y (α) that is comparable with the $d_{(011)}$ in PbI_2, consequently leading to similar peak positions (~25° 2θ) of their Bragg reflections [47]. In addition, the particular cell geometry of R_xPbI_y (α) also yielded an almost identical value of $d_{(21\bar{3})}$ and $d_{(013)}$. These two reflections thus superimpose, leading to an apparent single peak with a strong intensity, which rationalizes the observation of a prominent diffraction peak at $2\theta = 25.3°$ in the FIRA1.4 s XRD data (Figure 2.16a).

While the intense reflections from the (α) phase are centered within the 2θ range of 20–35°, the ones for the (β) phase are at lower 2θ angles, between 10° and 20°. The main reflections of both R_xPbI_y (α) and (β) structures resonate reasonably well with the experimental data of the $MAPbI_3$ precursor and intermediates reported previously in an in situ XRD study albeit with some discrepancies in their relative peak intensities and positions [89]. Note that because the structure coherence lengths in both structures are very small (~20 Å) (Figure 2.18b) relative to their unit cell dimensions and that their PDFs are always dominated by the atom-pair contribution from a very crystalline perovskite component, it is challenging to derive reliable long-range structure solutions of these intermediates solely from the PDF analysis. We do not exclude possibilities of a different space group/super cell and any intricate structure features that are common to the layered structures, for example, stacking defects. That being said, the corner-sharing $[PbI_6]$ 2D framework revealed by our PDF analysis offers critical insight into the full structure derivation of these intermediates.

The evolution of the phase fraction (wt%) of each component derived from the PDF refinement as a function of annealing time is shown in Figure 2.18a,

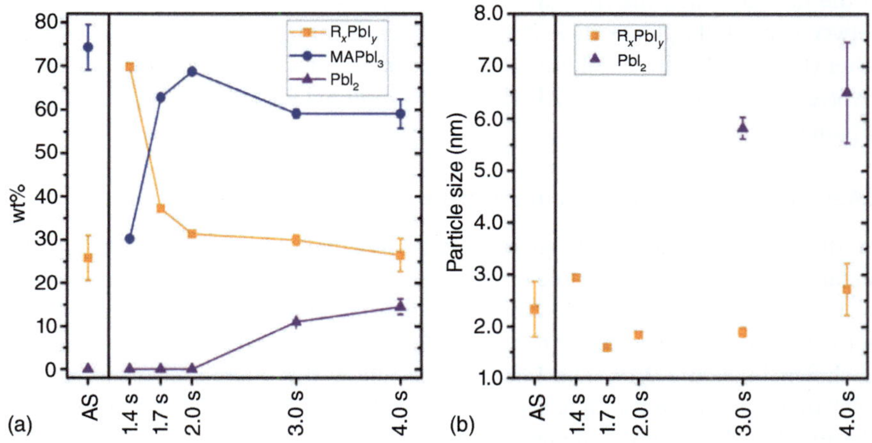

Figure 2.18 Evolution of the (a) weight fraction of each component and (b) particle sizes (reflecting the structure coherence lengths) of PbI_2 and R_xPbI_y derived from PDF refinements. Source: Adapted with permission from Sanchez et al. [80], © 2019 American Chemical Society.

along with the change of particle sizes of the intermediate R_xPbI_y and PbI_2 phases (Figure 2.18b). Based on the structure analysis of R_xPbI_y, we propose the (α) phase to form at the early stage of annealing. Its local structure is spatially related to PbI_2 whose solvated $[PbI_6]$ units in the precursor reconnect through vertices forming corner-sharing $[PbI_6]$ sheets. These sheets eventually stack, sandwiching MA (and/or other molecules, such as DMF) to form R_xPbI_y (α). Further incorporation of MA upon annealing triggers a significant tilting of the $[PbI_6]$ octahedra, giving rise to a conversion from R_xPbI_y (α) to (β), which eventually results in the transformation of the 2D building blocks to form the 3D perovskite network. Within about 2.0 seconds of FIRA annealing, this 2D-to-3D structure transition reaches its maximum conversion rate producing mixed phases that contain ~70 wt% perovskite and ~30 wt% R_xPbI_y (β). However, continued exposure to IR (>2.0 seconds) leads to a further increase in temperature, causing the degradation of both perovskite and intermediate phase.

The final decomposition products are PbI_2 for both phases, and no evident formation of any intermediate phase, such as R_xPbI_y (α), was observed. This implies that the degradation mechanism does not follow the reverse route of the perovskite formation via intermediate(s). This is consistent with an earlier in situ XRD study and can probably be explained using a layer-by-layer degradation mechanism proposed by in situ ED/TEM studies [88–90]. In contrast to the overall agreement on the degradation pathway, the perovskite formation mechanism may vary between different processing techniques. The pathway established from our structural analysis reflects a mechanism of a solution-based synthesis where solvated PbI_2 and MA^+ coexist in the system. It involves rearrangement of the $[PbI_6]$ unit, providing insight into the change of the local structure to support previous mechanistic study based on long-range structure characterization. This mechanism may differ from the general knowledge of the syntheses based on precoated PbI_2 films [91, 92], where the molecular components are believed to be simply inserted within the interlayers of PbI_2. In our proposed pathway, the $[PbI_6]$ rearrangement results in a reorientation of the interlayer spacing from that in between the (001) planes in PbI_2 to that in between the (001) planes in R_xPbI_y, which are essentially perpendicular to each other (Figure 2.17a,b). In addition, the transformation of the precursor to $MAPbI_3$ involves (at least) two additional steps via a sequential formation of R_xPbI_y (α) and (β) intermediates.

2.4.4 Internal Crystal Domain Structure

As discussed previously, the planar electrode architecture induces highly and preferentially ordered crystalline perovskite film, compared with films deposited on mesoporous substrates (which also induces oriented patterns). The grain domains in the SEM images correspond to the EBDS rings but exhibit an internal structure, as shown in Figure 2.19. The EBDS images show the internal structure by the color mapping of the multiple single particles (as can be defined in the contour). Figure 2.19a shows three different zoom-level top-view SEM images of a $MAPbI_3$ thin films, and Figure 2.19b shows the EBSD maps corresponding to the marked SEM areas. EBDS signal confirms the high degree of order in the FIRA-annealed

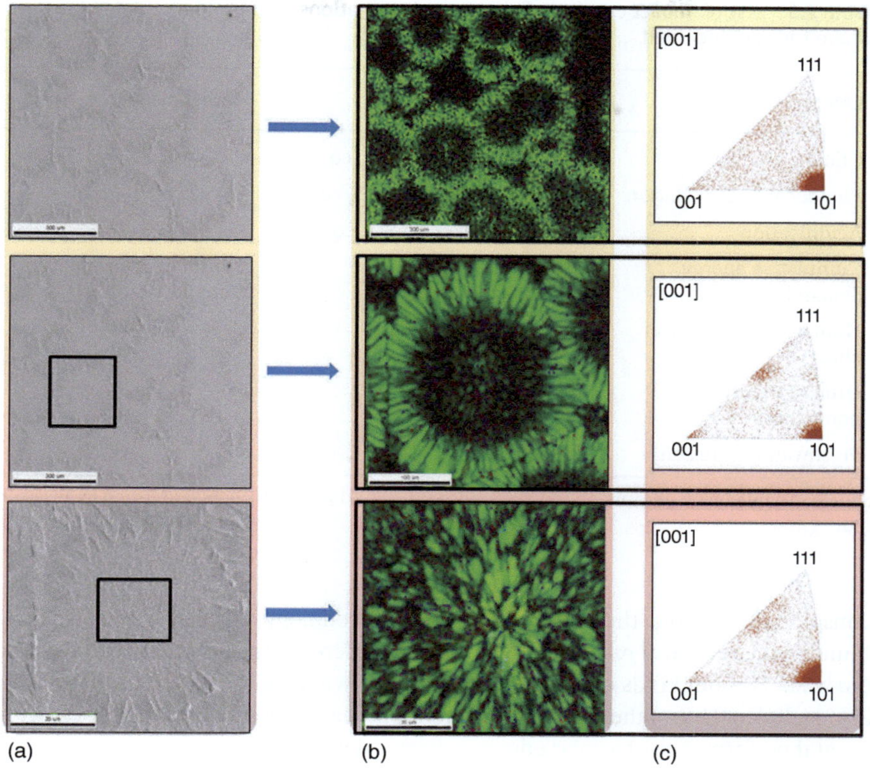

Figure 2.19 (a) SEM top view of aMAPbI$_3$ film synthesized with a 1.7 seconds FIRA pulse. (b) Electron backscattering diffraction spectroscopy (EBDS) from the marked areas in the SEM images. (c) Triangle plane denoting the preferential orientation of the single grains in the EBDS-mapped sample.

perovskite films on planar electrodes as previously described by the XRD analysis. The rings conform to single-crystal microrods with a minimum length of around 5 μm (with increasing size from the nucleation center to the ring border, reaching a maximum length of 20 μm). They have a preferential <101> in-plane orientation, showing the triangle plane orientation extracted from EBDS mapping in Figure 2.19c.

2.5 A Cost-Effective and Environmentally Friendly Method

2.5.1 Life-Cycle Assessment (LCA) of the Perovskite Film Synthesis Methods

LCA determines the potential of AS and FIRA processes for upscalability. To implement this, we have to consider that both devices can preserve the same efficiencies on larger substrates, although this will be more difficult for AS samples [93]. Further

Table 2.5 List of impact categories and their abbreviations, units, and methodologies where they are included.

Category	Abbreviation	Unit	Methodology
Climate change	GWP	kg CO_2 eq	CML baseline V3.02 [96]
Photochemical oxidation	POP	kg C_2H_4 eq	
Acidification	AP	kg SO_2 eq	
Cumulative energy demand	CED	MJ	Cumulative energy demand V1.09 [94]
Human toxicity, cancer effects	HTC	CTUh	Usetox V1.04 [95]
Human toxicity, non-cancer effects	HTNC	CTUh	
Freshwater ecotoxicity	ET	CTUe	

Source: Adapted with permission from reference [9], © (2019) The Authors. Published by Elsevier Ltd.

considerations about the assumptions and methodology adopted for LCA can be found in the section of materials and methods. Different impact categories have been analyzed for both kinds of samples; see Table 2.1 where abbreviations and units are also included. One of the most significant impact categories to measure the environmental performance of a solar energy collector device is the carbon footprint, since one of the main benefits of solar energy stems from a reduced greenhouse effect. Photochemical oxidation and acidification are also selected to have a broad panoply.

Energy is a fundamental aspect of perovskite modules as it is their only valuable output. Knowing the necessary production energy is a good practice to calculate how viable their production is. The cumulative energy demand method V1.09 (CED) [94] obtains the total cradle-to-gate energy invested in the production of the perovskite module. Externalities need to be included as well for a holistic assessment. Thus, the potential toxicity of the production and emission of chlorobenzene and nitrogen was performed with USEtox V1.04 method [95] including the impact categories, human toxicity (cancer), human toxicity (non-cancer), and freshwater ecotoxicity; see Table 2.5. CML, CED, and USEtox methods are incorporated within the SimaPro® 8.0.3.14 software.

2.5.2 Relative Cost and Environmental Impact of the AS and FIRA Methods

The environmental impacts and costs are normalized with respect to the corresponding highest value for FIRA and AS methods using the same functional unit (1 kWh). In this way, the results are independent of the solar cell lifetime. AS method presents the highest environmental impacts in each of the categories considered and also the highest cost. Hence, the value of 100% is assigned

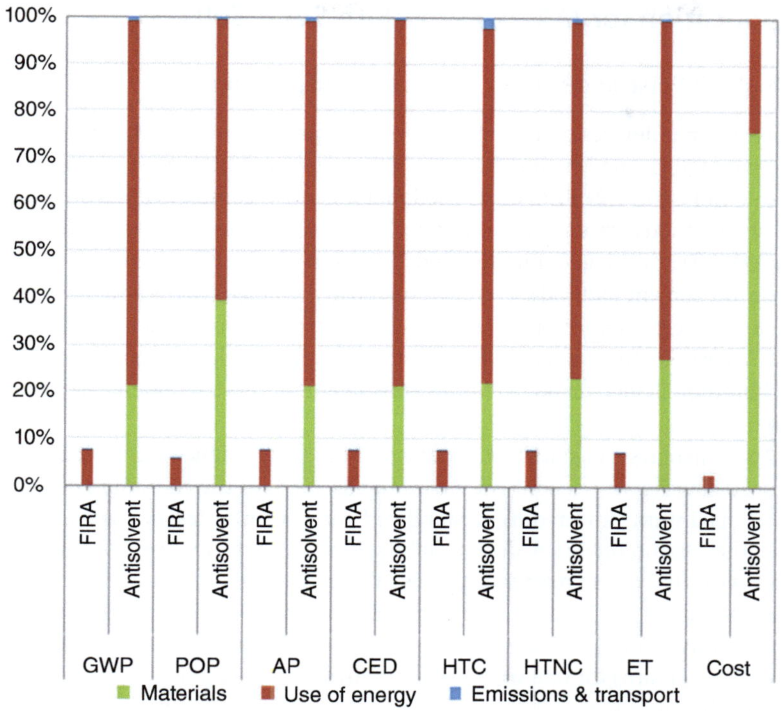

Figure 2.20 Relative costs and environmental impacts of the different process of the FIRA and antisolvent methods for perovskite layer deposition. Abbreviations of each impact category are explained in Table 2.5. Source: Adapted with permission from Sanchez et al. [9], © (2019) The Authors. Published by Elsevier Ltd.

in Figure 2.20, and the values for the FIRA method are relative to the values of AS method. Environmental impacts are disaggregated in the impacts produced by materials acquisition, energy consumption during the processing and emissions, and transport. Costs are broken down in materials and energy.

The environmental impacts of AS method are mainly incurred by the energy consumption during the annealing, an often-disregarded factor. The substitution of the long AS annealing by the fast FIRA process implies significantly lower electricity consumption and also the elimination of chlorobenzene consumption that also has adverse contributions. On the other hand, the environmental impacts of FIRA method are less than 8% compared with AS in all analyzed categories in Figure 2.20. Annealing was also identified to have one of the highest environmental impacts in the LCA of a preindustrial process based on a carbon stack architecture [97]. Moreover, the advantages of the FIRA method compared with AS are not just in terms of impact but in terms of cost. Materials prices have a significant influence (76%) in the cost of the antisolvent method. The reduction in nitrogen consumption, absence of chlorobenzene, and lower energy consumption in FIRA method imply that the cost is only 2% of the AS method.

2.6 Application for $MAPI_3$ Perovskite Solar Cells

2.6.1 Single IR Pulse and $MAPbI_3$ Perovskite Composition

To demonstrate that devices made with FIRA single pulse-annealed $MAPbI_3$ layers are as efficient as those made using the state-of-the-art AS method, two types of PSCs are compared in Figure 2.21 SEM cross-sectional images of complete PSCs prepared using the two methods are showed in Figure 2.21a. FIRA-annealed perovskite film is around 700–800 nm thick, compared with ~500 nm for the reference PSC (made by using the antisolvent method). The current density–voltage curves and the corresponding device performance parameters for the two champion cells are shown in Figure 2.21b. Also, FIRA-annealed PSCs exhibit a slightly higher FF and lower V_{oc} than the champion device made using the antisolvent method, while the J_{sc} values are similar.

Figure 2.21c displays the statistical analysis of the performance parameters collected from 20 devices, which confirm the trend observed for the champion devices. Finally, the maximum power output stabilized for five minutes in Figure 2.21d shows that the champion FIRA-device-stabilized PCE performance of ~18.3% is comparable with that of the reference that is among the highest values reported so far for $MAPbI_3$-based PSCs made by rapid thermal annealing. To support this, a long-term stability study has been carried out with FIRA and antisolvent devices by light exposure (1 sun) for one week under a nitrogen atmosphere. The results of this study are shown in Figure 2.22 and in Table 2.6, where the maximum PCE loss does not exceed 4% of the initial PCE, confirming stable PSCs made by the FIRA and antisolvent methods.

Moreover, the external quantum efficiency (EQE) of FIRA-annealed devices was measured to support the other PV parameters, as shown in Figure 2.23, explaining the high J_{sc} of 20.9 mA cm^{-2}. Note that the current density of FIRA devices decreases between short-circuit conditions and 0.6 V forward bias, while it is constant in the champion device. This is probably due to the fact the perovskite film is slightly thicker in the FIRA device than the ideal thickness of ~500 nm, as optimized for the antisolvent method. This results in slightly higher series resistance, the absence of which would enhance the fill factor even more. To support these arguments, an intensity-modulated photocurrent spectroscopy analysis is provided and discussed below (Figure 2.24).

The size of the perovskite crystals and the density of grain boundaries influence the ionic movement within the perovskite, which has been reported to affect the charge carrier collection efficiency [98]. As expected from the larger-crystalline domains, the device prepared using FIRA-annealing displays a faster ionic response than samples made by the antisolvent method, in agreement with the report by Correa-Baena et al. [77].

2.6.2 Large-Area Devices

Figure 2.25a shows the surface temperature of a perovskite film measured by an IR camera during heating. It shows that the IR radiation is mainly absorbed in the FTO

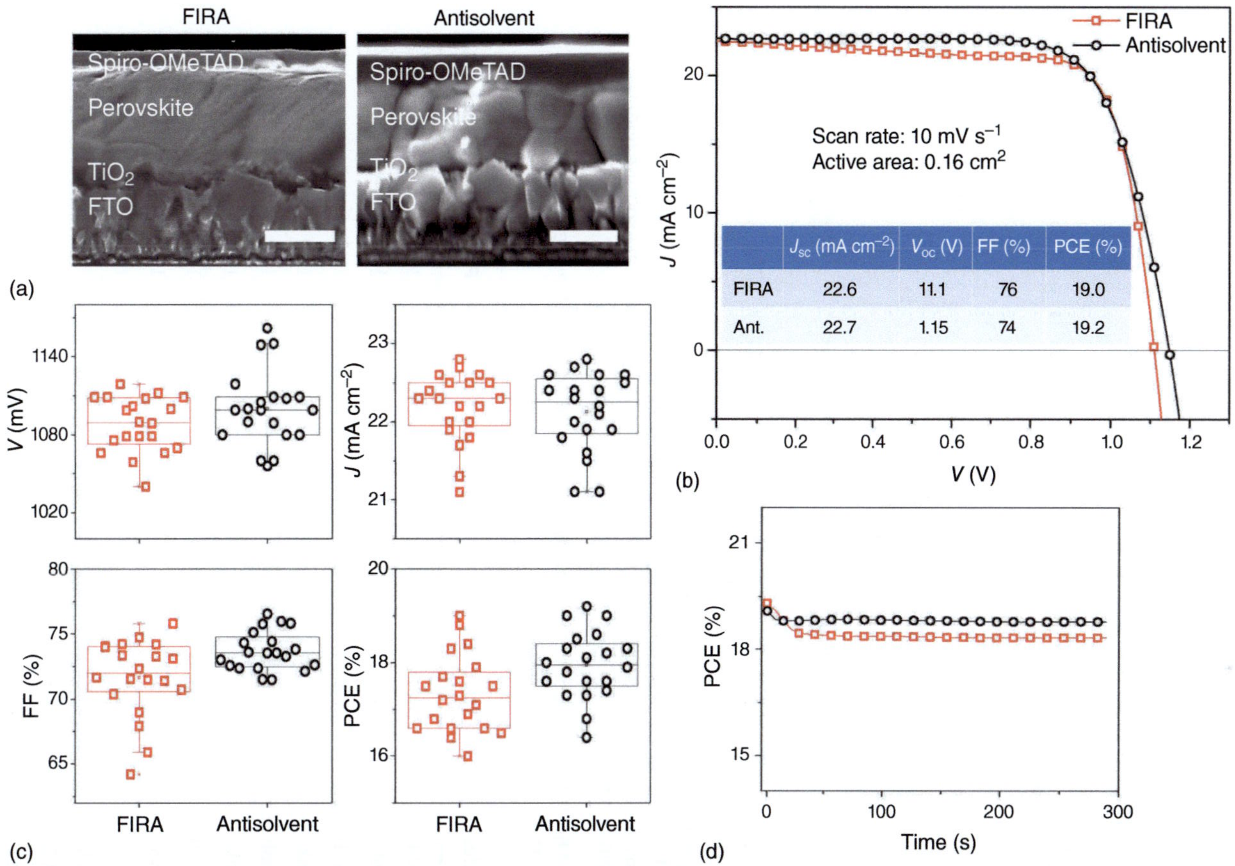

Figure 2.21 (a) SEM cross-sectional images of complete PSCs, scale bar of 500 nm. (b) Current density–voltage (J–V) scans collected at 10 mV s^{-1} of two champion devices under 1.5 AM irradiation and the corresponding photovoltaic performance parameters. (c) J_{sc}, V_{oc}, FF, and PCE statistics for 20 devices. (d) Maximum power point tracking of the champion cells (error: ±0.2 of the PCE). Source: Adapted with permission from Sanchez et al. [7], © (2018) WILEY-VCH Verlag GmbH & Co. KGaA, Weinheim.

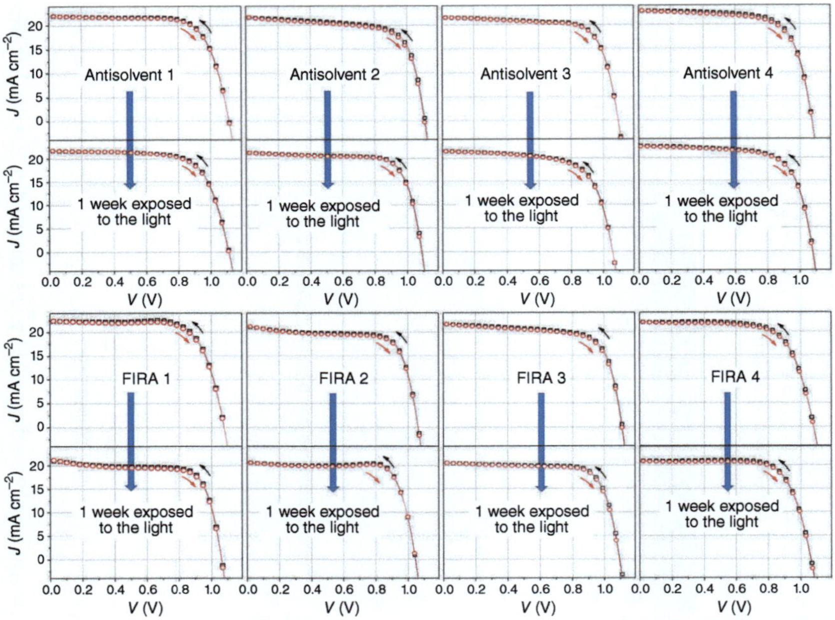

Figure 2.22 *J–V* characteristics during backward and forward scans of the four devices made by the FIRA and antisolvent methods, scanned at 10 mV s^{-1} with metal mask aperture of 0.16 cm^2. The devices were measured again after one week to 1 sun exposure under a nitrogen atmosphere.

Table 2.6 PV parameters of the stability study, including the PCE percent losses collected after one week of light exposure.

Samples	J_{sc} (mA cm^{-2})	V_{oc} (V)	FF (%)	PCE (%)	Losses (%)
Antisolvent 1	22.0	1.12	71.4	17.6	
1 week	21.7	1.11	70.6	17.0	3.4
Antisolvent 2	21.8	1.11	71.1	17.2	
1 week	21.2	1.09	72.3	16.9	2.9
Antisolvent 3	21.5	1.10	72.7	17.2	
1 week	21.5	1.06	72.8	16.6	3.5
Antisolvent 4	22.8	1.08	70.7	17.4	
1 week	22.5	1.08	68.7	16.7	4.0
FIRA 1	22.4	1.08	68.2	16.5	
1 week	21.3	1.06	70.9	16.0	3.0
FIRA 2	21.3	1.07	72.0	16.4	
1 week	20.7	1.05	73.6	16.0	2.4
FIRA 3	21.8	1.11	71.1	17.2	
1 week	20.7	1.09	74.0	16.9	2.9
FIRA 4	22.1	1.08	69.5	16.6	
1 week	20.9	1.06	72.7	16.1	3.0

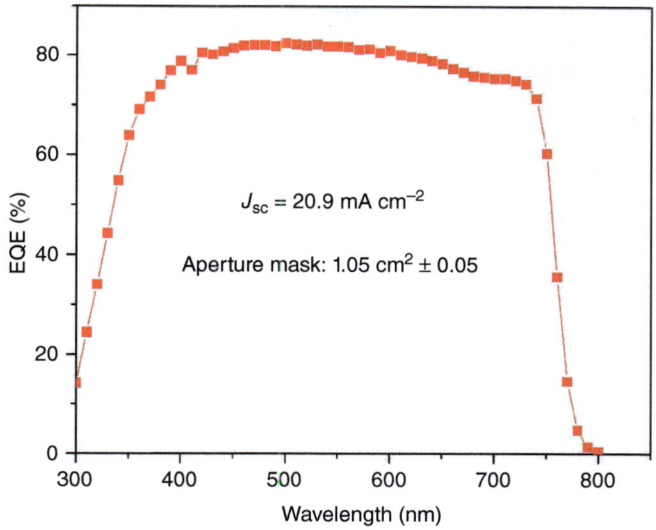

Figure 2.23 External quantum efficiency measured for an FIRA device with active area of 1.05 cm^2.

Figure 2.24 Intensity-modulated photocurrent spectroscopy, yielding the imaginary part of the current response in the low-frequency range for four devices.

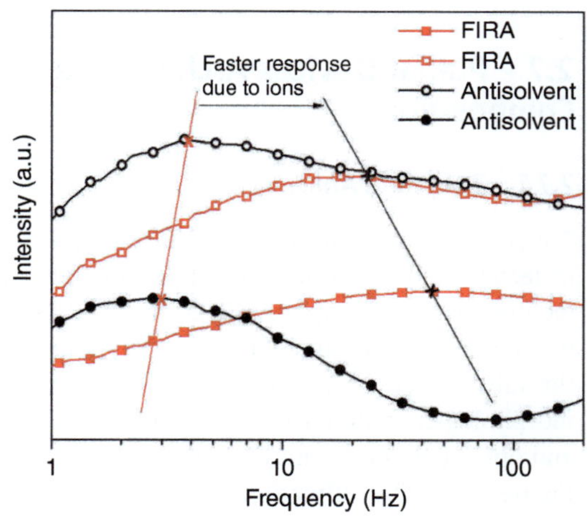

layer, which then heats the perovskite solution film to initiate the crystallization, while the substrate stays close to room temperature. This result indicates the compatibility of FIRA with the use of polymeric substrates, further enhancing its use in a roll-to-roll manufacturing process. Preliminary results indicating the feasibility of manufacturing uniform large-area perovskite films are shown in Figure 2.25b, both for glass and polymeric (PET) substrates (100 cm^2), documenting the utility of the FIRA method for large surface areas. To confirm the effectiveness of FIRA for large surfaces, devices with 1.4 cm^2 of active area are tested. Figure 2.25c shows the J–V curve for the champion large-area device with PCE of 14.6%, achieving a V_{oc} of more than 1 V. However, FF, J_{sc}, and V_{oc} are lower than the FIRA devices measured

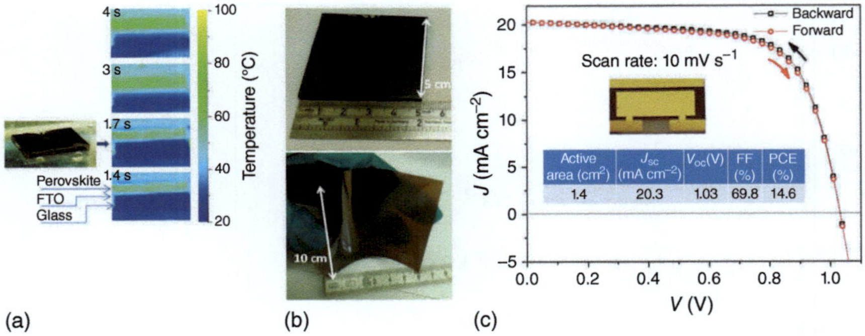

Figure 2.25 (a) Cross-sectional temperature variation of a perovskite film supported on a FTO glass substrate, measured in situ with an IR camera. (b) Images of large perovskite films made by FIRA on FTO-coated glass and ITO-coated PET (flexible) substrates. (c) Current density–voltage scans collected at 10 mV s^{-1} of the champion large-area device under 1.5AM irradiation and the corresponding photovoltaic performance parameters. Source: Adapted with permission from Sanchez et al. [7], © (2018) WILEY-VCH Verlag GmbH & Co. KGaA, Weinheim.

with an aperture area of 0.16 cm^2, but these PV parameters have nevertheless high enough values to be of interest for the scale-up of PSCs [99, 100].

2.7 Planar Devices Architecture and Mixed Perovskite Composition

2.7.1 Thin Film Analysis

The planar configuration has significant advantages from the upscaling point of view in terms of low cost and low environmental impact when compared with architectures with a mesoporous TiO_2 layer [101]. Using the FIRA method can result in a planar smooth and compact perovskite layer. The rapid crystal growth across the substrate can avoid the solution dewetting on the planar substrate. Despite the morphological differences, FIRA and AS perovskite films present similar structural and optical characteristics in terms of XRD and PL characterizations, as shown in Figure 2.26a. The diffraction pattern shows highly crystalline films with the tetragonal phase at room temperature corresponding to the mixed cations compounds with a large amount of FA (83%), which is stabilized by the addition of smaller cations such as MABr (17%) and CsI (5%), as previously reported [102, 103]. PL emission spectra correspond to the bandgap of the MAFACsPbIBr compound of about 1.6 eV and show in this case similar intensities and peak width [104–107]. Figure 2.26b shows a cross-sectional image of the manufactured devices comparing FIRA and AS. Both cases show the formation of pinhole-free and compact films.

2.7.2 PV Performance and Electronic Characteristic of the Devices

The regular architecture stack was used with FTO/TiO_2-planar/MAFACsPbIBr/spiro-OMeTAD/Au layers, as shown in the SEM cross-sectional image in

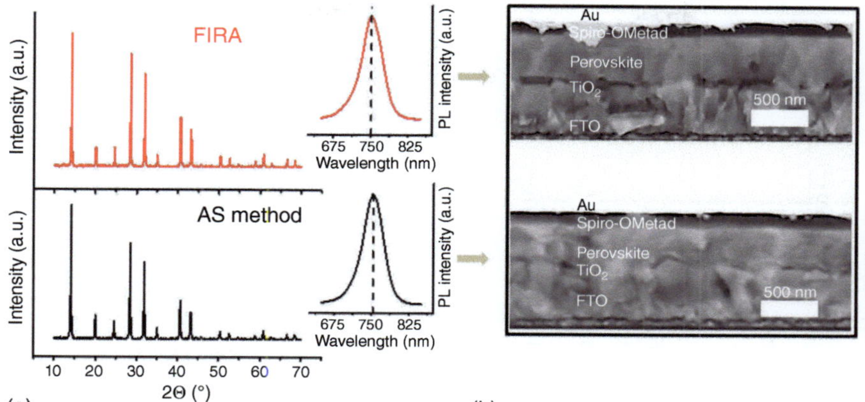

Figure 2.26 (a) XRD pattern (left) and photoluminescence spectra (right) of the FIRA and AS perovskite films. (b) Cross-sectional images of the devices with FIRA and AS perovskite films. Source: Adapted with permission from Sanchez et al. [9], © (2019) The Authors. Published by Elsevier Ltd.

Figure 2.26b. J–V characteristic of the champion devices in Figure 2.27a shows similar PV parameters for both studied methods, achieving a PCE of more than 18%. The forward scan displays little hysteresis [57, 108, 109] and has PCEs of >17%. Figure 2.27b compares FIRA and AS devices presenting statistics of the main PV parameters. Here, the standard deviation demonstrates a good reproducibility for both methods. Note that the data points are more spaced for FIRA than AS, as the median values are a bit lower in the FIRA case. We point out that all the FIRA processing was carried out in a standard fume hood, with a relative humidity of ~30%. Thus, the film formation can be improved even further by standardizing the atmospheric conditions [57].

Further insights on the device's working mechanisms can be obtained by means of impedance spectroscopy. The devices were measured under different light intensities at V_{oc} to evaluate charge recombination and charge storage behavior. The obtained Nyquist plot, with two distinguishable features at high and low frequencies, was adjusted to a simplified equivalent circuit, as previously reported [60, 110]. Although an in-depth model for PSCs is still under development, the results enable interesting observations regarding the operation mechanisms. High-frequency capacitance, which is associated with the perovskite geometric capacitance (C_g) [110], shows similar values for both FIRA and AS devices; see Figure 2.28a. These values are in good agreement with previous reports indicating a dielectric constant of $\varepsilon_r \sim 35$ regardless of the fabrication method. Similarly, low-frequency capacitance (C_{lf}) [111], with different interpretations in the literature but mostly pointing to an origin linked with ionic motion [98, 112–114], suggests that both analyzed preparation methods result in similar features in terms of ionic effect; see Figure 2.28. This observation is particularly interesting for the studied cases, as the grain size of the samples presents clear differences. As a direct result of the similar charge distribution of the perovskites, the recombination behavior of the analyzed samples does not present significant disparities. In particular, an estimation of the charge

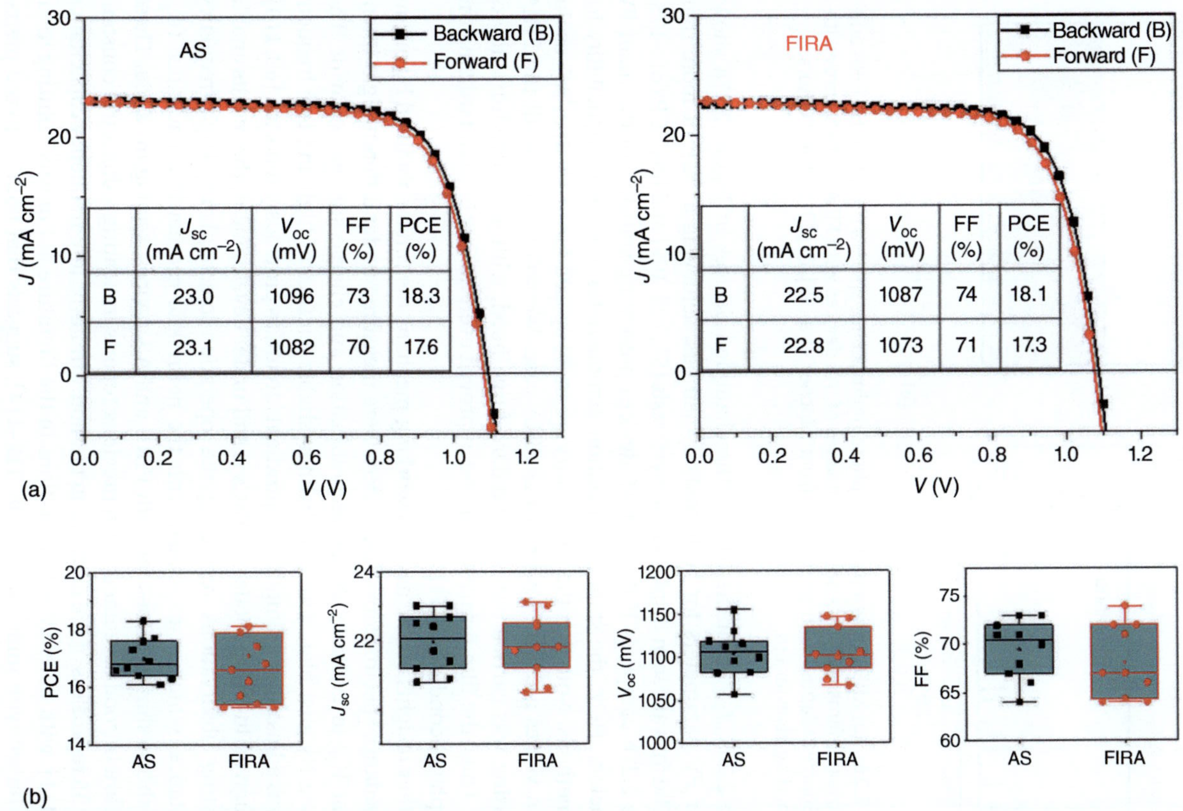

Figure 2.27 (a) Forward and backward JV scans of the champion devices made by FIRA and AS. The scan rate was 10 mV s^{-1} with an aperture area of 0.09 cm^2. (b) Statistic of the PV parameters for 10 devices made by FIRA and AS. Source: Adapted with permission from Sanchez et al. [9], © (2019) The Authors. Published by Elsevier Ltd.

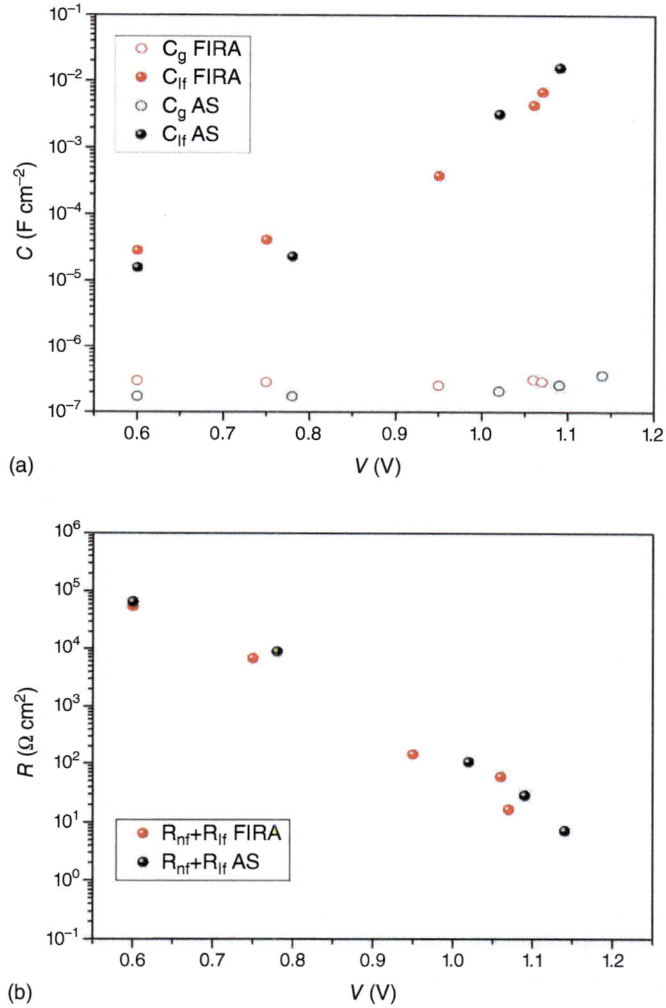

Figure 2.28 (a) Capacitive and (b) resistive parameters of FIRA- and AS-based perovskite solar cells measured under different light intensities at open circuit. The parameters are extracted from the fitting to the equivalent circuit discussed in the supporting information. Source: Adapted with permission from Sanchez et al. [9], © (2019) The Authors. Published by Elsevier Ltd.

recombination process can be extracted from the series addition of the resistances in both high- and low-frequency features [60, 115], displayed in Figure 2.28b).

2.8 Pulsed FIRA for Inorganic Perovskite Solar Cells

2.8.1 Thin Film Analysis

For all-inorganic perovskite lead iodide/bromide compositions, the boundary at 60% iodine and 40% bromine separates two different perovskite crystal structures.

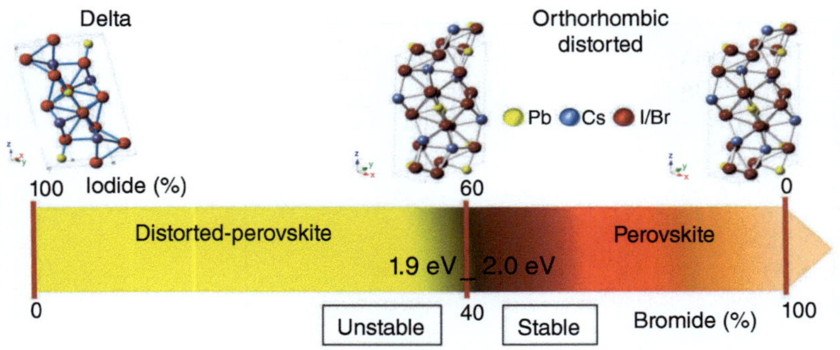

Figure 2.29 Illustration depicting the perovskite crystal structure as a function of the iodine/bromine ratio. Source: Adapted with permission from Sanchez et al. [7], © 2018 WILEY-VCH Verlag GmbH & Co. KGaA, Weinheim.

Compositions with iodine content lower than 60% and bromine content higher than 40% have a stable orthorhombic perovskite structure at room temperature [7]. $CsPbBr_3$ crystallizes in the orthorhombic (Pnma) space group. This adopts a distorted perovskite structure that occurs through two successive phase transitions at 88 and 130 °C, transforming the crystal structure to tetragonal (P4/mbm) and cubic (Pm-3m) [116–118]. However, compositions with bromine content lower than 40% and iodine higher than 60% evolve toward the delta phase, which is the characteristic of the 100% iodine composition. The phase stability sets the lower accessible bandgap for $CsPbI_xBr_{3-x}$ perovskite to 1.9–2 eV. The schematic found in Figure 2.29 summarizes the results of inorganic perovskite crystal structures as a function of the bromine/iodine ratio.

2.8.2 PV Performance

Figure 2.30a shows the *J–V* curves for champion devices prepared from $CsPbI_xBr_{(3-x)}$ with $x = 1.8$ (I60%, Br40%), 1.5 (I50%, Br50%), and 1.2 (I40%, Br60%). The insets show a PCE histogram for 10 devices of each composition. The composition with the highest content of iodide enabled higher PCE devices, as expected from the smaller bandgap. Devices prepared with the bromine-richer perovskite had the highest V_{oc} value of 1.3 V consistent with the largest bandgap [104]. EQE spectrum collected for the champion device is shown in Figure 2.30b, displaying a short-circuit current of 12 mA cm^{-2}, which is in close agreement with the value (12.3 mA cm^{-2}) extracted from the *JV* curve in Figure 2.30a. EQE plot also confirms the bandgap of roughly 2.0 eV (640 nm) for this particular halide composition.

The stability of PSCs with the three stoichiometries was tested in two ways. The first batch of devices was irradiated for 1000 hours with an AM1.5 spectrum under an inert atmosphere, followed by a *J–V* measurement. In a second stability test, perovskite layers were annealed at 200 °C for 1 hour before finalizing the device manufacturing with the deposition of the hole-selective layer and the gold contact. Figure 2.30c shows the PCE of freshly made (control) and stressed devices under light (100 hours AM1.5) or high temperature (1 hour at 200 °C), including the PCE

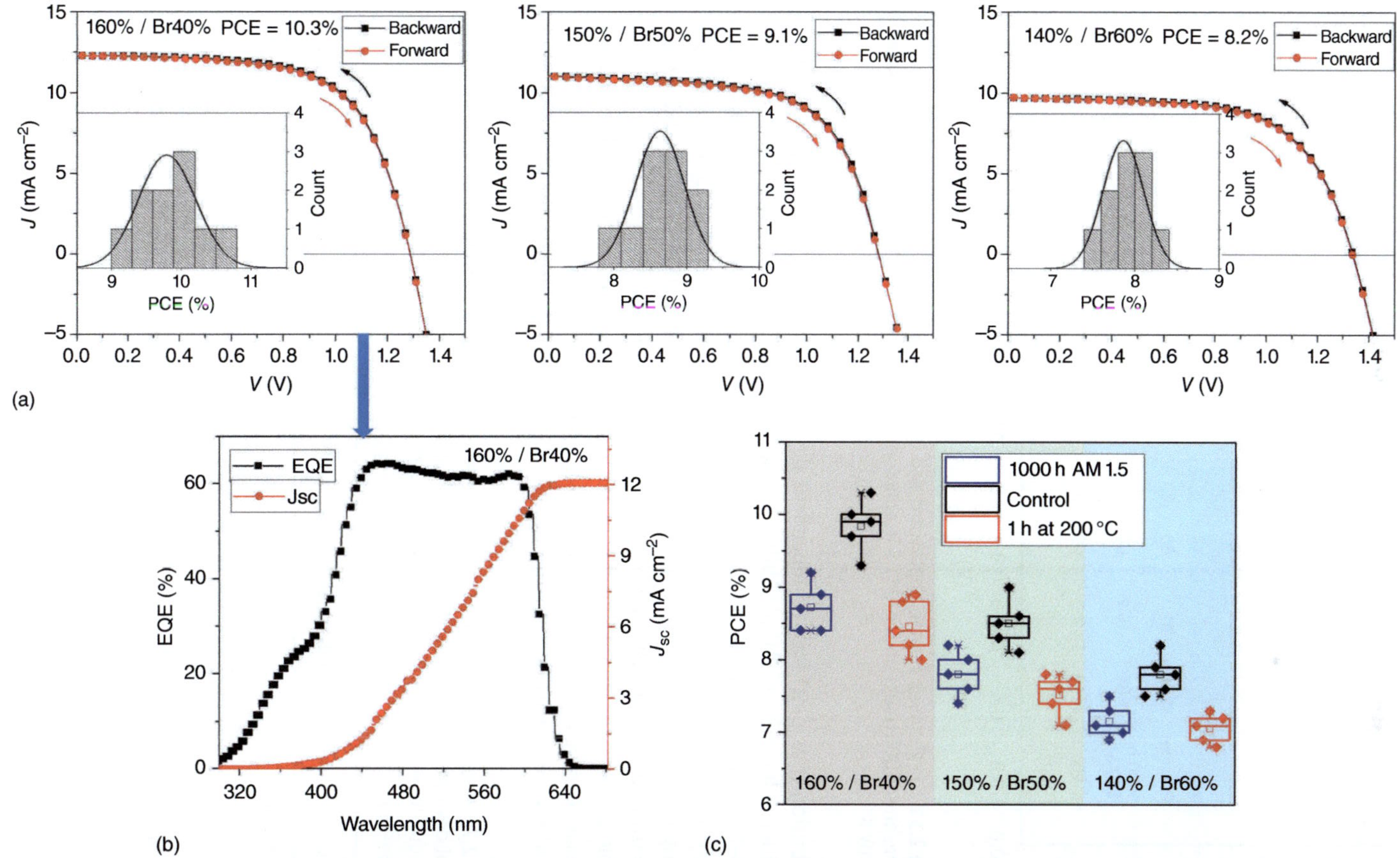

Figure 2.30 Performance and stability of PSCs prepared using 3 different inorganic $CsPbI_xBr_{(3-x)}$ compositions, with $x = 1.8$, 1.5, and 1.2. (a) *J–V* curves of the champion devices of each composition and the corresponding PCE histogram for 10 identically prepared devices. (b) EQE of the device prepared using the champion composition $CsPbI_{1.8}Br_{1.2}$. (c) PCE box plot comparing freshly made devices with devices stressed for 1000 hours under AM1.5 light and devices fabricated from perovskite films thermally stressed for 1 hour at 200 °C. Each box contains data from five independent devices. All *J–V* curves were acquired in ambient atmosphere with a humidity around 35% using a slow voltage scan rate of $10\,mV\,s^{-1}$ and AM1.5 simulated solar light without preconditioning the device. Source: Reproduced with permission from Sanchez et al. [7], © 2018 WILEY-VCH Verlag GmbH & Co. KGaA, Weinheim.

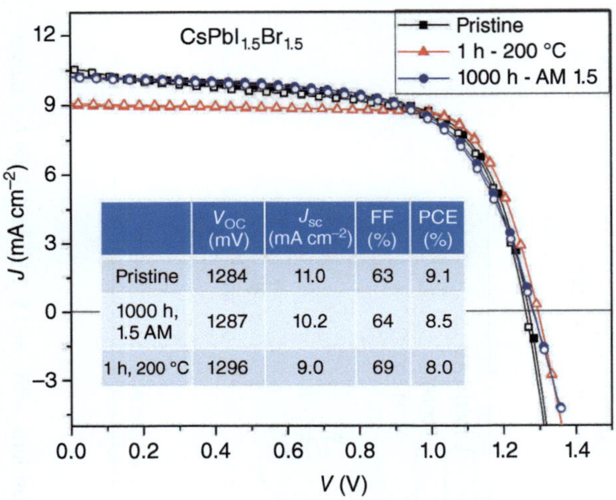

Figure 2.31 *J–V* curves scanned backward and forward of inorganic perovskite films, pristine, after 1000 hours under AM1.5 irradiation and heated to 200 °C for 1 hour; scan rate of 10 mV s^{-1}.

distribution (box plot) for 5 independent devices prepared for each stoichiometry. Despite the harshness of the stability measurements, PCE values deteriorated only by 10–16% compared with the control devices, with bromine-rich devices showing slightly higher stability (10–11% losses) than the iodide-rich devices (13–16% losses). A comparison of the *J–V* curves for 1000 hours AM1.5 and 200 °C-annealed devices made from the $CsPbI_{1.5}Br_{1.5}$ perovskite is shown in Figure 2.31.

Table 2.7 lists the champion device performance parameters collected for each chemical composition. Performance losses upon both light and thermal stress arise almost exclusively from the deterioration of the J_{sc} while the V_{oc} and the

Table 2.7 Photovoltaic performance parameters collected from the champion devices, for control devices, irradiated for 1000 hours with an AM1.5 spectrum and annealed at 200 °C for 1 hour, for each perovskite halide stoichiometry: $CsPbI_{1.8}Br_{1.2}$, $CsPbI_{1.5}Br_{1.5}$, and $CsPbI_{1.2}Br_{1.8}$.

Chem.	Record values	V_{oc} (mV)	J_{sc} (mA cm^{-2})	FF (%)	PCE (%)
I60% Br40%	Control	1288	12.3	65	10.3
	1000 h AM1.5	1287	10.9	66	9.2
	1 h at 200 °C	1295	10.4	66	8.9
I50% Br50%	Control	1282	11.0	65	9.1
	1000 h AM1.5	1281	10.2	63	8.2
	1 h at 200 °C	1294	9.1	66	7.8
I40% Br60%	Control	1332	9.7	64	8.2
	1000 h AM1.5	1291	8.5	68	7.5
	1 h at 200 °C	1296	8.0	70	7.3

FF remained unaffected or even slightly improved. Specifically, the light-soaking experiment for 1000 hours under 1 sun demonstrates a photostable material for all three compositions, which is consistent with the report by E. T. Hoke et al. [32] for mixed halide compositions where the reversibility of the photoinduced halide segregation is dominant [119–121].

2.9 Rapid Manufacturing of PSCs with an Adapted Perovskite Chemical Composition

2.9.1 Rapid Annealed TiO_2 Mesoscopic Film

Rapid thermal annealing through FIRA can also be applied for the sintering of TiO_2 mesoscopic layers. Figure 2.32a shows the temperature profile of the FIRA as a function of time, consisting of 10 cycles of 15 seconds of FIRA-on and 45 seconds of FIRA-off [59]. The consumed time was 10 minutes for the sintering of one batch (150 cm^2 of total area can be annealed in one time), plus 10 minutes more with a similar annealing procedure for the lithium treatment (see the section of methods). The maximum recorded temperature was 640 °C, and the median was roughly 620 °C. The highest temperature was kept for only a few seconds. It went down quickly due to the water-cooling system and gas flow. Note that this temperature was measured on the sample holder in the chamber, which is constantly under cooling. Figure 2.32b shows X-ray pattern of a TiO_2 film annealed by our regular procedure, i.e. on a hotplate at 450 °C during 30 minutes (the whole annealing time lasts 3 hours), and a TiO_2 film annealed by the FIRA cycles. As a blank reference we show an FTO substrate, for which the peak at 2θ 25° is representative of the anatase phase [47, 49, 50]. FIRA and conventional (hotplate) annealing results in a similar crystal structure as observed from the XRD.

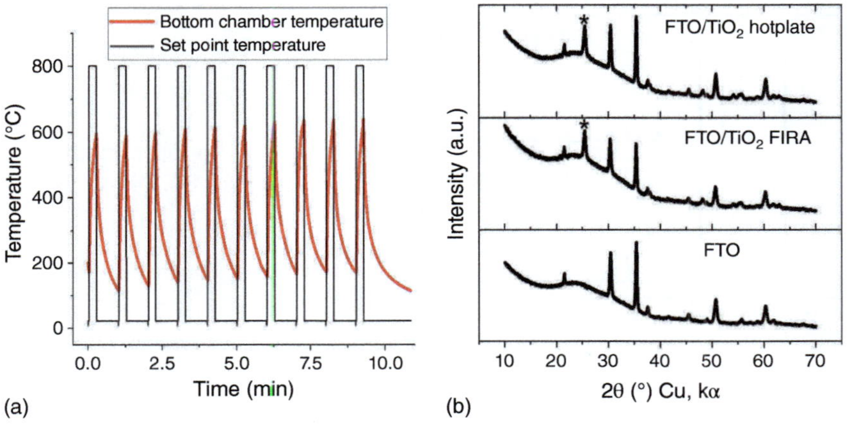

Figure 2.32 (a) Temperature as a function of time for TiO_2 mesoporous FIRA-annealed films. (b) X-ray pattern for FTO and TiO_2 films. Source: Adapted with permission from Sánchez et al. [59], © 2019 The Author(s). Published by Elsevier Ltd.

2.9.2 FCG Perovskite Stabilized with TBAI

After spin coating the perovskite solution in the electrodes, films were annealed during 1.6 seconds of pulse time and kept in the FIRA chamber for an additional 10 seconds (see the section of materials and methods).

Figure 2.33a shows the X-ray pattern for FA-based perovskite compositions, with cation stoichiometry variations. The patterns do not show a significant difference, and the α black phase in the desired three-dimensional cubic system matches all of them, being similar to the results reported by Qifeng Han et al. for $FAPbI_3$ single crystal [51]. A small difference is, however, observed in the patterns, i.e. for peaks at 2-theta 22° and 34°, which can be due to different cation stoichiometries. Therefore, there is no evidence of the yellow δ phase [34, 38, 40, 52]. The similarity in peak position and intensity suggests that the additional Cs and GA cations can be integrated into the lattice with the FIRA method [53–55]. The chosen annealing time is set to 1.6 seconds, giving the purest FA-based lead iodide perovskite phase. At this FIRA time, there is no evidence of a 2-theta peak at 12°, which is attributed to the remaining PbI_2 phase [56, 61, 122].

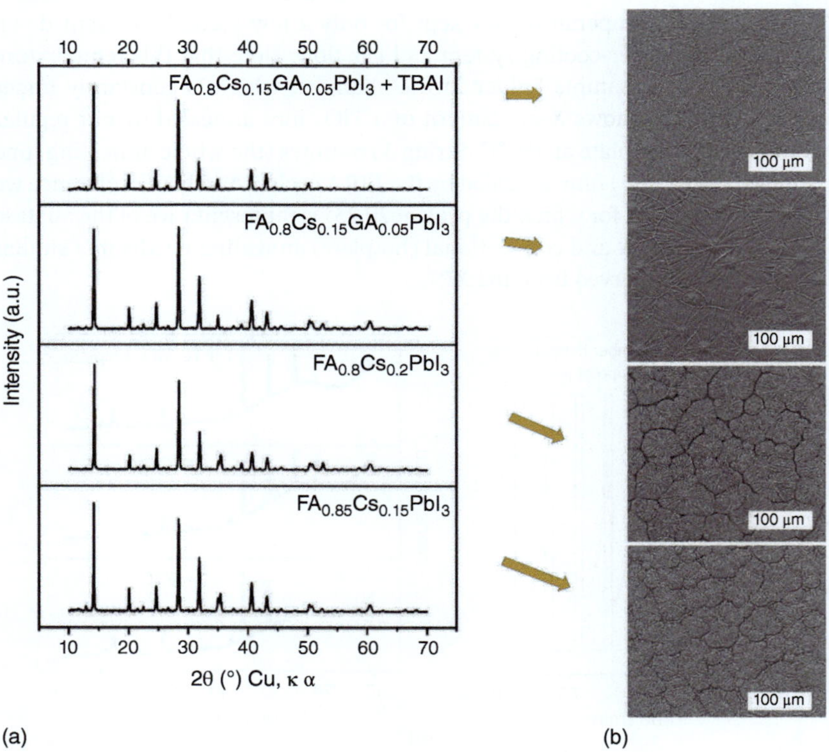

Figure 2.33 Formamidinium-based perovskite thin film. (a) X-ray pattern. (b) SEM top-view images. Source: Adapted with permission from Sánchez et al. [59], © 2019 The Author(s). Published by Elsevier Ltd.

To verify the film crystallization quality, Figure 2.33b shows SEM top-view images of the related pattern perovskite composition. Here, FA-Cs-based perovskite composition does not show a compact film, since gaps can be seen in the grain domain boundary. This can be attributed to the difference in chemical potential of the solutions (before annealing), from the ratios of the cation concentrations, where an excess of cesium can slow down the characteristic FIRA crystal growth, as previously reported [64]. To solve this, a large GA cation was added, which provides extra hydrogen bonding to increase the adhesive molecular force from solution to the substrate, decreasing the surface energy [34, 37, 62]. Therefore, we are able to grow a bigger grain domain, which collapses at the boundary with other neighborhood grains after FIRA annealing. This results in a compact thin film, as shown in Figure 2.33b).

Guanidinium cation addition contributes to planarize and smoothen out the perovskite films under FIRA synthesis, avoiding short circuits due to voids at the grain boundaries (i.e. cesium lead iodide tends to have an island growth mechanism; Figure 2.33b. Regarding the stability of the film and avoiding the phase polymorphism of $FAPbI_3$ [27, 52, 65], a posttreatment was included as a second step, with an additional FIRA annealing at low temperature. A few milligrams of TBAI molecule were dissolved in 1 ml of IPA to spin drop after the perovskite deposition. Afterward, the samples were annealed for 15 minutes at 100 °C in inert atmosphere (in FIRA furnace mode; see technical details in Section 2.10). To prove the efficacy of this treatment, two films with FCG perovskite, one with TBAI treatment and one without, were aged in ambient condition, as shown in Figure 2.34, an optical image taken in transmission mode.

One month later, FCG composition was degraded, as shown in the optical image and the respective collected spectrum found in Figure 2.34, top. Similar to the image, no optical absorption is displayed with respect to the pristine spectrum (estimated from optical transmittance). In contrast, FCG perovskite with TBAI treatment remains stable, and no significant differences can be seen with respect to the optical transmittance spectrum after one month (Figure 2.34, bottom).

2.9.3 PV Performance of the Manufactured PSCs

FCG perovskite is used in this study in a regular PSC architecture containing FTO/TiO_2/perovskite/spiro/Au. PV performances of the three devices with similar architecture but different perovskite compositions and manufacturing methods are also being studied. FCG perovskite is synthesized by FIRA and AS (both devices use TiO_2 mesoporous layer made by FIRA cycles), and the mixed (FA-MA-Cs) composition was made with regular (AS) procedure, which includes the TiO_2 mesoporous layer annealed in a standard hotplate. First, Figure 2.35 shows the *J–V* curves of the champion PSCs and their respective PV parameters, all of them with PCE over 20% denoting high-performance devices. The hysteresis of the all-FIRA FCG-processed device is slightly higher than the AS-processed perovskite,

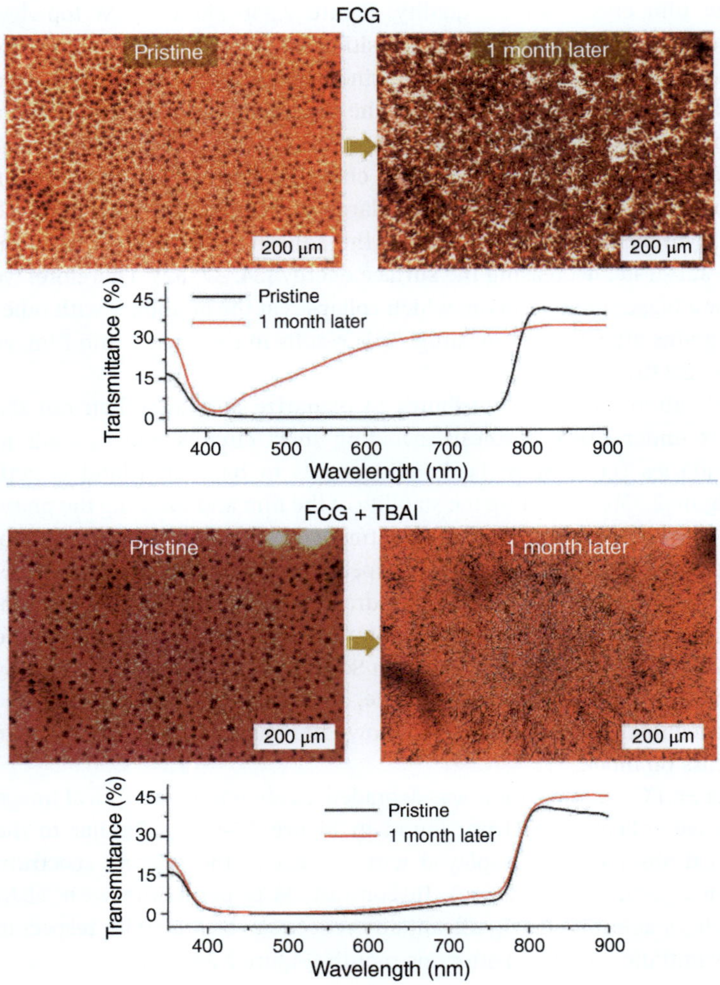

Figure 2.34 Optical images taken in transmission mode for FCG perovskite thin film with and without tetrabutylammonium (TBAI) treatment and optical transmittance spectrum for the chosen image pixels. The films were aged one month in ambient conditions, 40% of relative humidity and in shadow. Source: Adapted with permission from Sánchez et al. [59], © 2019 The Author(s). Published by Elsevier Ltd.

as observed by the difference in PCE for the forward scan, having a decrease of 5% with respect to the backward initial scan.

Figure 2.36a shows the IPCE of an all-FIRA-synthetized FCG device, showing a high EQE and the lowest bandgap energy with respect to the mixed composition, in agreement with its halide stoichiometry, which redshift the optical bandgap. In addition, Figure 2.36b shows the statistics of the PV parameters for 50 manufactured devices of the three chosen sets of PSCs, with its respective manufacturing methods

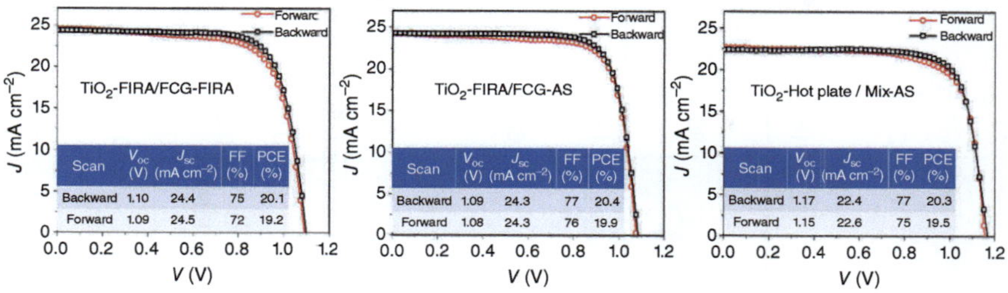

Figure 2.35 *J–V* curves and forward and backward scans of the champion devices, with scan speed of $10\,mV\,s^{-1}$ and mask aperture area of $0.1\,cm^2$. The measurements were carried out under constant light and by a *J–V* scan each for 10 minutes. Source: Adapted with permission from Sánchez et al. [59], © 2019 The Author(s). Published by Elsevier Ltd.

and perovskite composition. Interestingly, the results show a similar median average value of PCE, being 1% lower for the all-FIRA-FCG batch. This is directly related to the slightly lower FF for this batch of PSCs with respect to the AS-processed perovskite, which suggests that FIRA processing under ambient conditions (without inert atmosphere) may limit the quality of the perovskite film, as previously reported [14, 64]. However, with the FF median average of 69.8% ± 0.464%, there has been an improvement with respect to the previous related works. This can be attributed to the adapted FCG-FIRA perovskite composition and the TBAI posttreatment, reducing the series resistance of the devices. With respect to the V_{oc} and J_{sc}, the statistics show a good agreement between the perovskite optical bandgap energy and its higher V_{oc} for mixed perovskite and higher J_{sc} for FCG perovskite.

To prove the utility of FIRA-FCG process for larger-area devices, Figure 2.36c shows the *J–V* curve of a champion device, measured with $1.4\,cm^2$ of light aperture area. PCE achieved is up to 17.0% with 16.4% in the forward scan, with a median average of 15.9% ± 0.126% for 30 devices made for this experience. Finally, Figure 2.36d shows the stability measurement for FCG-FIRA and mixed devices with 24% and 20% losses, respectively, after 200 hours at 25 °C. The mixed device shows slightly smaller decays than the FCG-FIRA device. However, both compositions behave almost similarly after the first 50 hours. A reversible degradation component can be the cause of the PCE losses, but they remain stable after 200 hours of recorded data.

2.10 Outlook and Technical Details

2.10.1 Optimization of FIRA Process for Tandem Solar Cells

The control of the crystallization kinetics of the perovskite and metal oxide thin films serves to develop an industrial rapid thermal process. For this, it is necessary to carry out a series of optimizations of the FIRA thermodynamic parameters to obtain an optimum thin film quality for a large-scale process. The main parameters to be optimized are the intensity and the duration of the IR pulse, both combined with the pulse frequency for a series of cycles. For example, a tandem solar cell made with

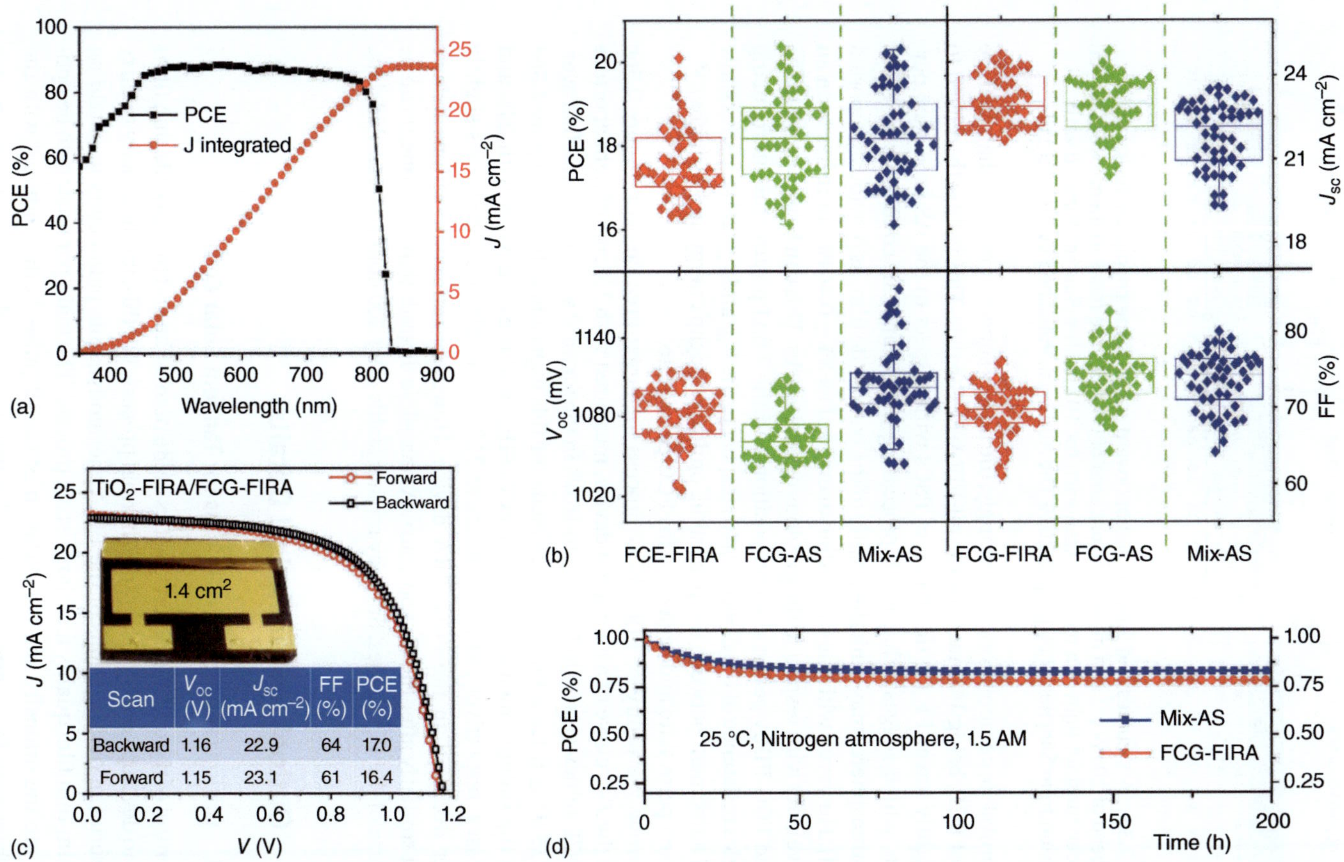

Figure 2.36 (a) IPCE and integrated photocurrent density of the champion FCG-FIRA device. (b) Statistics of the PV parameter for 50 samples of the three studied devices. (c) J–V curves for the champion large-area device with a mask aperture area of 1.4 cm^2 and scan speed of 10 mV s^{-1}. (d) Stability measurement recorded at 25 °C in nitrogen atmosphere during 200 hours at 1.5AM, for FCG and mixed devices. Source: Adapted with permission from Sánchez et al. [59], © 2019 The Author(s). Published by Elsevier Ltd.

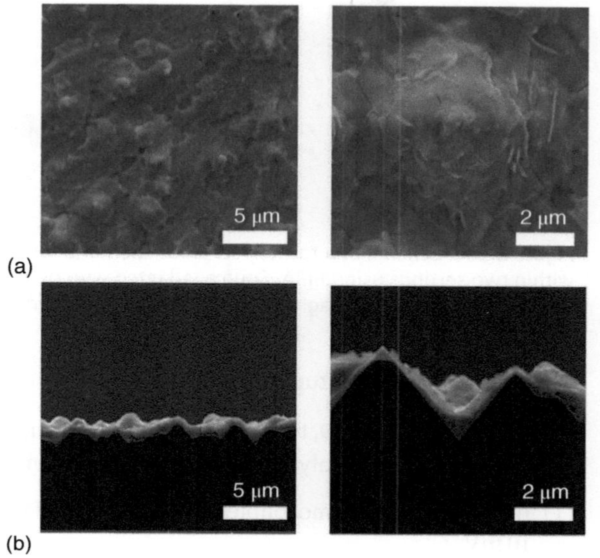

Figure 2.37 SEM images of the silicon-textured wafer/perovskite heterojunction. (a) Top-view SEM image. (b) Cross-sectional SEM image.

FIRA is one of the possible applications. Figure 2.37 shows a $MAPbI_3$ film annealed on top of the textured silicon wafer. The pulse time duration is 1.2 seconds, similar for FTO substrates, and the optimal solution molarity is 1.7 M in a ratio solvent of 2 v/v DMF: 1 v/v DMSO. Interestingly, perovskite film can conformally cover all the textured silicon area (Figure 2.37a). Figure 2.37b shows a cross-sectional image with pinhole-free films in the interface silicon/perovskite, which is fundamental for high-performance devices.

2.10.2 Automatic Roll-to-Roll System for the FIRA Manufacture of Perovskite Solar Cells

Predicting the commercialization of PSCs, Figure 2.38 shows a schematic of a possible roll-to-roll production method employing FIRA to crystallize the perovskite film. The perovskite precursor solution can be deposited onto the substrate using any of the roll-to-roll compatible methods available on the market [123], such as doctor blading, slot casting, spray coating, screen printing and inject printing. The deposited film is then passed through the FIRA box that enables rapid (within two seconds) perovskite film crystallization. By extending FIRA application to the coating of metal oxide hole conducting layers for solar cells, a single system can manufacture a complete perovskite solar cell [58, 124–126]. At this stage, hole conducting layers will be directly deposited and annealed on top of the perovskite film to create a working heterojunction solar cell. Electron and hole conducting layers such as NiO, SnO_2, and TiO_2 can form a planar inorganic architecture, all deposited from solution and rapid thermal processing.

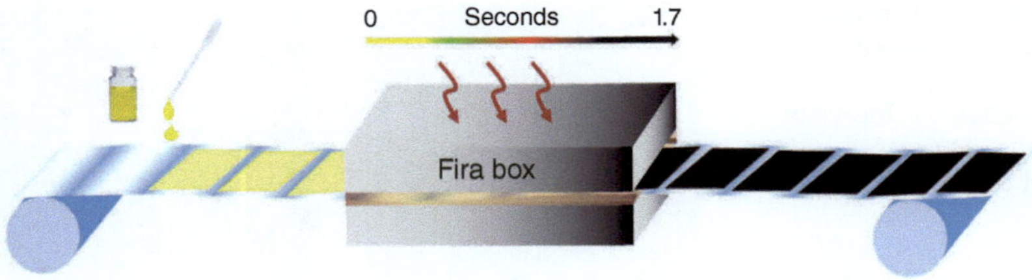

Figure 2.38 Scheme of a roll-to-roll production line for perovskite thin films crystallized within two seconds using FIRA. Source: Adapted with permission from Sanchez et al. [7], © (2018) WILEY-VCH Verlag GmbH & Co. KGaA, Weinheim.

2.10.3 Electronic Setup

As shown in Figure 2.39, to drive the high power output of the IR lamps, FIRA electric power setup is mainly composed of the following:

(1) Interface pour thermocouple K type, structured with an AD595 device to deliver $10\,mV\,°C^{-1}$.
(2) Module DAC Meilhaus to measure the temperature in degree Celsius.
(3) Power supply transformer: 230/5 V DC.
(4) Converter DC/DC Traco with voltage in +5 V and voltage out +12/−12 V.
(5) Three solid-state relays: 600 V AC/25 A.
(6) Three-phase power system: 400 V/16 A.

For the regulation of the heat power, a pulse width modulation signal (PWM) follows the PID signal output from 0 to 10. The network frequency is 50 Hz (with a period of 20 ms), and the three solid-state relays can be turned ON or OFF with a positive or negative alternation (10 ms). PWM max (100) corresponds to $100 \times 10\,ms = 1000\,ms$ (1 second). PWM value varies between 0 and 100. The particularity of solid-state relays is that the ON or OFF only occurs during the crossing of zero current (this fact avoids generating high peak disturbances in the power grid).

2.10.4 LabView Interface

The guide user interface is governed by a LabView script, as shown in Figure 2.40. Two modes of operation are represented in an intuitive interface. One mode is temperature proportional–integral–derivative (PID) regulation, which is similar to the temperature controller of any standard furnace. With this mode, any kind of temperature profile up to 600 °C can be programmed. The second mode is called full power and serves for turning on (at full power) the lamps. No PID controller is required to regulate the temperature. This mode is mainly used for rapid thermal annealing, having a pulse duration of 20 ms. A pyrometer is used as an extra sensor to record the surface temperature during fast annealing.

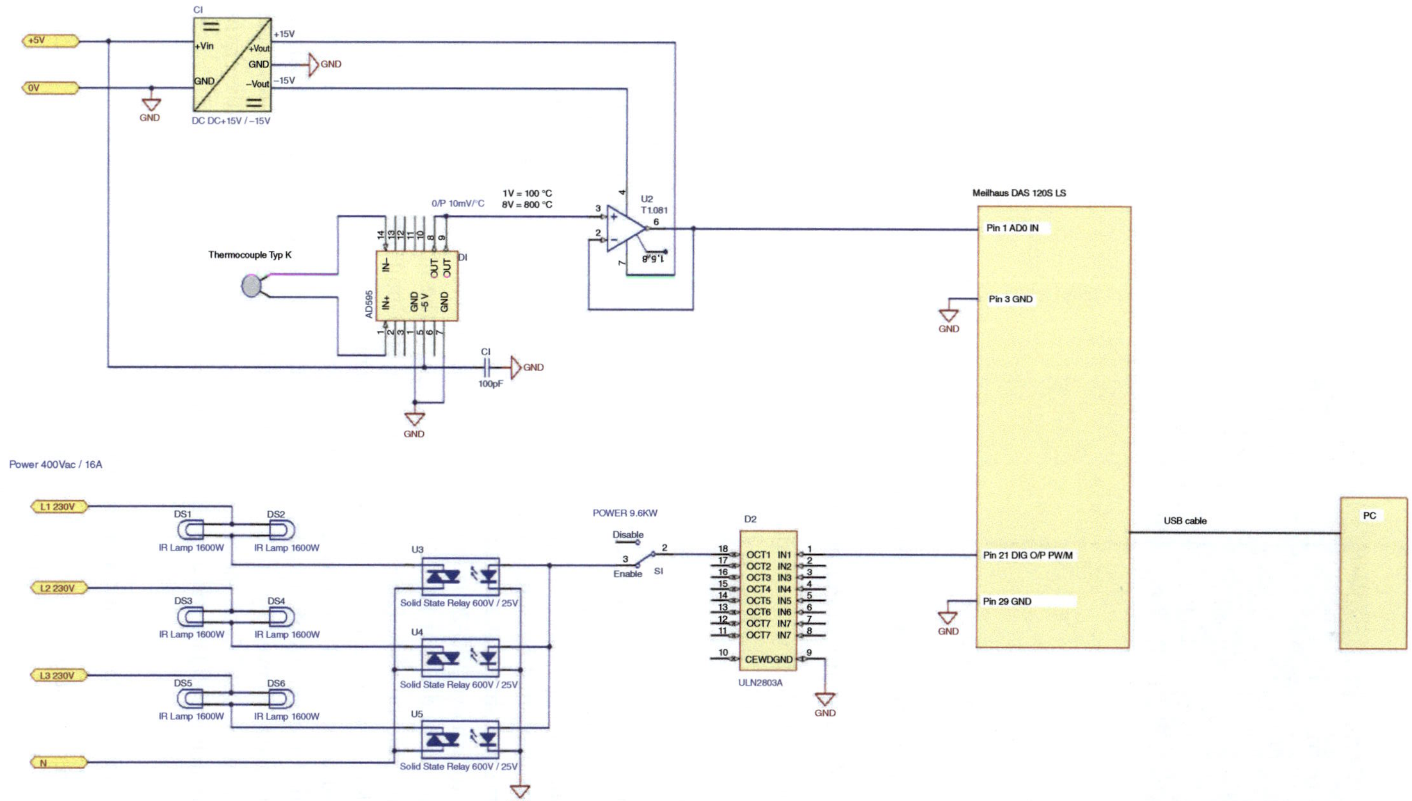

Figure 2.39 Schematic illustration of the electronic configuration of the FIRA oven.

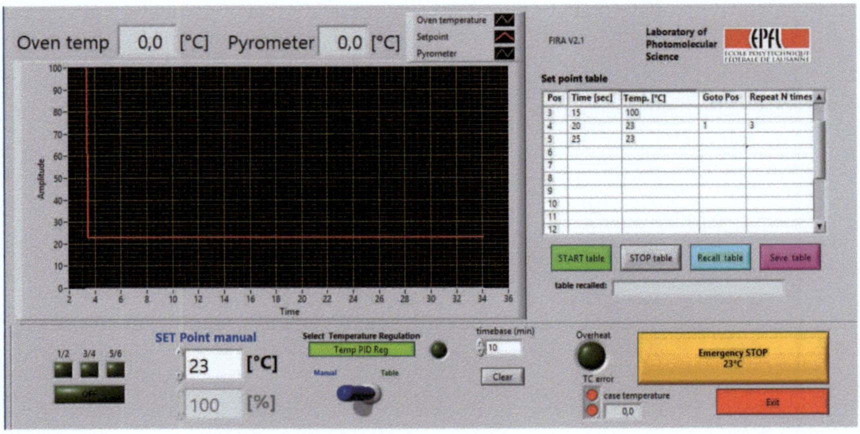

Figure 2.40 An image of the LabView interface of the FIRA oven.

2.11 Experimental Methods

2.11.1 Manufacture of Perovskite Solar Cells

Photovoltaic devices were fabricated on fluorine-doped tin oxide (FTO)-coated glass (Pilkington NSG **TEC™).** The substrates were cleaned with Hellmanex soap, followed by 30 minutes sonication in a 2% water solution of Hellmanex, 15 minutes sonication in IPA, and 5 minutes of oxygen plasma etching. Then, 30 nm-thick TiO_2 compact layers were deposited onto the FTO by spray pyrolysis at 450 °C from a precursor solution of titanium diisopropoxide bis(acetylacetonate) in anhydrous ethanol and acetylacetonate. After spraying, FTO substrates were left at 450 °C for 5 minutes before cooling to room temperature. Then, a mesoporous TiO_2 layer was deposited by spin coating for 10 seconds at 4000 rpm with a ramp of 2000 rpm s^{-1}, using a 30 nm particle size TiO_2 paste (Dyesol 30 NR-D) diluted in ethanol to achieve 150–200 nm-thick mesoporous layers. After spin coating, FTO substrates were dried at 100 °C for 10 minutes, and the films were annealed on a programmable hotplate (2000 W, Harry Gestigkeit GmbH) to crystallize TiO_2 at 450 °C for 30 minutes under dry airflow. Li doping of the mesoporous TiO_2, as described elsewhere [97], was accomplished by spin coating a 0.1 M solution of bis(trifluoromethane)sulfonimide lithium salt (Li-TFSI) in acetonitrile at 3000 rpm for 10 seconds followed by another sintering step at 450 °C for 30 minutes. After cooling down to 150 °C, all the substrates were kept in the fume hood (with relative humidity below 38%) for the deposition of the perovskite films.

2.11.2 Perovskite Solution Preparation

$MAPBI_3$ perovskite precursor solution contains MAI (1.2 M) and PbI_2 1.2 M in anhydrous 3 v/v DMF: 1 v/v DMSO. The organic salts were purchased from Greatcell: the cesium and lead compounds from TCI. The hybrid perovskite precursor solutions were deposited from a precursor solution containing CH_5N_2I (1 M), PbI_2 (1.1 M), CH_3NH_3Br (0.2 M), and $PbBr_2$ (0.2 M) in anhydrous DMF:DMSO 4 : 1 (v:v). Note that this composition contains a lead excess as reported elsewhere [98]. Then CsI, dissolved as a 1.5 M stock solution in DMSO, was added to the mixed perovskite precursor to achieve the desired triple-cation composition. The inorganic perovskite precursor solution was deposited from a precursor solution containing CsI (1.2 M). Moreover, to obtain the three final stoichiometries of $CsPbI_xBr_{3-x}$, a variation of PbI_2 (0.2, 0.3, and 0.4 M) and PbBr2 (1, 0.9, and 0.8 M) for $x = 1.2, 1.5$, and 1.8, respectively, was used.

2.11.3 Antisolvent Method

Perovskite films were deposited as previously reported [99, 122]. The perovskite solution was spin-coated in two steps at 1000 and 6000 rpm for 10 and 20 seconds, respectively. During the second step, 100 μl of chlorobenzene was poured onto the spinning substrate, 5 seconds before the end of the program. The substrates were then annealed at 100 °C for 1 hour in a nitrogen-filled glovebox. After perovskite annealing, the substrates were cooled down within a few minutes.

2.11.4 FIRA Method

The films made by FIRA method include the spin coating of the perovskite solution in a single step at 4000 rpm for 10 seconds. The substrates were then IR irradiated for 1.7 seconds in the FIRA oven and were kept there for ten additional seconds before the removal and cooling within several minutes. The samples made by pulsed FIRA were exposed to between 0 and 25 of 20 millisecond pulse duration followed by a 2 second pulse. All the substrates were kept in the oven for 10 additional seconds before the removal. FIRA processing was carried out in a standard fume hood. One of the most important parameters is the relative humidity during manufacture, which may well be the fundamental reason for the performance difference between antisolvent and FIRA devices. These devices were made in a fume hood with ≈30% relative humidity, which is a drawback that must be addressed in reaching the highest performances.

2.11.5 HTM Deposition and Back Contact Evaporation

Immediately after cooling the perovskite films, a Spiro-OMeTAD (Merck) solution (70×10^{-3} M in chlorobenzene) was deposited by spin coating at 4000 rpm for 20 seconds. Spiro-OMeTAD was doped with Li-TFSI (Sigma-Aldrich), tris(2-(1*H*-pyrazol-1-yl)-4-*tert*-butylpyridine)-cobalt(III) tris(bis(trifluoromethylsulfonyl)imide) (FK209, Dynamo), and 4-*tert*-butylpyridine (TBP) (Sigma-Aldrich). The molar ratios of the additives to Spiro-OMeTAD were 0.5, 0.03, and 3.3 for Li-TFSI, FK209, and TBP, respectively. For the inorganic perovskite devices, poly(triarylamine) (PTAA) (Merck) solution (35 mM in toluene) was deposited by spin coating at 4000 rpm for 20 seconds and was doped with bis(trifluoromethylsulfonyl)imide lithium salt (Li-TFSI) (Sigma-Aldrich),) and 4-*tert*-butylpyridine (TBP) (Sigma-Aldrich). The molar ratios of the additives to PTAA were 0.0825 and 3.9 for Li-TFSI and TBP, respectively. The inorganic HTM layers were deposited as follows. Bismuth triodide (Merck) solution (50 mM in tetrahydrofuran) and bis(methylcyclopentadienyl)nickel(II) (Merck) solution (0.1 M in isopropyl alcohol) were deposited by spin coating at 4000 rpm for 20 seconds. The inorganic HTM layers were synthesized by FIRA; the bismuth solution film was annealed for 3 seconds at maximum power, and the nickel metal–organic film was annealed from 10 to 40 minutes at temperatures between 150 and 450 °C. Finally, a two-step of thermal evaporation was used to evaporate first 1 nm of chromium, following a 80 nm-thick gold top electrode under high vacuum. The purpose of the thin chromium layer is to avoid gold migration into device and to improve the stability of the device as previously discussed [19].

2.11.6 Device Characterization

For photovoltaic measurements, a solar simulator from ABET Technologies (Model 11016 Sun 2000) with a xenon arc lamp was used, and the solar cell response was recorded using a Metrohm PGSTAT302N Autolab. The intensity of the solar simulator was calibrated to 100 mW cm^{-2} using a silicon reference cell from ReRa Solutions (KG5 filtered). J–V curves were measured in reverse and forward bias at a scan rate of 10 mV s^{-1}. A shadow mask was used to define an aperture of device active area of 0.16 cm^2. The aperture of devices termed "large area" was 1.4 cm^2. Maximum power point tracking was used to perform stability experiments. Incident-modulated photocurrent spectroscopy (IMPS) was performed according to the procedure described in the literature using a 625 nm LED driver at short-circuit conditions, a light intensity of 100 mW cm^{-2}, and a Metrohm PGSTAT302N Autolab. EQE setup was situated inside a glovebox filled with an inert atmosphere (nitrogen). It consisted of an Oriel Instruments QEPVSI-b system with a Newport 300 W Xenon arc lamp, a 35.5 Hz chopper, and a Newport Cornerstone 260 monochromator. A silicon diode was used to calibrate the lamp to subtract the background light noise. The monochromatic light spot was controlled such that the spot was located inside the active area of the device and the sample was approximately 10.5 mm away from the light source. EQE measurements were carried out at short circuit.

2.11.7 Material Characterization

Scanning electron microscopy and energy-dispersive X-ray spectroscopy were carried out on a Tescan MIRA 3 LMH with a field emission source operated at an acceleration voltage of 10 kV using an octane-pro EDS detector. Electron backscattering diffraction was performed with HKL Advanced EBSD System using a Nordlys nano camera. Powder X-ray diffraction was performed in transmission geometry with Cu target (λ = 1.5401 Å) using an STOE STADI P diffractometer.

Optical transmission measurements were performed using a Zeiss Axio-Scope A1 Pol using a Zeiss EC Epiplan-Apochromat 20× objective and a xenon light source (Ocean Optics HPX-2000). For spectroscopic measurements, an optical fiber (QP230-2-XSR, 230 μm core) collected the light reflected from the sample. The spectra were recorded by a spectrometer (Ocean Optics Maya2000 Pro), and a standard white diffuser was used as a reference.

Differential scanning calorimetry (DSC) measurements were performed using a Mettler-Toledo STAR system under a N_2 atmosphere. Experiments involving different heating and cooling cycles were performed on ~3 mg samples placed in standard DSC pans at 10 °C min^{-1}. Thermogravimetric analysis for powder materials was carried out in a PerkinElmer TGA 4000. Two to three milligram samples were placed in ceramic pans, and the weight loss was monitored between 25 and 600 °C at a rate of 10 °C min^{-1} under a nitrogen flow rate of 20 ml min^{-1}. Powder samples were carefully scraped off the substrates right after annealing. All spectra were obtained at room temperature in transmission mode on the as-prepared sample surfaces. During the measurements, the pressure in the chamber did not exceed 2×10^{-10} mbar.

2.11.8 Temperature Measurement

The temperature in the FIRA chamber was measured using a K-type thermocouple, and the surface temperature of the FTO was measured by a wire J-type thermocouple, both connected to an Analog Devices *AD595* amplifier interfaced with LabView software. The surface temperature of the perovskite film was measured in situ by a *FLIR T62101 IR Video Camera*.

List of Abbreviations

Au	gold
bkg	background
DSC	differential scanning calorimetry
DSSC	dye-sensitized solar cell
EBDS	electron back-scattering diffraction spectroscopy
EDS	energy-dispersive X-ray spectroscopy
EQE	external quantum efficiency
ESL	electron-selective layer
ETM	electron transporting material
eV	electron volt
FF	fill factor
FIRA	flash infrared annealing
FTO	fluorine-doped tin oxide
HI	hydrogen iodide
HTM	hole transporting material

IMPS	intensity-modulated photocurrent spectroscopy
IR	infrared
ITO	indium-doped tin oxide
J_{SC}	short-circuit current
MPPT	maximum power point tracking
NiO	nickel oxide
LDP	low-dimensional perovskite
Li-TFSI	bis(trifluoromethane)sulfonimide lithium salt
PCE	power conversion efficiency
PDF	pair distribution function
p-FIRA	pulsed flash infrared annealing
PSC	perovskite solar cell
PV	photovoltaic
SEM	scanning electron microscopy
spiro-OMeTAD	octakis(4-methoxyphenyl)-9,9′-spirobi[9*H*-fluorene]-2,2′,7,7′-tetramine
SnO_2	tin dioxide
TCO	transparent conducting oxide
TGA	thermogravimetric analysis
THF	tetrahydrofuran
TiO_2	titanium dioxide
UV	ultraviolet
V_{oc}	open-circuit potential
wp-FIRA	warmed-pulsed flash infrared annealing
XRD	X-ray diffraction

Acknowledgments

The EU Horizon 2020 project "WASP" with project number 825213 is gratefully acknowledged for its financial contribution to this study.

References

1 Chen, J.-S., Ober, C.K., Poliks, M.D. et al. (2004). *Polymer* 45: 1939–1950.
2 Abdelmageed, G., Mackeen, C., Hellier, K. et al. (2018). *Sol. Energy Mater. Sol. Cells* 174: 566–571.
3 Bae, S., Kim, S., Lee, S.-W. et al. (2016). *J. Phys. Chem. Lett.* 7: 3091–3096.
4 Leijtens, T., Hoke, E.T., Grancini, G. et al. (2015). *Adv. Energy Mater.* 5: 1500962.
5 Domanski, K., Correa-Baena, J.-P., Mine, N. et al. (2016). *ACS Nano* 10: 6306–6314.

6 Im, J.-H., Lee, C.-R., Lee, J.-W. et al. (2011). *Nanoscale* 3: 4088–4093.

7 Sanchez, S., Christoph, N., Grobety, B. et al. (2018). *Adv. Energy Mater.* 8: 1802060.

8 Sanchez, S., Hua, X., Phung, N. et al. (2018). *Adv. Energy Mater.* 8: 1702915.

9 Sanchez, S., Vallés-Pelarda, M., Alberola-Borràs, J.-A. et al. (2019). *Mater. Today* 31: 39–46.

10 Burschka, J., Pellet, N., Moon, S.-J. et al. (2013). *Nature* 499: 316.

11 Niu, G., Guo, X., and Wang, L. (2015). *J. Mater. Chem. A* 3: 8970–8980.

12 Niu, G., Li, W., Li, J. et al. (2017). *RSC Adv.* 7: 17473–17479.

13 Tian, M., Wang, J., Kurtz, J. et al. (2003). *Nano Lett.* 3: 919–923.

14 Jackson, K.A. (2010). *Kinetic Processes: Crystal Growth, Diffusion, and Phase Transitions in Materials*. Weinheim: Wiley-VCH Verlag GmbH & Co. KGaA.

15 Li, Z., Yang, M., Park, J.-S. et al. (2016). *Chem. Mater.* 28: 284–292.

16 Venables, J.A., Spiller, G.D.T., and Hanbucken, M. (1984). *Rep. Prog. Phys.* 47: 399.

17 Liang, P.-W., Liao, C.-Y., Chueh, C.-C. et al. (2014). *Adv. Mater.* 26: 3748–3754.

18 Schwartz, R.W. (1997). *Chem. Mater.* 9: 2325–2340.

19 Cacciuto, A., Auer, S., and Frenkel, D. (2004). *Nature* 428: 404–406.

20 De Yoreo, J.J. and Vekilov, P.G. (2003). *Rev. Mineral. Geochem.* 54: 57–93.

21 Sakai, N., Wang, Z., Burlakov, V.M. et al. (2017). *Small* 13: 1602808.

22 Zalewski, E.F. and Duda, C.R. (1983). *Appl. Opt.* 22: 2867–2873.

23 Pascoe, A.R., Gu, Q., Rothmann, M.U. et al. (2017). *Sci. China Mater.* 60: 617–628.

24 Venables, J.A. (1973). *Philos. Mag.: J. Theor. Exp. Appl. Phys.* 27: 697–738.

25 Wang, Y., Shi, Y., Xin, G. et al. (2015). *Cryst. Growth Des.* 15: 4741–4749.

26 Salim, T., Sun, S., Abe, Y. et al. (2015). *J. Mater. Chem. A* 3: 8943–8969.

27 Mukhopadhyay, M. and Dalvi, S.V. (2005). *J. Chem. Technol. Biotechnol.* 80: 445–454.

28 Doshi, P., Rohatgi, A., Ropp, M. et al. (1996). *Sol. Energy Mater. Sol. Cells* 41: 31–39.

29 Xiao, M., Huang, F., Huang, W. et al. (2014). *Angew. Chem. Int. Ed.* 53: 9898–9903.

30 Cölfen, H. and Antonietti, M. (2005). *Angew. Chem. Int. Ed.* 44: 5576–5591.

31 Saidaminov, M.I., Abdelhady, A.L., Murali, B. et al. (2015). *Nat. Commun.* 6: 7586.

32 Hoke, E. T., Daniel J. Slotcavage, Emma R. Dohner, et al.. Reversible photo-induced trap formation in mixed-halide hybrid perovskites for photovoltaics. *Chem. Sci.* 6, 613–617 (2015).

33 Ashkenazy, Y. and Averback, R.S. (2010). *Acta Mater.* 58: 524–530.

34 Kazuo, N., Michio, A., and Naozo, W. (1980). *Jpn. J. Appl. Phys.* 19: L563.

35 Terry, M.L., Straub, A., Inns, D. et al. (2005). *Appl. Phys. Lett.* 86: 172108.

36 Payne, D.N.R., Chan, C.E., Hallam, B.J. et al. (2016). *Sol. Energy Mater. Sol. Cells* 158 (Part 1): 102–106.

37 Xu, J., Hu, Z., Jia, X. et al. (2016). *Org. Electron.* 34: 84–90.
38 Troughton, J., Charbonneau, C., Carnie, M.J. et al. (2015). *J. Mater. Chem. A* 3: 9123–9127.
39 Troughton, J., Carnie, M.J., Davies, M.L. et al. (2016). *J. Mater. Chem. A* 4: 3471–3476.
40 Ye, J., Zheng, H., Zhu, L. et al. (2017). *Sol. Energy Mater. Sol. Cells* 160: 60–66.
41 Ninomaru, F.I.H. (1985). *JST Mater.* 69: 5.
42 Yue, L., Yan, B., Attridge, M., and Wang, Z. (2016). *Sol. Energy* 124: 143–152.
43 El-Bashir, S.M., Yahia, I.S., Binhussain, M.A., and AlSalhi, M.S. (2017). *Results Phys.* 7: 1852–1858.
44 Aouaj, M.A., Diaz, R., Belayachi, A. et al. (2009). *Mater. Res. Bull.* 44: 1458–1461.
45 Stair, R., Schneider, W.E., and Jackson, J.K. (1963). *Appl. Opt.* 2: 1151–1154.
46 Stair, R., Schneider, W.E., and Fussell, W.B. (1967). *Appl. Opt.* 6: 101–105.
47 Sanchez, S., Hua, X., Günzler, A. et al. (2020). *Cryst. Growth Des.* 2: 670–679.
48 Song, Z., Watthage, S.C., Phillips, A.B., and Heben, M.J. Pathways toward high-performance perovskite solar cells: review of recent advances in organo-metal halide perovskites for photovoltaic applications. *J. Photon. Energy* 6: 022001.
49 Yao, Z., Jones, T.W., Grigore, M. et al. (2018). *ACS Appl. Mater. Interfaces* 10: 14673–14683.
50 Aranda, C., Cristobal, C., Shooshtari, L. et al. (2017). *Sustainable Energy Fuels* 1: 540–547.
51 Gao, L.-L., Zhang, K.-J., Chen, N., and Yang, G.-J. (2017). *J. Mater. Chem. A* 5: 18120–18127.
52 Roldan-Carmona, C., Gratia, P., Zimmermann, I. et al. (2015). *Energy Environ. Sci.* 8: 3550–3556.
53 Xiao, M., Huang, F., Huang, W. et al. (2014). *Angew. Chem.* 126: 10056–10061.
54 Yang, W.S., Noh, J.H., Jeon, N.J. et al. (2015). *Science* 348: 1234–1237.
55 Bi, D., Yi, C., Luo, J. et al. (2016). *Nat. Energy* 1: 16142.
56 Hu, Q., Zhao, L., Wu, J. et al. (2017). *Nat. Commun.* 8: 15688.
57 Chen, B., Yang, M., Priya, S., and Zhu, K. (2016). *J. Chem. Phys. Lett.* 7: 905–917.
58 Jung, Y.-S., Hwang, K., Heo, Y.-J. et al. (2018). *Adv. Opt. Mater.* 6: 1701182.
59 Sánchez, S., Jerónimo-Rendon, J., Saliba, M., and Hagfeldt, A. (2020). *Mater. Today* 35: 9–15.
60 Zarazua, I., Han, G., Boix, P.P. et al. (2016). *J. Phys. Chem. Lett.* 7: 5105–5113.
61 Lee, J., Christopher Orilall, M., Warren, S.C. et al. (2008). *Nat. Mater.* 7: 222–228.
62 Nayak, P.K., Moore, D.T., Wenger, B. et al. (2016). *Nat. Commun.* 7: 13303.

63 Dualeh, A., Tétreault, N., Moehl, T. et al. (2014). *Adv. Funct. Mater.* 24: 3250–3258.

64 Fang, X., Wu, Y., Lu, Y. et al. (2017). *J. Mater. Chem. C* 5: 842–847.

65 Song, Z., Watthage, S.C., Phillips, A.B. et al. (2015). *Chem. Mater.* 27: 4612–4619.

66 Hao, F., Stoumpos, C.C., Liu, Z. et al. (2014). *J. Am. Chem. Soc.* 136: 16411–16419.

67 Miller, C.E. (1977). *J. Cryst. Growth* 42: 357–363.

68 Lovette, M.A. and Doherty, M.F. (2013). *Cryst. Growth Des.* 13: 3341–3352.

69 Zhou, X., Zhang, Y., Kong, W. et al. (2018). *J. Mater. Chem. A* 6: 3012–3021.

70 Yan, K., Long, M., Zhang, T. et al. (2015). *J. Am. Chem. Soc.* 137: 4460–4468.

71 Wang, Y., Guan, X., Li, D. et al. (2017). *Nano Res.* 10: 1223–1233.

72 Xia, B., Wu, Z., Dong, H. et al. (2016). *J. Mater. Chem. A* 4: 6295–6303.

73 Sangwal, K. (1998). *Prog. Cryst. Growth Charact. Mater.* 36: 163–248.

74 Eperon, G.E., Burlakov, V.M., Docampo, P. et al. (2014). *Adv. Funct. Mater.* 24: 151–157.

75 Bi, D., Luo, J., Zhang, F. et al. (2017). *ChemSusChem* 10: 1624–1630.

76 Tsai, H., Nie, W., Cheruku, P. et al. (2015). *Chem. Mater.* 27: 5570–5576.

77 Correa-Baena, J.-P., Anaya, M., Lozano, G. et al. (2016). *Adv. Mater.* 28: 5031–5037.

78 Huang, F., Pascoe, A.R., Wu, W.-Q. et al. (2017). *Adv. Mater.* 29: 1601715.

79 Yang, G.-J., Ding, B., Li, Y. et al. (2017). *J. Mater. Chem. A* 5: 6840–6848.

80 Sanchez, S., Steiner, U., and Hua, X. (2019). *Chem. Mater.* 31: 3498–3506.

81 Dualeh, A., Gao, P., Seok, S.I. et al. (2014). *Chem. Mater.* 26: 6160–6164.

82 Juarez-Perez, E.J., Ono, L.K., Maeda, M. et al. (2018). *J. Mater. Chem. A* 6: 9604–9612.

83 Saito, Y., Iihara, J., Yamaguchi, K. et al. (2008). *SEI Tech. Rev.*: 27.

84 Brivio, F., Frost, J.M., Skelton, J.M. et al. (2015). *Phys. Rev. B* 92: 144308.

85 Billing, D.G. and Lemmerer, A. (2007). *Acta Crystallogr. Sect. B* 63: 735–747.

86 Billing, D.G. and Lemmerer, A. (2008). *New J. Chem.* 32: 1736–1746.

87 Svane, K.L., Forse, A.C., Grey, C.P. et al. (2017). *J. Phys. Chem. Lett.* 8: 6154–6159.

88 Tubbs, M.R. (1975). *Phys. Status Solidi B* 67: 11–49.

89 Nenon, D.P., Christians, J.A., Wheeler, L.M. et al. (2016). *Energy Environ. Sci.* 9: 2072–2082.

90 Fan, Z., Xiao, H., Wang, Y. et al. (2017). *Joule* 1: 548–562.

91 Pellegrino, G., D'Angelo, S., Deretzis, I. et al. (2016). *J. Phys. Chem. C* 120: 19768–19777.

92 Jain, S.M., Philippe, B., Johansson, E.M.J. et al. (2016). *J. Mater. Chem. A* 4: 2630–2642.

93 Jiang, Y., Leyden, M.R., Qiu, L. et al. (2018). *Adv. Funct. Mater.* 28: 1703835.

94 Frischknecht, R., Jungbluth, N., Althaus, H.-J. et al. (2005). *Int. J. Life Cycle Assess.* 10: 3–9.
95 Rosenbaum, R.K., Bachmann, T.M., Gold, L.S. et al. (2008). *Int. J. Life Cycle Assess.* 13: 532.
96 Jolliet, O., Margni, M., Charles, R. et al. (2003). *Int. J. Life Cycle Assess.* 8: 324.
97 Alberola-Borràs, J.-A., Baker, J.A., De Rossi, F. et al. (2018). *iScience* 9: 542–551.
98 Bag, M., Renna, L.A., Adhikari, R.Y. et al. (2015). *J. Am. Chem. Soc.* 137: 13130–13137.
99 Lee, J., Kang, H., Kim, G. et al. (2017). *Adv. Mater.* 29: 1606363.
100 Chen, W., Wu, Y., Yue, Y. et al. (2015). *Science* 350: 944–948.
101 Alberola-Borràs, J.-A., Vidal, R., Juárez-Pérez, E.J. et al. (2017). *Sol. Energy Mater. Sol. Cells* 179: 169–177.
102 Saliba, M., Matsui, T., Seo, J.-Y. et al. (2016). *Energy Environ. Sci.* 9: 1989–1997.
103 McMeekin, D.P., Sadoughi, G., Rehman, W. et al. (2016). *Science* 351: 151–155.
104 Jeon, N.J., Noh, J.H., Yang, W.S. et al. (2015). *Nature* 517: 476–480.
105 Bi, D., Tress, W., Dar, M.I. et al. (2016). *Sci. Adv.* 2.
106 Draguta, S., Sharia, O., Yoon, S.J. et al. (2017). *Nat. Commun.* 8: 200.
107 Pellet, N., Gao, P., Gregori, G. et al. (2014). *Angew. Chem. Int. Ed.* 53: 3151–3157.
108 Unger, E.L., Hoke, E.T., Bailie, C.D. et al. (2014). *Energy Environ. Sci.* 7: 3690–3698.
109 Snaith, H.J., Abate, A., Ball, J.M. et al. (2014). *J. Phys. Chem. Lett.* 5: 1511–1515.
110 Guerrero, A., Garcia-Belmonte, G., Mora-Sero, I. et al. (2016). *J. Phys. Chem. C* 120: 8023–8032.
111 Juarez-Perez, E.J., Sanchez, R.S., Badia, L. et al. (2014). *J. Phys. Chem. Lett.* 5: 2390–2394.
112 Jacobs, D.A., Shen, H., Pfeffer, F. et al. (2018). *J. Appl. Phys.* 124: 225702.
113 Moia, D., Gelmetti, I., Calado, P. et al. (2019). *Energy Environ. Sci.* 12: 1296–1308.
114 Zarazua, I., Bisquert, J., and Garcia-Belmonte, G. (2016). *J. Phys. Chem. Lett.* 7: 525–528.
115 Zolfaghari, Z., Hassanabadi, E., Pitarch-Tena, D. et al. (2019). *ACS Energy Lett.* 4: 251–258.
116 Manser, J.S., Christians, J.A., and Kamat, P.V. (2016). *Chem. Rev.* 116: 12956–13008.
117 Green, M.A., Emery, K., Hishikawa, Y. et al. (2015). *Prog. Photovoltaics Res. Appl.* 23: 1–9.
118 Stoumpos, C.C., Malliakas, C.D., Peters, J.A. et al. (2013). *Cryst. Growth Des.* 13: 2722–2727.
119 Lemmens, M.J.P. (2007). *Handbook of Magnetism and Advanced Magnetic Materials*. Wiley.
120 Tang, H., He, S., and Peng, C. (2017). *Nanoscale Res. Lett.* 12: 410.
121 Xu, Q., Yang, D., Lv, J. et al. (2018). *Small Methods* 2: 1700316.
122 Watson, W.J. (2002). *Org. Process Res. Dev.* 6: 201–202.
123 Hwang, K., Jung, Y.-S., Heo, Y.-J. et al. (2015). *Adv. Mater.* 27: 1241–1247.

124 Zuo, C., Vak, D., Angmo, D. et al. (2018). *Nano Energy* 46: 185–192.
125 Dou, B., Whitaker, J.B., Bruening, K. et al. (2018). *ACS Energy Lett.* 3: 2558–2565.
126 Jung, H.S. and Park, N.-G. (2015). *Small* 11: 10–25.

3

Passivation of Hybrid/Inorganic Perovskite Solar Cells

Muhammad Akmal Kamarudin and Shuzi Hayase

i-Powered Energy System Research Center (i-PERC) The University of Electro-Communications, 1-5-1 Chofugaoka, Chofu, Tokyo, 182-8585, Japan

3.1 Introduction

Figure 3.1 shows the rapid progress of perovskite solar cells, reaching more than 20% efficiency just within 10 years [13]. Perovskite solar cells evolved from dye-sensitized solar cells where the use of mesoporous TiO_2 structure is needed for anchoring dye sensitizers. However, recent works have shown that high-efficiency devices can also be achieved using a planar structure. The earlier method of preparing perovskite layer using a one-step method in which the precursors, i.e. CH_3NH_3I and PbI_2, are dissolved in a common solvent and spin-coated onto a substrate resulted in poor morphology and formation of pinholes. Thus, a two-step method is introduced by preparing the PbI_2 layer first, followed by deposition of methylammonium iodide (MAI) solution. As a result, complete infiltration of perovskite into mesoporous TiO_2, excellent coverage, and high-quality perovskite is obtained. Perovskite fabrication method is further improved by introducing binary solvent, i.e. dimethylsulfoxide (DMSO) and dimethylformamide (DMF), forming an adduct to allow for slow crystal growth. The efficiency is further improved when an antisolvent method is introduced to promote homogenous nucleation of perovskite crystals. Significant improvement in the efficiency of perovskite solar cells is achieved through compositional engineering using various A site cations and mixed halide compositions. As the efficiency is reaching its limit, researchers start to move toward interfacial engineering, aiming to reduce interfacial recombination.

Passivation of perovskites is currently the research trend in improving the device performance. Through passivation technique, the intrinsic (lattice vacancies, Schottky, and Frenkel defects) and extrinsic (impurities) defects can be suppressed. Ball et al. have published a review article regarding defects in perovskites and discussed the impact of defects on various solar cell parameters [14]. Figure 3.2 shows the schematic representation of various types of defects and their location [15]. Trap-assisted non-radiative recombination is known to be detrimental to open-circuit voltage (V_{OC}), which limits the efficiency of perovskite solar cells [16].

Perovskite Solar Cells: Materials, Processes, and Devices, First Edition.
Edited by Shahzada Ahmad, Samrana Kazim, and Michael Grätzel.

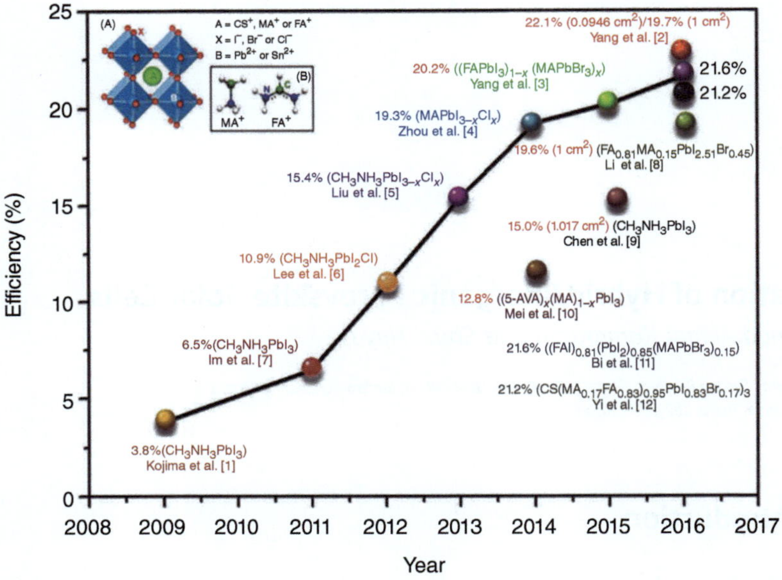

Figure 3.1 Progress of perovskite solar cells in recent years. Source: Fu et al. [13].

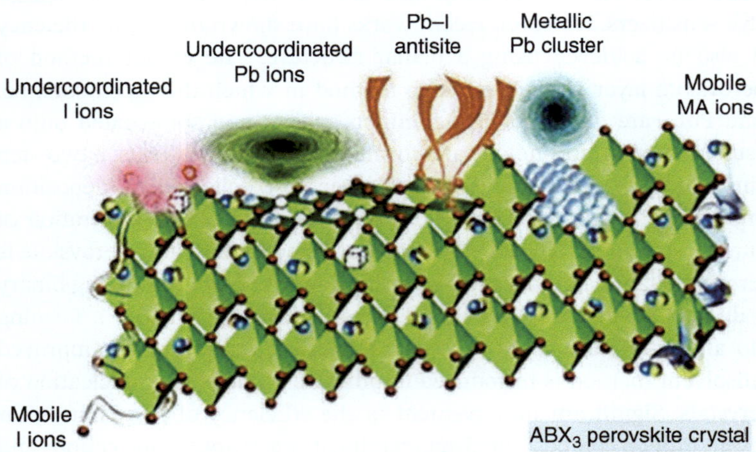

Figure 3.2 Schematic diagram showing various types of defects and passivation materials. Source: Chen et al. [15].

Passivation not only improved the V_{OC}, but it can also help to improve the air stability of perovskite solar cells as well as alleviating the hysteresis effect.

In this chapter, we will review recent progress in passivation of perovskite solar cells. As perovskite is a polycrystalline material, grain boundaries are extended into the perovskite bulk. Hence, we will compare both bulk and surface passivation techniques where the working mechanisms will be discussed. We will also review the materials and techniques that have been used to passivate the perovskite materials comparing both experimental and computational studies to further understand the progress of this field.

3.1.1 Types of Passivation

3.1.1.1 Bulk Passivation

Bulk passivation can be achieved by introducing the passivating agents directly into the perovskite films, and there are currently two common methods of introducing passivating material into perovskite bulk. In the first method, the passivating agent is added together in perovskite precursor solution. Meng et al. introduced poly(vinyl alcohol) (PVA) into $FASnI_3$ precursor solution where hydrogen bonding between –OH group in PVA and I^- in $FASnI_3$ solution control the perovskite nucleation rate and improve crystal quality [17]. As a result, the trap density measured from space charge limited current measurements is reduced from $3.5 \times 10^{17}\,cm^{-3}$ for the pristine device to $2.6 \times 10^{17}\,cm^{-3}$ for the polymer-modified device. They cited that the defects originating from iodide vacancy have been effectively suppressed by the formation of hydrogen bonds. This has been supported by Yin et al. through simulation study using density functional theory (DFT) model showed the changes of point defects with respect to various lattice vacancies [18]. Yang et al. then confirmed experimentally comparing perovskite samples with and without iodide deficiency [2]. Deep-level transient spectroscopy (DLTS) measurements showed that the trap density for the iodide-deficient device is $8.81 \times 10^{13}\,cm^{-3}$ compared to $5.28 \times 10^{14}\,cm^{-3}$ for the control device. The second method involves the dissolution of the passivating agent in anti-solvent, and upon dropping the antisolvent, the passivating material is incorporated within the perovskite layer. Guo et al. performed antisolvent treatment of $Cs_{0.05}FA_{0.81}MA_{0.14}PbI_{2.55}Br_{0.45}$ perovskite layer using 1*H*,1*H*-perfluorooctylamine (PFA) dissolved in chlorobenzene to suppress defects at grain boundaries and perovskite surface [19]. As a result, the trap density calculated from space charge limited current measurements reduced from $2.88 \times 10^{16}\,cm^{-3}$ to $1.36 \times 10^{15}\,cm^{-3}$ upon passivation. Niu et al. explored various organic halides to passivate the grain boundaries of $FAPbI_3$ [20]. In addition to reduced trap density, device stability in ambient air has also improved significantly where passivated device still show more than 90% of its original efficiency compared to 20% for control device after 80 days exposure to ambient environment.

3.1.1.2 Surface Passivation

Surface passivation has been gaining popularity in recent times as it effectively improves the performance of perovskite solar cells. In a perovskite solar cell, irrespective of the device structure, i.e. normal or inverted structure, there are three main interfaces.

1. Transparent conductive oxide (TCO)/electron transport layer (ETL) or hole transport layer (HTL)
2. ETL or HTL/perovskite
3. ETL or HTL/electrode

One of the commonly used ETLs in a normal structure device is TiO_2 due to its high electron mobility and suitable energy level. However, defect disorder, including Ti and O vacancies, may lead to the conversion of n-type behavior into p-type behavior [21]. Thus, surface passivation of TiO_2 is performed to suppress these defects, which could lead to electron–hole recombination. Zhen et al. have written an in-depth review regarding the surface modification of TiO_2 [22]. Li et al. showed that by passivating TiO_2 surface with PCBM, charge transfer from the perovskite layer into ETL has been enhanced [23]. A similar passivation technique can also be applied in other oxide-based ETLs. Surface passivation of SnO_2 with C60 fullerene derivatives showed a similar effect as a result of matching energy levels between ETL and perovskite aided by efficient charge extraction [24]. Energy barrier between conduction level of perovskite and ETL becoming smaller such that electron injection is more efficient.

poly(3,4-ethylenedioxythiophene):polystyrene sulfonate (PEDOT:PSS) is currently the most commonly used HTL for inverted structure perovskites solar cells (PSCs) [25, 26]. The best device performance reported using PEDOT:PSS is now reaching 22.3% [27]. However, due to the acidity of PEDOT:PSS, researchers are exploring inorganic HTLs [28, 29]. One candidate is NiO_x which has high charge carrier mobility and suitable work function [30]. The traps in NiO_x are originated from the presence of Ni^{3+}/Ni^{2+} together with nickel vacancies on the surface [31, 32]. Bromobenzoic acid has been used to passivate the surface of NiO_x, which helped to reduce the trap-assisted recombination at the NiO_x/perovskite interface [33]. This is confirmed by plotting the change of V_{OC} as a function of light intensity. The control device showed an ideality factor of 1.7, whereas the sample using bromobenzoic acid has an ideality factor of 1.5. A larger ideality factor suggests that the trap-assisted recombination is dominant [34]. Surface engineering via post-treatment with an organic basic solution of tetramethylammonium hydroxide significantly improves the device efficiency from 12.30% to 17.03% [35]. Tetramethylammonium hydroxide (TMAH) improves the interfacial contact between NiO_x and perovskite layer as suggested from atomic force microscopy (AFM) images showing lower roughness with TMAH treatment.

Commonly, passivation technique is performed on the perovskite surface. The passivation of the perovskite surface can be achieved by using organic halide salt phenethylammonium iodide to suppress the surface defects [36]. This results in the formation of a thin 2D PEA_2PbI_4 perovskite layer on the 3D perovskite. Wang et al. showed that the antisite defects originating from Pb are more prominent on the perovskite surface, and these can be effectively passivated with molecules possessing N—H and C=O groups [37]. The interaction between N—H and C=O groups with antisite Pb reduced the trap density of states by 1 order of magnitude. Apart from suppressing charge carrier recombination in perovskite, surface passivation could also improve the device stability [38]. Kim et al. compared the device stability with and without polyethylene oxide passivation under constant 1 sun illumination and found that the device with poly(ethylene oxide) (PEO) retained >95% of its initial efficiency while the one without PEO dropped down to 80% [38]. The thin PEO layer is proposed to encapsulate the perovskite layer while allowing charge transfer through tunneling effect. Additionally, there are reports that suggest simultaneous surface passivation and charge extraction using organic molecules [39, 40]. A π-conjugated Lewis base possessing C=O and CN groups is expected to coordinate with under-coordinated Pb, while the π-conjugated main chain allows for efficient charge transfer [41].

Degradation of the back electrode as a result of ion migration and metal oxidation in ambient air remains one of the challenges in achieving a stable device. To overcome these issues, the introduction of an interlayer between perovskite and back electrode has been proposed [42]. Tu et al. have synthesized a new type of quaternary ammonium salt, PTN–Br, to impede iodide diffusion from perovskite, which could react with the electrode [43]. PTN–Br also helped to passivate the presence of pinholes and avoid direct contact between perovskite and the electrode due to incomplete coverage of PCBM. Wu et al. investigated the effect of iodide and metal interaction by coating perovskite layer on Bi, Cu, Au, and Ag films and performed X-ray photoelectron spectroscopy (XPS) for fresh and aged samples [44]. Metallic Pb^0 peak has been observed for Cu and Ag films confirming the hypothesis that corrosion occurred when iodide reacts with the metal electrode. The perovskite solar cell prepared using Bi as interlayer showed almost no degradation even after 500 hours under constant light illumination at maximum power point.

3.1.2 Passivating Materials

3.1.2.1 Metal Halides

PbI_2, which is one of the precursors for perovskites, is used to passivate the defects at perovskite grain boundaries and surface (Figure 3.3a) [45]. Shi et al. compared the addition of excess PbI_2 in the two-step evaporation method and one-step antisolvent method to unravel the perovskite growth [48]. Depending on the fabrication method, the passivation effect differs such that in the case of the one-step antisolvent method, PbI_2 predominantly exists near the grain boundaries at the perovskite/HTL interface, while in the two-step evaporation-solution process, excess PbI_2 mainly present at the ETL/perovskite interface. Barbe et al. proposed that in order for this passivation technique to be effective, the thickness of PbI_2 should be around 20 nm [49]. The group used Raman spectroscopy and photocurrent mapping to evaluate locally the presence of excess PbI_2 and correlated it with the change in photogenerated current. In addition to lowering the defect density in perovskite, excess PbI_2 also helped to improve the growth orientation as observed from high-resolution transmission electron spectroscopy (HR-TEM), although the growth mechanism is not clear at this point [50]. Figure 3.3b shows passivation of halide vacancies by potassium. Potassium inhibits halide migration by forming a strong ionic bond and suppress non-radiative recombination originating from interstitial halides [46]. Passivation method using metal halide in tin perovskite solar cell has also been reported [51, 52]. According to Ito et al. bulk and surface XPS data revealed that the GeI_2 is forming a native oxide layer on tin perovskite surface rather than being incorporated into the perovskite lattice [53]. This has an effect on the device stability by maintaining 80% of its initial efficiency after 60 minutes in ambient air without encapsulation

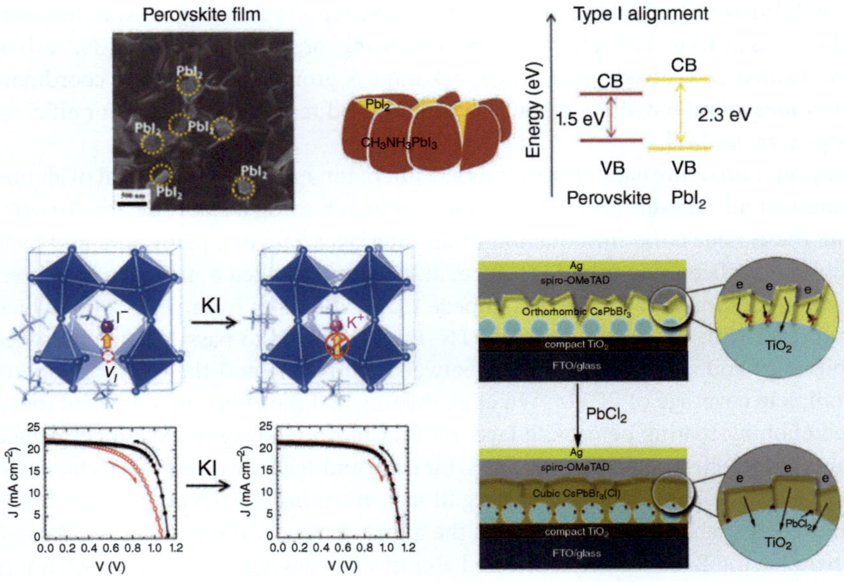

Figure 3.3 (a) SEM image and schematic diagram showing the location of excess PbI_2 at the perovskite grain boundary and the energy diagram comparing perovskite and PbI_2. Source: Reprinted with permission from Chen et al. [45]. (b) Potassium iodide passivating iodide vacancies by forming stronger K—I bonds, and as a result, *J*–*V* hysteresis is suppressed. Source: Son et al. [46]. (c) Surface passivation of TiO_2 using $PbCl_2$ suppressing TiO_2/perovskite interfacial defects. Source: Li et al. [47].

which is a remarkable performance for pure tin-based perovskite solar cell. Metal halide has also been used to passivate the surface of TiO_2, as shown in Figure 3.3c, where the small $PbCl_2$ grains filled the voids left by incomplete coverage of perovskite on mesoporous TiO_2, improving interfacial contact between perovskite/ETL [47]. This has been confirmed by Nakayashiki et al. using quartz crystal microbalance system showing the interaction between PbI_2 and TiO_2 forming Ti—O—Pb bonds [54]. Even after rinsing the TiO_2 surface with DMF, Pb still remains on the surface. Using a similar method, Zhang et al. performed surface passivation of carbon back electrode using PbI_2 and achieved a significant improvement of the efficiency from 2.21% for the reference device and 7.30% for the passivated device [55]. XPS revealed that the —OH groups on carbon surface react with PbI_2 to form C—O—Pb bonds and helped to suppress interfacial recombination of the perovskite/carbon electrode.

3.1.2.2 Organic Acids (—COOH, —SOOH, and —POOH)

The use of acids to passivate a metal oxide surface has been widely reported [56]. The carboxylic group (—COOH) chemically binds onto metal oxide surface, as shown in Figure 3.4a. Zu et al. have made a comparison of various self-assembled monolayers on perovskite solar cell performance [57]. Their work shows that a suitable choice of carboxylic group is important in achieving high-performing devices. The

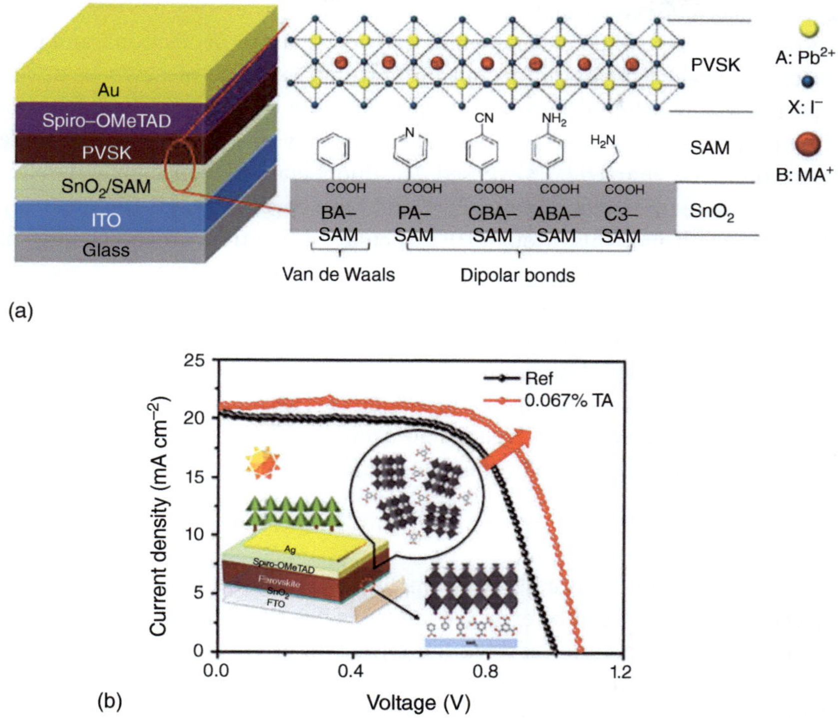

Figure 3.4 (a) Schematic diagram showing various molecules with carboxylic group functional group anchored onto SnO_2 surface. Source: Zuo et al. [57]. (b) –COOH can also passivate both TiO_2 surface and perovskite bulk. Source: Guan et al. [58].

effect of surface traps on TiO_2 has been clearly demonstrated in the case of tin-lead perovskite solar cells. Hamada et al. showed that by passivating TiO_2 surface with C_{60}–COOH, the undesirable formation of Ti–O–Sn can be avoided leading to significantly reduced trap density as measured from thermally stimulated current (TSC) spectroscopy [59]. The passivation not only affects the surface of TiO_2 but can also be extended deep into the perovskite bulk (Figure 3.4b) [58]. Furthermore, by introducing sulfonic acid into the perovskite bulk, Tai et al. achieved a high stability tin perovskite solar cell in ambient atmosphere [60]. The work suggested the interaction between the sulfonate group and Sn^{2+} hindered the oxidation of Sn^{2+}, and at the same time, the sulfonic acid molecules are forming a passivating layer on perovskite grains, protecting them from water molecules. Oxidation of I^- into I_2 increases trap density originating from iodide vacancy in perovskite solar cells, but this can be suppressed by introducing hypophosphorous acid (HPA) into perovskite precursor solution [61]. Zhang et al. proved that HPA could reduce I_2 into I^- and concurrently enlarging perovskite grains. Through Kelvin probe microscopy, they showed that the surface photovoltage is less in the case of HPA, signifying that the acceptor type (Pb^{2+}) surface traps have been passivated.

3.1.2.3 Organosulfur Compound

Organosulfur is a type of Lewis base which can donate a pair of electrons, and thus the reaction between organosulfur and Pb^{2+}, which is a type of Lewis acid, can be expected [62]. Noel et al. reported using thiophene to passivate the surface of $MAPbI_3$ [63]. They postulated that the iodide vacancy on the perovskite surface leaves the positively charged Pb^{2+} exposed. The thiophene will then donate free electron pairs to these under-coordinated Pb^{2+} and form covalent bonds (Figure 3.5a). Yu et al. have confirmed this through DFT calculation that the formation energy of surface defects is reduced upon passivation with S atoms where these S atoms bind with outermost Pb atoms [64]. This Lewis acid–base interaction can also be used on metal oxides, as shown by Wang et al., who used xanthate to passivate SnO_2 surface [65]. This is achieved by coating a thin layer of potassium O-hexylxathate on the SnO_2 surface, and upon annealing, it will decompose, leaving only sulfur anchored to the SnO_2 surface, as shown in Figure 3.5b. From the TSC result, the shallow trap has been significantly suppressed with sulfur treatment.

3.1.2.4 Amines

Amine is also a type of Lewis base; accordingly, the interaction between amine and under-coordinated Pb^{2+} can be expected, and this has been discussed in detail by Kerner et al. [66]. Xiao et al. reported the formation of 1D perovskite nanowires when butylamine is added into $MAPbI_3$ precursor solution [67]. The N atoms in butylamine react with under-coordinated Pb^{2+} and thus passivating the surface traps while the long alkyl chain of butylamine facilitates crystal growth in (110) facet orientation forming 1D nanowire. The use of bilateral alkylamine has also been reported to passivate the defects at perovskite grain boundaries and surface. Wu et al. compared various bilateral amines in their study where they introduced the amine molecules inside the perovskite precursor solution [68]. From the scanning electron microscopy (SEM) images, it is clear that a more compact perovskite film is observed as a result of grain boundaries passivation. The bilateral amine acts as a glue to bind neighboring perovskite grains. A similar passivation technique using bilateral amine has also been explored to suppress surface defects. Kamarudin et al. performed post-treatment on tin perovskite using ethylenediamine and achieve unprecedented high efficiency for pure tin perovskite solar cell reaching 10.18% (Figure 3.6a) [69]. The suppression of interfacial recombination increased the V_{OC} up to 0.63 V from a low 0.49 V for the control device, which is attributed to the passivation of under-coordinated Sn^{4+}. A similar passivation technique using amine can also be applied on the TiO_2 surface, as shown in Figure 3.6b, where dopamine is used to passivate the oxygen vacancies on the TiO_2 surface and crosslink TiO_2 with the perovskite layer [70]. Another interesting method of passivating the perovskite surface using amine is through post-treatment using vapor [71, 72]. Using triethyelenetetraamine (TETA) vapor treatment of a complete device increased the efficiency from 16.44% to 17.45% is obtained [73]. This simple post-treatment method has been suggested to modify the perovskite/charge transport layer (CTL) interface as seen from impedance spectroscopy.

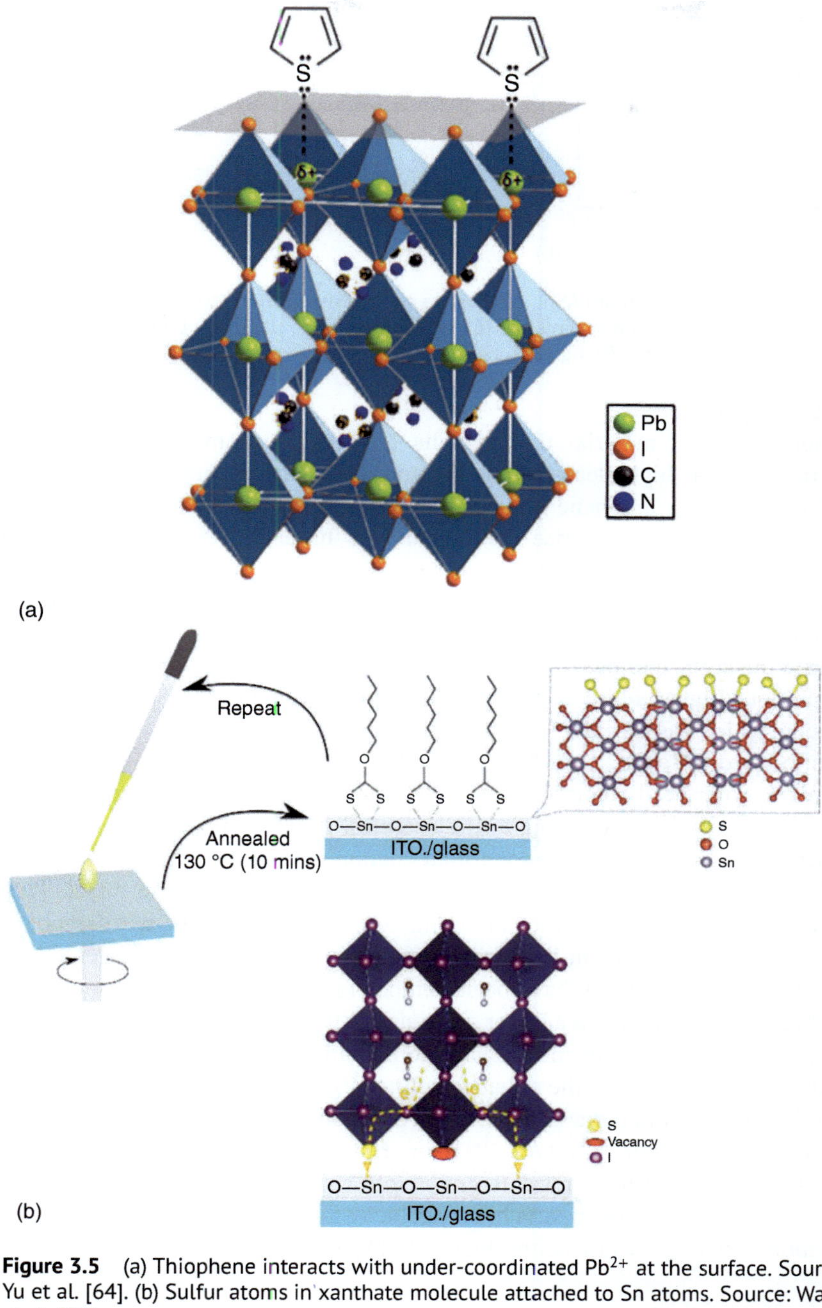

Figure 3.5 (a) Thiophene interacts with under-coordinated Pb^{2+} at the surface. Source: Yu et al. [64]. (b) Sulfur atoms in xanthate molecule attached to Sn atoms. Source: Wang et al. [65].

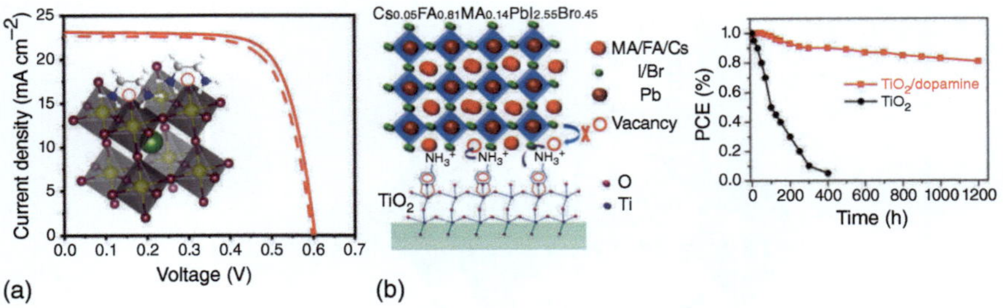

Figure 3.6 (a) Passivation of $FASnI_3$ perovskite surface using amine to suppress surface recombination. Source: Kamarudin et al. [69]. (b) Dopamine to passivate both TiO_2 and perovskite surface. Source: Zhang et al. [70].

3.1.2.5 Graphene

Graphene is a 2D material that exhibits remarkable chemical and electrical properties such as high electron mobility and mechanical strength. Due to these excellent attributes, graphene has been widely investigated for electronic applications. Thus, it is no surprise that researchers have also explored the use of graphene in perovskite solar cells. Figure 3.7a shows TiO_2 surface passivated with graphene nanoplatelets suppressing surface traps of TiO_2 and facilitating electron injection [74]. Wen et al. have reported the use of amino-function graphene to passivate the surface of perovskite (Figure 3.7b) [75]. The amine group acts as an electron-donating group interacting with under-coordinated Pb^{2+}. Simultaneously, graphene, which has excellent charge carrier mobility, improves the charge extraction into spiro-OMeTAD [78]. This method effectively reduces the V_{OC} loss by minimizing interfacial recombination and enhancing charge carrier extraction to improve J_{SC}. Additionally, owing to the chemical stability and hydrophobicity of graphene, Tavakoli et al. have shown that coating perovskite film with graphene could protect the film from thermal decomposition and moisture attack [79]. Even after 300 hours of continuous illumination, the device coated with graphene maintains more than 80% of its original efficiency. Furthermore, Nitrogen-doped graphene oxide nanosheets have been demonstrated to suppress the nucleation rate of perovskite, leading to a larger grain size [80]. Suppression of hole transport material (HTM)/perovskite interfacial recombination and improvement of hole selectivity are attributed to the enhancement of V_{OC}. Seo et al. introduced carbon nanotubes (CNTs) into perovskite precursor solution and prepared perovskite solar cells, as shown in Figure 3.7c [76]. CNTs improved the crystal growth mechanism and bridged neighboring perovskite grains to form bigger crystals. The presence of carbon atoms on CNTs imparts hydrophobicity to perovskite solar cells, making them more stable against moisture (Figure 3.7d) [77].

3.1.2.6 Metal Oxides

Metal oxides are commonly used as charge transport layers in perovskite solar cells [81]. These metal oxides include TiO_2 and SnO_2, which are used as ETLs and NiO_x as HTL. In addition to being used as CTL, they can also be used as passivating agents. Magnesium oxide (MgO) has been reported to reduce interfacial carrier

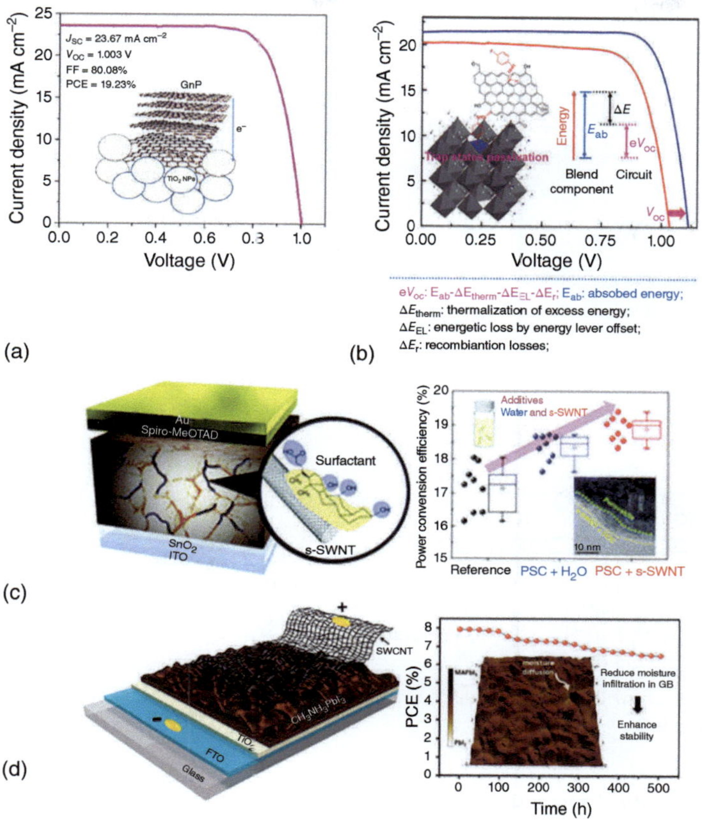

Figure 3.7 (a) Graphene nanoplatelets coated on TiO_2 surface to reduce surface defects and increase electron injection. Source: Sidhik et al. [74]. (b) Functionalized graphene oxide providing defect passivation on perovskite surface and reducing charge injection barrier into spiro-OMeTAD. Source: Wen et al. [75]. (c) Functionalized carbon nanotubes within perovskite bulk act as a crystal growth template and crosslinker between neighboring perovskite grains. Source: Reprinted with permission from Seo et al. [76]. (d) Perovskite surface passivation with carbon nanotube as hydrophobic coating against moisture. Source: Tran et al. [77].

recombination when coated on TiO_2 surface, as shown in Figure 3.8a [82]. The large bandgap of MgO making it a good hole blocking material due to its large hole injection barrier while allowing electrons to tunnel through to TiO_2. This thin layer of MgO can also be inserted between TCO and ETL, as shown in the works of Ma et al. [84]. Further to suppressing the interfacial charge recombination, coating the fluorine-doped tin oxide (FTO) surface with MgO nanoparticles allows for the growth of compact and dense SnO_2 layer and, in turn, facilitates the growth of a high-quality perovskite layer. Ultra-violet (UV) degradation in PSCs is most significant in normal structure devices employing TiO_2 ETL [85]. This is due to TiO_2 particles, which are known to promote photocatalysis under the presence of UV light. By passivating the surface of TiO_2 with strontium oxide (SrO_2), device

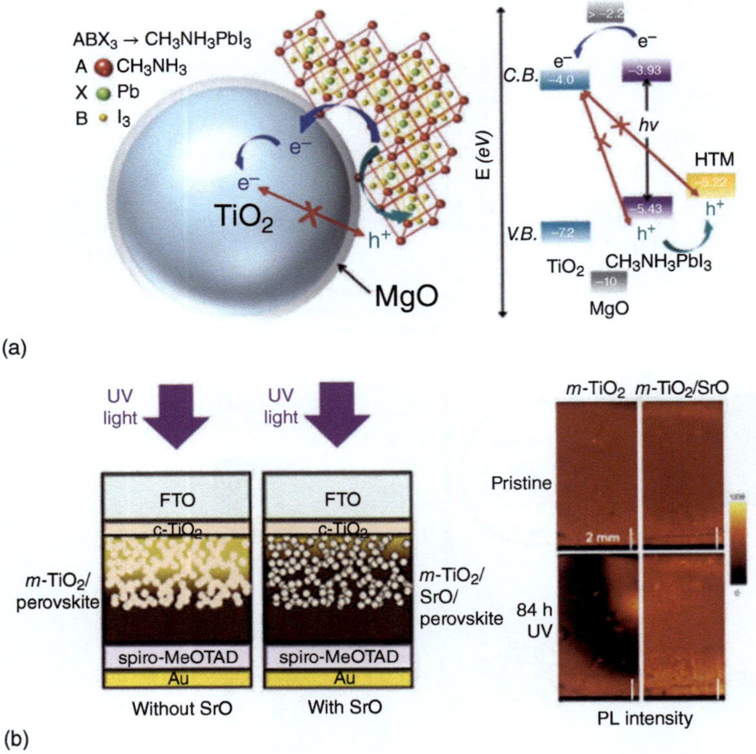

Figure 3.8 (a) Schematic diagram showing MgO-coated TiO_2 providing energy barrier to retard electron recombination. Source: Han et al. [82]. (b) Schematic diagram showing passivation of TiO_2 surface with SrO and comparison of photoluminescence (PL) intensity change after a prolonged UV light exposure. Source: Reprinted with permission from Lee et al. [83].

stability has been improved significantly (Figure 3.8b) [83]. The pristine device showed efficiency reduction of more than 65% after 100 hours under UV light illumination, while for the SrO-coated device, the reduction is around 40%.

3.1.2.7 Organic Halides

Perhaps the most investigated material for passivation of perovskite solar cells is organic halides, partly due to organic halides being one of the starting precursors. It is currently the most effective material to give passivation effect on perovskite solar cells either on metal oxide surface (Figure 3.9a) or within the perovskite bulk (Figure 3.9b) [86, 87]. Son et al. explored the effect of excess MAI on PSC performance [90]. Similar to the case of excess PbI_2, passivation of grain boundary defects is also observed with excess MAI. This is attributed to the formation of ionic conductivity pathways, which help charge carrier extraction. However, it is also important to note that there is a contrasting report suggesting that MAI-rich perovskite behaves as a p-type semiconductor [91]. This self-formed grain boundary

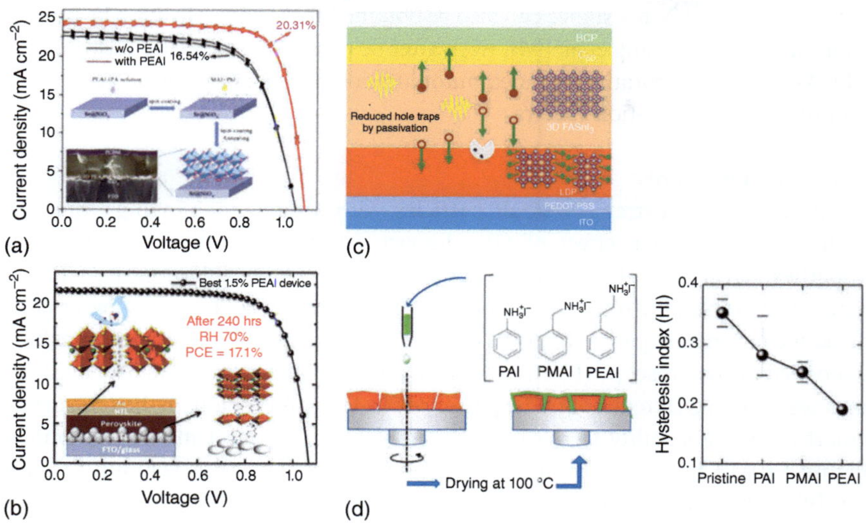

Figure 3.9 (a) PEAI coated onto nickel oxide surface. Source: Reprinted with permission from Liu et al. [86]. (b) Simultaneous TiO_2 surface passivation and perovskite grain boundary passivation with PEAI layer providing superior device stability at high humidity conditions. Source: Lee et al. [87]. (c) Coating the surface of PEDOT: PSS with PEABr assisting perovskite crystal growth together with hole trap passivation. Source: Chen et al. [88]. (d) Grain boundary and surface passivation with various organic halides resulting in lower hysteresis index. Source: Yoo et al. [89].

layer can also be achieved with MACl, in which case a significant change in the grain size is observed [92]. Moreover, Kelvin probe force microscopy reveals that with MACl treatment, electrons are initially trapped in a potential well which can easily be overcome upon light illumination. This is in stark contrast without MACl treatment, in which the potential barrier for electrons trapped at the grain boundaries is too large to overcome even with light illumination. The use of 2D precursors to passivate the surface of PEDOT:PSS has also proven effective in suppressing the electron–hole recombination reaction [88]. X-ray diffraction (XRD) analysis reveals that upon coating PEDOT:PSS surface with PEABr, the growth of a more preferential perovskite orientation is observed, suggesting that this 2D perovskite layer also acts as a crystal growth template (Figure 3.9c). Work by Yoo et al. suggests that surface passivation with phenylethyl ammonium iodide (PEAI) can also suppress hysteresis issues in PSCs [89]. They claimed that the dipole interaction and charge transfer mechanism at the perovskite/HTM interface are responsible for the hysteresis effect, and passivating perovskite surface with organic halide could alleviate this issue (Figure 3.9d). Excellent device stability against moisture has been demonstrated by He et al. using surface passivation of $MAPbI_3$ with triphenyl(9-ethyl-9*H*-carbazol-3-yl)-phosphonium bromide (TCPBr) [93]. After 1000 hours of storage in 60–80% relative humidity condition, the control device showed a sharp drop in efficiency to 18.2% of its initial efficiency while the TCPBr-coated device still retains 90.7% of its initial efficiency. Aside from stability against humidity, UV resistance can also be obtained through passivation by organic halide, as shown by Yang et al. [94]. Even after 60 minutes of exposure to strong UV light, $MAPbI_3$ incorporating pyreneammonium iodide did not show any change in color or UV–vis absorption spectra.

3.1.2.8 Quantum Dots

Quantum dots are extremely small particles with a diameter ranging from 2 to 10 nm, where the size effect of QDs impart various interesting optoelectronic properties. QDs have found applications in solar cells, light-emitting diodes, lasers, and single-electron transistors [95]. In perovskite solar cells, Zheng et al. introduced inorganic perovskite QDs as a passivating agent [96]. Fourier-transform photocurrent spectroscopy (FTPS) reveals that the tail and mid-gap states have been suppressed with QD treatment. The presence of a long alkyl chain on the QD surface impart moisture stability to the perovskite layer by making it more hydrophobic (Figure 3.10a). Chen et al. attempted a similar experiment by using lead sulfide (PbS) QDs [97]. The optimized concentration of PbS in anisole is 0.1 mg ml^{-1} enhancing the efficiency from 19.35% up to 21.07% with a small hysteresis index of 1.65%. From thermal admittance spectroscopy, the trap distribution can be understood where the shallow trap can be assigned to the trap at grain boundaries while deep traps are originated from surface traps (Figure 3.10b). Apart from inorganic-based QDs, it is also possible to use organic-based QDs as passivating agent [98]. DFT simulation study has confirmed that the C=O bond in QDs interacts with under-coordinated Pb. Compared to the pristine perovskite layer, carbon QD-passivated perovskite showed higher photoluminescence intensity and longer carrier lifetime, implying that the non-radiative recombination has been suppressed.

3.1.2.9 Polymers

$CsPbI_3$ are known to have stability issues in which the cubic α-phase $CsPbI_3$ undergo phase transition to orthorhomobic δ-phase [99]. Using polymer-induced passivation technique, a stable α-phase $CsPbI_3$ at room temperature has been realized. Zhang et al. introduced poly-vinylpyrrolidine (PVP) into the perovskite precursor solution in which the PVP–$CsPbI_3$ complex is formed (Figure 3.11a) [100]. The study claimed that acylamino groups in PVP attached to the surface of $CsPbI_3$, thus reducing the surface energy. As a result, a stable α-phase $CsPbI_3$ is maintained even after 80 days. Figure 3.11b shows the schematic diagram of perovskite surface passivation using insulating polymer, resulting in improved moisture stability [101]. Oxygen atoms in polyethylene oxide interact with under-coordinated defect sites at the surface, while the long alkyl chain of the polymer provides protection for the perovskite from water moisture. Even after 15 hours under light illumination, no degradation is observed for the polymer-passivated device. Zuo et al. further investigated the effect of polymers having various functional groups and investigated the strength of coordination bonds with Pb [104]. They discussed the effect of trap passivation from the viewpoint of electron pair sharing between the functional groups and under-coordinated Pb. However, it is worth to note that competing formation of coordination-type bonding (amino-Pb^{2+}) and ionic bonding ($CH_3NH_3{}^+I^-$) will finally determine the beneficial

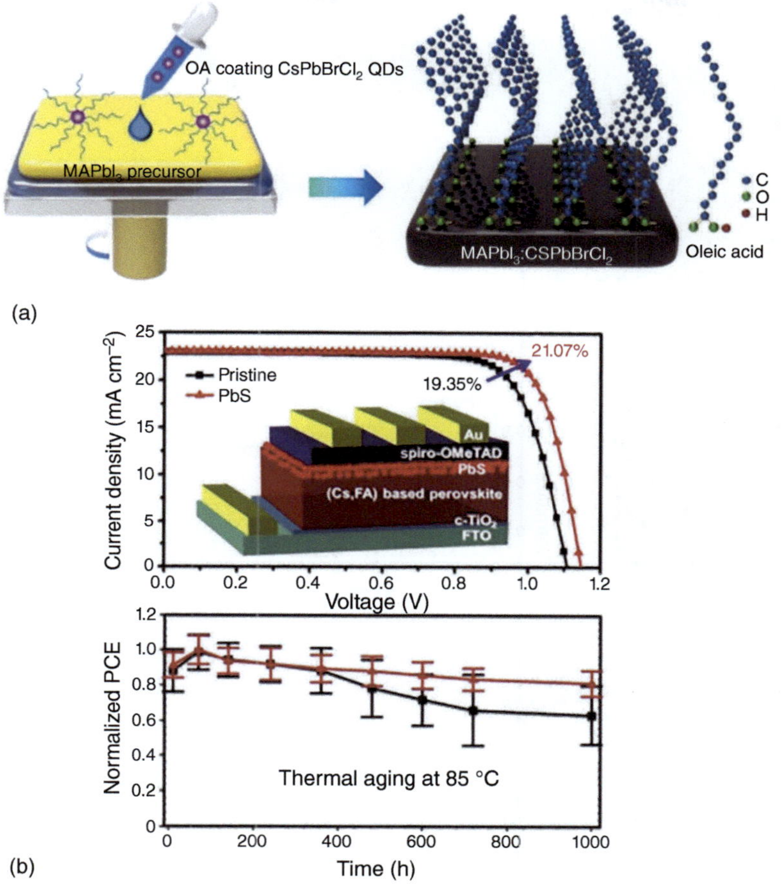

Figure 3.10 (a) Inorganic perovskite quantum dots coated onto $MAPbI_3$ perovskite layer protecting perovskite layer from high humidity. Source: Reprinted with permission from Zheng et al. [96]. (b) PbS quantum dots enhanced the V_{OC} as a result of reduced trap density at the perovskite/spiro-OMeTAD interface. Source: Chen et al. [97].

effect of surface passivation. This surface passivation method has also been adopted using semiconducting polymers [105, 106]. Figure 3.11c shows the schematic diagram of surface passivation using conjugated semiconducting polymer [102]. n-type conjugated polymer has been used to passivate defects at grain boundaries of perovskite [107, 108]. The authors conducted 2D-GIWAXS to confirm their hypothesis that the polymer is present at the grain boundaries. In addition to n-type polymers, p-type polymers have also been used to passivate defects [109]. Li et al. confirmed the passivation of undercoordinated Pb^{2+} through XPS, where the binding energy of 4f Pb shifted to a lower value suggesting an interaction between fluorine and Pb. This is in agreement with the calculated trap density where the reference device has a higher trap density of 1.91×10^{16} cm^{-3} while that for the polymer-passivated device is 1.45×10^{16} cm^{-3}. Further to defect passivation, the polymer is reported to enhance

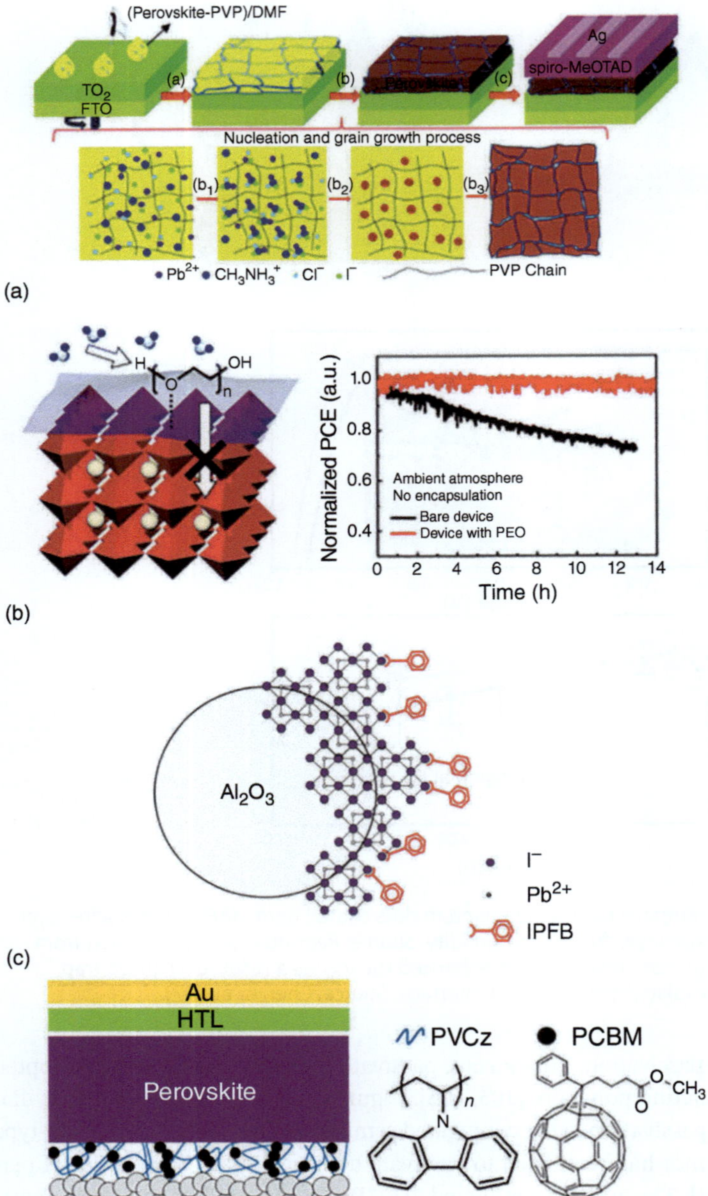

Figure 3.11 (a) Schematic diagram showing polymer-aided nucleation and grain growth of perovskite crystal. Source: Zhang et al. [100]. (b) Surface passivation of perovskite layer with superior stability in ambient atmosphere. Source: Kim et al. [101]. (c) Simultaneous trap passivation and charge extraction with π-conjugated semiconducting polymer. Source: Abate et al. [102]. (d) Passivation of TiO_2 surface using polyvinylcarbazole mixed with PCBM. Source: Okada et al. [103].

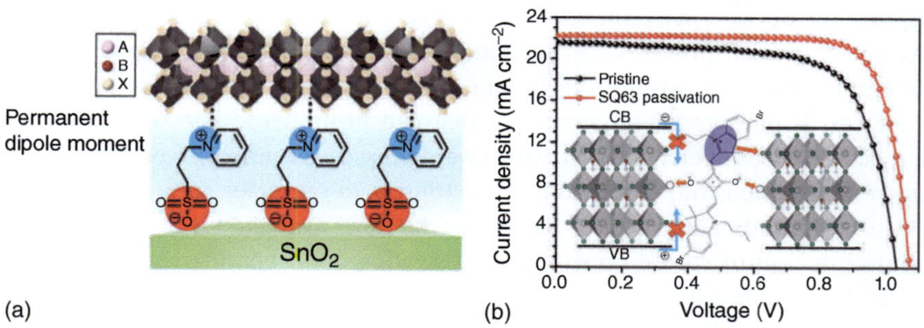

Figure 3.12 (a) Zwitterion passivated on both SnO_2 and perovskite surface. Source: Choi et al. [113]. (b) Squaraine-based zwitterion between neighboring perovskite grains. Source: Wang et al. [114].

the film quality by allowing for a more ordered crystallization [110]. The presence of many Lewis base species in the polymer, i.e. N and S atoms, will coordinate with Pb^{2+} forming an intermediate phase that allows for slow crystal growth, which will increase the grain size. Polyvinylcarbazole/PCBM mixture has been used to passivate the surface defect of TiO_2, as shown in Figure 3.11d [103]. In addition, insertion of this thin organic layer induced a spike structure energy level, which prevents from electron back recombination reaction.

3.1.2.10 Zwitterions

Zwitterions are molecules having both positive and negative charges at different locations within the same molecule [111]. Despite having both positive and negative charges, the overall net charge remains zero. Typical applications of zwitterions can be found in the biomedical field [112]. One example of using zwitterion in perovskite solar cells has been demonstrated by Choi et al., where the group passivated the surface of SnO_2 with 3-(1-pyridino)-1-propanesulfonate zwitterion (Figure 3.12a) [113]. They suggested that the hydroxy group on the SnO_2 surface react with the sulfonate group in the zwitterion while the positively charge N atom is anchored to the Pb–I antisite defects. Wang et al. further investigated the effect of zwitterion in the bulk of perovskite. Squaraine-based zwitterions are found to interact with under-coordinated Pb^{2+} and suppressed Pb–I antisite defects simultaneously (Figure 3.12b) [114]. The trap passivation effect is clearly evidenced from the TSC measurements, especially near the deep trap region. This has been supported by another work that used sulfobetaine zwitterion to stabilize $CsPbI_3$ perovskite to stabilize $CsPbI_3$ perovskite [115]. The mechanism of α-phase stabilization is explained by the interaction of sulfobetaine with ions and colloids in the precursor solution impeding uncontrolled fast crystallization.

3.2 Conclusion

In summary, we have reviewed some of the recent advances in passivation techniques in perovskite solar cells, focusing on the various passivation materials to improve the device's performance and stability. The beneficial effect of passivation has not only been proven experimentally but also supported by extensive computational studies. Passivation technique, either within the perovskite bulk or at interfaces, is one promising technique that can bring perovskite solar cells one step closer to commercialization. We have discussed the effect of passivation on suppressing various trap formations originating from vacancies, interstitial defects, and under-coordinated species. Recent works have also shown that passivating agents can control crystal growth and passivating grain boundaries by acting as linkers between neighboring grains. Excellent device stability under light illumination in ambient air further established the importance of this technique. Hysteresis issue in perovskite solar cells, which has long been associated with ion migration and charge build-up at the interface, has been eliminated through interfacial engineering. Passivation approach in perovskite solar cells has made full use of wide-ranging materials, including metal halides, organic halides, metal oxides, small molecules, and polymers which can either be insulators or semiconductors with various functional groups. The effectiveness of the passivation technique is undeniable, and we could expect similar approaches to be utilized on mixed SnPb and Sn-based perovskites, which have been reported to be unstable and possess high trap density.

References

1 Kojima, A., Teshima, K., Shirai, Y., and Miyasaka, T. (2009). *J. Am. Chem. Soc.* 131: 6050.

2 Yang, W.S., Park, B.W., Jung, E.H. et al. (2017). *Science (80-.)* 356: 1376.

3 Yang, W.S., Noh, J.H., Jeon, N.J. et al. (2015). *Science* 348: 1234–1237.

4 Zhou, H., Chen, Q., Li, G. et al. (2014). *Science* 345: 542–546.

5 Liu, M., Johnston, M.B., and Snaith, H.J. (2013). *Nature* 501: 395–398.

6 Lee, M.M., Teuscher, J., Miyasaka, T. et al. (2012). *Science* 338: 643–647.

7 Im, J.-H., Lee, C.-R., Lee, J.-W. et al. (2011). *Nanoscale* 3: 4088–4093.

8 Li, X., Bi, D., Yi, C. et al. (2016). *Science* 353: 58–62.

9 Chen, W., Wu, Y., Yue, Y. et al. (2015). *Science* 350: 944–948.

10 Mei, A., Li, X., Liu, L. et al. (2014). *Science* 345: 295–298.

11 Bi, D., Yi, C., Luo, J. et al. (2016). *Nat. Energy* 1: 16142.

12 Yi, C., Luo, J., Meloni, S. et al. (2016). *Energy Environ. Sci.* 9: 656–662.

13 Fu, Q., Tang, X., Huang, B. et al. (2018). *Adv. Sci.* 5: 1700387.

14 Ball, J.M. and Petrozza, A. (2016). *Nat. Energy* 1: 16149.

15 Chen, B., Rudd, P.N., Yang, S. et al. (2019). *Chem. Soc. Rev.* 48: 3842.

16 Wolff, C.M., Caprioglio, P., Stolterfoht, M., and Neher, D. (2019). *Adv. Mater.* 31: 1902762.

17 Meng, X., Lin, J., Liu, X. et al. (2019). *Adv. Mater.* 31: 1903721.

18 Yin, W.-J., Shi, T., and Yan, Y. (2014). *Adv. Mater.* 26: 4653.

19 Guo, P., Ye, Q., Yang, X. et al. (2019). *J. Mater. Chem. A* 7: 2497.

20 Niu, T., Lu, J., Munir, R. et al. (2018). *Adv. Mater.* 30: 1706576.

21 Bak, T., Nowotny, J., and Nowotny, M.K. (2006). *J. Phys. Chem. B* 110: 21560.
22 Zhen, C., Wu, T., Chen, R. et al. (2019). *ACS Sustain. Chem. Eng.* 7: 4586.
23 Li, H., Xue, Y., Zheng, B. et al. (2017). *RSC Adv.* 7: 30422.
24 Wang, J., Datta, K., Weijtens, C.H.L. et al. (2019). *Adv. Funct. Mater.* 29: 1905883.
25 Meng, L., You, J., Guo, T.-F., and Yang, Y. (2016). *Acc. Chem. Res.* 49: 155.
26 Liu, T., Chen, K., Hu, Q. et al. (2016). *Adv. Energy Mater.*: 6. https://doi.org/10.1002/aenm.201600457.
27 Zheng, X., Hou, Y., Bao, C. et al. (2020). *Nat. Energy* 5: 131.
28 Yu, Z. and Sun, L. (2018). *Small Methods* 2: 1700280.
29 Kung, P.-K., Li, M.-H., Lin, P.-Y. et al. (2018). *Adv. Mater. Interfaces* 5: 1800882.
30 Schulz, P., Whittaker-Brooks, L.L., MacLeod, B.A. et al. (2015). *Adv. Mater. Interfaces* 2: 1400532.
31 Roberts, M.W. and Smart, R.S.C. (1984). *J. Chem. Soc. Fraday Trans. 1 Phys. Chem. Condens. Phases* 80: 2957.
32 Marrani, A.G., Novelli, V., Sheehan, S. et al. (2014). *ACS Appl. Mater. Interfaces* 6: 143.
33 Wang, Q., Chueh, C.C., Zhao, T. et al. (2017). *ChemSusChem* 10: 3794.
34 Tress, W., Yavari, M., Domanski, K. et al. (2018). *Energy Environ. Sci.* 11: 151.
35 Lee, J.H., Noh, Y.W., Jin, I.S. et al. (2019). *ACS Sustain. Chem. Eng.* 7: 15495.
36 Jiang, Q., Zhao, Y., Zhang, X. et al. (2019). *Nat. Photonics* https://doi.org/10.1038/s41566-019-0398-2.
37 Wang, R., Xue, J., Wang, K.-L. et al. (2019). *Science (80-.)* 366: 1509–1513.
38 Chaudhary, B., Kulkarni, A., Jena, A.K. et al. (2017). *ChemSusChem* 10: 2473.
39 Wei, F., Jiao, B., Dong, H. et al. (2019). *J. Mater. Chem. A* 7: 16533.
40 Wu, T., Wang, Y., Li, X. et al. (2019). *Adv. Energy Mater.* 9: 1803766.
41 Lin, Y., Shen, L., Dai, J. et al. (2017). *Adv. Mater.* 29: 1604545.
42 Cho, A.-N. and Park, N.-G. (2017). *ChemSusChem* 10: 3687.
43 Tu, J., Liu, C., Fan, Y. et al. (2019). *J. Mater. Chem. A* 7: 15662.
44 Wu, S., Chen, R., Zhang, S. et al. (2019). *Nat. Commun.* 10: 1161.
45 Chen, Q., Zhou, H., Bin Song, T. et al. (2014). *Nano Lett.* 14: 4158.
46 Son, D.-Y., Kim, S.-G., Seo, J.-Y. et al. (2018). *J. Am. Chem. Soc.* 140: 1358.
47 Li, B., Zhang, Y., Zhang, L., and Yin, L. (2017). *J. Power Sources* 360: 11.
48 Chen, Y., Meng, Q., Xiao, Y. et al. (2019). *ACS Appl. Mater. Interfaces* 11: 44101.
49 Barbé, J., Newman, M., Lilliu, S. et al. (2018). *J. Mater. Chem. A* 6: 23010.
50 Park, B., Kedem, N., Kulbak, M. et al. (2018). *Nat. Commun.* 9: 3301.
51 Yang, F., Hirotani, D., Kapil, G. et al. (2018). *Angew. Chem. Int. Ed.* 57: 12745.
52 Chen, M., Ju, M.-G., Garces, H.F. et al. (2019). *Nat. Commun.* 10: 16.
53 Ito, N., Kamarudin, M.A., Hirotani, D. et al. (2018). *J. Phys. Chem. Lett.* 9: 1682.
54 Nakayashiki, S., Daisuke, H., Ogomi, Y., and Hayase, S. (2015). *J. Photonics Energy* 5: 057410.
55 Zhang, P., Kapil, G., Hamada, K. et al. (2018). *ACS Sustain. Chem. Eng.* 6: 10221.
56 Bin Kim, H., Im, I., Yoon, Y. et al. (2015). *J. Mater. Chem. A* 3: 9264.
57 Zuo, L., Chen, Q., De Marco, N. et al. (2017). *Nano Lett.* 17: 269.

58 Guan, L., Jiao, N., and Guo, Y. (2019). *J. Phys. Chem. C* 123: 14223.

59 Hamada, K., Tanaka, R., Kamarudin, M.A. et al. (2020). *ACS Appl. Mater. Interfaces* 12: 17776.

60 Tai, Q., Guo, X., Tang, G. et al. (2019). *Angew. Chem. Int. Ed.* 58: 806.

61 Zhang, W., Pathak, S., Sakai, N. et al. (2015). *Nat. Commun.* 6: 10030.

62 Bi, D., Li, X., Milić, J.V. et al. (2018). *Nat. Commun.* 9: 4482.

63 Noel, N.K., Abate, A., Stranks, S.D. et al. (2014). *ACS Nano* 8: 9815.

64 Yu, C.-J., Kye, Y.-H., Jong, U.-G. et al. (2020). *ACS Appl. Mater. Interfaces* 12: 1858.

65 Wang, Z., Kamarudin, M.A., Huey, N.C. et al. (2018). *ChemSusChem* 11: 3941.

66 Kerner, R.A., Schloemer, T.H., Schulz, P. et al. (2019). *J. Mater. Chem. C* 7: 5251.

67 Xiao, K., Cui, C., Wang, P. et al. (2018). *Nanotechnology* 29: 065401.

68 Wu, W.Q., Yang, Z., Rudd, P.N. et al. (2019). *Sci. Adv.* 5: 1.

69 Kamarudin, M.A., Hirotani, D., Wang, Z. et al. (2019). *J. Phys. Chem. Lett.* 10: 5277.

70 Zhang, Y., Liu, X., Li, P. et al. (2019). *Nano Energy* 56: 733.

71 Raga, S.R., Jiang, Y., Ono, L.K., and Qi, Y. (2017). *Energy Technol.* 5: 1750.

72 Yang, J., Qin, T., Xie, L. et al. (2019). *J. Mater. Chem. C* 7: 10724.

73 Yao, D., Mao, X., Wang, X. et al. (2020). *ACS Appl. Mater. Interfaces* 12: 6651.

74 Sidhik, S., Panikar, S.S., Pérez, C.R. et al. (2018). *ACS Sustain. Chem. Eng.* 6: 15391.

75 Wen, X., Wu, J., Gao, D., and Lin, C. (2016). *J. Mater. Chem. A* 4: 13482.

76 Seo, S., Jeon, I., Xiang, R. et al. (2019). *J. Mater. Chem. A* 7: 12987.

77 Tran, V.-D., Pammi, S.V.N., Dao, V.-D. et al. (2018). *J. Alloys Compd.* 747: 703.

78 Li, H., Tao, L., Huang, F. et al. (2017). *ACS Appl. Mater. Interfaces* 9: 38967.

79 Tavakoli, M.M., Tavakoli, R., Yadav, P., and Kong, J. (2019). *J. Mater. Chem. A* 7: 679.

80 Hadadian, M., Correa-Baena, J.-P., Goharshadi, E.K. et al. (2016). *Adv. Mater.* 28: 8681.

81 Shin, S.S., Lee, S.J., and Il Seok, S. (2019). *Adv. Funct. Mater.* 29: 1900455.

82 Han, G.S., Chung, H.S., Kim, B.J. et al. (2015). *J. Mater. Chem. A* 3: 9160.

83 Lee, S.-W., Kim, S., Bae, S. et al. (2018). *Org. Electron.* 63: 343.

84 Ma, J., Yang, G., Qin, M. et al. (2017). *Adv. Sci.* 4: 1700031.

85 Lee, S.-W., Kim, S., Bae, S. et al. (2016). *Sci. Rep.* 6: 38150.

86 Liu, Y., Duan, J., Zhang, J. et al. (2020). *ACS Appl. Mater. Interfaces* 12: 771.

87 Lee, D.S., Yun, J.S., Kim, J. et al. (2018). *ACS Energy Lett.* 3: 647.

88 Chen, K., Wu, P., Yang, W. et al. (2018). *Nano Energy* 49: 411.

89 Yoo, H.-S. and Park, N.-G. (2018). *Sol. Energy Mater. Sol. Cells* 179: 57.

90 Son, D.-Y., Lee, J.-W., Choi, Y.J. et al. (2016). *Nat. Energy* 1: 16081.

91 Wang, Q., Shao, Y., Xie, H. et al. (2014). *Appl. Phys. Lett.* 105: 163508.

92 Uribe, J.I., Ciro, J., Montoya, J.F. et al. (2018). *ACS Appl. Energy Mater.* 1: 1047.

93 He, Q., Worku, M., Xu, L. et al. (2020). *J. Mater. Chem. A* 8: 2039.

94 Yang, F., Zhang, P., Kamarudin, M.A. et al. (2018). *Adv. Funct. Mater.* 28: 1804856.

95 Litvin, A.P., Martynenko, I.V., Purcell-Milton, F. et al. (2017). *J. Mater. Chem. A* 5: 13252.

96 Zheng, X., Troughton, J., Gasparini, N. et al. (2019). *Joule* 3: 1963.

97 Chen, Y., Yang, J., Wang, S. et al. (2020). *iScience* 23: 100762.

98 Ma, Y., Zhang, H., Zhang, Y. et al. (2019). *ACS Appl. Mater. Interfaces* 11: 3044.

99 Eperon, G.E., Paternò, G.M., Sutton, R.J. et al. (2015). *J. Mater. Chem. A* 3: 19688.

100 Zhang, Y., Zhuang, X., Zhou, K. et al. (2017). *J. Mater. Chem. C* 5: 9037.

101 Kim, M., Motti, S.G., Sorrentino, R., and Petrozza, A. (2018). *Energy Environ. Sci.* 11: 2609.

102 Abate, A., Saliba, M., Hollman, D.J. et al. (2014). *Nano Lett.* 14: 3247.

103 Okada, W., Suga, T., Oyaizu, K. et al. (2019). *ACS Appl. Energy Mater.* 2: 2848.

104 Zuo, L., Guo, H., DeQuilettes, D.W. et al. (2017). *Sci. Adv.* 3: e1700106.

105 Cai, F., Cai, J., Yang, L. et al. (2018). *Nano Energy* 45: 28.

106 Chen, W., Shi, Y., Wang, Y. et al. (2020). *Nano Energy* 68: 104363.

107 Chen, W., Wang, Y., Pang, G. et al. (2019). *Adv. Funct. Mater.* 29: 1808855.

108 Yu, B., Zhang, L., Wu, J. et al. (2020). *J. Mater. Chem. A* 8: 1417.

109 Li, X., Li, W., Yang, Y. et al. (2019). *Sol. RRL* 3: 1900029.

110 Zhang, C.-C., Li, M., Wang, Z.-K. et al. (2017). *J. Mater. Chem. A* 5: 2572.

111 Cao, B., Tang, Q., and Cheng, G. (2014). *J. Biomater. Sci. Polym. Ed.* 25: 1502.

112 Wu, A., Gao, Y., and Zheng, L. (2019). *Green Chem.* 21: 4290.

113 Choi, K., Lee, J., Il Kim, H. et al. (2018). *Energy Environ. Sci.* 11: 3238.

114 Wang, Z., Pradhan, A., Kamarudin, M.A. et al. (2019). *ACS Appl. Mater. Interfaces* 11: 10012.

115 Wang, Q., Zheng, X., Deng, Y. et al. (2017). *Joule* 1: 371.

4

Tuning Interfacial Effects in Hybrid Perovskite Solar Cells

Rafael S. Sánchez[1], *Lionel Hirsch*[2], *and Dario M. Bassani*[3]

[1] *Universitat Jaume I, Institute of Advanced Materials (INAM), Physics Department, Av. de Vicent Sos Baynat, s/n, 12006 Castelló de la Plana, Spain*
[2] *Université de Bordeaux, CNRS, Bordeaux INP, IMS, UMR 5218, ENSCBP, 16, Av Pey Berland, F-33607 Pessac, France*
[3] *Université de Bordeaux, CNRS, Bordeaux INP, ISM, UMR 5255, 351, Cours de la Libération, F-33405 Talence, France*

4.1 Strategies for Interfacial Deposition and Analysis

Since the first report in 2009 [1], the performance of perovskite solar cells (PSCs) has evolved steadily, reaching recently a maximum certified power conversion efficiency (*PCE*) of 24.2% [2]. This is principally a direct consequence of the fundamental electronic properties of the active layer material, which include not only its light-harvesting properties but also its ability to separate and transport photogenerated charges to the electrodes. Several reasons account for the outstanding charge generation and extraction in PSC: an optical bandgap close to the optimal value (~1.5 eV) [3], high absorption coefficients (1.5×10^4 cm^{-1} at 550 nm) [4], large ambipolar (electron and hole) conductivity [5], long charge carrier diffusion lengths (>1 μm) [6], and low exciton binding energy [7]. However, although the properties of the active layer are essential factors to achieve high-performance devices, the nature of the interfaces within the stacked materials plays a crucial role for device efficiency and operational lifetime. In the last years, significant efforts have been devoted toward interface engineering for more efficient light harvesting [8, 9], enhanced charge separation and transport [10, 11], improved stability [12–14], and/or suppressed *current–voltage (I–V)* hysteresis [13–15]. The main interface engineering approaches that have been adopted broadly fall into two categories aimed at promoting enhanced device performance and stability: (i) exploitation of different deposition techniques, solvent treatment, and introduction of additives to improve control of the crystal morphology and quality, surface roughness, and coverage and (ii) incorporation of functional materials between the active layer and the charge selective contacts aimed at suppressing charge recombination processes or the diffusion of unwanted species such as lithium or moisture.

Perovskite Solar Cells: Materials, Processes, and Devices, First Edition.
Edited by Shahzada Ahmad, Samrana Kazim, and Michael Grätzel.

Most perovskite (PS) materials currently used in photovoltaic devices consist of organic and inorganic building blocks that form crystalline composite materials exhibiting a strong hybrid character. Although the crystallinity contributes to the desirable bulk electronic properties outlined above, it also introduces complications due to the finite macroscopic dimensions of the crystalline grains since every surface is, in essence, a defect. This leads to a wide variety of possible interfacial chemistries, in which dangling bonds, surface terminations, interface strain, and related physicochemical processes contribute in the alteration of the device performance. The complexity of the interfaces behind PSC has been identified to play a crucial role for device functionality and performance, including device stability [16]. Therefore, a thorough control over the interfacial properties and unraveling those optoelectronic mechanisms taking place at the interfaces, which in turn determine the overall device performance, is essential to further understand the working principles of PSC and improve their performance toward approaching the maximum theoretical limits of this technology.

4.1.1 Tailoring the PS Properties and Microstructural Interface Through Solvent Engineering

Over the last years, a series of strategies have been studied for the deposition of PS materials with the aim to control the morphological crystal features and their ensuing influence on the overall device performance. One of the most common deposition techniques for the preparation of PSC is based on solution processing (spin coating), which allows relatively simple and cost-efficient device production. The microstructural characteristics of PS films, i.e. grain size [17], grain boundaries [18], density of pinholes [19], crystal quality, and orientation of the crystalline growth [20], determine the performance of the PSC. Not surprisingly, numerous studies have focused on tailoring PS chemistry, grain size, and microstructure by modifying the deposition procedures and/or conditions. One- and two-step sequential depositions are two representative techniques that allow fabricating relatively high-quality films and devices. The one-step deposition is based on the spin-coating deposition of a precursor solution that contains all the necessary components for the PS formation. The crystalline film is formed using a thermal treatment for solvent evaporation, and this process generally suffers from a poor control over the crystallization process and, therefore, yields small-sized and shapeless crystals, rougher surface, and poorer substrate coverage [21, 22]. In the two-step sequential deposition process, a thin layer of lead precursor (e.g. PbI_2) is deposited on the substrate. Then, the organic cation, e.g. methylammonium iodide (MAI) or formamidinium iodide (FAI) in solution, is coated onto the pre-deposited PbI_2, and heat is applied to enable conversion into the PS phase. This process involves crystal nucleation and growth of the PS phase through a reaction between the PbI_2 in solid state and the organic cation in solution and therefore occurs from the surface toward the inner crystalline regions of the PbI_2 phase. Consequently, this method has been ineffective in producing high-performance perovskite films that are >500 nm in thickness due to incomplete conversion and uncontrolled

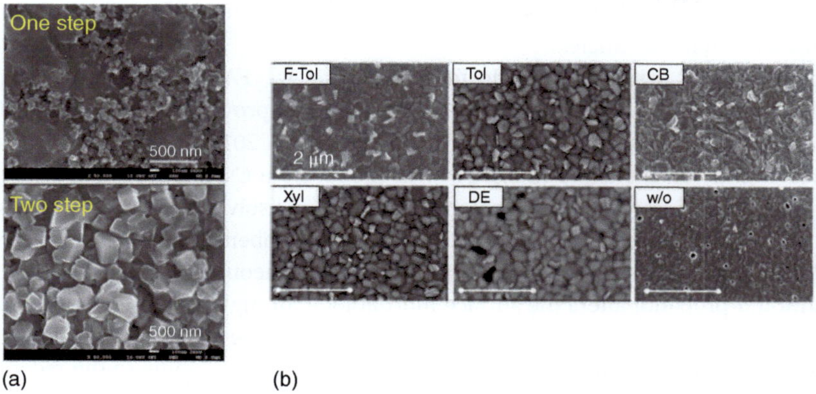

Figure 4.1 (a) Surface SEM images of $MAPbI_3$ films deposited by one-step and two-step procedures on an m-TiO_2 layer. Source: Adapted from Paek et al. [21] with permission from the American Chemical Society. (b) SEM images of PS films without (w/o) and with antisolvent treatment using trifluorotoluene (F-Tol), toluene (Tol), chlorobenzene (CB), xylene (Xyl), and diethyl ether (DE). Source: Adapted from Paek et al. [21] with permission from the American Chemical Society.

surface roughness [23]. An illustrative morphological comparison between PS films obtained through one-step and two-step deposition methods is shown in Figure 4.1a.

In their seminal work, in 2014, Seok and coworkers reported that antisolvent engineering is an effective method to enhance the efficiency of PSC [24]. Spin coating of PS precursors (e.g. PbI_2 and MAI) in a binary mixture of solvents (γ-butyrolactone and DMSO) followed by adding toluene during the end of the drop-casting process leads to highly uniform and dense perovskite layers via a MAI–PbI_2–DMSO intermediate phase. This enabled the fabrication of remarkably improved solar cells with a certified *PCE* value of 16.2% without observable hysteresis on the *I–V* curves. The process was later modified by Park and coworkers, who employed an equimolar mixture of MAI, PbI_2, and DMSO in DMF to deposit a PS precursor film. Next, diethyl ether was added as antisolvent during spin coating, which selectively removes the DMF, thus allowing the formation of an intermediate phase, which, after thermal treatment at low temperature, yields a smooth, uniform, and pinhole-free PS film. With this approach, *PCE* values of up to 19.7% were obtained with a *I–V* hysteresis factor of 2% at slow scan rates [25].

Several antisolvents such as diethyl ether [26], chlorobenzene [27], ethyl acetate [28], and others [29] have been reported in the literature (Figure 4.1b). High-quality PS films were recently obtained by using ethyl acetate instead of chlorobenzene for the preparation of mixed PS films that consist of a mixture of methylammonium (MA), formamidinium (FA) cations, and corresponding lead halides, with the formula $FA_{0.85}MA_{0.15}Pb(I_{0.85}Br_{0.15})_3$. These provide *PCE*s of up to 19.43% for laboratory-scale cells (0.16 cm^2) and 14.28% for module cells (5 × 5 cm^2) [28]. The enhanced performance of the devices prepared from ethyl acetate compared with the chlorobenzene counterparts was ascribed to an efficient inhibition of charge carrier recombination that arises from interfacial defects, although the reason for this difference remains unclear.

Triple-cation PS with the generic formula $Cs_x(MA_{0.17}FA_{0.83})_{1-x}Pb(I_{0.83}Br_{0.17})_3$ have been studied in the recent years due to their improved performance and enhanced stability in photovoltaic applications [30]. In 2017, Li and coworkers reported a study on the triple-cation PS with the formula $Cs_5FA_{79}MA_{16}PbI_{2.5}Br_{0.5}$ using a mixture of *i*-propanol and chlorobenzene as antisolvent [31]. The authors found that alcohol can efficiently remove the excess chlorobenzene from the surface and that this promotes the formation of more homogeneous and compact films. In addition, *i*-propanol alters the surface morphology through a slight solubility of PS that is proposed to allow healing of preliminary crystalline defects. Hence, larger PS grain size and crystallinity are obtained, which in turn contribute to the record efficiencies of 19.2%. Very recently, Chen and coworkers reported the preparation of highly efficient and stable PSC through using $FA_{0.85}Cs_{0.15}Pb(Br_{0.15}I_{2.85})$ films by employing chlorobenzene as an antisolvent with methylamine as an additive [27]. Methylamine-containing antisolvent strategy induced the formation of PS films with large grain size and preferential 101 crystal orientation. Steady-state photoluminescence (PL) and time-resolved photoluminescence (TRPL) experiments suggested that the interfacial charge transfer efficiency is improved by the presence of the additive. The devices produced via methylamine treatment achieved a *PCE* of 19.6%, but, more interestingly, the initial efficiency values decreased only by 5% after 500 hours under continuous light soaking in N_2 atmosphere. This represents an interesting improvement in the intrinsic stability of the PS active layer.

Film formation based on the antisolvent treatment in PS is indeed effective for promoting crystal nucleation and producing compact films, which is crucial for achieving efficient PSC. Despite this, the grain size of PS films prepared through standard antisolvent methods is relatively small compared with the traditional spin-coating methods, mainly due to the fast PS nucleation induced by the orthogonal solvent. As a consequence, more grain boundaries are present in the film, and these may act as charge recombination centers. Additionally, it has been generally accepted that charge mobilities are larger in the inner regions of the crystals rather than on the surface [32]. From these considerations, it appears necessary to target enlarged grain sizes while improving the crystal quality of PS films to further enhance the *PCE*. In 2015, Nie et al. reported the preparation of uniform films with millimeter-scale PS grains by casting a hot (~70 °C) mixture of lead iodide (PbI_2) and methylammonium chloride (MACl) solution onto a substrate maintained at a temperature of up to 180 °C [17]. The main difference between conventional spin coating/post-annealing and hot casting is the presence of solvent when heating the substrate. In the hot-casting conditions, there is an excess of solvent present when the substrate is maintained above the crystallization temperature for the formation of the PS, which allows for the prolonged growth of the perovskite crystals to yield large crystalline domains. The very significant increases in the short-circuit current (J_{sc}) from 3.5 to 22.4 mA cm^{-2}, open-circuit voltage (V_{oc}) from 0.4 to 0.92 V, and fill factor (*FF*) from 0.45 to 0.82 were observed when the grain size was enlarged from ~1 μm up to ~180 μm, which contributed to an increase in the overall *PCE* from 1% to ~18%. However, despite the large crystal sizes obtained from the hot melt process, *PCE* values reported are similar or even lower than those reported in more recent work exploiting the traditional antisolvent methodology, which proved that crystal size is not the sole parameter determining the quality of PS films.

4.1.2 Tailoring the PS Properties and Microstructural Interface Through Non-solvent Methods

Compared with solution processing, vacuum deposition techniques are probably one of the most prominent methodologies that offer advantageous possibilities to fine-tune the composition and purity, surface properties, crystallinity, and thickness of PS materials (Figure 4.2). Additionally, vacuum processing possesses several remarkable advantages compared with solution techniques, i.e. mature fabrication facilities, uniform and large-area deposition, low exposure to toxic solvents, and easy control of layered structures with nanometer resolution. Vapor deposition of $MAPbI_3$ based on either co-evaporation, sequential evaporation, or hybrid chemical vapor deposition (CVD) has been extensively explored [36–39], and the corresponding PSC has shown efficiencies that can compete with their solution-processed counterparts [33]. In 2016, a fully evaporated PSC with micrometer-sized crystal grains was reported (Figure 4.2a) with average V_{oc}, J_{sc}, *FF*, and *PCE* values of 1.1 V, 20.28 mA cm^{-2}, 80%, and 18%, respectively, and a record *PCE* value of 20.3% [34]. However, evaporation of the best performing and stable PS derivatives based on mixed cation/anion compositions had not been explored until recently. Bolink and coworkers reported the evaporation of double- and triple-cation PS derivatives, $Cs_{0.5}FA_{0.5}Pb(I_{0.83}Br_{0.17})_3$ and $Cs_{0.5}FA_{0.4}MA_{0.1}Pb(I_{0.83}Br_{0.17})_3$, but significantly lower *PCE* values (16%) compared with the *state-of-the-art* were achieved [35]. The double- and triple-cation PS films showed full surface coverage with compact and uniform morphology, but the crystal grain size ranged from 100 to 200 nm. Also, it is worth noting that large and irregular crystals appeared on top of the triple-cation

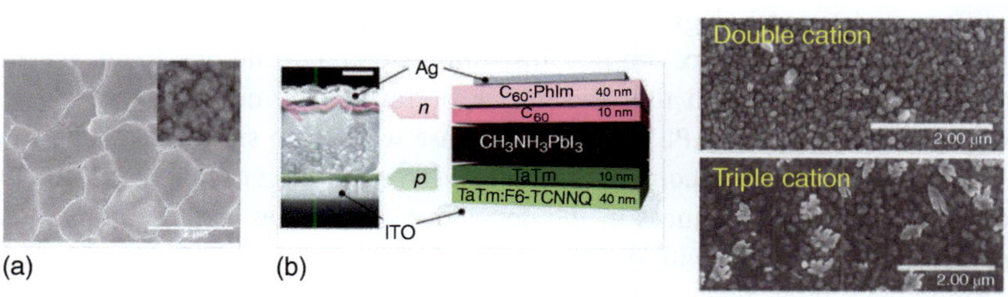

(a) (b) (c)

Figure 4.2 (a) Top-view SEM and inset AFM image (5 μm × 5 μm) of the PS thin film fabricated at 10^{-4} Torr partial pressure. Source: Adapted from Hsiao et al. [33] with permission from Wiley. (b) SEM cross section and schematics of a fully vacuum deposited p–i–n PSC (scale bar of 200 nm). Source: Adapted from Momblona et al. [34] with permission from the Royal Society of Chemistry. (c) Top-view SEM images of the double- and triple-cation PS layers. Source: Adapted from Gil-Escrig [35] with permission from Wiley.

PS film (Figure 4.2c). The relatively small grain size and the irregularities of the films could account for the limited performance.

Considering the importance of PS homogeneity, grain size, and crystallinity on the bulk and interface processes involved in the power conversion, further improvements might be obtained by a more precise control over the crystal and film morphology. Crystallinity and surface morphology of evaporated films can be modified by varying the PS composition, by controlling the evaporation rate and the substrate temperature in the evaporation chamber, and/or by thermal or chemical posttreatment of the deposited compounds. Additionally, evaporation of interlayer materials in between the PS and the charge selective layers is a feasible strategy to further improve the device performances through defect passivation mechanisms, doping, or energy level matching. Despite the thriving prospect of vapor-phase deposition, the crystal quality level of epitaxial growth may prove difficult to reach. Unfortunately, epitaxial crystal growth, which has been perfected for nearly all group III–V or II–VI semiconductors, has been seldom reported for PS compounds so far [40, 41]. In this regard, most common solvent and non-solvent processing techniques can rarely yield stoichiometric surfaces, thus usually showing defective regions [42]. Additional studies focused on the exploration of new PS compositions, enhancement of PS crystallinity, minimization of defect density, and reduction of dangling bonds at the crystal surface are required for further improvement.

4.2 Defect Formation in PS Films and Interfaces

The morphology and crystal quality of PS films are crucial parameters that contribute in determining the performance of the final devices. For this reason, topographic techniques, such as atomic force microscopy (AFM) and scanning tunneling microscopy (STM), proved to be valuable tools for characterizing PS thin films [43–45]. In particular, average grain size and root mean square (RMS) roughness are two parameters that provide information about the potential performance of solar cells. However, differences in performance are not always accompanied by a difference in morphology. In fact, the effects of grain boundaries or grain size on the solar cell performance are not fully understood. On the one hand, several research groups have found that the device performance and/or charge carrier lifetimes measured from TRPL generally increase with grain size, thus suggesting that non-radiative charge recombination is less likely for larger grains [17, 46]. On the other hand, recent advances in the deposition techniques have demonstrated high-performance devices with average grain size values in the sub-micron scale [28]. Along these lines, techniques such as Kelvin probe force microscopy (KPFM) suggest that grain boundaries can contribute to increasing V_{oc} by facilitating the charge separation processes [47]. Although still under debate, the most recent studies seem to converge toward the detrimental role of grain boundaries and the beneficial effect of increasing the PS grain size. Larger grains would result in improved carrier mobility and reduce the relative area of grain boundaries, which are often reported to yield higher concentrations of defects. It is worth noting that several types of defects coexist in PS films, characterized by their origin and spatial location: (i) film surface, (ii) grain bulk, and (iii) boundaries between neighboring grains. Therefore, attempts to tailor the properties of PS films may induce variations on the other parameters, e.g. variation of growth methodologies for increasing grain size could also affect the surface composition and/or bulk properties of PS grains, thus leading to ambiguous conclusions. In this sense, it is important to deconvolute and elucidate the contribution of these different microscopic factors on the electro-optical properties of polycrystalline PS films and on the performance of the final devices. In addition to the intrinsic defects of the as-grown PS material, the defects formed in the bulk material and at the interface with the other stacked layers during device operation must also be considered. Approaching the theoretical maximum device performance and improving the PSC stability are the two most important challenges that need to be tackled. Surface roughness, grain boundaries, and defect formation, both during the materials deposition and device functioning, affect the quality of the device. An additional question to be unveiled is whether the grain boundaries and/or the surface of PS films are the weakest regions where the degradation would first start.

4.2.1 Defect Formation in the PS Bulk and at the Surface During Film Crystallization

It is widely accepted that the presence of imperfections in the photoactive material may significantly influence device performance. Considering single point defects, there are three common types of intrinsic point defects that are formed in the crystalline structure of a semiconductor, i.e. vacancies, interstitials, and antisites, as shown in Figure 4.3.

These point defects can be formed intrinsically or through the presence of impurities introduced intentionally or unintentionally, and the density of a defect is governed by its formation energy. Depending on how efficiently the charge carriers (free electrons or holes) are trapped or detrapped, these defect states can be classified into

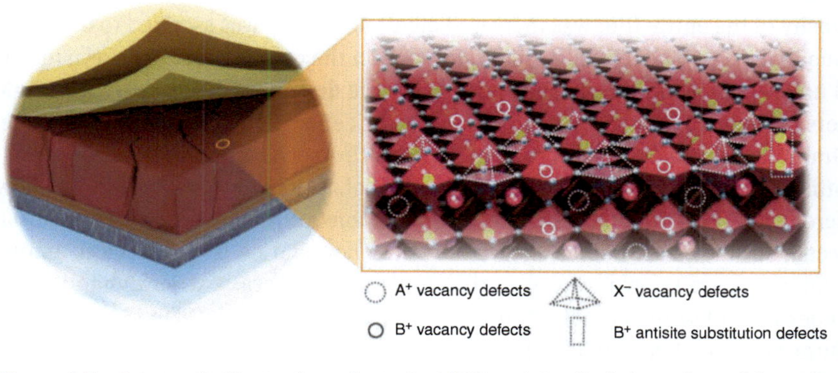

Figure 4.3 Schematic illustration of a typical PSC and detailed view of possible surface defects on PS crystals. Source: Hsiao et al. [33].

shallow or deep defects. Shallow defects account for those states that can potentially trap charge carriers with a low detrapping activation energy. Although these defects decrease the charge carrier mobilities, which in turn increases the probability of charge recombination, the low activation energy does not entirely shun the chance of charge carriers to be detrapped. Therefore, they can partially contribute to the charge extraction. In contrast, deep defects correspond to those states that irreversibly capture the charge carriers and act as non-radiative recombination centers. In this case, the high activation energy of detrapping impedes the charge carriers being extracted to the external circuit. Several theoretical and experimental studies indicate that the nature and density of defect states in PS are highly sensitive to the film deposition conditions.

In 2017, Heo et al. reported that defects originating from iodine atoms occupying a Pb site (I_{Pb}) lie at −0.60 eV below the conduction band (CB) minimum and are dominant in PS films prepared from the single-step spin-coating method. In contrast, defects from iodine atoms located at methylammonium positions (I_{MA}) are −0.75 eV below the CB and correspond to the main deep defects of PS films fabricated from a sequential deposition method [48]. Using density functional theory (DFT), Sargent et al. calculated the defect formation energies of $MAPbI_3$ using different conditions and precursors and predicted that PS grown under iodine-rich conditions are prone to show a high density of deep electronic traps, whereas low trap densities (below 10^{15} cm^{-3}) can be achieved in iodine-poor conditions. Considering the lack of stoichiometric compositions at the surfaces of crystal grains and the loss of organic molecules during the thermal annealing process, it is believed that most of the defects in PS films are mainly located at the grain boundaries or near the surface. Some studies have confirmed the presence of high density of defects at grain boundaries, since trap densities in solution-processed $MAPbI_3$ films (10^{16}–10^{17} cm^{-3}) are much higher than those found in PS single crystals (10^{9}–10^{10} cm^{-3}) [49, 50]. Ginger and coworkers identified the presence of bright and dark grains through fluorescence microscopy imaging and associated these to different densities of deep traps. In addition, lower PL intensities were observed at the grain boundaries compared with the inner part of grains, thus suggesting more defects and faster non-radiative kinetics at the grain boundaries [49]. After treatment with a solution of pyridine, PL intensity of the entire film was increased eightfold, which was ascribed to the passivation of defects. The role of grain boundaries on the overall device performance is still under intense debate: some studies point to a detrimental effect [51, 52], while others to a beneficial impact [47] or no significant effect on the device performance.

Grain boundaries are distributed over the entire polycrystalline film due to discontinuities in the crystal structure of the material. Neighboring crystal grains show different orientations that give rise to dislocations, misplaced atoms (interstitial defects), vacancies, and bond distortions at the boundaries. The interfaces among crystal grains are known to play a critical role in the charge carrier dynamics and photophysics of semiconductor thin films such as poly-Si [53], CdTe [54] or copper indium gallium selenide (CIGS) [55]. Although several models have been successful in explaining the charge carrier dynamics and the influence of grain boundaries in inorganic polycrystalline solar cells [56], PSC technology shows different behavior from other photovoltaic technologies. Hybrid PS are considered to be lightly doped materials [57]. Due to the presence of charged ionic defects, not only in the bulk but also at the grain boundaries and interfaces, it is also likely that the traps are electrically charged when empty and neutral when filled with photogenerated charges [58]. Unbalanced accumulation or depletion of charges between grain boundaries by charged ionic defects in PS may cause downward or upward band bending, which may then assist or impede the charge separation processes [45, 59]. For instance, a potential barrier for hole transport, i.e. downward bending, has been observed at the grain boundaries of $MAPbI_3$ on PEDOT:PSS substrates [60]. Therefore, the optoelectronic properties of grain boundary are sensitive to defect types, which in turn depend on the material composition, substrate properties, and preparation methods. Additionally, some researchers have found that $MAPbI_3$ shows a p-type behavior, while some others found n-type [61]. This means that PS films can be bipolarly doped and that different processing protocols can lead to opposite doping behaviors. In 2018, Garnett et al. reported a study of the effects of grain size and boundaries based on electron backscattered diffraction (EBSD) for characterizing $MAPbBr_3$ thin films [62]. They showed that, in general, larger grain size with less boundaries is beneficial for all PS properties due to the lower defect density. However, they also found that crystallinity at the grain boundaries may play a significant role. Longer PL lifetimes and higher PL intensity were found at grain boundaries with amorphous characteristics, thus suggesting that amorphous interfaces among grains can improve the optoelectronic properties through either grain boundary passivation or native defect gettering.

A different type of defect is caused by the presence of impurities, which may modulate the semiconductor properties by increasing the charge carrier concentration, i.e. semiconductor doping. The electrical, optical, and structural properties of semiconductors can be greatly altered by the presence of small amounts of certain impurities. This strategy has been extensively explored for classical inorganic and organic semiconductors [63–66]. It is generally accepted that PS films show a significant defect tolerance for the presence of certain impurities that only cause point defects and hardly influence the morphology of the film [67]. However, the presence of other impurities such as bismuth [68], even at very low concentrations, can form both shallow and deep trap states. Consequently, strong non-radiative recombination centers are created, thus causing a detrimental effect on the solar cell performance. Therefore, there are clear evidences that the defect tolerance, often associated with PS derivatives, does not extend to all kinds of impurities, and such species can be unintentionally introduced from external contaminants, solvents, or precursors during processing. Therefore, many intrinsic defects form shallow states, which is one of the major reasons for the defect tolerance of metal halide PS. Nevertheless, the formation of deep trap states and the resulting non-radiative losses are still significantly limiting the PSC device efficiency, thus impeding reaching the maximum theoretical efficiency (Shockley–Queisser limit) that lies beyond 30% [69].

The Shockley–Queisser limit is characterized by the condition that only radiative recombination prevails as the reciprocal process to absorption [70]; for PS devices, the photoluminescence quantum yield (PLQY) values are in the order of 1% at open-circuit conditions under illumination (AM1.5 G) [68]. In general, the *state-of-the-art* PSC shows a non-radiative voltage loss of 120–150 mV, thus showing V_{oc} values around 1.1 V [71–74]. These V_{oc} values are significant considering the polycrystalline nature of PS films, which are normally deposited via solution techniques at low temperature. However, they are still lower than their theoretical V_{oc} limit of 1.33 V (for $MAPbI_3$) under illumination conditions [75]. Very recently, PS devices with a record V_{oc} of 1.26 V have been reported [76], indicating that the non-radiative voltage losses are <100 mV in best cases. The anticipated defects originated from grain boundaries, crystal imperfections, and solvent remnants do not seem to act as effective recombination centers. This could be explained assuming that many of these intrinsic point defects are in fact shallow traps because of the strongly ionic bonds and the antibonding nature of the valence states as described by DFT simulations [77–79]. However, deep traps originating from methylammonium dissociation or charged iodine interstitial defects have also been predicted [80, 81]. The role of grain boundaries in charge recombination processes is still under considerable debate and probably depends on the chemical nature of the surface terminations on the crystallite surfaces and the respective surface states that are created.

Other sources of recombination may also be present if the interfaces to the charge transport contacts are not perfectly selective to the designated charge carrier. Chen et al. demonstrated that thermal deposition of a gold electrode under vacuum induces the formation of interfacial shallow traps located at an energy level of $E_T < 10$ meV above the valence band energy level (E_{VB}) with a concentration as high as 5.5×10^{19} cm^{-3}. These interfacial traps induce band bending by hole trapping that also facilitates electron injection [82]. Hence, tuning the charge selective materials and using interlayers are essential tools to transform the high chemical potential difference generated in the active layer upon excitation into a high V_{oc} at the external contacts. Despite the outstanding achievements, there is still a long way to go toward reaching the radiative efficiency limit considering that still around 99% of recombinations in the best devices is non-radiative.

4.2.2 Defect Formation and Dynamics of PSC Under Working Conditions

Apart from the intrinsic defects and imperfections generated in the PS bulk and at the interfaces during the deposition steps, it has been widely shown that PSC show a dynamic behavior under working conditions. Several physical mechanisms have been proposed to be acting during device operation, i.e. ion migration, charge accumulation, and/or chemical reactions at the interfaces or in the bulk, which may be either reversible or permanent and either beneficial or detrimental to the device performance. The occurrence of these processes can be strongly influenced by a great variety of external factors, e.g. fabrication methods, composition of the PS film, nature of the PS interfaces with the charge selective contacts, interlayers, light intensity, humidity, and oxygen concentration, and could account for the characteristic phenomena associated to PSC, e.g. I–V hysteresis, PL enhancement/quenching, capacitance at low frequencies, long transient photovoltage decays, or limited device stability. Several research groups have investigated the effects of light soaking on the performance of PSC [83–86]. For instance, Zhao et al. claimed that the defects present in the PS bulk and at the interfaces can be neutralized by the photogenerated charge carriers upon illumination, thus increasing the V_{OC} over the illumination time [83]. It is worth pointing out that these devices showed a reduced initial V_{oc} (0.51 V), and, therefore, such results may not be directly extrapolated to high performance devices. Hoke et al. observed a redshift of the PL signal upon light soaking of mixed-halide PSC, thus suggesting the formation of new sub-bandgap states due to photoinduced halide segregation. This would induce a reduction in the electronic bandgap and quasi-Fermi level splitting, and consequently a reduction of the V_{oc} could potentially be expected [85]. Tress et al. reported the observation of reversible V_{oc} reduction (from 1.2 V to less than 1 V) after light soaking that they assigned to the reversible photoinduced formation of narrow defect states situated 0.25 eV below the bandgap acting as dominant recombination centers [87]. Along these lines, Domanski et al. reported the formation of light-activated trap states upon light soaking that seemed to arise from ionic migration and trap accumulation, explaining the reversible loss of performance that is sometimes observed [86]. Ion migration, which has been observed in the past for alternative photovoltaic materials [88], has been proposed in the last years as a feasible mechanism to explain the anomalous phenomena observed in the PSC technology. Theoretical studies have focused on determining the activation energy (E_a) for the migration of the ions that constitute the PS materials. Eames et al. inferred that I^- ions migrate along the I^-–I^- edge of the PbI_6 octahedron showing an E_a of 0.58 eV; the migration of MA^+ was estimated to show a higher E_a of 0.84 eV and a much higher energy of 2.31 eV for Pb^{2+} migration. The authors measured an E_a of 0.60–0.68 eV from photocurrent relaxation at biased preconditioning at different voltages, thus suggesting the occurrence of I^- migration [89]. Haruyama et al. estimated lower E_a of 0.44 eV for I^- and 0.57 eV for MA^+ migration [90]. Although significantly different values have been reported, these studies generally indicate that the most mobile ions are I^-, followed by MA^+ and Pb^{2+}. However, all these precedents only considered ion migration through Schottky defects.

Yin et al. considered ion migration through interstitial (Frenkel) defects as well, such as three vacancies (V_{MA}, V_{Pb}, V_I), three interstitial species (MA_i, Pb_i, I_i), two cation substitutions (MA_{Pb}, Pb_{MA}), and four antisite substitutions (MA_I, Pb_I, I_{MA}, I_{Pb}). They obtained formation energies of the interstitial defects for I^- (0.23–0.83 eV) and MA^+ (0.20–0.93 eV) that are actually close to the E_a of I^- (e.g. 0.58 eV) and MA^+ (e.g. 0.84 eV) migration [91]. Additionally, Yuan et al. observed the electroinduced migration and redistribution of MA^+ ions at room temperature for the first time from photothermal induced resonance (PTIR) measurements [92]. Later, the same authors provided evidence for the macroscopic migration of I^- vacancies in $MAPbI_3$ perovskite films at 330 K, which had not been experimentally demonstrated before and suggested the main ion migration channels mediated by the defects present in the crystal structure (Figure 4.4) [93, 94].

Levine et al. deduced from their work that ion migration and accumulation at the interfaces of the perovskite layer determine the hysteresis in PSC and

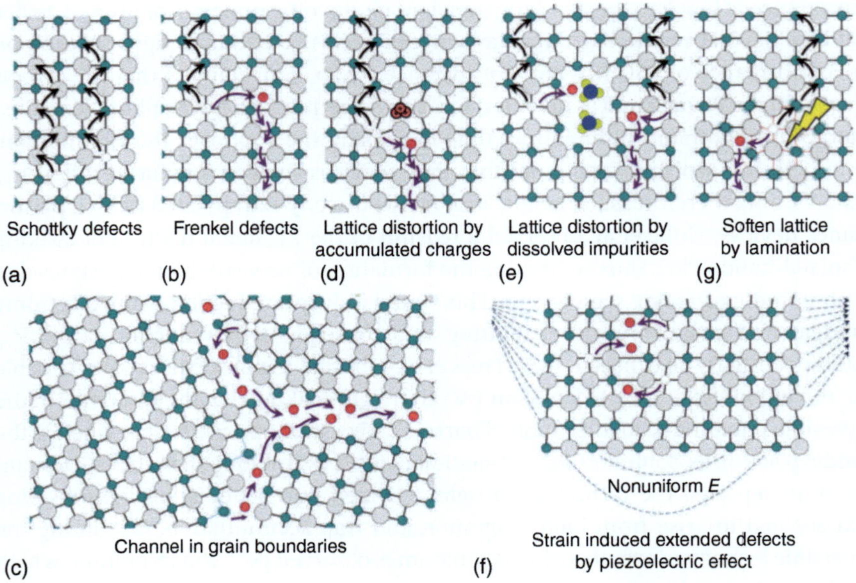

Figure 4.4 Illustration of the ion migration pathways enabled by (a) Schottky defects, (b) Frenkel defects, (c) open space and wrong bonds at grain boundaries, (d) lattice distortions due to accumulated charges, (e) dissolved impurities, and (f) nonuniform strain caused by piezoelectric effect, and (g) soften lattice caused by the light illumination induced bond weakening. Source: Yuan and Huang [93].

that the rate of migration is highly dependent on the contacting materials. This leads to different activation energies depending on the device architecture and interfaces employed [95]. In addition, the study of Levine et al. concludes that the variation of V_{oc} as a function of temperature shows that the main recombination mechanism is trap-assisted in the space charge region, thus suggesting that ion migration/accumulation hampers the extraction of charge carriers. Futscher et al. have recently reported that both MA^+ and I^- ions migrate at room temperature, but with very different diffusion coefficients of 10^{-9} and 10^{-12} cm^2 s^{-1}, respectively. Interestingly, they have observed that E_a of mobile I^- ions is 0.29 eV and is highly reproducible for different devices; instead, the activation energy of mobile MA^+ depends strongly on the device fabrication method and conditions [96]. Very recently, Lee et al. combined temperature-dependent conductivity and glow discharge-optical emission spectroscopy (GD-OES) under electrical bias to experimentally determine the I^- diffusion coefficient (1.3×10^{-12} cm^2 s^{-1}) and mobility (5×10^{-11} cm^2 V^{-1} s^{-1}) in $MAPbI_3$ [97]. They also measured an activation energy of about 0.25 eV, which agrees with that reported by Futscher et al. [96] motivated by the well-known proton-conducting phenomena in oxide perovskites [98, 99] and the protic nature of methylammonium-based perovskites, Egger et al. developed a theoretical study on the hydrogen motion in $MAPbI_3$ [100]. They calculated relatively low migration barriers for proton transfer and therefore concluded that protons are likely to be mobile under device working conditions and can contribute to defect formation/migration. Chen et al. studied the reaction kinetics of $MAPbI_3$ using non-deuterated and deuterated MA species, $CH_3NH_3^+$, $CH_3ND_3^+$, and $CD_3ND_3^+$, respectively [101]. Measurements of the dielectric constant of the devices as a function of the frequency and temperature were performed and allowed assigning the mid-frequency dielectric response in PSCs to proton motion through a tunneling mechanism. The authors measured E_a of 50 and 27 meV associated with proton migration for the tetragonal and orthorhombic phases, respectively, substantially decreasing upon deuteration of the ammonium group. Interestingly, these isotopic labeling experiments demonstrated that proton migration is principally localized at the interfacial region with the PEDOT:PSS layer, which indeed participates in the proton exchange with $MAPbI_3$. In agreement with these results, previous analysis of the dielectric response of $CsPbBr_3$, a PS that does not contain labile protons, did not revealed any mid-frequency dielectric response [102].

Other theoretical and experimental works suggest that *I–V* hysteresis is a combined effect of ion migration and variation of interfacial recombination induced by the accumulated ionic species or defects [103, 104]. Most of the research that focused on the elucidation of ion migration mechanisms only considered intrinsic ion migration, i.e. migration of ions that constitute the PS crystal structure. However, extrinsic ions that may be intentionally introduced into the PSC are also prone to migrate under working conditions and thus alter the device performance. For example, in 2017, Li et al. reported significant migration of ionic species employed as dopants in the charge transport layers [105]. The authors claimed that Li^+ ions, which are widely employed as dopant for the hole transport material (HTM), i.e. spiro-OMeTAD, can travel through the PS layer and accumulate at the electron transport material (ETM) interface, i.e. TiO_2. Their results suggest that Li^+ accumulation at PS/TiO_2 interface improved electron extraction from the PS to the TiO_2 and therefore lead to an enhancement of the V_{oc} and *FF* compared with Li^+-free devices. Additionally, it was suggested that intrinsic and extrinsic migration rates, which can be extrapolated to other ions such as H^+ or Na^+, would affect the device performance, stability, and device hysteresis. Alternatively, other reports focusing on elucidating the effect of light soaking in PS films and/or devices show that slow charge accumulation at the PS interface with the TiO_2 induces an upward band bending, thus creating an energetic barrier for electron extraction [106–108]. Moreover, crystal structure changes of the PS films upon illumination have been also suggested as feasible reasons that account for the changes in the electronic properties of PSC (e.g. charge carrier mobility) [109, 110]. Nevertheless, the real effect of light-induced structural variations on the device performance has not been fully demonstrated.

Considering the discrepancies surrounding the dynamic behavior of PSC under working conditions as suggested by a vast variety of research contributions, more conclusive studies are needed to understand the mechanisms behind device operation. One major concern is whether intrinsic ion migration and interfacial defect accumulation can induce severe performance losses upon prolonged operation at the *maximum power point* (*MPP*). For instance, Wang et al. suggested that photoinduced formation of iodine radicals can degrade the PS films, thus implying that $MAPbI_3$ devices suffer from intrinsic degradation pathways that cannot be avoided by sealing strategies [111]. In spite of this, much effort is being devoted not only on the search of new PS compositions toward defect-free PS films with enhanced stability but also on the exploration of passivation approaches that alleviate the defect impact on device performance and/or avoid extrinsic degradation reactions.

4.3 Passivation Strategies of PS

Up to now, a wide variety of organic and inorganic charge-selective contacts (HTM and ETM) with suitable orbital energy alignment and high charge mobility have been developed that can maximize the charge extraction [112, 113]. However, most PSC reported so far suffer from limited device stability, relatively low V_{oc} compared with the theoretical limit, and hysteresis on the I–V curves. As discussed in Section 4.2, many studies suggest that these limitations are closely related to ion migration and the presence of intrinsic or extrinsic defects and charge accumulation at the PS/ETM and/or PS/HTM interfaces [114, 115]. Therefore, interfacial charge recombination significantly affects the photovoltaic performance of PSC, thus leading to reduced charge carrier lifetimes and voltage losses. In addition, the combination of ion migration in the PS film and interfacial charge accumulation is believed to be responsible of the observed hysteretic behaviors in the I–V curves and lack of long-term device stability. Consequently, a straightforward approach to overcome these limitations may lie in the interface passivation to minimize detrimental interface phenomena and further improve the performance of PSC.

Three main approaches for reducing the impact of interfacial recombination in PSC have been explored. The first is based on reducing the defects density of the charge selective contacts and/or PS by adding dopants. For instance, Giordano et al. reported that Li^+ doping of TiO_2 passivates the electronic defect states, thus increasing the V_{oc} of the devices from 1.04 to 1.11 V. In addition, Li^+-doped devices showed negligible hysteresis [116]. Saliba et al. reported a PSC based on standard configuration with a stabilized *PCE* of 21.6% and V_{oc} of 1.18 V by incorporating 5% of Rb^+ into the PS film [117]. A more recent report by Zhang et al. based on inverted configuration devices exploits the introduction of NiO_x nanoparticles into the PS precursor solution that is deposited onto a NiO_x substrate (Figure 4.5a) [118]. By adding a suitable amount of nanoparticles, an increase of the grain size and crystallization quality is induced, which produces a better light-absorption capability. In addition, NiO_x is believed to play a significant role in passivating the boundaries between PS grains and improve the interface with the HTM. Therefore, a PS film with reduced trap states and enhanced charge separation in the PS bulk is obtained, which in turns produces a significant increase in the V_{oc}. Additionally, a significant increase in the *FF* also arises from the improved NiO_x/PS interface, which permits more efficient charge transfer and reduction of series resistance. It is worth mentioning that despite the fact that NiO_x shows a valence band that is properly aligned with PS for efficient hole extraction and a high conduction band that is able to efficiently block electrons, the interaction between these two materials is sometimes reported to be weak [120], which may hamper its long-term operation. However, the use of NiO_x nanoparticles addresses these limitations, since introduction of NiO_x into the

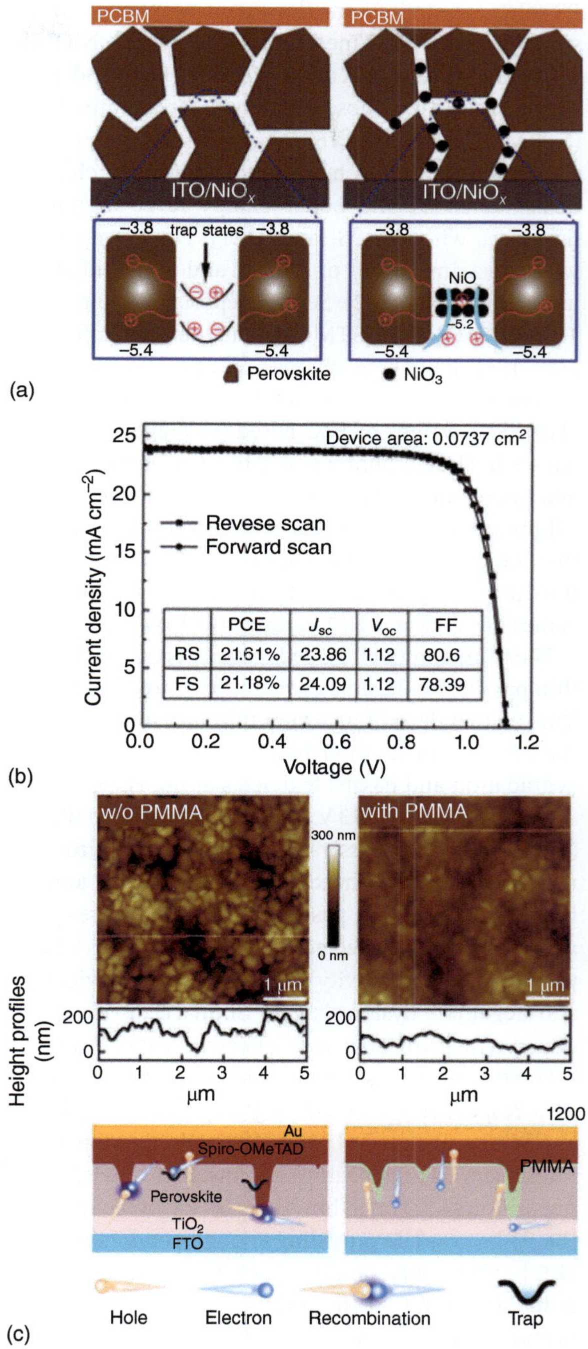

	PCE	J_{sc}	V_{oc}	FF
RS	21.61%	23.86	1.12	80.6
FS	21.18%	24.09	1.12	78.39

Figure 4.5 (a) Schematic illustration of PS films with and without passivating NiO nanoparticles incorporated into the PS layer. Source: Zhang et al. [118]. (b) I–V curve of the best device with an active area of 0.0737 cm^2 measured from reverse (RS) and forward scan (FS) under 1 sun conditions (AM1.5G, 100 mW cm^{-2}). The inset table shows the cell parameters in both scan directions. Source: Jiang et al. [119]. (c) AFM images of the perovskite layer surface without and with PMMA layer and height profiles of both samples along the dashed white line cut, respectively. The bottom part shows an illustration of the carrier dynamics model in PSC without and with PMMA layer. Source: Adapted with permission from Wang et al. [72]. Copyright (2017) American Chemical Society.

PS solutions strengthens the chemical interactions as demonstrated using infrared spectroscopy (FTIR) measurements. Overall, NiO_x-decorated PSC showed a V_{oc} of 1.12 V, J_{sc} of 23.01 mA cm^{-2}, *FF* of 75.06%, and *PCE* of 19.34%, which are significantly higher than those cell parameters obtained for devices without nanoparticles (1.08 V, 22.23 mA cm^{-2}, 68.97%, and 16.56%).

A different approach is to use alternative charge-selective contacts with lower defect densities and better energy level alignment that can form high-quality interfaces with the PS layer [121]. For example, Anaraki et al. reported devices with a maximum V_{oc} of 1.157 V and a *PCE* of 20.47% by using a Nb-doped SnO_2 layer as an ETM [122]. Several reports on the replacement of the widely employed TiO_2 by SnO_2 as an ETM in planar device configuration are found in the literature [123–125]. Jiang et al. showed that SnO_2 owns a deep conduction band, high electron mobility, and low defect densities, which can be processed at low temperature [119]. *PCE* values of PSC based on planar SnO_2 has been pushed beyond 21.6%, which is closely approaching the performance obtained with mesoporous ETM, and interestingly, these planar devices show negligible *I–V* hysteresis (Figure 4.5b). Additionally, Correa-Baena et al. showed that reducing the doping concentration of the hole-selective contact spiro-OMeTAD can suppress interfacial recombination, thus producing devices with maximum V_{oc} of 1.23 V and stabilized *PCE* and V_{oc} values of ~20% and 1.17 V, respectively [126].

The third approach for reducing interfacial recombination consists in introducing thin passivation layers between the PS and the charge transport layers. Zhang et al. showed that depositing a very thin fullerene derivative (α-bis-PCBM) layer through the antisolvent method between PS and HTM, spiro-OMeTAD can reduce cell degradation and passivate defects at the PS/HTM interface, thus yielding a *PCE* of 20.8% and a V_{oc} of 1.13 V [71]. Wang et al. used ultrathin poly(methyl methacrylate) (PMMA) layer to passivate the perovskite/spiro-OMeTAD interface (Figure 4.5c), resulting in an enhanced V_{oc} value of 1.06 V and reduced hysteresis [72]. PMMA provides exceptional passivation that increases the V_{oc} but conductivity and *FF* are detrimentally affected. However, mixtures of PMMA:PCBM allows tuning the passivation/conductivity (V_{oc}/*FF*) ratio, thus yielding a *PCE* of 20.4%, V_{oc} of 1.16 V, and negligible hysteresis [73]. Similarly, a very recent example reports on the use of poly-vinylcarbazole as an effective scaffold of PCBM to modify and passivate the TiO_2 surface. This strategy yielded devices with V_{oc} and *PCE* values of 1.16 V and 21%, respectively, and with high reproducibility [74].

Additionally, several examples exploiting the deposition of ultrathin metal oxide interlayers have appeared. For instance, Koushik et al. explored the atomic layer deposition (ALD) of a thin Al_2O_3 on top of the PS absorber, which effectively protects the PS layer while providing enough conductivity through a tunneling mechanism [127]. This study produced devices with a stabilized *PCE* of 18%, a significant reduction of the *I–V* hysteresis, and a device stability beyond 60 days. In a similar fashion, Kot et al. investigated the effect of an ALD-deposited Al_2O_3 film using different conditions [128], e.g. thickness, deposition temperature, or precursors. These experiments allowed them to conclude that the *PCE* of devices is connected to the chemical and electronic properties at the PS/Al_2O_3 interface, which promotes an interfacial shift of the valence band. Interestingly, Lv et al. explored the effects of functional ALD metal oxide layers on the device performance by employing the following device architecture ITO/NiO/$MAPbI_3$/PC61BM/Ag as reference [129]. A thin ALD-deposited TiO_2 layer was inserted between PC61BM and silver electrode, which induced an increase of the V_{oc} and *PCE* of the device due to the reduced interfacial charge recombination. In addition, a top Al_2O_3 layer was deposited to act as an encapsulating element, which significantly enhanced stability. The stability was assessed by monitoring the required time under working conditions to lose a 20% factor of the initial efficiency. Control devices reached that loss of performance in 16 hours, while those devices with TiO_2 and with $TiO_2 + Al_2O_3$ layers needed 40 and 600 hours, respectively. Stolterfoht et al. proposed that the suppression of interfacial recombination on high-efficiency inverted PSC, with general device architecture ITO/PTAA/PS/C_{60}/Cu, could be achieved by selective introduction of interlayers at the ETM(HTM)/PS interfaces [130]. They found significant non-radiative recombination losses in the bulk PS layer that reduce the potential V_{oc} by approximately 135 mV. It is worth mentioning that the radiative limit of the triple-cation PS-employed $CsPbI_{0.05}[(FAPbI_3)_{0.89}(MAPbBr_3)_{0.11}]_{0.95}$ corresponds to 1.345 V. Upon studying the p-type and n-type semi-cells independently through a combination of steady-state and time-resolved PL measurements, they calculated an additional free energy loss of 80 meV at each individual interface, which limits the V_{oc} of the complete cell to ~1.12 V. Therefore, interface recombination at each individual hybrid interface (PTAA/PS and PS/C_{60}) dominates the non-radiative loss current in the stack. The insertion of a thin layer of poly[(9,9-bis(30-((*N,N*-dimethyl)-*N*-ethylammonium)-propyl)-2,7-fluorene)-alt-2,7-(9,9-dioctylfluorene)] dibromide (PFN-P2) between PTAA/PS and an ultrathin layer of LiF between PS/C_{60} induced a significant suppression of the recombination losses by 65 and 35 meV at the p- and n-interfaces, respectively. By means of this strategy, PSCs with maximum *PCE* values above 20% and with a certified stabilized *PCE* of 19.83% were achieved.

A slightly different approach, based on depositing lower dimensionality PS derivatives on top of the commonly employed 3D-PS films, has been proposed for the enhancement of PSC performance. It has been generally observed that 2D-PS derivatives exhibit higher moisture resistance and photostability than the 3D counterparts [131, 132]. However, the device efficiencies obtained so far from pure 2D structures are still far from the *state of the art* in PSC [131–135]. Several research groups have focused on exploiting the enhanced stability of 2D-PS derivatives for passivating the 3D-PS films. Yoo et al. reported very recently a study using different alkylammonium bromide precursors for depositing a 2D-PS passivation layer on top of a 3D-PS film through solvent processing techniques (Figure 4.6a) [136]. A maximum *PCE* of 23.4% from a reverse *I–V* sweep was obtained with a certified stabilized *PCE* of 22.6%, which was attributed to the passivation of interfacial defects, thus minimizing charge carrier quenching, as the increase in V_{oc} of the 3D/2D devices suggests (Figure 4.6b,c).

Similarly, Liu et al. have recently reported the fabrication of highly efficient PSC based on 3D/2D PS bilayers with exceptional moisture stability by employing a perfluorinated ammonium derivative to produce a superhydrophobic, ultrathin,

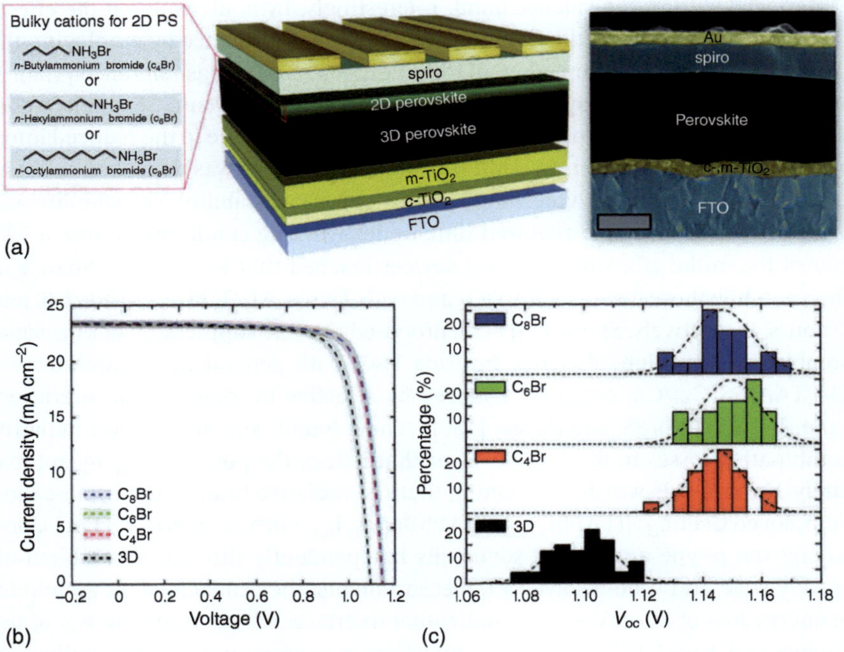

Figure 4.6 (a) Schematic illustration of a layered 3D/2D PSC with color-edited cross-sectional SEM (scale bar: 500 nm). (b) I–V curves of 3D and 3D/LP (C_4Br, C_6Br, C_8Br) PSCs fabricated using the selective precursor dissolution (SPD) strategy, with average and standard deviation shown as a dashed line and shaded area, respectively. (c) Histogram of V_{oc} for 3D and 3D/2D (C_4Br, C_6Br, C_8Br) PSCs. Source: Adapted from Yoo et al. [136] with permission from the Royal Society of Chemistry.

and highly uniform 2D PS material on top of the 3D absorber [137]. By means of this approach, V_{oc} was increased from 1.045 to 1.096 V, and PCE from 20.62% to 22.16% for the 3D and 3D/2D, respectively. Additionally, the stability of 3D/2D PSC was significantly improved upon prolonged dark storage in high relative humidity conditions (up to 90%).

All these studies evidence that the passivation of PS active layers is a successful strategy to further enhance the overall performance of devices and their long-term stability. Despite the valuable advances recently reached so far, there is ample room for improvement due to the large number of available materials with passivating properties and deposition techniques and the wide variety of approaches to be explored.

4.4 Measuring and Tuning the Work Function and Surface Potential in PSC

As noted in Sections 4.1–4.3, charge accumulation/recombination due to trap states in the bulk and at interfaces and energy level mismatch at the PS interfaces with

HTM and ETM play relevant roles in the performance of PCS. These are therefore key factors that can be tuned to further improve the photovoltaic functionality of this technology. Consequently, understanding fundamental charge carrier distribution within the devices is required to further enhance the device efficiency and long-term stability. It seems to be generally accepted that photogenerated charge carriers in PSC behave as free charges because the thermal energy is sufficient to overcome the exciton binding energy [7]. However, the mechanisms behind the charge carrier transportation toward the outer electrodes and the junction structures formed within PSC are still controversial issues under debate [82, 138–140]. For instance, Jiang et al. studied the electric field inside PSC with standard configuration using TiO_2 and spiro-OMeTAD as ETM and HTM, respectively [140]. They reported the observation of a p–n junction at the TiO_2/PS interface, whereas no junction was observed at the PS/spiro-OMeTAD interface. However, Bergmann et al. found an electrical potential barrier at the PS/spiro-OMeTAD interface through KPFM measurements performed both in dark and under illumination, thus suggesting a p–i–n junction structure [141]. The surface potential or work function of a sample can be measured through different techniques such as Kelvin probe (KP), KPFM, photoelectron spectroscopy (PES), or scanning electron microscopy (SEM) coupled with electron beam-induced current (EBIC) analysis [142]. Determining the surface potential/work function of the different materials employed in a complete device is crucial to obtain a proper energy level alignment among adjacent layers and, therefore, maximize the charge extraction efficiency.

Yue et al. studied systematically the impact of the work function of the cathode material on the performance of PSC with inverted architectures [143]. They exploited devices with architecture ITO/PEDOT:PSS/PS/PC61BM/buffer metal/Al, where different buffer layers, i.e. Ca, Mg, Ag, and Au, were studied. They found that lowering the work function of the cathode is of large benefit to performance improvement, including promoting the separation and conduction of charge carriers and even increasing the generation rate of excitons. This study revealed an increasing gradient of *PCE* values from 2.99% up to 17.09% for the highest and lowest work function buffer metals Au and Ca, respectively, indicating that cathodes with lower work function have a superior capability in suppressing the reverse transfer of charge carriers in devices with inverted structure.

One of the most studied device architectures in PSC is based on the use of TiO_2 as an ETM; however, its use is often accompanied by intrinsic limitations such as high processing temperature, presence of trap states, or photoinduced side reactions. Xie et al. reported the fabrication of high-efficiency PSC in planar configuration without introducing the commonly employed TiO_2 as ETM and using the following device architecture as benchmark devices: ITO/PCBM/PS/spiro-OMeTAD/Au [144]. Similarly to others, they measured a value of the ITO work function of −4.75 eV from ultraviolet photoelectron spectroscopy (UPS). It is worth mentioning that the LUMO level of PCBM lies at −3.8 eV below the vacuum, thus revealing that the energy levels at the ITO/PCBM interface are not optimally aligned for an efficient electron extraction. Interestingly, they then modified the ITO surface with a thin layer of poly [(9,9-bis(3′-(*N*,*N*-dimethylamino)propyl)-2,7-fluorene)-alt-2,7-(9,9-dioctylfluorene)] that induced a significant upward shift of the ITO work function due to the surface dipole mechanisms reaching −4.12 eV. Upon selective modification of the ITO surface, the PSC produced showed a significant increase of their performance compared to those made of bare ITO, showing *PCE* values of 12.2% and 7.9%, respectively. Despite the moderate *PCE* values reported with this approach, it is necessary to consider the relative simplicity of the device preparation procedure, which avoids the use of high-temperature processing that TiO_2 generally requires, thus broadening the compatibility with the use of other materials and substrates. In addition, this strategy reveals the significant role that surface engineering plays on the observed performance of devices.

Some reports in the literature show the formation of Schottky barrier PSC by substituting the charge-selective contacts by metal-like materials [145–148]. Possible benefits include avoiding some limiting issues related to the deposition of HTMs and/or ETMs, e.g. stability, energy level mismatching, processing complexity, or fabrication costs. Song et al. demonstrated the fabrication of HTM-free PSC based on the device architecture graphene/PS/PCBM/Ca/Al with relatively good efficiencies [149]. This approach is based on the selection of a metal-like contact with a suitable work function, in this case graphene. The performance of these devices was compared with those obtained from devices prepared with PEDOT:PSS as a HTM. These experiments revealed that the V_{oc} of the Schottky devices (1.05 V) was higher than that obtained for PEDOT:PSS-based devices, which was ascribed to the higher work function of graphene (−5.23 eV) compared with that of PEDOT:PSS (approximately −5.0 eV). The higher work function of graphene causes a more significant band bending at the graphene/PS interface, thus accounting for the higher V_{oc}. Despite this, moderate-to-low overall PCE were obtained mainly due to losses in the J_{sc} and *FF*. A different HTM-free device architecture that exploits mesoporous carbon counter electrodes has also been developed through printable techniques aiming at ultralow-cost and highly stable PSC. For instance, Mei et al. reported a fully printed stacked sequence of materials consisting of c-TiO_2/m-TiO_2/m-ZrO_2/m-carbon (c and m accounts for compact and mesoporous layers, respectively) onto which a PS precursor solution is drop cast [150]. Through this strategy, devices with J_{sc} of 22.8 mA cm^{-2}, V_{oc} of 0.858 V, *FF* of 66%, and *PCE* value of 12.84% were achieved. Remarkably, in the absence of the m-ZrO_2, the performance of devices is significantly poorer, having J_{sc} of 12.51 mA cm^{-2}, V_{oc} of 0.592 V, *FF* of 56%, and overall efficiency of 4.18%. This proves the important role of the ZrO_2 interlayer between the electron collector (TiO_2) and the *m*-carbon back contact in blocking the flow of photogenerated electrons to the back contact, thus preventing charge recombination. Later, Jiang et al. studied the influence of the work function of carbon electrodes on the performance of PSC with similar configuration c-TiO_2/m-TiO_2/m-ZrO_2/m-carbon/c-carbon [151]. The Fermi levels of pristine carbon materials such as carbon black or graphite are usually above the valence band of PS, thus inducing a misalignment of the energy levels that leads to inefficient hole extraction and low V_{oc}. In their work, the authors added different amounts of NiO to modify the work function of the carbon electrodes, as confirmed by Kelvin probe measurements. The work function of the *m*-carbon electrodes continuously increased from −4.94 to −5.04 eV as the ratio of NiO increased from 0% to 50%. They found that the V_{oc} and overall performance of the devices can be enhanced by tuning the work function of the *m*-carbon electrode without sacrificing J_{sc} and *FF*. Nevertheless, this approach showed drawbacks upon the analysis of devices with different ZrO_2 layer thicknesses. When the ZrO_2 layer is too thin (0.5–1 μm) and NiO is incorporated into the *m*-carbon layer, the extraction of holes is enhanced, but unfortunately the charge recombination at the TiO_2/carbon interface limits the device performance. Contrarily, when the thickness of the ZrO_2 is further increased (3 μm), the interfacial charge recombination at TiO_2/carbon interface is significantly reduced, due to the insulating properties of ZrO_2. However, the performance of the devices is sometimes sensitive to the addition of NiO and seems to be dominated by the increased charge diffusion length as a consequence of the thicker ZrO_2 layer.

The fabrication of ETM-free devices with architecture ITO/PEDOT:PSS/perovskite/Ti/Au, where the Ti interlayer avoids the direct contact between Au and PS was reported by Shi et al. [148]. Due to the nearly identical work functions of PEDOT:PSS (approximately −5.0 eV) and Au electrodes (−5.1 eV), the direct contact between Au and PS in such inverted architectures is generally considered to be unfavorable for charge collection, thereby favoring charge carrier recombination at the electrodes. However, since the work function of Ti is higher than that of PEDOT:PSS by about 1 eV and slightly lower than the conduction band edge of PS, inserting a layer of Ti thin film effectively raises the V_{oc} and blocks the holes without affecting the collection of electrons. Accordingly, Ti-based devices having averaged V_{oc}, J_{sc}, *FF*, and *PCE* values of 0.82 V, 21.96 mA cm^{-2}, 50.3% and 10.2%, respectively, were achieved. It was concluded that Ti layer can abate the effect of PS voids and inhibit the direct contact between Au and PS. This in turn passivates any charge traps in the PS films, thus minimizing the *I–V* hysteresis. A more recent report based on the same strategy further demonstrates the production of high-performance planar PSC by exploiting the Ti interlayer concept (Figure 4.7a,b) [152]. In this case, the effect of the Ti interlayer in combination with Au or Cu as a top cathode electrodes was tested. ITO/PTAA/PS/Ti/(Au or Cu) devices showed similar and outstanding cell parameters with J_{sc} of 22.5 mA cm^{-2}, V_{oc} of 1.03 V, and *FF* of 78% that yielded a *PCE* of 18.1%. Those devices without Ti interlayer showed significantly reduced performances, which again proves the beneficial effect of Ti not only on basis at its ability toward surface defect passivation of PS layers but also for its role as a diffusion barrier layer that prevents the diffusion of the cathode metal atoms. Deposition of the Ti forms a very smooth layer on the top of PS and therefore improves the top-metal film uniformity, as shown in Figure 4.7c–j.

Despite the fact that the stability of Cu-based devices was shown to be significantly limited compared with the Au counterparts, these results open new promising strategies for the development of high-performance and cost-efficient PSC. Unfortunately, the estimation of work function and surface potential of materials that are sensitive to environmental and processing conditions is not trivial, and this is often the case for PS materials. Zhang et al. have studied the contact potential difference and surface potential of a 3D-PS derivative ($MAPbBr_3$) and a 2D-PS based

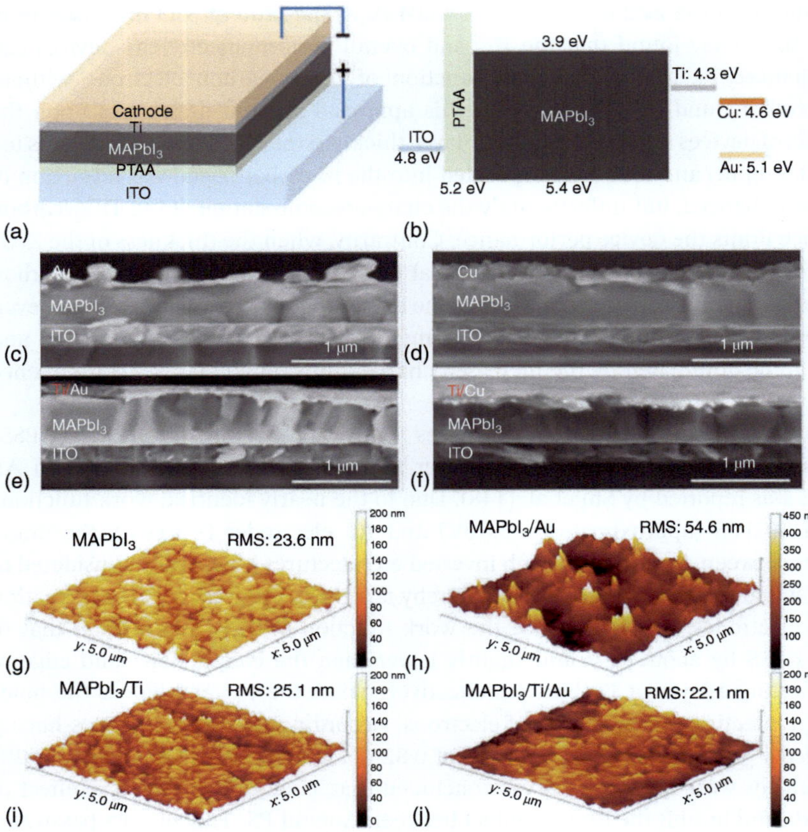

Figure 4.7 (a) Schematic and (b) energy level of the device structure using a Ti interlayer. Cross-sectional SEM images of devices with (c) Au, (d) Cu, (e) Ti/Au, and (f) Ti/Cu electrode, respectively. Topography AFM images and root-mean-square (RMS) surface roughness of (g) $MAPbI_3$, (h) $MAPbI_3$/Au (80 nm), (i) $MAPbI_3$/Ti (10 nm), and (j) $MAPbI_3$/Ti (10 nm)/Au (80 nm). Source: (a, b, j–g) Chen et al. [152], (c) Adapted from Chen et al. [152] with permission from Wiley.

on butylammonium cation (BA_2PbI_4) deposited on ITO substrates, respectively, in the dark and under illumination conditions through Kelvin probe measurements [153]. They observed a constant work function value of −4.43 eV for the 3D-PS in dark conditions and, upon illumination with green light (2.34 eV), an increase of the work function of the tip (from −4.10 to −4.24 eV) for the surface potential of the 3D-PS film (from −4.45 to −4.55 eV). After subsequent relaxation in the dark, the work function changes in the tip recovered the initial values. However, the changes suffered by the 3D-PS film were irreversible. Therefore, the non-reversible nature of the changes measured for the $MAPbBr_3$ film suggests the absence of band bending at the surface, strong Fermi level pinning, or fast degradation of $MAPbBr_3$. In contrast, BA_2PbI_4 exhibited a significant surface photovoltage upon illumination, which arises from a partial flattening of an upward surface band bending promoted by the accumulation of photogenerated carriers. After relaxation in the dark, a fast surface photovoltage recovery component ascribed to charge recombination and a slow quasi-irreversible component were observed. Although the non-reversible effect was significantly more notable for the 3D-PS, it was concluded that illumination of both 3D- and 2D-PS with wavelengths of higher energy than the bandgap leads to an irreversible halide loss from the surface of the PS, thus contaminating the KP tip. From an experimental point of view, the results show that Kelvin probe-based surface photovoltage measurements on PS surfaces should be carefully considered, since there is a high potential risk for erroneous conclusions that arise from halide-induced changes in the tip. Very recently, Sadhu et al. reported a systematic surface modification of $MAPbI_3$ with monolayers of perfluorinated molecules and studied its effect on the surface potential of the PS films by means of Kelvin probe measurements [154]. The surface of $MAPbI_3$ films deposited on ITO/PEDOT:PSS substrates was modified with a series of perfluorinated porphyrins, which induced a gradual tuning of the surface potential of the PS film from 5.02 to 5.17 eV depending on the porphyrin derivative. Pristine PS layers showed a surface potential value of 5.11 eV. These results could have a great impact on the performance of devices, including long-term stability, and unveil a suitable strategy toward a better matching of the PS work function with that of the adjacent layers.

Despite possible artifacts, KPFM has been implemented as a very useful technique to characterize PS films and PSC under varying conditions, since it is possible to identify the local work function at the nanoscale with high-resolution through imaging of surface topography. This makes it a valuable tool for exploring the surface composition and electronic states even at the microcrystal boundaries [155–159]. In fact, due to its nondestructive nature, KPFM constitutes a unique tool to probe surface energetics and charge dynamics of PS derivatives that are very often extremely sensitive to water, oxygen, and electron beam irradiation. Lan et al. disclosed a study on the cross-sectional surface potential profile of planar PSC with standard and inverted configuration (ITO/ZnO/PS/spiro-OMeTAD/Au and ITO/NiO/PS/ZnO/Ag, respectively) through KPFM [160]. As pointed out previously [161], this work assumes that biasing the device at a fixed DC potential in the dark induces a surface potential distribution within the device equivalent to that found under operating conditions. Their measurements revealed that for a standard configuration, a p–n junction is formed at the PS/ZnO interface with a depletion region of around 310 nm into the PS layer. Therefore, electron injection is affected by the electric field in the depletion region. Instead, holes can diffuse through the HTM layer with very low resistance. For inverted configuration devices, a p–i–n junction distribution is found, and the charge transport is instead determined by the electric field through the device. Hence, both hole and electron injection barriers were observed at the NiO_x/PS and PS/ZnO interfaces, respectively. These results are in agreement with the slightly lower efficiencies that are generally reported for inverted PSC [162, 163], compared with the standard PSC [117, 164]. The origin of such energy barriers may be related to an energy level misalignment or to a limited interface contact between PS and both metal oxides. Along similar lines, Yang et al. compared the surface potential of PS layer using a standard (fluorine tin oxide (FTO)/c-TiO_2/m-TiO_2/PS) and an inverted planar (FTO/c-NiO/PS) semi-cell configuration through KPFM [165]. The effect of top and below illumination was explored for both configurations. However, they observed an increase of the surface potential of 73 mV for the standard configuration between dark and illumination from the top (PS side). Contrarily, a decrease of 43 mV was observed for the inverted semi-cell configuration at the same illumination conditions. Upon illumination from the underneath (glass side), an increase of 280 mV and a decrease of 30 mV were obtained for the standard and inverted semi-cells, respectively. These results confirm that electrons and holes are injected at the PS/ETM(HTM) interfaces and that these induce the observed variations in the surface potential of PS. For standard architecture devices, most of the photoexcited electrons are generated at the TiO_2/PS interface followed by the electron extraction into the TiO_2 under the built-in electrical field when illuminated from the glass side. However, when the illumination is provided from the PS side, a large amount of the photoexcited electrons are generated on the surface of the absorber, and therefore, they need to diffuse throughout the PS thickness before being extracted to the ETM. This long-range diffusion process increases the chance of recombination within the PS, which accounts for the lower surface potential values measured. Instead, for the NiO/PS interface, the lower surface potential values indicate a smaller accumulated charge density than the TiO_2/PS samples, with the potential difference similar under both illumination conditions. These results highlight the different light sensitivity depending on the nature of the ETM(HTM)/PS interfaces and disclose the importance of the measuring conditions for a proper interpretation of surface photovoltage measurements through KPFM.

The role of the grain boundaries on the operation of PSC devices remains a controversial topic that is still under much debate. Adhikari et al. performed a study using KPFM measurements to quantify the barrier that prevents interfacial back-recombination between extracted electrons in TiO_2 and accumulated holes in the PS depending on the annealing time at a constant temperature (100 °C) [155]. At the same time, they also studied the surface potential at the grain boundaries within the PS. From XRD, topography, and surface potential KPFM images, it was concluded that the formation of a passivation layer of PbI_2 upon 15 minutes of thermal annealing increases the energetic barrier for the back-recombination of charges at the TiO_2/PS interface. Additionally, surface potential spatial mapping of PS layers shows higher surface potentials at the grain boundaries than that observed within the grains. However, the grain boundary potential was also found to decrease from the non-annealed samples to those annealed during 15 minutes but to increase for longer annealing times. These results suggest that formation of PbI_2 also passivates the PS grain boundaries, thus improving the charge transport properties, which in turn improves the device performance. Nevertheless, the device efficiencies of the samples studied in this work are far from the current *state of the art*, and the PS microscopy images reveal the presence of very irregular-shaped crystals and significant variations of the crystal size among samples. Yun et al. performed a related study focusing on the effects of ion migration and the role of grain boundaries on the photovoltaic performance of a mixed-cation/halide PS derivative $(FAPbI_3)_{0.85}(MAPbBr_3)_{0.15}$ [158]. They performed KPFM experiments on the top surface of PS in the dark under bias in noncontact mode to simulate open-circuit conditions and in contact mode to reproduce short-circuit conditions. They showed that biasing the PS induced changes in the contact potential difference (CPD) between the tip and the PS surface and that different bias induced variations in the CPD (with the effect being more significant for negative biasing). After removing the bias, CPD recovered the initial values but revealed slow kinetics, since the recovery time was about 720 seconds. Moreover, the change of CPD was larger at the grain boundaries than at the interior of the grains. The authors therefore concluded that these phenomena are directly related to ion migration processes, as energy-dispersive X-ray spectroscopy (EDX) results confirmed that MA^+ and/or FA^+ cations and I^- anions participate in such migration. Very importantly, these results point out that grain boundaries behave as fast ion migration channels and that, under illumination conditions, band bending is formed at the boundaries that facilitates charge carrier separation and contributes positively to the device operation. These results seem to agree with alternative reports based on KPFM and AFM measurements, where larger CPD values have been identified along the grain boundaries compared with the inner crystal regions upon illumination (Figure 4.8a–c) [47]. However, opposite conclusions are provided from a wide number of alternative scientific contributions that point out to a detrimental effect of grain boundaries

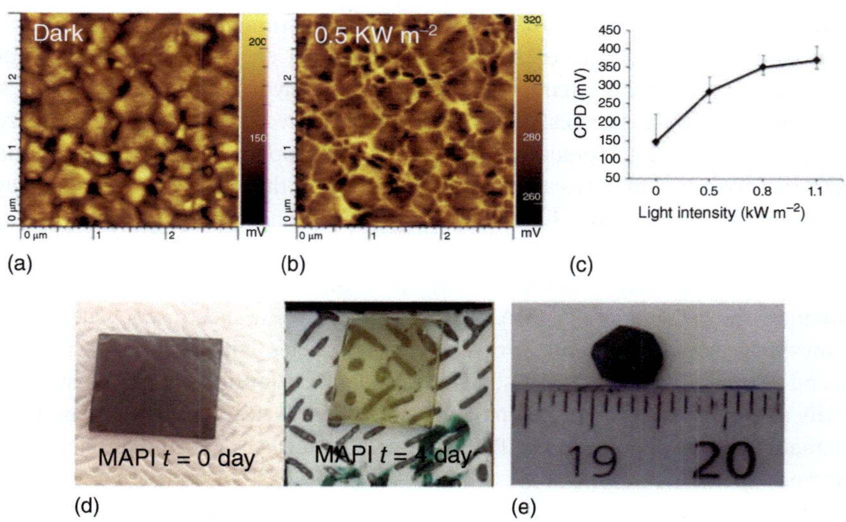

Figure 4.8 KPFM measurements performed on $MAPbI_3/TiO_2/FTO$/glass structure over an area of 3 μm^2. (a) CPD images taken in the dark. (b) CPD map under 500 nm laser illumination at an intensity of 0.5 $Kw\,m^{-2}$. (c) CPD at different illumination intensities (500 nm laser) measured from KPFM. Source: (a) Adapted with permission from Yun et al. [47]. Copyright (2005) American Chemical Society, (c) Yun et al. [47]. (d) Photography of an 800 nm-thick $MAPbI_3$ layer deposited by spin coating on a quartz substrate, just after the deposition ($t = 0$ day) and four days after the deposition ($t = 4$ days). (e) Photography of a millimeter-sized $MAPbI_3$ crystal, six months after its growth. Source: (d) Adapted from Jemli et al. [166].

[155, 167, 168]. In fact, the higher stability of PS single crystals than of PS thin films also demonstrates the important role of grain boundaries in the degradation mechanisms (Figure 4.8d,e) [166]. Although grain boundaries seem to be unanimously considered key elements that determine the performance, their role is still questionable. Thus, further development of more systematic and accurate studies is required for a full understanding of the photovoltaic working principles of PSC.

4.5 Tuning the Wettability and Compatibility Between Layers

Similar to other photovoltaic technologies, PSC consists of different vertically stacked materials, each assigned to accomplish specific functions, i.e. light absorption, charge transport and separation, inhibition of charge recombination, etc. Consequently, the compatibility between materials, deposition techniques, and processing conditions is a crucial factor in determining the viability of device preparation and their ensemble performance. In this sense, the wettability of the precursor solutions employed to deposit the different stacked materials on top of each other plays an important role in the formation of high-quality films and interfaces. PSCs are very often produced from wet deposition methods in which their solutions consist of ionic precursors, e.g. MAI and PbI_2, dissolved in high dielectric constant (ε) solvents, e.g. DMF and/or DMSO, thus providing a strong hydrophilic character. Consequently, PS layers are often successfully grown on hydrophilic organic or inorganic surfaces rather than on those with high hydrophobic characteristics. As pointed out above, the best performing PSC reported so far adopts a standard architecture where light enters from the ETL. The optical, morphological, and electrical properties of ETLs can significantly affect the PS crystal morphology and quality, as well as the light-harvesting and/or charge separation/transport properties of PSC. De Bastiani et al. investigated the crystallization of PS films on several metal oxide substrates with different composition and porosity [169]. In particular, they characterized $MAPbI_{3-x}Cl_x$ films by PL spectroscopy deposited on mesoporous substrates made of ZnO nanoparticles with different sizes (70, 50, and 20 nm), TiO_2, and Al_2O_3, respectively. They found that PS crystallizes rapidly at 100 °C when deposited on ZnO substrates. Instead, a one-hour thermal treatment is required to form the PS layer on TiO_2 and Al_2O_3 scaffolds. They also observed significant variations in surface coverage and crystal morphology among different-sized nanoparticle ZnO substrates. In addition, different PL characteristics were identified due to the presence of infiltrated PS into the scaffolds and capping layer on top. Infiltrated PS shows smaller crystal sizes due to volume limitations that provoke a faster PL decay kinetics, whereas the capping layer shows larger grains with slower characteristic decay times. Chen et al. reported a comparison between the performance of devices prepared from using quasi-compact TiO_2 films made of small nanoparticles (6 nm in diameter) and those obtained from spray pyrolysis compact TiO_2 films [169]. Contact angle measurements of DMF droplets on both substrates revealed significant differences. Quasi-compact nanoparticles and compact TiO_2 films showed contact angles of 7.2° and 36.7°, respectively. The smaller contact angle of the TiO_2 nanoparticles film implied an enhanced surface wettability, which yielded devices with remarkably higher efficiencies, 11.0% vs. 4.6%, due to an improved PS surface coverage. Following the concept of optimizing the wettability of the bottom layer films prior to the deposition of the next material, Li et al. explored the modification of the PS surface with graphene oxide (GO) toward improving the wettability of a spiro-OMeTAD solution for its deposition as an HTM [170]. After surface modification, a significant decrease of the contact angle of the HTM solution from 13.4° to 0° was observed. Interestingly, XPS results revealed that GO interacts with the perovskite by Pb—O bonding, and Raman spectroscopy analysis showed that C—C double bonds present in GO bind to spiro-OMeTAD through π–π interactions. The improved wettability between PS film and HTM solution, as well as the enhanced interface interaction, accounts for the 45% improvement of *PCE* observed, from 10% to 14.5%. These results prove the influence of the adjacent layers on the PS properties, which eventually will contribute to determining the ultimate performance of PSC.

Extensive work has been developed on PSC with inverted configuration, which has eventually nearly equaled the performance of standard architecture devices. PEDOT:PSS is an organic conductive polymer very often used as HTM in inverted architectures, partially due to its good compatibility with PS precursor solutions that arise from its hydrophilic nature. However, it shows intrinsic limitations due to its acidic nature, hygroscopicity, and limited ability to selectively block electrons. Therefore, several organic and inorganic alternatives, e.g. NiO, CuSCN, P3HT, and PTAA, have been proposed as alternative to PEDOT:PSS [171]. Bi et al. reported the fabrication of planar PSC with enlarged size crystal grains by exploiting a series of different non-wetting organic HTMs [172]. This approach relies on the use of smooth and non-wetting surfaces that prevent the formation of a too dense layer of crystal nucleation sites from heterogeneous nucleation, thus favoring crystal growth. Remarkably different water contact angles of 10°, 12°, 79°, 105°, and 108°, which can be extrapolated to DMF, were obtained for polyvinyl alcohol (PVA), PEDOT:PSS, cross-linked *N*4,*N*4′-bis(4-(6-((3-ethyloxetan-3-yl) methoxy)hexyl)phenyl)-*N*4,*N*4′-diphenylbiphenyl-4,4′-diamine (c-OTPD), poly(bis (4-phenyl)(2,4,6-trimethylphenyl)amine) (PTAA), and poly(*N*-9′-heptadecanyl-2,7-carbazole-alt-5,5-(4′,7′-di-2-thienyl-2′,1′,3′-benzothiadiazole)) (PCDTBT), respectively (Figure 4.9). Upon the optimization of the deposition methods to produce pinhole-free films on non-wetting substrates and device optimization to enhance the conductivity of organic HTM, those devices made with PTAA showed a *PCE* of 18.1% with V_{oc} of 1.07 V, J_{sc} of 22.0 mA cm^{-2}, and *FF* of 76.8% (Table 4.1). Despite not specifically studied in this example, the exploitation of HTM that do not suffer from the inherent limitations of PEDOT:PSS together with the production enlarged-size PS crystals was believed to yield devices with enhanced long-term stability.

Metal oxides are known to generally show better chemical, photochemical, and electrochemical stabilities than organic compounds. For instance, Du et al. [173] reported the fabrication of high-efficiency inverted PSC using NiO electrodes modified with PTAA. Despite untreated NiO previously reported to suffer from limited

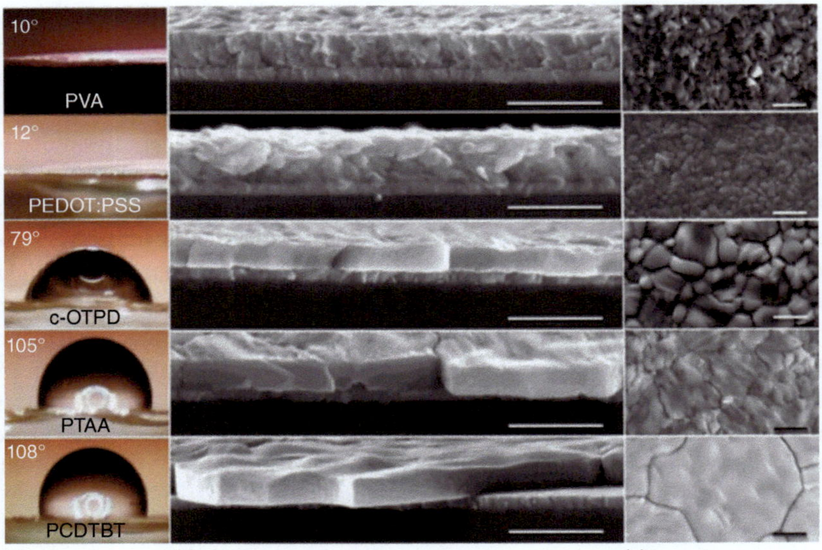

Figure 4.9 (a) Contact angle of water on a series of HTMs. (b) Cross-sectional and (c) top-view SEM. Scale bars are 1 μm. Source: Adapted from Bi et al. [172].

Table 4.1 Summary of the performance of the PSCs devices made with different HTMs.

HTL and active layer thickness	J_{sc} (mA cm^{-2})	V_{oc} (V)	FF (%)	PCE (%)
PEDOT:PSS 400 nm $MAPbI_3$	18.6	0.92	72.0	12.3
c-OTPD 500 nm $MAPbI_3$	22.4	1.05	75.6	17.8
PTAA 500 nm $MAPbI_3$	22.0	1.07	76.8	18.1

Source: Adapted from Bi et al. [172].

compatibility with PS for efficient charge extraction [174], its properties, i.e. energy levels and/or wettability, can be tuned by specific treatments to overcome these inherent limitations. In this direction, PTAA has been used to modify the surface properties of NiO_x films, which improves the interaction between NiO_x and PS and the PS quality. It is worth mentioning that both NiO_x and PTAA show good band alignment with PS, allowing the achievement of high V_{oc} devices with efficient hole extraction at the NiO_x/PS interface. Interestingly, NiO_x/PTAA substrates showed a water droplet contact angle that was higher than that observed for pristine NiO_x, 85.1° and 42.8°, respectively, thus suggesting a higher hydrophobic character for the PTAA-treated films. As previously mentioned, the difference in substrate wettability can alter and tailor the nucleation/growth process of PS crystals and therefore influence the charge transport/extraction properties [172]. As a result, planar PSC devices with a maximum efficiency of 17.1% and negligible I–V hysteresis were fabricated upon NiO_x surface modification.

Despite the best performing PSC being based on PS derivatives with 3D crystal structures [27, 28, 31, 164], the lack of stability hampers their commercialization and exploitation under working conditions. In contrast, their 2D-PS counterparts are known to provide much better long-term chemical and photochemical stabilities [175]. Although the performances obtained so far for low-dimensional PSC are still far below those of *state-of-the-art* 3D-PSC, 2D-PS derivatives have emerged as promising candidates to overcome stability issues. Significant efforts are being devoted toward enhancing the 2D-PSC efficiencies with high long-term stability that can effectively compete with the 3D counterparts. Based on this, Tian et al. recently developed a quasi-2D-PSC with a device structure of ITO/PTAA/PS/PC61BM/BCP/Ag based on the mixed-cation PS derivative $PEA_2MA_{3.5}FA_{0.5}Pb_5I_{16}$ (PEA: phenylethylammonium cation) [176]. These devices achieved a maximum *PCE* of 10.28% in reverse *I–V* scanning, with a J_{sc} of 13.31 mA cm^{-2}, V_{oc} of 1.14 V, and *FF* of 0.68. Moreover, forward scanning on the same device yielded a PCE of 9.75%, thus revealing a certain degree of hysteretic behavior. Upon depositing a thin layer of PFN (<5 nm) on the PTAA film, devices in reverse scanning direction showed a significant performance enhancement from 10.28% to 12.78% due to an increase in the J_{sc} and *FF*, and the hysteresis effect was partly alleviated. From SEM imaging of the PTAA/PS and PTAA/PFN/PS films, discontinuities and inhomogeneities of the 2D-PS film were observed for the sample without PFN due to the poor wettability. It is worth highlighting that water contact angles of 91.6° and 79.8° were measured. They found that the change in wettability and hydrophobicity upon PFN deposition not only improved the quality of PS in terms of enhancing the smoothness, crystal size, and surface coverage but also played a significant role on forming an unexpected distribution of crystal PS phases. Although the composition was nominally designed to produce a PS with $n = 5$, where n accounts for the number of 3D-PS sheets constituted by the small cations and separated by the bulky PEA organic layers, coexistence of multiple perovskite phases with different n values was detected, in agreement with the previous reports [177, 178]. It was confirmed that improving the device performance upon PFN treatment reduces the formation of the PS phase with $n = 1$, thus resulting in an enhanced light absorption over the entire spectral window and the presence of PFN minimizing the trap-assisted recombination without affecting the charge extraction.

In summary, tuning the wettability of sublayers prior to deposition of new materials is an important factor to be considered to ensure an enhanced surface coverage, compatibility between materials, crystal grain size, and/or distribution of PS crystal phases, among other important parameters. Additionally, optimizing and exploiting the compatibility among materials, e.g. PS/ETM(HTM) interfaces, interlayers, or 3D/2D PS combinations, significantly broadens the number of possibilities and strategies to be adopted for further development of PSC technology. Despite the general concepts mentioned above being able to be extrapolated to any kind of device architecture, deposition technique, and/or material, each specific case requires an individualized perspective, and different approaches can be potentially needed.

4.6 Effect on Device Efficiency and Lifetime

Despite the outstanding advances achieved so far in the field of PSC in the last few years, one of the main limitations of this technology is still the severe lack of long-term stability, especially under real-life working conditions [179, 180]. Currently, the benchmark technology in terms of performance and stability is inorganic semiconductor devices such as silicon-based technology that provides lifetimes of over 25 years. However, the longest durability achieved so far in PSC is of about six months [181]. PSCs suffer from different types of stability issues that arise from a variety of sources and induce distinct detrimental phenomena on the stacked materials and device operation. For instance, PS derivatives are known to be prone of degradation in humid conditions [182], showing sensitivity to UV irradiation [183] and heat [184].

The stability of PS crystal structures with the general formula ABX_3 can be determined by the tolerance factor t and an octahedral factor μ; $t = (r_A + r_X)/[\sqrt{2}(r_B + r_X)]$, $\mu = r_B/r_X$, where r_A, r_B, and r_X are the ionic radii of the corresponding ions. Generally, to form a stable PS structure at room temperature, the value of t must lie between 0.81 and 1.11. When t is between 0.9 and 1.0, it is expected to be ideal cubic perovskite, while distorted structures such as orthorhombic, rhombohedral, tetragonal, or hexagonal are likely to be formed when t is larger or smaller. Therefore, the properties of the PS films can be readily modified by adjusting the size of the ionic components, which can induce tremendous variations in the PS stability, among other characteristics.

However, the lack of stability does not uniquely arise from the degradation of the PS material, since charge-selective contacts and/or metallic electrodes also play a significant role on the long-term device reliability [112]. In this section, we will discuss the different sources of degradation more relevant in PSC and their effects on the physicochemical properties of materials and device functionality, as well as the main strategies that are reported toward overcoming such limitations.

4.6.1 Moisture Effects on PS Films and PSC

The presence of controlled amount of water at low levels, as observed for other Lewis bases [185], influences perovskite film formation and has even been proposed to be beneficial for enhancing device performance [179, 186]. However, exposure to moisture often produces detrimental effects on the functioning of devices due to the decomposition of the PS structure. This is especially significant for 3D-PS materials, which show higher sensitivity to water exposure than their 2D-PS counterparts [135, 187]. The insertion of water molecules into PS structures was detected and studied decades ago. Thus, two different crystal structures were identified, the monohydrate $MAPbI_3{\cdot}H_2O$ and the dihydrate $MA_4PbI_6{\cdot}2H_2O$ [188, 189]. There are several plausible mechanisms that would explain the PS decomposition triggered by moisture; a possible degradation mechanism was proposed based on the hygroscopic nature of MA^+ ion, which can lead to the formation of HI [190]. The effect of water in the degradation cycle is catalytic and proposed to occur upon incorporation of small quantities of water into the film, thus inducing the irreversible formation of PbI_2. An alternative mechanism that involves water molecules and the presence of oxygen and/or UV light exposure was also proposed [191, 192]. Very importantly, Leguy et al. showed that all the proposed degradation pathways are part of a multistep decomposition mechanism in which the initial step corresponds to the formation of the metastable monohydrate PS structure [193]; moisture in air induces a reversible hydration process of the PS films. However, the presence or absence of condensed water seems to play a significant role on the degradation phenomenon. In its absence, degradation does not lead to the formation of crystalline lead iodide, and the process can be reversed by blowing dry air over the samples. After prolonged exposure to moisture, decomposition of the dihydrate structure is accompanied by the formation of PbI_2 and H_2O. Because these materials are likely to phase-separate, the reaction is irreversible. Additionally, ellipsometry measurements showed that water is evenly distributed within a hydrated $MAPbI_3$ film, suggesting that H_2O molecules must diffuse along the grain boundaries. This would explain why compact, pinhole-free, and larger grain PS films show better stabilities against moisture [194].

Recently, Schlipf et al. investigated the water uptake into $MAPbI_3$ films depending on the relative humidity level through small-angle neutron scattering [182]. The monohydrate was observed for humidity levels ≥73%, while the dihydrate is only detected for those ≥93%. In all cases the films contain much more water in the form of water molecules adsorbed on the outside of the crystalline domains or amorphous regions. The uptake of water into the crystalline domains and the formation of hydrates were accompanied by an expansion of the grains. This showed to be reversible, although the grain shape changed from faceted to round-shaped particles, which could be detrimental for device functionality. It was claimed that even humidity levels of 41% lead to incorporation of ~10% volume of water into the PS film by adsorption on the crystals. These evidences the strong sensitivity of PS materials even at low humidity levels. However, completely different scenarios can be found when the complete device is considered, also depending on the PS composition, deposition technique, processing, nature of HTM/ETM, and/or storage conditions, among other parameters.

Christians et al. studied the stability of $MAPbI_3$ films and PSC with the architecture TiO_2/PS/spiro-OMeTAD under different humidity levels in the dark [192]. These PSC initially showed a J_{sc} of 20.6 mA cm^{-2}, a V_{oc} of 1.01 V, a *FF* of 63%, and a *PCE* of 13.1%. No significant variations were found for those cells stored in a dry environment, showing similar performance values after 21 days. However, the exposure to moisture caused a significant drop in the efficiency of devices. While the *PCE* of devices stored in 50% humidity conditions dropped to 60% of the initial value in 21 days, those stored in 90% relative moisture showed a null functionality upon six days. Interestingly, they noticed that the photovoltaic performance decreased much more rapidly than the UV–vis absorption and excited-state characteristics of the $MAPbI_3$ films. A bare PS film exposed to 90% relative humidity for three days only revealed a marginal decrease in absorbance and no change in excited-state dynamics on the picosecond timescale. However, the efficiency of a complete cell stored under the same conditions decreased to nearly 95%. This would indicate that alternative degradation pathways not uniquely related to PS layers may account for the loss of performance in PSC. Unfortunately, these studies were not conducted under illumination conditions, and the efficiency of the explored devices was far below the current *state of the art*.

The monovalent cation present in PS derivatives is one of the key factors that can determine the PS crystal structure and its stability. When the MA^+ cations are replaced by the slightly larger FA^+, there is a lattice expansion [195] and therefore a change in the tilt of the PbI_6 octahedra [196]. This leads to a slight decrease of the bandgap from ≈1.59 eV for $MAPbI_3$ to ≈1.49 eV for $FAPbI_3$ [197]. It was reported that the interaction of FA^+ with surrounding PbI_6 octahedra is stronger than that of MA^+ due to higher probability of forming hydrogen bonds [198]. Despite reports of the photostability being better for $FAPbI_3$ than for $MAPbI_3$, the incorporation of FA^+ is found to show poorer stability toward humidity than its MA^+ counterparts [199, 200]. Nevertheless, Wu et al. have recently reported on the fabrication of planar PSC with standard configuration based on mixing MA^+ and FA^+ cations that show high efficiency and significantly enhanced stability [201]. They exploited the two-step deposition method by adding dropwise a solution of MA^+/FA^+ onto a PbI_2 film. Upon optimization, closely packed, large, and flat PS crystals were formed (largest grain size of about 5 μm) that yielded devices with *PCE* values of 20.63%. Interestingly, PSC even without encapsulation exhibited excellent humidity stability, retaining 93% of its initial efficiency after 500 hours in air under 40% relative humidity conditions. It is worth highlighting that these stability measurements were carried out with the samples stored in the dark at room temperature. Therefore, the combination of MA^+ and FA^+ cations might not overcome photothermal and thermal instabilities. Leblanc et al. introduced large cations into 3D $[Pb_{1-x}I_{3-x}]$ α-FAPI-type perovskite network. They demonstrated that they could circumvent the Goldschmidt factor introducing hydroxyethylammonium (HEA^+) and thioethylammonium (TEA^+) with effective radii of 3.44 and 4.14 Å, respectively. More importantly, they found that thin films of $(TEA, FA)_{1+x}[Pb_{1-x}I_{3-x}]$ (x = 0.04 and 0.13) showed exceptional α-phase stability under ambient conditions: 1 order of magnitude higher than α-FAPI and α-(Cs, FA)PbI3 thin films [202].

The incorporation of inorganic cations, such as cesium (Cs^+), has been also explored in view of improving the device performance and stability. Therefore, all-inorganic PS derivatives such as $CsPbI_3$ have been prepared. Unfortunately, the cubic phase α-$CsPbI_3$, which shows a suitable bandgap for harvesting sunlight ($E_g \sim 1.7$ eV), is structurally unstable below 320 °C and undergoes a spontaneous phase transition to orthorhombic δ-$CsPbI_3$ ($E_g \sim 2.8$ eV) that possesses a poor photoresponse [203]. However, partial introduction of Cs^+ in $FAPbI_3$ PS has been reported as an effective method toward stabilizing the trigonal phase of $FAPbI_3$. Lee et al. reported the synthesis of unencapsulated $FA_{0.9}Cs_{0.1}PbI_3$ films and devices that were exposed to high moisture conditions (85% relative humidity) in the dark and to less humid conditions (<50% relative humidity) under illumination conditions (sulfur lamp at 100 mW cm^{-2}) [204]. The UV–vis absorption of the

$FA_{0.9}Cs_{0.1}PbI_3$ films was monitored in both conditions and compared with that of the controlled PS $FAPbI_3$. The photostability test revealed a decrease of the absorbance ($\lambda = 630$ nm) for both PS derivatives, although $FAPbI_3$ film showed an ~86% decrease of the absorbance while $FA_{0.9}Cs_{0.1}PbI_3$ showed a 65% lessening after 19 hours, thus suggesting a more severe degradation for the sample without Cs^+. Additionally, moisture stability test in the dark showed relatively better stability for $FA_{0.9}Cs_{0.1}PbI_3$ samples after seven hours, since an absorbance loss of 49.9% was observed compared with the 77.8% measured for the pure FA^+ samples. Nevertheless, complete devices prepared from $FAPbI_3$ and $FA_{0.9}Cs_{0.1}PbI_3$ showed limited stability since the *PCE* values were decreased by a fivefold factor in about 60 minutes of light exposure in the best case.

In addition to cation exchange, the introduction of different halides has allowed producing highly efficient PSC with enhanced stability. For instance, in 2016, McMeekin et al. reported a photostable PS with the formula $FA_{0.83}Cs_{0.17}PbI_{1.8}Br_{1.2}$ showing a bandgap of ~1.74 eV and allowing fabricating devices with V_{OC} of 1.2 V and $PCE > 17\%$ [205]. The photostability was judged based on the lack of phase segregation in the PS film upon irradiation with monochromatic light at 5 W cm^{-2} during 240 seconds. Also, Cai et al. demonstrated the fabrication of PSC based on $FA_{0.8}Cs_{0.2}PbI_{2.68}Br_{0.32}$ with relatively long-term moisture stability. A *PCE* of 17.0% was achieved, retaining 85% of the initial value in air at 60% of relative humidity after 45 days in the dark without encapsulation. Very recently, Yang et al. converted the surface of lead halide PS into a water-insoluble lead(II) oxysalt through reaction with sulfate or phosphate ions. With this approach they could obtain a PS single crystal that remained stable even after dipping it in water during four minutes [206]. They also made CsFAMA devices with efficiency of above 20%. Moreover, stability tests at the *MPP* of encapsulated devices (with and without oxysalt treatment) under irradiation were performed. This concludes that treated devices maintained stable for more than 1200 hours, whereas untreated devices showed an efficiency drop of 10% after only 300 hours.

Other approaches toward enhanced moisture stability have been proposed based on exploiting 2D-PS derivatives. Smith et al. reported the formation of the 2D-PS derivative $PEA_2MA_2Pb_3I_{10}$ and demonstrated that the XRD patterns after 46 days exposure at 52% relative humidity in the dark were unvaried, thus indicating its good stability to moisture [131]. However, moving from 3D to 2D-PS derivatives by introducing bulky cations limits charge transport, thus clamping the device efficiency. Interestingly, Liu et al. investigated the effect of adding small amounts (0.01%) of the bulky cation *tert*-butylammonium (*t*-BA^+) and reported that it induced a stabilization of $MAPbI_3$ [207]. They demonstrated a *PCE* of 19.3% and a *FF* of 80%. *PCE* decreased less than 20% after exposure to ambient environment at 50% of relative humidity in the dark at room temperature and without encapsulation.

Depositing hydrophobic layers on 3D-PSC has been also exploited toward improving the stability of films and devices while maintaining high efficiencies. Liu et al. demonstrated the formation of an ultrahydrophobic 2D capping layer $(FEA)_2PbI_4$ on top of a 3D-PS by treatment with pentafluorophenylethylammonium (FEA). This approach yielded devices with an initial *PCE* of 22.1% that retained 90% of their efficiency during photovoltaic operation upon 1000 hours in humid air under simulated sunlight [137]. Deposition of photocurable fluoropolymers has been also reported [181]. The authors proposed a luminescent fluoropolymer on top of the back contact that absorbs UV light and reemits visible light, thus protecting the PS layer from photodegradation while allowing absorption of the incident light. With this approach, devices with a *PCE* of 18.67% were achieved. More impressively, the performance of five devices that were kept for one month in a closed chamber in the presence of a beaker containing boiling water (~95% relative humidity) was evaluated. After one month, four of the five cells withstood the accelerated aging conditions and remarkably retained 96% of their initial *PCE*.

Despite the relatively good stabilities against moisture achieved through different approaches described in the literature, further advances for device stability under working conditions are required, since PSC suffers as well from severe photothermal and thermal degradations.

4.6.2 Photoinduced Degradation of PS Films and PSC

Photovoltaic devices are inevitably exposed to light during working conditions, and, despite the successful efficiency values obtained so far for PSC, these suffer from premature photothermally and thermally induced degradation. Stability improvements can be achieved by developing encapsulants and adding UV filters that alleviate the degradation of the different interlayers constituting the devices. Nevertheless, the intrinsic properties of each PS derivative play a significant role on the ultimate stability of films and PSC. As mentioned above, different stabilities and performances can be achieved by employing various types of organic and/or inorganic cations that fill the A^+ site or by modifying the halide at the site X^-. The most widely employed monovalent cation for the preparation of 3D-PS is probably MA^+, which has yielded high-efficiency devices. However, the stability of the resulting PS is among the lowest [208]. Therefore, significant efforts on elucidating the main degradation mechanisms and proposing new approaches are currently being devoted toward further improvement of PSC.

A material suitable for photovoltaic applications needs to be stable upon prolonged light exposure. Several groups have demonstrated that the optoelectronic properties of PS derivatives undergo significant variations under illumination. For instance, de Quilettes et al. demonstrated that the PL intensity and lifetimes of $MAPbI_3$ films without contacts increase over time under illumination [209]. This increase of the optical properties was correlated with migration of I^- species away from the illuminated area. Exploration of new PS derivatives with more complex composition aiming at overcoming stability issues has led to the observation of additional photoinduced phenomena. For instance, Hoke et al. reported on the reversible phase segregation into Br^-- and I^--rich phases when a $MAPb(Br_xI_{1-x})_3$ film was exposed to light [85]. Christians et al. studied the evolution of the device performance of PSC based on $(FA_{0.79}MA_{0.16}Cs_{0.05})_{0.97}Pb(I_{0.84}Br_{0.16})_{2.97}$ as an active layer and demonstrated that these devices retained 94% of their initial *PCE* after

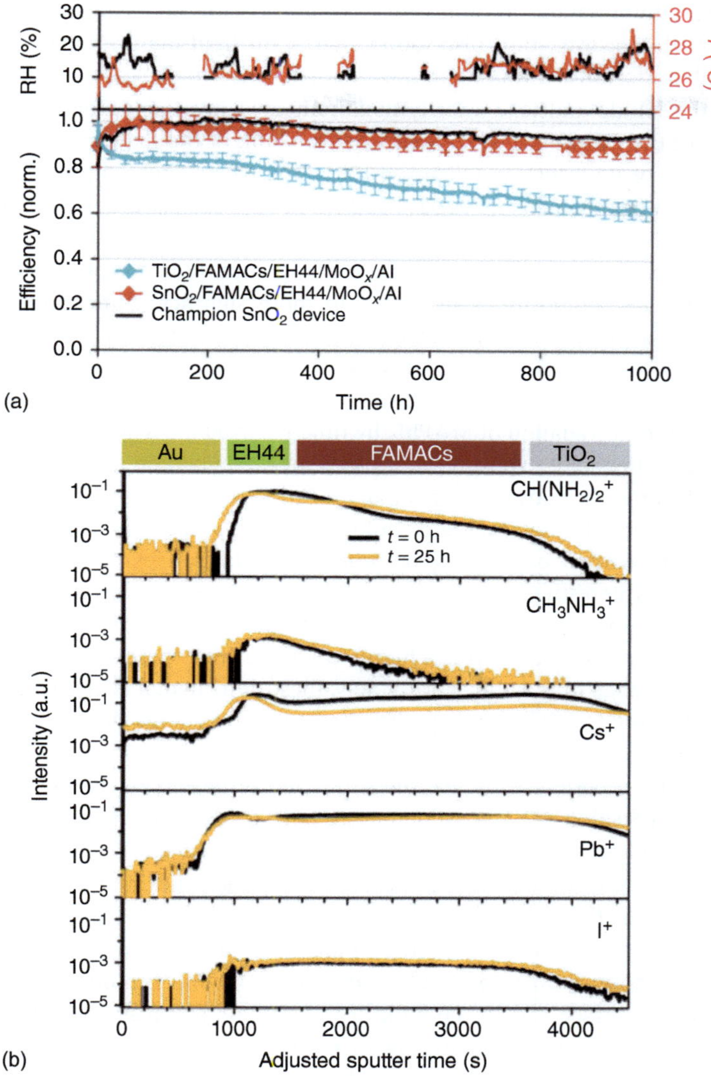

Figure 4.10 (a) Normalized average efficiency obtained from *I*–*V* scans over time for devices ETM/PS/EH44/MoO_x/Al in which the ETM layer is either TiO_2 or SnO_2, PS composition is $(FA_{0.79}MA_{0.16}Cs_{0.05})_{0.97}Pb(I_{0.84}Br_{0.16})_{2.97}$, and HTM employed is 9-(2-ethylhexyl)-*N*,*N*,*N*,*N*-tetrakis(4-methoxyphenyl) 9*H*-carbazole-2,7-diamine (EH44). The champion stability for a SnO_2/PS/EH44/MoO_x/Al device is shown as a black line. (b) Time-of-flight secondary ion mass spectrometry (ToF–SIMS) profiles of PS components measured in fresh and degraded devices after 25 hours of illumination with architecture TiO_2/PS/EH44/Au. Source: Christians et al. [210].

1000 hours under illumination in ambient conditions without encapsulation (Figure 4.10a) [210]. However, they also found significant light-induced changes in compositional uniformity throughout the depth of the mixed cation/anion PSC. Although the halide and lead distributions remained invariable, there was a rearrangement of the Cs^+ and FA^+ content from the bulk of the material to the HTM after 25 hours of illumination (Figure 4.10b). It is worth highlighting that these irreversible compositional changes occurred when TiO_2 was used as ETM but were not observed when SnO_2 was used instead.

Similar phenomena have been reported by other groups [86] concurring to prove that complex PS compositions may lead to more significant and unpredictable light-induced phenomena that can dramatically influence the ultimate performance of devices. Additionally, light soaking is known to accelerate the decomposition of materials in general, and PS derivatives are not an exception. Song et al. studied the photodegradation of $MAPbI_3$ by time-resolved mass spectrometry [211]. Upon ~20 minutes light exposure, the release of methylamine and HI was detected, thus indicating the start of the photodegradation process. The formation of H_2 and NH_3/H_2O was also detected and could be directly produced from direct photoinduced decomposition of $MAPbI_3$ or from the ionization by-products of methylamine species. Interestingly, an equilibrium between I_2 and I^- was detected in the PS films during photodegradation. Since iodine vapor can induce a series of chemical chain reactions [111] and HI is a highly corrosive acid, these species can likely cause severe damage to $MAPbI_3$. Fortunately, these detrimental effects could be mitigated by blocking UV light, as they identified a photon energy threshold of ~3 eV to trigger the decomposition of $MAPbI_3$.

Light has also been reported to induce the formation of defects on PS derivatives, as described in Section 4.2.2. Aristidou et al. have exhaustively studied the photodegradation reaction mechanism of $MAPbI_3$ in the presence of oxygen and highlighted the importance of defects, particularly iodide vacancies, in governing photoinduced decomposition [212, 213]. They found that diffusion of molecular oxygen into the bulk of $MAPbI_3$ films occurs immediately upon exposure and is complete within an hour, with oxygen adsorbing and diffusing through iodide vacancies both at the surface and in the bulk of the crystallites [213]. Kim et al. demonstrated that illumination dramatically increases the halide vacancy concentration, which results in increased ionic conductivity [214] as also reported in previous studies [215, 216]. Once adsorbed at vacancy sites, the molecular oxygen can effectively trap electrons from the conduction band of photoexcited PS and form highly reactive superoxide O_2^- species. Subsequently, an acid–base reaction with the acidic A-site cation can be initiated, resulting in the formation of water, methylamine gas, and lead iodide. The reaction rate was found to be influenced by both grain size and defect density, showing a significantly slower reaction rates for larger grains with passivated defects [213]. Additionally, Kim et al. proposed that I^- ions in the lattice can be oxidized by photogenerated holes, thus resulting in the concomitant formation of interstitial neutral iodine and iodide vacancies. Eventually, this process can induce decomposition to metallic lead, molecular iodine, and MAI. If the iodine remains in the lattice, this process may be reversible. However, if iodine is removed under illumination out of the PS bulk, irreversible decomposition occurs. In fact, several studies performing X-ray photoemission spectroscopy (XPS) have reported the appearance of Pb^0 upon illumination [217, 218].

Saliba et al. explored a quadruple-cation PS with the formula $FA_{0.75}MA_{0.15}Cs_{0.05}Rb_{0.05}Pb(I_{0.83}Br_{0.17})_3$ that achieved a *PCE* of 21.6% and a stability of 500 hours at 85 °C under full illumination under nitrogen [117]. The same research group later developed a MA-free PS derivative $Rb_{0.05}Cs_{0.1}FA_{0.85}PbI_3$ that yielded an efficiency of above 20% and excellent stability under illumination under nitrogen [219]. The good stability of the Rb-containing compounds is attributed to increased entropy of mixing and the removal of the volatile MA cation. However, the exact mechanisms behind the incorporation of Rb are still under study. It is worth highlighting that relatively recent works have suggested that Rb-containing PS are prone to form non-photoactive secondary phases [220, 221].

In summary, these results prove that even though encapsulation is crucial to guarantee an optimum insulation of devices, further improvement in the intrinsic photostability of PS derivatives is possible. Exploration of new compositions and maximization of the charge extraction rates and efficiencies seem to be crucial for an optimum long-term device functioning.

4.6.3 Thermal Degradation of PS Films and PSC

Degradation due to thermal stress is also an important issue to be exhaustively considered, since it is difficult to avoid the rise in temperature of the solar cells during operation. The effective operational temperature can range from −40 °C to +85 °C [222], and, typically, device temperatures can be over 45 °C. According to International Standards (IEC 61646 climatic chamber tests), long-term stability to at least 85 °C is required to compete with other solar cell technologies [223]. Therefore, the study of the thermal stability of PSC is currently a "hot topic" in the photovoltaic field. Although different degradation pathways have been proposed, this remains under discussion. It has been established that $MAPbI_3$ loses its light-harvesting properties over time due to the formation of PbI_2 resulting from the loss of MAI. The degradation mechanism is most likely accompanied by a release of gases, and, therefore, determining the exact composition of these volatile species upon thermal stress is of interest to shed light into the degradation processes and consequently suggest new approaches toward enhancing the PS stability. Different interpretations can be found in the literature regarding the chemical nature of the gases released during thermal degradation, and several studies have analyzed the effect of intrinsic degradation and thermal stress on PSC. In general, it is accepted that $MAPbI_3$ decomposes via sublimation of the organic and halide components (Eq. (4.1)); however, further evolution of the reaction is still under debate, and several alternative mechanisms have been proposed (Eqs. (4.2, 4.3)) [224, 225]:

$$CH_3NH_3PbI_{3(s)} \rightarrow PbI_2 + [CH_3NH_3^+ I^-] \tag{4.1}$$

$$CH_3NH_3^+ + I^- \rightarrow CH_3I_{(g)} + NH_{3(g)} \tag{4.2}$$

$$CH_3NH_3^+ + I^- \rightarrow CH_3NH_{2(g)} + HI_{(g)} \tag{4.3}$$

Conings et al. studied the effect of temperature on the performance of PSC based on $MAPbI_3$ under different storage conditions in the dark [223]. They treated the devices at 85 °C for 24 hours under inert atmosphere, pure oxygen, and ambient air and compared the resulting cell parameters to those obtained prior to the thermal aging. A severe degradation of the samples stored in ambient air was observed due to the contribution of heat and humidity. In addition, samples stored in pure oxygen suffered from slightly greater degradation than that those stored under inert atmosphere. Nevertheless, this experiment revealed that, even in the absence of oxygen and moisture, PSC degrades from thermal stress. This is due, at least partially, to the volatile character of some of the building blocks, i.e. methylamine and hydrogen iodide. FA^+ cation was proposed as an alternative to MA^+ toward the fabrication of high-efficiency PSC with enhanced stability. However, pure FA^+ PS shows the coexistence of two different crystal phases, black α-$FAPbI_3$ and yellow δ-$FAPbI_3$. The α-phase, which possesses photovoltaic activity, is metastable at temperatures below 160 °C and tends to convert into the thermodynamically more stable δ-phase that shows limited light-harvesting behavior [226]. In spite of this, Eperon et al. demonstrated that $FAPbI_3$ was fully stable and did not show any film discoloration at 150 °C in air, whereas its $MAPbI_3$ counterpart lost its characteristic dark color after 30 minutes [227]. Unfortunately, both derivatives showed similar instability against moisture, thus limiting their exploitation for commercial applications. Outstanding device efficiency values have been achieved by mixing different cations such as FA/MA, Cs/MA, FA/MA/Cs, etc. [23, 117, 205]. For instance, Niu et al. used Cs^+ doping in the $MAPbI_3$ lattice to form $Cs_xMA_{1-x}PbI_3$ to attempt to overcome the stability issues [228]. A *PCE* of as high as 18.1% was achieved for the $Cs_xMA_{1-x}PbI_3$ devices ($x = 0.09$). These devices showed good thermal stability with $x = 0.09$ and 0.02, but further increasing the amount of Cs^+ leads to a severe decrease of the efficiency and stability of devices. Stability tests under air exposure at 85 °C revealed a 60% and 20% reduction of performance for the devices with $x = 0$ and $x = 0.09$, respectively, after 60 minutes. The enhanced thermal stability of the Cs^+-containing devices was ascribed to the higher stability of Cs^+ than the MA halide species.

4.6.4 Other Sources of Degradation in PSC

Despite the intrinsic stability of the PS derivatives being a key factor toward improving the long-term stability of devices, additional degradation mechanisms associated to the other components of PSC exist. For instance, the interface of PS with charge-selective contacts, i.e. HTM/ETM, or even the metallic electrodes, was demonstrated to have a significant impact on the lifetime of devices. In addition, PS derivatives are mostly produced through low-temperature solution processing techniques that lead to the formation of polycrystalline domains with high surface areas and inherent defects and local non-stoichiometric compositions. Grain boundaries are considered to be significantly active sites for initiating the degradation process [229]. Therefore, interfaces between crystal grains must be considered as a key factor for achieving long-term stable devices. Interfacial engineering approaches can be divided into two general groups: those focused on the PS/HTM(ETM)/metal interfaces and those involving the PS grain boundaries. It is worth mentioning that both are intimately related since adjacent layers can strongly influence the crystallization, passivation, and/or permeability of PS films and hence affect their stability.

As previously mentioned, TiO_2 and spiro-OMeTAD materials are commonly employed as ETM and HTM, respectively, for the construction of PSC. However, several studies indicate that these materials suffer from limited stability and/or induce a premature loss of device performance under functioning conditions. For instance, it was reported that encapsulated devices containing m-TiO_2 suffered a faster drop in performance under UV light than unencapsulated devices, which was attributed to the formation of oxygen desorption states at the TiO_2 surface that are normally passivated in the presence of oxygen [230]. This effect was circumvented when Al_2O_3 was employed instead. Similarly, several groups have reported on the limited stability of PS/spiro-OMeTAD interface [231–233], although high efficiencies have been achieved. Therefore, tuning the device interfaces and searching new charge-selective contacts and/or buffer materials have become important topics within the PSC field over the last years.

To minimize the detrimental effects of TiO_2 on the device stability, doping is an effective approach to tune the electrical properties of the TiO_2 and improve the stability of PSC. Chen et al. successfully fabricated devices with Nb-doped TiO_x electrodes to improve rapid carrier extraction, whose devices maintained 97% of the initial *PCE* after 1000 hours of light soaking [234]. As an alternative to doping, insertion of interlayers between the PS/TiO_2 interface can block the UV-induced degradation of PS. Along these lines, Ito et al. inserted a Sb_2S_3 interlayer and demonstrated a beneficial effect on the stability. While control devices showed null functionality after 12 hours of exposure to light, the modified devices maintained 65% of the initial *PCE* value. Alternative materials that can substitute TiO_2 have been also proposed. For instance, Shin et al. reported the use of an ETM based on La-doped $BaSnO_3$ that shows high electrical mobility for the preparation of PSC. The best performing devices exhibited *PCE* values of 21.2% with an efficiency drop of less than 10% after 1000 hours under illumination at 1 sun [235].

Remarkable efforts have been also made to replace spiro-OMeTAD as HTM due to its high cost, relatively low hole mobility, and poor conductivity. Lithium bis(trifluoromethanesulfonyl)imide salt (Li-TFSI) and 4-*tert*-butylpyridine (TBP) are added to spiro-OMeTAD films with the aim of enhancing the hole mobility and suppressing charge recombination. Unfortunately, these additives promote oxidative reactions and may migrate into the PS layer, thus limiting its functionality over prolonged periods. Several dopant-free HTM have been proposed that provide good performance and stability [236–239]. Also, trux-OMeTAD, new organic HTM possessing high charge carrier mobility and suitable surface energy, was prepared by introducing triarylamine and aliphatic side chains onto a truxene core. This HTM did not require doping processes, and good *PCE* values of 18.6% with better stability than the spiro-OMeTAD counterparts were obtained

[240]. Nevertheless, inorganic materials seem to be an appealing choice toward enhancing the device stability while providing high efficiency. Materials such as NiO_x [241], CuO_x [242], CuSCN [243], CuI [244], or Ta-WO_x [245], among others, have been investigated and sometimes led to high-efficiency devices with enhanced stability. Arora et al. added a conductive reduced graphene oxide spacer layer between a CuSCN layer and the Au electrode, which resulted in a *PCE* of 20.4% with negligible hysteresis. Through this interface engineering approach, the device retained more than 95% of its initial *PCE* after aging under illumination at 60 °C for 1000 hours [246]. Likewise, a thin layer of tantalum-doped tungsten oxide (Ta-WO_x) inserted between a poly[5,5′-bis(2-butyloctyl)-(2,2′-bithiophene)-4,4′-dicarboxylate-alt-5,5′-2,2′-bithiophene] (PDCBT) as HTM and Au electrode yielded devices with a *PCE* of 21.2% for a planar PSC [245]. Interestingly, PDCBT/Ta-WO_x-based device retained almost 95% of its *PCE* after 1000 hours of exposure to 1 sun illumination under N_2.

An additional concern for the long-term stability of PSC is the reactivity of the metal contacts that are evaporated on top of the devices. For instance, it has been demonstrated that aluminum and silver electrodes suffer from rapid degradation and corrosion to form AlO_x and AgI, respectively [247, 248]. Although metallic gold is more stable, it has also been observed to corrode in the presence of the highly reactive iodide species formed from PS decomposition under illumination with visible light [249]. In this sense, while some metals are more stable in contact with PS structures, most are prone to react with the PS decomposition products such as MAI, HI, CH_3I, and/or I_2 that can be produced even in an encapsulated device. There are three main mechanisms that account for the premature degradation related to metal electrodes: (i) halide/halogen species that arise from PS decomposition diffused to the metal electrode, thus corroding it and causing halide deficiency in the PS absorber layer (Figure 4.11a) [250]; (ii) metal contacts form a redox couple with Pb^{2+} at the PS interface, thus promoting the loss of halide species and forming metallic Pb (Figure 4.11b) [251]; and (iii) the diffusion of the metal electrode into the PS layer upon photothermal and/or thermal activation, which can potentially create insulating metal halide species or defect states at the PS interface and/or bulk [252]. The use of compact diffusion barriers has shown to alleviate the detrimental effects of the metal diffusion (Figure 4.11c).

Metallic electrode degradation has been mitigated through either alternative electrode layers or interlayers between the PS and the metal electrode. Metal electrode-free PSC using carbon electrodes are promising architectures as an alternative to conventional metallic electrode devices. A combination of m-TiO_2 and m-ZrO_2 and carbon electrodes in which the PS is infiltrated has demonstrated an invariable performance for unencapsulated devices after one year of storage in 54% relative humidity in ambient air [253] and after 1000 hours under 1 sun in ambient air [150]. Despite the maximum *PCE* values of these devices still slightly below the *state of the art* of conventional architectures, recent advances have demonstrated efficiencies beyond 15% [254, 255].

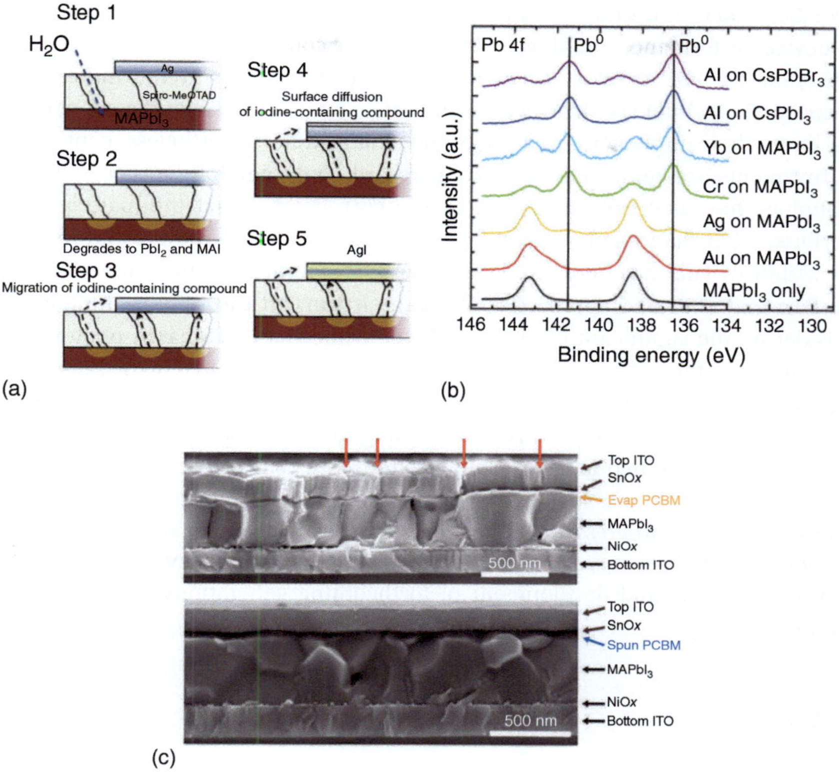

Figure 4.11 (a) Schematic illustration of the proposed mechanism of AgI formation based on XPS and XRD data. (Step 1) H_2O in air enters pinholes in the spiro-MeOTAD layer. (Step 2) Decomposition of $MAPbI_3$ gives rise to iodine-based volatile compounds (MAI and/or HI). (Step 3) Migration of the iodine-containing volatile compound from the $MAPbI_3$ layer corroding the electrode from both top and bottom surfaces of Ag layer. (Step 4) Surface diffusion of the iodine-containing volatile compound. (Step 5) AgI formation. Source: Kato et al. [250]. (b) Pb 4f XPS spectra of samples with Al, Yb, Cr, Ag, and Au metals directly on $CsPbBr_3$, $CsPbI_3$, or $MAPbI_3$ perovskite films. All PS/metal interfaces show evidence of redox chemistry via the presence of a Pb^0 peak except the $MAPbI_3$/Au sample that shows evidence of partial charge transfer. Source: Zhao et al. [251]. (c) Cross-sectional SEM images of ITO barrier layers on $MAPbI_3$ cells with evaporated (top) and spun PCBM (bottom) interlayers. Red arrows show the diffusion channels in the evaporated PCBM sample that propagate along the rough perovskite grain boundaries. These channels are not present in the spun PCBM sample. Source: Adapted with permission from Boyd et al. [252]. Copyright (2018) American Chemical Society.

4.7 Conclusions and Prospects

In this chapter, we presented a general overview on the representative interface engineering being currently explored toward further development of PSC. We also introduced the main limitations encountered in this photovoltaic technology and showed the most representative strategies reported in the literature for materials deposition and tuning of the electro-optical interface properties aimed at optimizing the device performance and durability. The power conversion efficiencies that the *state-of-the-art* PSC exhibit are high enough to compete with the much more advanced Si technology. However, long-term stability is a significant concern that limits their implementation for commercialization. Nevertheless, continuous improvements are being achieved in this regard, and, therefore, a success in this technology has been predicted. It is clearly corroborated that the materials and interfaces other than PS play a significant role on the ultimate device performance and stability. This has motivated the scientific community to search for new interface materials and approaches that can circumvent the stability limitations. Considering the significant efforts worldwide dedicated to PS-based photovoltaic technology and the outstanding advances and knowledge acquired in the last decade since the first reports, it is a matter of time until PSC become a real alternative for solar energy conversion. It is worth mentioning that PS materials can be exploited not only for single-junction devices but also for more complex device architectures involving multijunctions, such as tandem solar cells. In summary, PSC technology represents a huge progress in photovoltaics that can not only contribute to the production of clean energy with reduced costs but also provide new insights toward the understanding and development of forthcoming generations of photovoltaic technologies.

References

1 Kojima, A., Teshima, K., Shirai, Y., and Miyasaka, T. (2009). Organometal halide perovskites as visible-light sensitizers for photovoltaic cells. *J. Am. Chem. Soc.* 131 (17): 6050–6051. https://doi.org/10.1021/ja809598r.

2 (NREL), N.R.E.L. (2019). Efficiency Chart. https://www.nrel.gov/pv/cell-efficiency.html.

3 Kim, H.-S., Lee, C.-R., Im, J.-H. et al. (2012). Lead iodide perovskite sensitized all-solid-state submicron thin film mesoscopic solar cell with efficiency exceeding 9%. *Sci. Rep.* 2: 591. https://doi.org/10.1038/srep00591.

4 Park, N.-G. (2015). Perovskite solar cells: an emerging photovoltaic technology. *Mater. Today* 18 (2): 65–72. https://doi.org/10.1016/j.mattod.2014.07.007.

5 Li, F., Ma, C., Wang, H. et al. (2015). Ambipolar solution-processed hybrid perovskite phototransistors. *Nat. Commun.* 6: 8238.

6 Stranks, S.D., Eperon, G.E., Grancini, G. et al. (2013). Electron-hole diffusion lengths exceeding 1 micrometer in an organometal trihalide perovskite absorber. *Science (80-.)* 342 (6156): 341–344. https://doi.org/10.1126/science.1243982.

7 Miyata, A., Mitioglu, A., Plochocka, P. et al. (2015). Direct measurement of the exciton binding energy and effective masses for charge carriers in organic–inorganic tri-halide perovskites. *Nat. Phys.* 11: 582.

8 Chen, Z., Dong, Q., Liu, Y. et al. (2017). Thin single crystal perovskite solar cells to harvest below-bandgap light absorption. *Nat. Commun.* 8 (1): 1890. https://doi.org/10.1038/s41467-017-02039-5.

9 Zhang, Y.-N., Li, B., Fu, L. et al. (2019). Enhanced optical absorption and efficient cascade electron extraction based on energy band alignment double absorbers perovskite solar cells. *Sol. Energy Mater. Sol. Cells* 194: 168–176. https://doi.org/10.1016/j.solmat.2019.02.014.

10 Shin, S.S., Suk, J.H., Kang, B.J. et al. (2019). Energy-level engineering of the electron transporting layer for improving open-circuit voltage in dye and perovskite-based solar cells. *Energy Environ. Sci.* 12 (3): 958–964. https://doi.org/10.1039/C8EE03672A.

11 Wei, K., Jiang, T., Xu, Z. et al. (2018). Ultrafast carrier transfer promoted by interlayer coulomb coupling in 2D/3D perovskite heterostructures. *Laser Photonics Rev.* 12 (10): 1800128. https://doi.org/10.1002/lpor.201800128.

12 Cao, J., Yin, J., Yuan, S. et al. (2015). Thiols as interfacial modifiers to enhance the performance and stability of perovskite solar cells. *Nanoscale* 7 (21): 9443–9447. https://doi.org/10.1039/C5NR01820J.

13 Wei, J., Li, H., Zhao, Y. et al. (2016). Suppressed hysteresis and improved stability in perovskite solar cells with conductive organic network. *Nano Energy* 26: 139–147. https://doi.org/10.1016/j.nanoen.2016.05.023.

14 Li, P., Zhang, Y., Liang, C. et al. (2018). Phase pure 2D perovskite for high-performance 2D–3D heterostructured perovskite solar cells. *Adv. Mater.* 30 (52): 1805323. https://doi.org/10.1002/adma.201805323.

15 Tress, W., Correa Baena, J.P., Saliba, M. et al. (2016). Inverted current–voltage hysteresis in mixed perovskite solar cells: polarization, energy barriers, and defect recombination. *Adv. Energy Mater.* 6 (19): 1600396. https://doi.org/10.1002/aenm.201600396.

16 Schulz, P., Cahen, D., and Kahn, A. (2019). Halide perovskites: is it all about the interfaces? *Chem. Rev.* 119 (5): 3349–3417. https://doi.org/10.1021/acs.chemrev.8b00558.

17 Nie, W., Tsai, H., Asadpour, R. et al. (2015). High-efficiency solution-processed perovskite solar cells with millimeter-scale grains. *Science (80-.)* 347 (6221): 522–525. https://doi.org/10.1126/science.aaa0472.

18 Wen, X., Sheng, R., Ho-Baillie, A.W.Y. et al. (2014). Morphology and carrier extraction study of organic–inorganic metal halide perovskite by one- and two-photon fluorescence microscopy. *J. Phys. Chem. Lett.* 5 (21): 3849–3853. https://doi.org/10.1021/jz502014r.

19 Huang, Z., Wang, D., Wang, S., and Zhang, T. (2018). Highly efficient and stable $MAPbI_3$ perovskite solar cell induced by regulated nucleation and Ostwald recrystallization. *Materials (Basel, Switzerland)* 11 (5): 778. https://doi.org/10.3390/ma11050778.

20 Passarelli, J.V., Fairfield, D.J., Sather, N.A. et al. (2018). Enhanced out-of-plane conductivity and photovoltaic performance in $n = 1$ layered perovskites through organic cation design. *J. Am. Chem. Soc.* 140 (23): 7313–7323. https://doi.org/10.1021/jacs.8b03659.

21 Im, J.-H., Kim, H.-S., and Park, N.-G. (2014). Morphology-photovoltaic property correlation in perovskite solar cells: one-step versus two-step deposition of $CH_3NH_3PbI_3$. *APL Materials* 2 (8): 81510. https://doi.org/10.1063/1.4891275.

22 Williams, S.T., Zuo, F., Chueh, C.-C. et al. (2014). Role of chloride in the morphological evolution of organo-lead halide perovskite thin films. *ACS Nano* 8 (10): 10640–10654. https://doi.org/10.1021/nn5041922.

23 Yang, W.S., Noh, J.H., Jeon, N.J. et al. (2015). High-performance photovoltaic perovskite layers fabricated through intramolecular exchange. *Science (80-.)* 348 (6240): 1234–1237. https://doi.org/10.1126/science.aaa9272.

24 Jeon, N.J., Noh, J.H., Kim, Y.C. et al. (2014). Solvent engineering for high-performance inorganic–organic hybrid perovskite solar cells. *Nat. Mater.* 13: 897.

25 Ahn, N., Son, D.-Y., Jang, I.-H. et al. (2015). Highly reproducible perovskite solar cells with average efficiency of 18.3% and best efficiency of 19.7% fabricated via Lewis base adduct of lead(II) iodide. *J. Am. Chem. Soc.* 137 (27): 8696–8699. https://doi.org/10.1021/jacs.5b04930.

26 Wang, H., Zeng, W., and Xia, R. (2018). Antisolvent diethyl ether as additive to enhance the performance of perovskite solar cells. *Thin Solid Films* 663: 9–13. https://doi.org/10.1016/j.tsf.2018.07.041.

27 Huang, Y., Wu, S., Chen, R. et al. (2019). Efficient methylamine-containing antisolvent strategy to fabricate high-efficiency and stable $FA_{0.85}Cs_{0.15}Pb(Br_{0.15}I_{2.85})$ perovskite solar cells. *ACS Appl. Mater. Interfaces* 11 (20): 18415–18422. https://doi.org/10.1021/acsami.9b03323.

28 Bu, T., Wu, L., Liu, X. et al. (2017). Synergic interface optimization with green solvent engineering in mixed perovskite solar cells. *Adv. Energy Mater.* 7 (20): 1700576. https://doi.org/10.1002/aenm.201700576.

29 Paek, S., Schouwink, P., Athanasopoulou, E.N. et al. (2017). From nano- to micrometer scale: the role of antisolvent treatment on high performance perovskite solar cells. *Chem. Mater.* 29 (8): 3490–3498. https://doi.org/10.1021/acs.chemmater.6b05353.

30 Saliba, M., Matsui, T., Seo, J.-Y. et al. (2016). Cesium-containing triple cation perovskite solar cells: improved stability, reproducibility and high efficiency. *Energy Environ. Sci.* 9 (6): 1989–1997. https://doi.org/10.1039/C5EE03874J.

31 Wang, Y., Wu, J., Zhang, P. et al. (2017). Stitching triple cation perovskite by a mixed anti-solvent process for high performance perovskite solar cells. *Nano Energy* 39: 616–625. https://doi.org/10.1016/j.nanoen.2017.07.046.

32 Wang, F., Bai, S., Tress, W. et al. (2018). Defects engineering for high-performance perovskite solar cells. *npj Flex. Electron.* 2 (1): 22. https://doi.org/10.1038/s41528-018-0035-z.

33 Hsiao, S.-Y., Lin, H.-L., Lee, W.-H. et al. (2016). Efficient all-vacuum deposited perovskite solar cells by controlling reagent partial pressure in high vacuum. *Adv. Mater.* 28 (32): 7013–7019. https://doi.org/10.1002/adma.201601505.

34 Momblona, C., Gil-Escrig, L., Bandiello, E. et al. (2016). Efficient vacuum deposited P–i–n and n–i–p perovskite solar cells employing doped charge transport layers. *Energy Environ. Sci.* 9 (11): 3456–3463. https://doi.org/10.1039/C6EE02100J.

35 Gil-Escrig, L., Momblona, C., La-Placa, M.-G. et al. (2018). Vacuum deposited triple-cation mixed-halide perovskite solar cells. *Adv. Energy Mater.* 8 (14): 1703506. https://doi org/10.1002/aenm.201703506.

36 Liu, M., Johnston, M.B., and Snaith, H.J. (2013). Efficient planar heterojunction perovskite solar cells by vapour deposition. *Nature* 501: 395.

37 Leyden, M.R., Ono, L.K., Raga, S.R. et al. (2014). High performance perovskite solar cells by hybrid chemical vapor deposition. *J. Mater. Chem. A* 2 (44): 18742–18745. https://doi.org/10.1039/C4TA04385E.

38 Luo, P., Liu, Z., Xia, W. et al. (2015). Uniform, stable, and efficient planar-heterojunction perovskite solar cells by facile low-pressure chemical vapor deposition under fully open-air conditions. *ACS Appl. Mater. Interfaces* 7 (4): 2708–2714. https://doi.org/10.1021/am5077588.

39 Ávila, J., Momblona, C., Boix, P.P. et al. (2017). Vapor-deposited perovskites: the route to high-performance solar cell production? *Joule* 1 (3): 431–442. https://doi.org/10.1016/j.joule.2017.07.014.

40 Li, Y., Chen, Y., Han, L. et al. (2019). Epitaxial growth of horizontally aligned single-crystal arrays of perovskite. *Sci. China Mater.* 62 (1): 59–64. doi: 10.1007/s40843-018-9304-7.

41 Wei, Y., Li, K., Cheng, Z. et al. (2019). Epitaxial growth of $CsPbX_3$ (X = cl, Br, I) perovskite quantum dots via surface chemical conversion of Cs_2GeF_6 double perovskites: a novel strategy for the formation of leadless hybrid perovskite phosphors with enhanced stability. *Adv. Mater.* 31 (16): 1807592. https://doi.org/10.1002/adma.201807592.

42 Hieulle, J., Stecker, C., Ohmann, R. et al. (2018). Scanning probe microscopy applied to organic–inorganic halide perovskite materials and solar cells. *Small Methods* 2 (1): 1700295. https://doi.org/10.1002/smtd.201700295.

43 Chen, Q., Zhou, H., Song, T.-B. et al. (2014). Controllable self-induced passivation of hybrid lead iodide perovskites toward high performance solar cells. *Nano Lett.* 14 (7): 4158–4163. https://doi.org/10.1021/nl501838y.

44 Zhou, N., Shen, Y., Zhang, Y. et al. (2017). CsI pre-intercalation in the inorganic framework for efficient and stable FA1−x Cs_xPbI_3(cl) perovskite solar cells. *Small* 13 (23): 1700484. https://doi.org/10.1002/smll.201700484.

45 Ono, L.K. and Qi, Y. (2016). Surface and interface aspects of organometal halide perovskite materials and solar cells. *J. Phys. Chem. Lett.* 7 (22): 4764–4794. https://doi.org/10.1021/acs.jpclett.6b01951.

46 Yang, B., Dyck, O., Poplawsky, J. et al. (2015). Perovskite solar cells with near 100% internal quantum efficiency based on large single crystalline grains and vertical bulk heterojunctions. *J. Am. Chem. Soc.* 137 (29): 9210–9213. https://doi.org/10.1021/jacs.5b03144.

47 Yun, J.S., Ho-Baillie, A., Huang, S. et al. (2015). Benefit of grain boundaries in organic–inorganic halide planar perovskite solar cells. *J. Phys. Chem. Lett.* 6 (5): 875–880. https://doi.org/10.1021/acs.jpclett.5b00182.

48 Heo, S., Seo, G., Lee, Y. et al. (2017). Deep level trapped defect analysis in $CH_3NH_3PbI_3$ perovskite solar cells by deep level transient spectroscopy. *Energy Environ. Sci.* 10 (5): 1128–1133. https://doi.org/10.1039/C7EE00303J.

49 de Quilettes, D.W., Vorpahl, S.M., Stranks, S.D. et al. (2015). Impact of microstructure on local carrier lifetime in perovskite solar cells. *Science (80-.)* 348 (6235): 683–686. https://doi.org/10.1126/science.aaa5333.

50 Shi, D., Adinolfi, V., Comin, R. et al. (2015). Low trap-state density and long carrier diffusion in organolead trihalide perovskite single crystals. *Science (80-.)* 347 (6221): 519–522. https://doi.org/10.1126/science.aaa2725.

51 Reid, O.G., Yang, M., Kopidakis, N. et al. (2016). Grain-size-limited mobility in methylammonium lead iodide perovskite thin films. *ACS Energy Lett.* 1 (3): 561–565. https://doi.org/10.1021/acsenergylett.6b00288.

52 Ren, X., Yang, Z., Yang, D. et al. (2016). Modulating crystal grain size and optoelectronic properties of perovskite films for solar cells by reaction temperature. *Nanoscale* 8 (6): 3816–3822. https://doi.org/10.1039/C5NR08935B.

53 DiStefano, T.H. and Cuomo, J.J. (1977). Reduction of grain boundary recombination in polycrystalline silicon solar cells. *Appl. Phys. Lett.* 30 (7): 351–353. https://doi.org/10.1063/1.89396.

54 Li, C., Wu, Y., Poplawsky, J. et al. (2014). Grain-boundary-enhanced carrier collection in CdTe solar cells. *Phys. Rev. Lett.* 112 (15): 156103. https://doi.org/10.1103/PhysRevLett.112.156103.

55 Gloeckler, M., Sites, J.R., and Metzger, W.K. (2005). Grain-boundary recombination in Cu(In, Ga)Se_2 solar cells. *J. Appl. Phys.* 98 (11): 113704. doi: 10.1063/1.2133906.

56 Landsberg, P.T. and Abrahams, M.S. (1984). Effects of surface states and of excitation on barrier heights in a simple model of a grain boundary or a surface. *J. Appl. Phys.* 55 (12): 4284–4293. https://doi.org/10.1063/1.333038.

57 Leijtens, T., Eperon, G.E., Barker, A.J. et al. (2016). Carrier trapping and recombination: the role of defect physics in enhancing the open circuit voltage of metal halide perovskite solar cells. *Energy Environ. Sci.* 9 (11): 3472–3481. https://doi.org/10.1039/C6EE01729K.

58 Sherkar, T.S., Momblona, C., Gil-Escrig, L. et al. (2017). Recombination in perovskite solar cells: significance of grain boundaries, Interface traps, and defect ions. *ACS Energy Lett.* 2 (5): 1214–1222. https://doi.org/10.1021/acsenergylett.7b00236.

59 Dymshits, A., Henning, A., Segev, G. et al. (2015). The electronic structure of metal oxide/organo metal halide perovskite junctions in perovskite based solar cells. *Sci. Rep.* 5: 8704.

60 Li, J.-J., Ma, J.-Y., Ge, Q.-Q. et al. (2015). Microscopic investigation of grain boundaries in organolead halide perovskite solar cells. *ACS Appl. Mater. Interfaces* 7 (51): 28518–28523. https://doi.org/10.1021/acsami.5b09801.

61 Frolova, L.A., Dremova, N.N., and Troshin, P.A. (2015). The chemical origin of the P-type and n-type doping effects in the hybrid methylammonium–lead iodide ($MAPbI_3$) perovskite solar cells. *Chem. Commun.* 51 (80): 14917–14920. https://doi.org/10.1039/C5CC05205J.

62 Adhyaksa, G.W.P., Brittman, S., Āboliņš, H. et al. (2018). Understanding detrimental and beneficial grain boundary effects in halide perovskites. *Adv. Mater.* 30 (52): 1804792. https://doi.org/10.1002/adma.201804792.

63 Cui, Y., Duan, X., Hu, J., and Lieber, C.M. (2000). Doping and electrical transport in silicon nanowires. *J. Phys. Chem. B* 104 (22): 5213–5216. https://doi.org/10.1021/jp0009305.

64 Li, X., Fan, L., Li, Z. et al. (2012). Boron doping of graphene for graphene–silicon p–n junction solar cells. *Adv. Energy Mater.* 2 (4): 425–429. https://doi.org/10.1002/aenm.201100671.

65 Yu, B., Zebarjadi, M., Wang, H. et al. (2012). Enhancement of thermoelectric properties by modulation-doping in silicon germanium alloy nanocomposites. *Nano Lett.* 12 (4): 2077–2082. https://doi.org/10.1021/nl3003045.

66 Tietze, M.L., Pahner, P., Schmidt, K. et al. (2015). Doped organic semiconductors: trap-filling, impurity saturation, and reserve regimes. *Adv. Funct. Mater.* 25 (18): 2701–2707. https://doi.org/10.1002/adfm.201404549.

67 Poindexter, J.R., Hoye, R.L.Z., Nienhaus, L. et al. (2017). High tolerance to iron contamination in lead halide perovskite solar cells. *ACS Nano* 11 (7): 7101–7109. https://doi.org/10.1021/acsnano.7b02734.

68 Yavari, M., Ebadi, F., Meloni, S. et al. (2019). How far does the defect tolerance of lead-halide perovskites range? The example of Bi impurities introducing efficient recombination centers. *J. Mater. Chem. A* https://doi.org/10.1039/C9TA01744E.

69 Pazos-Outón, L.M., Xiao, T.P., and Yablonovitch, E. (2018). Fundamental efficiency limit of lead iodide perovskite solar cells. *J. Phys. Chem. Lett.* 9 (7): 1703–1711. https://doi.org/10.1021/acs.jpclett.7b03054.

70 Shockley, W. and Queisser, H.J. (1961). Detailed balance limit of efficiency of p–n junction solar cells. *J. Appl. Phys.* 32 (3): 510–519. https://doi.org/10.1063/1.1736034.

71 Zhang, F., Shi, W., Luo, J. et al. (2017). Isomer-pure Bis-PCBM-assisted crystal engineering of perovskite solar cells showing excellent efficiency and stability. *Adv. Mater.* 29 (17): 1606806. https://doi.org/10.1002/adma.201606806.

72 Wang, F., Shimazaki, A., Yang, F. et al. (2017). Highly efficient and stable perovskite solar cells by interfacial engineering using solution-processed polymer layer. *J. Phys. Chem. C* 121 (3): 1562–1568. https://doi.org/10.1021/acs.jpcc.6b12137.

73 Peng, J., Wu, Y., Ye, W. et al. (2017). Interface passivation using ultrathin polymer–fullerene films for high-efficiency perovskite solar cells with negligible hysteresis. *Energy Environ. Sci.* 10 (8): 1792–1800. https://doi.org/10.1039/C7EE01096F.

74 Okada, W., Suga, T., Oyaizu, K. et al. (2019). Perovskite/TiO2 interface passivation using poly(Vinylcarbazole) and fullerene for the photovoltaic conversion efficiency of 21%. *ACS Appl. Energy Mater.* 2 (4): 2848–2853. https://doi.org/10.1021/acsaem.9b00162.

75 Tress, W., Marinova, N., Inganäs, O. et al. (2015). Predicting the open-circuit voltage of CH3NH3PbI3 perovskite solar cells using electroluminescence and photovoltaic quantum efficiency spectra: the role of radiative and non-radiative recombination. *Adv. Energy Mater.* 5 (3): 1400812. https://doi.org/10.1002/aenm.201400812.

76 Liu, Z., Krückemeier, L., Krogmeier, B. et al. (2019). Open-circuit voltages exceeding 1.26 V in planar methylammonium lead iodide perovskite solar cells. *ACS Energy Lett.* 4 (1): 110–117. https://doi.org/10.1021/acsenergylett.8b01906.

77 Kim, J., Lee, S.-H., Lee, J.H., and Hong, K.-H. (2014). The role of intrinsic defects in methylammonium lead iodide perovskite. *J. Phys. Chem. Lett.* 5 (8): 1312–1317. https://doi.org/10.1021/jz500370k.

78 Buin, A., Pietsch, P., Xu, J. et al. (2014). Materials processing routes to trap-free halide perovskites. *Nano Lett.* 14 (11): 6281–6286. https://doi.org/10.1021/nl502612m.

79 Brandt, R.E., Stevanović, V., Ginley, D.S., and Buonassisi, T. (2015). Identifying defect-tolerant semiconductors with high minority-carrier lifetimes: beyond hybrid lead halide perovskites. *MRS Commun.* 5 (2): 265–275. https://doi.org/10.1557/mrc.2015.26.

80 Delugas, P., Filippetti, A., and Mattoni, A. (2015). Methylammonium fragmentation in amines as source of localized trap levels and the healing role of cl in hybrid lead-iodide perovskites. *Phys. Rev. B* 92 (4): 45301. https://doi.org/10.1103/PhysRevB.92.045301.

81 Park, J.-S., Calbo, J., Jung, Y.-K. et al. (2019). Accumulation of deep traps at grain boundaries in halide perovskites. *ACS Energy Lett.* 4 (6): 1321–1327. https://doi.org/10.1021/acsenergylett.9b00840.

82 Chen, Y.-F., Tsai, Y.-T., Bassani, D.M. et al. (2016). Evidence of band bending induced by hole trapping at MAPbI3 perovskite/metal interface. *J. Mater. Chem. A* 4 (44): 17529–17536. https://doi.org/10.1039/C6TA08979H.

83 Zhao, C., Chen, B., Qiao, X. et al. (2015). Revealing underlying processes involved in light soaking effects and hysteresis phenomena in perovskite solar cells. *Adv. Energy Mater.* 5 (14): 1500279. https://doi.org/10.1002/aenm.201500279.

84 Ahn, N., Kwak, K., Jang, M.S. et al. (2016). Trapped charge-driven degradation of perovskite solar cells. *Nat. Commun.* 7: 13422.

85 Hoke, E.T., Slotcavage, D.J., Dohner, E.R. et al. (2015). Reversible photo-induced trap formation in mixed-halide hybrid perovskites for photovoltaics. *Chem. Sci.* 6 (1): 613–617. https://doi.org/10.1039/C4SC03141E.

86 Domanski, K., Roose, B., Matsui, T. et al. (2017). Migration of cations induces reversible performance losses over day/night cycling in perovskite solar cells. *Energy Environ. Sci.* 10 (2): 604–613. https://doi.org/10.1039/C6EE03352K.

87 Tress, W., Yavari, M., Domanski, K. et al. (2018). Interpretation and evolution of open-circuit voltage, recombination, ideality factor and subgap defect states during reversible light-soaking and irreversible degradation of perovskite solar cells. *Energy Environ. Sci.* 11 (1): 151–165. https://doi.org/10.1039/C7EE02415K.

88 Theelen, M., Hans, V., Barreau, N. et al. (2015). The impact of alkali elements on the degradation of CIGS solar cells. *Prog. Photovolt. Res. Appl.* 23 (5): 537–545. https://doi.org/10.1002/pip.2610.

89 Eames, C., Frost, J.M., Barnes, P.R.F. et al. (2015). Ionic transport in hybrid Lead iodide perovskite solar cells. *Nat. Commun.* 6: 7497.

90 Haruyama, J., Sodeyama, K., Han, L., and Tateyama, Y. (2015). First-principles study of ion diffusion in perovskite solar cell sensitizers. *J. Am. Chem. Soc.* 137 (32): 10048–10051. https://doi.org/10.1021/jacs.5b03615.

91 Yin, W.-J., Shi, T., and Yan, Y. (2014). Unusual defect physics in CH3NH3PbI3 perovskite solar cell absorber. *Appl. Phys. Lett.* 104 (6): 63903. https://doi.org/10.1063/1.4864778.

92 Yuan, Y., Chae, J., Shao, Y. et al. (2015). Photovoltaic switching mechanism in lateral structure hybrid perovskite solar cells. *Adv. Energy Mater.* 5 (15): 1500615. https://doi.org/10.1002/aenm.201500615.

93 Yuan, Y. and Huang, J. (2016). Ion migration in organometal trihalide perovskite and its impact on photovoltaic efficiency and stability. *Acc. Chem. Res.* 49 (2): 286–293. https://doi.org/10.1021/acs.accounts.5b00420.

94 Yuan, Y., Wang, Q., Shao, Y. et al. (2016). Electric-field-driven reversible conversion between methylammonium lead triiodide perovskites and lead iodide at elevated temperatures. *Adv. Energy Mater.* 6 (2): 1501803. https://doi.org/10.1002/aenm.201501803.

95 Levine, I., Nayak, P.K., Wang, J.T.-W. et al. (2016). Interface-dependent ion migration/accumulation controls hysteresis in MAPbI3 solar cells. *J. Phys. Chem. C* 120 (30): 16399–16411. https://doi.org/10.1021/acs.jpcc.6b04233.

96 Futscher, M.H., Lee, J.M., McGovern, L. et al. (2019). Quantification of ion migration in CH3NH3PbI3 perovskite solar cells by transient capacitance measurements. *Mater. Horiz.* https://doi.org/10.1039/C9MH00445A.

97 Lee, H., Gaiaschi, S., Chapon, P. et al. (2019). Effect of halide ion migration on the electrical properties of methylammonium lead tri-iodide perovskite solar cells. *J. Phys. Chem. C* 123 (29): 17728–17734. https://doi.org/10.1021/acs.jpcc.9b04662.

98 Kreuer, K.D. (2003). Proton-conducting oxides. *Annu. Rev. Mater. Res.* 33 (1): 333–359. https://doi.org/10.1146/annurev.matsci.33.022802.091825.

99 Kreuer, K.-D., Paddison, S.J., Spohr, E., and Schuster, M. (2004). Transport in proton conductors for fuel-cell applications: simulations, elementary reactions, and phenomenology. *Chem. Rev.* 104 (10): 4637–4678. https://doi.org/10.1021/cr020715f.

100 Egger, D.A., Kronik, L., and Rappe, A.M. (2015). Theory of hydrogen migration in organic–inorganic halide perovskites. *Angew. Chem. Int. Ed.* 54 (42): 12437–12441. https://doi.org/10.1002/anie.201502544.

101 Chen, Y.-F., Tsai, Y.-T., Hirsch, L., and Bassani, D.M. (2017). Kinetic isotope effects provide experimental evidence for proton tunneling in methylammonium lead triiodide perovskites. *J. Am. Chem. Soc.* 139 (45): 16359–16364. https://doi.org/10.1021/jacs.7b09526.

102 Fabregat-Santiago, F., Kulbak, M., Zohar, A. et al. (2017). Deleterious effect of negative capacitance on the performance of halide perovskite solar cells. *ACS Energy Lett.* 2 (9): 2007–2013. https://doi.org/10.1021/acsenergylett.7b00542.

103 van Reenen, S., Kemerink, M., and Snaith, H.J. (2015). Modeling anomalous hysteresis in perovskite solar cells. *J. Phys. Chem. Lett.* 6 (19): 3808–3814. https://doi.org/10.1021/acs.jpclett.5b01645.

104 Calado, P., Telford, A.M., Bryant, D. et al. (2016). Evidence for ion migration in hybrid perovskite solar cells with minimal hysteresis. *Nat. Commun.* 7: 13831.

105 Li, Z., Xiao, C., Yang, Y. et al. (2017). Extrinsic ion migration in perovskite solar cells. *Energy Environ. Sci.* 10 (5): 1234–1242. https://doi.org/10.1039/C7EE00358G.

106 Hu, J., Gottesman, R., Gouda, L. et al. (2017). Photovoltage behavior in perovskite solar cells under light-soaking showing photoinduced interfacial changes. *ACS Energy Letters* 2 (5): 950–956. https://doi.org/10.1021/acsenergylett.7b00212.

107 Lopez-Varo, P., Jiménez-Tejada, J.A., García-Rosell, M. et al. (2018). Device physics of hybrid perovskite solar cells: theory and experiment. *Adv. Energy Mater.* 8 (14): 1702772. https://doi.org/10.1002/aenm.201702772.

108 Hu, J., Gouda, L., Kama, A. et al. (2019). Radiative recombination changes under light-soaking in CsPbBr3 films on TiO2 and insulating glass contacts: interface versus bulk effects. *ACS Appl. Energy Mater.* 2 (5): 3013–3016. https://doi.org/10.1021/acsaem.9b00335.

109 Gottesman, R., Haltzi, E., Gouda, L. et al. (2014). Extremely slow photoconductivity response of CH3NH3PbI3 perovskites suggesting structural changes under working conditions. *J. Phys. Chem. Lett.* 5 (15): 2662–2669. https://doi.org/10.1021/jz501373f.

110 Kedem, N., Brenner, T.M., Kulbak, M. et al. (2015). Light-induced increase of electron diffusion length in a p–n junction type CH3NH3PbBr3 perovskite solar cell. *J. Phys. Chem. Lett.* 6 (13): 2469–2476. https://doi.org/10.1021/acs.jpclett.5b00889.

111 Wang, S., Jiang, Y., Juarez-Perez, E.J. et al. (2016). Accelerated degradation of methylammonium lead iodide perovskites induced by exposure to iodine vapour. *Nat. Energy* 2: 16195.

112 Roose, B., Wang, Q., and Abate, A. (2019). The role of charge selective contacts in perovskite solar cell stability. *Adv. Energy Mater.* 9 (5): 1803140. https://doi.org/10.1002/aenm.201803140.

113 Wang, S., Sakurai, T., Wen, W., and Qi, Y. (2018). Energy level alignment at interfaces in metal halide perovskite solar cells. *Adv. Mater. Interfaces* 5 (22): 1800260. https://doi.org/10.1002/admi.201800260.

114 Garcia-Belmonte, G. and Bisquert, J. (2016). Distinction between capacitive and noncapacitive hysteretic currents in operation and degradation of perovskite solar cells. *ACS Energy Letters* 1 (4): 683–688. https://doi.org/10.1021/acsenergylett.6b00293.

115 Lee, J.-W., Kim, S.-G., Yang, J.-M. et al. (2019). Verification and mitigation of ion migration in perovskite solar cells. *APL Mater.* 7 (4): 41111. https://doi.org/10.1063/1.5085643.

116 Giordano, F., Abate, A., Correa Baena, J.P. et al. (2016). Enhanced electronic properties in mesoporous TiO2 via lithium doping for high-efficiency perovskite solar cells. *Nat. Commun.* 7: 10379.

117 Saliba, M., Matsui, T., Domanski, K. et al. (2016). Incorporation of rubidium cations into perovskite solar cells improves photovoltaic performance. *Science (80-.)* 354 (6309): 206–209. https://doi.org/10.1126/science.aah5557.

118 Zhang, F., Guang, S., Ye, M. et al. (2018). Chemical decoration of perovskites by nickel oxide doping for efficient and stable perovskite solar cells. *ACS Appl. Mater. Interfaces* 10 (43): 36841–36850. https://doi.org/10.1021/acsami.8b08658.

119 Jiang, Q., Chu, Z., Wang, P. et al. (2017). Planar-structure perovskite solar cells with efficiency beyond 21%. *Adv. Mater.* 29 (46): 1703852. https://doi.org/10.1002/adma.201703852.

120 Bai, Y., Chen, H., Xiao, S. et al. (2016). Effects of a molecular monolayer modification of NiO nanocrystal layer surfaces on perovskite crystallization and interface contact toward faster hole extraction and higher photovoltaic performance. *Adv. Funct. Mater.* 26 (17): 2950–2958. https://doi.org/10.1002/adfm.201505215.

121 Stolterfoht, M.; Caprioglio, P.; Wolff, C. M., et al. (2018). The Perovskite/Transport Layer Interfaces Dominate Non-Radiative Recombination in Efficient Perovskite Solar Cells. *arXiv Prepr. arXiv1810.01333.*

122 Halvani Anaraki, E., Kermanpur, A., Mayer, M.T. et al. (2018). Low-temperature Nb-doped SnO2 electron-selective contact yields over 20% efficiency in planar perovskite solar cells. *ACS Energy Lett.* 3 (4): 773–778. https://doi.org/10.1021/acsenergylett.8b00055.

123 Correa Baena, J.P., Steier, L., Tress, W. et al. (2015). Highly efficient planar perovskite solar cells through band alignment engineering. *Energy Environ. Sci.* 8 (10): 2928–2934. https://doi.org/10.1039/C5EE02608C.

124 Jiang, Q., Zhang, X., and You, J. (2018). SnO2: a wonderful electron transport layer for perovskite solar cells. *Small* 14 (31): 1801154. https://doi.org/10.1002/smll.201801154.

125 Luan, Y., Yi, X., Mao, P. et al. (2019). High-performance planar perovskite solar cells with negligible hysteresis using 2,2,2-trifluoroethanol-incorporated SnO2. *iScience* 16: 433–441. https://doi.org/10.1016/j.isci.2019.06.004.

126 Correa-Baena, J.-P., Tress, W., Domanski, K. et al. (2017). Identifying and suppressing interfacial recombination to achieve high open-circuit voltage in perovskite solar cells. *Energy Environ. Sci.* 10 (5): 1207–1212. https://doi.org/10.1039/C7EE00421D.

127 Koushik, D., Verhees, W.J.H., Kuang, Y. et al. (2017). High-efficiency humidity-stable planar perovskite solar cells based on atomic layer architecture. *Energy Environ. Sci.* 10 (1): 91–100. https://doi.org/10.1039/C6EE02687G.

128 Kot, M., Kegelmann, L., Das, C. et al. (2018). Room-temperature atomic-layer-deposited Al2O3 improves the efficiency of perovskite solar cells over time. *ChemSusChem* 11 (20): 3640–3648. https://doi.org/10.1002/cssc.201801434.

129 Lv, Y., Xu, P., Ren, G. et al. (2018). Low-temperature atomic layer deposition of metal oxide layers for perovskite solar cells with high efficiency and stability under harsh environmental conditions. *ACS Appl. Mater. Interfaces* 10 (28): 23928–23937. https://doi.org/10.1021/acsami.8b07346.

130 Stolterfoht, M., Wolff, C.M., Márquez, J.A. et al. (2018). Visualization and suppression of interfacial recombination for high-efficiency large-area pin perovskite solar cells. *Nat. Energy* 3 (10): 847–854. https://doi.org/10.1038/s41560-018-0219-8.

131 Smith, I.C., Hoke, E.T., Solis-Ibarra, D. et al. (2014). A layered hybrid perovskite solar-cell absorber with enhanced moisture stability. *Angew. Chem. Int. Ed.* 53 (42): 11232–11235. https://doi.org/10.1002/anie.201406466.

132 Cao, D.H., Stoumpos, C.C., Farha, O.K. et al. (2015). 2D homologous perovskites as light-absorbing materials for solar cell applications. *J. Am. Chem. Soc.* 137 (24): 7843–7850. https://doi.org/10.1021/jacs.5b03796.

133 Quan, L.N., Yuan, M., Comin, R. et al. (2016). Ligand-stabilized reduced-dimensionality perovskites. *J. Am. Chem. Soc.* 138 (8): 2649–2655. https://doi.org/10.1021/jacs.5b11740.

134 Liao, Y., Liu, H., Zhou, W. et al. (2017). Highly oriented low-dimensional tin halide perovskites with enhanced stability and photovoltaic performance. *J. Am. Chem. Soc.* 139 (19): 6693–6699. https://doi.org/10.1021/jacs.7b01815.

135 Tsai, H., Nie, W., Blancon, J.-C. et al. (2016). High-efficiency two-dimensional Ruddlesden–popper perovskite solar cells. *Nature* 536: 312.

136 Yoo, J.J., Wieghold, S., Sponseller, M.C. et al. (2019). An interface stabilized perovskite solar cell with high stabilized efficiency and low voltage loss. *Energy Environ. Sci.* 12 (7): 2192–2199. https://doi.org/10.1039/C9EE00751B.

137 Liu, Y., Akin, S., Pan, L. et al. (2019). Ultrahydrophobic 3D/2D fluoroarene bilayer-based water-resistant perovskite solar cells with efficiencies exceeding 22%. *Sci. Adv.* 5 (6): eaaw2543. https://doi.org/10.1126/sciadv.aaw2543.

138 Edri, E., Kirmayer, S., Mukhopadhyay, S. et al. (2014). Elucidating the charge carrier separation and working mechanism of CH3NH3PbI3−xClx perovskite solar cells. *Nat. Commun.* 5: 3461.

139 Gonzalez-Pedro, V., Juarez-Perez, E.J., Arsyad, W.-S. et al. (2014). General working principles of CH3NH3PbX3 perovskite solar cells. *Nano Lett.* 14 (2): 888–893. https://doi.org/10.1021/nl404252e.

140 Jiang, C.-S., Yang, M., Zhou, Y. et al. (2015). Carrier separation and transport in perovskite solar cells studied by nanometre-scale profiling of electrical potential. *Nat. Commun.* 6: 8397.

141 Bergmann, V.W., Weber, S.A.L., Javier Ramos, F. et al. (2014). Real-space observation of unbalanced charge distribution inside a perovskite-sensitized solar cell. *Nat. Commun.* 5: 5001.

142 Melitz, W., Shen, J., Kummel, A.C., and Lee, S. (2011). Kelvin probe force microscopy and its application. *Surf. Sci. Rep.* 66 (1): 1–27. https://doi.org/10.1016/j.surfrep.2010.10.001.

143 Yue, S., Lu, S., Ren, K. et al. (2017). Insights into the influence of work functions of cathodes on efficiencies of perovskite solar cells. *Small* 13 (19): 1700007. https://doi.org/10.1002/smll.201700007.

144 Xie, X., Liu, G., Xu, C. et al. (2017). Tuning the work function of indium-tin-oxide electrodes for low-temperature-processed, titanium-oxide-free perovskite solar cells. *Org. Electron.* 44: 120–125. https://doi.org/10.1016/j.orgel.2017.02.011.

145 Liu, D., Yang, J., and Kelly, T.L. (2014). Compact layer free perovskite solar cells with 13.5% efficiency. *J. Am. Chem. Soc.* 136 (49): 17116–17122. https://doi.org/10.1021/ja508758k.

146 Yan, K., Wei, Z., Li, J. et al. (2015). High-performance graphene-based hole conductor-free perovskite solar cells: Schottky junction enhanced hole extraction and electron blocking. *Small* 11 (19): 2269–2274. https://doi.org/10.1002/smll.201403348.

147 Zhang, L., Liu, T., Liu, L. et al. (2015). The effect of carbon counter electrodes on fully printable mesoscopic perovskite solar cells. *J. Mater. Chem. A* 3 (17): 9165–9170. https://doi.org/10.1039/C4TA04647A.

148 Shi, T., Chen, J., Zheng, J. et al. (2016). Ti/au cathode for electronic transport material-free organic-inorganic hybrid perovskite solar cells. *Sci. Rep.* 6: 39132.

149 Song, Y., Osherov, A., Bulović, V., and Kong, J. (2018). Graphene–perovskite Schottky barrier solar cells. *Adv. Sustain. Syst.* 2 (4): 1700106. https://doi.org/10.1002/adsu.201700106.

150 Mei, A., Li, X., Liu, L. et al. (2014). A hole-conductor–free, fully printable mesoscopic perovskite solar cell with high stability. *Science (80-.)* 345 (6194): 295–298. https://doi.org/10.1126/science.1254763.

151 Jiang, P., Xiong, Y., Xu, M. et al. (2018). The influence of the work function of hybrid carbon electrodes on printable mesoscopic perovskite solar cells. *J. Phys. Chem. C* 122 (29): 16481–16487. https://doi.org/10.1021/acs.jpcc.8b02163.

152 Chen, T., Shi, T., Li, X. et al. (2018). Efficient perovskite solar cells with titanium cathode interlayer. *Sol. RRL* 2 (11): 1800167. https://doi.org/10.1002/solr.201800167.

153 Zhang, F., Ullrich, F., Silver, S. et al. (2019). Complexities of contact potential difference measurements on metal halide perovskite surfaces. *J. Phys. Chem. Lett.* 10 (4): 890–896. https://doi.org/10.1021/acs.jpclett.8b03878.

154 Sadhu, S., Aqueche, K., Buffeteau, T. et al. (2019). Unexpected surface interactions between fluorocarbons and hybrid organic inorganic perovskites evidenced by PM-IRRAS and their application towards tuning the surface potential. *Mater. Horiz.* 6 (1): 192–197. https://doi.org/10.1039/C8MH01119B.

155 Adhikari, N., Dubey, A., Khatiwada, D. et al. (2015). Interfacial study to suppress charge carrier recombination for high efficiency perovskite solar cells. *ACS Appl. Mater. Interfaces* 7 (48): 26445–26454. https://doi.org/10.1021/acsami.5b09797.

156 Adhikari, N., Dubey, A., Gaml, E.A. et al. (2016). Crystallization of a perovskite film for higher performance solar cells by controlling water concentration in methyl ammonium iodide precursor solution. *Nanoscale* 8 (5): 2693–2703. https://doi.org/10.1039/C5NR06687E.

157 Kang, Z., Si, H., Shi, M. et al. (2019). Kelvin probe force microscopy for perovskite solar cells. *Sci. China Mater.* 62 (6): 776–789. https://doi.org/10.1007/s40843-018-9395-y.

158 Yun, J.S., Seidel, J., Kim, J. et al. (2016). Critical role of grain boundaries for ion migration in formamidinium and methylammonium lead halide perovskite solar cells. *Adv. Energy Mater.* 6 (13): 1600330. https://doi.org/10.1002/aenm.201600330.

159 Lee, J.-W., Bae, S.-H., De Marco, N. et al. (2018). The role of grain boundaries in perovskite solar cells. *Mater. Today Energy* 7: 149–160. https://doi.org/10.1016/j.mtener.2017.07.014.

160 Lan, F., Jiang, M., Tao, Q., and Li, G. (2018). Revealing the working mechanisms of planar perovskite solar cells with cross-sectional surface potential profiling. *IEEE J. Photovolt.* 8 (1): 125–131. https://doi.org/10.1109/JPHOTOV.2017.2762525.

161 Chen, Q., Mao, L., Li, Y. et al. (2015). Quantitative operando visualization of the energy band depth profile in solar cells. *Nat. Commun.* 6: 7745.

162 Lin, H.-S., Jeon, I., Xiang, R. et al. (2018). Achieving high efficiency in solution-processed perovskite solar cells using C60/C70 mixed fullerenes. *ACS Appl. Mater. Interfaces* 10 (46): 39590–39598. https://doi.org/10.1021/acsami.8b11049.

163 Xu, C., Liu, Z., and Lee, E.-C. (2019). High-performance inverted planar perovskite solar cells using a pristine fullerene mixture as an electron-transport layer. *J. Mater. Chem. C* 7 (23): 6956–6963. https://doi.org/10.1039/C9TC01741K.

164 Yang, W.S., Park, B.-W., Jung, E.H. et al. (2017). Iodide management in formamidinium-lead-halide–based perovskite layers for efficient solar cells. *Science (80-.)* 356 (6345): 1376–1379. https://doi.org/10.1126/science.aan2301.

165 Yang, C., Du, P., Dai, Z. et al. (2019). Effects of illumination direction on the surface potential of CH3NH3PbI3 perovskite films probed by kelvin probe force microscopy. *ACS Appl. Mater. Interfaces* 11 (15): 14044–14050. https://doi.org/10.1021/acsami.8b21774.

166 Jemli, K., Diab, H., Lédée, F. et al. (2016). Using low temperature photoluminescence spectroscopy to investigate CH3NH3PbI3 hybrid perovskite degradation. *Molecules* 21 (7): 885.

167 Zong, Y., Zhou, Y., Zhang, Y. et al. (2018). Continuous grain-boundary functionalization for high-efficiency perovskite solar cells with exceptional stability. *Chem* 4 (6): 1404–1415. https://doi.org/10.1016/j.chempr.2018.03.005.

168 Ham, S., Choi, Y.J., Lee, J.-W. et al. (2017). Impact of excess CH3NH3I on free carrier dynamics in high-performance nonstoichiometric perovskites. *J. Phys. Chem. C* 121 (5): 3143–3148. https://doi.org/10.1021/acs.jpcc.6b11984.

169 De Bastiani, M., D'Innocenzo, V., Stranks, S.D. et al. (2014). Role of the crystallization substrate on the photoluminescence properties of organo-lead mixed halides perovskites. *APL Mater.* 2: 81509–81509. https://doi.org/10.1063/1.4889845.

170 Li, W., Dong, H., Guo, X. et al. (2014). Graphene oxide as dual functional Interface modifier for improving wettability and retarding recombination in hybrid perovskite solar cells. *J. Mater. Chem. A* 2 (47): 20105–20111. https://doi.org/10.1039/C4TA05196C.

171 Bakr, Z.H., Wali, Q., Fakharuddin, A. et al. (2017). Advances in hole transport materials engineering for stable and efficient perovskite solar cells. *Nano Energy* 34: 271–305. https://doi.org/10.1016/j.nanoen.2017.02.025.

172 Bi, C., Wang, Q., Shao, Y. et al. (2015). Non-wetting surface-driven high-aspect-ratio crystalline grain growth for efficient hybrid perovskite solar cells. *Nat. Commun.* 6: 7747. https://doi.org/10.1038/ncomms8747.

173 Du, Y., Xin, C., Huang, W. et al. (2018). Polymeric surface modification of NiO_x-based inverted planar perovskite solar cells with enhanced performance. *ACS Sustain. Chem. Eng.* 6: 16806–16812. https://doi.org/10.1021/acssuschemeng.8b04078.

174 Jeng, J.-Y., Chen, K.-C., Chiang, T.-Y. et al. (2014). Nickel oxide electrode interlayer in CH3NH3PbI3 perovskite/PCBM planar-heterojunction hybrid solar cells. *Adv. Mater.* 26 (24): 4107–4113. https://doi.org/10.1002/adma.201306217.

175 Yang, Y., Gao, F., Gao, S., and Wei, S.-H. (2018). Origin of the stability of two-dimensional perovskites: a first-principles study. *J. Mater. Chem. A* 6 (30): 14949–14955. https://doi.org/10.1039/C8TA01496E.

176 Tian, S., Chen, J., Lian, X. et al. (2019). Engineering the underlying surface to manipulate the growth of 2D perovskites for highly efficient solar cells. *J. Mater. Chem. A* 7 (23): 14027–14032. https://doi.org/10.1039/C9TA03022K.

177 Liu, J., Leng, J., Wu, K. et al. (2017). Observation of internal photoinduced electron and hole separation in hybrid two-dimensional perovskite films. *J. Am. Chem. Soc.* 139 (4): 1432–1435. https://doi.org/10.1021/jacs.6b12581.

178 Zhou, N., Shen, Y., Li, L. et al. (2018). Exploration of crystallization kinetics in quasi two-dimensional perovskite and high performance solar cells. *J. Am. Chem. Soc.* 140 (1): 459–465. https://doi.org/10.1021/jacs.7b11157.

179 Petrus, M.L., Schlipf, J., Li, C. et al. (2017). Capturing the sun: a review of the challenges and perspectives of perovskite solar cells. *Adv. Energy Mater.* 7 (16): 1700264. https://doi.org/10.1002/aenm.201700264.

180 Rong, Y., Hu, Y., Mei, A. et al. (2018). Challenges for commercializing perovskite solar cells. *Science (80-.)* 361 (6408): eaat8235. https://doi.org/10.1126/science.aat8235.

181 Bella, F., Griffini, G., Correa-Baena, J.-P. et al. (2016). Improving efficiency and stability of perovskite solar cells with photocurable fluoropolymers. *Science (80-.)* 354 (6309): 203–206. https://doi.org/10.1126/science.aah4046.

182 Schlipf, J., Bießmann, L., Oesinghaus, L. et al. (2018). In situ monitoring the uptake of moisture into hybrid perovskite thin films. *J. Phys. Chem. Lett.* 9 (8): 2015–2021. https://doi.org/10.1021/acs.jpclett.8b00687.

183 Lee, S.-W., Kim, S., Bae, S. et al. (2016). UV degradation and recovery of perovskite solar cells. *Sci. Rep.* 6: 38150.

184 Ava, T., Al Mamun, A., Marsillac, S., and Namkoong, G. (2019). A review: thermal stability of methylammonium lead halide based perovskite solar cells. *Appl. Sci.* 9 (1): 188.

185 Zhu, L., Xu, Y., Zhang, P. et al. (2017). Investigation on the role of Lewis bases in the ripening process of perovskite films for highly efficient perovskite solar cells. *J. Mater. Chem. A* 5 (39): 20874–20881. https://doi.org/10.1039/C7TA05378A.

186 You, J., Yang, Y., Hong, Z. et al. (2014). Moisture assisted perovskite film growth for high performance solar cells. *Appl. Phys. Lett.* 105 (18): 183902.

187 Grancini, G., Roldán-Carmona, C., Zimmermann, I. et al. (2017). One-year stable perovskite solar cells by 2D/3D interface engineering. *Nat. Commun.* 8: 15684.

188 Vincent, B.R., Robertson, K.N., Cameron, T.S., and Knop, O. (1987). Alkylammonium lead halides. Part 1. Isolated PbI64– ions in (CH3NH3) 4PbI6• 2H2O. *Can. J. Chem.* 65 (5): 1042–1046.

189 Poglitsch, A. and Weber, D. (1987). Dynamic disorder in methylammoniumtrihalogenoplumbates (II) observed by millimeter-wave spectroscopy. *J. Chem. Phys.* 87 (11): 6373–6378.

190 Frost, J.M., Butler, K.T., Brivio, F. et al. (2014). Atomistic origins of high-performance in hybrid halide perovskite solar cells. *Nano Lett.* 14 (5): 2584–2590. https://doi.org/10.1021/nl500390f.

191 Niu, G., Guo, X., and Wang, L. (2015). Review of recent progress in chemical stability of perovskite solar cells. *J. Mater. Chem. A* 3 (17): 8970–8980.

192 Christians, J.A., Miranda Herrera, P.A., and Kamat, P.V. (2015). Transformation of the excited state and photovoltaic efficiency of CH3NH3PbI3 perovskite upon controlled exposure to humidified air. *J. Am. Chem. Soc.* 137 (4): 1530–1538. https://doi.org/10.1021/ja511132a.

193 Leguy, A.M.A., Hu, Y., Campoy-Quiles, M. et al. (2015). Reversible hydration of CH3NH3PbI3 in films, single crystals, and solar cells. *Chem. Mater.* 27 (9): 3397–3407. https://doi.org/10.1021/acs.chemmater.5b00660.

194 Ma, Y., Hangoma, P.M., Park, W.I. et al. (2019). Controlled crystal facet of MAPbI3 perovskite for highly efficient and stable solar cell via nucleation modulation. *Nanoscale* 11 (1): 170–177. https://doi.org/10.1039/C8NR08344D.

195 Jacobsson, T.J., Correa-Baena, J.-P., Pazoki, M. et al. (2016). Exploration of the compositional space for mixed lead halogen perovskites for high efficiency solar cells. *Energy Environ. Sci.* 9 (5): 1706–1724.

196 D'innocenzo, V., Grancini, G., Alcocer, M.J.P. et al. (2014). Excitons versus free charges in organo-lead tri-halide perovskites. *Nat. Commun.* 5: 3586.

197 Lee, J., Seol, D., Cho, A., and Park, N. (2014). High-efficiency perovskite solar cells based on the black polymorph of HC (NH2) 2PbI3. *Adv. Mater.* 26 (29): 4991–4998.

198 Amat, A., Mosconi, E., Ronca, E. et al. (2014). Cation-induced band-gap tuning in organohalide perovskites: interplay of spin–orbit coupling and octahedra tilting. *Nano Lett.* 14 (6): 3608–3616.

199 Koh, T.M., Fu, K., Fang, Y. et al. (2013). Formamidinium-containing metal-halide: an alternative material for near-IR absorption perovskite solar cells. *J. Phys. Chem. C* 118 (30): 16458–16462.

200 Binek, A., Hanusch, F.C., Docampo, P., and Bein, T. (2015). Stabilization of the trigonal high-temperature phase of formamidinium lead iodide. *J. Phys. Chem. Lett.* 6 (7): 1249–1253.

201 Wu, R., Yao, J., Wang, S. et al. (2019). Ultracompact, well-packed perovskite flat crystals: preparation and application in planar solar cells with high efficiency and humidity tolerance. *ACS Appl. Mater. Interfaces* 11 (12): 11283–11291. https://doi.org/10.1021/acsami.8b17300.

202 Leblanc, A., Mercier, N., Allain, M. et al. (2019). Enhanced stability and band gap tuning of α-[HC(NH2)2]PbI3 hybrid perovskite by large cation integration. *ACS Appl. Mater. Interfaces* 11 (23): 20743–20751. https://doi.org/10.1021/acsami.9b00210.

203 Zhao, B., Jin, S.-F., Huang, S. et al. (2018). Thermodynamically stable orthorhombic γ-CsPbI3 thin films for high-performance photovoltaics. *J. Am. Chem. Soc.* 140 (37): 11716–11725. https://doi.org/10.1021/jacs.8b06050.

204 Lee, J.-W., Kim, D.-H., Kim, H.-S. et al. (2015). Formamidinium and cesium hybridization for photo- and moisture-stable perovskite solar cell. *Adv. Energy Mater.* 5 (20): 1501310. https://doi.org/10.1002/aenm.201501310.

205 McMeekin, D.P., Sadoughi, G., Rehman, W. et al. (2016). A mixed-cation lead mixed-halide perovskite absorber for tandem solar cells. *Science (80-.)* 351 (6269): 151–155. https://doi.org/10.1126/science.aad5845.

206 Yang, S., Chen, S., Mosconi, E. et al. (2019). Stabilizing halide perovskite surfaces for solar cell operation with wide-bandgap lead oxysalts. *Science (80-.)* 365 (6452): 473–478. https://doi.org/10.1126/science.aax3294.

207 Liu, X., He, J., Wang, P. et al. (2019). Fabrication of efficient and stable perovskite solar cells in high-humidity environment through trace-doping of large-sized cations. *ChemSusChem* 12 (11): 2385–2392. https://doi.org/10.1002/cssc.201900106.

208 Han, Y., Meyer, S., Dkhissi, Y. et al. (2015). Degradation observations of encapsulated planar CH3NH3PbI3 perovskite solar cells at high temperatures and humidity. *J. Mater. Chem. A* 3 (15): 8139–8147. https://doi.org/10.1039/C5TA00358J.

209 de Quilettes, D.W., Zhang, W., Burlakov, V.M. et al. (2016). Photo-induced halide redistribution in organic–inorganic perovskite films. *Nat. Commun.* 7: 11683.

210 Christians, J.A., Schulz, P., Tinkham, J.S. et al. (2018). Tailored interfaces of unencapsulated perovskite solar cells for >1,000 hour operational stability. *Nat. Energy* 3 (1): 68–74. https://doi.org/10.1038/s41560-017-0067-y.

211 Song, Z., Wang, C., Phillips, A.B. et al. (2018). Probing the origins of photodegradation in organic–inorganic metal halide perovskites with time-resolved mass spectrometry. *Sustain. Energy Fuels* 2 (11): 2460–2467. https://doi.org/10.1039/C8SE00358K.

212 Aristidou, N., Sanchez-Molina, I., Chotchuangchutchaval, T. et al. (2015). The role of oxygen in the degradation of methylammonium lead trihalide perovskite photoactive layers. *Angew. Chem. Int. Ed.* 54 (28): 8208–8212. https://doi.org/10.1002/anie.201503153.

213 Aristidou, N., Eames, C., Sanchez-Molina, I. et al. (2017). Fast oxygen diffusion and iodide defects mediate oxygen-induced degradation of perovskite solar cells. *Nat. Commun.* 8: 15218.

214 Kim, G.Y., Senocrate, A., Yang, T.-Y. et al. (2018). Large tunable photoeffect on ion conduction in halide perovskites and implications for photodecomposition. *Nat. Mater.* 17 (5): 445–449. https://doi.org/10.1038/s41563-018-0038-0.

215 Zhao, Y.-C., Zhou, W.-K., Zhou, X. et al. (2017). Quantification of light-enhanced ionic transport in lead iodide perovskite thin films and its solar cell applications. *Light Sci. Appl.* 6: e16243.

216 Mosconi, E., Meggiolaro, D., Snaith, H.J. et al. (2016). Light-induced annihilation of Frenkel defects in organo-lead halide perovskites. *Energy Environ. Sci.* 9 (10): 3180–3187. https://doi.org/10.1039/C6EE01504B.

217 Li, Y., Xu, X., Wang, C. et al. (2017). Light-induced degradation of CH3NH3PbI3 hybrid perovskite thin film. *J. Phys. Chem. C* 121 (7): 3904–3910. https://doi.org/10.1021/acs.jpcc.6b11853.

218 Cappel, U.B., Svanström, S., Lanzilotto, V. et al. (2017). Partially reversible photoinduced chemical changes in a mixed-ion perovskite material for solar cells. *ACS Appl. Mater. Interfaces* 9 (40): 34970–34978. https://doi.org/10.1021/acsami.7b10643.

219 Turren-Cruz, S.-H., Hagfeldt, A., and Saliba, M. (2018). Methylammonium-free, high-performance, and stable perovskite solar cells on a planar architecture. *Science (80-.)* 362 (6413): 449–453. https://doi.org/10.1126/science.aat3583.

220 Zhang, M., Yun, J.S., Ma, Q. et al. (2017). High-efficiency rubidium-incorporated perovskite solar cells by gas quenching. *ACS Energy Lett.* 2 (2): 438–444. https://doi.org/10.1021/acsenergylett.6b00697.

221 Hu, Y., Aygüler, M.F., Petrus, M.L. et al. (2017). Impact of rubidium and cesium cations on the moisture stability of multiple-cation mixed-halide perovskites. *ACS Energy Lett.* 2 (10): 2212–2218. https://doi.org/10.1021/acsenergylett.7b00731.

222 Cheacharoen, R., Boyd, C.C., Burkhard, G.F. et al. (2018). Encapsulating perovskite solar cells to withstand damp heat and thermal cycling. *Sustain. Energy Fuels* 2 (11): 2398–2406. https://doi.org/10.1039/C8SE00250A.

223 Conings, B., Drijkoningen, J., Gauquelin, N. et al. (2015). Intrinsic thermal instability of methylammonium lead trihalide perovskite. *Adv. Energy Mater.* 5 (15): 1500477. https://doi.org/10.1002/aenm.201500477.

224 Williams, A.E., Holliman, P.J., Carnie, M.J. et al. (2014). Perovskite processing for photovoltaics: a spectro-thermal evaluation. *J. Mater. Chem. A* 2 (45): 19338–19346. https://doi.org/10.1039/C4TA04725G.

225 Juarez-Perez, E.J., Hawash, Z., Raga, S.R. et al. (2016). Thermal degradation of CH3NH3PbI3 perovskite into NH3 and CH3I gases observed by coupled thermogravimetry–mass spectrometry analysis. *Energy Environ. Sci.* 9 (11): 3406–3410. https://doi.org/10.1039/C6EE02016J.

226 Qiu, W., Ray, A., Jaysankar, M. et al. (2017). An Interdiffusion method for highly performing cesium/formamidinium double cation perovskites. *Adv. Funct. Mater.* 27 (28): 1700920. https://doi.org/10.1002/adfm.201700920.

227 Eperon, G.E., Stranks, S.D., Menelaou, C. et al. (2014). Formamidinium lead trihalide: a broadly tunable perovskite for efficient planar heterojunction solar cells. *Energy Environ. Sci.* 7 (3): 982–988. https://doi.org/10.1039/C3EE43822H.

228 Niu, G., Li, W., Li, J. et al. (2017). Enhancement of thermal stability for perovskite solar cells through cesium doping. *RSC Adv.* 7 (28): 17473–17479. https://doi.org/10.1039/C6RA28501E.

229 Wang, Q., Chen, B., Liu, Y. et al. (2017). Scaling behavior of moisture-induced grain degradation in polycrystalline hybrid perovskite thin films. *Energy Environ. Sci.* 10 (2): 516–522. https://doi.org/10.1039/C6EE02941H.

230 Leijtens, T., Eperon, G.E., Pathak, S. et al. (2013). Overcoming ultraviolet light instability of sensitized TiO2 with meso-superstructured organometal tri-halide perovskite solar cells. *Nat. Commun.* 4: 2885.

231 Domanski, K., Correa-Baena, J.-P., Mine, N. et al. (2016). Not all that glitters is gold: metal-migration-induced degradation in perovskite solar cells. *ACS Nano* 10 (6): 6306–6314. https://doi.org/10.1021/acsnano.6b02613.

232 Sanchez, R.S. and Mas-Marza, E. (2016). Light-induced effects on spiro-OMeTAD films and hybrid lead halide perovskite solar cells. *Sol. Energy Mater. Sol. Cells* 158: 189–194. https://doi.org/10.1016/j.solmat.2016.03.024.

233 Jena, A.K., Numata, Y., Ikegami, M., and Miyasaka, T. (2018). Role of spiro-OMeTAD in performance deterioration of perovskite solar cells at high temperature and reuse of the perovskite films to avoid Pb-waste. *J. Mater. Chem. A* 6 (5): 2219–2230. https://doi.org/10.1039/C7TA07674F.

234 Chen, W., Wu, Y., Yue, Y. et al. (2015). Efficient and stable large-area perovskite solar cells with inorganic charge extraction layers. *Science (80-.)* 350 (6263): 944–948. https://doi.org/10.1126/science.aad1015.

235 Shin, S.S., Yeom, E.J., Yang, W.S. et al. (2017). Colloidally prepared La-doped BaSnO3 electrodes for efficient, photostable perovskite solar cells. *Science (80-.)* 356 (6334): 167–171. https://doi.org/10.1126/science.aam6620.

236 Liu, J., Wu, Y., Qin, C. et al. (2014). A dopant-free hole-transporting material for efficient and stable perovskite solar cells. *Energy Environ. Sci.* 7 (9): 2963–2967. https://doi.org/10.1039/C4EE01589D.

237 Wang, Y.-K., Yuan, Z.-C., Shi, G.-Z. et al. (2016). Dopant-free spiro-triphenylamine/fluorene as hole-transporting material for perovskite solar cells with enhanced efficiency and stability. *Adv. Funct. Mater.* 26 (9): 1375–1381. https://doi.org/10.1002/adfm.201504245.

238 Ni, J.-S., Hsieh, H.-C., Chen, C.-A. et al. (2016). Near-infrared-absorbing and dopant-free heterocyclic quinoid-based hole-transporting materials for efficient perovskite solar cells. *ChemSusChem* 9 (22): 3139–3144. https://doi.org/10.1002/cssc.201600923.

239 Li, Z., Zhu, Z., Chueh, C.-C. et al. (2016). Rational design of dipolar chromophore as an efficient dopant-free hole-transporting material for perovskite solar cells. *J. Am. Chem. Soc.* 138 (36): 11833–11839. https://doi.org/10.1021/jacs.6b06291.

240 Huang, C., Fu, W., Li, C.-Z. et al. (2016). Dopant-free hole-transporting material with a C3h symmetrical truxene core for highly efficient perovskite solar cells. *J. Am. Chem. Soc.* 138 (8): 2528–2531. https://doi.org/10.1021/jacs.6b00039.

241 Cao, J., Yu, H., Zhou, S. et al. (2017). Low-temperature solution-processed NiOx films for air-stable perovskite solar cells. *J. Mater. Chem. A* 5 (22): 11071–11077. https://doi.org/10.1039/C7TA02228J.

242 Yu, Z.-K., Fu, W.-F., Liu, W.-Q. et al. (2017). Solution-processed CuOx as an efficient hole-extraction layer for inverted planar heterojunction perovskite solar cells. *Chin. Chem. Lett.* 28 (1): 13–18. https://doi.org/10.1016/j.cclet.2016.06.021.

243 Yang, I.S., Lee, S., Choi, J. et al. (2019). Enhancement of open circuit voltage for CuSCN-based perovskite solar cells by controlling the perovskite/CuSCN Interface with functional molecules. *J. Mater. Chem. A* 7 (11): 6028–6037. https://doi.org/10.1039/C8TA12217B.

244 Sun, W., Ye, S., Rao, H. et al. (2016). Room-temperature and solution-processed copper iodide as the hole transport layer for inverted planar perovskite solar cells. *Nanoscale* 8 (35): 15954–15960. https://doi.org/10.1039/C6NR04288K.

245 Hou, Y., Du, X., Scheiner, S. et al. (2017). A generic interface to reduce the efficiency-stability-cost gap of perovskite solar cells. *Science (80-.)* 358 (6367): 1192–1197. https://doi.org/10.1126/science.aao5561.

246 Arora, N., Dar, M.I., Hinderhofer, A. et al. (2017). Perovskite solar cells with CuSCN hole extraction layers yield stabilized efficiencies greater than 20%. *Science (80-.)* 358 (6364): 768–771. https://doi.org/10.1126/science.aam5655.

247 Jiang, Z., Chen, X., Lin, X. et al. (2016). Amazing stable open-circuit voltage in perovskite solar cells using AgAl alloy electrode. *Sol. Energy Mater. Sol. Cells* 146: 35–43. https://doi.org/10.1016/j.solmat.2015.11.026.

248 Besleaga, C., Abramiuc, L.E., Stancu, V. et al. (2016). Iodine migration and degradation of perovskite solar cells enhanced by metallic electrodes. *J. Phys. Chem. Lett.* 7 (24): 5168–5175. https://doi.org/10.1021/acs.jpclett.6b02375.

249 Shlenskaya, N.N., Belich, N.A., Grätzel, M. et al. (2018). Light-induced reactivity of gold and hybrid perovskite as a new possible degradation mechanism in perovskite solar cells. *J. Mater. Chem. A* 6 (4): 1780–1786. https://doi.org/10.1039/C7TA10217H.

250 Kato, Y., Ono, L.K., Lee, M.V. et al. (2015). Silver iodide formation in methyl ammonium lead iodide perovskite solar cells with silver top electrodes. *Adv. Mater. Interfaces* 2 (13): 1500195. https://doi.org/10.1002/admi.201500195.

251 Zhao, L., Kerner, R.A., Xiao, Z. et al. (2016). Redox chemistry dominates the degradation and decomposition of metal halide perovskite optoelectronic devices. *ACS Energy Lett.* 1 (3): 595–602. https://doi.org/10.1021/acsenergylett.6b00320.

252 Boyd, C.C., Cheacharoen, R., Bush, K.A. et al. (2018). Barrier design to prevent metal-induced degradation and improve thermal stability in perovskite solar cells. *ACS Energy Lett.* 3 (7): 1772–1778. https://doi.org/10.1021/acsenergylett.8b00926.

253 Hu, Y., Si, S., Mei, A. et al. (2017). Stable large-area (10 × 10 Cm2) printable mesoscopic perovskite module exceeding 10% efficiency. *Sol. RRL* 1 (2): 1600019. https://doi.org/10.1002/solr.201600019.

254 Tian, C., Zhang, S., Mei, A. et al. (2018). A multifunctional bis-adduct fullerene for efficient printable mesoscopic perovskite solar cells. *ACS Appl. Mater. Interfaces* 10 (13): 10835–10841. https://doi.org/10.1021/acsami.7b18945.

255 Hu, Y., Zhang, Z., Mei, A. et al. (2018). Improved performance of printable perovskite solar cells with bifunctional conjugated organic molecule. *Adv. Mater.* 30 (11): 1705786. https://doi.org/10.1002/adma.201705786.

5

All-inorganic Perovskite Solar Cells

Yaowen Li[1] *and Yongfang Li*[1,2]

[1] *Soochow University, College of Chemistry, Chemical Engineering and Materials Science, Laboratory of Advanced Optoelectronic Materials, Suzhou Key Laboratory of Novel Semiconductor-optoelectronics Materials and Devices, Suzhou, Jiangsu 215123, China*
[2] *Beijing National Laboratory for Molecular Sciences, Institute of Chemistry, Chinese Academy of Sciences, Beijing 100190, China*

5.1 Introduction

With the excessive development and overuse of chemical energy, environmental pollution is becoming an increasingly serious problem, and the shortage of energy an urgent global issue. As a source of green and sustainable energy, solar energy is expected to replace traditional energy sources and may play an important role in solving energy shortage and environmental pollution problems. Recently, organic–inorganic hybrid perovskite solar cells (pero-SCs) are prospective powerful candidates for the commercial application due to their feasible manufacturing process, high power conversion efficiency (PCE), ultralow cost, and other appealing merits [1–3]. Since 2012, the hybrid pero-SCs have made unprecedented progress and the certified PCE has recently reached 25.2%, which is comparable to commercial polysilicon, CdTe and CIGS solar cells. However, the instability at higher temperature of the hybrid perovskites is an urgent issue for its future commercialization, which is mainly induced by volatile and hygroscopic organic cations in the hybrid perovskites [4].

In recent years, all-inorganic perovskites with completely replacing organic cations in the organic–inorganic hybrid perovskites by inorganic cerium (Cs) cations have attracted great interests in the field of pero-SCs. Due to their enhanced inherent stability and photoelectric properties compared to those of the organic–inorganic hybrid perovskites [5], Cs-based inorganic perovskites can show great potential and act as stable photovoltaic absorbers. The initial work on the all-inorganic pero-SCs based on $CsSnI_3$ perovskite only exhibited a low PCE of 0.88% [6]. Subsequently, various inorganic perovskites were investigated and applied to inorganic pero-SCs. Different methodologies such as device structure optimization, crystal growth control, and interface engineering were explored to improve the efficiency and stability of the inorganic pero-SCs. Among them,

Perovskite Solar Cells: Materials, Processes, and Devices, First Edition.
Edited by Shahzada Ahmad, Samrana Kazim, and Michael Grätzel.

$CsPbX_3$ (where X = Br or I) perovskites were mostly studied and the highest PCE has reached 18.4% [7] recently for the all-inorganic pero-SCs.

Pure $CsPbBr_3$ perovskite is found to exhibit outstanding long-term stability against heat, moisture [8], and ultraviolet irradiation [9], while its wide bandgap excludes the possibility of harvesting light in the long-wavelength spectrum, hence limiting the PCE of the $CsPbBr_3$-based pero-SCs. In contrast, pure $CsPbI_3$ is the most promising inorganic perovskite for photovoltaic application. However, $CsPbI_3$ is easily converted from black to yellow phase upon exposure under ambient condition, which is a consequence of a low Goldschmidt tolerance factor (t) for α-phase $CsPbI_3$ [10]. Decreasing crystal size [11], ionic incorporation [12], and organic compounds capping protection [13] can effectively suppress the phase-transferring process. Mixed halide perovskites, such as $CsPbI_2Br$ and $CsPbIBr_2$, have been developed, and these Br-rich perovskites show better stability than $CsPbI_3$ perovskite and can generate high open-circuit voltage (V_{oc}) without sacrificing excessive photocurrent, which is beneficial for tandem devices and water splitting systems [14, 15]. To date, the highest PCEs for the pero-SCs based on $CsPbI_2Br$ and $CsPbIBr_2$ are 16.37% [16] and 10.88%, respectively [17]. In addition, lead-free inorganic perovskite such as Sn-based, Ge-based, Sb-based, and Bi-based perovskites and double perovskites were also investigated, with only relatively low PCEs when applied to solar cells. Meanwhile, Sn^{2+} is known for the facile oxidation to Sn^{4+} under ambient condition, which induces near-short-circuit diode condition and seriously hampers the device performance [18]. Despite some stability-related problems among partial perovskites, inorganic perovskites still show bright application prospects, and these problems will certainly be fully addressed in future development.

In this chapter, we discuss and summarize recent advances of the all-inorganic pero-SCs. This chapter is divided into five parts: in the first section, we briefly introduce the basic knowledge of the all-inorganic perovskite; in the second section, we discuss the specific performance and development of lead-based inorganic pero-SCs in detail, including $CsPbI_3$-based, $CsPbI_2Br$-based, $CsPbIBr_2$-based, and $CsPbBr_3$-based pero-SCs; in the third section, we illustrate the progress and strategies applied to Sn-based pero-SCs; in the fourth section, we briefly introduce some other inorganic pero-SCs; and finally, we present a brief conclusion of the recent advances in the all-inorganic pero-SCs.

5.2 Basic Knowledge of All-inorganic Pero-SCs

5.2.1 Crystalline Structure

The basic crystal structure of inorganic perovskite is $AB^{2+}X_3$, where A site is generally occupied by Cs monovalent cations, B site the Pb divalent cations (rarely Sn, Ge, Sb, and Bi ion), and X site the halogen monovalent anions, such as $CsPbI_3$ [19]. In addition, some other derivative structures have also been explored, such as $A_2B^{4+}X_6$ and double perovskite structure $A_2B^{1+}B^{3+}X_6$, (Figure 5.1a) [5, 20]. $A_2B^{4+}X_6$ crystal

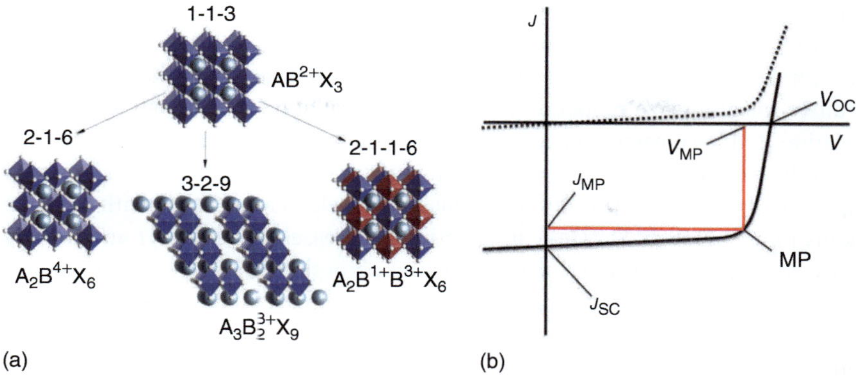

Figure 5.1 (a) Schematic relation between the crystal structures of Pb-based perovskites and Pb-free perovskite derivatives. Source: Giustino and Snaith [20]. (b) Current density–voltage (*J*–*V*) curves of a solar cell. Source: Mazzio and Luscombe [21].

can be considered obtained from $AB^{2+}X_3$ by removing half of the B site cations in a checkerboard pattern, which requires B site cation to be in its +4 oxidation state, such as Cs_2SnI_6 [22, 23]. Similarly, double perovskite structure $A_2B^{1+}B^{3+}X_6$ can be considered obtained by replacing every pair of adjacent B^{2+} cations by one B^{1+} and one B^{3+} cation, such as $Cs_2BiAgBr_6$ [24].

5.2.2 Stability

Except for the continuously rising efficiency, perovskite stability plays an equally important role toward commercial application. Organic–inorganic hybrid perovskite decomposes upon long-term heat stress, moisture attack, and illumination. By contrast, all-inorganic perovskite exhibits better stability under these conditions.

5.2.2.1 Thermal Stability

Industrialization requires solar cells to operate well at 85 °C, which is the temperature of the outdoor solar cells and can easily reach in some geographic locations under high solar irradiance [25]. Organic–inorganic hybrid perovskites are susceptible to degrade at a slightly high temperature (above 85 °C) due to the presence of organic cations, thereby damaging the long-term stability of the pero-SCs. In contrast, all-inorganic perovskites can easily withstand temperatures above 400 °C with no degradation, exhibiting their long-term excellent thermal stability [26].

5.2.2.2 Phase Stability

The Goldschmidt tolerance factor is a criterion for evaluating the phase stability of perovskites. When the Goldschmidt tolerance factor is between 0.9 and 1, the ABX_3 structure perovskite can maintain a stable cubic phase. It is expected that the moisture will introduce vacancies to the crystal lattice and lower the nucleation free-energy barrier [27]. For $CsPbX_3$ perovskite, when the I content is too high, its tolerance factor will drop to close to 0.8. Therefore it is easy to undergo a phase transition to a non-perovskite material under the attack of water vapor. To solve this issue, methodologies involving organic compound layers to prevent moisture [13], additive for grain size reduction [11], and encapsulation technology for moisture isolation [28] have been applied to effectively improve the phase stability. Notably, when the Br content is rich, the phase stability of the inorganic perovskite itself is significantly enhanced [8]. Besides, replacing Pb element at B site with a smaller radius metal (such as Sn, Ge) can effectively increase the value of t and improve phase stability. In addition, for the Sn-based perovskite, divalent Sn is easily oxidized to tetravalent Sn, thus leading to near short-circuit diode condition, which is detrimental for device performance. Usually, adding some excess of divalent Sn ions can effectively suppress the oxidation process [18].

5.2.2.3 Light Stability

Photoinduced phase separation contributes to the decline in perovskite performance, which may be affected by lattice strain, traps, light intensity, pressure, or compositional uniformity, but the origin is still unclear [29]. The photoinduced phase separation is a common phenomenon in organic–inorganic hybrid perovskites, as well as in $CsPb(I_xBr_{1-x})_3$ ($0.4 < x < 1$) all-inorganic perovskites [14]. Luckily, it is proved that high-quality inorganic perovskite films can effectively suppress this phase separation. Chen et al. found that high-quality inorganic perovskite films could maintain good stability under ultraviolet irradiation [9], which is also a major advantage of inorganic perovskites compared with organic–inorganic hybrid perovskites.

5.2.3 Working Principle

An all-inorganic pero-SC is usually composed of a highly transparent conductive bottom electrode, a perovskite bulk film as a light absorption layer sandwiched between an electron transporting layer (ETL) and a hole transporting layer (HTL) with an opaque metal layer as a top electrode. When the perovskite film absorbs sunlight, electrons and holes are generated inside the perovskite film, separated by the build-in electric field, and then transferred to the transporting layers and collected by the top/bottom electrodes to produce photocurrent and photovoltage. PCE is one of the most important parameters to evaluate the performance of a pero-SC, which can be obtained from the current density–voltage (J–V) curve (Figure 5.1b) [21]. The PCE value is defined as the ratio of the maximum power output (P_{out}) to the incident light power (P_{in}) or determined by open-circuit voltage (V_{oc}), short-circuit current density (J_{sc}), and fill factor (FF) as follows:

$$\mathrm{PCE} = \frac{P_{out}}{P_{in}} = \frac{V_{oc} * J_{sc} * \mathrm{FF}}{P_{in}}$$

Recently, the PCE of all-inorganic pero-SC has reached 18.4%. Theoretical PCE value for single-junction pero-SC is 33.3% [30], and there is a huge boost for PCE improvement in the future.

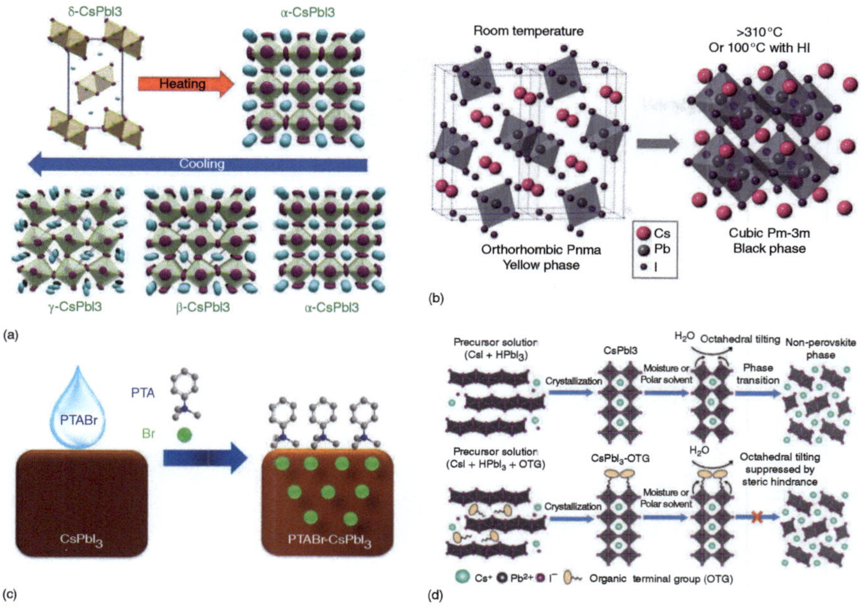

Figure 5.2 (a) Structural phase transitions in $CsPbI_3$ vs. temperature. Source: Marronnier et al. [32]. (b) Crystal structures of yellow and black phase $CsPbI_3$. Source: Eperon et al. [33]. (c) Schematic illustration of gradient Br doping and PTA organic cation surface passivation on $CsPbI_3$ perovskite thin film. Source: Wang et al. [13]. (d) Mechanism of the OTG-induced phase stabilization of $CsPbI_3$ perovskite. Source: Wu et al. [34].

5.3 Lead-Based Inorganic Pero-SCs

5.3.1 $CsPbI_3$

$CsPbI_3$ is an inorganic perovskite formed by substituting organic cation (MA^+) in $MAPbI_3$ completely with Cs^+. Due to a small Goldschmidt tolerance factor of 0.81 [31], it usually shows a yellow non-perovskite phase (denoted as δ-phase) at room temperature, which exhibits poor photovoltaic performance. As shown in Figure 5.2a, upon heating the δ-phase sample to 360 °C, the yellow phase will convert to black cubic α-phase, which has a bandgap of 1.73 eV, suitable for photovoltaic active layer. During the cooling process, α-phase $CsPbI_3$ tends to convert at 260 °Cto tetragonal β-phase and at 175 °C to orthorhombic γ-phase. Both are black phases and will convert to δ-phase after a few days [32]. Among them, α-phase and γ-phase $CsPbI_3$ are commonly used as perovskite layers for fabricating high-performance inorganic pero-SCs. Basic properties of lead-based all-inorganic perovskites are summarized in Table 5.1. The PCE of the $CsPbI_3$-based pero-SCs recently increased from 2.9% [33] to 18.4% [7]. Several strategies have been applied for fabricating high-performance $CsPbI_3$ pero-SCs. Here, we will discuss the strategies as follows.

Table 5.1 Basic properties of lead-based all-inorganic perovskites.

<table>
<tr><th rowspan="2">Composition</th><th rowspan="2">Phase transition</th><th rowspan="2">Bandgap (eV)</th><th colspan="4">Stability</th></tr>
<tr><th>Light</th><th>Heat</th><th>Moisture</th><th>Others</th></tr>
<tr><td>$CsPbI_3$</td><td>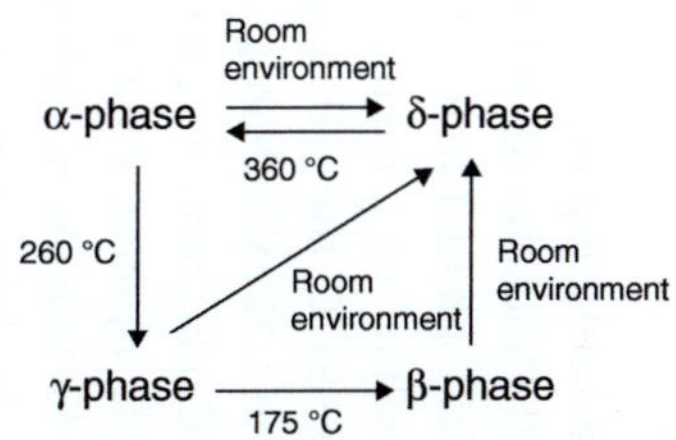
</td><td>~1.73</td><td>Superior white light stability</td><td>Withstand temperature above 400 °C</td><td>Poor</td><td>—</td></tr>
<tr><td>$CsPbI_2Br$</td><td>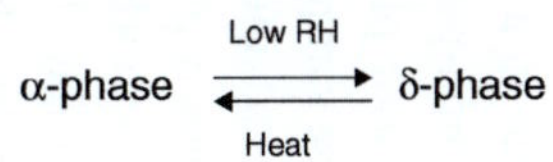
</td><td>~1.92</td><td>Superior UV light, white light stability</td><td></td><td>Normal</td><td>Superior stability against electron beam</td></tr>
<tr><td>$CsPbIBr_2$</td><td>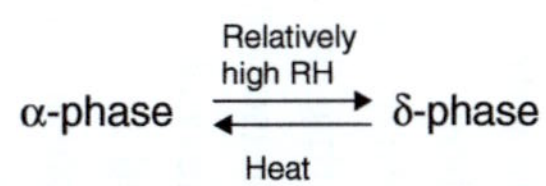
</td><td>~2.05</td><td>Phase-segregate upon illumination</td><td></td><td>Good</td><td>—</td></tr>
<tr><td>$CsPbBr_3$</td><td>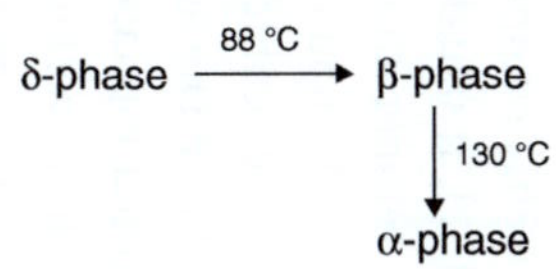
</td><td>~2.3</td><td>Excellent UV light stability</td><td></td><td>Excellent</td><td>Superior temperature cycling stability</td></tr>
</table>

5.3.1.1 Additive Engineering

α-Phase $CsPbI_3$ is usually obtained above 360 °C. Such a high temperature restricts the operability and application in flexible devices, thus requiring a lower post annealing temperature of $CsPbI_3$ perovskite. Additive engineering is a convenient and efficient way to solve this issue.Eperon et al. were the first to add hydroiodic acid (HI) into a $CsPbI_3$ precursor, and the thermal annealing temperature was successfully lowered to 100 °C (Figure 5.2b) [33]. It may be attributed to the smaller crystal size, which effectively facilitated phase transition at a low temperature. Based on its planar heterojunction architecture, the device exhibited a PCE of 2.9% [33]. Subsequently, as a commonly used additive, HI has been widely used in $CsPbI_3$-based solar cells [35, 36]. Jena et al. introduced Eu^{3+} into $CsPbI_3$ perovskite and successfully stabilized the α-$CsPbI_3$ at a lower annealing temperature (85 °C), which is ascribed to the crystal structure distortion and the reduced grain size. A PCE above 6% was achieved [31]. Except for additive engineering, dual-source evaporation method could also produce black phase $CsPbI_3$ at 50 °C when CsI was optimally rich [37]. The resultant γ-phase $CsPbI_3$ exhibited large carrier lifetime and outstanding charge carrier mobilities [38]. A *p–i–n* pero-SC with a maximum PCE exceeding 12% and high self-stability of more than 1200 hours was demonstrated. The performance parameters of the $CsPbI_3$-based inorganic pero-SCs are summarized in Table 5.2.

5.3.1.2 Organic Compound Treatment

α-Phase $CsPbI_3$ is unstable under ambient condition because of its self-instability and the attacking external water vapor. Several studies explored the influence of organic compounds on the improvement of stability of $CsPbI_3$ perovskite. The organic species usually form a capping layer above the perovskite film, preventing the perovskite from damage by moisture [13, 39] or forming steric hindrance [34, 36] to suppress phase transition. Both effects improve the stability of the black phase. In addition, organic compounds possess an outperformed advantage, i.e. trap density passivation effect. Most of the organic compounds used in the $CsPbI_3$ pero-SCs have lone unpaired electrons, which are donated to the empty orbitals of Pb^{2+} in the perovskite, therefore passivating the defect states. This passivation strategy is also commonly used in organic–inorganic hybrid pero-SCs and has been widely applicated in the $CsPbI_3$ pero-SCs [51, 52].

Different organic compounds have been applied to the inorganic pero-SCs. Wang et al. found that phenyltrimethylammonium bromide (PTABr) post-treatment can lead to a void-free $CsPbI_3$ film. PTA organic cations terminated on the surface of $CsPbI_3$ perovskite prevent $CsPbI_3$ film from moisture (Figure 5.2c). As a result, a high PCE of 17.06% was obtained by this strategy [13]. It could maintain 91% of its initial PCE after continuous white light LED illumination for 500 hours. Zhang et al. found that a small quantity of 2D $EDAPbI_4$ perovskite containing the ethylenediamine (EDA) cation could stabilize the black phase of $CsPbI_3$ by reducing the crystal size [40]. Moreover, the obtained α-phase $CsPbI_3$ could stabilize at room temperature for months and at 100 °C for over 150 hours because the $EDAPbI_4$ may connect the crystallites for efficient carrier transfer and passivate the trap

Table 5.2 Photovoltaic parameters of $CsPbI_3$-based pero-SCs.

Composition	Configuration	V_{oc} (V)	J_{sc} (mA cm^{-2})	FF (%)	PCE (%)	References
$CsPbI_3$	FTO/TiO_2/$CsPbI_3$/spiro/Au	0.8	12	30.2	2.9	[33]
	FTO/TiO_2/$CsPbI_3$-PTABr/spiro/Ag	1.104	19.15	80.6	17.06	[13]
	FTO/TiO_2/$CsPbI_3$:$EuCl_3$/spiro/Au	0.898	11.1	68	6.8	[31]
	ITO/PTAA/$CsPbI_3$/C_{60}/BCP/Cu	0.96	17.8	73	12.5	[37]
	FTO/SnO_2/$CsPbI_3$/HTM/Ag	1.00	13.0	68	8.8	[38]
	FTO/TiO_2/$CsPbI_3$/PTAA/Au	1.059	18.95	74.9	15.07	[36]
	FTO/TiO_2/$CsPbI_3$-PVP/spiro/Au	1.16	14.88	65	10.74	[39]
	FTO/PTAA/$CsPbI_3$-OTG3/PCBM/BCP/Ag	1.12	15.81	75.19	13.32	[34]
	FTO/TiO_2/$CsPbI_3$:0.025$EDAPbI_4$/spiro/Ag	1.15	14.53	71	11.86	[40]
	FTO/TiO_2/$Cs_{0.9}PEA_{0.1}PbI_3$/spiro/Au	0.838	10.96	62	5.7	[41]
	ITO/PTAA/$CsPbI_3$: zwitterion /PCBM/C_{60}/BCP/cathode	1.08	14.9	70	11.4	[42]
	FTO/TiO_2/$CsPbI_3$:0.05$DETAI_3$/P3HT/Au	1.06	12.21	61	7.89	[43]
	FTO/TiO_2/$BA_2CsPb_2I_7$/spiro/Au	0.957	8.88	57	4.84	[44]
	FTO/TiO_2/$CsPb_{0.96}Bi_{0.04}I_3$/CuI/Au	0.97	18.76	72.59	13.21	[12]
	FTO/TiO_2/$CsPbI_3$/P3HT/Au	1.04	16.53	65.7	11.30	[11]
	FTO/TiO_2/$CsPb_{1-x}Ca_xI_3$/spiro/Au	0.84	16.37	66.91	9.20	[45]
	FTO/TiO_2/$CsPb_{0.95}Ca_{0.05}I_3$/P3HT/Au	0.945	17.9	80.0	13.5	[46]
	FTO/TiO_2/$CsPbI_3$:$Cl_{0.03}$/PTAA/Au	1.084	19.58	75.7	16.07	[47]
	ITO/SnO_2/$CsPbI_3$/spiro/Au	1.08	18.41	79.32	15.71	[48]
	FTO/TiO_2/$CsPbI_3$/CHI/spiro/Ag	1.11	20.23	82	18.4	[7]
	FTO/TiO_2/$CsPbI_3$/PTAA:UCNPs/Au	1.113	19.17	74.33	15.86	[49]
	N-CQDs EDS/Glass/FTO/TiO_2/$CsPbI_3$/PTAA/Au	1.106	19.15	75.6	16.02	[50]

density. A PCE of 11.8% of the pero-SCs based on $CsPbI_3·0.025EDAPbI_4$ perovskite film was realized. Similarly, phenylethylammonium anion (PEA^+) [41], PEA [53], sulfobetaine zwitterions [42], triple cation $NH_3^+C_2H_4NH_2^+C_2H_4NH_3^+$ [43], polymer poly-vinylpyrrolidone (PVP) [39], and 2D cesium lead iodide perovskite $BA_2CsPb_2I_7$ [44] were also introduced into $CsPbI_3$ perovskite to reduce crystal size and/or to form a defect-passivating organic cation terminated surface. The PCEs in these reports were mostly high of 5.7% [41], 11.4% [42], 7.89% [43], 10.74% [39], and 4.84% [44]. In addition, by adding organic terminal group (OTG) molecules into the precursor solution, the steric hindrance of surface OTGs could be systematically investigated to suppress the octahedral tilting of $[PbI_6]^{4-}$ units in cubic $CsPbI_3$ [34]. Steric hindrance caused by OTGs could maintain the linearity of the Pb—I—Pb bonds and therefore suppress the octahedral tilting of adjacent $[PbI_5]^{4-}$ units, which effectively prevent the lattice distortion and thus increase the energy barrier for phase transition (Figure 5.2d). This resulted in a stable PCE of 13.2% for the inverted planar $CsPbI_3$ pero-SCs with improved air stability. Phenylethyl ammonium iodide (PEAI) could also optimize the crystallization and stabilize the distorted black phase by hindering phase transition via its steric effects [36]. A PCE as high as 15.07% was obtained by this strategy and the device could keep 92% of its initial efficiency after being stored for two months under ambient condition.

5.3.1.3 Crystal Size Reduction and Morphology Optimization

The stabilization of the black phase $CsPbI_3$ perovskite can be improved when the grain sizes are reduced. Except for crystal size reduction due to incorporation of organic species, the presence of other compounds may contribute in the fabrication of stable $CsPbI_3$ films. Ionic incorporation is a commonly used method to increase the Goldschmidt tolerance factor by partial substitution. For example, Bi^{3+} was introduced into $CsPbI_3$ to partially substitute Pb^{2+}. The radium of Bi^{3+} is 1.03 Å, smaller than that of Pb^{2+} (1.19 Å), thus increasing the Goldschmidt tolerance factor and stabilizing α-phase $CsPbI_3$ [12]. When the amount of Bi^{3+} is increased, the crystal size of perovskite gradually decreases. Specially, the average grain size in the $CsPb_{0.96}Bi_{0.04}I_3$ film was the smallest (<100 nm), resulting in a high PCE of 13.21% with the improved air stability. Zhao et al. introduced a small amount of H_2O into the perovskite precursor to reduce crystal size, obtaining a high-performance γ-$CsPbI_3$ thin film without introducing any organic ligands mixed cations/anions (Figure 5.3a) [11]. Theoretical calculations coupled with experiments indicated that γ-$CsPbI_3$ with a lower surface free energy has become thermodynamically preferred over δ-$CsPbI_3$ at surface areas >8600 $m^2\ mol^{-1}$ and exhibited comparable optoelectronic properties to α-$CsPbI_3$, resulting in 11.3% efficiency. Crystal size reduction does benefit the stabilization of the black phase $CsPbI_3$. Another different route for enhanced phase stability is to improve the crystal quality, thus reducing the voids and trap density of the perovskite films. For example, an optimal 0.25% Ca^{2+}-doped $CsPbI_3$ film exhibited the least pinholes and the best compactness. At the same time, a strongest peak intensity indicated the best crystallization, which was beneficial for photovoltaic performance [45]. Another report clarified that a 5% Ca^{2+}-doped $CsPbI_3$ exhibited the best performance due to the size reduction of

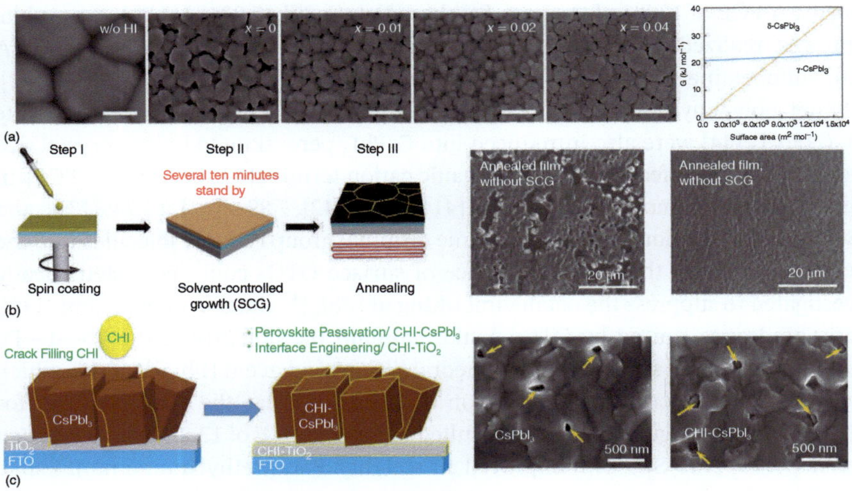

Figure 5.3 (a) Left: top-view SEM images with different H_2O concentrations. Right: calculated Gibbs free energy G of $CsPbI_3$ polymorphs relative to bulk-$CsPbI_3$ as a function of surface area. Source: Zhao et al. [11]. (b) Left: Schematic illustration of $CsPbI_3$-perovskite crystallization procedures via solvent-controlled growth (SCG). Right: SEM images of annealed $CsPbI_3$ perovskite precursor films without and with SCG, respectively. Source: Wang et al. [48]. (c) Left: Schematic illustration of crack-filling process. Right: SEM images of $CsPbI_3$ and CHI-$CsPbI_3$ thin films. Source: Wang et al. [7].

colloidals in the precursor solution containing Ca^{2+} [46]. Surface passivation caused by Ca-rich oxide layer on the surface of the $CsPbI_3$ film leads to a remarkable PCE of 13.5%. Encapsulated device also showed a remarkable stability, maintaining 85% of its initial PCE for over two months. Wang et al. incorporated optimal chlorine ions and obtained black orthorhombic γ-phase $CsPbI_3$ [47]. The chlorine incorporation improves the crystallization dynamics for favorable surface morphology and crystalline orientation. Consequently, a 3% Cl-treated γ-$CsPbI_3$ pero-SC give a PCE as high as 16.07%. Meanwhile, the optimized device shows slight PCE degradation after continuous light soaking or long-term air exposure. Without adding any compounds, Wang et al. proposed a solvent-controlled growth (SCG) method to deposit $CsPbI_3$ films [48]. They put the as-prepared film for several 10 minutes before thermal annealing and the resultant film exhibited pinholes-free morphology with up to 5 μm grain size (Figure 5.3b), resulting in a remarkable PCE of 15.75% as well as outstanding light stability. Recently, Wang et al. mitigated the effects of cracks and pinholes in the β-phase $CsPbI_3$ perovskite layer by surface treating with choline iodide (Figure 5.3c) [7]. It was proved that CHI homogeneously distributed on the surface of perovskite film and the pinhole locations were filled with CHI, which effectively increased the charge carrier lifetime and improved the energy-level alignment between the β-$CsPbI_3$ layer and carrier-selective contacts. As expected, a significantly improved PCE of 18.4% was realized, which is the highest PCE for the all-inorganic pero-SCs till now. More importantly, the CHI-treated perovskite film exhibited good thermal stability and greater resistance toward high humidity. In other words, a small grain size stabilizes black phase $CsPbI_3$ by reducing surface free energy, and high-quality crystal can also be responsible for void-free and stable black phase film.

5.3.1.4 Current Density Increase

Compared with organic–inorganic hybrid pero-SC, inorganic pero-SCs output lower current density, owing to their shallower light absorption range. Some technologies were used to increase current density of the $CsPbI_3$ pero-SCs. Liu's group introduced core–shell-structured $NaLuF_4$:Yb,Er@$NaLuF_4$ up-conversion nanoparticles (UCNPs) into the HTL of γ-$CsPbI_3$ pero-SC to improve device performance [49]. Combined with scattering and reflecting sunlight, an enhanced current density was achieved from 18.81 to 19.17 mA cm^{-2}. A high PCE of 15.86% was obtained simultaneously. Almost at the same time, they blended nitrogen-doped graphene quantum dots (N-GQDs) into γ-$CsPbI_3$ pero-SC [50]. N-GQDs was advantageous at converting harmful UV light into visible photons for improved solar cell efficiency. The resultant current density increased by 0.5 mA cm^{-2} and the PCE exceeded 16%.

Above all, black phase $CsPbI_3$ owes a bandgap of ~1.73 eV, which is suitable for single-junction solar cell or for tandem solar cell. The main issue needs to be solved is the poor phase stability of $CsPbI_3$ itself. Though metal ion replacement, organic compound and 2D perovskite incorporation, as well as encapsulation can effectively improve the stability, the intrinsic instability still hinders its future application. More efforts need to be devoted to solve the stability issue of the $CsPbI_3$-based pero-SCs.

5.3.2 $CsPbI_2Br$

In considering that partial substitution of I^- with smaller Br^- can modulate the Goldschmidt tolerance factor and stabilize the black phase, there are increasing attention focused on another inorganic perovskite $CsPbI_2Br$. $CsPbI_2Br$ possesses an optical E_g of about 1.9 eV [54], which is also suitable for application in photovoltaic device or tandem solar cell [14]. Unlike $CsPbI_3$ film, the $CsPbI_2Br$ film will not degrade quickly from black phase to yellow phase when stored in a dry environment. However, it also tends to transfer to orthorhombic δ-phase upon moisture attack. The orthorhombic δ-phase is a non-perovskite yellow phase (2.85 eV) and is not suitable for a photovoltaic material. Fortunately, this transferring process is reversible when reheating the $CsPbI_2Br$ perovskite film to 350 °C under a dry condition, and the basic PV performance is almost unchanged (Figure 5.4a) [55]. The PCE of the $CsPbI_2Br$ pero-SCs has been increased to 16.37% recently [16]. Though still lagging behind the PCE value of the $CsPbI_3$ pero-SC, $CsPbI_2Br$ pero-SC owns a much more brighter prospect due to its better stability.

5.3.2.1 Fabrication Methods

Vapor Evaporation Vapor evaporation is a commonly used method for fabricating high-quality film. Ma et al. investigated the effect of stoichiometry on the stability of $CsPbI_2Br$ pero-SCs [55]. It is found that the air stability of $CsPbI_2Br$ perovskite

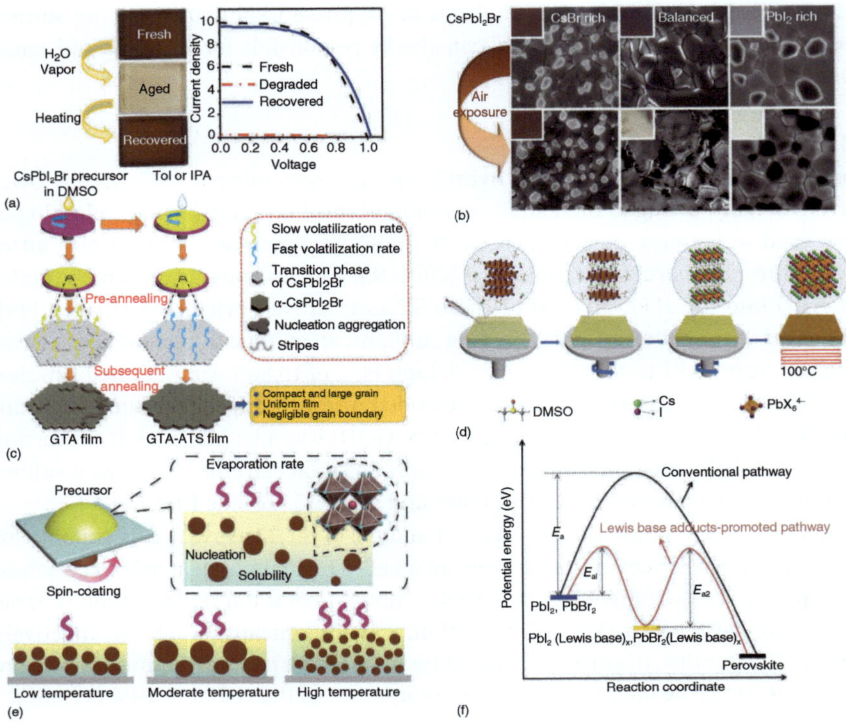

Figure 5.4 (a) Degraded/recovered process for $CsPbI_2Br$ films and related $J-V$ curves. Source: Mariotti et al. [55]. (b) Top-view SEM images of unencapsulated fresh (top row) and air-exposed (bottom row) CsBr-rich, stoichiometrically balanced and PbI_2-rich $CsPbI_2Br$ films. Source: Ma et al. [56]. (c) Schematic illustration of $CsPbI_2Br$ perovskite crystallization process via GTA or GTA-ATS processing. Source: Chen et al. [57]. (d) Schematic for the evolution of the $CsPbI_2Br$ film with introducing DMSO. Source: Zai et al. [58]. (e) Growth mechanism of thin $CsPbI_2Br$ film with spin-coating processing and schematic illustration of the film growth at different temperatures. Source: Bai et al. [59] (f) The reaction coordinate diagram of the inorganic perovskite formation via the conventional pathway and DSMO-adduct promoted pathway. Source: Jiang et al. [60].

film is closely related to the ratio of precursors during dual-sources evaporation process. As a result, the CsBr-rich films have the best stability under moisture attacking, while the stoichiometrically balanced film displayed the best photovoltaic performance (Figure 5.4b). Subsequently, Chen et al. also verified the importance of precursor stoichiometry and studied the influence of different post annealing times on the film morphology and device performance [61]. A high PCE of 11.8% for the $CsPbI_2Br$-based solar cell as well as superior long-term storage stability was realized (summarized in Table 5.3).

Solvent Engineering Among all the fabrication methods, a one-step solution method is a facile and effective way for fabricating high-performance film, where antisolvent is always applied during the spinning process. Antisolvent treatment plays a critical role in perovskite nuclei formation and crystal growth. Chlorobenzene (CB) is a

Table 5.3 Photovoltaic parameters of $CsPbI_2Br$-based pero-SCs.

Composition	Configuration	V_{oc} (V)	J_{sc} ($mA\,cm^{-2}$)	FF (%)	PCE (%)	References
$CsPbI_2Br$	ITO/TiO_2/$CsPbI_2Br$/spiro/Au	1.05	12.7	68.4	9.08	[55]
	ITO/SnO_2/PN4N/$CsPbI_2Br$/PDCBT/MoO_3/Ag	1.30	15.3	81.5	16.2	[62]
	FTO/TiO_2/$CsPbI_2Br$/P3HT/Au	1.005	11.5	67	7.7	[56]
	ITO/Ca/C_{60}/$CsPbI_2Br$/TAPc/TAPc:MoO_3/Ag	1.13	15.2	68	11.8	[61]
	ITO/TiO_2/$CsPbI_2Br$/carbon	1.15	13.54	64.2	10.0	[63]
	ITO/TiO_2/$CsPbI_2Br$/spiro/Au	1.23	16.79	77.81	16.07	[57]
	FTO/TiO_2/$CsPbI_2Br$/carbon	1.15	13.87	64	10.21	[64]
	FTO/NiMgLiO/$CsPbI_2Br$/PCBM/BCP/Ag	0.98	14.18	66	9.14	[65]
	ITO/SnO_2/$CsPbI_2Br$/PTAA/Au	1.162	15.565	19.06	14.30	[58]
	FTO/TiO_2/$CsPbI_2Br$/FA^+ $CsPbX_3$/FA^+ QDs/PTAA/Au	1.221	15.1	80.3	14.81	[59]
	ITO/NiO_x/$CsPbI_2Br$/C_{60}/BCP/Ag	1.05	12.6	78.7	10.4	[66]
	FTO/TiO_2/$CsPbI_2Br$/spiro/Au	1.18	14.89	77.23	13.54	[60]
	PET/FTO/TiO_2/$CsPbI_2Br$/spiro/Au	1.19	14.6	67.3	11.7	
	FTO/TiO_2/$CsPbI_2Br$/spiro/Au	1.22	15.33	78.70	14.78	[67]
	ITO/SnO_2/$CsPbI_2Br$/PTAA/MoO_3/Al	1.17	15.04	71.0	12.5	[68]
	FTO/TiO_2/$CsPbI_2Br$/PTAA/Au	1.23	12.0	73	10.7	[69]
	FTO/TiO_2/$CsPbI_2Br$/spiro/Ag	1.243	13.56	74.3	12.52	[70]
	ITO/SnO_2/$CsPbI_2Br$/spiro/Ag	1.06	15.99	77.12	13.09	[71]
	FTO/TiO_2/$CsPb_{0.9}Zn_{0.1}I_2Br$/spiro/Ag	1.18	15.8	72.7	13.6	[72]
	FTO/TiO_2/$CsPbI_2Br$/PTAA/Au	1.227	14.55	78.7	14.05	[73]
	FTO/TiO_2/$CsPb_{0.095}Eu_{0.05}I_2Br$/spiro/Au	1.22	14.63	76.6	13.71	[74]
	FTO/SnO_2/$CsPb_{0.8}Ge_{0.2}I_2Br$/P3HT/spiro/Au	1.27	12.15	70.1	10.8	[75]
	FTO/TiO_2/$CsPb_{0.98}Sr_{0.02}I_2Br$/P3HT/Au	1.043	15.3	69.9	11.2	[76]
	ITO/SnO_2/$CsPbI_2Br$: Nb/carbon	1.20	12.06	72	10.42	[77]
	FTO/TiO_2/$CsPbI_2Br$:$MnCl_2$/QDs/PTAA/Au	1.172	14.37	80.0	13.47	[78]
	FTO/TiO_2/$Cs_{1-x}K_xPbI_2Br$/spiro/Au	1.18	11.6	73	10.0	[79]
	FTO/TiO_2/$CsPbI_{1.78}F_{0.22}Br$/spiro/Ag	1.01	14.94	68	10.26	[80]
	FTO/NiO_x/$CsPbI_2Br$:$InCl_3$/ZnO@C_{60}/Ag	1.15	15.1	78	13.74	[81]
	FTO/TiO_2/Al_2O_3/NiO_x/carbon/$CsPbI_2Br$	0.944	14.33	62.50	8.44	[82]
	ITO/TiO_2/$CsPbI_2Br$/P3HT/Au	1.30	13.13	70.4	12.02	[83]
	FTO/SnO_2/$CsPbI_2Br$/spiro/Ag	1.29	12.41	76.9	12.34	[84]
	FTO/NiO_x/$CsPbI_2Br$/ZnO@C_{60}/Ag	1.14	15.2	77	13.3	[85]

(Continued)

Table 5.3 (Continued)

Composition	Configuration	V_{oc} (V)	J_{sc} (mA cm^{-2})	FF (%)	PCE (%)	References
	ITO/SnO_2/$CsPbI_2Br$/PMMA/carbon	1.202	12.64	71	10.95	[86]
	ITO/SnO_2/$CsPbI_2Br$/CsBr/spiro/Au	1.271	16.72	77.18	16.37	[16]
	ITO/SnO_2/$CsPbI_2Br$/Co_3O_4/carbon	1.187	13.09	72.12	11.21	[87]
	FTO/TiO_2/$CsPbI_2Br$/HTL/Au	1.31	14.55	78.58	14.86	[88]
	ITO/ZnO/SnO_2/$CsPbI_2Br$/HTL/ MoO_3/Ag	1.28	14.7	76.4	14.45	[89]
	FTO/TiO_2/$CsPbI_2Br$/HTL/Au	1.20	14.65	76	13.36	[90]
	ITO/SnO_2/ZnO/$CsPbI_2Br$/spiro/ MoO_3/Ag	1.23	15.0	78.7	14.6	[91]
	ITO/NiO_x/$CsPbI_2Br$/c-Nb_2O_5/ PCBM/Bphen/Ag	1.06	14.13	78.4	11.74	[92]
	FTO/TiO_2/$CsPbI_2Br$/PTAA/Au	1.19	12.93	80.5	12.39	[93]
	FTO/TiO_2/$CsPbI_2Br$/FA^+ $CsPbX_3$/FA^+ QDs/PTAA/Au	1.223	14.51	79.6	14.12	[94]
	ITO/SnO_2/$CsPbI_2Br$/spiro/MoO_3/Ag	1.12	81.5	15.4	14.05	[95]

commonly used antisolvent in hybrid organic–inorganic pero-SCs, but its toxicity undoubtedly restricts the further fabrication in ambient environment. Therefore, several green antisolvents were introduced to replace CB for preparing $CsPbI_2Br$ film. Dong et al. used a green antisolvent ethyl acetate (EA) to treat the spinning film [63]. EA-treated films exhibited a larger grain size and smoother morphology than CB-treated films. As a result, a PCE of 10.0% was obtained, much higher than that of the CB-treated device (8.07%). Recently, Chen et al. applied two green antisolvents (toluene and isopropyl alcohol [IPA]) on $CsPbI_2Br$ wet film [57]. Both could extract the residual solvent dimethylsulphoxide (DMSO) molecules without washing away perovskite precursor. Excitingly, IPA treatment could fully eliminate the solvent edge effect and result in compact and large grains (Figure 5.4c). Efficiencies of 13.82% and 16.07% were achieved for toluene-treated and IPA-treated $CsPbI_2Br$ pero-SCs, respectively. Simultaneously, moisture and UV light resistance were significantly improved. There was also a solution method combined with antisolvent and multistep deposition. Dong et al. utilized green ethanol (EtOH)-treated PbI_2 films with high porosity and randomized orientation and then deposited CsBr on the porous PbI_2 layer to form a high-thickness and enhanced crystalline $CsPbI_2Br$ film [64]. Finally, when applying carbon as electrode, the $CsPbI_2Br$ pero-SC exhibited a PCE of 10.21% and excellent long-term ambient stability.

Except for antisolvent treatment, solvent composition plays an important role in crystalline process and morphology control [96]. Zhang et al. optimized solvent of *N,N*-dimethylformamide (DMF)/DMSO mixture with the volume ratio of 3 : 2 and then obtained a void-free and compact $CsPbI_2Br$ film [65]. The resultant device exhibited a PCE of 9.14% and showed great stability upon heat stress as well as continuous light soaking. Recently, similar strategy was applied and the PCE of the $CsPbI_2Br$ pero-SC was further raised to 14.30% (Figure 5.4d) [58]. Besides solvent composition, solvent temperature has a great influence on nuclei formation and crystal growth. Bai et al. carefully optimized the solvent temperature, thus suppressing too much nuclei formation and increasing the crystallization rate simultaneously when the temperature was locked at 100 °C(Figure 5.4e) [59]. As expected, a stabilized PCE of 14.81% for the $CsPbI_2Br$ pero-SC was achieved. Solvent annealing method was also proposed, which could lower the thermal annealing temperature and contribute to a high-quality perovskite film with 10.4% efficiency [66]. In addition, DMSO is expected to significantly promote the perovskite grain growth during annealing process [97]. It can coordinate with PbX_2 to form an intermediate phase PbX_2(DMSO), therefore retarding the reaction between precursors, which is beneficial for crystal growth [98]. Here, DMSO adducts was used to substitute PbX_2 precursor. They could effectively lower the activation energy to form $CsPbI_2Br$ perovskite films (Figure 5.4f) [60]. Finally, a 120 °C post thermal annealing realized a PCE of 13.54% for rigid device and a 11.73% PCE for flexible device. DMSO adducts precursors could also slow the crystallization of $CsPbI_2Br$ film, thus realizing large grain size, flat morphology, low density defects, and long carrier lifetimes [67]. The resultant device exhibited a high PCE of 14.78%.

Temperature Control Temperature is another important factor influencing the crystal quality, including solvent temperature (mentioned above), substrate temperature, and post annealing temperature. Wang et al. cast $CsPbI_2Br$ perovskite precursor solution on a hot substrate (55 °C) followed by a post annealing at 120 °C, and they obtained a compact and pinhole-free $CsPbI_2Br$ film with a PCE of 12.5% [68]. Nam et al. unveiled the relationship between the post annealing temperature and crystal formation [69]. They optimized the post annealing temperature at 280 °C and obtained a void-free and phase-stable $CsPbI_2Br$ film. A PCE of 10.7% was realized. Zhang et al. controlled the nucleation and crystallization process with the assistance of hot air flow (HAF) method [70]. Combined with DMSO solvent, a wide solution-processing window including a wide range of solvent components and solute concentrations had been realized. A maximum PCE of 12.52% was realized for the $CsPbI_2Br$ pero-SC. Another annealing strategy was flash annealing method, where a 240 °C post annealing for 10 seconds resulted in a high PCE of 13.09% [71]. In fact, besides the temperature control mentioned above, other factors like environment temperature will undoubtedly affect the nucleation process, which may be difficult to precisely control.

5.3.2.2 Ionic Incorporation

B Site Incorporation B site incorporation strategy is commonly used in $CsPbI_2Br$ pero-SCs. Partially substituting Pb^{2+} with other metal ions can effectively improve the lattice stability and film morphology (Figure 5.5a). Sun et al. found that the Pb^{2+}-reduced $CsPb_{0.9}Zn_{0.1}I_2Br$ promoted the PCE from 11.8% to 13.6% [72]. It is expected that Zn has higher chemical activity than Pb, thus coordinating

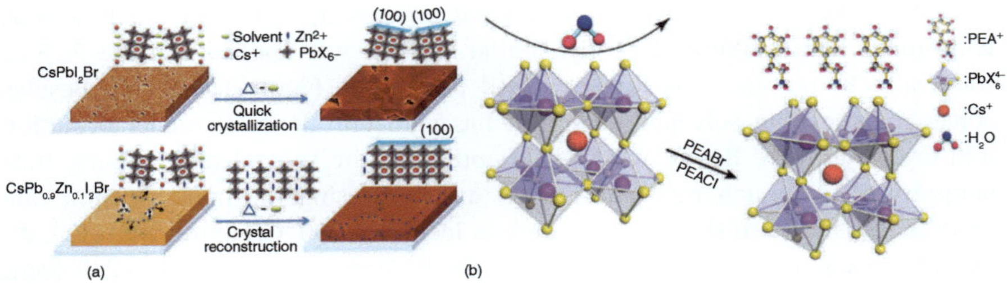

Figure 5.5 (a) Scheme of the nucleation and crystalline growth mechanism. Source: Sun et al. [72]. (b) Schematic of the PEAX-stabilized $CsPbI_2Br$ structure. Source: Wang et al. [73].

stronger with halides and resulting in enhanced growth orientation and enlarged crystal size. Moreover, the formed Cs–Zn–I/Br compound could passivate grain boundaries and facilitate the charge transport. Similarly, Eu-doped $CsPbI_2Br$ exhibited a high PCE of 13.71% and it is proved that Eu is incorporated into the perovskite lattice on the atomic level [74]. Ge^{2+}-doped $CsPbI_2Br$ showed a better effective recombination lifetime and low trap densities as well as an ultrahigh V_{oc} of 1.34 V [75]. Strontium-doped $CsPbI_2Br$ device had a Sr-enrich surface, which provided a passivation effect, therefore contributing to a higher lifetime and a high PCE of 11.2% [76]. Niobium ionic incorporation successfully reduced the charge recombination in the $CsPbI_2Br$ perovskite, leading to a V_{oc} of 1.27 V and negligible hysteresis [77]. Besides B site substitution, interstitial Mn^{2+} in $CsPbI_2Br$ lattice could suppress the nucleation process as well as a decreased growth rate. Assisted by trap density passivation effect, a high PCE of 13.47% was obtained [78].

Incorporation of Other Sites For A site incorporation, Nam et al. investigated that potassium cation incorporation (A site) lead to the contraction of PbX_6 octahedral volume, therefore improving the $CsPbI_2Br$ stability. A PCE of 10% was obtained [79]. In addition, X site incorporation is usually associated with halogen doping. Fu et al. introduced fluorine to X site of $CsPbI_2Br$ to modulate bulk-phase heterostructure and tolerance factor. An improved PCE of 10.26% was realized due to the matched energy band offsets and retarded charge recombination [80]. Besides, a co-doping strategy was proposed where In^{3+} and Cl^- partly substituted Pb^{2+} and X^- in the $CsPbI_2Br$ perovskite structure, respectively [81]. Through increasing the tolerance factor and decreasing grain size, the α-phase $CsPbI_2Br$ was effectively stabilized, which was associated with spatial symmetry with a closely packed atom arrangement. The resultant $InCl_3$:$CsPbI_2Br$ pero-SC exhibited an improved PCE of 13.74% for small area device (0.09 cm^2) and 11.4% for large area device (1 cm^2). Organic compounds were also incorporated into $CsPbI_2Br$ pero-SCs. Phenylethylammonium chlorine (PEACl) incorporation realized a co-doping effect (Figure 5.5b) [73]. PEA^+ could form on the $CsPbI_2Br$ surface in the edge-on orientation, and Cl^- would be incorporated into $CsPbI_2Br$ lattice and benefit crystal growth. A PCE of 14.05% was obtained with much improved moisture resistance. A nonvolatile additive 4-aminobenzoic acid (ABA) was introduced and they could bridge the adjacent $CsPbI_2Br$ to form a grain-interconnected film during thermal annealing process

[82], resulting in a highly crystalline perovskite film with a PCE of 8.44%. More importantly, the phase stability and humidity stability were significantly improved.

5.3.2.3 Interface Engineering

Passivation Effect Organic HTL can be utilized as passivation layer to reduce energy loss so that to get a high V_{oc}. A thin poly(3-hexylthiophene) (P3HT) layer is deposited on the$CsPbI_2Br$ film, providing an effective interface passivation [83]. Here, the sulfur (S) atoms in P3HT would bond with surface Cs atoms and passivate the interfacial positively charged under-coordinated Cs^+ ions. In addition, antisite defects of Pb were passivated by S atoms (Figure 5.6a), resulting in a low energy loss (E_{loss}) with high V_{oc}. Tian et al. applied an amino-functionalized polymer (PN4N) as cathode interlayer and a dopant-free hole-transporting polymer poly[5,5′-bis(2-butyloctyl)-(2,2′-bithiophene)-4,4′-dicarboxylate-alt-5,5′-2,2′-bithiophene] (PDCBT) as an HTL [62]. A similar passivation effect was observed

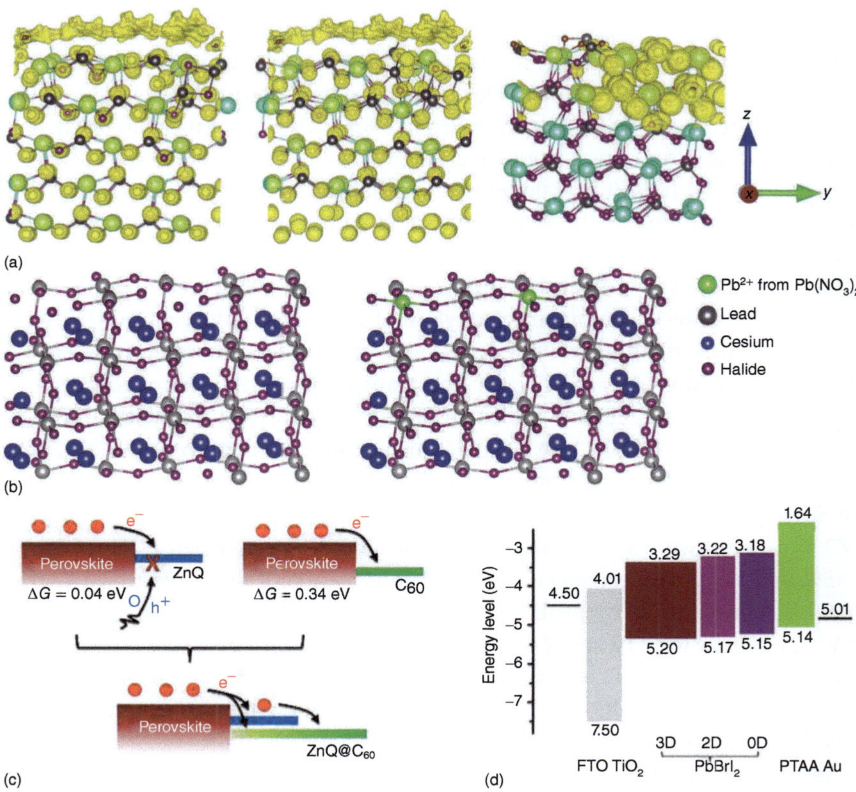

Figure 5.6 (a) Density distribution plots of $CsPbI_2Br$ 110-surface. Source: Zeng et al. [83]. (b) Schematic diagram for the defects of the perovskite surface passivated by Pb^{2+} ions. Source: Yuan et al. [84]. (c) Energy-level diagram of multi-graded $CsPbBrI_2$ PSCs. Source: Liu et al. [85]. (d) Energy-level diagram of multigraded $CsPbI_2Br$ pero-SCs. Source: Zhang et al. [93].

and the photovoltaic performance and photostability were significantly improved. A remarkable PCE of 16.2% and a V_{oc} of 1.3 V were achieved owing to the dual interfacial passivation strategy. In addition, interfacial passivation is also applied to reduce the trap density of $CsPbI_2Br$ perovskite and $Pb(NO_3)_2$ solution post-processing strategy to passivate the deep trap states (Pb vacancy (V_{Pb}) and I interstitial (I_i)) of $CsPbI_2Br$ films (Figure 5.6b) [84]. A high V_{oc} of 1.29 V was obtained with a PCE of 12.34%. Zhang et al. used a solution-processed poly(methyl methacrylate) (PMMA) layer to modify the $CsPbI_2Br$ film [86]. It is found that nonradiative recombination in $CsPbI_2Br$ film is suppressed, owing to the passivation effect of PMMA, thus resulting in the promoted charge extraction. More recently, CsBr was evaporated on the surface of $CsPbI_2Br$ film, indicating that the evaporated CsBr gathered at grain boundaries of $CsPbI_2Br$ layers and the Br-enriched perovskites were formed, which provided a passivation effect, therefore reducing the defects and nonradiative recombination loss of perovskites. A remarkable PCE of 16.37% was realized, which is the highest PCE for the $CsPbI_2Br$-based pero-SCs up to now [16].

Smooth Transferring It is expected that nonradiative recombination was largely governed by the mismatched energy levels at the interfaces between the perovskite and the hole/ETLs [99]. Recently, energy level-matched transporting layer was produced to passivate the trap density of perovskite and effectively extract the charge carriers. Inorganic oxide Co_3O_4 was applied as an HTL to fabricate high-performance $CsPbI_2Br$ pero-SC [87]. It was demonstrated that the inorganic oxide Co_3O_4 HTL could markedly promote the photo-generated carrier separation and extraction, leading to an improved photovoltaic performance with a maximum PCE of 11.21%. Besides, organic HTLs were also synthesized and utilized in the $CsPbI_2Br$ pero-SCs. For example, poly-3- methyl methacrylate (P3HT), poly-triarylamine (P-TAA), poly-fluoren-8-triarylamine (PF8-TAA), and poly-indenofluoren-8-triarylamine (PIF8-TAA) functioned as an HTL for hole transportation [88]. Through matching proper the highest occupied molecular orbital (HOMO) level with $CsPbI_2Br$ perovskite, the PIF8-TAA HTM (hole transporting material)-based mesoscopic pero-SC exhibited the highest efficiency of 14.86% and an ultrahigh V_{oc} of 1.31 V. Similarly, polytriphenylamines-based hole transport polymer derivatives (H-Z1, H-Z2 and H-Z3) were synthesized and the related devices yielded PCEs of 14.1%, 14.45%, and 11.1%, respectively [89]. Zhang and coworkers also synthesized a series of 9,9-bis(4-methoxyphenyl)-substituted fluorene with different terminal groups (H for YT1, methoxy for YT2 and methoxyphenyl-fluorene for YT3) [90]. The $CsPbI_2Br$ device based on YT3 HTL exhibited the best performance (a PCE of 13.36%) due to the lower lying HOMO level, which contributes to better hole extraction and transfer capability.

Additionally, double transporting layers are commonly used to minimize the energy-level difference therefore smooth transferring. ZnO@C_{60} bilayer was utilized as an ETL in an inverse structure solar cell, which demonstrated high carrier extraction efficiency and low leakage loss (Figure 5.6c) [85]. The resultant $CsPbI_2Br$ pero-SC exhibited a PCE of 13.3%, one of the highest PCEs in inverse structure $CsPbI_2Br$ pero-SCs. Recently, Yan et al. applied a SnO_2/ZnO bilayer

ETL for low energy loss and large V_{oc} [91]. It was found that a desirable cascade energy-level alignment between the $CsPbI_2Br$ perovskite and SnO_2/ZnO ETLs was formed due to the higher-lying conduction band minimum of ZnO. As a result, the optimized $CsPbI_2Br$ pero-SC produced a PCE of 14.6%. Another metal oxide Nb_2O_5 nanosheets was prepared and combined with [6,6]-phenyl-C61-butyric acid methyl ester ($PC_{61}BM$) as bilayer ETL, which promoted the PCE of $CsPbI_2Br$ pero-SC from 9.10% to 11.74% [92]. Zhang et al. firstly designed a 3D-2D-0D multiple graded interface based on $CsPbI_2Br$ bulk, nanosheets, and QDs via a facile solution process, which reduced the charge recombination at interface by the graded hole channel. A PCE of 12.39% was achieved by this strategy (Figure 5.6d) [93].

Some other works focused on the interface modification for enhancing the carrier extraction ability. Zhang et al. fabricated an efficient $CsPbI_2Br$ pero-SC by applying a dimension-grading heterojunction based on a quantum dots (QDs)/bulk film structure following with organic iodine salt treatment. This generated a proper band edge bending, which contributed to charge carrier collection, resulting in a PCE of 14.12% [94]. Zhou found that an interfacial layer MoO_3 between spiro and Ag electrode could suppress the carrier recombination and lead to highly efficient charge extraction, achieving a PCE of 14.05% [95].

5.3.3 $CsPbIBr_2$

$CsPbIBr_2$ is formed when substituting two-thirds of the iodine in $CsPbI_3$ with bromine, which owns a large bandgap of ~2.05 eV [100]. It can also stabilize well at room temperature and present in orange red for cubic phase. When exposed in ambient condition, $CsPbIBr_2$ also transfers to δ-phase (non-perovskite phase) in the atmosphere with higher relative humidity (RH). $CsPbIBr_2$ exhibits much better stability than $CsPbI_3$ and $CsPbI_2Br$ due to the high content of Br ions, but the resultant device PCE is relatively lower than the I-rich devices due to its narrow absorption range. When the Br content is too high (>40%), the perovskite tends to phase-segregate upon illumination [14]. Li et al. also found $CsPbIBr_2$ would segregate into I^- and Br^- rich phases under light and electron beam illumination [101]. "I-rich" phases form at grain boundaries as well as segregate as clusters inside the film. Phase segregation generates a high density of mobile ions moving along grain boundaries as ion migration "highways." Finally, these mobile ions can pile up at the perovskite/TiO_2 interface resulting in the formation of larger injection barriers, hampering electron extraction and leading to strong current density–voltage hysteresis in the $CsPbIBr_2$ pero-SCs. However, most researchers did not mention phase segregation in the $CsPbIBr_2$ pero-SCs, and they improved the device PCE from 4.7% [100] to 10.88% [17] recently (summarized in Table 5.4). Further decreasing the bandgap to 1.79 and 1.64 eV by partially incorporating Sn ions, the PCEs of devices raised to 11.33% [102] and 11.53% [103], respectively, owing to their broad light absorption range.

Researchers also found the $CsPbIBr_2$ pero-SCs could undergo thermally-driven, moisture-mediated reversible transitions between a transparent non-perovskite phase (81.7% visible transparency) and a deeply colored perovskite phase (35.4%

Table 5.4 Photovoltaic parameters of $CsPbIBr_2$-based pero-SCs.

Composition	Configuration	V_{oc} (V)	J_{sc} (mA cm^{-2})	FF (%)	PCE (%)	References
$CsPbIBr_2$	$FTO/TiO_2/CsPbIBr_2$/carbon	0.959	8.7	56	4.7	[100]
	$FTO/TiO_2/SmBr_3/CsPbIBr_2$/spiro/Au	1.17	12.75	73	10.88	[17]
	$FTO/TiO_2/CsPb_{0.9}Sn_{0.1}IBr_2$/carbon	1.26	14.30	63	11.33	[102]
	$ITO/SnO_2/C_{60}/CsPb_xSn_{1-x}IBr_2$/spiro/Au	1.21	12.57	75.8	11.53	[103]
	$FTO/NiO_x/CsPbIBr_2$/ZnO/AZO/ITO	1.00	7.92	59.2	4.69	[104]
	$FTO/TiO_2/CsPbIBr_2$/carbon	1.245	10.66	69	9.16	[105]
	$FTO/TiO_2/CsPbIBr_2$/carbon	1.142	9.11	63	6.55	[106]
	$FTO/TiO_2/CsPb_{0.995}Mn_{0.005}I_{1.01}Br_{1.99}$/carbon	0.99	13.15	57	7.36	[107]
	$FTO/TiO_2/CsPb(Ba)IBr_2$/spiro/Au	1.19	11.91	74	10.51	[108]
	$FTO/TiO_2/CsPbIBr_2$: PEG/spiro/Au	1.28	8.8	64.9	7.31	[109]
	$ITO/SnO_2/CsPbIBr_2$: CsXth/P3HT/Au	1.3	10.19	73.81	9.78	[110]
	$FTO/In_2S_3/CsPbIBr_2$/spiro/Au	1.09	7.76	65.94	5.59	[111]
	$FTO/NiO_x/CsPbIBr_2/MoO_x$/Au	0.85	10.56	62	5.52	[112]
	$ITO/TiO_2/CsPbIBr_2$/carbon	1.23	8.50	67	7.00	[113]
	$ITO/SnO_2/CsPbIBr_2$/YD2-o-C8/spiro/Au	1.37	12.05	61	10.13	[114]

visible transparency) (Figure 5.7a) [104]. It was suitable for smart photovoltaic windows utilizing this phase transition process. An excellent device stability over repeated phase transition cycles without color fade or performance degradation was achieved. PCEs of 4.69% and 0.15% for two phases-based semitransparent devices were obtained. We can speculate that by optimizing the ratio of I/Br in the inorganic pero-SCs, various colors-smart windows can be realized. Here, some technologies for fabricating high-performance $CsPbIBr_2$ pero-SCs will be introduced as follows.

5.3.3.1 Crystal Growth

In terms of crystal growth, Zhu et al. demonstrated a facile intermolecular exchange strategy for $CsPbIBr_2$ crystal growth, i.e. spin-coating a CsI solution on the $CsPbIBr_2$ precursor film for intermolecular exchange process (Figure 5.7b) [105]. Compared with conventional one-step solution route, this strategy produced a more compact and pinhole-free film as well as enhanced crystallinity, resulting in an optimized PCE of 9.16%. It could also maintain 90% and 97% of its initial PCEs under ambient condition (45% humidity) for over 60 days and at 85 °C for over 7 days. Zhu et al. stood by the wet film for three weeks and then annealed the film on a 100 °C hotplate [106]. This strategy successfully lowered the post thermal annealing temperature. The resultant device yielded a PCE of 6.55%, and the long-term operational stability against moisture and heat were improved.

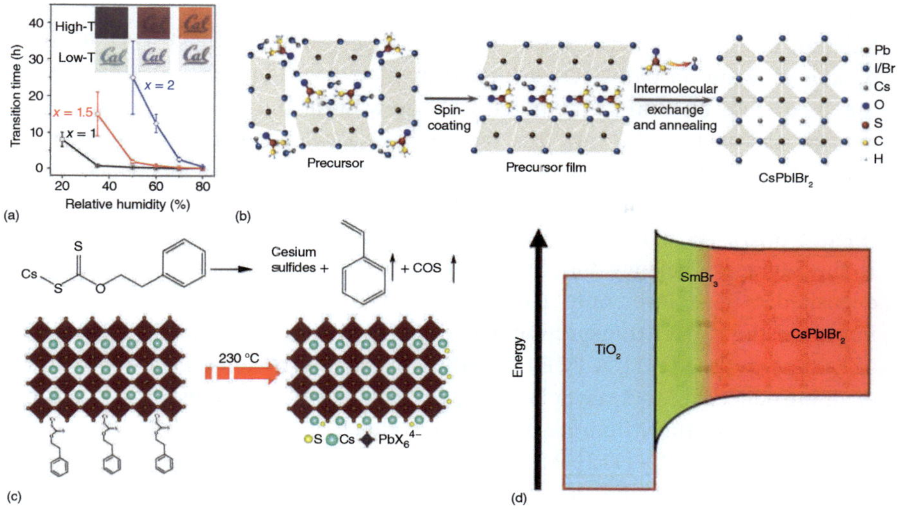

Figure 5.7 (a) Variation of the high-T to low-T phase transition time of $CsPbI_{3-x}Br_x$ films measured at different humidity conditions with RH = 20–80%. Source: Lin et al. [104]. (b) Illustration of intermolecular exchange strategy. Source: Zhu et al. [105]. (c) Illustration of CsXth decomposition and sulfur doping in perovskite. Source: Wang et al. [110]. (d) Schematic of the SIM-induced gradient band structure. Source: Subhani et al. [17].

5.3.3.2 Ionic Incorporation

The strategy of ionic incorporation was also applied to the $CsPbIBr_2$ pero-SCs. Liang et al. added an amount of Mn^{2+} into $CsPbIBr_2$ film [107]. Vertical branches could be observed at the surface of $CsPbIBr_2$ film, which would protrude into the upper carbon electron, increasing the contact area for smooth hole transferring. A high PCE of 7.36% was obtained based on $CsPb_{0.995}Mn_{0.005}I_{1.01}Br_{1.99}$ solar cells. The encapsulated device showed only 8% PCE decrease after >300 hours storage at ambient condition. Recently, Subhani et al. introduced barium ion (Ba^{2+}) into $CsPbIBr_2$, resulting in increased grain size and enhanced crystallinity of the film [108]. Therefore, trap density was reduced, and non-combination was effectively suppressed, thus leading to a remarkable PCE of 10.51% as well as improved stability. Besides metal ion substitution, organic compounds were also added into $CsPbIBr_2$ precursor for performance improvement. Lu et al. introduced polyethylene glycol (PEG) into $CsPbI_2Br$ film [109]. PEG, acted as a supporting matrix, restrained aggregation of perovskite crystals and slowed the rate of crystal growth, thus improving the film morphology and coverage. The best-performance $CsPbIBr_2$ pero-SC exhibited a PCE of 7.31% with a high V_{oc} of 1.28 V, owing to the defect passivation. Wang et al. applied xanthate (CsXth) to perovskite to realize a sulfur-doped $CsPbIBr_2$ for superior phase stability [110]. The additive CsXth could be decomposed into cesium sulfides during annealing process which could strongly coordinate with neighboring grains closer (Figure 5.7c). The resultant PCE and V_{oc} reached to 9.78% and 1.3 V, respectively, and the unencapsulated device exhibited almost no PCE decay in 10 hours under ambient air with 65% RH.

5.3.3.3 Interface Engineering

Subhani et al. employed a $SmBr_3$ interface layer on a TiO_2 ETL [17]. The perovskite growing on bare TiO_2 exhibited many exposed cracks and pinholes, but the perovskite on $SmBr_3$ modified TiO_2 grew well and covered completely. It may be ascribed to the interface layer ($SmBr_3$), which acted as a nucleus for better crystallinity. Encouragingly, a gradient energy band was created at the interface for electron smooth transferring and an outstanding hole blocking effect (Figure 5.7d). As a result, a remarkable PCE of 10.88% was obtained, which is the highest efficiency for the $CsPbIBr_2$ based pero-SCs to date. Yang et al. deposited a In_2S_3 layer onto TiO_2 ETL to form a gradient energy band, which improved the carrier transporting and suppressed recombination at the interface of ETL/perovskite [111]. A PCE of 5.59% with reduced hysteresis was realized. Liu et al. introduced a MoO_3 layer between $CsPbIBr_2$ layer and Au electrode to form a new device structure (fluorine-doped tin-oxide (FTO)/NiO_x/$CsPbIBr_2$/MoO_x/Au) of inorganic pero-SCs [112]. MoO_x could independently serve as an outstanding cathode buffer layer, and the lower work-function MoO_x leaded to decreases of the Schottky barrier, the contact resistance and the interface trap-state density. Above all, the interface layer can effectively contribute to the charge transferring and reduce recombination at interface. Except for energy-level matching, interface layer also plays an important role in passivating adjacent layers to reduce trap density. Guo et al. employed a $SnCl_2$ solution onto SnO_2 ETL, providing a surface passivation effect and suppressing the non-recombination process [113]. By this strategy, the V_{oc} of $CsPbIBr_2$ device dramatically improved from 1.07 to 1.31 V. Recently, Yang et al. introduced a dye molecule 5, 15-bis (2, 6-dioctoxyphenyl)-10-(bis(4-hexylphenyl)-amino-20-4-carboxyphenyl ethynyl) porphyrinato]zinc(II) (YD2-o-C8) between $CsPbIBr_2$ layer and HTL [114]. The results demonstrated that YD2-o-C8 had a bifunctional effect, not only acting as a photoactive layer for $CsPbIBr_2$ broadening the absorption spectrum but also reducing the E_{loss} by interface passivation effect. As a result, the J_{sc} and V_{oc} dramatically improved from 9.37 mA cm^{-2} and 1.17 V to 12.05 mA cm^{-2} and 1.37 V, respectively. By employing this dye molecule, the PCE of the pero-SC was markedly improved from 7.02% to 10.13%.

5.3.4 $CsPbBr_3$

$CsPbBr_3$ perovskite are featured with outstanding long-term stability upon high humidity, heat [8], and illumination (even under UV illumination) [9] and it possesses a remarkably high carrier mobility of ~4500 cm^2 V^{-1} s^{-1}. $CsPbBr_3$ perovskite exhibits three different phases: orthorhombic, tetragonal, and cubic phases. Orthorhombic δ-phase $CsPbBr_3$ tends to convert at 88 °C to tetragonal β-phase and at 130 °C to cubic α-phase. A large direct bandgap of about 2.3 eV seriously restricts its light response range and PCE of the corresponding pero-SCs. However, $CsPbBr_3$ pero-SC is suitable for a top cell in tandem solar cells. Chen et al. used two-sources evaporation method to fabricate a high-quality $CsPbBr_3$ film, which could almost fully absorb, utilize, and filter short wavelength light (<550 nm), especially UV light (Figure 5.8a) [9]. The transmittance above 550 nm was even up to 80%. The

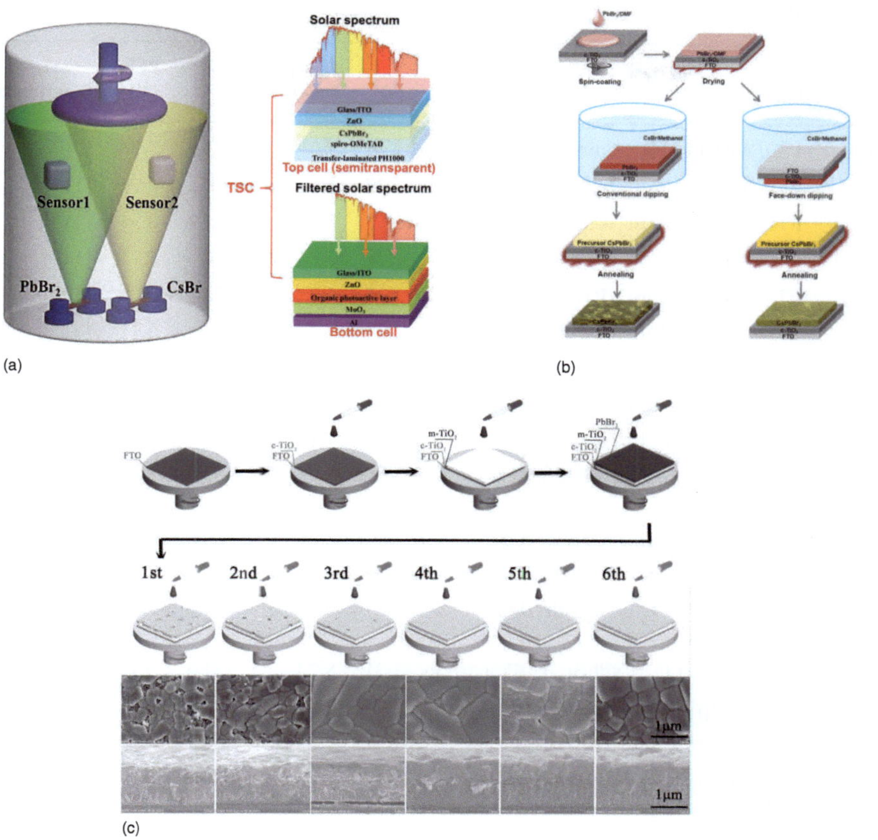

Figure 5.8 (a) Left: schematic diagram of the vacuum chamber for the DSVE method. Right: device structure of four-terminal tandem solar cells. Source: Chen et al. [9]. (b) Schematic process for the preparation of $CsPbBr_3$ films obtained by face-up dipping process and the face-down dipping process. Source: Teng et al. [115]. (c) The illumination of Multi-step solution-processed deposition of $CsPbBr_3$ perovskite film and corresponding top-view and cross four-sectional SEM images. Source: Duan et al. [116].

opaque $CsPbBr_3$ pero-SC yielded a PCE of 7.73% with a high V_{oc} of 1.44 V. After replacing Au electrode with semitransparent PH1000 electrode, a PCE of 6% was realized (summarized in Table 5.5) for the semitransparent pero-SCs. In addition, when combined with bottom organic solar cell, a four terminal tandem solar cell exhibited a promising PCE of 14.03%. Encouragingly, the UV light stability of the organic solar cell and the whole tandem solar cell was significantly improved.

5.3.4.1 Fabrication Method

The fabrication methods of $CsPbBr_3$ perovskite film are limited by the poor solubility of CsBr in commonly used polar solvents [131], which verifies the infeasibility of preparing $CsPbBr_3$ by one-step solution method. Here, vapor evaporation method, dipping method and multistep spin-coating method are introduced as follows.

Table 5.5 Photovoltaic parameters of $CsPbBr_3$-based pero-SCs.

Composition	Configuration	V_{oc} (V)	J_{sc} (mA cm^{-2})	FF (%)	PCE (%)	References
$CsPbBr_3$	ITO/ZnO/$CsPbBr_3$/spiro/Au	1.44	7.01	77.11	7.78	[9]
	FTO/TiO_2/$CsPbBr_3$/spiro/Au	1.27	6.97	78.5	6.95	[117]
	FTO/TiO_2/$CsPbBr_3$-$CsPb_2Br_5$/spiro/Ag	1.296	8.48	75.9	8.34	[118]
	FTO/TiO_2/$CsPbBr_3$/spiro/Au	1.28	6.24	74	5.95	[119]
	FTO/TiO_2/$CsPbBr_3$/PTAA/Au	1.25	6.7	73	6.2	[120]
	FTO/TiO_2/$CsPbBr_3$/carbon	1.24	7.4	73	6.7	[8]
	FTO/TiO_2/$CsPbBr_3$/carbon	1.34	6.46	68.04	5.86	[115]
	FTO/TiO_2/$CsPbBr_3$/carbon	1.458	8.12	82.1	9.72	[116]
	FTO/TiO_2/$CsPbBr_3$/carbon	1.522	7.35	84.3	9.42	[121]
	FTO/TiO_2/$Cs_{1-x}R_xPbBr_3$/carbon	1.552	7.73	82.2	9.86	[122]
	FTO/TiO_2/$CsPb_{0.97}Sm_{0.03}Br_3$/carbon	1.594	7.48	85.1	10.14	[123]
	FTO/TiO_2/$CsPbBr_3$/P3HT: ZnPc/carbon	1.578	7.652	83.06	10.03	[124]
	FTO/TiO_2/$CsPb_{0.97}Tb_{0.03}Br_3$/SnS: ZnS/NiO_x/ carbon	1.57	8.21	79.6	10.26	[125]
	FTO/TiO_2/SnO_2/$CsPbBr_3$/CuPc/carbon	1.310	8.24	81.4	8.79	[126]
	FTO/TiO_2/$CsPbBr_3$/HTM/Ag	1.02	8.47	71.6	6.21	[127]
	FTO/GQDs/$CsPbBr_3$/PQDs/carbon	1.21	5.08	66.7	4.10	[128]
	FTO/TiO_2/CQDs/$CsPbBr_3$/RPQDs/ carbon	1.470	7.33	76	8.2	[129]
	FTO/TiO_2/$CsPbBr_3$/CuPc/carbon	1.26	6.62	74.4	6.21	[130]

Vapor Evaporation Chen et al. firstly investigated the importance of stoichiometric precursors and low evaporation rate to achieve high-quality $CsPbBr_3$ crystal [9]. After that, following the rule of stoichiometric, Lei et al. studied that the optimal temperatures of substrate (300 °C) and post thermal annealing (500 °C) would contribute to a high-quality $CsPbBr_3$ film by using co-evaporation method. Stabilized PCEs of 6.95% for a 0.09 cm^2 device and 5.37% for a 1 cm^2 device were obtained [117]. In addition to stoichiometric precursor, Li et al. found that excessive $PbBr_2$ could form a perovskite derivative phase $CsPb_2Br_5$, which lay at the interface of $CsPbBr_3$ crystals, effectively passivating grain defects and reducing the carrier recombination [118]. The resultant device yielded a PCE of 8.34% and improved air stability.

Dipping Method Dipping method was firstly proposed by Kulbak et al. They firstly produced a $PbBr_2$ layer by spin-coating method followed by a thermal annealing process. Then they put the samples face-up into a hot CsBr (methanol) solution for a complete reaction. After several minutes, the samples were washed by IPA, dried by nitrogen and annealed on a hotplate [119]. This method could produce high-performance $CsPbBr_3$ pero-SC with a PCE of 5.95%. After that, they further improved the PCE to 6.2% [120]. Liang et al. increased the PCE to 6.7% [8] using carbon electrode. Based on above studies, Teng et al. found that during face-up dipping process, $PbBr_2$ would react rapidly with CsBr, but the formed $CsPbBr_3$ precursor film would be easily decomposed before complete reaction with $PbBr_2$ [115]. So they inversed the sample and made it face-down to grow, which could effectively inhibit the decomposition of $CsPbBr_3$ precursor films by liquid-space restriction (Figure 5.8b). This dipping method solved the problem of precursor solubility and pushed the PCE increasing of $CsPbBr_3$ pero-SCs. However, the dipping and post-treatment processes are uncontrollable, which limit the further application of the $CsPbBr_3$ pero-SCs.

Multistep Spin-coating Method Duan et al. invented a novel method to fabricate $CsPbBr_3$ film, i.e. multistep spin-coating method [116]. They also firstly deposited a $PbBr_2$ layer and then spin-coated CsBr solution on the formed $PbBr_2$ layer, followed by a thermal annealing process. After repeating the process for four cycles, a high-purity $CsPbBr_3$ perovskite film was formed with a large grain size (Figure 5.8c). After interface modification with GQDs, a remarkable PCE of 9.72% was achieved, which was significantly improved in comparison with the efficiencies reported previously. In addition, the $CsPbBr_3$ pero-SC exhibited outstanding thermal stability (at 80 °C) and moisture stability (90% humidity). Recently, they fully utilized this technology and produced a series of high-performance $CsPbBr_3$ pero-SCs [121–125]. Liu et al. also ameliorated the coating process. They combined the dipping method and multistep spin-coating method, thus obtaining a crystallinity enhanced film [126]. The PCE of the planar device increased to 8.79% with the assistance of double ETLs.

5.3.4.2 Ionic Incorporation

Li et al. incorporated various alkali metal cations into $CsPbBr_3$ crystal lattice to partially substitute Cs ion [122]. This A site doping strategy could suppress nonradiative losses and radiative recombination. The device based on $Cs_{0.91}Rb_{0.09}PbBr_3$ delivered a PCE of 9.86% and it could remain 97% of its initial PCE when suffering persistent attack under 80% RH under ambient condition for over 700 hours. Except for A site incorporation, B site incorporation also plays an important role in fabricating high-performance $CsPbBr_3$ pero-SCs. The lattice doping of lanthanide ions into $CsPbBr_3$ was presented for enlarged grain size and prolonged carrier lifetimes [123]. As a result, the Ln^{3+}-doped device yielded the highest PCE of 10.14% and an ultrahigh V_{oc} of 1.594 V. Halogen atom was also introduced into $CsPbBr_3$ for X site incorporation. Cl incorporation could effectively enhance carrier lifetime, diffusion length, and extraction rates and suppress nonradiative recombination [127].

5.3.4.3 Interface Engineering

QDs are usually applied as interface layer for smooth transferring and band energy alignment. A device with a structure of FTO/$CsPbBr_3$/Carbon only delivered a PCE of 2.35%, due to serious recombination at interface and poor charge carrier extraction ability. Upon inserting QD layers at interfaces, an increased PCE of 4.1% was achieved with enhanced charge extraction as well as the reduced recombination reactions [128]. Similar modification strategies were also reported [116, 129]. The PCEs in these reports were mostly high of 8.2% [129] and 9.72% [116].

Except for QDs interface modification, Liu et al. used Cu-phthalocyanine (CuPc) as an HTL to fabricate $CsPbBr_3$ pero-SC, proving a simple and effective route to facilitate charge transfer and suppress charge recombination [130]. A PCE of 6.21% was obtained with an outstanding durability and a promising thermal stability. Similarly, poly(3-hexylthiophene)/zinc phthalocyanine (P3HT/ZnPc) composite was employed as an HTL with tunable energy level [124]. Through optimizing the mass ratios of P3HT and ZnPc, the energy level of composite could match well with adjacent layers, thus promoting charge separation and reducing charge recombination at $CsPbBr_3$/carbon interface. A remarkable PCE of 10.03% and an excellent humidity stability were realized. Yuan et al. introduced an inorganic SnS: ZnS photoactive layer as interface to broaden spectral absorption from 550 to 700 nm [125]. The interface layer could reduce perovskite surface defects and improve charge extraction simultaneously, yielding a high PCE of 10.26%.

5.4 Tin-Based Inorganic Pero-SCs

Although Pb-based all-inorganic pero-SCs have achieved a rapid development right now, their intrinsic toxicity of Pb ion will inevitably pollute the environment and bring potential hazards [132]. Therefore, some Pb-free inorganic perovskite are also developed in recent years. Among them, Sn-based perovskite has a great potential for photovoltaic application.

5.4.1 $CsSnI_3$

$CsSnI_3$ is an unusual perovskite, which has an ideal bandgap of 1.4 eV with a higher theoretical short-circuit photocurrent (J_{sc}) density of 34.3 mA cm^{-2} [5], a low exciton binding energy of 18 meV [133] and a high hole mobility (~585 cm^2 V^{-1} s^{-1}) with high conductivity (~200 S cm^{-1}) [134, 135]. It exhibits four polymorphs, in which two polymorphs can exist independently at room temperature: one is the one-dimensional double-chain yellow phase (Y-$CsSnI_3$) and the other is a three-dimensional black phase (B-γ-$CsSnI_3$). When heated to 500–708 K, the Y-$CsSnI_3$ tends to transform to a black cubic phase (B-α-$CsSnI_3$). When cooled, the cubic phase B-α-$CsSnI_3$ transforms to a tetragonal phase (B-β-$CsSnI_3$) at 380 K and orthorhombic phase (B-γ-$CsSnI_3$) at 300 K, respectively (Figure 5.9a). In addition, B-γ-$CsSnI_3$ will easily undergoes a phase transition to Y-$CsSnI_3$ upon air exposure, which is not suitable for light absorption [134]. It is proved that black $CsSnI_3$ is a p-type metal under a Sn-poor growth condition, which will lead to a high concentration of acceptor defects, such as Sn or Cs vacancies. These defects become electron-hole recombination centers with high energy. In contrast, when Sn

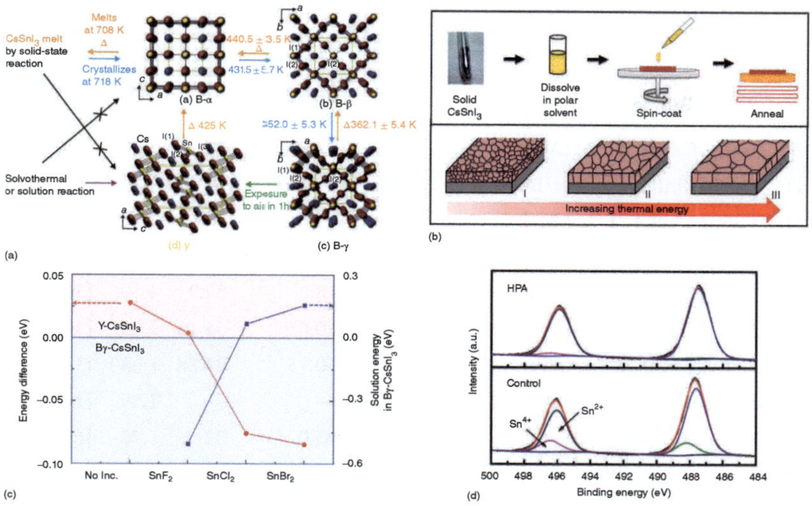

Figure 5.9 (a) Illustration of phase change details of four $CsSnI_3$ polymorphs and their crystal structures. Source: Chung et al. [134]. (b) Top: Schematic representation of the one-step solution fabrication of B-γ-$CsSnI_3$ thin films. Bottom: Schematic representation of the grain-coarsening driven morphology evolution of as-deposited B-γ-$CsSnI_3$ thin film with increasing heat-treatment temperature. Source: Wang et al. [136]. (c) Total energy difference between Y-$CsSnI_3$ and B-γ-$CsSnI_3$ per formula unit (left *y*-axis) and the solution energies (right *y*-axis) with regard to SnX_2 inclusion in the supercell. Source: Heo et al. [137]. (d) High-resolution XPS spectra (Sn 3d) of the $CsSnIBr_2$ thin films. Source: Li et al. [138].

becomes rich, the acceptor defects will be reduced, therefore, the p-type conductivity of perovskites may drop to intrinsic and even weakly n-type, which make $CsSnI_3$ more suitable for light absorption. Meanwhile, the dropped p-type conductivity may also cause deep-level donor defect SnI, which is more like to trigger electron trapping and nonradiative electron-hole recombination [139]. All these issues will undoubtedly result in a low PCE of $CsSnI_3$ pero-SCs. Considering the above facts, black $CsSnI_3$ can be used as both HTL and light absorption layer in pero-SCs. Among them, B-γ-$CsSnI_3$ is the most studied and applied as light absorption layer for photovoltaic devices. Various methods are produced for improving the performance of B-γ-$CsSnI_3$ pero-SCs.

5.4.1.1 Fabrication Methods

In 2012, Chen et al. prepared a $CsSnI_3$ thin film by alternately depositing layers of $SnCl_2$ and CsI followed by a thermal annealing in ambient condition. The resultant device only yielded a PCE of 0.9% (summarized in Table 5.6) [6]. After that, Wang et al. reported a solution method to fabricate B-γ-$CsSnI_3$ films. It was found that coarsening of the grains became larger as the thermal annealing temperatures were raised (Figure 5.9b). An optimal device architecture was applied, where NiO_x acted as an HTL, providing energetically favorable band alignments. As a result, a PCE up to 3.31% was obtained without using any additives [136]. Zhang et al. prepared a p-type B-γ-$CsSnI_3$ thin film using a simple solution-processed method, and

Table 5.6 Photovoltaic parameters of Sn-based and other inorganic pero-SCs.

Composition	Configuration	V_{oc} (V)	J_{sc} (mA cm^{-2})	FF (%)	PCE (%)	References
Sn-based	FTO/TiO_2/$CsSnI_3$/HTM/Au	0.24	22.70	37	2.02	[18]
	ITO/CuI/$CsSnI_3$/fullerene/BCP/Al	0.55	—	—	1.5	[140]
	ITO/$CsSnI_3$: $SnCl_2$/PC_{61}BM/BCP/Al	—	—	—	3.56	[141]
	FTO/TiO_2/$CsSnI_3$/PTAA/Au	0.381	25.71	49.05	4.81	[142]
	FTO/TiO_2/$CsSnI_3$/PTAA/Au	0.44	18.5	52.9	4.30	[137]
	FTO/TiO_2/$CsSnI_3$/PTAA/Au	0.338	20.63	54.18	3.83	[143]
	FTO/TiO_2/$CsSnI_3$/PTAA/Au	0.17	30.75	34.88	1.83	[144]
	ITO/CuSCN/$CsSnI_3$/PC_{61}BM/BCP/Al	—	—	—	2.59	[145]
	ITO/$CsSnI_3$/Au/Ti	0.42	4.80	22	0.88	[6]
	ITO/NiO_x/$CsSnI_3$/PC_{61}BM/Al	0.52	10.21	62.5	3.31	[136]
	ITO/$CsSnI_3$/Cs_2SnI_6/Au	0.63	6.1	28.4	1.1	[146]
	FTO/TiO_2/$CsSnI_3$/spiro/Au	0.27	17.26	47.98	2.23	[147]
	FTO/TiO_2/$CsSnBr_3$/spiro/Au	0.41	3.99	58	0.95	[148]
	FTO/TiO_2/Al_2O_3/$CsSnIBr_2$/carbon	0.31	17.4	56	3.2	[138]
Ge-based	FTO/TiO_2/$CsGeI_3$/spiro/Au	0.074	5.7	27	0.11	[149]
Sb-based	FTO/TiO_2/$Cs_3Sb_2I_9$: HCl/Au	0.61	3.55	55.85	1.21	[150]
	ITO/PEDOT: PSS/PC_{71}BM/C_{60}/BCP/Al	0.72	5.31	38.97	1.5	[151]
Bi-based	FTO/TiO_2/$AgBi_2I_7$/P_3HT/Au	0.56	3.30	67.41	1.22	[152]
	FTO/TiO_2/Ag_3BiI_6/PTAA/Au	0.63	10.7	64	4.3	[153]
	FTO/TiO_2/Ag_3BiBr_6/P3HT/Ag	0.61	0.12	44.83	0.033	[154]
Double perovskite	FTO/TiO_2/$Cs_2AgBiBr_6$/spiro/Au	0.98	3.93	63	2.43	[155]
	ITO/SnO_2/$Cs_2AgBiBr_6$/P3HT/Au	1.04	1.78	78	1.44	[156]
	ITO/TiO_2/$Cs_2AgBiBr_6$/spiro/Au	1.06	1.55	74	1.22	[157]
	FTO/TiO_2/$Cs_2AgBiBr_6$/spiro/MoO_3/Ag	1.01	3.82	65	2.51	[158]

Source: Based on Chen et al. [6].

the film thickness was dependent on the concentration of precursor solution [146]. What's more, they produced a Cs_2SnI_6 upper layer on B-γ-$CsSnI_3$ film, which not only served as an n-type material to form a heterojunction but also could effectively prevent the degradation of B-γ-$CsSnI_3$ perovskite in air. Finally, a PCE of 1.1% was achieved and the device could maintain 90% of its initial PCE after being exposed in air for 20 hours. Recently, Zhu et al. produced a uniform, dense and pinhole-free $CsSnI_3$ film via an evaporation-assisted solution method [147]. They firstly deposited a SnI_2/SnF_2 layer by spin-coating method, then a CsI layer was evaporated onto the film surface, followed with a thermal annealing process. The resultant $CsSnI_3$ device demonstrated a PCE of 2.23%.

5.4.1.2 Additive Engineering

Though being regarded as a semiconductor, Sn-based perovskites often exhibit metallic behavior because of a strong "self-doping" effect. A p-type metal will form due to the easy oxidation of Sn^{2+} to Sn^{4+} and the low formation energy of Sn vacancies [134, 159], which will induce near short-circuit diode condition and seriously hamper the device performance [18]. Kumar et al. firstly introduced SnF_2 into $CsSnI_3$ perovskite films to control the metallic conductivity [18]. They found that the optimal SnF_2 played a critical role in the reduction of Sn vacancies, thereby rendering the $CsSnI_3$ less conducting in nature, yielding a high current density of 22.7 mA cm^{-2} and a PCE of 2.02%. SnF_2 was also proved to significantly improve the chemical stability of B-γ-$CsSnI_3$ [160]. In addition to the SnF_2 additive, Marshall et al. reported an effective strategy for the PCE increasement and air stability of $CsSnI_3$ device [140]. They fabricated the $CsSnI_3$ film using excessive SnI_2 precursor. Based on a CuI/$CsSnI_3$/fullerene planar layer architecture, a higher V_{oc} of 0.55 V was obtained owing to alignment between the perovskite conduction band edge and the lowest occupied molecular orbital (LUMO) in the fullerene ETL. Later, they also incorporated $SnCl_2$ into $CsSnI_3$ films and prepared perovskite photovoltaic devices without an HTL [141]. A higher PCE of 3.56% was realized. It was shown that the improved performance and tolerance to pinholes in the perovskite film stemmed from n-doping of the fullerene electron-transport layer by $SnCl_2$. Song et al. found that excess SnI_2 would disperse uniformly into the perovskite films and functioned as a compensator as well as a suppressor of Sn^{2+} vacancies, therefore effectively reducing the p-type conductivity [142]. When optimizing the ratio of CsI/SnI_2 to 0.4, a higher PCE of 4.81% was successfully obtained, which was one of the highest PCEs among the Pb-free pero-SCs. Recently, Heo et al. systematically investigated the roles of SnX_2 (X = F, Cl, Br) additives in Sn-based halide perovskites and the resultant photovoltaic performance [137]. Following the first-principles-based theoretical calculations, they found that these additives could effectively passivate the perovskite surface and stabilize the black phase, thereby promoting the stability of $CsSnI_3$, instead of annihilating the defects (Figure 5.9c). The results indicated that $SnBr_2$ was the best choice for preventing the phase transformation of black $CsSnI_3$ perovskite to the yellow phase. A promising PCE of 4.3% was obtained with $SnBr_2$ additive. Except the SnX_2 additives, piperazine was also introduced into precursor solution to suppress the undesirable p-doping of $CsSnI_3$ films [143]. The conductivity of $CsSnI_3$ films was significantly reduced and the film coverage was improved. With the assistance of SnF_2 incorporation, a best-performing $CsSnI_3$ pero-SC yielded a PCE of 3.83%. Song et al. incorporated hydrazine into $CsSnI_3$ perovskite film, which could reduce vapor atmosphere during the perovskite film preparation process, thereby suppressing oxidized Sn^{4+} impurities and lowering levels of defect states. PCEs of 1.83% and 3.04% were achieved for the $CsSnI_3$-based and $CsSnBr_3$-based devices, respectively [144].

5.4.1.3 Substrate Control

Marshall et al. elucidated the role of the different hole-extracting electrodes on the stability and efficiency of B-γ-$CsSnI_3$ pero-SCs [161]. They found that $CsSnI_3$ film stability depended strongly on the film quality, such as the density of pinholes and grain boundaries, not on the preferred $CsSnI_3$ crystallite orientation. B-γ-$CsSnI_3$ devices using Au or ITO as hole-extracting electrodes with no HTLs were more stable than that using the poly(3,4-ethylenedioxythiophene) (PEDOT) HTL, which indicated that this simplification in device structure could be extended to other stable hole-extracting electrodes. In addition, solution-processed CuI showed unsuitability as an HTL for an inverted $CsSnI_3$ pero-SC, because it was almost completely displaced by the $CsSnI_3$ precursor solution. Wijesekara et al. proved similar situation would also appear when applying CuSCN as an HTL, though the up-layer $CsSnI_3$ exhibited a high uniformity and coverage morphology [145]. As a result, a PCE of 2.59% was achieved.

5.4.2 $CsSnI_xBr_{3-x}$

By chemical doping of $CsSnI_3$ with Br, a mixed perovskite $CsSnI_xBr_{3-x}$ was formed, and the V_{oc} was improved even without the addition of SnF_2 [148]. It is attributed to the decrease of Sn vacancies, which is reflected by the lower charge carrier densities of 10^{15} cm^{-3} and a high resistance to charge recombination. When adding SnF_2 as an additive, the device based on $CsSnBr_3$ exhibited a higher V_{oc} of 0.41 V, far greater than that of $CsSnI_3$ (V_{oc} of 0.201 V). SnF_2 was also investigated for the long-term stability of the $CsSnBr_3$ pero-SCs [162]. The operational stability of the $CsSnBr_3$ pero-SCs with SnF_2 in an inert atmosphere was tested. The device kept nearly stable upon light illumination and could maintained 96% of its initial photocurrent over five hours. X-ray photoelectron spectroscopy (XPS) spectrum also indicated the important role of SnF_2 to the stability of $CsSnBr_3$, which is possibly due to reduction of the background carrier density by neutralizing traps, or possibly *via* filling of cation vacancies. Hypophosphorous acid (HPA) was also applied to the $CsSnIBr_2$ perovskite, which could speed up the nucleation process while inhibiting the formation of Sn^{4+} during the formation process of $CsSnIBr_2$ thin films (Figure 5.9d) [138]. The $CsSnIBr_2$ device presented a PCE of 3.2%, and the encapsulated devices exhibited negligible PCE loss in 77 days and remarkably stable power output within nine hours at high temperatures up to 200 °C. Moghe et al. produced an all vapor-deposited lead-free doped $CsSnBr_3$ planar pero-SC. The unencapsulated $CsSnBr_3$ solar cells in ambient condition showed no change over the span of hours [163]. While other $CsSnX_3$ (e.g. X = Cl, I) were extremely air sensitive and decomposed within seconds.

5.5 Other Inorganic Pero-SCs

Except for Sn-based perovskite, other Pb-free inorganic perovskites including Ge-, Sb-, and Bi-based perovskites as well as double perovskite also exhibit some prospect for fabricating solar cells.

5.5.1 Ge-Based Inorganic Pero-SCs

Ge is revealed as a candidate element to replace Pb in halide perovskite compounds, which is suitable for light harvesting. Krishnamoorthy et al. had synthesized $CsGeI_3$, which could maintain stability at 150 °C [149]. The $CsGeI_3$-based device could only display a photocurrent of about 6 mA cm^{-2}, because the poor film forming abilities and oxidizing tendencies limited the device performance. According to defect calculations, Ming et al. showed that $CsGeI_3$ was a p-type semiconductor and its hole density could be modified by varying the chemical potentials of the constituent elements [164]. Long-range Coulomb attraction can be reduced by strong screening, but the strong Ge–Ge covalent bonding in $CsGeI_3$ led the iodine vacancy to a deep electron trap, which limited electron transporting efficiency in p-type $CsGeI_3$.

5.5.2 Sb-Based Inorganic Pero-SCs

Sb is also a possible replacement of Pb in the inorganic pero-SCs with the structure of $A_3B_2X_9$. $Cs_3Sb_2I_9$ has two known structural modifications, the 0-D dimer form and the 2-D layered form. However, the former, which has a large indirect bandgap of 2.4 eV, is not so suitable for application in pero-SCs as an active layer, compared to the latter that exhibits a nearly direct 2.06 eV bandgap. Some fabrication methods were proposed to prepare $Cs_3Sb_2I_9$ perovskite films. Saparov et al. employed two approaches to fabricate large grain and continuous 2-D $Cs_3Sb_2I_9$ thin films [165]. For one approach, highly oriented $Cs_3Sb_2I_9$ thin films were fabricated through co-evaporation of CsI and SbI_3. For another approach, a CsI film was first deposited through evaporation and, subsequently, annealed in SbI_3 vapor, which could yield randomly oriented $Cs_3Sb_2I_9$ thin films. The resultant thin films exhibited enhanced stability in ambient condition, an optical bandgap value of 2.05 eV and a weakly p-type character, suggesting that $Cs_3Sb_2I_9$ exhibited promising optical properties, comparable to that of $CH_3NH_3PbI_3$. Harikesh et al. used a simple solution processable methodology to develop layered Sb-based perovskite with the structure of $A_3Sb_2I_9$ (A = Cs and Rb) (Figure 5.10a) [166]. They demonstrated that a smaller cation like Sb in A position could effectively stabilize the structure in its layered form. Umar et al. developed a HCl-assisted solution method to fabricate layered phase $Cs_3Sb_2I_9$ perovskite (Figure 5.10b) [150]. They utilized HCl to pronouncedly decrease the layer phase reaction temperature and obtained a compact and uniform antimony iodide perovskite (AIP) film with the assistance of antisolvent. As a result, the device yielded a PCE value of 1.21%, while the dimer one only achieved a PCE of 0.43%. The AIP-based devices also demonstrated superior ambient storage and thermal stability compared with the dimer one. Singh et al. also prepared a layered polymorph of $Cs_3Sb_2I_9$ in both layer and dimer form through solution-processing method (Figure 5.10c) [151]. They obtained PCEs of 1.5% and 0.89%, respectively.

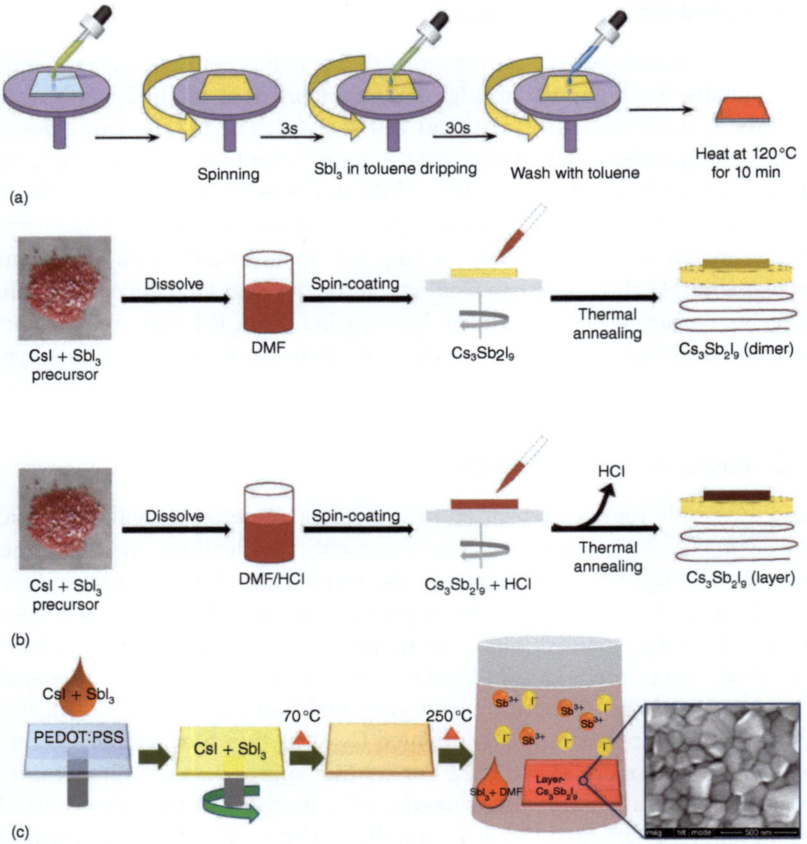

Figure 5.10 (a) Procedure adopted to form $Rb_3Sb_2I_9$ films. Source: Harikesh et al. [166]. (b) Schematic representation of the antisolvent solution method of w/o HCl incorporated $Rb_3Sb_2I_9$-dimer film and w/HCl-based $Rb_3Sb_2I_9$-layer film. Source: Umar et al. [150]. (c) Schematic representation of the sequential introduction of precursor for the preparation of the stoichiometric layer-form $Rb_3Sb_2I_9$, accompanied by an SEM image of the layered perovskite morphology. Source: Singh et al. [151].

5.5.3 Bi-Based Inorganic Pero-SCs

5.5.3.1 $A_3B_2I_9$ Structure

Bi-based perovskite is another candidate for construction of lead-free and air-stable perovskites solar cells. Among these categories of perovskites, the ternary metal halides ($A_3B_2I_9$) have attracted considerable interest because they have good air stability and reduced toxicity compared with Pb-based halide perovskites. Park et al. prepared $Cs_3Bi_2I_9$ perovskite using a one-step spin-coating method [167]. The resultant $Cs_3Bi_2I_9$ perovskite showed a high PL intensity, which suggested suppressed nonradiative recombination. As a result, the $Cs_3Bi_2I_9$ pero-SCs achieved a PCE over 1%. In addition, Gu et al. had successfully doped Ru^{3+} into $Cs_3Bi_2I_9$ perovskite and yielded a PCE up to 4.3% using a hydrothermal method [168].

UV-Vis spectra indicated that the Ru-doped $Cs_3Bi_2I_9$ showed a decreased bandgap as the doping levels increased, suggesting that the doping strategy promoted a wide optoelectronic application. Stability tests further confirmed the good environmental and thermal stability of $Cs_3Bi_{2-x}Ru_xI_9$. Pal et al. synthesized colloidal $Cs_3Bi_2I_9$ nanocrystals by using the hot injection method [169]. Absorption spectrum at 10 K confirmed that $Cs_3Bi_2I_9$ nanocrystals has a very high exciton binding energy. However, the effective phonon energy was significantly smaller than the exciton binding energy, probably suppressing the rate of the carrier relaxation process from conduction band minimum to exciton states. The main issue with $A_3X_2I_9$ is their bandgap, which is relatively large for application in solar cells. To solve the problem, Hong et al. found that the bandgap of $Cs_3Bi_2I_9$ could be successfully modulated by using dual metal cations, i.e. by forming Cs_3BiMI_9 (M: trivalent metal cation) through theoretical study [170]. They investigated various trivalent atoms and found that In and Ga showed a promising bandgap modulation behavior for realizing lower bandgap trivalent metal-based perovskites.

5.5.3.2 Other Structures

Except for $Cs_3Bi_2I_9$, there are some other Bi-based perovskites. Johansson et al. synthesized $CsBi_3I_{10}$ perovskite by using solution-processing method [171]. A bandgap of 1.77 eV was obtained for the $CsBi_3I_{10}$ film, which is smaller than that of $Cs_3Bi_2I_9$ (2.03 eV), as well as an extended visible light absorption spectrum (up to 700 nm). Kim et al. prepared solution-processed $AgBi_2I_7$ thin films by spin-coating silver and bismuth precursors dissolved in *n*-butylamine [152]. X-ray diffraction analysis revealed that the pure cubic structure was formed. The resultant $AgBi_2I_7$ thin films exhibited a lower bandgap of 1.87 eV, which was suitable for photovoltaic applications. The resultant device exhibited a PCE of 1.22% and excellent air stability. Tu et al. prepared Ag_3BiBr_6 by one-step spin-coating process at room temperature [154]. The resultant pero-SC produced a preliminary efficiency of 0.033% with a relatively higher V_{oc} of 0.61 V. They demonstrated that Ag_3BiBr_6 films were stable in ambient condition. In addition, Turkevych et al. studied the structure and optoelectronic properties of Ag_3BiI_6, Ag_2BiI_5, $AgBiI_4$, $AgBi_2I_7$ [153]. All the studied Ag-Bi-I rudorffite compositions featured direct optical bandgaps in the range of 1.79–1.83 eV. They demonstrated that the pero-SC with the structure of FTO/c-m-TiO_2/Ag_3BiI_6/PTAA/Au fabricated in ambient conditions exhibited a PCE of 4.3%.

5.5.4 Double B site Cation Perovskite

Recently, trivalent cations like Bi^{3+}, together with monovalent cations like Ag^+, have concurrently been introduced into the B-sites of lead-free perovskites, thus leading to double B site cation perovskites with the general chemical formula of $A_2B^{1+}B^{3+}X_6$. These double perovskites have been reported to present promising optoelectronic properties including long carrier recombination lifetime, good stability against air and moisture, and low carrier effective masses [24, 172, 173]. Within this new family of semiconductors, the bandgap of $Cs_2BiAgBr_6$ is about

1.95–2.2 eV, which is a bit smaller than those in the analogous lead halide perovskites and in favor of PV application. McClure et al. synthesized $Cs_2BiAgBr_6$ from both solid state and solution routes and demonstrated that $Cs_2BiAgBr_6$ was stable when exposed to air but degraded over several weeks when exposed to both light and ambient condition [24]. Greul et al. prepared $Cs_2BiAgBr_6$ films and applied them to solar cells for the first time [155]. The fabrication process of $Cs_2BiAgBr_6$ film was shown in Figure 5.11a. Both the precursor solution and the substrate were preheated to 75 °C before spin-coating. Then the hot precursor solution was spin-coated onto the TiO_2 substrate. Finally, the substrate underwent a post-processing in ambient condition to form a $Cs_2BiAgBr_6$ film. It is worth to mention that a high-annealing-temperature of at least 250 °C was required to fully convert the precursors into $Cs_2BiAgBr_6$. The resultant device yielded a PCE close to 2.5% and a V_{oc} exceeding 1 V. Wu et al. produced a planar heterojunction solar cell using a high-quality $Cs_2BiAgBr_6$ film, fabricated by low-pressure assisted solution processing in ambient condition (Figure 5.11b) [156]. The device presented a PCE of 1.44% and the HTL-free device without any encapsulation showed no obvious PCE loss after storage in ambient condition (20–30 °C and 40–60% humidity) for 30 days. Ning et al. fabricated high-quality $Cs_2BiAgBr_6$ films, which contributed to the minimized grain boundary length and carrier recombination [157]. The $Cs_2BiAgBr_6$ films showed long electron-hole diffusion lengths 100 nm, enabling the fabrication of planar structure double pero-SCs (Figure 5.11c). As a result, the solar cells based on planar TiO_2 exhibited an average PCE over 1%. Igbari et al. fabricated $Cs_2BiAgBr_6$ thin films using two approaches of the deposition methods of vacuum sublimation and solution processing (Figure 5.11d) [158]. Compared with the vacuum sublimation based one, the solution-processing $Cs_2BiAgBr_6$ film exhibited higher crystal quality. The devices based on solution- and vacuum-processed

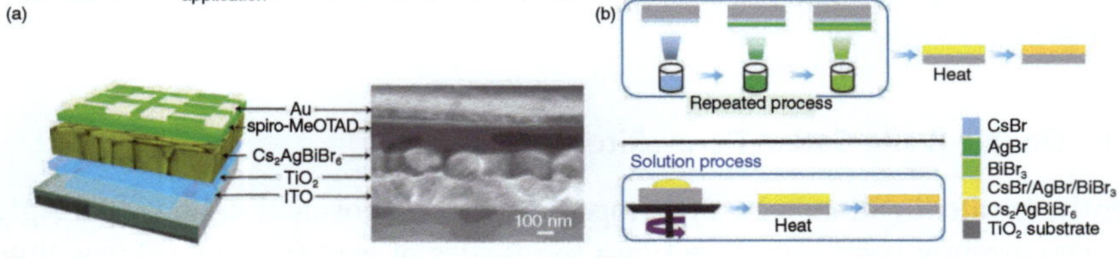

Figure 5.11 (a) Schematic of the synthesis route for $Cs_2AgBiBr_6$ thin films. Source: Greul et al. [155]. (b) The film fabrication process diagram. Source: Wu et al. [156]. (c) Planar $Cs_2AgBiBr_6$ solar cell structure, ITO/compact TiO_2/$Cs_2AgBiBr_6$/spiro-OMeTAD/Au. Source: Ning et al. [157]. (d) Schematic representation of the preparation of $Cs_2AgBiBr_6$ thin films by vacuum-sublimation and solution-processing. Source: Igbari et al. [158].

films obtained PCEs of 2.51% and 1.41%, respectively. These results indicate that solution-processing could guarantee a precise composition stoichiometry to form high-quality multicomponent perovskite films.

In addition, some other works were carried out for further understanding of the double perovskites. Xiao et al. demonstrated that the stable chemical potential region for pure $Cs_2AgBiBr_6$ was narrow, which was proved by density functional theory calculations [174]. For this reason, $Cs_2BiAgBr_6$ was less stable than the Pb-based single-perovskite counterpart $CH_3NH_3PbBr_3$. They also indicated that Ag vacancies easily form shallow acceptors, leading to intrinsic p-type conductivity, while Bi vacancies and AgBi antisites were deep acceptors that should be the dominant defects under the Br-rich growth conditions. The results suggested that $Cs_2AgBiBr_6$ prepared under Br-poor/Bi-rich conditions would be beneficial for suppressing the formation of deep defects. Savory et al. demonstrated that the combination of Ag(I) and Bi(III) led to the wide indirect bandgaps owing to a mismatch in angular momentum of the frontier atomic orbitals [175]. Replacing Ag with In or Tl could overcome the limitation on the PCE of less than 10%. In addition, Du et al. synthesized the substituted double perovskites, $Cs_2Ag(Bi_{1-x}In_x)Br_6$ ($x = 0$, 0.25, 0.5, and 0.75) and $Cs_2Ag(Bi_{1-x}Sb_x)Br_6$ ($x = 0$, 0.125, and 0.375), using In^{III}/Sb^{III} to partly replace Bi^{3+} to modulate the bandgap [176]. They found that the In content enlarged the bandgap, whereas increasing Sb content reduced the bandgap. Apart from above reports, Feng et al. studied the Cs_2BBiX_6 (B = Ag, Cu; X = Br, Cl)/TiO_2 interfaces [177]. They found that Cs, Ag, and Bi tend to bond with the oxygen from TiO_2 side with a shorter bond length compared with that of metal-halide ions. A clear charge flow pathway for the Cs_4X_4 (X = Br and Cl)/TiO_2-mediated interfaces facilitated the photo absorption and charge transfer/separation.

Above all, although these Pb-free inorganic perovskites are friendly to the environment, there are also some problems impeding their optoelectronic applications. As for the Ge-based inorganic pero-SCs, it is hard to fabricate high-performance perovskite films due to their poor film forming abilities and oxidizing tendencies. Solution-processed Sb-based perovskite-like structures tend to form the dimer phase, which show poor photovoltaic properties. Bi-based inorganic perovskite materials showed wide indirect bandgaps and deep defect states, which severely limit their photoelectronic conversion efficiency when applied in photovoltaic devices. For double perovskite materials, their inherently indirect and large bandgap are caused by the mismatch of Ag d- and Bi s-orbitals at the band edges, thus leading to the low voltages and efficiencies. All above issues are great obstacles that impedes their further development.

5.6 Conclusion

In this chapter, we summarized recent advances of the all-inorganic pero-SCs. Significant progress has been made toward high-performance all-inorganic pero-SCs. For lead-based perovskite, $CsPbI_3$ owns a suitable bandgap and the corresponding pero-SCs have a great potential to further increase the PCE value. The problem of phase instability needs to be completely solved and effective phase stabilization methods should be further developed. $CsPbBr_3$ perovskites exhibit excellent stability against heat, moisture, and UV light, and the PCE of the $CsPbBr_3$-based pero-SCs has already reached 10.26% with an ultrahigh V_{oc}, which has application prospects in the transparent field. The mixed halide perovskites $CsPbIBr_2$ and $CsPbI_2Br$ also have a potential in fabricating tandem and semitransparent solar cells due to their relatively wide bandgaps. All of these lead-based perovskites exhibited good thermal stability. For Sn-based perovskite, it is urgent to decrease the intrinsic defect density for decreased V_{oc} loss. To do this, the oxidation process from Sn^{2+} to Sn^{4+} should be considerably suppressed. For other inorganic perovskites, to improve the device PCE value, various fabrication methodologies need to be developed for obtaining high-performance crystalline perovskite, and the energy level-matched transporting layers should be developed at the same time.

References

1 Burschka, J., Pellet, N., Moon, S.-J. et al. (2013). Sequential deposition as a route to high-performance perovskite-sensitized solar cells. *Nature* 499: 316.

2 Lee, M.M., Teuscher, J., Miyasaka, T. et al. (2012). Efficient hybrid solar cells based on meso-superstructured organometal halide perovskites. *Science* 338: 643–647.

3 Yang, W.S., Park, B.-W., Jung, E.H. et al. (2017). Iodide management in formamidinium-lead-halide-based perovskite layers for efficient solar cells. *Science* 356: 1376–1379.

4 Conings, B., Drijkoningen, J., Gauquelin, N. et al. (2015). Intrinsic thermal instability of methylammonium lead trihalide perovskite. *Adv. Energy Mater.* 5: 1500477.

5 Stoumpos, C.C., Malliakas, C.D., and Kanatzidis, M.G. (2013). Semiconducting tin and lead iodide perovskites with organic cations: phase transitions, high mobilities, and near-infrared photoluminescent properties. *Inorg. Chem.* 52: 9019–9038.

6 Chen, Z., Wang, J., Ren, Y. et al. (2012). Schottky solar cells based on $CsSnI_3$ thin-films. *Appl. Phys. Lett.* 101: 093901.

7 Wang, Y., Dar, M.I., Ono, L.K. et al. (2019). Thermodynamically stabilized β-$CsPbI_3$-based perovskite solar cells with efficiencies >18%. *Science* 365: 591–595.

8 Liang, J., Wang, C., Wang, Y. et al. (2016). All-inorganic perovskite solar cells. *J. Am. Chem. Soc.* 138: 15829–15832.

9 Chen, W., Zhang, J., Xu, G. et al. (2018). A semitransparent inorganic perovskite film for overcoming ultraviolet light instability of organic solar cells and achieving 14.03% efficiency. *Adv. Mater.* 30: 1800855.

10 Zhou, Y. and Zhao, Y. (2019). Chemical stability and instability of inorganic halide perovskites. *Energy Environ. Sci.* 12: 1495–1511.

11 Zhao, B., Jin, S.-F., Huang, S. et al. (2018). Thermodynamically stable orthorhombic γ-$CsPbI_3$ thin films for high-performance photovoltaics. *J. Am. Chem. Soc.* 140: 11716–11725.

12 Hu, Y., Bai, F., Liu, X. et al. (2017). Bismuth incorporation stabilized α-$CsPbI_3$ for fully inorganic perovskite solar cells. *ACS Energy Lett.* 2: 2219–2227.

13 Wang, Y., Zhang, T., Kan, M., and Zhao, Y. (2018). Bifunctional stabilization of all-inorganic α-$CsPbI_3$ perovskite for 17% efficiency photovoltaics. *J. Am. Chem. Soc.* 140: 12345–12348.

14 Beal, R.E., Slotcavage, D.J., Leijtens, T. et al. (2016). Cesium lead halide perovskites with improved stability for tandem solar cells. *J. Phys. Chem. Lett.* 7: 746–751.

15 Zhang, H., Yang, Z., Yu, W. et al. (2018). A sandwich-like organolead halide perovskite photocathode for efficient and durable photoelectrochemical hydrogen evolution in water. *Adv. Energy Mater.* 8: 1800795.

16 Zhang, Y., Wu, C., Wang, D. et al. (2019). High efficiency 16.37% of cesium bromide-passivated all-inorganic $CsPbI_2Br$ perovskite solar cells. *Sol. RRL* https://doi.org/10.1002/solr.201900254.

17 Subhani, W.S., Wang, K., Du, M. et al. (2019). Interface-modification-induced gradient energy band for highly efficient $CsPbIBr_2$ perovskite solar cells. *Adv. Energy Mater.* 9: 1803785.

18 Kumar, M.H., Dharani, S., Leong, W.L. et al. (2014). Lead-free halide perovskite solar cells with high photocurrents realized through vacancy modulation. *Adv. Mater.* 26: 7122–7127.

19 Lignos, I., Morad, V., Shynkarenko, Y. et al. (2018). Exploration of near-infrared-emissive colloidal multinary lead halide perovskite nanocrystals using an automated microfluidic platform. *ACS nano* 12: 5504–5517.

20 Giustino, F. and Snaith, H.J. (2016). Toward lead-free perovskite solar cells. *ACS Energy Lett.* 1: 1233–1240.

21 Mazzio, K.A. and Luscombe, C.K. (2014). The future of organic photovoltaics. *Chem. Soc. Rev.* 44: 78–90.

22 Lee, B., Stoumpos, C.C., Zhou, N. et al. (2014). Air-stable molecular semiconducting iodosalts for solar cell applications: Cs_2SnI_6 as a hole conductor. *J. Am. Chem. Soc.* 136: 15379–15385.

23 Maughan, A.E., Ganose, A.M., Bordelon, M.M. et al. (2016). Defect tolerance to intolerance in the vacancy-ordered double perovskite semiconductors Cs_2SnI_6 and Cs_2TeI_6. *J. Am. Chem. Soc.* 138: 8453–8464.

24 McClure, E.T., Ball, M.R., Windl, W., and Woodward, P.M. (2016). Cs_2AgBiX_6 (X = Br, Cl): new visible light absorbing, lead-free halide perovskite semiconductors. *Chem. Mater.* 28: 1348–1354.

25 Sutton, R.J., Eperon, G.E., Miranda, L. et al. (2016). Bandgap-tunable cesium lead halide perovskites with high thermal stability for efficient solar cells. *Adv. Energy Mater.* 6: 1502458.

26 Zeng, Q., Zhang, X., Liu, C. et al. (2019). Inorganic $CsPbI_2Br$ perovskite solar cells: the progress and perspective. *Sol. RRL* 3: 1800239.

27 Ning, W., Zhao, X., Klarbring, J. et al. (2019). Thermochromic lead-free halide double perovskites. *Adv. Funct. Mater.* 29: 1807375.

28 Akbulatov, A.F., Luchkin, S.Y., Frolova, L.A. et al. (2017). Probing the intrinsic thermal and photochemical stability of hybrid and inorganic lead halide perovskites. *J. Phys. Chem. Lett.* 8: 1211–1218.

29 Xiao, Z., Zhao, L., Tran, N.L. et al. (2017). Mixed-halide perovskites with stabilized bandgaps. *Nano Lett.* 17: 6863–6869.

30 Alharbi, F.H. and Kais, S. (2015). Theoretical limits of photovoltaics efficiency and possible improvements by intuitive approaches learned from photosynthesis and quantum coherence. *Renew Sustain. Energy Rev.* 43: 1073–1089.

31 Jena, A.K., Kulkarni, A., Sanehira, Y. et al. (2018). Stabilization of α-$CsPbI_3$ in ambient room temperature conditions by incorporating Eu into $CsPbI_3$. *Chem. Mater.* 30: 6668–6674.

32 Marronnier, A., Roma, G., Boyer-Richard, S. et al. (2018). Anharmonicity and disorder in the black phases of cesium lead iodide used for stable inorganic perovskite solar cells. *ACS Nano* 12: 3477–3486.

33 Eperon, G.E., Paternò, G.M., Sutton, R.J. et al. (2015). Inorganic caesium lead iodide perovskite solar cells. *J. Mater. Chem. A* 3: 19688–19695.

34 Wu, T., Wang, Y., Dai, Z. et al. (2019). Efficient and stable $CsPbI_3$ solar cells via regulating lattice distortion with surface organic terminal groups. *Adv. Mater.* 39: 1900605.

35 Luo, P., Xia, W., Zhou, S. et al. (2016). Solvent engineering for ambient-air-processed, phase-stable $CsPbI_3$ in perovskite solar cells. *J. Phys. Chem. Lett.* 7: 3603–3608.

36 Wang, K., Jin, Z., Liang, L. et al. (2018). All-inorganic cesium lead iodide perovskite solar cells with stabilized efficiency beyond 15. *Nat. Commun.* 9: 4544.

37 Becker, P., Márquez, J.A., Just, J. et al. (2019). Low temperature synthesis of stable γ-$CsPbI_3$ perovskite layers for solar cells obtained by high throughput experimentation. *Adv. Energy Mater.* 9: 1900555.

38 Hutter, E.M., Sutton, R.J., Chandrashekar, S. et al. (2017). Vapour-deposited cesium lead iodide perovskites: microsecond charge carrier lifetimes and enhanced photovoltaic performance. *ACS Energy Lett.* 2: 1901–1908.

39 Li, B., Zhang, Y., Fu, L. et al. (2018). Surface passivation engineering strategy to fully-inorganic cubic $CsPbI_3$ perovskites for high-performance solar cells. *Nat. Commun.* 9: 1076.

40 Zhang, T., Dar, M.I., Li, G. et al. (2017). Bication lead iodide 2D perovskite component to stabilize inorganic α-$CsPbI_3$ perovskite phase for high-efficiency solar cells. *Sci. Adv.* 3: e1700841.

41 Li, F., Pei, Y., Xiao, F. et al. (2018). Tailored dimensionality to regulate the phase stability of inorganic cesium lead iodide perovskites. *Nanoscale* 10: 6318–6322.

42 Wang, Q., Zheng, X., Deng, Y. et al. (2017). Stabilizing the α-phase of $CsPbI_3$ perovskite by sulfobetaine zwitterions in one-step spin-coating films. *Joule* 1: 371–382.

43 Ding, X., Chen, H., Wu, Y. et al. (2018). Triple cation additive $NH_3^+C_2H_4NH_2^+C_2H_4NH_3^+$-induced phase-stable inorganic α-$CsPbI_3$ perovskite films for use in solar cells. *J. Mater. Chem. A* 6: 18258–18266.

44 Liao, J., Rao, H., Chen, B. et al. (2017). Dimension engineering on cesium lead iodide for efficient and stable perovskite solar cells. *J. Mater. Chem. A* 5: 2066–2072.

45 Zhao, H., Xu, J., Zhou, S. et al. (2019). Preparation of tortuous 3D γ-$CsPbI_3$ films at low temperature by CaI_2 as dopant for highly efficient perovskite solar cells. *Adv. Funct. Mater.* 29: 1808986.

46 Lau, C.F.J., Deng, X., Zheng, J. et al. (2018). Enhanced performance via partial lead replacement with calcium for a $CsPbI_3$ perovskite solar cell exceeding 13% power conversion efficiency. *J. Mater. Chem. A* 6: 5580–5586.

47 Wang, K., Jin, Z., Liang, L. et al. (2019). Chlorine doping for black γ-$CsPbI_3$ solar cells with stabilized efficiency beyond 16%. *Nano Energy* 58: 175–182.

48 Wang, P., Zhang, X., Zhou, Y. et al. (2018). Solvent-controlled growth of inorganic perovskite films in dry environment for efficient and stable solar cells. *Nat. Commun.* 9: 2225.

49 Liang, L., Liu, M., Jin, Z. et al. (2019). Optical management with nanoparticles for a light conversion efficiency enhancement in inorganic gamma-$CsPbI_3$ solar cells. *Nano Lett.* 19: 1796–1804.

50 Bian, H., Wang, Q., Yang, S. et al. (2019). Nitrogen-doped graphene quantum dots for 80% photoluminescence quantum yield for inorganic γ-$CsPbI_3$ perovskite solar cells with efficiency beyond 16%. *J. Mater. Chem. A* 7: 5740–5747.

51 Li, X., Zhang, W., Wang, Y. et al. (2018). In-situ cross-linking strategy for efficient and operationally stable methylammoniun lead iodide solar cells. *Nat. Commun.* 9: 3806.

52 Li, N., Tao, S., Chen, Y. et al. (2019). Cation and anion immobilization through chemical bonding enhancement with fluorides for stable halide perovskite solar cells. *Nat. Energy* 4: 408–415.

53 Shpatz Dayan, A., Cohen, B.-E., Aharon, S. et al. (2018). Enhancing stability and photostability of $CsPbI_3$ by reducing its dimensionality. *Chem. Mater.* 30: 8017–8024.

54 Zhang, J., Hodes, G., Jin, Z., and Liu, S. (2019). All-inorganic $CsPbX_3$ perovksite solar cells: progress and prospects. *Angew. Chem. Int. Ed.* https://doi.org/10.1002/anie.201901081 10.1002/anie.201901081.

55 Mariotti, S., Hutter, O.S., Phillips, L.J. et al. (2018). Stability and performance of $CsPbI_2Br$ thin films and solar cell devices. *ACS Appl. Mater. Interfaces* 10: 3750–3760.

56 Ma, Q., Huang, S., Chen, S. et al. (2017). The effect of stoichiometry on the stability of inorganic cesium lead mixed-halide perovskites solar cells. *J. Phys. Chem. C* 121: 19642–19649.

57 Chen, W., Chen, H., Xu, G. et al. (2019). Precise control of crystal growth for highly efficient $CsPbI_2Br$ perovskite solar cells. *Joule* 3: 191–204.

58 Zai, H., Zhang, D., Liang, L. et al. (2018). Low-temperature-processed inorganic perovskite solar cells via solvent engineering with enhanced mass transport. *J. Mater. Chem. A* 6: 23602–23609.

59 Bai, D., Bian, H., Jin, Z. et al. (2018). Temperature-assisted crystallization for inorganic $CsPbI_2Br$ perovskite solar cells to attain high stabilized efficiency 14.81%. *Nano Energy* 52: 408–415.

60 Jiang, H., Feng, J., Zhao, H. et al. (2018). Low temperature fabrication for high performance flexible $CsPbI_2Br$ perovskite solar cells. *Adv. Sci.* 5: 1801117.

61 Chen, C., Lin, H., Chiang, K.-M. et al. (2017). All-vacuum-deposited stoichiometrically balanced inorganic cesium lead halide perovskite solar cells with stabilized efficiency exceeding 11%. *Adv. Mater.* 29: 1605290.

62 Tian, J., Xue, Q., Tang, X. et al. (2019). Dual interfacial design for efficient $CsPbI_2Br$ perovskite solar cells with improved photostability. *Adv. Mater.* 31: 1901152.

63 Dong, C., Han, X., Zhao, Y. et al. (2018). Green anti-solvent process for high performance carbon-based $CsPbI_2Br$ all-inorganic perovskite solar cell. *Sol. RRL* 2: 1800139.

64 Dong, C., Han, X., Li, W. et al. (2019). Anti-solvent assisted multi-step deposition for efficient and stable carbon-based $CsPbI_2Br$ all-inorganic perovskite solar cell. *Nano Energy* 59: 553–559.

65 Zhang, S., Wu, S., Chen, W. et al. (2018). Solvent engineering for efficient inverted perovskite solar cells based on inorganic $CsPbI_2Br$ light absorber. *Mater. Today Energy* 8: 125–133.

66 Rao, H., Ye, S., Gu, F. et al. (2018). Morphology controlling of all-inorganic perovskite at low temperature for efficient rigid and flexible solar cells. *Adv. Energy Mater.* 8: 1800758.

67 Yin, G., Zhao, H., Jiang, H. et al. (2018). Precursor engineering for all-inorganic $CsPbI_2Br$ perovskite solar cells with 14.78% efficiency. *Adv. Funct. Mater.* 28: 1803269.

68 Wang, Z., Liu, X., Lin, Y. et al. (2019). Hot-substrate deposition of all-inorganic perovskite films for low-temperature processed high-efficiency solar cells. *J. Mater. Chem. A* 7: 2773–2779.

69 Nam, J.K., Jung, M.S., Chai, S.U. et al. (2017). Unveiling the crystal formation of cesium lead mixed-halide perovskites for efficient and stable solar cells. *J. Phys. Chem. Lett.* 8: 2936–2940.

70 Zhang, L., Li, B., Yuan, J. et al. (2018). High-voltage-efficiency inorganic perovskite solar cells in a wide solution-processing window. *J. Phys. Chem. Lett.* 9: 3646–3653.

71 Gao, Y., Dong, Y., Huang, K. et al. (2018). Highly efficient, solution-processed $CsPbI_2Br$ planar heterojunction perovskite solar cells via flash annealing. *ACS Photonics* 5: 4104–4110.

72 Sun, H., Zhang, J., Gan, X. et al. (2019). Pb-reduced $CsPb_{0.9}Zn_{0.1}I_2Br$ thin films for efficient perovskite solar cells. *Adv. Energy Mater.* 9: 1900896.

73 Wang, H., Bian, H., Jin, Z. et al. (2018). Synergy of hydrophobic surface capping and lattice contraction for stable and high-efficiency inorganic $CsPbI_2Br$ perovskite solar cells. *Sol. RRL* 2: 1800216.

74 Xiang, W., Wang, Z., Kubicki, D.J. et al. (2019). Europium-doped $CsPbI_2Br$ for stable and highly efficient inorganic perovskite solar cells. *Joule* 3: 205–214.

75 Yang, F., Hirotani, D., Kapil, G. et al. (2018). All-inorganic $CsPb_{1-x}Ge_xI_2Br$ perovskite with enhanced phase stability and photovoltaic performance. *Angew. Chem. Int. Ed.* 57: 12745–12749.

76 Lau, C.F.J., Zhang, M., Deng, X. et al. (2017). Strontium-doped low-temperature-processed $CsPbI_2Br$ perovskite solar cells. *ACS Energy Lett.* 2: 2319–2325.

77 Guo, Z., Zhao, S., Liu, A. et al. (2019). Niobium incorporation into $CsPbI_2Br$ for stable and efficient all-inorganic perovskite solar cells. *ACS Appl. Mater. Interfaces* 11: 19994–20003.

78 Bai, D., Zhang, J., Jin, Z. et al. (2018). Interstitial Mn^{2+}-driven high-aspect-ratio grain growth for low-trap-density microcrystalline films for record efficiency $CsPbI_2Br$ solar cells. *ACS Energy Lett.* 3: 970–978.

79 Nam, J.K., Chai, S.U., Cha, W. et al. (2017). Potassium incorporation for enhanced performance and stability of fully inorganic cesium lead halide perovskite solar cells. *Nano Lett.* 17: 2028–2033.

80 Fu, L., Zhang, Y., Chang, B. et al. (2018). A fluorine-modulated bulk-phase heterojunction and tolerance factor for enhanced performance and structure stability of cesium lead halide perovskite solar cells. *J. Mater. Chem. A* 6: 13263–13270.

81 Liu, C., Li, W., Li, H. et al. Structurally reconstructed $CsPbI_2Br$ perovskite for highly stable and square-centimeter all-inorganic perovskite solar cells. *ACS Appl. Mater. Interfaces* 11: 19994–20003.

82 Zhang, T., Li, H., Liu, S. et al. (2018). Low-temperature stable α-phase inorganic perovskite compounds via crystal cross-linking. *J. Phys. Chem. Lett.* 10: 200–205.

83 Zeng, Q., Zhang, X., Feng, X. et al. (2018). Polymer-passivated inorganic cesium lead mixed-halide perovskites for stable and efficient solar cells with high open-circuit voltage over 1.3 V. *Adv. Mater.* 30: 1705393.

84 Yuan, J., Zhang, L., Bi, C. et al. (2018). Surface trap states passivation for high-performance inorganic perovskite solar cells. *Sol. RRL* 2: 1800188.

85 Liu, C., Li, W., Zhang, C. et al. (2018). All-inorganic $CsPbI_2Br$ perovskite solar cells with high efficiency exceeding 13%. *J. Am. Chem. Soc.* 140: 3825–3828.

86 Zhang, X., Zhou, Y., Li, Y. et al. (2019). Efficient and carbon-based hole transport layer-free $CsPbI_2Br$ planar perovskite solar cells using PMMA modification. *J. Mater. Chem. C* 7: 3852–3861.

87 Zhou, Y., Zhang, X., Lu, X. et al. (2019). Promoting the hole extraction with Co_3O_4 nanomaterials for efficient carbon-based $CsPbI_2Br$ perovskite solar cells. *Sol. RRL* 3: 1800315.

88 Kim, D.H., Heo, J.H., and Im, S.H. (2019). Hysteresis-less $CsPbI_2Br$ mesoscopic perovskite solar cells with high open circuit voltage exceeding 1.3 V and 14.86% of power conversion efficiency. *ACS Appl. Mater. Interfaces* 11: 19123–19131.

89 Zhang, B., Zhou, Y., Xue, Q. et al. (2019). The energy-alignment engineering in polytriphenylamines-based hole transport polymers realizes low energy loss and high efficiency for all-inorganic perovskite solar cells. *Sol. RRL.* https://doi.org/10.1002/solr.201900265.

90 Hua, Y., Zhang, D.Y., Wu, T. et al. (2019). Importance of terminated groups in 9,9-bis(4-methoxyphenyl)-substituted fluorene-based hole transport materials for highly efficient organic-inorganic hybrid and all-inorganic perovskite solar cells. *J. Mater. Chem. A* 7: 10319–10324.

91 Yan, L., Xue, Q., Liu, M. et al. (2018). Interface engineering for all-inorganic $CsPbI_2Br$ perovskite solar cells with efficiency over 14%. *Adv. Mater.* 30: 1802509.

92 Han, D., Xin, Y., Yuan, Q. et al. (2019). Solution-processed 2D $Nb_2O_5(001)$ nanosheets for inverted $CsPbI_2Br$ perovskite solar cells: interfacial and diffusion engineering. *Sol. RRL* 3: 1900091.

93 Zhang, J., Bai, D., Jin, Z. et al. (2018). 3D-2D-0D interface profiling for record efficiency all-inorganic $CsPbBrI_2$ perovskite solar cells with superior stability. *Adv. Energy Mater.* 8: 1703246.

94 Zhang, J., Jin, Z., Liang, L. et al. (2018). Iodine-optimized interface for inorganic $CsPbI_2Br$ perovskite solar cell to attain high stabilized efficiency exceeding 14%. *Adv. Sci.* 5: 1801123.

95 Zhou, L., Guo, X., Lin, Z. et al. (2019). Interface engineering of low temperature processed all-inorganic $CsPbI_2Br$ perovskite solar cells toward PCE exceeding 14%. *Nano Energy* 60: 583–590.

96 Liu, D., Yang, C., Bates, M., and Lunt, R.R. (2018). Room temperature processing of inorganic perovskite films to enable flexible solar cells. *iScience* 6: 272–279.

97 Ahn, N., Son, D.-Y., Jang, I.-H. et al. (2015). Highly reproducible perovskite solar cells with average efficiency of 18.3% and best efficiency of 19.7% fabricated via Lewis base adduct of lead (II) iodide. *J. Am. Chem. Soc.* 137: 8696–8699.

98 Yang, W.S., Noh, J.H., Jeon, N.J. et al. (2015). High-performance photovoltaic perovskite layers fabricated through intramolecular exchange. *Science* 348: 1234–1237.

99 Bhosale, S.S., Jokar, E., Fathi, A. et al. (2018). Functionalization of graphene oxide films with Au and MoO_x nanoparticles as efficient p-contact electrodes for inverted planar perovskite solar cells. *Adv. Funct. Mater.* 28: 1803200.

100 Ma, Q., Huang, S., Wen, X. et al. (2016). Hole transport layer free inorganic $CsPbIBr_2$ perovskite solar cell by dual source thermal evaporation. *Adv. Energy Mater.* 6: 1502202.

101 Li, W., Rothmann, M.U., Liu, A. et al. (2017). Phase segregation enhanced ion movement in efficient inorganic $CsPbIBr_2$ solar cells. *Adv. Energy Mater.* 7: 1700946.

102 Liang, J., Zhao, P., Wang, C. et al. (2017). $CsPb_{0.9}Sn_{0.1}IBr_2$ based all-inorganic perovskite solar cells with exceptional efficiency and stability. *J. Am. Chem. Soc.* 139: 14009–14012.

103 Li, N., Zhu, Z., Li, J. et al. (2018). Inorganic $CsPb_{1-x}Sn_xIBr_2$ for efficient wide-bandgap perovskite solar cells. *Adv. Energy Mater.* 8: 1800525.

104 Lin, J., Lai, M., Dou, L. et al. (2018). Thermochromic halide perovskite solar cells. *Nat. Mater.* 17: 261–267.

105 Zhu, W., Zhang, Q., Chen, D. et al. (2018). Intermolecular exchange boosts efficiency of air-stable, carbon-based all-inorganic planar $CsPbIBr_2$ perovskite solar cells to over 9%. *Adv. Energy Mater.* 8: 1802080.

106 Zhu, W., Zhang, Q., Zhang, C. et al. (2018). Aged precursor solution toward low-temperature fabrication of efficient carbon-based all-inorganic planar $CsPbIBr_2$ perovskite solar cells. *ACS Appl. Energy Mater.* 1: 4991–4997.

107 Liang, J., Liu, Z., Qiu, L. et al. (2018). Enhancing optical, electronic, crystalline, and morphological properties of cesium lead halide by Mn substitution for high-stability all-inorganic perovskite solar cells with carbon electrodes. *Adv. Energy Mater.* 8: 1800504.

108 Subhani, W.S., Wang, K., Du, M., and Liu, S. (2019). Goldschmidt-rule-deviated perovskite $CsPbIBr_2$ by barium substitution for efficient solar cells. *Nano Energy* 61: 165–172.

109 Lu, J., Chen, S., and Zheng, Q. (2018). Defect passivation of $CsPbIBr_2$ perovskites for high-performance solar cells with large open-circuit voltage of 1.28 V. *ACS Appl. Energy Mater.* 1: 5872–5878.

110 Wang, Z., Baranwal, A.K., kamarudin, M.A. et al. (2019). Xanthate-induced sulfur doped all-inorganic perovskite with superior phase stability and enhanced performance. *Nano Energy* 59: 258–267.

111 Yang, B., Wang, M., Hu, X. et al. (2019). Highly efficient semitransparent $CsPbIBr_2$ perovskite solar cells via low-temperature processed In_2S_3 as electron-transport-layer. *Nano Energy* 57: 718–727.

112 Liu, C., Li, W., Chen, J. et al. (2017). Ultra-thin MoO_x as cathode buffer layer for the improvement of all-inorganic $CsPbIBr_2$ perovskite solar cells. *Nano Energy* 41: 75–83.

113 Guo, Z., Teo, S.h.H., Xu, Z. et al. (2019). Achievable high V_{oc} of carbon based all-inorganic $CsPbIBr_2$ perovskite solar cells through interface engineering. *J. Mater. Chem. A* 7: 1227–1232.

114 Yang, S., Guo, Z., Gao, L. et al. (2019). Bifunctional dye molecule in all-inorganic $CsPbIBr_2$ perovskite solar cells with efficiency exceeding 10%. *Sol. RRL* https://doi.org/10.1002/solr.201900212.

115 Teng, P., Han, X., Li, J. et al. (2018). Elegant face-down liquid-space-restricted deposition of $CsPbBr_3$ films for efficient carbon-based all-inorganic planar perovskite solar cells. *ACS Appl. Mater. Interfaces* 10: 9541–9546.

116 Duan, J., Zhao, Y., He, B., and Tang, Q. (2018). High-purity inorganic perovskite films for solar cells with 9.72 % efficiency. *Angew. Chem. Int. Ed.* 57: 3787–3791.

117 Lei, J., Gao, F., Wang, H. et al. (2018). Efficient planar $CsPbBr_3$ perovskite solar cells by dual-source vacuum evaporation. *Sol. Energy Mater. Sol. Cells* 187: 1–8.

118 Li, H., Tong, G., Chen, T. et al. (2018). Interface engineering using a perovskite derivative phase for efficient and stable $CsPbBr_3$ solar cells. *J. Mater. Chem. A* 6: 14255–14261.

119 Kulbak, M., Cahen, D., and Hodes, G. (2015). How important is the organic part of lead halide perovskite photovoltaic cells? Efficient $CsPbBr_3$ cells. *J. Phys. Chem. Lett.* 6: 2452–2456.

120 Kulbak, M., Gupta, S., Kedem, N. et al. (2016). Cesium enhances long-term stability of lead bromide perovskite-based solar cells. *J. Phys. Chem. Lett.* 7: 167–172.

121 Duan, J., Hu, T., Zhao, Y. et al. (2018). Carbon-electrode-tailored all-inorganic perovskite solar cells to harvest solar and water–vapor energy. *Angew. Chem. Int. Ed.* 57: 5746–5749.

122 Li, Y., Duan, J., Yuan, H. et al. (2018). Lattice modulation of alkali metal cations doped $Cs_{1-x}R_xPbBr_3$ halides for inorganic perovskite solar cells. *Sol. RRL* 2: 1800164.

123 Duan, J., Zhao, Y., Yang, X. et al. Lanthanide ions doped $CsPbBr_3$ halides for HTM-free 10.14%-efficiency inorganic perovskite solar cell with an ultrahigh open-circuit voltage of 1.594 V. *Adv. Energy Mater.* 8: 1802346.

124 Liu, Y., He, B., Duan, J. et al. (2019). Poly(3-hexylthiophene)/zinc phthalocyanine composites for advanced interface engineering of 10.03%-efficiency $CsPbBr_3$ perovskite solar cells. *J. Mater. Chem. A* 7: 12635–12644.

125 Yuan, H., Zhao, Y., Duan, J. et al. (2018). All-inorganic $CsPbBr_3$ perovskite solar cell with 10.26% efficiency by spectra engineering. *J. Mater. Chem. A* 6: 24324–24329.

126 Liu, X., Tan, X., Liu, Z. et al. (2019). Boosting the efficiency of carbon-based planar $CsPbBr_3$ perovskite solar cells by a modified multistep spin-coating technique and interface engineering. *Nano Energy* 56: 184–195.

127 Li, B., Zhang, Y., Zhang, L., and Yin, L. (2017). $PbCl_2$-tuned inorganic cubic $CsPbBr_3(Cl)$ perovskite solar cells with enhanced electron lifetime, diffusion length and photovoltaic performance. *J. Power Sources* 360: 11–20.

128 Duan, J., Zhao, Y., He, B., and Tang, Q. (2018). Simplified perovskite solar cell with 4.1% efficiency employing inorganic $CsPbBr_3$ as light absorber. *Small* 14: 1704443.

129 Liao, G., Duan, J., Zhao, Y., and Tang, Q. (2018). Toward fast charge extraction in all-inorganic $CsPbBr_3$ perovskite solar cells by setting intermediate energy levels. *Sol. Energy* 171: 279–285.

130 Liu, Z., Sun, B., Liu, X. et al. (2018). Efficient carbon-based $CsPbBr_3$ inorganic perovskite solar cells by using Cu-phthalocyanine as hole transport material. *Nano-Micro Lett.* 10: 34.

131 Hoffman, J.B., Zaiats, G., Wappes, I., and Kamat, P.V. (2017). $CsPbBr_3$ solar cells: controlled film growth through layer-by-layer quantum dot deposition. *Chem. Mater.* 29: 9767–9774.

132 Zhang, Q., Hao, F., Li, J. et al. (2018). Perovskite solar cells: must lead be replaced-and can it be done? *Sci. Technol. Adv. Mater.* 19: 425–442.

133 Chen, Z., Yu, C., Shum, K. et al. (2012). Photoluminescence study of polycrystalline $CsSnI_3$ thin films: determination of exciton binding energy. *J. Lumin.* 132: 345–349.

134 Chung, I., Song, J.-H., Im, J. et al. (2012). $CsSnI_3$: semiconductor or metal? High electrical conductivity and strong near-infrared photoluminescence from a single material. High hole mobility and phase-transitions. *J. Am. Chem. Soc.* 134: 8579–8587.

135 Rajendra Kumar, G., Kim, H.-J., Karupannan, S., and Prabakar, K. (2017). Interplay between iodide and tin vacancies in $CsSnI_3$ perovskite solar cells. *J. Phys. Chem. C* 121: 16447–16453.

136 Wang, N., Zhou, Y., Ju, M. et al. (2016). Heterojunction-depleted lead-free perovskite solar cells with coarse-grained B-γ-$CsSnI_3$ thin films. *Adv. Energy Mater.* 6: 1601130.

137 Heo, J.H., Kim, J., Kim, H. et al. (2018). Roles of SnX_2 (X = F, Cl, Br) additives in tin-based halide perovskites toward highly efficient and stable lead-free perovskite solar cells. *J. Phys. Chem. Lett.* 9: 6024–6031.

138 Li, W., Li, J., Li, J. et al. (2016). Additive-assisted construction of all-inorganic $CsSnIBr_2$ mesoscopic perovskite solar cells with superior thermal stability up to 473 K. *J. Mater. Chem. A* 4: 17104–17110.

139 Xu, P., Chen, S., Xiang, H. et al. (2014). Influence of defects and synthesis conditions on the photovoltaic performance of perovskite semiconductor $CsSnI_3$. *Chem. Mater.* 26: 6068–6072.

140 Marshall, K.P., Walton, R.I., and Hatton, R.A. (2015). Tin perovskite/fullerene planar layer photovoltaics: improving the efficiency and stability of lead-free devices. *J. Mater. Chem. A* 3: 11631–11640.

141 Marshall, K.P., Walker, M., Walton, R.I., and Hatton, R.A. (2016). Enhanced stability and efficiency in hole-transport-layer-free $CsSnI_3$ perovskite photovoltaics. *Nat. Energy* 1: 16178.

142 Song, T.-B., Yokoyama, T., Aramaki, S., and Kanatzidis, M.G. (2017). Performance enhancement of lead-free tin-based perovskite solar cells with reducing atmosphere-assisted dispersible additive. *ACS Energy Lett.* 2: 897–903.

143 Song, T.-B., Yokoyama, T., Logsdon, J. et al. (2018). Piperazine suppresses self-doping in $CsSnI_3$ perovskite solar cells. *ACS Appl. Energy Mater.* 1: 4221–4226.

144 Song, T.-B., Yokoyama, T., Stoumpos, C.C. et al. (2017). Importance of reducing vapor atmosphere in the fabrication of tin-based perovskite solar cells. *J. Am. Chem. Soc.* 139: 836–842.

145 Wijesekara, A., Varagnolo, S., Dabera, G.D.M.R. et al. (2018). Assessing the suitability of copper thiocyanate as a hole-transport layer in inverted $CsSnI_3$ perovskite photovoltaics. *Sci. Rep.* 8: 15722.

146 Zhang, J., Li, S., Yang, P. et al. (2018). Enhanced stability of lead-free perovskite heterojunction for photovoltaic applications. *J. Mater. Sci.* 53: 4378–4386.

147 Zhu, P., Chen, C., Gu, S. et al. (2018). $CsSnI_3$ solar cells via an evaporation-assisted solution method. *Sol. RRL* 2: 1700224.

148 Sabba, D., Mulmudi, H.K., Prabhakar, R.R. et al. (2015). Impact of anionic Br-substitution on open circuit voltage in lead free perovskite ($CsSnI_{3-x}Br_x$) solar cells. *J. Phys. Chem. C* 119: 1763–1767.

149 Krishnamoorthy, T., Ding, H., Yan, C. et al. (2015). Lead-free germanium iodide perovskite materials for photovoltaic applications. *J. Mater. Chem. A* 3: 23829–23832.

150 Umar, F., Zhang, J., Jin, Z. et al. (2019). Dimensionality controlling of $Cs_3Sb_2I_9$ for efficient all-inorganic planar thin film solar cells by HCl-assisted solution method. *Adv. Optical Mater.* 7: 1801368.

151 Singh, A., Boopathi, K.M., Mohapatra, A. et al. (2018). Photovoltaic performance of vapor-assisted solution-processed layer polymorph of $Cs_3Sb_2I_9$. *ACS Appl. Mater. Interfaces* 10: 2566–2573.

152 Kim, Y., Yang, Z., Jain, A. et al. (2016). Pure cubic-phase hybrid iodobismuthates $AgBi_2I_7$ for thin-film photovoltaics. *Angew. Chem. Int. Ed.* 55: 9586–9590.

153 Turkevych, I., Kazaoui, S., Ito, E. et al. (2017). Photovoltaic rudorffites: lead-free silver bismuth halides alternative to hybrid lead halide perovskites. *ChemSusChem* 10: 3754–3759.

154 Tu, J., Kou, C., Liu, M. et al. (2019). The preparation of Ag_3BiBr_6 films and their preliminary use for solution processed photovoltaics. *SN Appl. Sci.* 1: 620.

155 Greul, E., Petrus, M.L., Binek, A. et al. (2017). Highly stable, phase pure $Cs_2AgBiBr_6$ double perovskite thin films for optoelectronic applications. *J. Mater. Chem. A* 5: 19972–19981.

156 Wu, C., Zhang, Q., Liu, Y. et al. (2018). The dawn of lead-free perovskite solar cell: highly stable double perovskite $Cs_2AgBiBr_6$ film. *Adv. Sci.* 5: 1700759.

157 Ning, W., Wang, F., Wu, B. et al. (2018). Long electron-hole diffusion length in high-quality lead-free double perovskite films. *Adv. Mater.* 30: 1706246.

158 Igbari, F., Wang, R., Wang, Z. et al. (2019). Composition stoichiometry of $Cs_2AgBiBr_6$ films for highly efficient lead-free perovskite solar cells. *Nano Lett.* 19: 2066–2073.

159 Iefanova, A., Adhikari, N., Dubey, A. et al. (2016). Lead free $CH_3NH_3SnI_3$ perovskite thin-film with p-type semiconducting nature and metal-like conductivity. *AIP Adv.* 6: 085312.

160 Kontos, A.G., Kaltzoglou, A., Siranidi, E. et al. (2017). Structural stability, vibrational properties, and photoluminescence in $CsSnI_3$ perovskite upon the addition of SnF_2. *Inorg. Chem.* 56: 84–91.

161 Marshall, K.P., Walker, M., Walton, R.I., and Hatton, R.A. (2017). Elucidating the role of the hole-extracting electrode on the stability and efficiency of inverted $CsSnI_3/C_{60}$ perovskite photovoltaics. *J. Mater. Chem. A* 5: 21836–21845.

162 Gupta, S., Bendikov, T., Hodes, G., and Cahen, D. (2016). $CsSnBr_3$, a lead-free halide perovskite for long-term solar cell application: insights on SnF_2 addition. *ACS Energy Lett.* 1: 1028–1033.

163 Moghe, D., Wang, L., Traverse, C.J. et al. (2016). All vapor-deposited lead-free doped $CsSnBr_3$ planar solar cells. *Nano Energy* 28: 469–474.

164 Ming, W., Shi, H., and Du, M. (2016). Large dielectric constant, high acceptor density, and deep electron traps in perovskite solar cell material $CsGeI_3$. *J. Mater. Chem. A* 4: 13852–13858.

165 Saparov, B., Hong, F., Sun, J. et al. (2015). Thin-film preparation and characterization of $Cs_3Sb_2I_9$: a lead-free layered perovskite semiconductor. *Chem. Mater.* 27: 5622–5632.

166 Harikesh, P.C., Mulmudi, H.K., Ghosh, B. et al. (2016). Rb as an alternative cation for templating inorganic lead-free perovskites for solution processed photovoltaics. *Chem. Mater.* 28: 7496–7504.

167 Park, B.-W., Philippe, B., Zhang, X. et al. (2015). Bismuth based hybrid perovskites $A_3Bi_2I_9$ (A: methylammonium or cesium) for solar cell. *Appl. Adv. Mater.* 27: 6806–6813.

168 Gu, J., Yan, G., Lian, Y. et al. (2018). Bandgap engineering of a lead-free defect perovskite $Cs_3Bi_2I_9$ through trivalent doping of Ru^{3+}. *RSC Adv.* 8: 25802–25807.

169 Pal, J., Bhunia, A., Chakraborty, S. et al. (2018). Synthesis and optical properties of colloidal $M_3Bi_2I_9$ (M = Cs, Rb) perovskite nanocrystals. *J. Phys. Chem. C* 122: 10643–10649.

170 Hong, K.-H., Kim, J., Debbichi, L. et al. (2017). Band gap engineering of $Cs_3Bi_2I_9$ perovskites with trivalent atoms using a dual metal cation. *J. Phys. Chem. C* 121: 969–974.

171 Johansson, M.B., Zhu, H., and Johansson, E.M.J. (2016). Extended photo-conversion spectrum in low-toxic bismuth halide perovskite solar cells. *J. Phys. Chem. Lett.* 7: 3467–3471.

172 Slavney, A.H., Hu, T., Lindenberg, A.M., and Karunadasa, H.I. (2016). A bismuth-halide double perovskite with long carrier recombination lifetime for photovoltaic applications. *J. Am. Chem. Soc.* 138: 2138–2141.

173 Filip, M.R., Hillman, S., Haghighirad, A.A. et al. (2016). Band gaps of the lead-free halide double perovskites $Cs_2BiAgCl_6$ and $Cs_2BiAgBr_6$ from theory and experiment. *J. Phys. Chem. Lett.* 7: 2579–2585.

174 Xiao, Z., Meng, W., Wang, J., and Yan, Y. (2016). Thermodynamic stability and defect chemistry of bismuth-based lead-free double perovskites. *ChemSusChem* 9: 2628–2633.

175 Savory, C.N., Walsh, A., and Scanlon, D.O. (2016). Can Pb-free halide double perovskites support high-efficiency solar cells? *ACS Energy Lett.* 1: 949–955.

176 Du, K., Meng, W., Wang, X. et al. (2017). Bandgap engineering of lead-free double perovskite $Cs_2AgBiBr_6$ through trivalent metal alloying. *Angew. Chem. Int. Ed.* 56: 8158–8162.

177 Feng, H.-J., Deng, W., Yang, K. et al. (2017). Double perovskite Cs_2BBiX_6 (B = Ag, Cu, X = Br, Cl)/TiO_2 heterojunction: an efficient Pb-free perovskite interface for charge extraction. *J. Phys. Chem. C* 121: 4471–4480.

6

Tin Halide Perovskite Solar Cells

Thomas Stergiopoulos[1,2]

[1] *National Center for Scientific Research "Demokritos", Institute of Nanoscience and Nanotechnology, 15310, Ag. Paraskevi Athens, Greece*
[2] *Aristotle University of Thessaloniki, Laboratory of Physical Chemistry, Department of Chemistry, 54124 Thessaloniki, Greece*

6.1 Introduction

Lead halide perovskite solar cells, currently affording efficiencies close to crystalline silicon, represent a technology on the road of commercialization. However, these materials are highly ionic compounds with questionable stability in humid air. Therefore, given that water-soluble lead compounds can accumulate into the human body, their seemingly imminent commercialization could create a new pathway of toxic lead invasion into our society. A solution to these issues comes from tin, which sits just above lead, belonging to group 14 of the periodic table, and is generally considered as a metal of higher environmental compatibility. Tin perovskite solar cells already demonstrate acceptable stability and efficiencies approaching those of other emerging technologies such as dye solar cells and amorphous silicon. Here, we review the progress in this technology. In particular, we focus on fabrication methods of tin-based materials and devices, highlighting the key role of additives to compensate tin vacancies, control film morphology, passivate traps, and enhance tolerance toward oxidation. We also discuss ongoing challenges facing tin perovskites and give an outlook on future research avenues that might enable the transition for emerging optoelectronics from lead to tin halide perovskites.

6.2 Why Tin Halide Perovskites?

6.2.1 Tin as the Sole Viable Alternative

A large number of lead-free halide perovskites (or perovskite-like compounds) based on different metal ions has been reported [1]. For instance, bismuth [2] or d-block metals [3] are able to form halides of variable structural dimensionality.

Perovskite Solar Cells: Materials, Processes, and Devices, First Edition.
Edited by Shahzada Ahmad, Samrana Kazim, and Michael Grätzel.

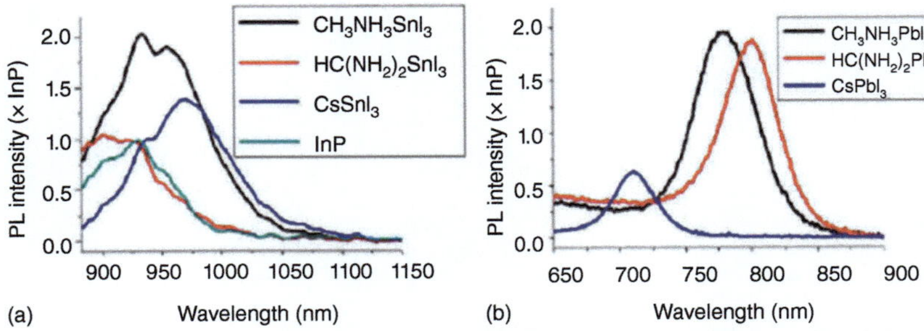

Figure 6.1 Photoluminescence (PL) spectra of AMI_3 perovskites (where A = Cs, MA, and FA and M = Sn and Pb) demonstrating the clear difference in optical bandgap between Sn and Pb compounds (shown in part (a) and (b), respectively). The data are normalized toward a single crystalline sample of InP that is used as a benchmark. Source: Stoumpos et al. [9].

However, the corresponding solar cells have not presented appreciable efficiencies thus far [4]. A special case came from germanium that also belongs to group 14 but is much lighter than lead. Viable solar cells of around 7% were possible for $CsGeI_3$ perovskites but only when $CsSnI_3$ was combined at 50% molar ratio in the structure [5]. On the other hand, Sn-only perovskite devices already present efficiencies of around 10% [6]. Therefore, it looks like there is currently a failure to suggest a viable alternative for toxic lead perovskites except for the case of tin perovskites.

6.2.2 Favorable Optoelectronic Properties of Tin Perovskites

6.2.2.1 Low Bandgap

Record lead halide perovskites possess an optical bandgap of 1.5–1.6 eV. Their photovoltaic efficiency now lies at over 25% [7], implying that they already approach their thermodynamic limit that is around 30% according to the Shockley–Queisser (S–Q) model. An even higher theoretical maximum is predicted for materials of lower bandgap, i.e. 32.5% for a 1.4 eV semiconductor [8]. Notably, tin can form halide perovskites such as $CsSnI_3$, $MASnI_3$ ($MA{=}CH_3NH_2$), and $FASnI_3$ (FA stands for $NH_2CH{=}NH_2$), which have a bandgap of ~1.3–1.4 eV [9]. Figure 6.1 clearly shows the difference between the two metals (Sn and Pb), depicting the emission wavelength maximum (from photoluminescence spectra) that matches well with the absorption energy edge of the perovskites. Therefore, these materials seem even more suitable from their lead analogues, primarily as a result of different relativistic interactions.

6.2.2.2 High Charge Carrier Mobility

Tin halide perovskites have demonstrated high charge carrier mobilities in comparison with those of lead perovskites [10]. These mobilities result from exceptionally low effective masses of both electrons and holes due to the large dispersion of the conduction and valence band [11]. Spectacular THz mobilities of up to $170\,cm^2\,V^{-1}\,s^{-1}$ for $MASnI_3$ polycrystalline films have been reported [12]. Even more impressive Hall mobilities of $200–585\,cm^2\,V^{-1}\,s^{-1}$ have been determined for single crystals or polycrystals of $MASnI_3$, $FASnI_3$, and $CsSnI_3$ perovskites. The above values are significantly larger than those attained by pure lead perovskites [13]. These unique mobilities enable the materials to sustain an affordable charge carrier diffusion length and therefore ensure an efficient charge extraction. For instance, a diffusion length of 0.55 μm for polycrystalline $MASnI_3$ films [14] or 0.93 μm for $CsSnI_3$ ingots, containing high-quality single crystals, has been determined [15].

6.2.2.3 Similar Properties with Lead Perovskites

Sn perovskites have nearly comparable optoelectronic properties with Pb perovskites. Both semiconductors have large absorption coefficients in the visible region (10^4–10^5 cm^{-1}) due to their direct band transition [16]. As in the case of $MAPbI_3$ perovskite, no Stokes shift was found between the photoluminescence (PL) peak and the absorption onset for $MASnI_3$ [17]. Additionally, the observed steep absorption edge of the material implies a sharp Urbach tail [18]. The corresponding energy at 300 K was determined to be as low as 12 meV, which is comparable with that of the $MAPbI_3$ perovskite. All these contribute to a low number of localized states within the bandgap [19].

Despite the fact that high mobilities have been demonstrated, electron–phonon interactions, spin–orbit coupling [20], and slow charge carrier cooling effects [21] have been also observed in tin perovskites that could open the possibility for solar cells beyond the S–Q limit [22]. Furthermore, the majority of Sn perovskites exhibit binding energies at room temperature within a range similar to that of the lead perovskites (10–50 meV), facilitating free charge carriers formation [23]. Interestingly, Milot et al. determined an exciton binding energy of only 3.1 meV for $FASnI_3$, which is even lower than values typically observed for lead iodide perovskites and compares well with values for traditional inorganic semiconductors of similar bandgap energy [24]. The combination of these unique photophysical properties prove that tin perovskites have the potential to match the optoelectronic performance of lead perovskites. To do so, serious issues should be adequately addressed.

6.3 Concerns About Tin-Based Perovskites

6.3.1 Severe Non-radiative Recombination

In solar cells made with lead halide perovskite films of excellent optoelectronic quality (i.e. well passivated), total recombination is ultraslow, as photocarrier lifetimes of almost 10 μs under solar illumination fluences have been reported [25]. Notably, even in the case of films with a high trap-state density (10^{14}–10^{17} cm^{-3}), comparable with the photoexcited charge carrier density in an operating solar cell under AM1.5 illumination (~10^{14}–10^{16} cm^{-3}), the lifetime is still long (over 100 ns) and even longer than that of conventional direct gap semiconductors [26]. In above case, steady-state photocarrier densities are significantly larger than the intrinsic carrier densities (~10^5 cm^{-3}) and doping densities [27]. However, for Sn-based perovskites, the photoexcitation densities compare well with the hole

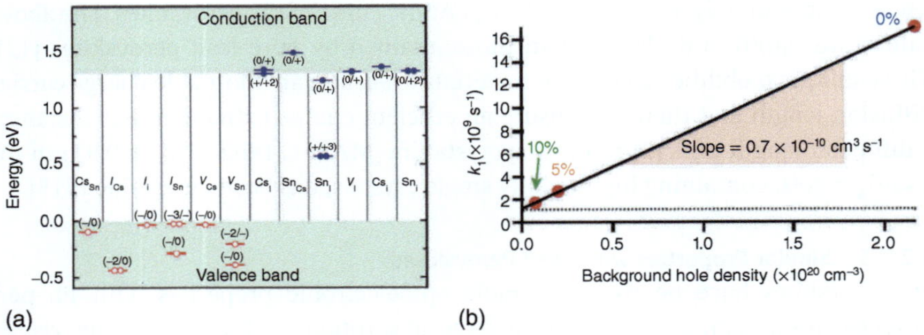

Figure 6.2 (a) Calculated transition energy levels for various intrinsic defects including tin vacancies (V_{Sn}). The region under valence band maximum (VBM) is in shaded color. One hole can be released during the transition of the defect charge state. Source: Xu et al. [29]. (b) Linear dependence between hole doping density and monomolecular charge carrier recombination rate, k_1. Source: Milot et al. [24].

doping densities, derived to be more than 10^{18} cm^{-3} [28]. This creates an additional channel for fast recombination of electrons with the high background density of holes, leading to the relatively poor lifetimes of picoseconds up to only 10–20 ns [14]. First-principles calculations have shown that the high hole concentration results from the very low formation energy of Sn^{2+} vacancies (V_{Sn}). Remarkably, under Sn-poor conditions during perovskite growth, V_{Sn} become the dominant intrinsic defects as they present even negative formation energy (Figure 6.2a) [29].

Certainly, hole (h^+) doping can also result from the ease oxidation of Sn^{2+} to Sn^{4+} [30]:

$$Sn^{2+} + O_2 \rightarrow SnO_2 + 2h^+ \tag{6.1}$$

Heavy p-doping (~10^{17} cm^{-3}) and metal-like conductivity (hole mobility measured as high as 585 cm^2 V^{-1} s^{-1}) of $CsSnI_3$ perovskite has been attributed to V_{Sn} [31]. Massive hole doping densities of up to 10^{20} cm^{-3} have been also determined for $FASnI_3$ films. Notably, the authors suggested a linear increase of monomolecular charge recombination with the doping density in $FASnI_3$ (Figure 6.2b) [24]. Similar hole doping densities have been observed in other Sn^{2+} (non-perovskite) semiconductors such as SnTe and SnS [32]. Based on the discussion above, it is evident that perovskites with missing Sn^{2+} ions can present metal-like conductivity and the corresponding solar cells will thus suffer from severe recombination, demonstrating a bad diode rectifying behavior or even a complete short circuit.

6.3.2 Poor Stability

Grancini et al. have proved that lead perovskite solar cells can stay operational for one year [33], which is still shorter than expected for commercialized photovoltaic (PV) technologies [34]. Thus, the stability of perovskite solar cells is still an open issue. Now, one should bear in mind that Sn perovskites have significantly worse stability than their lead analogues. After exposure to ambient air, Sn^{2+}-based materials are prone to oxidation, degrading to Sn^{4+}-based compounds such as SnO_2 and other relevant hydroxides. Air stability should not be a primary concern for real solar panels since they are always encapsulated to ensure water- and oxygen-free conditions during the whole lifetime of the device [1]. However, here, intrinsic factors, which are directly correlated with the propensity of tin to equally appear in both 2+ and 4+ oxidation states, impede the stability of the materials even with encapsulation [35].

6.4 Control of Hole Doping

6.4.1 Sn^{2+} Compensation/Necessity of Adding SnF_2

Xu et al. have suggested that Sn-rich growth conditions should be used to control the formation of tin vacancies and thus regulate the concentration of hole carriers to a moderate level – but not too rich since extra Sn could occupy I sites and Sn_I antisite defects also act as recombination centers [29]. To do so, a "Sn self-compensation" strategy should be employed, and the most obvious way to compensate missing Sn^{2+} is to add more of it. For this, excess of SnX_2 (X = I, Br, Cl) was added to $CsSnI_3$ or $MASnI_3$ precursor solutions to improve both efficiency and stability of the devices [36–39]. In a special case, Yokoyama et al. observed that a large excess of SnO and $Sn(OH)_2$ secondary phases, in situ generated by a reaction of methylamine gas with SnI_2 films, also resulted in low hole carrier concentration [40].

Notably an alternative tin source, SnF_2 being a powerful reducing agent due to the presence of fluoride that has the greatest electronegativity in the periodic table, seems the more adequate additive to effectively lower the doping densities in Sn-based perovskites [41]. For instance, when 5–20% m of SnF_2 was added in $CsSnI_3$ precursor solution, the initial high carrier density ($>10^{19}$ cm^{-3}), with holes as the majority carriers, was severely reduced down to 10^{17} cm^{-3} for the 20% tin fluoride (Figure 6.3a) [42]. Strikingly, the above carrier density was further reduced to 6×10^{15} cm^{-3} when Br^- ions were added in the solution [43]. The beneficial role of the simultaneous presence of Br^- ions and SnF_2 was also confirmed in the case of $FASnI_3$ perovskite [44]. When the doping density is to be controlled down to the 10^{15} cm^{-3} range, then a long diffusion length of over 1 μm will be achieved, ensuring the effective extraction and collection of all the carriers photogenerated within a half micron-thick polycrystalline perovskite film (Figure 6.3b) [28]. Currently, the incorporation of (around 10%) SnF_2 in Sn-based perovskites has become a universal method to fabricate devices with high performance and reproducibility.

6.4.2 Additives to Improve SnF_2 Dispersion

Kumar et al. indicated that, above a specific molar ratio, SnF_2 could not properly mix with $CsSnI_3$. Instead, SnF_2 tends to exist in the grain boundaries, forming a detrimental secondary phase [42]. Phase separation takes place because F^- is too small to be incorporated in the crystal lattice, when its ratio exceeds 10–20% [45]. However,

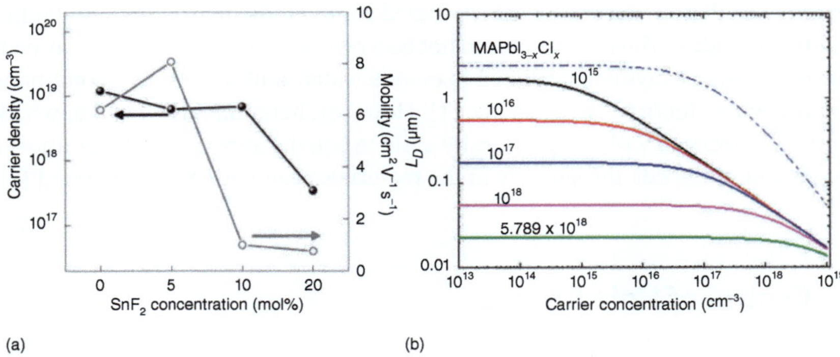

Figure 6.3 (a) Charge carrier density (and Hall mobility) of $CsSnI_3$ films for various SnF_2 concentrations. Source: Kumar et al. [42]. (b) Theoretical values of diffusion length (L_D) of $MASnI_3$ perovskite against the carrier concentration for different hole doping. A decrease in the background level of doped holes down to 10^{15} cm^{-3} corresponds to an increase in the diffusion length up to over 1 μm. Source: Noel et al. [28].

it seems that fine-tuning of SnF_2 molar ratio can result in phase separation-free films only under certain conditions. There are specific cases where other additives, should be incorporated in the precursor solution to assist SnF_2 to uniformly spread in the perovskite and consequently reduce tin vacancies without deteriorating film morphology. These additives are usually Lewis bases, with an ability to donate electrons to SnF_2 that is a typical Lewis acid. These Lewis acid–base interactions result in the formation of stable adducts such as that of SnI_2·3DMSO already identified by X-ray diffraction [46]. A characteristic example is pyrazine that restricted phase separation for tin fluoride through forming an adduct with Sn^{2+}. This mediator could be easily removed during film annealing process due to its low boiling point of 115 °C. Remarkably, a champion efficiency of 4.8% was achieved by devices that also maintained the 98% of their initial efficiency after 100 days' storage in the dark under ambient conditions [47]. Similar effects have been observed with other Lewis bases such as trifluoroethylamine [48] trimethylamine [49], and triethylphosphine [50]. Another example of additives, acting as Lewis bases to donate electrons to Sn^{2+}, is sulfonate. More specifically, SO_3^-, from the potassium salt of hydroquinone sulfonic acid, was found to interact with excessive Sn^{2+} (in the form of $SnCl_2$) in $FASnI_3$ precursor solution via coordination interactions and electrostatic attraction. This forms a $SnCl_2$-additive complex and suppresses $SnCl_2$ aggregation (Figure 6.4). With this, as-fabricated un-encapsulated devices attained appreciable stabilized efficiencies of 6.6% while simultaneously retaining the 70% of their initial efficiency after 600 hours storage in ambient [51].

Similarly positive effects have been shown by SCN^- (added as FASCN in the precursor solution) that improved the crystallographic and morphological quality of PEA-$FASnI_3$ perovskite films (where PEA stands for phenethylammonium cation), resulting in functional devices of stabilized efficiencies up to 7.8% [52]. Even higher efficiencies were attained by the same perovskite with NH_4SCN as additive [53].

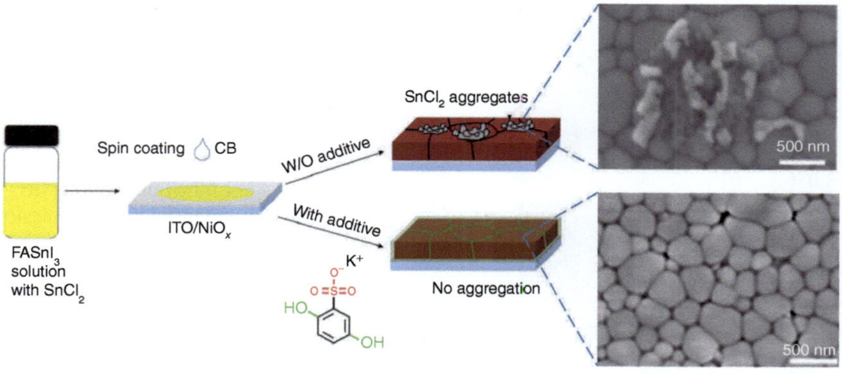

Figure 6.4 Schematic illustration of the preparation of the perovskite films and their morphologies with or without potassium salt of hydroquinone sulfonic acid as the additive. Source: Tai et al. [51]. Adapted with permission from John Wiley & Sons.

6.4.3 Elimination of Sn^{4+} Impurities

Besides the formation of Sn^{2+} vacancies during film processing, hole doping may take place due to undesirable oxidation to Sn^{4+}. Currently, there are three different ways to prevent this.

6.4.3.1 SnI_2 Purification

Considering the inevitable oxidation, Sn^{4+} (mostly as SnI_4 or even SnO_2) can be even formed during the synthesis of SnI_2 that is the precursor molecule for Sn-based perovskites. Even if advanced purification methods are adopted to produce high-purity, commercially available SnI_2, it should be noted that both SnI_2 and SnI_4 are synthesized through the addition of (1 or 2 m of) iodine in Sn powder. Therefore, different purity from the same commercial source (99%, 99.99%, 99.999% SnI_2) may result in a different SnI_4 content [42]. Gu et al. have already demonstrated that even 0.5 mol% addition of SnI_4 in commercial SnI_2 can deteriorate device performance [54].

To inhibit the presence of harmful Sn^{4+}, Kanatzidis group have fabricated self-developed SnI_2 and further purified it in vacuum, discovering that fluorescence lifetimes of films, deposited out of it, were significantly longer than those of films made with 99.999% commercial SnI_2 [9]. For this, Gu et al. recommend to always sublime commercial tin halide powders under inert atmospheres before any use [45].

6.4.3.2 Reaction of Sn Powder with Sn^{4+} Residuals

A more straightforward idea is to add Sn powder (or granules) in SnI_2 solution. If this solution contains SnI_4 impurities, then solid Sn will react with liquid Sn^{4+} to form SnI_2 following the reaction below:

$$Sn + SnI_4 \rightarrow 2SnI_2 \tag{6.2}$$

The reverse reaction is a disproportionation when Sn^{2+} is simultaneously oxidized and reduced. However it is not spontaneous ($\Delta G > 0$). Therefore, loss of Sn^{2+} cannot be due to the disproportionation. This case was confirmed by Zhang et al. [55].

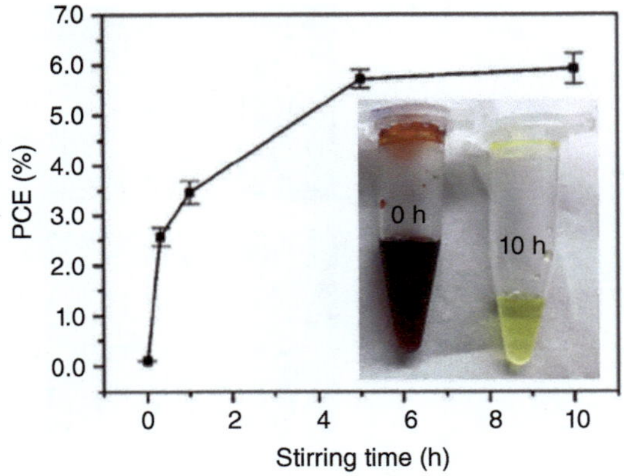

Figure 6.5 The evolution of efficiency of devices prepared from the precursor solutions stirring with Sn powder for different time. Inset: Solutions of SnI_2 (99%) + FAI + SnF_2 in DMSO without (left) and with Sn powder (right) after filtration. Source: Gu et al. [54]. Adapted with permission from John Wiley & Sons.

On the contrary, controlling reaction (6.2) can result in a complete conversion of any present SnI_4 back to SnI_2. Indeed, a dark red solution of 99% non-sublimed SnI_2 (along with SnF_2 and FAI) is progressively becoming orange/yellowish when excess of tin powder was added. After sufficient reaction time (defined by the stirring time during which the powder interacts with the solution), the solution turned into light yellow, implying that most of the SnI_4 disappeared. With this solution after filtration, compact and smooth $FASnI_3$ films, with crystal grains of similar size, were deposited on poly(3,4-ethylenedioxythiophene)–poly(styrenesulfonate) substrates (i.e. PEDOT:PSS (poly(3,4-ethylenedioxythiophene)-poly(styrenesulfonate))/FTO (fluorine-doped tin oxide). On the contrary films, without the addition of Sn powder, they were coarse and full of pinholes. Evidently, the stirring time controls the film quality and thus the efficiency of the devices. Remarkably, nonoperational devices without Sn powder were able to achieve reproducible efficiencies of 6%, when Sn powder reacted with SnI_4 residues for 10 hours (Figure 6.5) [54]. The approach was recently followed by others, leading to the record efficiencies of 21% for mixed lead–tin perovskites [56].

6.4.3.3 Addition of Reducing Agents

To further suppress undesirable oxidation reactions, very powerful reducing agents are commonly added in the precursor solutions along with SnF_2. For instance, hypophosphorous acid (HPA) was added in a $SnF_2/CsSnI_2Br$ precursor solution. X-ray photoelectron spectroscopy (XPS) measurements have clearly indicated that the amount of Sn^{4+} in HPA-treated $CsSnIBr_2$ (4%) was much lower than that (23%) of the non-treated sample. With this, the charge carrier density decreased from 1.6×10^{16} down to 1.5×10^{15} cm^{-3}. More notably, the as-fabricated devices presented stabilized efficiencies of 3.3% and an impressive stability for 77 days in ambient environment (humidity ~20%) after encapsulation. The enhanced tolerance toward oxidation was attributed to the presence of HPA, reacting slowly with O_2 and thus acting as an oxygen sink [57].

In a similar approach, Song et al. have incorporated hydrazine vapors during crystallization of SnF_2-doped $CsSnI_3$ perovskite films, suppressing the formation of Sn^{4+}. This was confirmed by XPS and further justified by UV–vis and transient PL spectroscopies that have shown fewer defects and severely reduced trap-mediated recombination. Strikingly, a short-circuit current (J_{sc}) of more than 30 mA cm^{-2} could be extracted from the devices, resulting in the efficiency increase from 0.2% to 1.8% in the presence of hydrazine [58]. Following the above concept and given that N_2H_4 is a highly toxic and explosive reagent, very similar molecules have been incorporated in $MASnI_3$ or $FASnI_3$ perovskites such as hydrazinium chloride [59] or hydrazinium iodide [60, 61]. In a very recent report, trihydrazine dihydriodide additive in $FAMASnI_3$ perovskite films led to an impressive stabilized efficiency of 8.1% [62].

6.5 Films Deposition

6.5.1 Crystallization Tuning

Besides controlling hole doping and thus recombination in tin perovskites, smooth and pinhole-free polycrystalline films with large grains are imperative to fabricate highly efficient solar cells. In the case of lead, an effective methodology for depositing high-quality perovskite layers is the Lewis acid–base adduct approach where PbI_2 forms a stable adduct with Lewis bases when dissolved in typical polar aprotic solvents such as dimethyl sulfoxide [63]. The weak chemical interaction in the adduct regulates crystallization kinetics/growth and thus controls morphology of perovskite films. Then, an antisolvent is added to "break" the adduct by extracting the solvent and precipitate a fully crystalline perovskite [64]. For tin perovskites, one should consider that SnI_2 is a stronger Lewis acid than PbI_2. Therefore, it reacts faster, resulting in the uncontrollable crystallization and discontinuous film formation. This is most probably why two additional molecules of DMSO are needed to stabilize SnI_2 through the formation of $SnI_2{\cdot}3DMSO$ intermediate compared with the 1 : 1 $PbI_2{\cdot}DMSO$ adduct [46].

6.5.1.1 Solvent Engineering

As anticipated, the most prominent solvent to stabilize SnI_2 intermediates is pure DMSO (so excess of DMSO is present to easily react with SnI_2), which promoted homogeneous nucleation and enabled adjustable perovskite film growth rate [62]. Alternatively, ~3.2 mol of DMSO are added in 1 mol of dimethylformamide (DMF) to form the $SnI_2{\cdot}3DMSO$ intermediate again [65]. After the adduct formation, various antisolvents have been employed effectively, such as toluene [53], diethyl ether [66], and, more commonly, chlorobenzene [67]. With this, a flowerlike film with low surface coverage turns into a film with uniform surface, full coverage, and clear features of grains (Figure 6.6). After an additional annealing step at 70–100 °C, $FASnI_3$ devices with efficiencies over 7–8% are commonly attained. The approach was effective for $FASnI_3$ also for typical mixed-cation perovskites such as $Cs_{0.08}FA_{0.92}SnI_3$ [68] and $FA_{0.75}MA_{0.25}SnI_3$ [69]. Interestingly in the first case, cesium enhances the structural stability of $FASnI_3$, resulting in the retention of 70% of the initial efficiency of the device after exposure for 300 minutes in air [68]. In the latter case, hot (65 °C)

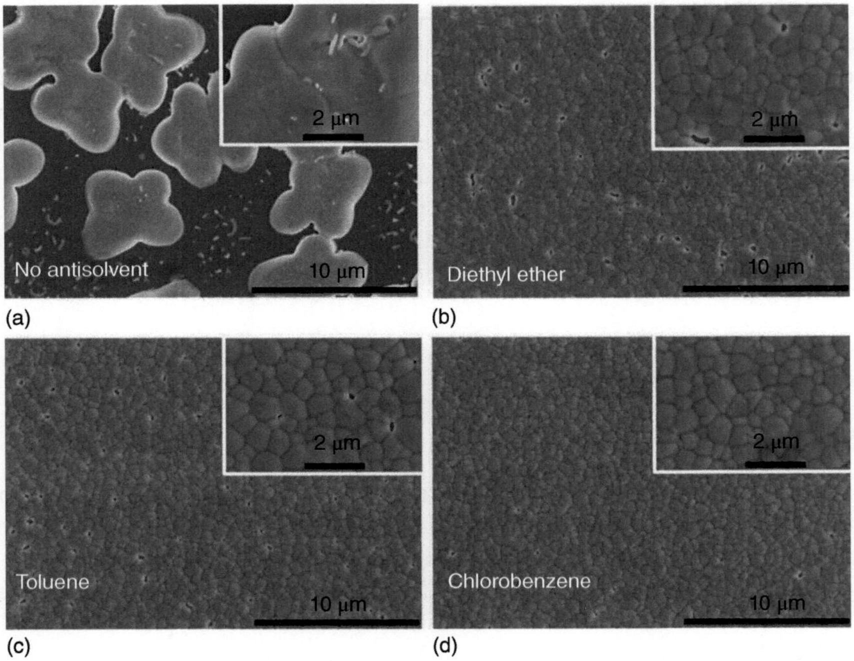

Figure 6.6 Top-view scanning electron microscopy (SEM) images of $FA_{0.75}MA_{0.25}SnI_3$ films deposited on indium-doped tin oxide (ITO)/PEDOT:PSS substrates without (a) or with antisolvents (b, c, and d). Source: Liu et al. [67]. Adapted with permission from American Chemical Society.

chlorobenzene was added as antisolvent, improving the density of nucleation sites, while additional DMSO vapor atmosphere, during annealing, led to grain coarsening. The synergy of both effects resulted in stabilized efficiencies of 6.5% [69]. In a similar approach, methylammonium chloride vapor annealing resulted in *en*-doped $FASnI_3$ films (where *en* stands for ethylenediammonium cation) with large grains and low crystal defects over an aperture area of ~1 cm^2, leading to a device efficiency of 6.3% [70].

6.5.1.2 Additives to Slow Down Crystallization Kinetics

Even if solvent engineering has proven effective thus far, various additives with functional groups may also alter the crystallization dynamics through interactions with species in precursor solutions. A characteristic example is a small or bigger molecule, bearing hydroxyl groups, which can form hydrogen bonds with iodide ions, retarding the crystal growth of $FASnI_3$ perovskite.

For instance, the incorporation of 5-ammonium valeric acid iodide (AVAI) in $FASnI_3$ precursor solution led to the formation of a hydrogen bond (O—H$\cdots$$I^-$ and N—H$\cdots$$I^-$) between the carboxylic acid (-COOH) and the ammonium (-NH_3^+) end groups with I^- from $[SnI_6]^{4-}$ octahedra. The as-deposited films presented better crystallinity and enhanced carrier lifetime. With this, a significant increase in device efficiency from 3.4% to 7.0% was realized [71].

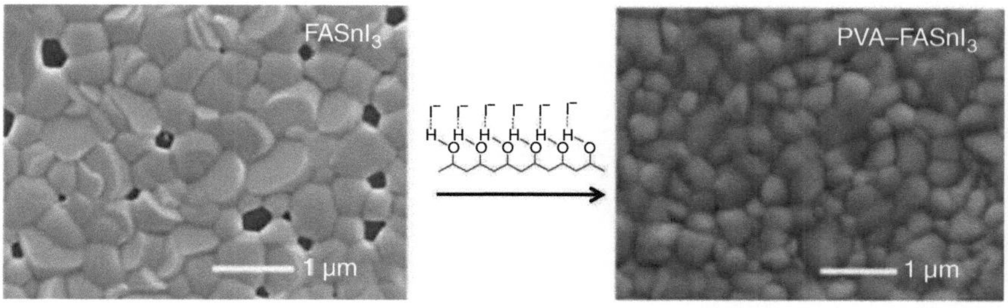

Figure 6.7 Top-view SEM images of pristine $FASnI_3$ perovskite films and after the PVA–$FASnI_3$ interaction through hydrogen bonding. Source: Meng et al. [65]. Adapted with permission from John Wiley & Sons.

In the same direction, by adding polyvinyl alcohol (PVA), ^1H nuclear magnetic resonance (NMR) and density functional theory (DFT) calculations have demonstrated hydrogen bonding interactions (O—H· · ·I) between hydroxyl group and iodide. This interaction slowed down crystal growth by increasing the number of nucleation sites, resulting in the homogeneous and pinhole-free uniform films (Figure 6.7). The PVA–$FASnI_3$ devices attained a severely enhanced efficiency of 8.7% (from 6.4% without PVA additive) [65].

6.5.2 Posttreatment Strategies/Surface Trap Passivation

After annealing, additional posttreatment of the as-fabricated films with appropriate molecules is a usual practice for lead perovskites to heal traps (e.g. dangling bonds) inevitably forming at the surface of the film [72]. A typical category of materials being used is that of halide compounds that bear large hydrophobic cations such as diammonium and phenethylammonium (PEA^+) [73]. Notably, in the case of PEAI, a record efficiency of 23.3% was achieved for $FAMAPbI_3$ perovskite-based devices [74].

Following the above concept, $FASnI_3$ films were posttreated with PEABr solution in isopropanol. Space charge-limited current measurements have shown a significantly reduced defect density at the surface upon modification. Simultaneously, Br^- ions contributed to a severe enlargement of grain size. The combination of these effects led to an increase of the device efficiency from 4.8% to 7.9%. At the same time, PEABr-treated devices retained approximately 80% of their initial efficiency after 350 hours of continuous illumination in air [75]. In another study, PEABr was deposited on top of PEDOT:PSS, acting as the crystallization substrate for $FASnI_3$ perovskite (Figure 6.8a). PL measurements verified that this interlayer reduced trap states close to the interface (Figure 6.8b). With this, stabilized efficiencies of 6.3% were achieved [76].

A small amount of ethylenediamine cation has been found to stabilize the α-phase of $CsPbI_3$ and passivate surface defects [77]. In the case of *en*-$FASnI_3$ perovskite, a posttreatment with ethylenediamine was found to effectively passivate the surface defects through stabilization of undercoordinated Sn^{2+}, as suggested by XPS measurements. Remarkably, the efficiency of the devices improved from 8.1% to 9.4%, which became 10.2% after one-week storage [78].

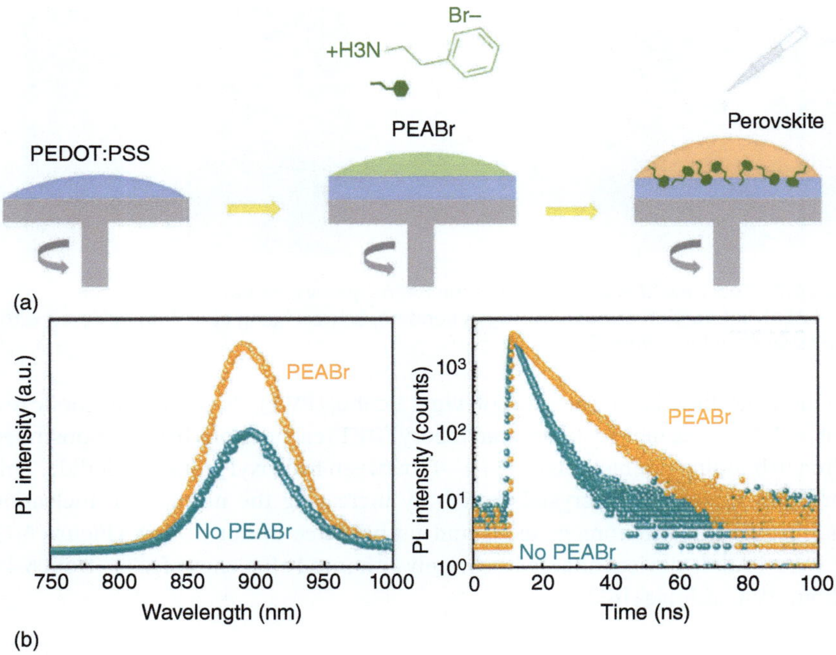

Figure 6.8 (a) Schematic diagram depicting the deposition of PEABr between PEDOT:PSS/ITO substrate and perovskite. (b) Steady-state and time-resolved PL spectra (measured at 890 nm) for PEDOT:PSS/$FASnI_3$ samples with and without PEABr interlayer (excitation of 540 nm from the ITO side). Source: Chen et al. [76].

6.6 Contacts/Interface Engineering

When having a perovskite film of optimum optoelectronic quality in hand, then proper contacting ensures efficient extraction and collection of the photogenerated carriers at the electrodes. As in lead perovskites, n–i–p (regular) and p–i–n (inverted) structures are employed also in tin perovskites.

In the first case, typical electron acceptors such as inorganic TiO_2 and SnO_2, usually in combination with a thin film of mesoporous TiO_2, are used while poly(triaryl)amine polymer is adopted as hole acceptor. Since the conduction band minimum of $FASnI_3$ film lies approximately 0.8–0.9 eV higher than that of TiO_2, interfacial layers such as ZnS [79] or C_{60} pyrrolidine tris-acid [80] are deposited on top to assist the more facile cascade of photoexcited electrons. In the latter case, a record open-circuit voltage (V_{oc}) of 0.72 V, accompanied with a stabilized efficiency of 6.0%, was attained for *en*-$FASnI_3$ devices [80]. In another work, benzodithiophene-based organic small molecules containing tetra- and di-triphenyl amine donors have been shown to be more energetically suitable than conventional hole acceptors for *en*-$FASnI_3$ solar cells, delivering efficiencies up to 7.6% [81].

However, the most common solar cell stratification is the inverted structure. This is most probably why the exposed surface of a tin perovskite easily oxidizes

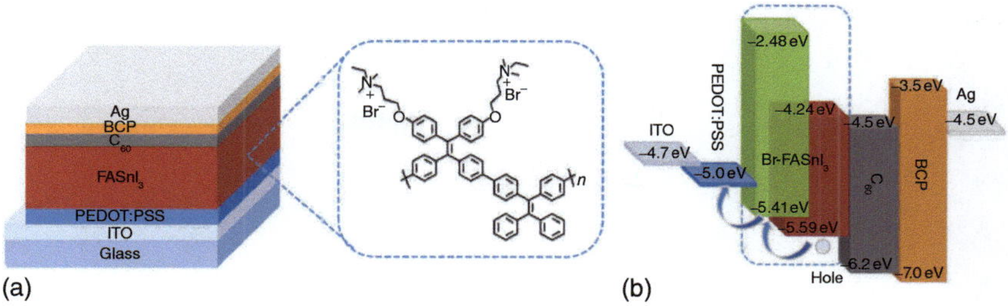

Figure 6.9 (a) Inverted device architecture of solar cells based on organic semiconductor-incorporated $FASnI_3$ perovskite (the chemical structure of the organic semiconductor is also shown). (b) Energy level diagram of the components of the device. It is clear that hole transfer between perovskite grains and between perovskite and PEDOT:PSS is enhanced through the incorporation of the Br-containing organic semiconductor into the $FASnI_3$ perovskite precursor solution. Source: Liu et al. [84].

to SnO_2 that is an n-type material. Then C_{60} and bathocuproine (BCP) are typically deposited (usually out of evaporation) on top to extract electrons. For hole extraction, PEDOT:PSS polymer is adopted. Due to its easy oxidizing ability, $FASnI_3$ usually exhibits a deeper valence band, creating a mismatch with the energy levels of the hole transport layer. For this, LiF [82] and polyethylene glycol [83] have been introduced as discrete interfacial layers to improve hole extraction or as modifiers to control the work function of PEDOT:PSS, respectively. In a special case, a semiconducting molecule poly[tetraphenylethene 3,3-(((2,2-diphenylethene-1,1-diyl)bis(4,1-phenylene))bis(oxy))bis(*N*,*N*-dimethylpropan-1-amine)tetraphenylethene] has been introduced into $FASnI_3$ perovskite precursor to serve as a hole transport medium between grains and improve band alignment with PEDOT:PSS (Figure 6.9). With this, a significant increase in efficiency from 4.8 (without the molecule) to 7.3% was achieved [84].

6.7 Ongoing Challenges

6.7.1 Efficiency

Most of the devices with champion stabilized efficiencies of 7.0–8.7% are based on $FASnI_3$ perovskite [67]. Recently, following the progress of lead halide perovskites [85], more focus has been devoted on multidimensional (2D/3D), quasi-2D, or even purely 2D perovskites via the incorporation of large organic cations, which cannot fit in the 3D cage of the $FASnI_3$ perovskite, violating the geometric tolerance factor rules.

For instance, the addition of 8–15% PEAI in $FASnI_3$ resulted in oriented 2D/3D perovskite grains of high crystallinity, leading to the lower trap density and high device efficiencies ranging from 9.0% [86] to 9.4% with negligible hysteresis [53]. Further addition of ethylammonium cations in PEA-$FASnI_3$ films increased the size of grains, which also had better orientation in the out-of-plane direction [66]. Likewise,

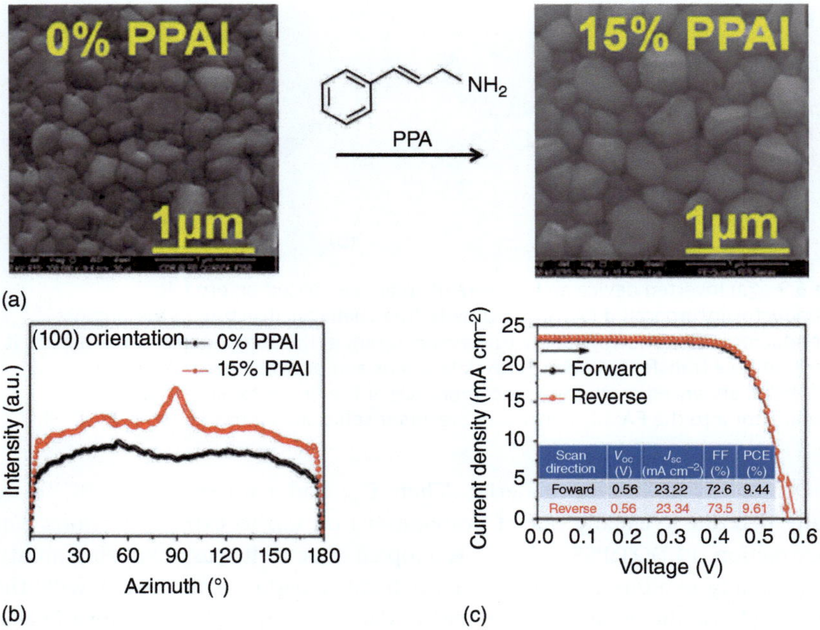

Figure 6.10 (a) SEM images of pristine $FASnI_3$ (0% PPAI) and 15% PPAI-modified $FASnI_3$ films, demonstrating the increase of grain size from around 250–400 nm. (b) Radially integrated intensity plots of the grazing incidence wide-angle X-ray scattering profiles along the ring at $q = 1\,\text{Å}^{-1}$ corresponding to the (100) orientation of the $FASnI_3$ structure. (c) J–V curves of PPA-$FASnI_3$ devices with different scanning directions. Source: Ran et al. [89]. Adapted with permission from Elsevier.

butylammonium [87] and hydroxyethylammonium cations have been employed, reaching efficiencies of up to 5.5% [88].

Since all above cations are insulating, Ran et al. thought of using a cation with a more conjugated and polarizable backbone to facilitate charge transport. For this, they incorporated the iodide salt of 3-phenyl-2-propen-1-amine (PPAI) in $FASnI_3$ precursor solutions (in 15% M ratio) and prepared films with enlarged grain size and preferential (100) orientation in comparison with the additive-free films. The as-fabricated inverted devices exhibited enhanced charge transport, resulting in the very high reverse scan efficiencies of 9.6% with unimportant hysteresis (Figure 6.10) [89].

A unique concept was introduced by the group of Kanatzidis when the dication of ethylene diamine (*en*) was incorporated into the 3D $MASnI_3$ lattice. Surprisingly, this small cation becomes part of the 3D structure; hence, the dimensionality is not altered. However, a high density of SnI_2 vacancies is created, forming so-called "hollow" perovskites. This enlarged unit cell volume and bandgap. However, the films have shown improved film morphology with lower trap-state density and much longer carrier lifetime, leading to the efficiencies of 6.6% for *en*-$MASnI_3$ devices of n–i–p structure [90].

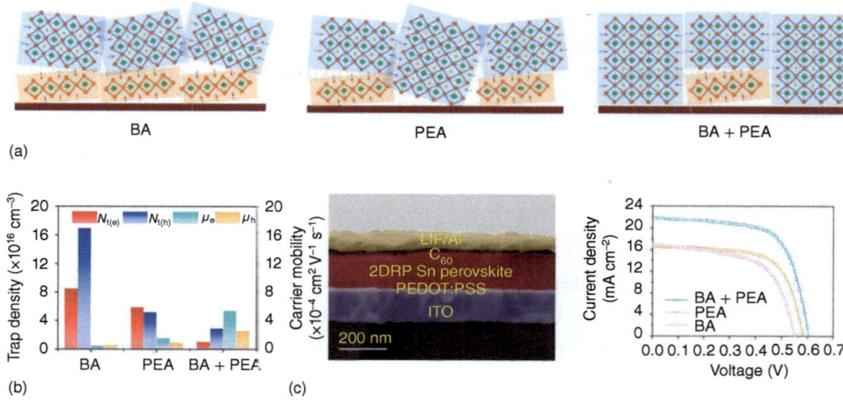

Figure 6.11 (a) Schematic illustrations of crystal orientation of 2D $FASnI_3$ perovskite films with BA, PEA and BA+PEA insulating cations (derived from grazing-incidence wide-angle X-ray scattering [GIWAXS] profiles) and (b) Carrier (electron and hole) trap density and mobility of different perovskite films (determined from dark current measurements on electron-only and hole-only devices) (c) Inverted device structure (from cross-sectional SEM image) and *J–V* curves of 2D perovskite devices based on BA, PEA, and BA + PEA cations. Source: Qiu et al. [94]. Adapted with permission from American Chemical Society.

The insulating cations, such as butylammonium, can be also utilized as organic separating interlayers (spacers) between the $[SnI_6]^{4-}$ octahedra, forming purely 2D perovskites [91]. If these materials can be vertically grown on a rigid substrate, then a continuous carrier transport pathway, free of confinement barrier impediment, could form, enhancing photovoltaic performance. Indeed, highly oriented perovskite films, perpendicular to the substrate, could be deposited when phenethylammonium [92], 5-ammonium valeric acid [93], or mixed butylammonium and phenylethylammonium [94] were incorporated in the precursor solutions. Notably in the latter case, almost all crystals grew perpendicular to the substrate (Figure 6.11a). This resulted in the significant increase in carrier mobility and decrease of trap density in the BA + PEA system compared with those of individual BA and PEA perovskite films, proved through grazing incidence wide-angle X-ray scattering characterization (Figure 6.11b). The obtained films were incorporated in a typical inverted device structure, attaining very high efficiencies (8.8%), surpassing those achieved by BA (5.5%) or PEA (6.3%), and approaching those of 3D perovskites (Figure 6.11c) [94].

Unquestionably, the above values are far from being optimum, approaching only the ~30–40% of their theoretical maximum [8]. This is attributed to both J_{sc} and V_{oc} values that fall short in comparison with their upper limit. More specifically, a J_{sc} of 23–24 mA cm^{-2} is typically attained, representing only the 55–65% of the maximum that can be determined by the bandgap of the material (e.g. $FASnI_3$ perovskite). Even more seriously, huge energy losses of more than 0.65–0.85 eV in comparison with the optical bandgap diminish the measurable V_{oc} down to 0.5–0.7 V [95]. Therefore, improving device performance remains an ongoing challenge. We stress that unveiling the nature of residual traps after SnF_2 and additives incorporation, controlling the doping level of the exposed surface, and understanding the reason behind the significant increase of efficiency upon storage in the dark are keys for future developments.

6.7.2 Stability

Sn-based perovskites are thought to be extremely unstable and degrade rapidly even if processed in an inert atmosphere. Recently, an outstanding result was published when *en*-$FASnI_3$ devices have shown an impressive increase of their initial efficiency from 6.3% up to 8.9% (stabilized at 8%) when stored for 2000 hours in N_2 glovebox [87]. Even more importantly, these perovskites have shown also significant air stability. For instance, encapsulated devices based on *en*-$FASnI_3$ perovskite films retained 96% of their initial efficiency upon aging for 1000 hours in the dark [96]. Likewise, *en*-$FASnI_3$ devices, when incorporating an additional nonpolar guanidinium cation, were able to preserve more than 90% of their efficiency after storage of six to seven days in humid air (20% or 60% relative humidity) without any encapsulation (Figure 6.12a). Interestingly, the efficiency of these devices was recorded at 8.3%, which is currently the only recorded efficiency for a Sn-only perovskite solar cell [97].

Outstanding stability was also demonstrated upon continuous illumination of Sn-based devices. Un-encapsulated AVA-$FASnI_3$ solar cell devices maintained their initial performance at 50% relative humidity conditions for 100 hours under 1 sun illumination [71], while PVA–$FASnI_3$ devices retained their entire initial efficiency at the maximum power point in air after 400 hours of operation (Figure 6.12b) [65].

The most pressing hurdle for tin perovskite solar cells, which is the inherent instability of the material even in an inert atmosphere, seems to be resolved by the incorporation of large hydrophobic cations that block moisture ingress at the boundaries of the perovskite films. However, for the technology to reach the standards of lead perovskites, thermal and operational stabilities under proper aging conditions should be demonstrated [98]. The key issue is the stabilization of the Sn^{2+} ion to prevent its oxidation to Sn^{4+}. The possible solution for this is alloying with other metal ions. Indeed, 1 : 1 solid solutions of $CsSnI_3$ with $CsGeI_3$ formed a perovskite that delivered a promising efficiency of ~7%, with less than 10% decay in efficiency after 100 hours of continuous operation in air under 1-sun illumination. This striking performance is attributed to the formation of a native-oxide layer (GeO_2), which fully encapsulates the perovskite surfaces [5].

6.7.3 Performance over the S–Q Limit/Toward Multijunction Solar Cells

One effective way to aid in the advancement of solar cell technology is to push the efficiency for single-junction solar cells close to their thermodynamic (S–Q) limit. When appropriate, a concept, parallel to the above main strategy, is to overcome this theoretical limit (which applies only for single-junction solar cells). Among various

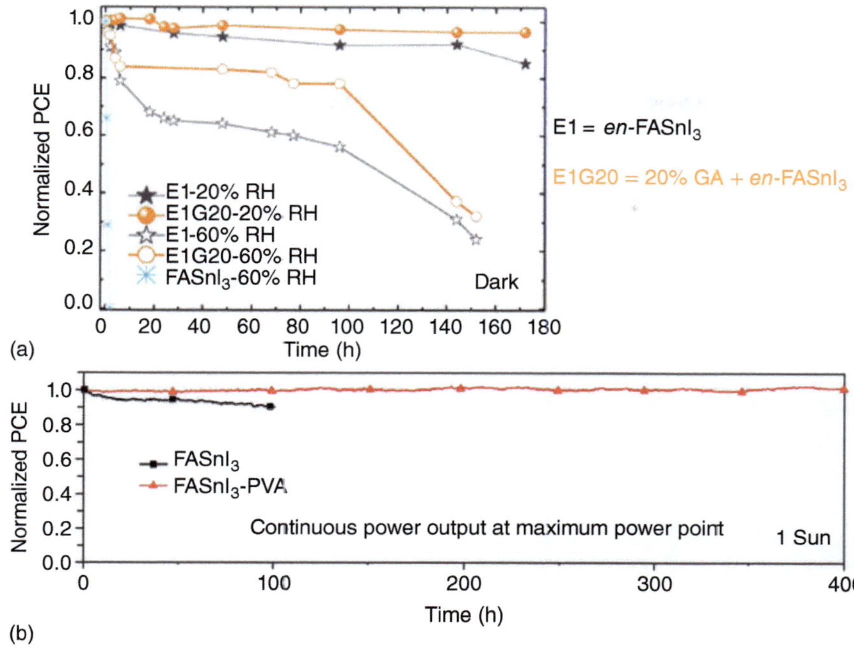

Figure 6.12 (a) Stability of *en*-$FASnI_3$ devices with or without guanidinium (GA) additive when stored in ambient air with relative humidity 20% and 60% without encapsulation. Source: Jokar et al. [97]. (b) Device stability under maximum power point tracking and continuous AM1.5G 1 sun illumination (with a 420 nm cutoff UV filter) at 25 °C in air. Source: Based on Meng et al. [65].

concepts, tandems, or multijunctions, cell design has successfully been proven to overcome the S–Q limit.

Fortunately, halide perovskites can serve as ideal top and bottom cells in a tandem solar cell. While 1.7–1.8 eV lead halide perovskites appear ideal as top cells [99], bandgaps lower than 1.4 eV are not possible for these materials. On the contrary, tin perovskites have an ideal bandgap for a bottom cell in a tandem configuration. However, the instability of pure tin perovskites impeded the fabrication of a tandem solar cell thus far.

Even lower bandgaps (1.1–1.2 eV) can be attained through mixtures of lead with tin due to an observed anomaly of Vegard's law [100]. After a lot of optimization studies, current mixed FA-MA perovskite films, containing 40–60% tin (along with lead), possess an optimum bandgap of 1.2–1.3 eV, accompanied with very low defect density. This leads to record carrier lifetimes of greater than 1 μs and diffusion lengths of 2.5 μm. With this, single-junction solar cells and four-terminal perovskite-on-perovskite tandems could be fabricated with efficiencies of 20.5% and 25.0%, respectively [101]. These results were further verified by similar single-junction devices attaining an efficiency of 21.5% (certified at 19.8%) accompanied with a record J_{sc} of around 31 mA cm^{-2}. Remarkably a V_{oc} of more than 0.82 V was also achieved, implying energy losses of less than 0.45 eV

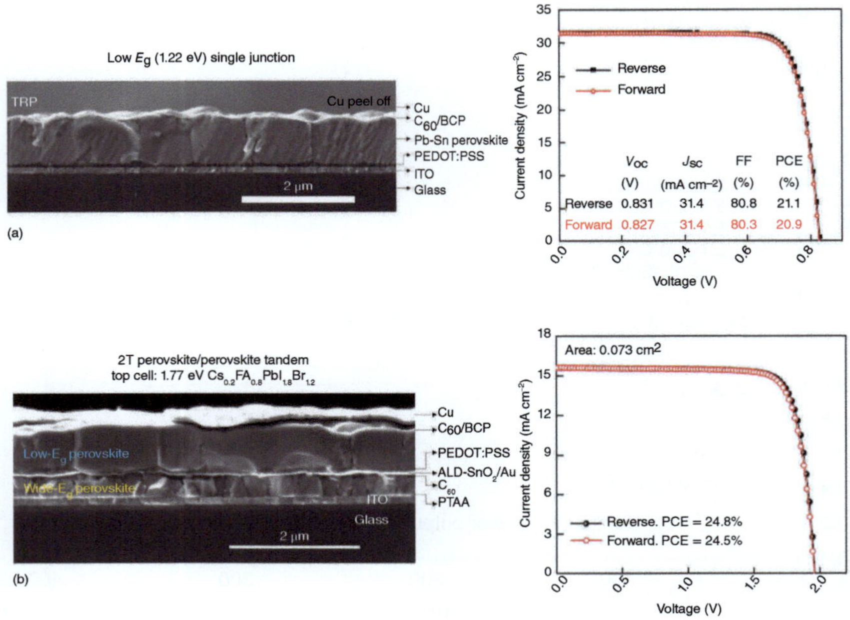

Figure 6.13 (a) Cross-sectional SEM image of a narrow-bandgap Sn–Pb (1,1) perovskite solar cell and corresponding J–V curves under 1 sun AM1.5G illumination. Source: Lin et al. [56]. Adapted with permission from Springer Nature. (b) Cross-sectional SEM image of a two-terminal perovskite/perovskite tandem solar cell and corresponding J–V curves of the best performing small-area tandem solar cell (0.073 cm^2) under forward and reverse scans.

(Figure 6.13a). Even more importantly, using the above perovskite as bottom cell in a two-terminal tandem configuration, an impressive efficiency of 24.8% (certified) was demonstrated (Figure 6.13b) [56].

In terms of stability, low-bandgap mixed lead–tin perovskite devices (without any hole transport material) have retained full initial efficiency under operation near the maximum power point upon 1 sun illumination for over 1000 hours in air. Impressively, they also preserved the 95% of their initial efficiency after 1000 hours at 85 °C in air in the dark and survived from a damp heat test when stored at 85 °C with 85% relative humidity [102]. We should highlight also here that a mixed Sn/Pb perovskite can also serve as a top cell in a tandem configuration. For instance, a mixed Br–I perovskite of the type $MAPb_{0.75}Sn_{0.25}(I_{0.4}Br_{0.6})_3$ was synthesized without presenting significant phase segregation over time. This material had a bandgap of 1.73 eV and presented an appreciable efficiency of 12% accompanied with a V_{oc} of 1 V [103].

Even though the efficiencies of perovskite/perovskite tandems are really impressive, they still fall behind those attained by silicon/perovskite tandems (certified at 28%) [7] and are definitely far from the theoretical maximum that is more than 44% [104]. Besides the inevitable increase of the efficiency of low-bandgap perovskites, other challenges such as energy losses in the wide-bandgap front cell and optical losses caused by the tunnel recombination junction should be mitigated. Further advances in current matching between subcells and lattice matching for maximum crystal quality should enable efficiencies exceeding 30% in the near future [105].

6.7.4 Sustainability

Tin suitability over lead is under debate. It is 15 times more expensive than lead [106], being the third most critical metal in strategic energy technologies with a production almost exclusively concentrated in countries with high political risks [107].

In terms of toxicity, it is commonly accepted that lead (along with mercury, cadmium, chromium, and arsenic) has the greatest potential to cause long-term environmental damage and poses threats to human health when compared with tin. However, there are reports on inorganic Sn or organotin causing hematological changes in animals, hemolytic anemia, or even problems in the central nervous system [108, 109].

In the specific case of perovskite solar cells, there are also serious concerns regarding the suitability of tin perovskites to manifest as viable alternatives over lead analogues. For instance, Babayagit et al. compared the environmental impact of PbI_2 and SnI_2 unveiling the main degradation products of the corresponding perovskite solar cells and concluded that, at equal concentration for both compounds, a higher lethal response was found in zebra fishes exposed to SnI_2 than for PbI_2. The authors rationalized this in terms of the higher acidification of the water due to easier release of HI [110]. Likewise, Serrano-Lujan et al. discovered that actually the main contributors to environmental impact are compounds other than PbI_2 or SnI_2 [111]. Additionally, SnI_2 easily hydrolyzes, forming the almost insoluble SnO_2 that is an inert and biocompatible compound. Thus, it is obvious that toxicity is not directly related to the heavy metal nature. Also, more attention should be paid to key factors such as control of pH reached by the soil [30].

6.8 Conclusion

In 2014, when tin or tin–lead perovskite solar cells have presented their first appreciable efficiencies, people was skeptical about the reproducibility and lifetime of the materials themselves and the corresponding devices. Only a few years later, we have witnessed unprecedented progress in the development of the technology when appropriate means have been discovered to optimize the optoelectronic quality of the film, focusing mainly on controlling the extensive hole doping and thus mitigating recombination. The presence of additives in the precursor solutions (such as SnF_2 and polyvinyl alcohol) not only reduced tin vacancies but also produced extra advantages for the as-prepared films such as grain size enlargement, improved surface coverage, and stability toward oxidation. It appears that the community is converging on the p–i–n inverted structure based on $FASnI_3$ perovskite. Organic, hydrophobic molecules (such as guanidinium iodide or ethylenediamine) seem imperative to passivate traps and seal the grains from water and oxygen ingress. There has been also a huge progress in the case of tin–lead perovskites with efficiencies approaching those of record lead halide perovskites. These low-bandgap materials are crucial to fabricate perovskite-on-perovskite tandems that would be otherwise impossible in the case of pure lead perovskites.

Despite these striking results, one could understand that there is still room for further improvements in the technology as intense research has only just begun. The main areas that require special attention are understanding and mitigation of recombination after tin vacancies compensation, proper contacting to efficiently extract carriers from perovskite, and inhibition of material/device degradation. If the above issues are addressed, then it is very possible that, within the next few years, a tin-based optoelectronic material with a probability of lower toxicity will exhibit as a viable alternative to lead halide perovskites.

Acknowledgments

We acknowledge the financial support from the European Research Council (ERC) through Consolidator Grant (818615-MIX2FIX) and Hellenic Foundation for Research and Innovation (HFRI) under the "First Call for H.F.R.I. Research Projects to support Faculty members and Researchers and the procurement of high-cost research equipment grant" (Project Number: 1027).

References

1 Abate, A. (2017). *Joule* 1: 659.
2 Slavney, A.H., Hu, T., Lindenberg, A.M., and Karunadasa, H.I. (2016). *J. Am. Chem. Soc.* 1387: 2138.
3 Theofylaktos, L., Kosmatos, K.O., Giannakaki, E. et al. (2019). *Dalton Trans.* 48: 9516.
4 Ke, W. and Kanatzidis, M.G. (2019). *Nat. Commun.* 10: 965.
5 Chen, M., Ju, M.-G., Garces, H.F. et al. (2019). *Nat. Commun.* 10: 16.
6 Wei-Guang Diau, E., Jokar, E., and Rameez, M. (2019). *ACS Energy Lett.* 48: 1930.
7 Green, M.A., Dunlop, E.D., Levi, D.H. et al. (2019). *Prog. Photovoltaics* 27: 565.
8 Park, N.-G. and Segawa, H. (2018). *ACS Photonics* 58: 2970.
9 Stoumpos, C.C., Malliakas, C.D., and Kanatzidis, M.G. (2013). *Inorg. Chem.* 52: 9019.
10 Kagan, C.R., Mitzi, D.B., and Dimitrakopoulos, C.D. (1999). *Science* 286: 945.
11 Stoumpos, C. and Kanatzidis, M.G. (2016). *Adv. Mater.* 28: 5778.
12 Zhao, B., Abdi-Jalebi, M., Tabachnyk, M. et al. (2017). *Adv. Mater.* 29: 1604744.
13 Herz, L.M. (2017). *ACS Energ. Lett.* 2: 1539.
14 Ma, L., Hao, F., Stoumpos, C.C. et al. (2016). *J. Am. Chem. Soc.* 138: 14750.
15 Wu, B., Zhou, Y., Xing, G. et al. (2017). *Adv. Funct. Mater.* 27: 1604818.
16 Yokoyama, T., Cao, D.H., Stoumpos, C.C. et al. (2016). *J. Phys. Chem. Lett.* 7: 776.

17 Zhu, L., Akiyama, H., and Kanemitsu, Y. (2018). *Sci. Rep.* 8: 11704.
18 Wakamiya, A. and Kanemitsu, Y. (2019). *APL Mater.* 7: 080903.
19 Handa, T., Aharen, T., Wakamiya, A., and Kanemitsu, Y. (2018). *Phys. Rev. Mater.* 2: 075402.
20 Stroppa, A., Di Sante, D., Barone, P. et al. (2014). *Nat. Commun.* 5: 5900.
21 Fang, H.-H., Adjokatse, S., Shao, S. et al. (2018). *Nat. Commun.* 9: 243.
22 Yang, Y., Ostrowski, D.P., France, R.M. et al. (2016). *Nat. Photonics* 10: 53.
23 Manser, J.S., Christians, J.A., and Kamat, P.V. (2016). *Chem. Rev.* 116: 12956.
24 Milot, R.L., Klug, M.T., Davies, C.L. et al. (2018). *Adv. Mater.* 30: 1804506.
25 de Quilettes, D.W., Koch, S., Burke, S. et al. (2016). *ACS Energy Lett.* 1: 438.
26 Walker, A.W., Heckelmann, S., Karcher, C. et al. (2016). *J. Appl. Phys.* 119: 155702.
27 Staub, F., Hempel, H., Hebig, J.-C. et al. (2016). *Phys. Rev. Appl.* 6: 044017.
28 Noel, N.K., Stranks, S.D., Abate, A. et al. (2014). *Energy Environ. Sci.* 7: 3061.
29 Xu, P., Chen, S., Xiang, H.-J. et al. (2014). *Chem. Mater.* 26: 6068.
30 Nasti, G. and Abate, A. (2019). *Adv. Energy Mater.* 10: 1902467. https://doi.org/10.1002/aenm.201902467.
31 Chung, I., Song, J.-H., Im, J. et al. (2012). *J. Am. Chem. Soc.* 134: 8579.
32 Tan, G., Zhao, L.-D., Shi, F. et al. (2014). *J. Am. Chem. Soc.* 136: 7006.
33 Grancini, G., Roldán-Carmona, C., Zimmermann, I. et al. (2017). *Nat. Commun.* 8: 15684.
34 Meng, L., You, J., and Yang, Y. (2018). *Nat. Commun.* 9: 5265.
35 Konstantakou, M. and Stergiopoulos, T. (2017). *J. Mater. Chem. A* 5: 11518.
36 Marshall, K.P., Walker, M., Walton, R.I., and Hatton, R.A. (2016). *Nat. Energy* 1: 16178.
37 Tsai, C.-M., Mohanta, N., Wang, C.-Y. et al. (2017). *Angew. Chem. Int. Ed.* 56: 13819.
38 Heo, J.H., Kim, J., Kim, H. et al. (2018). *J. Phys. Chem. Lett.* 9: 6024.
39 Song, T.-B., Yokoyama, T., Aramaki, S., and Kanatzidis, M.G. (2017). *ACS Energy Lett.* 24: 897.
40 Yokoyama, T., Song, T.-B., Cao, D.H. et al. (2017). *ACS Energy Lett.* 2: 22.
41 Koh, T.M., Krishnamoorthy, T., Yantara, N. et al. (2015). *J. Mater. Chem. A* 3: 14996.
42 Kumar, M.H., Dharani, S., Leong, W.L. et al. (2014). *Adv. Mater.* 26: 7122.
43 Sabba, D., Mulmudi, H.K., Prabhakar, R.R. et al. (2015). *J. Phys. Chem. C* 119: 1763.
44 Lee, S.J., Shin, S.S., Im, J. et al. (2018). *ACS Energy Lett.* 3: 46.
45 Gu, F., Zhao, Z., Wang, C. et al. (2019). *Solar RRL* 3: 1900213.
46 Hao, F., Stoumpos, C.C., Guo, P. et al. (2015). *J. Am. Chem. Soc.* 137: 11445.
47 Lee, S.J., Shin, S.S., Kim, Y.C. et al. (2016). *J. Am. Chem. Soc.* 138: 3974.
48 Yu, B.-B., Xu, L., Liao, M. et al. (2019). *Solar RRL* 3: 1800290.
49 Zhu, Z., Chueh, C.-C., Li, N. et al. (2017). *Adv. Mater.* 30: 1703800.
50 Cao, D.H., Stoumpos, C.C., Yokoyama, T. et al. (2017). *ACS Energy Lett.* 2: 982.
51 Tai, Q., Guo, X., Tang, G. et al. (2018). *Angew. Chem. Int. Ed.* 58: 806.
52 Kim, H., Lee, Y.H., Lyu, T. et al. (2018). *J. Mater. Chem. A* 6: 18173.

53 Wang, F., Jiang, X., Chen, H. et al. (2018). *Joule* 2: 2732.
54 Gu, F., Ye, S., Zhao, Z. et al. (2018). *Solar RRL* 2: 1800136.
55 Zhang, Y.-Y., Chen, S., Xu, P. et al. (2018). *Chin. Phys. Lett.* 35: 036104.
56 Lin, R., Xiao, K., Qin, Z. et al. (2019). *Nat. Energy* 4: 864.
57 Li, W., Li, J., Li, J. et al. (2016). *J. Mater. Chem. A* 4: 17104.
58 Song, T.-B., Yokoyama, T., Stoumpos, C.C. et al. (2017). *J. Am. Chem. Soc.* 139: 836.
59 Kayesh, M.E., Chowdhury, T.H., Matsuishi, K. et al. (2018). *ACS Energy Lett.* 3: 1584.
60 Tsarev, S., Boldyreva, A.G., Luchkin, S.Y. et al. (2018). *J. Mater. Chem. A* 6: 21389.
61 Li, F., Zhang, C., Huang, J.-H. et al. (2019). *Angew. Chem. Int. Ed.* 58: 6688.
62 Li, F., Fan, H., Zhang, J. et al. (2019). *Solar RRL* 3: 1900285.
63 Jeon, N.J., Noh, J.H., Kim, Y.C. et al. (2014). *Nat. Mater.* 13: 897.
64 Konstantakou, M., Perganti, D., Falaras, P., and Stergiopoulos, T. (2017). *Crystals* 7: 291.
65 Meng, X., Lin, J., Liu, X. et al. (2019). *Adv. Mater.* 31: 1903721. https://doi.org/10.1002/adma.201903721.
66 Shao, S., Dong, J., Duim, H. et al. (2019). *Nano Energy* 60: 810.
67 Liu, X., Yan, K., Tan, D. et al. (2018). *ACS Energy Lett.* 3: 2701.
68 Gao, W., Ran, C., Li, J. et al. (2018). *J. Phys. Chem. Lett.* 9: 6999.
69 Zhao, Z., Gu, F., Li, Y. et al. (2017). *Adv. Sci.* 4: 1700204.
70 Chowdhury, T.H., Kayesh, M.E., Lee, J.-J. et al. (2019). *Solar RRL* 3: 1900245. (7 pages) https://doi.org/10.1002/solr.201900245.
71 Kayesh, M.E., Matsuishi, K., Kaneko, R. et al. (2019). *ACS Energy Lett.* 4: 278.
72 Noel, N.K., Abate, A., Stranks, S.D. et al. (2014). *ACS Nano* 8: 9815.
73 Zhao, T., Chueh, C.C., Chen, Q. et al. (2016). *ACS Energy Lett.* 1: 757.
74 Jiang, Q., Zhao, Y., Zhang, X. et al. (2019). *Nat. Photonics* 13: 460.
75 Liao, M., Yu, B.-B., Jin, Z. et al. (2019). *ChemSusChem* 12: 5007–5014. https://doi.org/10.1002/cssc.201902000.
76 Chen, K., Wu, P., Yang, W. et al. (2018). *Nano Energy* 49: 411.
77 Zhang, T., Dar, M.I., Li, G. et al. (2017). *Sci. Adv.* 3: 1700841.
78 Kamarudin, M.A., Hirotani, D., Wang, Z. et al. (2019). *J. Phys. Chem. Lett.* 10: 5277.
79 Ke, W., Stoumpos, C.C., Logsdon, J.L. et al. (2016). *J. Am. Chem. Soc.* 138: 14998.
80 Yang, Z., Zhong, M., Liang, Y. et al. (2019). *Adv. Funct. Mater.* 29: 1903621. (6 pages) https://doi.org/10.1002/adfm.201903621.
81 Vegiraju, S., Ke, W., Priyanka, P. et al. (2019, https://doi.org/10.1002/adfm.201905393). *Adv. Funct. Mater.* 29: 1905393. (7 pages).
82 Ran, C., Xi, J., Gao, W. et al. (2018). *ACS Energy Lett.* 3: 713.
83 Liu, X., Wang, Y., Xie, F. et al. (2018). *ACS Energy Lett.* 35: 1116.
84 Liu, C., Tu, J., Hu, X. et al. (2019). *Adv. Funct. Mater.* 29: 1808059.
85 Kosasih, F.U. and Ducati, C. (2018). *ChemSusChem* 11: 4193.
86 Shao, S., Liu, J., Portale, G. et al. (2018). *Adv. Energy Mater.* 8: 1702019.

87 Jokar, E., Chien, C.-H., Fathi, A. et al. (2018). *Energy Environ. Sci.* 11: 2353.

88 Tsai, C.-M., Lin, Y.-P., Krishna Pola, M. et al. (2018). *ACS Energy Lett.* 3: 2077.

89 Ran, C., Gao, W., Li, J. et al. (2019, https://doi.org/10.1016/j.joule.2019.08.023). *Joule* 3: 3072–3087.

90 Ke, W., Stoumpos, C.C., Spanopoulos, I. et al. (2017). *J. Am. Chem. Soc.* 139: 14800.

91 Qiu, J., Xia, Y., Chen, Y., and Huang, W. (2018). *Adv. Sci.* 6: 1800793.

92 Liao, Y., Liu, H., Zhou, W. et al. (2017). *J. Am. Chem. Soc.* 139: 6693.

93 Xu, H., Jiang, Y., He, T. et al. (2019 https://doi.org/10.1002/adfm.201807696). *Adv. Funct. Mater.* 29: 1807696. (pages 11).

94 Qiu, J., Xia, Y., Zheng, Y. et al. (2019). *ACS Energy Lett.* 47: 1513.

95 Du, H.-J., Wang, W.-C., and Zhu, J.-Z. (2016). *Chin. Phys. B* 25: 108802.

96 Ke, W., Stoumpos, C.C., Zhu, M. et al. (2017). *Sci. Adv.* 3: 1701293.

97 Jokar, E., Chien, C.-H., Tsai, C.-M. et al. (2018). *Adv. Mater.* 31: 1804835.

98 Saliba, M. (2018). *Science* 359: 388.

99 Wang, Z., Song, Z., Yan, Y. et al. (2019). *Adv. Sci.* 6: 1801704.

100 Hao, F., Stoumpos, C.C., Chang, R.P.H., and Kanatzidis, M.G. (2014). *J. Am. Chem. Soc.* 136: 8094.

101 Tong, J., Song, Z., Hoe Kim, D. et al. (2019). *Science* 364: 475.

102 Prasanna, R., Leijtens, T., Dunfield, S.P. et al. (2019, https://doi.org/10.1038/s41560-019-0471-6). *Nat. Energy* 4: 939–947.

103 Yang, Z., Rajagopal, A., Jo, S.B. et al. (2016). *Nano Lett.* 16: 7739.

104 Araújo, G.L. and Martí, A. (1994). *Sol. Energy Mater. Sol. Cells* 33: 213.

105 Hörantner, M.T., Leijtens, T., Ziffer, M.E. et al. (2017). *ACS Energy Lett.* 210: 2506.

106 LME Official Prices. www.lme.com (accessed 28 May 2021).

107 Moss, R.L., Tzimas, E., Kara, H. et al. (2011). *Critical Metals in Strategic Energy Technologies: Assessing Rare Metals as Supply-Chain Bottlenecks in Low-Carbon Energy Technologies*. Luxembourg: European Commission.

108 Chmielnicka, J., Zareba, G., Polkowska-Kulesza, E. et al. (1993). *Biol. Trace Elem. Res.* 36: 73.

109 de Carvalho Oliveira, R. and Santelli, R.E. (2010). *Talanta* 82: 9.

110 Babayigit, A., Thanh, D.D., Ethirajan, A. et al. (2016). *Sci. Rep.* 6: 18721.

111 Serrano-Lujan, L., Espinosa, N., Larsen-Olsen, T.T. et al. (2015). *Adv. Energy Mater.* 5: 1501119.

7

Low-Temperature and Facile Solution-Processed Two-Dimensional Materials as Electron Transport Layer for Highly Efficient Perovskite Solar Cells

Shao Hui[1], Najib H. Ladi[1], Han Pan[2], Yan Shen[2], and Mingkui Wang[2]

[1] *China-EU Institute for Clean and Renewable Energy, Huazhong University of Science and Technology, School of Energy and Power Engineering, Luoyu Road 1037, Wuhan 430074, China*
[2] *Huazhong University of Science and Technology, Wuhan National Laboratory for Optoelectronics, Luoyu Road 1037, Wuhan 430074, China*

7.1 Introduction

Organic–inorganic hybrid perovskites have attracted worldwide attention as solar irradiance harvesting materials with remarkable photoelectric properties such as broad visible light capture capability, high charge mobility, and long diffusion lengths of charge carriers (both electrons and holes) [1–3]. The appealing advantages along the intensive research around the world have resulted in unprecedented power conversion efficiency (PCE) values for this emerging technology. To date, PCE of organic–inorganic halide perovskite-based solar cells have rapidly increased from 3.8% to 25.6% within a decade of research investigation [4, 5].

Usually an efficient perovskite solar cell (PSC) device is composed of a transparent conductive oxide (TCO), a perovskite light absorption layer sandwiched between electron transporting layer (ETL) and hole transporting layer (HTL), and a back-contact electrode [6, 7]. In order to enhance the performance of PSC devices, buffer layers can be also inserted at the interface between perovskite layer and ETL or HTL or between back-contact electrode and ETL or HTL. Furthermore, PSCs can be divided into mesoporous structure and flat heterojunction structure according to the materials' formation in devices. Depending on the incidence of light they can be further divided into n–i–p and p–i–n architectures, where n- and p- refer to the n-type and p-type charge carrier transporting materials and i refer to the perovskite optical absorption layer. Their definition is also based on the fact whether the ETL or HTL is designed to contact with the transparent conductive substrates. For instance, in a conventional planar PSC device, an ETL can be deposited on the TCO layer such as fluorine-doped tin oxide (FTO) or indium tin oxide (ITO),

Perovskite Solar Cells: Materials, Processes, and Devices, First Edition.
Edited by Shahzada Ahmad, Samrana Kazim, and Michael Grätzel.

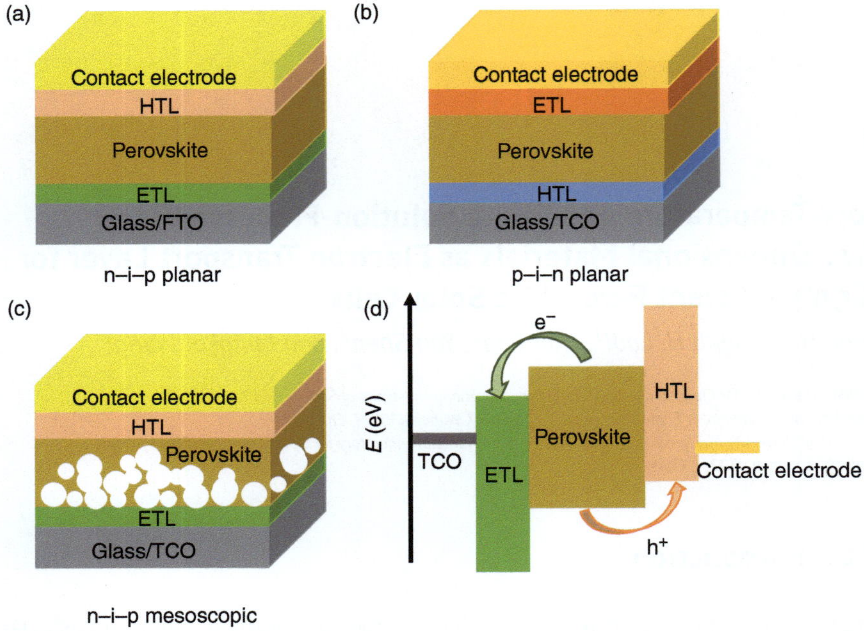

Figure 7.1 Typical perovskite solar cells with (a) n–i–p planar structure. Source: Based on Yoo et al. [8]; Solanki et al. [9]; Lin et al. [10]; Dong et al. [11]; Liu et al. [12]. (b) p–i–n planar structure and (c) n–i–p mesoporous structure. (d) Energy level scheme for transporting direction of electron and hole carriers in n–i–p-structured perovskite solar cells. Source: (d) Based on Ren et al. [13] and Varo et al. [14].

which will function as the cathode. Then, a perovskite layer is prepared on the top of this ETL using spin-coating or vacuum evaporation process, followed by the successive deposition of an HTL and a metal electrode, such as Au or Ag, to form an n–i–p-structured device (Figure 7.1a) [8–12]. However in an inverted PSC design, the positions of ETL and HTL are simply opposite (see Figure 7.1b).

Compared with the planar n–i–p-structured PSCs, the mesoporous PSCs utilize a compact TiO_2 layer formed on FTO or ITO substrates, and then additional mesoporous TiO_2 layer is deposited as both ETL and structural scaffold [15–17]. Afterward, the perovskite layer is deposited to form the absorber layer followed by fabrication of the HTL and metal electrode (Figure 7.1c). A special feature for the mesoporous structured PSCs is that the mesoscopic transport layer offers large contact areas and interfacial roughness for perovskite layer deposition. Despite these different structures, the working mechanism of PSC photovoltaics follows the same principle. Upon irradiated by sunlight, the electron–hole pairs (i.e., excitons) are generated in the light-harvesting perovskite layer. After separation of electron–hole pairs (induced by thermal energy due to a low exciton binding energy), photo-generated electrons are transported to the conduction band of ETL, and holes are transported to the highest occupied molecular orbital (HOMO) energy level of HTL, which are then collected by cathode and anode, respectively (Figure 7.1d) [13, 14].

7.2 Charge Transport in Perovskite Solar Cells

For a PSC device, both ETL and HTL play a crucial role not only in extracting and transporting the charge carriers but also in controlling the growth of perovskite thin film. Moreover, these charge transporting layers are of great importance to the stability and scalability of the devices [18, 19]. The ETL (usually highly doped n-type metal oxide, donated as n^+) must possess a high electron concentration to conduct the electrons to the cathode with a small voltage drop. Under a given bias voltage across the device terminals, the electrical current through the ETL/perovskite and perovskite/HTL contact interfaces must be the same as those of the rest of device. Therefore, the competition between charge carrier recombination and charge transport is critical for the PSC devices having a typical total thickness of about submicrometer [20]. It clearly becomes critical on how to further reduce charge carrier recombination at interfaces between ETL/perovskite and choose the best contact electrode for the PSC devices.

Following the above discussion, highly efficient PSCs generally possess a common feature of two interfaces, such as ETL/perovskite interface and HTL/perovskite heterojunction interface, both of which enable efficient charge separation and collection. Figure 7.2a illustrates various charge carrier kinetics at these two interfaces, including charge transfer, charge carrier collection, and charge carrier recombination. Processes 1–4 show the radiation recombination of charge carriers in the PSC devices, while the rest are non-radiative recombination processes. To achieve a high-performance PSCs, we must enhance the charge generation and extraction processes while suppressing the non-radiative recombination based on one assumption of well-crystallization of perovskite thin films. The interfacial recombination (non-radiative process) occurs at the ETL/perovskite or perovskite/HTL interface, which also is often called the free-carrier recombination (bimolecular recombination). It will greatly affect the open-circuit voltage (V_{oc}) of the photovoltaic devices and can be experimentally verified according to the equation [21]

$$V_{oc} = \frac{E_g}{q} - \frac{K_B T}{q} \text{In}\left(\frac{\gamma_L N_C N_v}{G}\right) \tag{7.1}$$

where γ_L is Langevin recombination constant, K_B is the Boltzmann constant, G is the non-radiative recombination rate, and N_c and N_v are the effective density of states. Therefore, for the ETL, it is necessary to reduce the interfacial charge accumulation by enabling a long diffusion length and a balanced charge transport. As a result, this allows for less non-radiative recombination opportunities at the ETL/perovskite interface and could be contributed to the augmented V_{oc}. Subsequently, Tress et al. [22, 23] rewrote Eq. (7.1) into Eqs. (7.2) and (7.3) to represent the Langevin recombination and the Shockley–Read–Hall (SRH) recombination, respectively:

$$qV_{oc} = E_g - K_B T \text{In}\left(\frac{\gamma_L N_C N_V}{1\,\text{m}^3\text{s}}\right) + K_B T\left(\frac{G}{1\,\text{m}^{-3}\text{s}^{-1}}\right) \tag{7.2}$$

$$qV_{oc} = E_g - K_B T \text{In}\left(\frac{C_{SRH} N_C N_V}{1\,\text{m}^3\text{s}}\right) + 2K_B T\left(\frac{G}{1\,\text{m}^{-3}\text{s}^{-1}}\right) \tag{7.3}$$

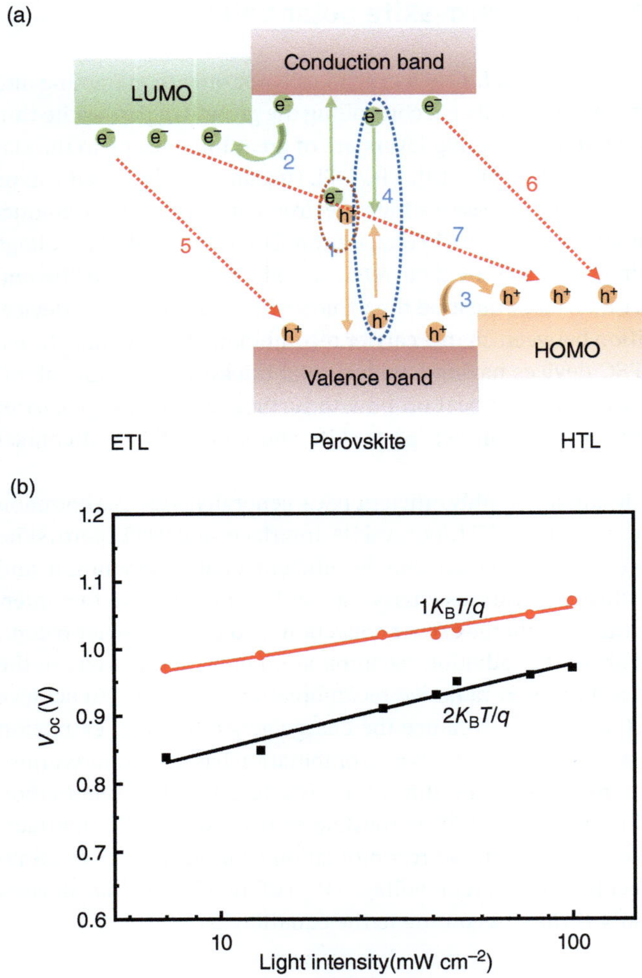

Figure 7.2 (a) Diagram for charge transfer, collection, and recombination of planar PSC devices with structure of ITO/ETL/perovskite/HTL/Au. Photon exciton (1), charge transfer (2 and 3), collection, and recombination (4, 5, 6, and 7) steps take place at the interfaces. (b) Schematic diagram of V_{oc} dependence upon different light intensity based on the same architecture.

where C_{SRH} is the SRH recombination constant. Given that the generation rate is proportional to the incident light, Kuil and Tress et al. proposed the V_{oc} – $\log(P_{in})$ to describe the relationship between the V_{oc} and the incident light intensity (as illustrated in Figure 7.2b). The results can be used to the recombination mechanism that affects the device physical performance [24]. In principle, a slope of $1\,K_BT/q$ means Langevin recombination, while a large slope of $2\,K_BT/q$ means that additional interfacial trap-assisted SRH recombination is involved. For efficient planar PSC devices, a slope of lower than $1\,K_BT/q$ has been frequently observed [25].

ETLs in PSC devices play important roles in collecting and transferring charge carriers after the injection of electrons from perovskite active layer and, more importantly, realizing effective charge separation and suppressing charge carrier recombination. An inefficient charge transport could cause inhomogeneous charge accumulation and serious interfacial recombination at the ETL/perovskite interface. Therefore, it is highly desirable to design and fabricate high-quality ETLs to guarantee an efficient charge transport and thus effective photovoltaic performance of devices. As a matter of fact, more efforts have made tremendous progress in the research of the ETLs.

7.3 Brief Development of Perovskite Solar Cells

In 2009, Miyasaka and coworkers reported the first sensitized solar cell using $MAPbI_3$ instead of dyes as a light-absorbing material [4]. Despite a low efficiency (~3.8%), this pioneer work has attracted most of attention in the photovoltaic field. Park and coworkers improved the efficiency by optimizing TiO_2 mesoporous layer and $MAPbI_3$ quantum dot of up to 6.5% [26]. In 2012, Snaith and coworkers reported that utilization of Al_2O_3 mesoporous layer as the scaffold for the PSCs could achieve much faster electron transport with the perovskite layer, if compared with that using the n-type TiO_2 with the perovskite layer [27]. Subsequently, extensive work has been devoted to improving the performance of PSCs. However, the most commonly used ETL TiO_2 in PSC devices has a low electron mobility (bulk mobility: $1\ cm^2\ V^{-1}\ s^{-1}$ [28]) and requires high-temperature annealing (>450 °C [29]), which limit their development. In this sense, several other metal oxides have been tested and optimized as ETL. In 2010, Snaith and Ducati reported for the first time that SnO_2 was used as an ETL in dye-sensitized solar cells [30], making the ETL type diversified. Subsequently, Kumar et al. successfully used ZnO as an ETL and achieved a PCE of 8.90% [31]. In contrast to mesoporous TiO_2, ZnO layer is substantially thinner and requires no sintering, which also make it applicable to flexible devices. Furthermore, Gong et al. reported the use of bilayer F-doped SnO_2 (F:SnO_2) nanocrystals as the ETL in a planar PSCs [32]. Compared with the widely used single-layer ETL, it has a better energy alignment to perovskite and electrode, which can effectively decrease the recombination of carrier. Consequently, the devices using this bilayer ETLs have achieved a PCE of 20.2%, which also expands the structure of ETL.

Figure 7.3 shows the representative progress of ETLs in the n–i–p architecture. Up to date, the perovskite photovoltaics have offered excellent performance. It showed that the PCE of PSCs rapidly increased to above 25% in a short period time. Although the efficiency has grown fast enough to rival that of silicon solar cell technologies (noting that this comparison is not made at a same device size for both), performance issues related to device stability, scalability, and flexibility prevent them from achieving their market potential. To further enhance the device' stability and scalability in operation conditions and reduce the production costs, key factors controlling the charge carrier collection, such as the charge transport layers and

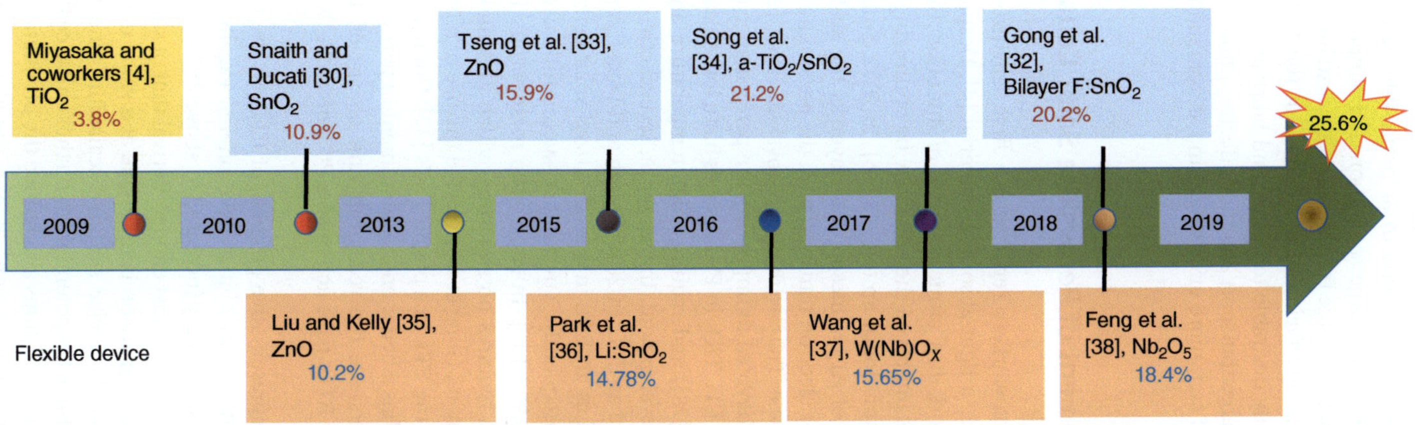

Figure 7.3 The representative progress of ETLs in the n–i–p architecture.

their contact to absorber layers, have now been taken into serious considerations [14, 39]. With this in mind, the research of new materials that address the obstacles to practical application of PSCs is crucial for its further development. In this case, two-dimensional (2D) materials have been extensively studied in recent years as robust and effective alternatives to the corresponding components owing to their significant photoelectric, chemical, and mechanical properties [40, 41]. For example, due to its unique morphological, structural, and optical properties, 2D layered perovskite has demonstrated appealing photoelectric properties and translational possibilities from its parent three-dimensional (3D) perovskite [42]. It is therefore important to highlight the latest advances in this rapidly developing field of research.

Here, we summarize the application of 2D materials in ETL of PSCs, as well as their recent progress. Furthermore, we evaluate the challenges and requirements for the future development of efficient and low-cost PSCs.

7.4 Functions and Requirements of Electron Transport Layer

Recently, there are vast studies on finding the effective ETLs to be used in PSC devices to further increase charge carrier transport. The main reason is that the V_{oc} values of most PSC devices are not large enough compared with the typical optical bandgap of the light absorber, considering that the short-circuit current (J_{SC}) is almost close to its limit (~−26 mA cm^{-2}) for a bandgap of around 1.55 eV [43]. However, the improvement of V_{oc} could be a very effective strategy to further increase PCE of PSCs. Figure 7.4 illustrates the open-circuit losses in PSC devices, in which the non-radiation loss occurring in the perovskite bulk layer and at either the perovskite/transport layer interfaces or the transport layers is key to high efficiency. Therefore, reducing the non-radiative recombination loss is crucial in the improvement of PSC performance. Effective ETLs can not only enhance the

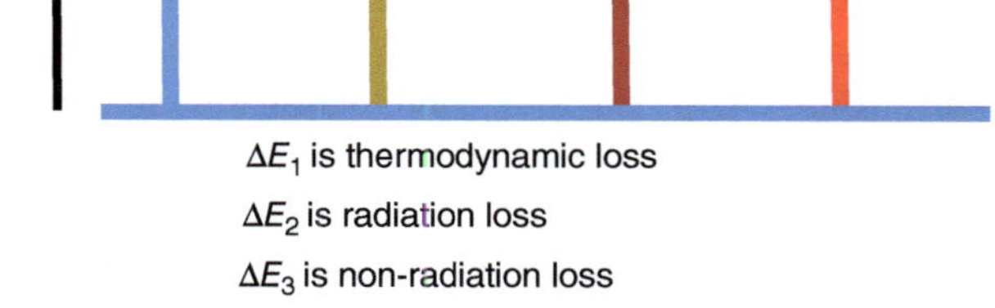

Figure 7.4 The principle of open-circuit voltage loss.

transport of charge carriers and reduce the recombination of interface charge but also play a vital role in increasing the PCE and lifetime of the device.

High-quality ETLs should have the following properties: (i) an appropriate energy alignment to other components, which can ensure the effective transfer of electrons, (ii) high electron mobility and electrical conductivity, and (iii) high transparency under light irradiation and stability under UV light [29].

For high-efficiency optoelectronic device, the electron mobility (μ_n) in the ETL must be high enough to effectively transport the charge carriers, which can be expressed by the following equation:

$$\mu_n = \frac{e\tau}{m_e^*} \tag{7.4}$$

where e is the elementary charge, τ is the average free time, and m_e^* is the electron effective mass. Electron free time greatly affects the electron mobility, which is inseparable from the features of the material and the scattering mechanism of carriers [44]. In addition, the electrical conductivity of semiconductor materials is affected by the electron mobility and is directly proportional to the number of free electrons in the conduction band, e.g.

$$\sigma = nDE \tag{7.5}$$

where σ is the electrical conductivity, n is the concentration of free electrons, D is the diffusion coefficient of electrons, and E is the electric field. An increase of either charge carriers (n) or diffusion coefficient of electrons (D) can increase the electrical conductivity of the ETLs. Also, we can know from Einstein equation that an increase in electron mobility (μ_n) can also enhance the electrical conductivity. We can also roughly estimate the μ_n and σ for the ETL in PSCs. Here, σ could be calculated as $\sigma = (d \times J_{SC})/V$, with the voltage drop (V) equal or less than the thermal voltage V_T [45]. Given the J_{sc} value of 25 mA cm^{-2} for the PSC with a bandgap of 1.6 eV [46] and thickness of 40 nm (d) for the ETL, the obtained σ should be larger than 4×10^{-6} S cm^{-1}. Moreover, μ_n can be quantified as $\sigma = q\mu_n N_D$, where N_D is the doping density of donor and q is the elementary charge. The electron mobility μ_n is estimated to be higher than ~2.5×10^{-5} cm^2 V^{-1} s^{-1} using the charge density $N_D = 1.0 \times 10^{18}$ cm^{-3} for most of efficient solar cell devices [47].

A suitable ETL should be also optically transparent to guarantee maximum light absorption by the perovskite layer in the n–i–p PSC. This makes choosing the right materials difficult. For example, some organic materials might have problems with good transparency. The most popular inorganic ETLs are transition metal oxides such as TiO_2 [48], ZnO [49], and SnO_2 [50], which can be modeled as a network of junctions formed from the connected nanoparticles. Figure 7.5a shows the mechanism for charge transfer through metal oxide semiconductor nanocrystalline particles and surface charge carrier recombination among various oxide particles and/or between oxides and perovskite layers, along with a charge injection–accumulation–diffusion model for charge transfer dynamics of those semiconducting nanocrystals illustrated in Figure 7.5b. We have firstly discussed this model in our previous study [51]. It is clear that defect states in these nanocrystals play a crucial role on the charge transport process (i.e. capture and release

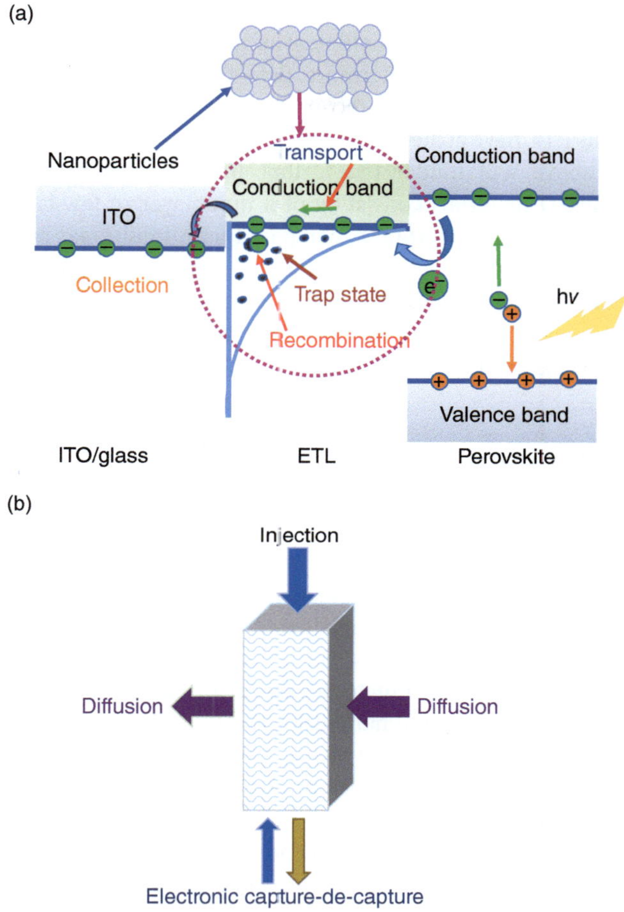

Figure 7.5 (a) Schematic illustrated of mechanism for charge transfer through metal oxide nanocrystalline particles and surface charge carrier recombination. (b) Charge injection–accumulation–diffusion model for charge transfer dynamics on a nanocrystalline particle.

processes). The local variation of free electron concentration can be affected by the trap–detrap from defect states, which slow down the transfer of electrons. One of the evidences for this phenomenon is that the conductivity of most of metal oxide nanocrystals is generally lower than that of the crystalline film. Consequently, a low charge transport in metal oxide nanocrystals significantly influences the maximum conversion efficiency of thermal radiation energy into electricity via such types of photovoltaic devices.

For the traditional nanoporous metal oxide ETLs such as TiO_2 and SnO_2, the charge transport in these films is strongly dependent on the properties of the junction linking these nanoparticles [52]. Owing to the intrinsic disorder of the semiconductor material between the nanocrystalline particles, the contact junctions of these nanoparticles may act as effective potential energy barriers for transferring electrons. The existence of potential barriers will slow down the transfer of charge between nanoparticles, so that the diffusion coefficient of charge decreases, thus reducing the diffusion lengths of charge. Furthermore, defect states on the surface of the nanoparticles may induce non-radiative recombination. Even though PCS devices using metal oxides as ETL have achieved a relatively high efficiency (>20%), the drawbacks described above will hinder the further improvement of their performance. The emerging 2D materials consisting of a monolayer of few-layer nanosheets can achieve atomic-level smoothness, compared with traditional bulk materials, which can effectively reduce defect states and suppress non-radiative recombination. When used as charge selective transport layers, they can effectively enhance the charge extraction and transmission. Therefore, they have been one of the most extensively studied classes of materials.

7.5 Features and Advantages of Two-Dimensional Electron Transport Materials

2D materials represent an emerging class of materials that are typically composed of a single layer or few layers of nanosheets bonded together by van der Waals (vdW) forces, whose optoelectronic properties can be effectively tailored by controlling the number of exfoliated layers [53]. Since the discovery of mechanically exfoliated graphene in 2004 [54], many research on 2D materials has grown exponentially in the fields of materials science, chemistry, and nanotechnology [55]. The intriguing feature of 2D materials endues them with unique physical and chemical characteristics, of which the electrical and optical properties present specific opportunities in electronic and optoelectronic devices including transistors, photodetectors, light-emitting diodes, and solar cells. In Sections 7.10 and 7.11, we briefly describe the application of 2D materials in the ETL of PSC devices and give a direction for its further development in the future.

7.6 Van der Waals Heterojunctions

One of the most exciting prospects for 2D materials is the ability to stack them in atomically abrupt heterojunctions bonded solely vdW interactions. Unlike covalently bonded heterojunctions in bulk semiconductors, vdW heterojunctions can avoid dangling bonds and trapped charges. This relaxation of the lattice matching condition allows vast permutations of 2D materials heterojunctions [56, 57]. Moreover, the surface of 2D semiconductor material has atomic-level smoothness, which offers the platform for free movement of electrons and a highly possibility of fabrication of large-area thin film devices. The excellent electrical properties are considered advantages even at a very thin thickness.

Therefore, lattice matching and material types have less effects on the performance of vdW heterojunctions because there is no covalent bond between the layers. Consequently, a perfect heterojunction can be obtained by continuously stacking

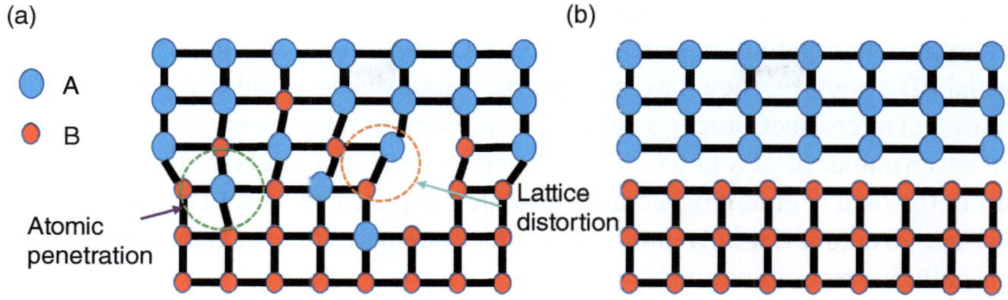

Figure 7.6 Schematic diagram of heterojunction interface among different layers. (a) Three-dimensional semiconductor heterojunction. (b) Van der Waals heterojunction. A and B represent elemental atoms from different materials.

different materials. Figure 7.6 illustrates a schematic diagram of the interface for various heterojunctions, in which the left side shows the heterojunction obtained by traditional methods. Due to the influence of covalent bonds, it is difficult to avoid defects such as interfacial stress, dislocations, and atomic penetration. In contrast, the vdW heterostructure layers are connected by weak vdW forces, which makes the atoms firmly fixed in the plane bonded by covalent bonds and prevents them from permeating into other layers and forming lattice distortion due to uneven forces (as illustrated in Figure 7.6a). Thus, defects such as interface stress are formed. Therefore, compared with traditional heterojunctions, vdW heterojunctions have many excellent photoelectric properties. For example, the interface defects of the metal–semiconductor heterostructure obtained by stacking are very few, and the pinning of the Fermi level is released, so that the height of the interface barrier tends to reach the Schottky–Mott limit. Consequently, the barrier height can be adjusted according to the work function of metals [58]. More importantly, devices processed with such 2D-based heterojunctions may have very good ductility and bendability and thus possess broad application prospects in the fields of flexible electronics. In this case, epitaxial crystal growth techniques such as molecular beam epitaxy and metal organic chemical vapor deposition (CVD) have become popular methods to have monolayer (few angstroms thick) control in the chemical composition of the growing crystal. Nearly most of semiconductors extending from small bandgap to wide bandgap materials have been preciously grown onto 2D materials-based substrates by epitaxial techniques. This allows special growth of heterojunctions where electronic properties can be altered.

In addition, based on the pure 2D layered materials, van der Waals heterojunction formed by artificially stacking different 2D materials in a selected order is an emerging research field in the past few years [59]. By subsequently stacking 2D materials with different properties (electrical, optical, etc.), the features of the combined "new" material can be artificially adjusted. Due to the weak van der Waals force between the layers, the adjacent layers are no longer restricted by the lattice matching in crystal growth. Moreover, since there is no composition transition, the heterojunction has a sharp gradient of carrier (potential field) at the atomic level in the vicinity of junctions. Owing to the transition metal represented by the double chalcogenide, non-graphene 2D layered materials can usually form two types of energy band relationships, so the heterojunctions built on them have very strong carrier separation capabilities. Furthermore, as a result of their ultrathin thickness and special 2D structure, they have strong grid responsiveness and compatibility with traditional microelectronics processing technology and flexible substrate. The excellent performance of vdW heterojunctions has made 2D materials the preferred materials in the field of emerging technologies such as photovoltaics. In short, its perfect interface structure not only enhances charge transport but also greatly reduces the non-radiative recombination.

7.7 Quantum Confinement Effect in Two-Dimensional Electron Transport Materials and Its Application

The quantum confinement effect is a unique feature of low-dimensional nanostructured materials that are of great concern since they provide a bridge between bulk and molecular domain [60]. This offers a new way to control optical and electronic properties through size [61]. In general, when a material exhibits significant changes in its electronic or optical properties due to the limitation of at least one dimension (1D), it is called a quantum confinement structure. Here, one can distinguish these structures from the amount of confinement where 2D materials like thin films and quantum wells (Q-wells) are confined in 1D, 1D materials like quantum wires (Q-wires) are confined in 2D, and zero-dimensional (0D) materials like quantum dots (Q-dots) are confined in all three dimensions. The size of a quantum confinement material may vary from material to material and also depends on the quantum mechanical properties of the electrons and holes in the material [62]. Figure 7.7 illustrates the density of states in semiconductors with the bulk, 2D, 1D, and 0D structures. In Q-wells, charge carriers (electrons and holes) are confined to move in

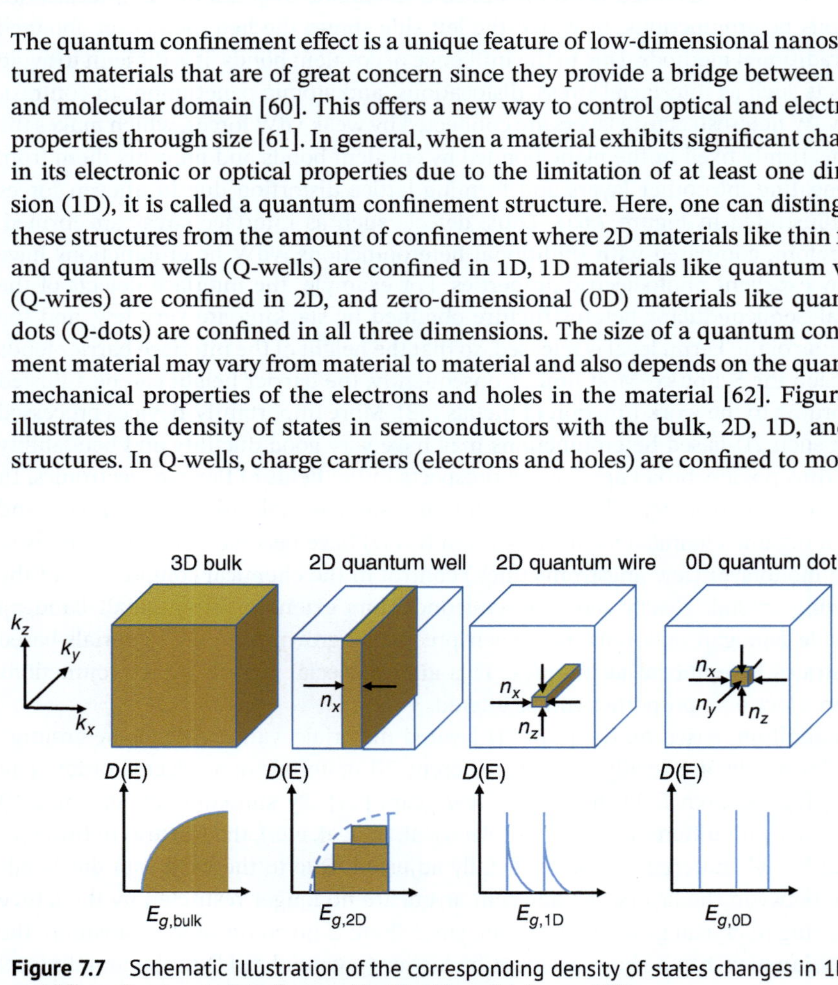

Figure 7.7 Schematic illustration of the corresponding density of states changes in 1D, 2D, and 3D confined materials. Source: Modified from Edvinsson [62].

a plane and are free to move in a 2D semiconductor. Also the energy level of one of the quantum numbers changes from continuous to discrete. Compared with bulk semiconductors, Q-wells have a higher density of electronic states near the edges of the conduction and valence bands, and therefore a higher concentration of carriers can contribute to the band-edge emission.

Owing to their strong quantum confinement in the vertical direction, the electronic structure of 2D materials can change significantly with the numbers of layers [63]. Size-dependent bandgaps and plasmon frequencies have been quintessential features of most low-dimensional materials. For instance, group VI 2D materials (such as MoS_2, WS_2, and WSe_2) in 2H phases show an evolution from an indirect bandgap of ~1.3, 1.4, and 1.2 eV in the bulk to a direct bandgap of ~1.9, 2.1, and 1.7 eV in the monolayer limit, respectively [64, 65]. As the thickness is increased from monolayer to bilayer, the direct bandgap at the K point of the Brillouin zone remains relatively unchanged due to the dominance of transition metal d-orbital that is relatively isolated from interlayer coupling [66]. However, bands at the Γ point change significantly due to the increased contribution of chalcogen p_z orbitals [67]. This emergence of a smaller indirect gap between Γ and Q points in bilayers decreases photoluminescence (PL) quantum yields in MoS_2 by up to a factor of $\sim 10^4$ between monolayers and bilayers [68]. In addition, the ultrathin body of 2D materials implies incomplete screening of externally applied electric and magnetic fields, enabling field-dependent tuning of electronic structure and thus novel field-effect devices [66, 69]. In brief, the diversity of electronic properties in 2D materials, including metals and insulators, combined with the tunability of semiconducting band structure with thickness and fields offers numerous opportunities for electronic devices.

7.8 Other Physical Properties of Two-Dimensional Electron Transport Materials

Besides the above features, 2D materials have many other advantages. For instance, due to the strong in-plane covalent bonds and atomic thickness, ultrathin 2D nanomaterials usually exhibit excellent mechanical strength and flexibility. Moreover, the atomic thickness of these materials also present an exceptional optical transparency, which gradually increases as the thickness decreases. Also, high exposure of surface atoms provides 2D materials with conditions for the easy control of material properties through surface modification/functionalization, element doping and/or defects, strain, phase engineering, etc. A summary of comparison highlighting the outstanding properties of different 2D materials is presented in Table 7.1.

Such fascinating properties discussed above have provided an intrinsic motivation to the research community to develop facile, effective, rapid, and reproducible synthetic methods that are suitable for large-scale production. These great developments for the production of 2D materials have also paved the way for their effective use in PSCs.

Table 7.1 Overview of fundamental properties of various 2D materials.

2D materials	Bandgap (eV)	Carrier mobility ($cm^2 V^{-1} s^{-1}$)	Thermal conductance ($Wm^{-1} K^{-1}$)	Young's modulus (GPa)	Fracture strain (%)	References
Graphene	0	$\approx 2 \times 10^5$	$\approx 5 \times 10^3$	1000	27–38	[70]
Phophorene	0.3–2.0	≈1000	10–36	35–166	24–32	[70]
MoS_2	1.2–1.8	10–200	34.5–52	270 ± 100	25–33	[70]
WSe_2	1.2–1.7	140–500	9.7	75–195	26–37	[70]
SnS_2	2.18–2.44	50–230	6.41	145	—	[71, 72]
TiS_2	0.02–2.5	7.24	11.76	220–270	—	[73, 74]

7.9 Synthesis of Various Two-Dimensional Materials

To characterize the layer-dependent changes in properties and to provide pathways for their integration into a multitude of applications, it is important to explore the reliable syntheses of monolayer and few-layer 2D materials. Therefore, many strategies, such as exfoliation, hydrothermal method, CVD, and so on, have been developed to synthesized high-quality and ultrathin nanosheets exhibiting their own advantages and disadvantages in preparing 2D nanomaterials.

The preparation of 2D materials by exfoliation is cost-efficient and versatile because it can be combined with other chemical treatments to produce a variety of functionalized 2D materials [75]. The mechanical exfoliation method is widely used to prepare single- or few-layer 2D materials by the application of mechanical forces to solutions of layered materials via stirring, shaking, or ultrasonication. Surfactants are commonly employed to reduce van der Waals interactions that hold layers together and keep the exfoliation products suspended in the solvent [76]. Lee and coworkers exfoliated bulk-layered materials (graphene, h-boron nitride, MoS_2, etc.) by controlling the temperature of the sonication bath to address the issue that 2D materials are difficult to be dissolved directly in common solvents (such as water, ethanol, and acetone) without prior surface treatment [77]. The exfoliated 2D platelets remain well dispersed in water without using any surfactants for over a month. Han and coworkers developed a modified way of mechanical exfoliation for making 2D materials by introducing a homebuilt exfoliation machine [78]. The device is capable of reliably producing 2D materials that depend on the speed of the exfoliation and can also accommodate a variety of samples and simultaneously cleave multiple samples. In addition, this method can be applied in various environments, such as inert glove box. Mechanical exfoliation is a straightforward method to obtain only one- or few-layer nanosheets, which well maintains the crystal structure and properties. Beside this, its versatility and low cost makes it highly popular for synthesizing 2D materials and extremely convenient for fundamental research. Even so, this method is limited to the laboratory research and seems impossible to scale up for industrial production. Although large-size and high-quality ultrathin nanosheets can be obtained in this way, this method is only applicable for layered van der Waals solids. However, nanosheets of layered ionic solids and nonlayered materials cannot be obtained using this method.

Hydro/solvothermal method is one of the wet chemical methods that is promising for the preparation of all types of 2D materials owing to their high yield, low cost, and mass production. The advantages of hydro/solvothermal method over other types of crystal growth include low temperature (generally in a temperature range of 100–240 °C) and convenience of adjusting reaction conditions. Many factors play the key roles in the synthesis of ultrathin nanostructure, which include temperature, reaction time, reactant ratio, and so on. This method is usually used for the synthesis of inorganic materials, such as MoS_2, TiO_2, ZnO, and Rh. Li and coworkers successfully synthesized ultrathin Rh nanosheets through solvothermal method [79]. Atomic force microscopy (AFM) image shows that Rh nanosheets have a single-crystalline nature, which possesses a hexagonal structure, with a lattice parameter of c. 2.6 Å, closer to the atomic distance of Rh (2.69 Å) in the (111) plane of bulk Rh phase. Xue and coworkers reported that a novel composite consisting of MoS_2 nanosheets and ultrasmall Fe_3O_4 nanoparticles was synthesized as an anode for lithium ion battery [80]. In the composite anode, MoS_2 nanosheets provide flexible substrates and high reversible capacity, which effectively enhances the performance of the device. In addition to the hydro/solvothermal method, template synthesis is also a common wet chemical synthesis method that can nearly prepare all types of nanostructure including quantum dot (0D), nanowire (1D), and nanosheets (2D).

Recently, CVD technique has been recognized as a popular method to generate high-quality 2D materials with scalable size, controllable thickness, and excellent electronic properties [81]. This technique is a high-temperature chemical synthesis process by which a desired material is deposited on substrates. CVD processes have been extensively studied for synthesizing thin film coatings of a wide range of materials, including metals, semiconductors, and insulators. CVD growth of 2D materials involves activated chemical reactions of precursors in a specially controlled environment. It uses transition metal (e.g. Pt, Ni) as catalysts. It is possible to change the process of 2D material production by adjusting the catalyst to prepare films with different qualities. Figure 7.8 summarizes the general experimental setup of CVD method [82]. Its main advantages are represented by the access to high-quality, high-purity 2D nanomaterials with controlled properties, which allow one to control the morphology, crystallinity, and defects of 2D nanostructures by tuning the

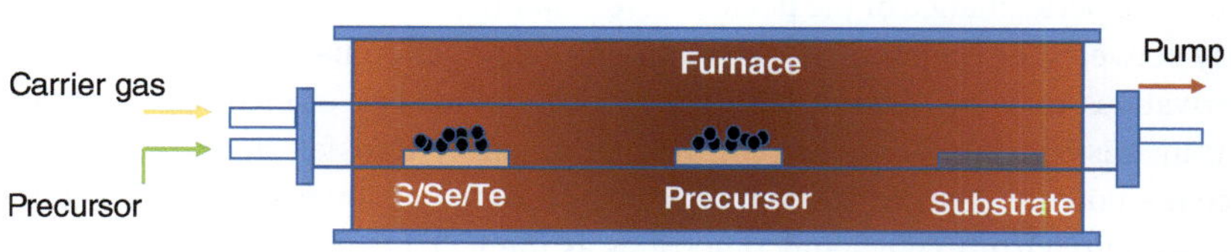

Figure 7.8 Schematic illustration of experimental setup for the 2D materials synthesis by a noncatalytic vapor deposition process. Source: Based on Butler et al. [82].

process parameters. Owing to these superiorities, many 2D nanomaterials have been prepared by this method conveniently, such as hexagonal boron and MoS_2 [83, 84].

CVD method is also another annealing strategy for the growth of high-quality single-crystalline 2D sheet on substrates, which have been the cutting-edge research area for the synthesis of transition metal dichalcogenides in recent years. In addition to the synthetic methods discussed above, there are many other synthetic methods such as ion-change exfoliation [85], microwave-assisted chemical [86], and ultrasonic exfoliation [87] emerging over the past decade. All of them have made great contributions to the practical application.

7.10 Application of Two-Dimensional Material as an Electron Transport Layer in Perovskite Solar Cells

As the most popular third-generation solar cell, PSCs have always attracted the attention of the research community. Introducing 2D materials into PSCs as ETL has several benefits. For example, the tunable bandgap and smooth surface of the 2D materials can effectively solve the problems such as energy level mismatch and potential barriers faced by nanoparticles in the process of charge transfer of the ETLs in PSC devices, thus resulting in less non-radiative recombination. In addition, the ultrathin nature of 2D materials also makes it ideal for flexible devices. In the past few years, many 2D materials have been successfully used as effective ETL in PSC devices.

As a representative of graphene derivatives, graphene oxides (GOs) and the reduced graphene oxides (rGOs) are bipolar conducting materials for both electrons and holes. They show significantly high carrier mobility, wide range of spectral absorption, and high optical transparency, along with tunable energy levels (by functionalization) [88, 89]. Recently, graphene-based derivatives of ETLs have achieved great success in scalable fabrication processes by optimizing charge carrier dynamics, showing significant advantages over conventional ETLs. Adding graphene to ETL can enhance the layers' conductivity and reduce the series resistance of the device, thereby improving the overall performance the solar cells [90].

As a widely used ETL, TiO_2 has suitable energy levels of conduction band minimum (conduction band minimum [CBM] = −3.9 eV) and valence band maximum (valence band maximum [VBM] = −7.2 eV) [91]. However, the poor electron mobility results in an unbalanced charge transfer between it and the perovskite photoactive layer, which causes carriers to accumulate at the interface and severely destroys the performance of the device. Given that the work function of graphene is exactly between the typical FTO glass and TiO_2 conduction band, therefore adding its derivatives to TiO_2 can not only improve the efficiency of electron extraction and transmission but also play the role of electron transfer intermediary and reduce the potential barrier at the interface [92]. Han and coworkers fabricated the device using rGO/mp-TiO_2 nanocomposites as the ETL (as illustrated Figure 7.9a) [93]. Since the energy level of rGO is lower than the TiO_2 conduction band, the photoexcited electrons on perovskite and TiO_2 can be transferred to rGO effectively.

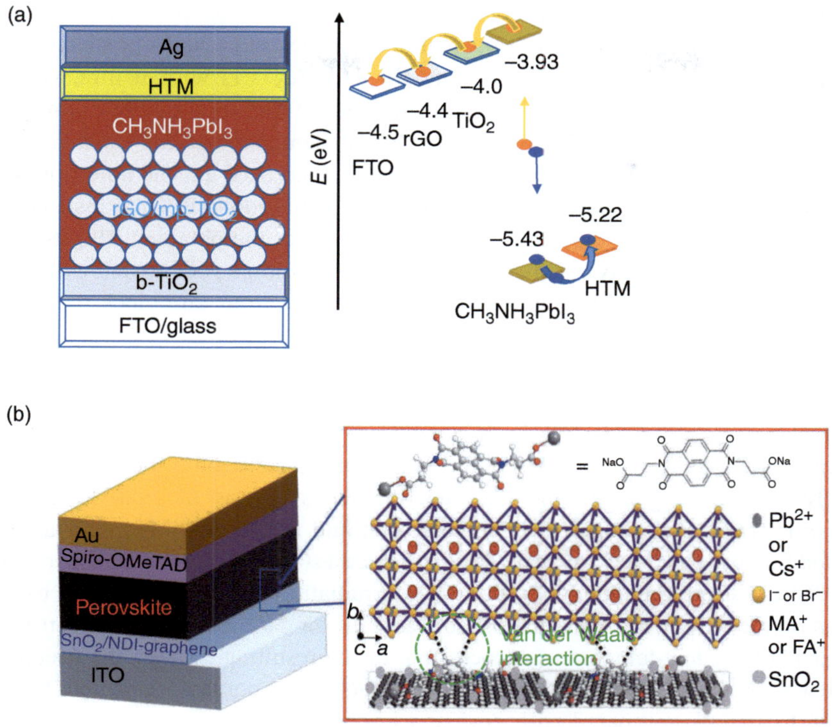

Figure 7.9 (a) Diagram of rGO/mp-TiO_2 nanocomposite-based perovskite solar cells. Source: Based on Yang et al. [93]. (b) Schematic architecture of planar perovskite solar cells. The expanded scheme shows the chemical structure of *N,N′*-bis-[2-(ethanoic acid sodium)]-1,4,5,8-naphthalene diimide (NDI) and the bonding condition of NDI–graphene with perovskite films. Source: Zhao et al. [94].

Therefore, PSCs based on rGO/mp-TiO_2 nanocomposite ETL can enhance electron transport and collection efficiency. After optimization, the resistivity of the device was reduced from 3×10^5 to 4.94×10^4 Ω cm with the addition of rGO, along with a decreased internal resistance and an enhanced charge collection.

Subsequently, Zhao et al. reported the use of electronic conducting graphene to form the graphene/SnO_2 composites, which was then used as an ETL in the PSC (Figure 7.9b) [94]. vdW interactions between perovskite materials and graphene have been shown to be able to immobilize octahedron $[PbI_6]^{4-}$ at the perovskite/ETL interface. After optimization, the device with a high PCE of 20.2% and a high fill factor (FF) of 82% can be obtained by using the $FA_{0.75}MA_{0.15}Cs_{0.1}PbI_{2.65}Br_{0.35}$ perovskite and an optimization graphene addition amount of 5 v/v%. This good performance is attributed to the improved electron mobility, extraction ability, and reduced carrier recombination.

Phosphorene is a material with a unique two-dimensional structure. As a monolayer 2D analogue of semiconducting black phosphorus derivative introduced to the family of 2D materials in 2014 [95], it has many appealing features, such as high carrier mobility and orientation-dependent (anisotropic) properties and tunable bandgap, which make it a promising material for electronic and optoelectronic devices. For instance, Lau and coworkers modified the ZnO ETL with BP film, which enhanced the electron extraction of the excited active film and greatly improved the performance of the device [96]. This study also demonstrated the feasibility of BP as an electronic transmission material. Fu et al. then used a facile liquid-phase exfoliation method to prepare black phosphorus quantum dots (BPQDs) instead of the ETL used in the PSCs [97]. The results showed that the obtained BPQDs not only showed the bipolarity conductivity but also could form a good match with the required energy band, which can effectively extract electrons and block the hole injection. In addition, these could contribute to the formation of dense, uniform, and orderly oriented perovskite films, reducing defects and suppressing interfacial recombination, which thus greatly enhance the device efficiency. The application of BPQDs as electron-selective layer has attracted a lot of attention. However, obtaining high-quality and reproducible black phosphorus nanosheets depends heavily on exploring a feasible and reproducible synthesis method. Recently, Batmunkh et al. [98] proposed a simple, novel, and fast liquid-phase exfoliation method that used a vortex fluidic device to produce a stable phosphorene dispersion under shear stress generated by pulsed laser irradiation. Subsequently, they used the as-produced phosphorene nanosheets as ETL for TiO_2-based photoelectrode planar PSCs for low-temperature processing. In the resulting devices, the average efficiency of PSCs with phosphorene nanosheets incorporated increased from 14.32% to 16.53%, compared with devices without phosphorene nanosheets, and the best devices produced a PCE of 17.85%.

Although graphene derivatives can improve the performance of PSCs, the performance of devices based on the ETL of whole graphene derivatives is still somewhat inadequate compared with the conventional ETL, except as additives. The transition metal dichalcogenides (TMDs), generally with a formula of MX_2 (M is generally a transition metal such as W and Mo, while X is usually a chalcogen element such as Se and S), have attracted a lot of research interests in recent years and been applied in many different fields. Unlike the carbon monatomic layer graphene, TMDs, such as tin disulfide (SnS_2), titanium disulfide (TiS_2), tungsten diselenide (WSe_2), and molybdenum disulfide (MoS_2), are composed of two sulfur group layers sandwiched with a transition metal layer. This unique structure endows them with appealing optical and electronic properties and chemical stability [99]. For instance, we can tune its electronic properties by changing transition metals or chalcogens into semiconductor or metal behavior [100], which makes this the best material for various energy-related applications. In addition, the properties of TMDs will also change greatly when the number of exfoliated layers of bulk crystals to make it a single layer or few-layers is changed. For example, when exfoliated to a single layer, the bulk TMD material will change from indirect bandgap to direct bandgap. In addition, studies have shown that single-layer TMDs with a thickness of less than 1 nm can effectively absorb 5–10% of incident sunlight, which is an order of magnitude higher than the absorption of sunlight by GaAs and Si [101]. Owing to their excellent characteristics, TMD materials are widely used especially in the application of high-efficiency PSCs, which has attracted a lot of attentions and opened a new path for the research of high-efficiency and stable PSCs.

Singh et al. [102] employed microwave-assisted low-temperature liquid-phase synthesis to fabricate a thin and uniform MoS_2 film directly on glass/FTO substrates and used it as ETL in PSCs. The transport layer forms a thermally stable interface with the perovskite photoactive layer, so no thermal annealing is required. Due to the high transparency of MoS_2 and high electrical conductivity in the vertical direction, the device achieved a high current density and PCE, and its performance was almost the same as the device based on TiO_2 and SnO_2.

Huang et al. [103] prepared 2D TiS_2 by room-temperature and aqueous solution process. TiS_2 was used as the ETL to prepare perovskite solar cells, and ETL-free devices were fabricated as the reference. The highest efficiency of the control device without ETL is 13.42%. Primary testing TiS_2 as the ETL showed the highest efficiency of the device being only 9.44%. Detailed investigation showed that when 2D TiS_2 was deposited on ITO as an ETL, the work function increased from 4.75 to 4.79 eV, which was much lower than the perovskite conduction band (approximately 4.4 eV), seriously deteriorating device performance. Interestingly, after treating TiS_2 with ultraviolet ozone (UVO) for 15 minutes, the work function of TiS_2-coated ITO film could be shifted up to 4.64 eV (Figure 7.10a). The highest efficiency of the device was 18.79% when the UVO-treated TiS_2 was used as the ETL. This performance improvement may be attributed to the decreased work function after UVO treatment and the enhanced electron transport capability of the ETL. Meanwhile, UVO treatment may induce an oxidation of the surface of TiS_2 and passivate structural defects such as sulfur vacancies by interaction with lithium [104] (as shown in Figure 7.10b).

Zhao et al. recently showed a new approach of utilizing 2D SnS_2 as the ETL in efficient PSCs via a self-assembly stacking deposition method (Figure 7.10c) [105]. The

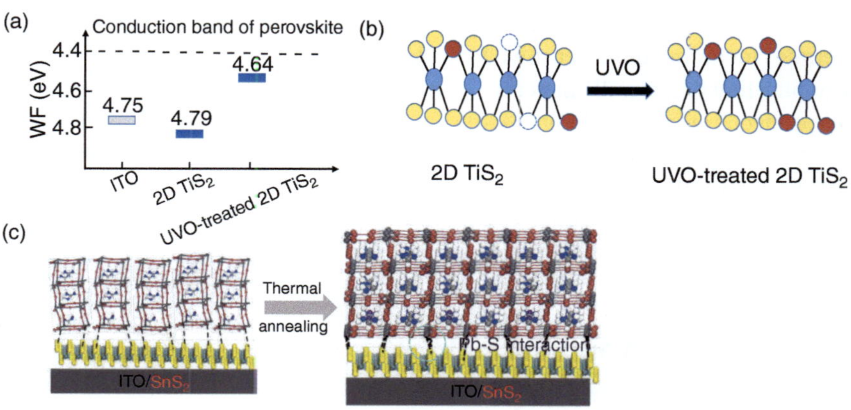

Figure 7.10 (a) Work functions (WFs) of ITO with the modification of 2D TiS_2 and UVO-treated 2D TiS_2. (b) Schematic diagram of the oxidation surface of 2D TiS_2 (yellow, blue, and red circles and dotted circle represent S, Ti, and O atoms and S vacancy, respectively). Source: Based on Kim et al. [104]. (c) Illustration of the perovskite films deposited on the SnS_2 substrate. The green dashed circle depicts the chemical reaction of Pb and S. Source: Zhao et al. [105].

large-scale 2D multilayer SnS_2 sheet structure promoted a heterogeneous nucleation over the perovskite precursor film. In addition, intermolecular Pb–S interactions could promote the formation of a micro-mosaic structure at the interface between 2D SnS_2 and perovskite, which suppresses charge carrier recombination and thus facilitates electron extraction for balanced charge transport at interfaces between ETL/perovskite and HTL/perovskite. Based on this structure, a high PCE of 20.12% and a high J_{SC} of 23.55 mA cm^{-2} device were achieved for the SnS_2-based PSC.

In addition, the application of metal oxide nanosheets is also a new trend, which can effective solve the challenges faced by metal oxide nanoparticles. For example, when exposed to UV irradiation, the holes generated in TiO_2 easily react with the oxygen adsorbed on the surface oxygen vacancies, thereby forming deep defects, causing great charge recombination and degradation of the films. This will greatly affect the performance of the PSCs [106]. In view of this, Chen et al. [107] used multilayer atomic thin titanium dioxide ($Ti_{0.92}O_2$) as the ETL in PSCs. They used Langmuir–Blodgett deposition process to deposit $Ti_{0.92}O_2$ on the FTO substrate at room temperature. Although the ETL was only 5 nm thick, much thinner than the conventional compact TiO_2 layer (~30–80 nm), its photoelectric performance (~14.5%) was comparable with its counterpart devices. Moreover, the deposition process could form a more uniform and better coverage film, which could effectively reduce the recombination of electron and holes, thereby reducing leakage current. In addition, since this is a low-temperature-processed method, it is also suitable for the assembly of flexible photovoltaic devices. Subsequently, Chen et al. [108] used self-assembled atomic stacking of 2D titanium atomic sheets as ETL to hybrid perovskite solar cells. Because the 2D titanium atomic plate has a high ultraviolet light transparent height, almost no oxygen vacancies is produced in the films. Therefore, compared with the traditional c-TiO_2, PSCs based on this ETL can greatly inhibit the ultraviolet-induced degradation, thereby enhancing the stability of the devices. In addition, since it is a full-solution process, ETL can be deposited on a variety of substrates including those sensitive to heat. This has also opened up a new idea for the application of many 2D materials based on solution-processed homogeneous or heterogeneous atomic stacking in large-area photovoltaic or photovoltaic devices.

From the studies above, we can easily see that 2D metal oxide nanocrystals have more advantages than traditional metal oxide ETL, such as ultrathin, low-temperature solution processing, etc. All these illustrate its huge potential as an ETL for efficient photovoltaic application.

7.11 Conclusion and Outlook

In this overview, we summarize the recent developments of ETLs based on 2D materials for highly efficient PSCs. We explored the advantages of using these emerging 2D materials as efficient and stable ETLs in PSCs. Compared with the traditional three-dimensional nanoparticle ETL, the 2D-based ETL can extract and transport electrons effectively and reduce interface recombination, thus guaranteeing an effective photo-to-electricity conversion by such devices. Indeed, ultrathin thickness and nearly perfect surface show great potential in realizing flexible and large-scale PSCs with high efficiency and long-term stability. Although a high efficiency of 21.73% has been achieved by using TiS_2, there are still some issues to be solved, including low yield, long time consumption of the synthesis process, and poor film coverage due to the single deposition method. Therefore, to develop high-quality PSCs based on 2D ETLs in the future, one should focus on the following directions.

At present, we can produce 2D materials through many methods, such as mechanical exfoliation, CVD, and liquid phase exfoliation. However, low yield, high cost, and low reproducibility prevent its further development. Therefore, it is important for us to explore new synthetic methods so that it can be universally applied to various 2D materials and be applied to the assembly of large-scale PSCs.

The high surface area-to-volume ratio of 2D materials means that chemical functionalization greatly affects the properties of unshared electron pairs on sulfur atoms in TMD. However, the challenge is to determine the covalent functionalization chemistry that can be regulated electronically without interfering with spectral band delocalization and charge carrier mobility. In terms of photoelectricity, the photon-to-electron conversion efficiency in photodetectors and solar cells and the electron-to-photon conversion efficiency in light-emitting diodes ultimately depend on the relative time scales of radiation and non-radiation processes. Therefore, the chemical route of passivating non-radiation composite sites holds great promise for enhancing 2D material-based optoelectronic technology.

Most of 2D materials with hydrophobic surfaces are difficult to dissolve directly in polar solvents. We can improve its solubility through covalent functionalization, since the covalent adsorption of functional groups is usually achieved by reacting with unsaturated atoms at the edges or defect sites, which changes the characteristics of the 2D materials. This also results in the improvement of their solubility in common solvents.

Altogether, the exploration of 2D materials provides a new direction for the development of PSCs. Also, more efforts need to be done to further improve and optimize them to develop toward large-scale modules for commercialization.

List of Abbreviations

0D	zero-dimension
1D	one dimension
2D	two-dimension
BPQDs	black phosphorus quantum dots
CVD	chemical vapor deposition
ETL	electron transport layer
FTO	fluorine-doped tin oxide
GOs	graphene oxides
HOMO	highest occupied molecular orbital
HTL	hole transport layer
ITO	indium tin oxide

J_{SC}	short-circuit current
MoS_2	molybdenum disulfide
PCE	power conversion efficiency
PL	photoluminescence
PSCs	perovskite solar cells
PVK	perovskite
rGOs	reduced graphene oxides
SnS_2	tin disulfide
SRH	Shockley–Read–Hall
TCO	transparent conductive oxide
TiS_2	titanium disulfide
TMDs	transition metal dichalcogenide
UVO	ultraviolet ozone
vdW	van der Waals
V_{oc}	open circuit voltage
WSe_2	tungsten diselenide

References

1 Eperon, G., Leijtens, T., Bush, K. et al. (2016). *Science* 354: 861.
2 Lin, Q., Armin, A., Nagiri, R. et al. (2015). *Nat. Photonics* 9: 106.
3 Wehrenfennig, C., Eperon, G., Johnston, M. et al. (2014). *Adv. Mater.* 26: 1584.
4 Kojima, A., Teshima, K., Shirai, Y., and Miyasaka, T. (2009). *J. Am. Chem. Soc.* 131: 6050.
5 Jeong, J., Kim, M., Seo, J. et al. (2021). *Nature* 592: 381–385.
6 Yu, J., Lee, C., Joh, H. et al. (2017). *Nanoscale* 9: 17167.
7 Agresti, A., Pescetelli, S., Palma, A. et al. (2017). *ACS Energy Lett.* 2: 279.
8 Yoo, J., Wieghold, S., Sponseller, M. et al. (2019). *Energy Environ. Sci.* 12: 2192–2199.
9 Solanki, A., Yadav, P., Turren-Cruz, S. et al. (2019). *Nano Energy* 58: 604–611.
10 Lin, D., Zhang, T., Wang, J. et al. (2019). *Nano Energy* 59: 619–625.
11 Dong, C., Han, X., Li, W. et al. (2019). *Nano Energy* 59: 553–559.
12 Liu, Y., Chen, Q., Duan, H. et al. (2015). *J. Mater. Chem. A* 3: 11940–11947.
13 Ren, X., Wang, Z., and Choy, W. (2019). *Adv. Opt. Mater.* 7: 1900407.
14 Varo, P., Tejada, J., Rosell, M. et al. (2018). *Adv. Energy Mater.* 8: 1702772.
15 Yang, T., Jeon, N., Shin, H. et al. (2019). *Adv. Sci.* 6: 1900528.
16 Li, F., Zhang, C., Huang, J. et al. (2019). *Angew. Chem. Int. Ed.* 58: 6688–6692.
17 Shi, X., Ding, Y., Zhou, S. et al. (2019). *Adv. Sci.* 6: 1901213.
18 Snaith, H., Abate, A., Ball, J. et al. (2014). *J. Phys. Chem. Lett.* 5: 1511.
19 Baena, J., Steier, L., Tress, W. et al. (2015). *Energy Environ. Sci.* 8: 2928.
20 Pan, H., Zhao, X., Gong, X. et al. (2020). *Mater. Horiz.* 7: 2276.
21 Koster, L., Smits, E., Mihailetchi, V., and Blom, P. (2005). *Phys. Rev. B* 72: 085205.
22 Tress, W., Merten, A., Furno, M. et al. (2013). *Adv. Energy Mater.* 3: 631–638.

23 Tress, W., Leo, K., and Riede, M. (2013). *Appl. Phys. Lett.* 102: 163901.
24 Zhang, H., Xue, L., Han, J. et al. (2016). *J. Mater. Chem. A* 4: 8724–8733.
25 Shai, X., Wang, J., Sun, P. et al. (2018). *Nano Energy* 48: 117–127.
26 Im, J., Lee, C., Lee, J. et al. (2011). *Nanoscale* 3: 4088.
27 Lee, M., Teuscher, J., Miyasaka, T. et al. (2012). *Science* 338: 643–647.
28 Yang, G., Tao, H., Qin, P. et al. (2016). *J. Mater. Chem. A* 4: 3970–3990.
29 Bati, A., Batmunkh, M., and Shapter, J. (2019). *Adv. Energy Mater.* 10: 1902253.
30 Snaith, H. and Ducati, C. (2010). *Nano Lett.* 10 (4): 1259–1265.
31 Kumar, M., Yantara, N., Dharani, S. et al. (2013). *Chem. Commun.* 49: 11089–11091.
32 Gong, X., Sun, Q., Liu, S. et al. (2018). *Nano Lett.* 18: 3969–3977.
33 Tseng, Z.-L., Chiang, C.-H., and Wu, C.-G. (2015). *Sci. Rep.* 5: 13211.
34 Song, S., Kang, G., Pyeon, L. et al. (2017). *ACS Energy Lett.* 2: 12, 2667–2673.
35 Liu, D. and Kelly, T.L. (2014). *Nature Photon* 8: 133–138.
36 Park, M., Kim, J.-Y., Son, H.J. et al. (2016). *Nano Energy* 26: 208.
37 Wang, K., Shia, Y., Gao, L. et al. (2017). *Nano Energy* 31: 424.
38 Feng, J., Yang, Z., Yang, D. et al. (2017). *Nano Energy* 36: 1–8.
39 Zhou, H., Chen, Q., Li, G. et al. (2014). *Science* 345: 542–546.
40 Acik, M. and Darling, S. (2016). *J. Mater. Chem. A* 4: 6185.
41 Das, S., Pandey, D., Thomas, J., and Roy, T. (2019). *Adv. Mater.* 31: 1802722.
42 Li, H., Wang, X., Zhang, T. et al. (2019). *Adv. Funct. Mater.* 29: 1903293.
43 Jiang, Q., Zhao, Y., Zhang, X. et al. (2019). *Nat. Photonics* 13: 460–466.
44 Reshak, A. (2014). *RSC Adv.* 4: 39565–39571.
45 Corre, V., Stolterfoht, M., Toro, L. et al. (2019). *ACS Appl. Energy Mater.* 2: 6280–6287.
46 Daboczi, M., Hamilton, I., Xu, S. et al. (2019). *ACS Appl. Mater. Interfaces* 11: 46808–46817.
47 Chang, W., Tian, H., Fang, G. et al. (2019). *Sol. Energy* 186: 323–327.
48 Zhang, Y., Liu, X., Li, P. et al. (2019). *Nano Energy* 56: 733–740.
49 Zheng, D., Wang, G., Huang, W. et al. (2019). *Adv. Funct. Mater.* 29: 1900265.
50 Tu, B., Shao, Y., Chen, W. et al. (2019). *Adv. Mater.* 31: 1805944.
51 Xu, X., Zhang, H., Cao, K. et al. (2014). *ChemSusChem* 7: 3088–3094.
52 Piotrowiak, P. (2013). *Solar Energy Conversion: Dynamics of Interfacial Electron and Excitation Transfer*. Cambridge, UK: RSC Publishing.
53 Nicolosi, V., Chhowalla, M., Kanatzidis, M. et al. (2013). *Science* 340: 1226419.
54 Novoselov, K., Geim, A., Morozov, S. et al. (2004). *Science* 306: 666.
55 Tan, C., Cao, X., Wu, X. et al. (2017). *Chem. Rev.* 117: 6225–6331.
56 Grigorieva, I. and Geim, A. (2013). *Nature* 499: 419–425.
57 Lotsch, B. (2015). *Annu. Rev. Mater. Res.* 45: 85–109.
58 Liu, Y., Guo, J., Zhu, E. et al. (2018). *Nature* 557: 696–700.
59 Novoselov, K., Mishchenko, A., Carvalho, A., and Neto, A. (2016). *Science* 353: 6298.
60 Brus, L. (1984). *J. Chem. Phys.* 80: 4403–4409.
61 Jacobon, T. and Edvinsson, T. (2012). *J. Phys. Chem. C* 116: 15692–15701.
62 Edvinsson, T. (2018). *R. Soc. Open Sci.* 5: 180387.

63 Bandurin, D., Tyurnina, A., Yu, G. et al. (2017). *Nat. Nanotechnol.* 12: 223–227.

64 Wang, Q., Zadeh, K., Kis, A. et al. (2012). *Nat. Nanotechnol.* 7: 699–712.

65 Gong, C., Zhang, H., Wang, W. et al. (2013). *Appl. Phys. Lett.* 103: 053513.

66 Kuc, A. and Heine, T. (2015). *Chem. Soc. Rev.* 44: 2603–2614.

67 Sangwan, V. and Hersam, M. (2018). *Annu. Rev. Phys. Chem.* 69: 299–325.

68 Mak, K., Lee, C., Hone, J. et al. (2010). *Phys. Rev. Lett.* 105: 136805.

69 Klein, J., Wierzbowski, J., Regler, A. et al. (2016). *Nano Lett.* 16: 1554–1559.

70 Batmunkh, M., Erdene, M., and Shapter, J. (2016). *Adv. Mater.* 28: 8586.

71 Huang, Y., Sutter, E., Sadowski, J. et al. (2014). *ACS Nano* 8: 10743.

72 Lee, M., Ahn, J., Sung, J. et al. (2016). *Nat. Commun.* 7: 12011.

73 Li, G., Yao, K., and Gao, G. (2018). *Nanotechnology* 29: 015204.

74 Wan, C., Gu, X., Dang, F. et al. (2015). *Nat. Mater.* 14: 622.

75 Shao, H., Ladi, N., Pan, H. et al. (2021). *Solar RRL* 5, 2000566.

76 Cai, M., Thorpe, D., Adamson, D., and Schniepp, H. (2012). *J. Mater. Chem.* 22: 24992–25002.

77 Kim, J., Kwon, S., Cho, D. et al. (2015). *Nat. Commun.* 6: 8294.

78 Yuan, L., Ge, J., Peng, X. et al. (2016). *AIP Adv.* 6: 125201.

79 Duan, H., Yan, N., Yu, R. et al. (2014). *Nat. Commun.* 5: 3093.

80 Chen, Y., Song, B., Tang, X. et al. (2014). *Small* 10: 1536–1543.

81 Lang, F., Gluba, M., Albrecht, S. et al. (2015). *J. Phys. Chem. Lett.* 6: 2745–2750.

82 Butler, S., Hollen, S., Cao, L. et al. (2013). *ACS Nano* 7: 2898–2926.

83 Kim, K., Hsu, A., Jia, X. et al. (2011). *Nano Lett.* 12: 161–166.

84 Lee, Y., Zhang, X., Zhang, W. et al. (2012). *Adv. Mater.* 24: 2320–2325.

85 Ma, R., Takada, K., Fukuda, K. et al. (2008). *Angew. Chem. Int. Ed.* 47: 86–89.

86 Wang, H., Yang, X., Shao, W. et al. (2015). *J. Am. Chem. Soc.* 137: 11376–11382.

87 Liu, Y., Wang, F., Xiao, Y. et al. (2015). *Sci. Rep.* 4: 5998.

88 Jo, G., Na, S., Oh, S. et al. (2010). *Appl. Phys. Lett.* 97: 213301.

89 Dubacheva, G., Liang, C., and Bassani, D. (2012). *Coord. Chem. Rev.* 256: 2628–2639.

90 Ahmed, M., Hussain, Z., Mujahid, M. et al. (2016). *AIP Adv.* 6: 065303.

91 Ke, W., Fang, G., Wang, J. et al. (2014). *ACS Appl. Mater. Interfaces* 6: 15959–15965.

92 Snaith, H. and Gratzel, M. (2006). *Adv. Mater.* 18: 1910–1914.

93 Yang, P., Hu, Z., Zhao, X. et al. (2017). *ChemistrySelect* 2: 9433–9437.

94 Zhao, X., Tao, L., Li, H. et al. (2018). *Nano Lett.* 18: 2442–2449.

95 Liu, H., Neal, A., Zhu, Z. et al. (2014). *ACS Nano* 8: 4033.

96 Lin, S., Liu, S., Yang, Z. et al. (2016). *Adv. Funct. Mater.* 26: 864.

97 Fu, N., Huang, C., Lin, P. et al. (2018). *J. Mater. Chem. A* 6: 8886.

98 Batmunkh, M., Vimalanathan, K., Wu, C. et al. (2019). *Small Methods* 3: 1800521.

99 Wang, Q., Zadeh, K., Kis, A. et al. (2012). *Nat. Nanotechnol.* 7: 699.

100 Li, H., Shi, Y., Chiu, M., and Li, L. (2015). *Nano Energy* 18: 293.

101 Bernardi, M., Palummo, M., and Grossman, J. (2013). *Nano Lett.* 13: 3664.

102 Singh, R., Giri, A., Pal, M. et al. (2019). *J. Mater. Chem. A* 7: 7151.

103 Huang, P., Yuan, L., Zhang, K. et al. (2018). *ACS Appl. Mater. Interfaces* 10: 14796.

104 Kim, C., Nguyen, T., Le, Q. et al. (2015). *Adv. Funct. Mater.* 25: 4512–4519.

105 Zhao, X., Liu, S., Zhang, H. et al. (2019). *Adv. Funct. Mater.* 29: 1805168.

106 Berhe, T., Su, W., Chen, C. et al. (2016). *Energy Environ. Sci.* 9: 323.

107 T. Chen, S. Li, C. Lin, et al., 2017, 49rd International Conference on Solid State Devices and Materials, Sendai, Japan, Japanese Journal of Applied Physics, 1039–1040.

108 Chen, T., Lin, C., Li, S. et al. (2018). *Adv. Energy Mater.* 8: 1701722.

8

Metal Oxides in Stable and Flexible Halide Perovskite Solar Cells: Toward Self-Powered Internet of Things

Carlos Pereyra[1], Haibing Xie[1], Amir N. Shandy[1], Vanessa Martínez[1,2], Henck Pierre[1], Elia Santigosa[1,3,4], Daniel A. Acuña-Leal[1,5], Laia Capdevila[1], Quentin Billon[1], Löis Mergny[1], María Ramos-Payán[4], Mónica Gomez[2], Bindu Krishnan[5], Mario Muñoz[3], David M. Tanenbaum[1,6], Anders Hagfeldt[7], and Monica Lira-Cantu[1]

[1] *Catalan Institute of Nanoscience and Nanotechnology (ICN2), CSIC and the Barcelona Institute of Science and Technology (BIST), Building ICN2, Campus UAB E-08193, Bellaterra, Barcelona, Spain*
[2] *Universidad Nacional de Ingeniería, Faculty of Science, Av. Túpac Amaru 210, Lima, Peru*
[3] *Universitat Autonoma de Barcelona, Department of Analytical Chemistry, Carrer Til·lers, 08193 Bellaterra, Spain*
[4] *University of Seville, Department of Analytical Chemistry, Faculty of Chemistry, c/Prof. García González s/n, 41012, Seville, Spain*
[5] *Universidad Autónoma de Nuevo León, Facultad de Ingeniería Mecánica y Eléctrica, San Nicolás de los Garza, Nuevo León 66455, México*
[6] *Department of Physics and Astronomy, Pomona College, Claremont, CA 91711, USA*
[7] *Laboratory of Photomolecular Science (LSPM), Institute of Chemical Sciences and Engineering, School of Basic Sciences, Ecole Polytechnique Fédérale de Lausanne, CH-1015 Lausanne, Switzerland*

8.1 Introduction

Halide perovskite solar cells (PSCs) have revolutionized the photovoltaic (PV) arena providing power conversion efficiencies (PCE) currently above 25%, low cost, and ease of fabrication. With theoretical PCEs of ~33% for single, ~42% for tandem, and ~49% triple junction devices, PSC-based PV technologies will permit building the terawatt-scale energy production required for a low-carbon economy shaping our society's energy future. Moreover, their low weight, semitransparency, and flexibility, together with the possibility of being integrated into arbitrary shapes and architectures, make them ideal energy sources for applications in self-powered devices required for the future Internet of Things (IoT). The imminent development of IoT networks, integrated by many smart devices, is expected to grow from the current estimated seven billion connected devices to one trillion by 2035, as predicted by IBM [1]. Powering billions of devices is one of the most challenging bottlenecks in achieving the future vision of IoT. Based on PSCs, photovoltaic technology is a potential alternative to satisfy the requirements of power energy, especially for wearable electric devices, due to their high power-per-area and the ubiquity of light harvested from natural solar and artificial sources.

Perovskite Solar Cells: Materials, Processes, and Devices, First Edition.
Edited by Shahzada Ahmad, Samrana Kazim, and Michael Grätzel.

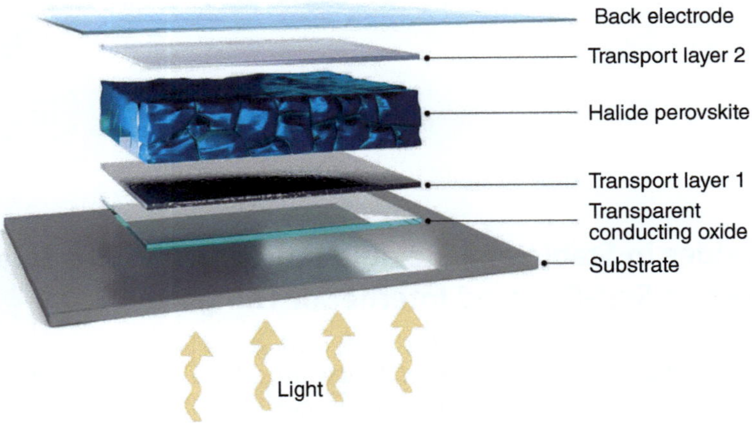

Figure 8.1 Schematic representation of a PSC. When the transport layer 1 (TL1), next to the TCO, is a hole transport oxide (HTO), the solar cell is in its inverted configuration (or p–i–n). If the TL1 is an electron transport oxide (ETO), it is called normal configuration (or e–i–p). When both the TL1 and TL2 are metal oxides, the PSC is in an "Oxide-Sandwiched" PSC configuration. If the back electrode is carbon, it is a carbon-based PSC (C-PSC).

Metal oxides have been applied as transport layers in emerging solar cells, such as hybrid [2], dye-sensitized [3], organic [4–6], and PSCs [7, 8]. Their many advantages include chemical compatibility, good band alignment, ease of fabrication, low cost, no toxicity, and good stability if compared to organic transport layers. Among the most efficient PSCs, we can find metal oxides as part of the constituent materials of the device (Figure 8.1). Among the most stable PSCs are carbon-based PSCs, but most interesting is that flexible PSCs can utilize metal oxides to confer moisture resistance without compromising device flexibility and transparency. This review highlights the use of metal oxides as transport layers in PSCs to enhance solar cell efficiency and device stability. We also underline their importance for the fabrication of flexible PSCs by solution processing. For this, we have divided this review work into different sections depending on the PSC configuration. Considering the PSC schematics shown in Figure 8.1, the configuration name will depend on the oxide placed at the transport layer 1 (TL1). If TL1 is an electron transport oxide (ETO), the PSC is in its normal configuration (also known as the n–i–p). If the TL1 is a hole transport oxide (HTO), the PSC is in its inverted configuration (or p–i–n). If both TL1 and TL2 are metal oxides, we will refer to the "oxide-sandwich" configuration. We will also describe the advantages observed when oxide multilayers are applied in PSCs. A separate section describes the Carbon-based PSC where a bilayer of mesoporous oxides, usually an ETO (like TiO_2 or ZnO) and another insulating oxide (usually ZrO_2 or Al_2O_3), work together with a mesoporous thin film of a conducting carbon. The PSC is known as the triple-layer Carbon-based PSC. This triple mesoporous layer of TiO_2/ZrO_2/Carbon is used as the skeleton or the "shell," where an infiltrated halide perovskite/organic additive mixture acts as the absorber.

The information in this review contemplates research results of PSCs applying metal oxides as part of their architecture with a focus on device stability. The data was obtained from published articles where highly efficient PSCs were reported applying metal oxides, and stability studies were considered for the devices. We only considered works published around the years 2018, 2019, and the first months of 2020.

8.2 Metal Oxides in Normal (n–i–p), Inverted (p–i–n) and "Oxide-Sandwich" Halide Perovskite Solar Cells

The properties of metal oxides (MO) permit their application as the hole and/or electron transport layer (hole transport layer [HTL] or electron transport layer [ETL], respectively) in the PSC. The three configurations, the normal (or n–i–p), the inverted (or p–i–n), and the "oxide-sandwiched" architectures, can be found in the literature. Among them, the normal configuration is the most studied and currently delivers the highest efficiency. For example, to ensure highly efficient devices with good thermal stability, Seo and coworkers applied a fluorene-terminated hole-transporting material with a fine-tuned energy level and a high glass transition temperature. They used TiO_2 as the ETL. As a result, photovoltaic devices with 23.2% PCE were obtained. Their devices observed a steady-state efficiency of 22.85% for small-area (~0.094 cm^2) cells and 21.7% efficiency for large-area (~1 cm^2) cells. Stability analysis following International Summit on Organic Photovoltaic Stability (ISOS) protocols demonstrated a highly stable device that retained almost 95% of its initial performance after more than 500 hours [9].

One of the most efficient and stable PSC applying metal oxides in the normal configuration was developed by Qi Jiang's group. They applied SnO_2 as ETL and used an organic halide salt, the phenethylammonium iodide (PEAI) on the $HC(NH_2)_2$–CH_3NH_3 mixed halide perovskite for surface defect passivation. They found that the PEAI occupies the surface of the halide perovskite, reducing defects and suppressing non-radiative recombination. As a result, planar PSCs were obtained with a certified efficiency of 23.32% (quasi-steady-state). Stability analysis revealed highly stable devices which can sustain more than 500 hours under continuous light irradiation of 1 sun [10].

The Inverted (p–i–n) configuration PSC was initially developed to enhance device stability. It emerged as an alternative to the unstable normal configuration PSCs. In this configuration, the HTL, such as poly(3,4-ethylenedioxythiophene) polystyrene sulfonate (PEDOT:PSS) or 2,2′,7,7′-Tetrakis[*N*,*N*-di(4-methoxyphenyl)amino]-9,9′-spirobifluorene (spiro-OMeTAD), can be replaced by HTL made of semiconductor oxides like NiO, V_2O_5, or MoO_3. The hole transport oxide is an essential part of PSCs. It provides important functions such as extracting the holes (and blocking electrons) from the perovskite layer, conducting the holes through the HTL layer, and isolating the perovskite layer from the oxygen and moisture from the atmosphere, in the case of semiconductor oxides. A standard HTL applied in inverted PSCs is PEDOT:PSS, a conducting polymer that possesses high optical transmittance in the visible light wavelength range. It is acidic, which is a drawback for the long-term stability of devices. As an alternative to PEDOT:PSS, inorganic p-type semiconductors have been employed in Inverted (p–i–n) PSCs. Moreover, it was initially reported that the placement of the hole transport layer next to the transparent conducting electrode resulted in highly stable devices that sustain ambient conditions, especially high humidity.

Currently, the best PCE reported for the inverted configuration PSCs are not above 21%. However, this is mostly due to larger research efforts (and thus more fabrication experience) invested in the Normal (e–i–p) configuration. Among the most efficient and stables devices are those developed by the Group of Prof. Wei Chen. They developed a PSC applying room-temperature solution-processed NiO_x as the HTL and compared their PSC devices applying different halide perovskites as absorbers. Specifically, $MAPbI_3$ and mixed cation $(FA_{0.85}MA_{0.15}Pb(Br_{0.15}I_{0.85})_3$ based solar cells were fabricated in the inverted device structure of the type ITO/NiO_x/halide perovskite/$PC_{61}BM$/zirconium (IV) acetylacetonate/Ag. Devices applying both halide perovskites exhibited excellent performance, with champion cells showing PCE of 19.45 and 17.77% for the formamidinium and methyl ammonium (FAMA) and the pure MA, respectively. The solution-processed NiO_x thin film serves as an excellent HTL in the inverted planar PSCs. That may be attributed to the reduced recombination of electrons and holes at the anode side owing to the good hole extraction ability of NiO_x [11]. This same research group developed a novel NiO_x HTL with molecular doping. The Fermi level of NiO_x HTLs was increased, and the valence band maximum decreased after doping with 2,20-(perfluoro-naphthalene-2,6-diylidene) dimalononitrile (F6TCNNQ). The champion cell conversion efficiency of CsFAMA mixed cations and $MAPbI_3$-based devices resulted in 20.86% and 19.75% efficiency, respectively. Unlike the passivation effect, the results offer an extremely promising molecular doping method to fabricate high-quality inorganic charge transport layers (CTLs) in PSCs [12].

The solar cell with the highest efficiency "oxide-sandwich configuration" that showed a PCE of 18.85% was developed by Yaxiong Guo's group. They developed an organic–inorganic hole transport layer made of the 5,6-difluorobenzothiadiazole-based conjugated polymer FBT-TH_4 and copper oxide, FBT-Th_4/Cu_xO. In this hybrid HTL, the organic part provides high hole mobility, good thermal and chemical stability, and compatibility with the halide perovskite absorber. Simultaneously, the inorganic oxide efficiently blocks the transport of electrons (which is a difficult task for the organic part). The study also revealed an optimal Cu_xO thin film layer thickness of 12 nm. Long-term stability was observed after 500 hours of aging, and the Th_4-FBT/Cu_xO HTM device retained 90% of its initial performance [13]. Another interesting result for the "oxide-sandwich" PSC configuration was developed by Zhou Xing's group. They applied a Cu-doped NiO (Cu:NiO_x) oxide as HTL. The thin film was also engineered at the interface between the metal cathode and the ETLs, improving device performance and stability. The Inverted (p–i–n) PSCs with a $PC_{61}BM$/CeO_x ETL showed a maximum PCE of 17.35%, which is superior to those devices applying single layers of either the $PC_{61}BM$ or the CeO_x thin films. Moreover, the PSCs applying the $PC_{61}BM$/CeO_x exhibit excellent stability under light soaking, attributed to the chemical stability effect shown by the CeO_x interlayer [14].

Figure 8.2 and Table 8.1 summarize the different metal oxides applied as transport layers in PSCs with normal (n–i–p), inverted (p–i–n), and "oxide-sandwiched" configurations. The PSC's efficiency (Figure 8.2, left *y*-axis) and the efficiency loss resulted after the stability test (Figure 8.2, right *y*-axis) demonstrate that highly efficient and highly stable devices can be obtained when metal oxides are applied as transport layers. For the three PSC architectures, we observe higher efficiency for the normal configuration, followed by the inverted and "oxide-sandwiched" architecture, respectively. Most importantly, the lifetime of all PSC configurations shows less than 20% performance loss after stability analysis, an indication of the good stability metal oxides provide to the complete solar cells.

Figure 8.3 summarizes the metal oxides most applied in PSCs. The graph is made independently of the types of PSC configurations. We can observe that electron transport oxides, such as TiO_2 and SnO_2, are the most applied and result in the highest efficiencies for the PSC. On the lowest efficiency range are the solar cells employing NiO as the hole transport oxide. In general, we can observe several trends: the application of simple binary oxides, doped oxides, bilayers of metal oxides, or oxide multilayers. The application of multilayers provides extra benefits such as enhanced humidity resistance or the passivation of surface defects at interfaces that reduce charge recombination. As shown in Section 8.3, the application of mesoporous bilayers of oxides (TiO_2/ZrO_2 or similar) results in one of the most stable PSC configurations: The carbon-based PSC.

8.3 Mesoporous Metal Oxide Bilayers in Highly Stable Carbon-Based Perovskite Solar Cells

In general, classical PSCs employ metal electrodes, such as Au, Cu, or Ag, as the back electrode together with expensive organic hole conductors such as spiro-OMeTAD. The application of such materials can limit the long-term stability of the device due to their degradation at temperatures above 25 °C and their high reactivity (metal migration under light irradiation). As a result, the hole-conductor-free PSCwas initially reported by Etgar et al. [41]. It was followed by the carbon-based PSC (C-PSC), which employs carbon as a counter electrode and also eliminates the use of the HTM [42]. Moreover, C-PSCs are currently one of the most stable PSCs configurations. This is partially due to the presence of oxide bilayers (mesoscopic TiO_2 and ZrO_2), which can protect the device from ambient atmospheres and humidity. This C-PSC configuration is usually fabricated by screen-printing method. A "shell," made of a triple layer of the mesoporous TiO_2, ZrO_2, and Carbon, is first fabricated as the skeleton of the device (see schematic representation in Figure 8.4). The C-PSC involves using organic modifiers or organic additives mixed with the halide perovskite absorber and are introduced into the "shell" by infiltration. The additives have the purpose of passivating traps or defects found in the halide perovskite material and at the perovskite/transport layer interface.

We distinguish different groups of devices for the C-PSCs depending on the ETM applied in the architecture (see Figure 8.5 and Table 8.2). For example, using a

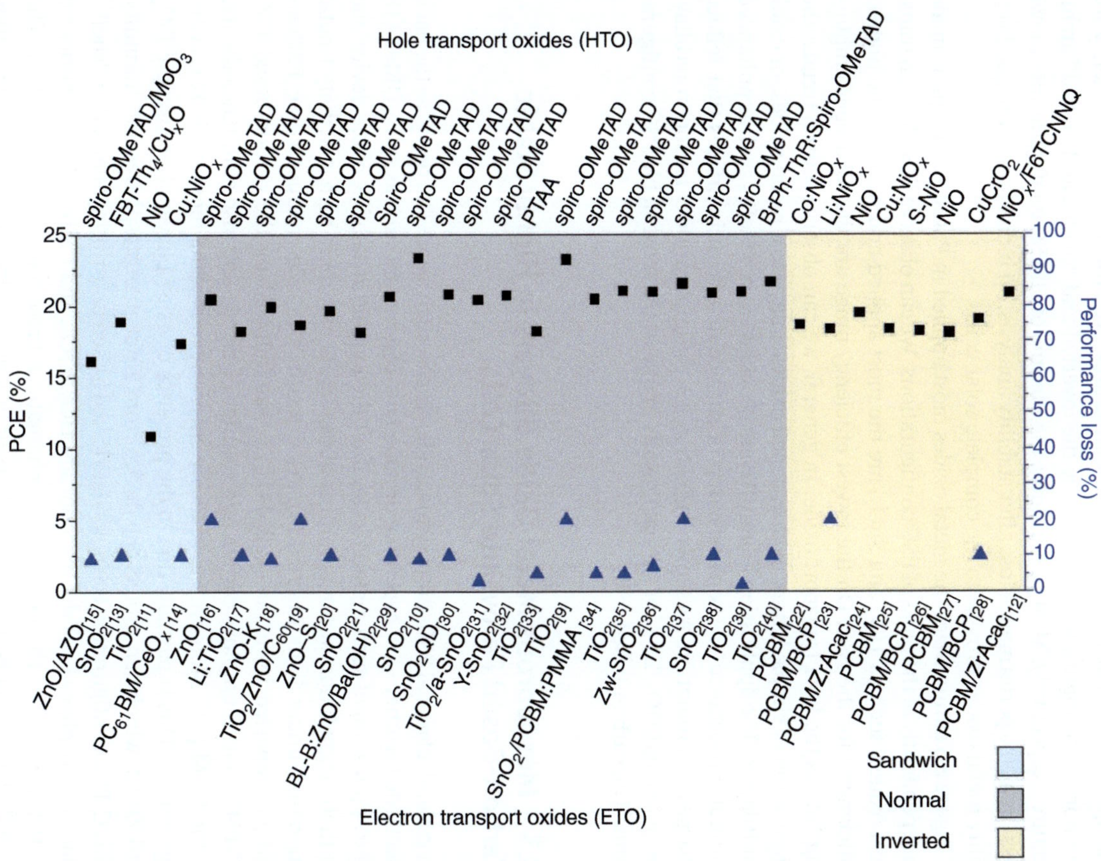

Figure 8.2 Halide perovskite solar cells applying metal oxides as transport layers in the normal (e–i–p, grey), Inverted (p–i–n, yellow), and "Oxide-Sandwiched" (blue) configurations. Power conversion efficiency of solar cells (lefty *y*-axis) and performance loss after stability analysis (right *y*-axis). Performance loss is the difference between the initial efficiency and the final efficiency value after analysis. The type of stability analysis varies for each device. All devices are fabricated on rigid substrates. Letter in parentheses corresponds to the PSC described in Table 8.1.

Table 8.1 Perovskite solar cells with oxide-based transport layers on a rigid substrate.

Device structure	Absorber (Perovskite)	Photovoltaics parameters (reverse) V_{oc} (V)	J_{sc} (mA cm^{-2})	*FF* (%)	*PCE* (%)	ETM	ETM method	HTM	HTM method	CE	CE method	Stability	References
FTO/ZnO/AZO/Perovskite/spiro/MoO_3/Ag	$MAPbI_3$	1.09	20.60	71.60	16.10	ZnO/AZO	SC	spiro/MoO_3	SC	Ag	TE	ST/1080/T_{91}	[15]
FTO/SnO_2/Perovskite/FBT-Th_4/Cu_xO/Au	$MAPbI_3$	1.09	22.35	75.4	18.85	SnO_2	SC	FBT-Th4/Cu_xO	SC	Au	TE	ST/500/T_{90}	[13]
FTO/TiO_2/Perovskite/NiO/Au	$MAPbI_3$	0.97	18.20	62.00	10.89	TiO_2	SP	NiO	SC	Au	TE	—	[11]
FTO/Cu:NiO_x/Perovskite/$PC_{61}BM$/CeO_x/Ag	$MAPbI_3$	1.06	21.52	76.07	17.35	$PC_{61}BM$/CeO_x	SC	Cu:NiO_x	SC	Ag	TE	ST/1000/T_{90}	[14]
ITO/ZnO/Perovskite/spiro/Ag	$MA_{0.1}FA_{0.75}Cs_{0.15}PbI_{2.9}Br_{0.1}$	1.13	23.49	77.00	20.44	ZnO	SC	spiro-OMeTAD	SC	Ag	TE	AC/100/T_{80}	[16]
FTO/c-TiO_2/Li:m-TiO_2/Perovskite/spiro/Au	$MAFACsPbI_3$	1.14	23.30	69.78	18.20	c-TiO_2/Li:m-TiO_2	SP	spiro-OMeTAD	SC	Au	TE	D/360/T_{90}	[17]
ITO/ZnO-K/Perovskite/spiro/Au	$MAPbI_3$	1.13	23.00	75.40	19.90	ZnO-K	SC	spiro-OMeTAD	SC	Au	TE	ST/800/T_{91}	[18]
FTO/TiO_2/ZnO/C_{60}/Perovskite/spiro/Au	$MAPbI_{3-x}Cl_x$	1.07	22.06	79.00	18.63	TiO_2/ZnO/C_{60}	SC	spiro-OMeTAD	SC	Au	TE	ST/336/T_{80}	[19]
ITO/ZnO-S/Perovskite/spiro/Au	$MAPbI_3$	1.10	23.15	79.00	19.65	ZnO–S	SC	spiro-OMeTAD	SC	Au	TE	ST/960/T_{90}	[20]
ITO/SnO_2/Perovskite/spiro/Ag	$MAPbI_3$	1.06	22.24	76.94	18.10	SnO_2	SC	spiro-OMeTAD	SC	Ag	TE		[21]
ITO/Co:NiO_x/Perovskite/PCBM/Ag	$MAPbI_3$	1.05	22.30	79.00	18.60	PCBM	SC	Co:NiO_x	SC	Ag	TE		[22]
FTO/Li-NiO_x/Perovskite/PCBM/BCP/Ag	$MAPbI_3$	1.12	21.35	76.74	18.35	PCBM/BCP	SC	Li-NiO_x	SP	Ag	TE	ST/792/T_{80}	[23]

(Continued)

Table 8.1 (Continued)

Device structure	Absorber (Perovskite)	Photovoltaics parameters (reverse) V_{oc} (V)	J_{sc} (mA cm^{-2})	*FF* (%)	*PCE* (%)	ETM	ETM method	HTM	HTM method	CE	CE method	Stability	References
ITO/NiO_x/Perovskite/PCBM/ZrAcac/Ag	$MAFABrPbI_3$	1.09	22.58	79.10	19.45	PCBM/ZrAcac	SC	NiO	SC	Ag	TE		[24]
ITO/Cu:NiO_x/Perovskite/PCBM/Al	$MAPbI_3$	1.11	23.60	70.00	18.30	PCBM	SC	Cu:NiO_x	SC	Al	TE		[25]
FTO/S-NiO/Pereovskite/PCBM/BCP/Ag	$MAPbI_3$	0.99	22.92	80.30	18.15	PCBM/BCP	SC	S-NiO	SC	Ag	TE		[26]
ITO/NiO/Perovskite/PCBM/Ag	$MAPbI_{3-x}Cl_x$	1.09	22.70	73.00	18.06	PCBM	SC	NiO	SC	Ag	TE		[27]
ITO/$CuCrO_2$/Perovskite/PCBM/BCP/Ag	$MAPbI_3$	1.07	21.94	81.00	19.00	PCBM/BCP	SC	$CuCrO_2$	SC	Ag	TE	ST/1000/T_{92}	[28]
ITO/NiO_x/F6TCNNQ/Perovskite/PCBM/ZrAcac/Ag	$MAFACsPbI_3$	1.12	23.18	80.30	20.86	PCBM/ZrAcac	SC	NiO_x/F6TCNNQ	SC	Ag	TE		[12]
ITO/BL-B:ZnO&$Ba(OH)_2$/Perovskite/spiro/Au	$MAPbI_3$	1.12	23.50	78.00	20.62	BL-B:ZnO/$Ba(OH)_2$	ES/SC	spiro-OMeTAD	SC	Au	TE	ST/2400/T_{90}	[29]
ITO/SnO_2/Perovskite/PEAI/spiro/Au	$FA_{1-x}MA_xPbI_3$	1.18	25.20	78.40	23.32	SnO_2	SC	spiro-OMeTAD	SC	Au	TE	ST/500/T_{91}	[10]
FTO/SnO_2QD/Perovskite/spiro/Au	$MAFACsBrPbI_3$	1.13	23.05	79.80	20.79	SnO_2QD	SC	spiro-OMeTAD	SC	Au	TE	ST/2/T_{91}	[30]
FTO/c-TiO_2/mp-TiO_2/a-SnO_2/Perovskite/spiro/Au	$MAFABrPbI_3$	1.16	22.51	77.60	20.40	c-TiO_2/mp-TiO_2/a-SnO_2	SP/SC	spiro-OMeTAD	SC	Au	TE	ST/60/T_{97}	[31]
FTO/Y-SnO_2/Perovskite/spiro/Au	$MAFACsBrPbI_3$	1.13	23.56	77.78	20.71	Y-SnO_2	SC	spiro-OMeTAD	SC	Au	TE		[32]
ITO/TiO_2/Perovskite/PTAA/Au	$FAPbI_{3-x}Br_x$	1.04	22.90	76.60	18.20	TiO_2	SC	PTAA	SC	Au	TE	ST/1000/T_{95}	[33]

FTO/bl-TiO_2/mp-TiO_2/ Perovskite/spiro/Au	$(FAPbI_3)_{0.95}$ $(MAPbBr_3)_{0.05}$	24c9114	81.29	23.20	bl-TiO_2/ mp-TiO_2	SP/SC	spiro-OMeTAD	SC	Au	TE	ST/120/T_{80}	[9]
FTO/SnO_2/PCBM:PMMA/ Perovskite/PMMA/spiro/ Au	$Rb_xCs_yFA_{(100-x-y)}$ PbI_3	25c0608	75.50	20.44	SnO_2/ PCBM: PMMA	ADL/SC	spiro-OMeTAD	SC	Au	TE	ST/1000/T_{95}	[34]
ITO/TiO_2-Cl/Perovskite/ spiro/Au	MAFACsPI$_3$-Cl	22c2016	81.30	21.00	TiO_2	SC	spiro-OMeTAD	SC	Au	TE	ST/720/T_{95}	[35]
FTO/Zw-SnO_2/Perovskite/ spiro/Au	MAFACsBrPbI$_3$	23c3014	78.80	20.91	Zw-SnO_2	SC	spiro-OMeTAD	SC	Au	TE	ST/140/T_{93}	[36]
FTO/TiO_2/Perovskite/ spiro/Au	(Cs,MA,FA) $Pb(I_{0.85}Br_{0.15})_3$	23c2017	79.00	21.50	TiO_2	SC	spiro-OMeTAD	SC	Au	TE	ST/300/T_{80}	[37]
FTO/SnO_2/Perovskite/ spiro/Au	KCsFAMAPbI$_3$	22c9415	79.10	20.86	SnO_2	CBD	spiro-OMeTAD	SC	Au	TE	ST/1000/T_{90}	[38]
FTO/c-TiO_2/mp-TiO_2/ Perovskite/spiro/Au	$(FAI)_{0.9}Cs_{0.1}$ $(PbI_2)_{1.05}$:SN[a]	24c0015	75.00	20.90	TiO_2	SC	spiro-OMeTAD	SC	Au	TE	ST/1000/T_{98}	[39]
FTO/c-TiO_2/mp-TiO_2/ Perovskite/spiro/Au	$(FAI)_{0.81}(PbI_2)_{0.85}$ $(MABr)_{0.15}$ $(PbBr_2)_{0.15}$	23c8813	77.00	21.60	TiO_2	SC	BrPh-ThR: spiro-OMeTAD	SC	Au	TE	ST/1500/T_{90}	[40]

a) SN: bifunctional 3-(5-mercapto-1*H*-tetrazol-1-yl)benzenaminium iodide.

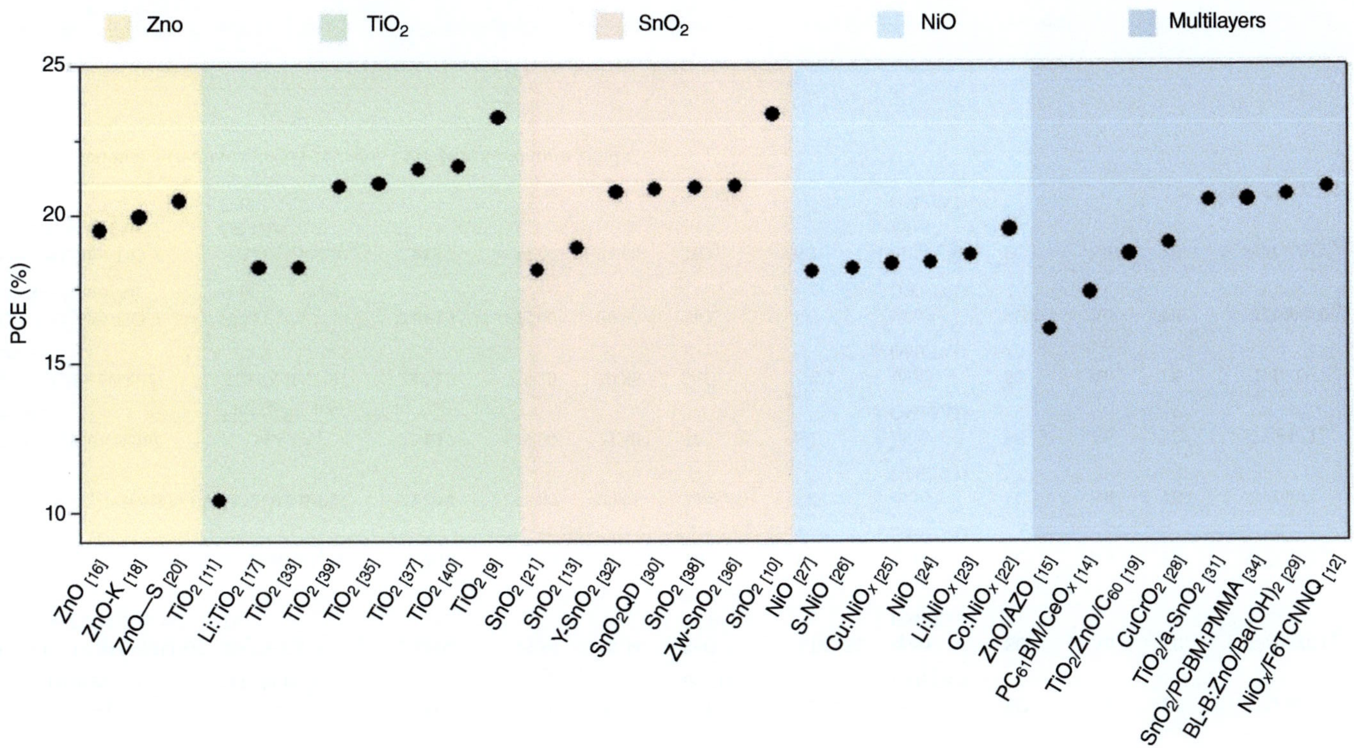

Figure 8.3 Metal oxides most employed as transport layers in PSC. We can find about five categories based on ZnO (yellow), TiO_2 (green), SnO_2 (orange), NiO (blue), and Multioxide layers (gray). The number in parentheses corresponds to the PSC described in Table 8.1.

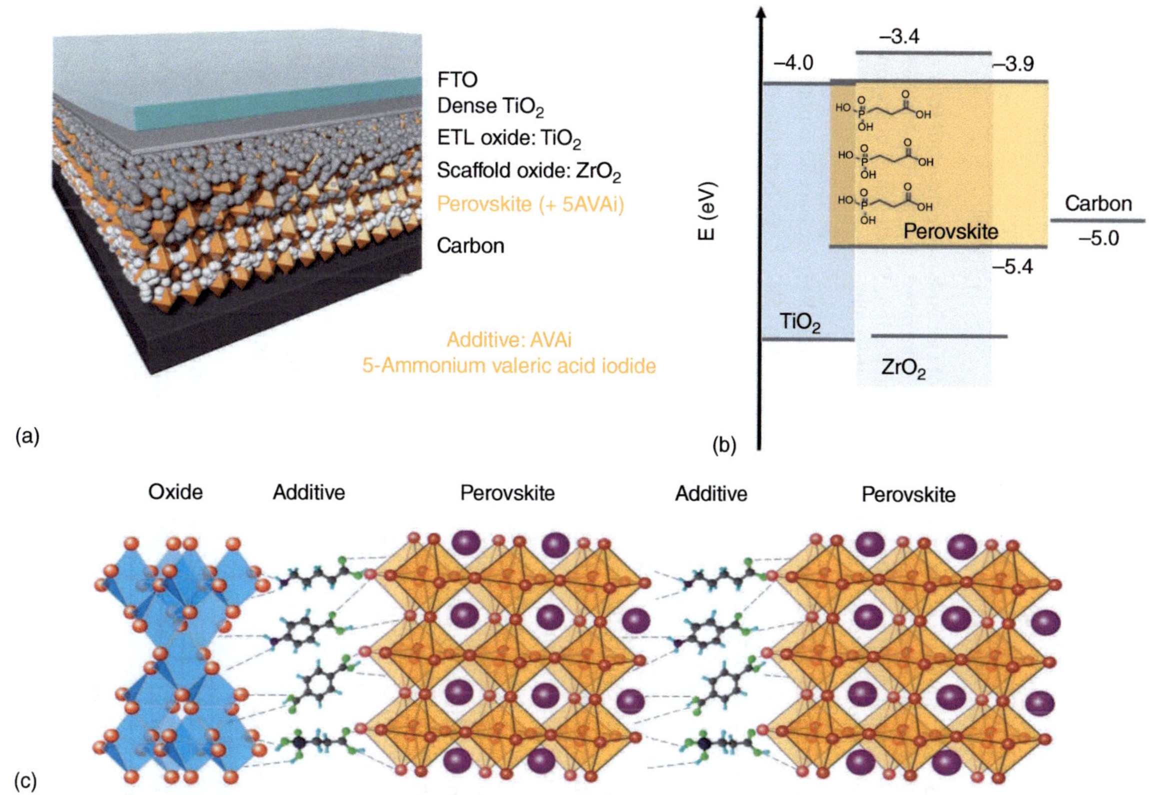

Figure 8.4 Carbon-based PSC (C-PSC). (a) Schematic representation of the C-PSC; (b) Band alignment of the different layers of the device; (c) Schematic representation of the interaction between an additive, the oxide/perovskite, and the perovskite/perovskite interfaces.

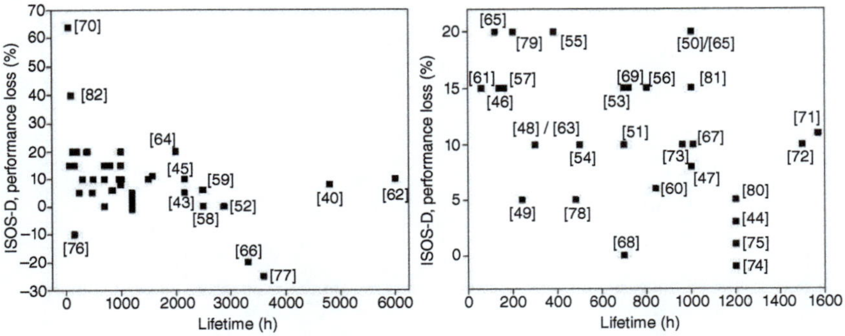

Figure 8.5 Left: Summary of the performance loss of carbon-based perovskite solar cells that apply metal oxides as transport layers (on rigid substrates) after stability analyses (ISOS-D). Right: Zoom of the cluster of data points on the left. The cluster of black squares without references (left) corresponds to the graph in the right.

bilayer of compact (c-) and mesoporous (m-) TiO_2 results in an excellent performing device with a PCE of 16.37%. The cell loses only 5% of its performance after 2160 hours under storage in the dark. The incorporation of an ultrathin ferroelectric oxide, $PbTiO_3$, layer effectively suppresses the non-radiative recombination. Due to this effect, the V_{oc} and the total photovoltaic performance were improved compared to PSCs applying the standard TiO_2 [43]. Another example employs a Ni-doped c-TiO_2 thin film. Compared to standard c-TiO_2, the doped layer facilitates the charge carrier transfer and reduces the recombination rate through trap states. Therefore, all solar cell parameters become enhanced, leading to a solar cell with a PCE of 17.46%. When stored in the dark, the cell loses 3% of its initial performance after 1200 hours. One reason for this high stability is the use of CuPc as HTM [44].

In terms of efficiency and stability, one of the best performing PSC applies a SnO_2 nanoparticle modified TiO_2 as the ETL, obtaining a champion efficiency of 18.1% and a performance loss in the dark of only 3% after 1600 hours of stability analysis. A new HTM made of P3HT/graphene bilayer is introduced and compared with a standard P3HT as HTM. It is reported that the hole mobility is greatly improved, leading to an improvement in hole extraction. Under continuous light illumination, the PSC shows an 11% performance loss from its initial efficiency after 600 hours of analysis [99].

Other PSCs worth mentioning with a hexamethonium bromide (HMB)-doped C_{60} layer as the ETM. It offers a smoother surface for the perovskite layer, faster electron transport, lower electron recombination, enhanced electron mobility, and decreased electron trap state density compared to the standard C_{60}. The best device shows a PCE of 16.03%, with a performance loss of 10% when stored under illumination for 338 hours [100]. However, the extraordinary stability reported by Grancini et al. for carbon-based PSC has not been improved in the past two years. In Grancini's work, a 2D/3D interface engineering at the perovskite layer is made, resulting in reducing the humidity and leading to a lifetime of more than 10 000 hours under continuous illumination without performance loss. The sample showed an initial PCE of 11.9% [101].

Table 8.2 Carbon-based perovskite solar cells.

Mesoporous triple-layer ("Shell") ETL/scaffold/ carbon-based electrode	Perovskite	Additive	HTM	V_{OC} (V)	J_{sc} (mA cm^{-2})	FF (%)	PCE (%)	Stability	References
c-/m-TiO_2/$PbTiO_3$/CNT				0.93	23.47	75	16.37	D/2160/T_{95}	[43]
c-/m-TiO_2/ZrO_2/C	$CH_3NH_3PbI_3$			0.95	24.19	68.5	15.77		[83]
c-/m-TiO_2/SWC NT/C	$CH_3NH_3PbI_3$	SWCNT		0.97	22.36	72	15.73	D/2160/T_{90}	[45]
c-/m-TiO_2/ZrO_2/C	$CH_3NH_3PbI_3$	5 AVAi		0.98	23.20	69.10	15.70	D/140/T_{85}	[46]
c-/m-TiO_2/C	$(Cs_{0.05}MA_{0.16}FA_{0.79}Pb(I_{0.84}Br_{0.16})_3$			1.07	21.44	68	15.66	D/1000/T_{92}	[47]
c-/m-TiO_2/C	$CH_3NH_3PbI_3$	AB		0.94	23.40	71.00	15.60	L/100/T_{90}	[84]
c-/m-TiO_2/ZrO_2/C	$MAI{\cdot}Pb_xZn_{1-x}I_2$			1.01	23.56	64.6	15.37		[85]
c-/m-TiO_2/C	$CH_3NH_3PbI_3$			0.99	21.55	70.83	15.21	D/4800/T_{92}	[86]
c-/m-TiO_2/C	$Cs_{0.1}FA_{0.9}PbI_3$			0.92	23.63	69	15	D/300/T_{90}	[48]
c-/m-TiO_2/C	$Cs_{0.15}(MA_{0.7}FA_{0.3})_{0.85}PbI_3$			0.93	22.8	62.95	15	D/240/T_{95}	[49]
c-/m-TiO_2/C	$FA_{0.8}Cs_{0.2}PbI_{2.64}Br_{0.36}$/ poly(ethylene oxide)			1.08	21.2	64.7	14.9		[87]
c-/m-TiO_2/ZrO_2/C	$CH_3NH_3PbI_3$	MACl		1.03	20.14	70	14.5	D/1000/T_{80}	[50]
c-/m-TiO_2/C	$(FAPbI_3)_{1-x}(MAPbBr_3)_x (x = 0.15)$		BDT2-FMeDPA	1.00	21.4	66	14.5	D/700/T_{90}	[51]
c-/m-TiO_2 NYFS NP/C	$CH_3NH_3PbI_3$			0.99	22.02	65.96	14.47	D/2880/T_{100}	[52]
c-/m-TiO_2/C	$(MA)_{1-x}PbI_3$	5-AVAi		0.93	23.30	66	14.26		[53]
c-/m-TiO_2/ZrO_2/C	$MAPbI_{2.95}(BF4)_{0.05}$			0.99	20.55	70	14.26		[88]
c-/m-TiO_2/ZrO_2	$Cs_{0.1}FA_{0.9}PbBr_{0.1}I_{2.9}$			1.01	22.16	62.7	14.14	D/500/T_{90}	[54]
c-/m-TiO_2/C	$(MA)_{1-x}PbI_3$	5-AVAi		0.88	23.12	69.00	14.04		[89]
c-/Au@SiO_2 NPs: m-TiO_2/C	$CH_3NH_3PbI_3$			0.99	21.43	64.3	13.85		[90]
c-/m-TiO_2/C	$CH_3NH_3PbI_3$		NiO	0.97	22.49	62	13.6	D/>384/T_{80}	[55]

(Continued)

Table 8.2 (Continued)

Mesoporous triple-layer ("Shell") ETL/scaffold/ carbon-based electrode	Perovskite	Additive	HTM	V_{OC} (V)	J_{sc} (mA cm^{-2})	FF (%)	PCE (%)	Stability	References
c-/m-TiO_2/C	$CH_3NH_3PbI_3$			1.05	20.25	63.00	13.50	D/960/T_{90}	[91]
c-/m-TiO_2/C	$CH_3NH_3PbI_3$			1,05	20.25	63.00	13.50	D/800/T_{85}	[56]
c-/m-TiO_2/C	$CH_3NH_3PbI_3$			0.98	21.69	63.09	13.46	D/160/T_{85}	[57]
c-/m-TiO_2/ZrO_2/C	$CH_3NH_3PbI_{3-x}Cl_x$			0.95	22.10	64.00	13.44		[92]
c-/m-TiO_2/ZrO_2/C	$CH_3NH_3PbI_3$			0.90	21.4	69.1	13.3		[93]
c-/m-TiO_2/C	$CH_3NH_3PbI_3$	5-AVAi	Co_3O_4	0.88	23.43	64.00	13.27	D/2500/T_{100}	[58]
c-/m-TiO_2/C	$(MA)_{1-x}PbI_3$	5-AVAi		0.90	22.63	64.26	13.09	D/2496/T_{94}	[59]
c-/m-TiO_2/ZrO_2/C	$CH_3NH_3PbI_xCl_{3-x}$			0.91	22.39	64	13.03	L/175/T_{80}	[94]
c-/m-TiO_2/C	ILPF6: $CH_3NH_3PbI_3$			0.97	21.49	62	13	D/840/T_{94}	[60]
c-/m-TiO_2/C	$CH_3NH_3PbI_3$	PTAA		1.01	21.60	59	13	D/60/T_{85}	[61]
c-/m-TiO_2/ZrO_2/C	$CH_3NH_3PbI_3$	5-AVAi	Cu:NiO_x	9.57	2.18	59	12.1	D/>6000/T_{90}	[62]
c-/m-TiO_2/C	$CH_3NH_3PbI_3$			0.98	18.41	63.05	11.49	D/300/T_{90}	[63]
c-/m-TiO_2/C	$CH_3NH_3PbI_3$			1.01	17.74	64	11.37	D/2000/T_{80}	[64]
c-/m-TiO_2/C	$CH_3NH_3PbI_3$			0.98	18.26	63.23	11.28	D/700/T_{85}	[95]
c-/m-TiO_2/C	$CH_3NH_3PbI_3$			0.84	21.39	60	10.87	D/>120/T_{80}	[65]
c-/m-TiO_2/C	$(MA)_{1-x}PbI_3$	5-AVAi		0.88	21.40	57	10.70	D/3312/T_{120}	[66]
c-/m-TiO_2/C	$CH_3NH_3PbI_3$			0.87	19.33	63.54	10.66		[96]
c-/m-TiO_2/ZrO_2/C	$CH_3NH_3PbI_3$			1.00	21.72	49	10.55	D/>1008/T_{90}	[67]
c-/m-TiO_2/ZrO_2/C	$CH_3NH_3PbI_3$		CNT	0.97	22.02	45.87	10.39	D/1000/T_{80}	[97]
c-/m-TiO_2/C	$CH_3NH_3PbI_3$			0.95	22.86	46	10.11	D/700/T_{100}	[68]
Ni:c-TiO_2/C	CsMAFAPb$(IBr)_3$		CuPc	1.07	22.41	72.60	17.46	D/1200/T_{97}	[44]

Mg:TiO_2/C	CsMAFAPb$(IBr)_3$		CuPc	1.02	22.26	68.90	15.73	D/720/T_{85}	[69]
m-TiO_2/C	$CH_3NH_3PbI_3$		—	0.91	18.69	75.00	12.80	D/48/T_{36}	[70]
c-TiO_2/C	$CH_3NH_3PbI_3$		spiro	0.89	20.25	65.00	11.90	D/1570/T_{89}	[71]
c-TiO_2/C	$CH_3NH_3PbI_2C_l$			0.84	22.05	63.15	11.7	D/1500/T_{90}	[72]
c-TiO_2(CsBr)/C	$CsPbIBr_2$			1.26	11.80	72	10.71		[98]
c-TiO_2/C	$CsPbI_2Br$			1.15	13.87	64	10.21	D/960/T_{90}	[73]
SnO_2@TiO_2/C	$(NHCHNH_3)_{0.3}(NH_3CH_3)_{0.7}PbI_3$		P3HT/ graphene	1.06	22.3	74	18.1	L/600/T_{90}	[99]
m-TiO_2/c-SnO_2/C	CsMAFAPb$(IBr)_3$		CuPc	0.98	23.28	67.00	15.39	D/1200/T_{101}	[74]
Zn:SnO_2/C	FAMAPb(IBr)	Cs	CuPc	1.08	23.4	69.2	17.78	D/1200/T_{99}	[75]
SnO_2/C	PbI_2/FAI:MABr:MACl			1.08	21.67	63.17	14.77	D/150/T_{110}	[76]
SnO_2/C	$(FAPbI_3)_x(MAPbBr_3)_{1-x}$			1.07	22.50	59.00	14.50	D/3600/T_{125}	[77]
SnO_2/C	$CsPbI_2Br$		PMMA	1.20	12.64	71	10.95	D/480/T_{95}	[78]
SnO_2/C	$CsPbI_2Br$			1.15	12.91	66.1	10.13	D/200/T_{80}	[79]
HMB:C_{60}/C	$CH_3NH_3PbI_3$			1.06	22.68	66.74	16.03	L/338/T_{90}	[100]
ZnO/C	$CH_3NH_3PbI_3$	NH_4C_l		0.98	20.31	68.43	13.62	D/>1200/T_{95}	[80]
SAM/C	$CH_3NH_3PbI_3$		CuSCN	1.00	17.21	71	12.31	D/1000/T_{85}	[81]
Nb_2O_5/C	$Cs_{0.99}Rb_{0.01}PbI_2Br$			1.24	14.02	69	12	D/96/T_{60}	[82]
c-/m-TiO_2/ZrO_2/C	2D/3D $CH_3NH_3PbI_3$			0.85	23.60	58.7	11.9	L/10000	[101]

A triple layer called the "shell," made of ETL/Scaffold/Carbon-based electrode, is fabricated and used as the device's skeleton where the halide perovskite absorber is infiltrated by drop-casting. The absorber usually applies an additive to enhance crystal growth and reduce defects.

8.4 Solution-Processable Metal Oxides for Flexible Halide Perovskite Solar Cells

Metal Oxides are a good alternative for the fabrication of flexible PSCs (Figure 8.6 and Table 8.3). The applied flexible substrates can be Polymer/ITO, a metallic thin film, or even an ultrathin flexible quartz substrate. This review distinguishes flexible PSC in the normal (n–i–p) and inverted (p–i–n) architectures. The best performing normal (n–i–p) PSC applies ZnO as the ETL, resulting in an efficiency of 17.11% and a performance loss of only 10% after 700 hours stored in the dark [15]. The perovskite film contains the three cations MA, FA, and Cs. Furthermore, a thin layer of TiO_2 can be deposited on top of the ZnO layer to reduce the recombination losses via passivation of the ZnO layer. The improvement of the photovoltaic performance depending on the TiO_2-layer was proven by comparison of PSC produced on glass/fluor tin oxide (FTO)-substrates with different material thicknesses starting at zero. The recipe is applied on polyethylene terephthalate (PET) substrate and achieves similar photovoltaic performances [106].

Figure 8.6 Flexible halide perovskite solar cells. (a–c) Solar cells fabricated by roll-to-roll method [102]. (d–e) Large-area flexible PSCs fabricated by TNO [103]. Source: Reprinted with permission from Galagan [103].

Table 8.3 Flexible perovskite solar cells applying metal oxides as transport layers.

"Oxide-sandwich" configuration													
Substrates	**HTM**	**MET**	**Perovskite**	**MET**	**ETM**	**MET**	**CE**	**Stability**	J_{sc} (mA cm^{-2})	V_{OC} (V)	**FF (%)**	**PCE (%)**	**References**
PEN/ITO	NiO	Spun cast	$CH_3NH_3PbI_3$	SC	PCBM/ZnO	Spun cast/ SC	Al (TE)	L/1000/T_{85}	20.6	1.02	73	15,3	[104]
PEN/ITO	NiO	Spun cast	$Cs_{0.05}(MA_{0.17}FA_{0.83})0.95Pb(_{I2.7}Br_{0.3})$	SC	PCBM/ZnO	Spun cast	Al (TE)	L/500/T_{92}	21.2	1.02	76	16.6	[104]
PET/Ag (TE)	MoO_3/ PEDOT:PSS	TE/SC	$CH_3NH_3PbI_3$	SC	PCBM	SC	Cu/Ag/ MoO_3	D/1680/T_{95}	17.57	1.01	75.18	12.81	[105]

Normal configuration													
Substrates	**HTM**	**MET**	**Perovskite**	**MET**	**ETM**	**MET**	**CE**	**Stability**	J_{sc} (mA cm^{-2})	V_{OC} (V)	**FF (%)**	**PCE (%)**	**References**
PET/AgNW (SC)	ZnO/TiO_2	SC/ALD	$MA_{0.1}FA_{0.75}Cs_{0.15}PbI_{2.9}Br_{0.1}$	SC	spiro-OMeTAD	SC	MoO_3/ Ag(TE)	D/700/T_{90}	22.45	1.03	74	17.11	[106]
PET/ITO	ZnO/p-(F)-PO-TAZ	Spun cast/SC	$CH_3NH_3PbI_3$	Mixed stirring heated	PB(NAP-Th) TBT	SC	Ag (TE)	D/480/T_{85} L/180/T_{75} [AM1.5G illumination (100 mW cm^{-2})]	19.94	1.04	69.56	14.51	[107]
PEN/ITO	ZnO	SC	$CH_3NH_3PbI_3$	SC	spiro-OMeTAD	SC	Au (TE)	L/7200/T_{100}	17.4	1.05	74.4	13.56	[108]
PET/ITO	ZnO	Spun cast	$CH_3NH_3PbI_3$	Mixed stirring heated	PB(NAP-Th) TBT	SC	Ag (TE)	D/480/T_{60} L/180/T_{45} [AM1.5G illumination (100mW cm^{-2})]	18.68	1.02	66.26	12.70	[107]

(Continued)

Table 8.3 (Continued)

Normal configuration													
Substrates	HTM	MET	Perovskite	MET	ETM	MET	CE	Stability	J_{sc} (mA cm^{-2})	V_{OC} (V)	FF (%)	PCE (%)	References
PET/ITO	ZnO (+IL-BF_4)	SC	$CH_3NH_3PbI_3$	SC	spiro-OMeTAD	SC	Au (TE)		22.9	0.94	55	12.1	[109]
PES/AZO/AgNW/a-AZO	ZnO	SC	$CH_3NH_3PbI_3$	SC	spiro-OMeTAD	SC	Au (TE)		18.9	0.99	59.7	11.23	[108]
PET/PEDOT:PSS	ZnO	SC	$CH_3NH_3PbI_3$	SC	spiro-OMeTAD	SC	Au (TE)		17.3	0.99	60	10.3	[110]
PET/ITO	m-TiO_2	SC	NH_4Cl: Cs/MA/FA	SC	spiro-OMeTAD	SC	Au (TE)	D/168/T_{95}	22.45	1.12	70.6	17.69	[111]
PEN/ITO	TiO_2	Laser dep	$FA_{0.85}MA_{0.15}PbI_{2.55}$-$Br_{0.45}$	SC	spiro-OMeTAD	SC	Au		23.2	0.97	75.6	17.1	[112]
Ti-foil	TiO_2	T.O	$CH_3NH_3PbI_3$	SC	spiro-OMeTAD	SC	Cu/Au (e-B)		17.9	1.09	74	14.9	[113]
PET/ITO	Nb_2O_5-TiO_2	SC	$CH_3NH_3PbI_3$	SC	spiro-OMeTAD	SC	Au (TE)		20.04	0.99	69	13.60	[114]
PET/ITO	Nb:TiO_2	Sol–gel	$Cs_{0.05}(MA_{0.15}FA_{0.85})_{0.95}Pb(I_{0.85}Br_{0.15})_3$	SC	spiro-OMeTAD	SC	Au		19.82	1.11	61.4	13.51	[115]
PET/graphene	TiO_2/PCBM	SC	$CH_3NH_3PbI_3$	SC	spiro-OMeTAD	SC	CSCNTS	L/>1200/T_{90}	20.25	0.89	65	11.9	[71]
PET/ITO	HfO_2/SnO_2	ALD/SC	$Rb_{5-x}K_x[Cs_{0.05}(FA_{0.83}MA_{0.17})_{0.95}(I_{0.83}Br_{0.17})_3]$	SC	spiro-OMeTAD	SC	Au (TE)	D/1400/T_{90}	21.24	1.13	79.2	19.11	[116]
PET/ITO	SnO_2 + Al_2O_3	SC	$[(FA_{0.83}MA_{0.17})_{0.95}Cs_{0.05}PbI_2Br_{0.5}]$		spiro-OMeTAD		Au		22.8	1.05	76	18.1	[117]
PEN/ITO	SnO_2 QD	SC	$CH_3NH_3PbI_3$	SC	Li-spiro-OMeTAD	SC	Au		19.7	1.13	79.6	17.7	[118]
PEN/ITO	SnO_2	SC	$Cs_{0.05}(MA_{0.17}FA_{0.83})_{0.95}Pb(I_{0.83}Br_{0.17})_3$	SC	spiro-OMeTAD	SC	Au (TE)	D/840/T_{95}	20.1	1.16	72.8	17.1	[119]
PET/ITO	SnO_2 NCs	SC	KCsFAMA	SC	spiro-OMeTAD	SC	Au (TE)	D/500/T_{80}	19.89	1.07	77	16.5	[120]

PET/ITO	SnO_2	Sol–gel	KCsFAMA PbI_xBr_{3-x}	SC	spiro-OMeTAD	SC	Au	D/1600/T_{80}	20.03	1.07	71	15.27	[121]
PEN/ITO	$SnO_2 + SnCl_2$	SC	$((FAPbI_3)_{0.97}(MAPbBr_3)_{0.03}$	SC	spiro-OMeTAD	Coated	Au (TE)	D/2664/T_{85} (without encapsulation) D/3648/T_{74}	21.023	1.01	71.29	15.21	[122]
PET/ITO	SnO_2	Slot die coating	CsFAMA	SC	spiro-OMeTAD	SC	Au (TE)	D/1200/T_{80}	3.28	6.7	69	15.2	[123]
PET/ITO	EDTA complexed-SnO_2	SC	$FA_{0.95}Cs_{0.05}PbI_3$	SC	spiro-OMeTAD	SC	Au (TE)		23.42	1.09	71.6	18.28	[124]
PEN/ITO	c-C_{60}/C_{60}-NP	SC	$CH_3NH_3PbI_3$	SC	spiro-OMeTAD	SC	Au (TE)	D/≫ 1200/T_{90} L/720/T_{80}	22.12	1.07	73	17.28	[125]
PET/ITO	PEI	SC	$CH_3NH_3PbI_3$	Flash meth	spiro-OMeTAD	SC	Ag (TE)	D/≫ 1920/T_{80}	21.68	1.04	75	16.81	[126]
Flex/IZO (Sput/SC)	PEIE/C_{60}	SC/TE	$CH_3NH_3PbI_3$	TE/SC/TE	spiro-OMeTAD	SC	Au (TE)		19.7	1.06	74	15.8	[127]
AZO/(Sputt)	LiF/C_{60}	TE	$CH_3NH_3PbI_3$	SC	spiro-OMeTAD	SC	Au		18.2	1.07	74	14.8	[128]
Flex/IZO (Sput/SC)	PEIE/C_{60}	SC/TE	$CH_3NH_3PbI_3$	TE/SC/TE	spiro-OMeTAD/MoOx/IZO	SC/TE/TE	NI/Al (e-beam Eva)		18.7	1.06	69	14	[127]
PET/ITO	C_{60}	TE	$CH_3NH_3PbI_3$	SC	spiro-OMeTAD	SC	MoO_3/Au (TE)	D/≫ 1440/T_{100}	18.1	1.01	70	12.6	[129]
PEN	PEDOT:PSS/PCBM	SC	$CH_3NH_3PbI_3$	TE/SC	spiro-OMeTAD	SC	Au (TE)	D/150/T_{90}	17.4	1	65.5	11.4	[130]

(Continued)

Table 8.3 (Continued)

Inverted configuration													
Substrates	**HTM**	**MET**	**Perovskite**	**MET**	**ETM**	**MET**	**CE**	**Stability**	**J_{sc} (mA cm^{-2})**	**V_{OC} (V)**	**FF (%)**	**PCE (%)**	**References**
PET/ITO	PEDOT:PSS/ NiO		$CH_3NH_3PbI_{3-x}Cl_x$		PCBM, a-TIPD		Al	D/1440/T_{80}	23.3	0.98	77.4	18	[131]
PEN/AgNW (SC)	PEDOT:PSS		$CH_3NH_3PbI_3$		PCBM/C60/ BCP		Al		15.88	0.93	54.2	8	[132]
Text/ polyurethane (lamination)	PEDOT:PSS/ CNT PEDOT:PSS	Bar coating	$CH_3NH_3PbI_3$	Bar coating	PCBM	Bar coating	Ag (TE)	D/300/T_{85}	12.4	0.88	49	5.72	[133]
PET/ITO	NiO_x/PDA	SC	$CH_3NH_3PbI_3$	SC	PCBM/BCP	SC	Ag	D/450/T_{90}	22.43	1.05	78.1	18.35	[134]
PEN/ITO	Cu:NiO	SC	Perovskite	SC	C_{60}/PCBM		Ag	D/1000/T_{95}	21.34	1.10	73.5	17.41	[135]
PEN/ITO	NiO	SC	Perovskite	SC	C_{60}/PCBM		Ag		21.24	1.08	71.5	16.41	[135]
PEN/ITO	NiO	Spun cast	$CH_3NH_3PbI_3$	SC	PCBM	Spun cast	Al (TE)		20.5	0.98	75	15	[104]
PET/ITO	NiO_x	SC	$MAPbI_{3-x}Cl_x$	SC	PCBM		Ag (TE)	D/1200/T_{90}	21.28	1.02	65	14.11	[136]
PET/ITO	NiO_x	SC	$CsPbI_2Br$	SC	C_{60}/BCP	TE	Ag (TE)	D/13/T_7	11.5	0.97	65	7.3	[137]
PEN/ITO	Reduced graphene oxides	SC	$CH_3NH_3PbI_3$	SC	PCBM	SC	Ag (TE)	D/1000/T_{65}	20.5	0.94	74.5	14.5	[138]
PET/ITO	B-HTL	S-G/ev.	$CH_3NH_3PbI_3$	SC	C60/BCP		Ag		21.21	1.00	68	14.39	[139]
PEN	PTAA	SC	$CH_3NH_3PbI_3$	SC	PCBM	SC	Ag (TE)	D/150/T_{85}	18.37	0.97	59.2	10.51	[140]

SC = spin-coating, TE = thermal evaporation, DC = DC reactive magnetron sputtering, SG = sol–gel, I = immersion, ALD = atomic layer deposition, CBD = chemical bath deposition, USD = ultrasonic spray deposition, SP = spray-pyrolisis, ECD = electrochemical deposition, SPT = sputtering, HT = hydrothermal method, EB = electron beam, DB = doctor blade, G = surface growth.

For solar cells applying TiO_2 as the ETM in an n–i–p configuration, the best PCE reported is 17.69%. When stored in the dark, this cell loses only 5% of its PCE in 168 hours. The solar cell efficiency was increased about 12% compared to standard m-TiO_2 using NH_4Cl-modified mesoporous TiO_2 on rigid substrates. Furthermore, a reduction in the J–V hysteresis was achieved. The recipe was deposited on PET/ITO substrates [111].

The best performing cell in this review corresponds to the KRbCsMAFA-based PSC with an efficiency of 19.11%. The sample is SnO_2-based with an ultrathin layer of HfO_2, enhancing the charge carrier collection by reducing interfacial recombination at the ITO surface. It shows promising stability properties, with a loss of performance of about 10% after storage in the dark for 1400 hours [116]. Another example is doping agents that can lead to PCEs of 18.28% [10]. For example, the use of EDTA-SnO_2 for the ETL results in decreased PSC's hysteresis and increased charge carrier mobility. The effect is attributed to the defect passivation effect of the EDTA on the SnO_2 oxide. Moreover, the Fermi level's successful matching between the EDTA-SnO_2 ETM and perovskite increased the open-circuit voltage, enhancing device efficiency [124].

Flexible PSCs fabricated with inverted architecture have also been reported. The best performing solar cell applies PEDOT:PSS as HTM on top of the TCO results in a PCE of 18% with a performance loss in the dark of 20% after 1440 hours of stability analysis [35]. The cell is supplied with an HTM made of a double layer of PEDOT:PSS and NiO_x. It has been reported that the double transport layer structure reduces the surface roughness on which the perovskite layer will be deposited. The amount of leakage currents in solar cells is minimized, which increases the FF and the PCE of the sample [131].

Another interesting example applies NiO_x as HTM, with an efficiency of 18.35% [38]. The cell loses 10% of its performance within 400 hours of stability analysis in the dark. The HTM is cross-linked polydopamine (PDA), which reduces the brittleness of the layer, improving the solar cell's mechanical stability during bending compared to the classical NiO_x [134]. A flexible PSC with efficiency as high as 14.5% and performance losses of 35% after 1000 hours of storage in the dark [45] applies graphene-oxide (GO) and reduced graphene-oxide (rGO) as HTM. Although the hole extraction rate for standard GO is much higher than for rGO, the retardation of charge recombination by the presence of delocalized holes in benzene rings of the rGO fully compensates this rate disadvantage and even improve their performance compared to standard GO [138].

We should mention that stability analysis carried out for PSC, including flexible devices, should follow the recent protocols recommended by the PSC community [141, 142]. These protocols are highly focused on harsh environments of light intensity, humidity, and/or temperature. However, PSCs for the future IoT (wearables, portable, etc.) will require less aggressive environments, and a deeper understanding of the need for flexible devices should be considered. Especially, stability reports under different light intensities and light sources for indoor (or artificial lighting) are still in their infancy.

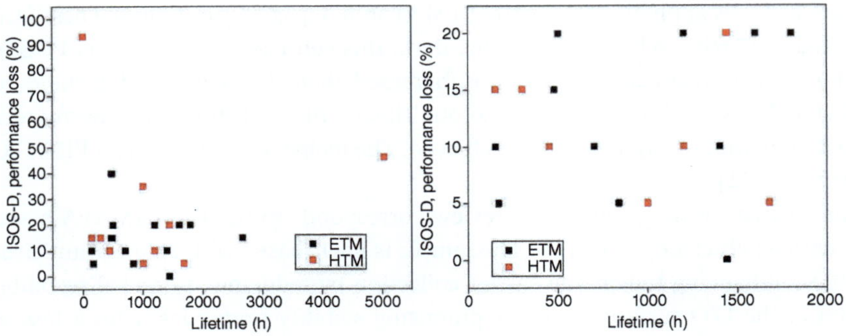

Figure 8.7 Left: Summary of the performance loss reported for flexible perovskite solar cell (ISOS-D) for devices applying oxides as HTM (red) and ETM (black). The cluster of squares without reference (left) corresponds to the graph on the right.

Although the efficiency and stability summarized in this work for flexible PSCs indicate a bright future for the technology, there is still a need for improvement, especially for developing more stable devices. It is expected that more sophisticated encapsulation methods will be developed in the future, which can contribute to the deployment of the technology. Currently, PSCs can already be fabricated on flexible substrates by large-area printing methods such as inkjet, roll-to-roll, or slot die coating (Figure 8.6). The fabrication of large active area devices and modules is also a reality (Figure 8.7).

8.5 Characterization of PSC by Electrochemical Impedance Spectroscopy (EIS)

In this section, we wanted to include recently published works where the EIS technique has been applied to characterize PSC applying metal oxides (Table 8.4). Among many other techniques, we believe this characterization method can contribute to understanding the stability of flexible PSCs, especially under constant irradiation conditions. However, in this section, we would like to review the different approaches found for applying electrochemical impedance spectroscopy (EIS). The method can be described as a stationary technique measuring the current response when an AC voltage is applied in a frequency range. An important advantage of EIS over other techniques is using low amplitude voltages to obtain very low disturbances in the system.

Dong et al. use an ITO/ZnO/perovskite/spiro-OMeOTAD/Au solar cell with an initial 20% PCE [16]. An impedance study was performed for which measurements were made under dark conditions. Only one semicircle is observed under dark conditions, which corresponds to the resistance of the interface between HTL and the perovskite. It offered information about the charge transfer resistance and (R_{ct}) and/or resistance to recombination (R_{rec}). This is since charge transfer resistance and/or resistance to recombination of the interface between ETL and

Table 8.4 Electrochemical impedance spectroscopy (EIS).

Device structure (the best)	Perovskite	Photovoltaics parameters (reverse)					Impedance parameters (in dark)			References
		V_{OC} (V)	J_{sc} (mA cm^{-2})	FF (%)	PCE (%)	VV)	R_s (Ω)	R_{rec}/R_{tr} (Ω)	CPE (F)	
FTO/TiO_2-c/TiO_2-m/perovskite/ZrO_2/Cu:NiO_x/C	$MAPbI_3$	1	22.15	0.57	12.79	0.2		4000		[62]
ITO/SnO_2/perovskite/spiro-OMeTAD/Au	$MAPbI_3$	1.06	23.14	0.75	18.34		46	215	5E-8	[143]
ITO/ZnO/perovskite/spiro-OMeTAD/Au	$MA_{0.1}FA_{0.75}Cs_{0.15}PbI_{2.9}Br_{0.1}$	1.13	23.49	77.00	20.44	1.1	23	1116		[16]
ITO/Cu:NiO_x/perovskite/PCBM/Al	$MAPbI_3$	1.11	23.60	70.00	18.30	1.0	22	425		[25]
FTO/S-NiO/pereovskite/PCBM/BCP/Ag	$MAPbI_3$	0.99	22.92	80.30	18.15			5000		[26]

Device structure (the best)	Perovskite	Photovoltaics parameters (reverse)										References
		V_{OC} (V)	J_{sc} (mA cm^{-2})	FF (%)	PCE (%)	VV)	R_s (Ω)	R_{tr} (Ω)	CPE1 (F)	R_{rec} (Ω)	CPE2 (F)	
FTO/TiO_2-c/TiO_2-m/perovskite/ZrO_2/Cu:NiO_x/C	$MAPbI_3$	1.00	22.2	0.57	12.8	0.2		22		24		[62]
FTO/TiO_2c/TiO_2m/perovskite/P3HT:PCBM/Au	$MAPbI_3$	0.98	19.1	0.7	13.1			474	1.6E-7	642	2.2E-5	[144]
FTO/SnO_2/perovskite/spiro-OMeTAD/Au	Cs5FAMA	1.10	22.7	0.77	19.4	1	18	70		110		[38]
FTO/TiO_2c/perovskite/PMMA/spiro-OMeTAD/Ag	$MAPbI_3$	1.0	21.86	0.78	17.15	−1		1200				[145]
ITO/PEDOT:PSS/perovskite/PCBM/Ag	$MAPbI_3$	0.98	19.2	0.73	14.2			500		3000		[146]
FTO/TiO_2c/m-$Zn_2Ti_3O_8$/perovskite/spiro-OMeTAD/Au	$MAPbI_3$	1.02	22.0	0.78	17.2	0.7	2.5	92.8		486.7		[147]

(Continued)

Table 8.4 (Continued)

Device structure (the best)	Perovskite	Photovoltaics parameters (reverse)										References
		V_{OC} (V)	J_{sc} (mA cm^{-2})	FF (%)	PCE (%)	VV)	R_s (Ω)	R_{tr} (Ω)	CPE1 (F)	R_{rec} (Ω)	CPE2 (F)	
FTO/SnO_2/Cs-perovskite/WO_3/spiro-OMeTAD/Au	Cs-$MAPbl_2$	1.17	23.70	0.77	21.44	1		4250				[148]
ITO/ZnO/perovskite/P(NDI2DT-TTCN)/Au	$MAPbI_3$	1.00	22.0	0.77	17.0					25 000		[149]
FTO/dye-NIO_x(M)/perovskite/BSO/PCBM/Ag	$MAPbI_3$	1.04	21.5	0.71	16.2	0.7	2.4	4794		9251		[150]
FTO/NB:TIO_2/perovskite/spiro-OMeOTAD/Au	$MAPbI_3$	1.13	23.50	0.78	19.1		33	2.51E6				[151]
FTO/TiO_2c/TiO_2m/perovskite/spiro-OMeTAD/Au	$MAPbI_3$	1.01	20.4	0.72	15.5	0.4	58.3	8541		2.87E4		[152]
FTO/ZnO/perovskite/P3TAA (42%)/Ag	$MAPbI_3$	0.95	15.7	0.67	9.98	0	9.43	44.8	1E-6	788	49E-6	[153]
ITO/PVO/perovskite/PC_{71}BM/AG	$MAPbI_3$	0.88	22.7	0.74	15.0	0.9	12.6	16.9	61E-9	60	17E-9	[154]
ITO/ZnO/perovskite/spiro-OMeOTAD/Au	$MA_{0.1}FA_{0.75}Cs_{0.15}PbI_{2.9}Br_{0.1}$	1.13	23.49	0.77	20.44	0.9	33	331		5361		[16]
FTO/S-NiO/pereovskite/PCBM/BCP/Ag	$MAPbI_3$	0.99	22.92	80.30	18.15			5000				[36]
FTO/Zw-SnO_2/perovskite/spiro-OMeTAD/Au	MAFACs $BrPbI_3$	1,14	23,30	78.80	20.91		8.1	255				[14]
FTO/Cu:NiO_x/perovskite/PC_{61}BM/CeO_x/Ag	$MAPbI_3$	1.06	21.52	76.07	17.35	0.8	23.92	650				[17]
FTO/c-TiO_2/Li:m-TiO_2/perovskite/spiro-OMeTAD/Au	$MAFACsPbI_3$	1.14	23.30	69.78	18.20			525				[155]

Device structure (the best)	Perovskite	Photovoltaics parameters (reverse)					Impedance parameters (1 sun)					Refs.
		V_{OC} (V)	J_{sc} (mA cm^{-2})	FF (%)	PCE (%)	VV)	R_s (Ω)	R_{tr} (Ω)	CPE1 (F)	R_{rec} (Ω)	CPE2 (F)	
a-FTO/bI-SnO_2/perovskite/ spiro-OMeTAD/Au	$MAPbI_3$	1.11	22.8	0.76	19.2		4.88	28				[156]
b-FTO/SnO_2/perovskite/ spiro-OMeTAD/Au	$MAPbI_3$	1.05	23.03	0.75	18.2			1200				[157]
b-FTO/SnO_2/perovskite/ spiro-OMeTAD/Au	$MAPbI_3$	1.05	23.03	0.75	18.2	0.6		800				[157]

PSC applying metal oxides which have been characterized by the impedance spectroscopy technique.

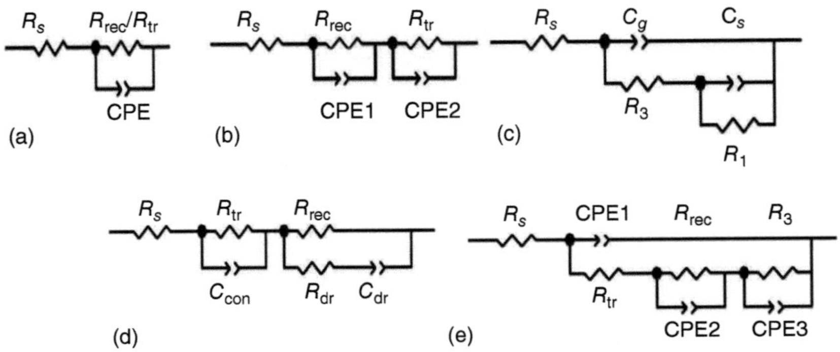

Figure 8.8 Application of the electrochemical impedance spectroscopy (EIS) for the study of PSCs. Most applied equivalent circuits (a–e) employed for the characterization of PSCs.

the perovskite is small to be detected due to the rapid electron injection, large contact area, and better interface contact. [16, 25, 26, 62, 158] Kim et al. employed a-FTO/SnO_2/Cs-perovskite/WO_3/spiro-OMeTAD/Au solar cell with an initial 21.44% PCE. The series resistance of the high-frequency region is attributed to the wiring and FTO interface. In this study, the Nyquist plots' main arc corresponds to the recombination resistance and the chemical capacitance (CPE) of the ETLs/$CH_3NH_3PbI_3$/spiro-OMeTAD interfaces. A secondary semicircle appears at the low-frequency region, which is associated with the counter electrode interface [148]. In general, under the conditions of 1 Sun, a semicircle is the high-frequency region observed. This semicircle is attributed to the charge transfer resistance between the HTL/perovskite interface. A second semicircle appears in the region of low frequencies attributed to resistance to recombination. On the other hand, the capacitance of high frequency (CPE1) is generally attributed to the polarization capacity of perovskite and the bulk dielectric capacitance of the HTM. In contrast, the capacitance at low frequency (CPE2) is much related to the ion accumulation in the device interval and the slow movement of the charges (ions) inside the perovskite [14, 16, 17, 36, 38, 144–147, 149–155, 159, 160].

Another group directed by Ngo et al. and Pham et al. suggest another type of equivalent circuit (see Figure 8.8) to explain the different processes between a PSC interface. The authors determine a capacitive element (C_g) at high frequencies associated with the dielectric properties of the perovskites, and a resistive element (R_3) associated with the transport resistance of the electron of the perovskite interface. At low frequencies, we obtain the elements attributed to the properties of the perovskite, C_s, and R_1. Cs refers to ionic or electronic accumulation capacitance at electrode interfaces under dark or illuminated conditions. Meanwhile, R_1 determinate to recombination resistance [161, 162]. Other authors such as Xiong et al. and Ulfa et al. suggest other equivalent circuits (Figure 8.8d,e) based on previous studies by Bisquert et al. and Guerrero et al., respectively. Both authors adapt the equivalent circuits depending on the interfaces of their devices, slightly with respect to the circuits discussed above [142, 156, 157, 160].

8.6 Conclusions

PSCs are one of the most important players in the Solar Energy arena. The increasing demand for energy to power electronic devices connected in a future network (IoT) will require energy sources such as PSC to be compatible with light-weight, flexible, semitransparency, low cost, and scalability. Flexible PSCs have already demonstrated high efficiency. However, their long-term stability remains unsolved. This work presented a brief overview (including published works from 2018 to the first months of 2020) of the most recent developments observed for PSC fabricated with metal oxides acting as transport layers and scaffolds. Metal oxides are recognized as a source of stability for many solar cells, providing moisture and temperature resistance to the device. They interact with the PSC additives reducing the defect density and enhancing device efficiency, also controlling also ion mobility which is a known source of instability. Many metal oxides can be fabricated as dense thin films at low temperatures, which benefits the fabrication of flexible substrates. This has also benefited their scale-up fabrication by applying printing methods such as roll-to-roll or slot die (among others). Thus, large and module fabrication of flexible PSCs is now a reality for research centers like TNO and companies like SauleTech. Among the different PSC configurations on rigid substrates (inverted, normal, oxide-sandwiched, and C-based), we have observed that oxides provide stability in any of these configurations. It is the Carbon-based PSC configuration the one with the record in stability (under continuous illumination). For flexible PSCs, the technology is still limited by stability studies. Mainly because it depends on encapsulation technology which should also pass permeation analysis. These tests are still under investigation. Nevertheless, we have observed that highly stable flexible PSC devices can sustain many hours in the dark following the ISOS-D protocol.

Acknowledgments

We thank the Spanish MINECO through the Severo Ochoa Centers of Excellence Program under Grant No. SEV-2017-0706 for the postdoctoral contract of H.X., as well as for the project RTI2018-101974-B-C22 and PCI2020-112185. We also thank the Spanish State Research Agency for the grant Self-Power (PID2019-104272RB-C54/AEI/10.13039/501100011033) and the OrgEnergy Excellence Network (CTQ2016-81911-REDT). The Agència de Gestió d'Ajuts Universitaris i de Recerca (AGAUR) also provided support to the consolidated Catalonia research group 2017 SGR 329 and the *Xarxa d'R+D+I Energy for Society (XRE4S)*. Part of this work is under the Ph.D. programme in Chemistry of C.P. in the Universitat Autonoma de Barcelona (UAB, Spain), and we are grateful for CONACYT for the scholarship of C.P. ICN2 is supported by the Severo Ochoa program from Spanish MINECO (Grant No. SEV-2017-0706) and is funded by the CERCA Programme/Generalitat de Catalunya.

References

1 AI is the future of the IoT – IBM Internet of Things blog (2020) Available: https://www.ibm.com/blogs/internet-of-things/ai-future-iot (accessed 21 February 2020).

2 Lira-Cantu, M., Chafiq, A., Faissat, J. et al. (2011). Oxide/polymer interfaces for hybrid and organic solar cells: anatase vs. rutile TiO_2. *Sol. Energy Mater. Sol. Cells* 95 (5): 1362–1374.

3 Gonzalez-Valls, I. and Lira-Cantu, M. (2009). Vertically-aligned nanostructures of ZnO for excitonic solar cells: a review. *Energy Environ. Sci.* 2 (1). Royal Society of Chemistry: 19–34.

4 Terán-Escobar, G., Pampel, J., Caicedo, J.M., and Lira-Cantú, M. (2013). Low-temperature, solution-processed, layered V_2O_5 hydrate as the hole-transport layer for stable organic solar cells. *Energy Environ. Sci.* 6 (10): 3088–3098.

5 Lima, F.A.S., Beliatis, M.J., Lira-Cantu, M. et al. (2016). Flexible ITO-free organic solar cells applying aqueous solution-processed V_2O_5 hole transport layer: an outdoor stability study. *APL Mater.* 4 (2): 026104.

6 Gonzalez-Valls, I., Angmo, D., Gevorgyan, S.A. et al. (2013). Comparison of two types of vertically aligned ZnO NRs for highly efficient polymer solar cells. *J. Polym. Sci. Part B Polym. Phys.* 51 (4): 272–280.

7 Pérez-Tomás, A., Mingorance, A., Tanenbaum, D., and Lira-Cantú, M. (2018). Metal oxides in photovoltaics: all-oxide, ferroic, and perovskite solar cells. In: *The Future of Semiconductor Oxides in Next-Generation Solar Cells*, 267–356. Elsevier.

8 Lira-Cantú, M. (2017). *The Future of Semiconductor Oxides in Next-Generation Solar Cells*. Elsevier.

9 Jeon, N.M., Na, H., Jung, E.H. et al. (2018). A fluorene-terminated hole-transporting material for highly efficient and stable perovskite solar cells. *Nat. Energy* 3 (8): 682–689.

10 Jiang, Q., Zhao, Y., Zhang, X. et al. (2019). Surface passivation of perovskite film for efficient solar cells. *Nat. Photonics* 13: 460–466. https://doi.org/10.1038/s41566-019-0398-2.

11 Icli, K.C. and Ozenbas, M. (2018). Fully metal oxide charge selective layers for n–i–p perovskite solar cells employing nickel oxide nanoparticles. *Electrochim. Acta* 263: 338–345.

12 Chen, W., Zhou, Y., Wang, L. et al. (2018). Molecule-doped nickel oxide: verified charge transfer and planar inverted mixed cation perovskite solar cell. *Adv. Mater.* 30 (20): 1–9.

13 Guo, Y., Lei, H., Xiong, L. et al. (2018). An integrated organic-inorganic hole transport layer for efficient and stable perovskite solar cells. *J. Mater. Chem. A* 6 (5): 2157–2165.

14 Xing, Z., Wu, B.-S., Wang, X. et al. (2018). Photovoltaic performance and stability of fullerene/cerium oxide double electron transport layer superior to single one in p–i–n perovskite solar cells. *J. Power Sources* 389 (August 2017): 13–19.

15 Wu, S.H., Lin, M.Y., Chang, S.H. et al. (2018). A design based on a charge-transfer bilayer as an electron transport layer for improving the performance and stability in planar perovskite solar cells. *J. Phys. Chem. C* 122 (1): 236–244.

16 Dong, X., Chen, D., Zhou, J. et al. (2018). High crystallization of a multiple cation perovskite absorber for low-temperature stable ZnO solar cells with high-efficiency of over 20%. *Nanoscale* 10 (15): 7218–7227.

17 Zhang, R., Liu, D., Wang, Y. et al. (2018). Theoretical lifetime extraction and experimental demonstration of stable cesium-containing tri-cation perovskite solar cells with high efficiency. *Electrochim. Acta* 265: 98–106.

18 Azmi, R., Hwang, S., Yin, W. et al. (2018). High efficiency low-temperature processed perovskite solar cells integrated with alkali metal doped ZnO electron transport layers. *ACS Energy Lett.* 3 (6): 1241–1246.

19 Xu, J., Fang, M., Chen, J. et al. (2018). ZnO-assisted growth of $CH_3NH_3PbI_{3-x}$ Cl_x film and efficient planar perovskite solar cells with a $TiO_2/ZnO/C_{60}$ electron transport trilayer. *ACS Appl. Mater. Interf.* 10 (24): 20578–20590.

20 Azmi, R., Lee, C.L., Jung, I.H., and Jang, S.Y. (2018). Simultaneous improvement in efficiency and stability of low-temperature-processed perovskite solar cells by interfacial control. *Adv. Energy Mater.* 8 (14): 1–7.

21 Wang, W.T., Sharma, J., Chen, J-W. et al. (2018). Nanoparticle-induced fast nucleation of pinhole-free PbI_2 film for ambient-processed highly-efficient perovskite solar cell. *Nano Energy* 49 (April): 109–116.

22 Xie, Y., Lu, K., Duan, J. et al. (2018). Enhancing photovoltaic performance of inverted planar perovskite solar cells by cobalt-doped nickel oxide hole transport layer. *ACS Appl. Mater. Interf.* 10 (16): 14153–14159.

23 Guo, Y., Ma, J., Lei, H. et al. (2018). Enhanced performance of perovskite solar cells: via anti-solvent nonfullerene Lewis base IT-4F induced trap-passivation. *J. Mater. Chem. A* 6 (14): 5919–5925.

24 Chen, W., Wu, Y., Tu, B. et al. (2018). Inverted planar organic–inorganic hybrid perovskite solar cells with NiO_x hole-transport layers as light-in window. *Appl. Surf. Sci.* 451: 325–332.

25 He, J., Xiang, Y., Zhang, F. et al. (2018). Improvement of red light harvesting ability and open circuit voltage of Cu:NiO_x based p–i–n planar perovskite solar cells boosted by cysteine enhanced interface contact. *Nano Energy* 45 (November 2017): 471–479.

26 Tang, L.J., Chen, X., Wn, T.Y. et al. (2018). A solution-processed transparent NiO hole-extraction layer for high-performance inverted perovskite solar cells. *Chem. A Eur. J.* 24 (12): 2845–2849.

27 Zhao, P., Liu, X., Lin, Z. et al. (2018). Device simulation of inverted $CH_3NH_3PbI_{3-x}$ Cl_x perovskite solar cells based on PCBM electron transport layer and NiO hole transport layer. *Sol. Energy* 169 (April): 11–18.

28 Zhang, H., Wang, H., Zhu, H. et al. (2018). Low-temperature solution-processed $CuCrO_2$ hole-transporting layer for efficient and photostable perovskite solar cells. *Adv. Energy Mater.* 8 (13): 1–9.

29 Rehman, F., Mahmood, K., Khalid, A. et al. (2019). Solution-processed barium hydroxide modified boron-doped ZnO bilayer electron transporting

materials: toward stable perovskite solar cells with high efficiency of over 20.5%. *J. Colloid Interface Sci.* 535: 353–362.

30 Yang, G., Chen, C., Yao, F. et al. (2018). Effective carrier-concentration tuning of SnO_2 quantum dot electron-selective layers for high-performance planar perovskite solar cells. *Adv. Mater.* 30 (14): 1–9.

31 Tavakoli, M.M., Giordano, F., Zakeeruddin, S.M., and Grätzel, M. (2018). Mesoscopic oxide double layer as electron specific contact for highly efficient and UV stable perovskite photovoltaics. *Nano Lett.* 18 (4): 2428–2434.

32 Song, J., Zhang, W., Wang, D. et al. (2019). Colloidal synthesis of Y-doped SnO_2 nanocrystals for efficient and slight hysteresis planar perovskite solar cells. *Sol. Energy* 185 (April): 508–515.

33 You, M.S., Heo, J.H., Park, J.K. et al. (2019). Low temperature solution processable TiO_2 nano-sol for electron transporting layer of flexible perovskite solar cells. *Sol. Energy Mater. Sol. Cells* 194 (February): 1–6.

34 Turren-Cruz, S.-H., Hagfeldt, A., and Saliba, M. (2018). Methylammonium-free, high-performance, and stable perovskite solar cells on a planar architecture. *Science (80-.)* 362 (6413): 449–453.

35 Saidaminov, M.I., Kim, S., Jain, A. et al. (2018). Suppression of atomic vacancies via incorporation of isovalent small ions to increase the stability of halide perovskite solar cells in ambient air. *Nat. Energy* 3 (8): 648–654.

36 Choi, K., Lee, J., Kim, H.I. et al. (2018). Thermally stable, planar hybrid perovskite solar cells with high efficiency. *Energy Environ. Sci.* 11 (11): 3238–3247.

37 Abdi-Jalebi, M., Andaji-Garmaroudi, Z., Cacovich, S. et al. (2018). Maximizing and stabilizing luminescence from halide perovskites with potassium passivation. *Nature* 555 (7697): 497–501.

38 Bu, T., Liu, X., Li, W. et al. (2019). Sub-sized monovalent alkaline cations enhanced electrical stability for over 17% hysteresis-free planar perovskite solar mini-module. *Electrochim. Acta* 306: 635–642.

39 Bi, D., Li, X., Milić, J.V. et al. (2018). Multifunctional molecular modulators for perovskite solar cells with over 20% efficiency and high operational stability. *Nat. Commun.* 9: 4482 (1–10). https://doi.org/10.1038/s41467-018-06709-w.

40 Zhang, F., Bi, D., Pellet, N. et al. (2018). Suppressing defects through the synergistic effect of a Lewis base and a Lewis acid for highly efficient and stable perovskite solar cells. *Energy Environ. Sci.* 11: 3480–3490.

41 Etgar, L., Gao, P., Xue, Z. et al. (2012). Mesoscopic $CH_3NH_3PbI_3/TiO_2$ heterojunction solar cells. *J. Am. Chem. Soc.* 134 (42): 17396–17399.

42 Chen, H. and Yang, S. (2017). Carbon-based perovskite solar cells without hole transport materials: the front runner to the market? *Adv. Mater.* 29 (24). Wiley-VCH Verlag: 1603994.

43 Yang, Y., Liu, Z., Ng, W.K. et al. (2019). An ultrathin ferroelectric perovskite oxide layer for high-performance hole transport material free carbon based halide perovskite solar cells. *Adv. Funct. Mater.* 29 (1): 1806506.

44 Liu, X., Liu, Z., Sun, B. et al. (2018). 17.46% efficient and highly stable carbon-based planar perovskite solar cells employing Ni-doped rutile TiO_2 as electron transport layer. *Nano Energy* 50: 201–211.

45 Wang, Y., Zhao, H., Mei, Y. et al. (2019). Carbon nanotube bridging method for hole transport layer-free paintable carbon-based perovskite solar cells. *ACS Appl. Mater. Interfaces* 11 (1): 916–923.

46 Tian, C., Mei, A., Zhang, S. et al. (2018). Oxygen management in carbon electrode for high-performance printable perovskite solar cells. *Nano Energy* 53 (June): 160–167.

47 Lee, K., Kim, J., Yu, H. et al. (2018). A highly stable and efficient carbon electrode-based perovskite solar cell achieved: via interfacial growth of 2D PEA 2 PbI_4 perovskite. *J. Mater. Chem. A* 6: 24560–24568.

48 Hou, X., Xu, M., Tong, Ch. et al. (2019). High performance printable perovskite solar cells based on $Cs_{0.1}$ $FA_{0.9}$ PbI_3 in mesoporous scaffolds. *J. Power Sources* 415 (1): 105–111.

49 Liu, P., Gong, Y., Xiao, Y. et al. (2019). Highly efficient and stable air-processed hole-transport-material free carbon based perovskite solar cells with caesium incorporation. *Chem. Commun.* 55: 218–221.

50 Jiang, H., Liu, X., Chai, N. et al. (2018). Alleviate the *J–V* hysteresis of carbon-based perovskite solar cells *via* introducing additional methylammonium chloride into $MAPbI_3$ precursor. *RSC Adv.* 8 (61): 35157–35161.

51 Ren, J., Qu, J., Chen, J. et al. (2018). Fluorinated dopant-free hole-transporting material for efficient and stable perovskite solar cells with carbon cathode. *J. Power Sources* 401 (May): 29–36.

52 Zhang, H., Xiao, Y., Qi, F. et al. (2019). Near-infrared light-sensitive hole-transport-layer free perovskite solar cells and photodetectors with hexagonal NaYF 4: Yb^{3+}, Tm^{3+} @SiO_2 upconversion nanoprism-modified TiO_2 scaffold. *ACS Sustain. Chem. Eng.* 7 (9): 8236–8244.

53 Xiong, Y., Zhu, Z., Mei, A. et al. (2018). Bifunctional Al_2O_3 interlayer leads to enhanced open-circuit voltage for hole-conductor-free carbon-based perovskite solar cells. *Sol. RRL* 2 (5): 1800002.

54 Wang, P., Chai, N., Wang, C. et al. (2019). Enhancing the thermal stability of the carbon-based perovskite solar cells by using a Cs_x FA_{1-x} $PbBr_x$ I_{3-x} light absorber. *RSC Adv.* 9 (21): 11877–11881.

55 Cai, C., Zhou, K., Guo, H. et al. (2019). Enhanced hole extraction by NiO nanoparticles in carbon-based perovskite solar cells. *Electrochim. Acta* 312: 100–108.

56 Chu, L., liu, L., Qin, Z. et al. (2018). Boosting efficiency of hole conductor-free perovskite solar cells by incorporating p-type NiO nanoparticles into carbon electrodes. *Sol. Energy Mater. Sol. Cells* 178: 164–169.

57 Xiao, Y., Wang, C., Kondamareddy, K.K. et al. (2018). Efficient electron transport scaffold made up of submicron TiO_2 spheres for high-performance hole-transport material free perovskite solar cells. *ACS Appl. Energy Mater.* 1 (10): 5453–5462.

58 Bashir, A., Shukla, S., Lew, J.H. et al. (2018). Spinel Co_3O_4 nanomaterials for efficient and stable large area carbon-based printed perovskite solar cells. *Nanoscale* 10 (5): 2341–2350.

59 Liu, H., Yang, B., Chen, H. et al. (2018). Efficient and stable hole-conductor-free mesoscopic perovskite solar cells using SiO_2 as blocking layer. *Org. Electron.* 58: 69–74.

60 Zhou, X., Wang, Y., Li, C., and Wu, T. (2019). Doping amino-functionalized ionic liquid in perovskite crystal for enhancing performances of hole-conductor free solar cells with carbon electrode. *Chem. Eng. J.* 372: 46–52.

61 Han, J., Yin, X., Zhou, Y. et al. (2018). Perovskite/poly[bis(4-phenyl)(2,4,6-trimethylphenyl)amine] bulk heterojunction for high-efficient carbon-based large-area solar cells by gradient engineering. *ACS Appl. Mater. Interfaces* 10 (49): 42328–42334.

62 Bashir, A., Lew, J.H., Shukla, S. et al. (2019). Cu-doped nickel oxide interface layer with nanoscale thickness for efficient and highly stable printable carbon-based perovskite solar cell. *Sol. Energy* 182: 225–236.

63 Zhang, Y., Zhuang, X., Zhou, K. et al. (2018). Vibration treated carbon electrode for highly efficient hole-conductor-free perovskite solar cells. *Org. Electron.* 52: 159–164.

64 Chen, W., Yin, X., Que, M. et al. (2019). A comparative study of planar and mesoporous perovskite solar cells with printable carbon electrodes. *J. Power Sources* 412 (October 2018): 118–124.

65 Meng, F., Gao, L., Yan, Y. et al. (2019). Ultra-low-cost coal-based carbon electrodes with seamless interfacial contact for effective sandwicsh-structured perovskite solar cells. *Carbon N. Y.* 145: 290–296.

66 Raminafshar, C., Dracopoulos, V., Mohammadi, M.R., and Lianos, P. (2018). Carbon based perovskite solar cells constructed by screen-printed components. *Electrochim. Acta* 276: 261–267.

67 Zhang, B., Fu, W., Meng, X. et al. (2019). Improved crystallization of perovskite films using $PbTiO_3$-decorated mesoporous scaffold layers for high stable carbon-counter-electrode solar cells. *Org. Electron. Phys. Mater. Appl.* 69 (March): 164–173.

68 Cheng, J., Quiang, Y., Zhou, C. et al. (2019). Effective improvement of the photovoltaic performance of carbon-based perovskites solar cells by grinding process and its capacitor model. *J. Power Sources* 422 (March): 131–137.

69 Liu, X., liu, Z., Sun, B. et al. (2018). All low-temperature processed carbon-based planar heterojunction perovskite solar cells employing Mg-doped rutile TiO_2 as electron transport layer. *Electrochim. Acta* 283: 1115–1124.

70 Huang, G., Wang, C., Zhang, H. et al. (2018). Post-healing of defects: an alternative way for passivation of carbon-based mesoscopic perovskite solar cells: via hydrophobic ligand coordination. *J. Mater. Chem. A* 6: 2449–2455.

71 Luo, Q., Ma, H., Hou, Q. et al. (2018). All-carbon-electrode-based endurable flexible perovskite solar cells. *Adv. Funct. Mater.* 28: 1706777.

72 Sajid, S., Elseman, A.M., wei, D. et al. (2019). NiO@carbon spheres: a promising composite electrode for scalable fabrication of planar perovskite solar cells at low cost. *Nano Energy* 55: 470–476.

73 Dong, C., Han, X., Li, W. et al. (2019). Anti-solvent assisted multi-step deposition for efficient and stable carbon-based $CsPbI_2$ Br all-inorganic perovskite solar cell. *Nano Energy* 59: 553–559.

74 Liu, Z., Sun, B., Liu, X. et al. (2018). 15% efficient carbon based planar-heterojunction perovskite solar cells using a TiO_2/SnO_2 bilayer as the electron transport layer. *J. Mater. Chem. A* 6 (17): 7409–7419.

75 Ye, H., Liu, Z., Liu, X. et al. (2019). 17.78% efficient low-temperature carbon-based planar perovskite solar cells using Zn-doped SnO_2 electron transport layer. *Appl. Surf. Sci.* 478: 417–425.

76 Yan, J., lin, S., Qiu, X. et al. (2019). Accelerated hole-extraction in carbon-electrode based planar perovskite solar cells by moisture-assisted post-annealing. *Appl. Phys. Lett.* 114 (10): 103503.

77 Lin, S., Yang, B., Qiu, X. et al. (2018). Efficient and stable planar hole-transport-material-free perovskite solar cells using low temperature processed SnO_2 as electron transport material. *Org. Electron.* 53: 235–241.

78 Zhang, X., Zhou, Y., Li, Y. et al. (2019). Efficient and carbon-based hole transport layer-free $CsPbI_2$ Br planar perovskite solar cells using PMMA modification. *J. Mater. Chem. C* 7: 3852–3861.

79 Meng, X., Wang, Z., Qian, W. et al. (2019). Excess cesium iodide induces spinodal decomposition of $CsPbI_2$ Br perovskite films. *J. Phys. Chem. Lett.* 10 (2): 194–199.

80 Yang, M., Li, J., Yuan, Z. et al. (2019). High efficient and long-time stable planar heterojunction perovskite solar cells with doctor-bladed carbon electrode. *J. Power Sources* 424: 61–67.

81 Mashhoun, S., Hou, Y., Chen, H. et al. (2018). Resolving a critical instability in perovskite solar cells by designing a scalable and printable carbon based electrode-interface architecture. *Adv. Energy Mater.* 8: 1802085.

82 Guo, Y., Zhao, F., Tao, J. et al. (2019). Efficient and hole-transporting-layer-free $CsPbI_2$ Br planar heterojunction perovskite solar cells through rubidium passivation. *ChemSusChem* 12 (5): 983–989.

83 Li, D., Tong, C., Ji, W. et al. (2019). Vanadium oxide post-treatment for enhanced photovoltage of printable perovskite solar cells. *ACS Sustain. Chem. Eng.*

84 Hu, Y., Zhang, Z., Mei, A. et al. (2018). Improved performance of printable perovskite solar cells with bifunctional conjugated organic molecule. *Adv. Mater.* 30 (11): 1705786.

85 Zhu, X.-Y., Chen, M.-W., Wang, B. et al. (2018). Improved photovoltaic properties of nominal composition CH_3 NH_3 $Pb_{0.99}$ $Zn_{0.01}$ I_3 carbon-based perovskite solar cells. *Opt. Express* 26 (26): A984–A995. https://doi.org/10.1364/OE.26.00A984.

86 Xiao, Y., Wang, C., Kondamareddy, K.K. et al. (2019). Enhancing the performance of hole-conductor free carbon-based perovskite solar cells through rutile-phase passivation of anatase TiO_2 scaffold. *J. Power Sources* 422 (March): 138–144.

87 Wu, Z., Liu, Z., Hu, Z., Hawash, Z. et al. (2019). Highly efficient and stable perovskite solar cells via modification of energy levels at the perovskite/carbon electrode interface. *Adv. Mater.* 31 (11): 1–7.

88 Guan, Y., Mei, A., Rong, Y. et al. (2018). Fullerene derivative as an additive for highly efficient printable mesoscopic perovskite solar cells. *Org. Electron. Phys. Mater. Appl.* 62 (May): 653–659.

89 Jiang, P., Jones, T.W., Duffy, N.W. et al. (2018). Fully printable perovskite solar cells with highly-conductive, low-temperature, perovskite-compatible carbon electrode. *Carbon N. Y.* 129: 830–836.

90 Qi, F., Wang, C., Chang, N. et al. (2018). Improving the performance through SPR effect by employing Au@SiO_2 core–shell nanoparticles incorporated

TiO_2 scaffold in efficient hole transport material free perovskite solar cells. *Electrochim. Acta* 282: 10–15.

91 Chu, Q.-Q., Ding, B., Qiu, Q. et al. (2018). Cost effective perovskite solar cells with a high efficiency and open-circuit voltage based on a perovskite-friendly carbon electrode. *J. Mater. Chem. A* 6 (18): 8271–8279.

92 Kim, B., Ko, S.G., Sonu, K.S. et al. (2018). Effects of adding PbI_2 on the performance of hole-transport material-free mesoscopic perovskite solar cells with carbon electrode. *J. Electron. Mater.* 47 (10): 6266–6271.

93 Mishra, A., Ahmad, Z., Zimmermann, I. et al. (2019). Effect of annealing temperature on the performance of printable carbon electrodes for perovskite solar cells. *Org. Electron. Phys. Mater. Appl.* 65 (October 2018): 375–380.

94 Ko, S.G., Kim, B., Cha, G. et al. (2019). Effects of thiourea on the perovskite crystallization for fully printable solar cells. *Sol. Energy Mater. Sol. Cells* 196 (October 2018): 105–110.

95 Hu, R., Zhang, R., Ma, Y. et al. (2018). Enhanced hole transfer in hole-conductor-free perovskite solar cells via incorporating CuS into carbon electrodes. *Appl. Surf. Sci.* 462 (May): 840–846.

96 Barichello, J., Vesce, L., Matteocci, F. et al. (2019). The effect of water in carbon-perovskite solar cells with optimized alumina spacer. *Sol. Energy Mater. Sol. Cells* 197: 76–83.

97 Chen, D., Lu, Y., Wu, J. et al. (2019). Perovskite solar cells-TiO_2 tandem assembly for photoelectrocatalytic degradation of organic pollutants. *J. Phys. Chem. Solids* 132 (September 2018): 204–212.

98 Zhu, W., Zhang, Z., Chai, W. et al. (2019). Band alignment engineering towards high efficiency carbon-based inorganic planar $CsPbIBr_2$ perovskite solar cells. *ChemSusChem* 12: 2318–2325.

99 Chu, Q.Q., Ding, B., Peng, J. et al. (2019). Highly stable carbon-based perovskite solar cell with a record efficiency of over 18% via hole transport engineering. *J. Mater. Sci. Technol.* 35 (6): 987–993.

100 Zhou, J., Hou, J., Tao, X. et al. (2019). Solution-processed electron transport layer of n-doped fullerene for efficient and stable all carbon based perovskite solar cells. *J. Mater. Chem. A* 7 (13): 7710–7716.

101 Grancini, G., Roldan-Carmona, C., Zimmermann, I. et al. (2017). One-year stable perovskite solar cells by 2D/3D interface engineering. *Nat. Commun.* 8: 15684.

102 Sun, H., Deng, K., Zhu, Y. et al. (2018). A novel conductive mesoporous layer with a dynamic two-step deposition strategy boosts efficiency of perovskite solar cells to 20%. *Adv. Mater.* 30 (28): 1801935.

103 Galagan, Y. (2018). Perovskite solar cells: toward industrial-scale methods. *J. Phys. Chem. Lett.* 9 (15). American Chemical Society: 4326–4335.

104 Najafi, M., Di Giacomo, F., Zhang, D. et al. (2018). Highly efficient and stable flexible perovskite solar cells with metal oxides nanoparticle charge extraction layers. *Small* 14 (12): 1–10.

105 Hanmandlu, C., Liu, C.-C., Chaen, C.-Y. et al. (2018). Top illuminated hysteresis-free perovskite solar cells incorporating microcavity structures on metal electrodes: a combined experimental and theoretical approach. *ACS Appl. Mater. Interfaces* 10 (21): 17973–17984.

106 Jin, T.Y., Li, W., Li, Y.-Q. et al. (2018). High-performance flexible perovskite solar cells enabled by low-temperature ALD-assisted surface passivation. *Adv. Opt. Mater.* 6 (24): 1–10.

107 Chakravarthi, N., Park, H.-Y., Kranthiraja, K. et al. (2018). Substituent position engineering of phosphine oxide functionalized triazine-based cathode interfacial materials for flexible organic and perovskite solar cells. *Org. Electron. physics, Mater. Appl.* 54 (October 2017): 54–63.

108 Lee, E., Ahn, J., Kwon, H.-C. et al. (2018). All-solution-processed silver nanowire window electrode-based flexible perovskite solar cells enabled with amorphous metal oxide protection. *Adv. Energy Mater.* 8 (9): 1–11.

109 Chu, W., Yang, J., Jiang, Q. et al. (2018). Enhancement of photovoltaic performance of flexible perovskite solar cells by means of ionic liquid interface modification in a low temperature all solution process. *Appl. Surf. Sci.* 440: 1116–1122.

110 Zhang, Y., Wu, Z., Li, P. et al. (2018). Fully solution-processed TCO-free Semi-transparent perovskite solar cells for tandem and flexible applications. *Adv. Energy Mater.* 8 (1): 1–10.

111 Jiang, J., Jia, X., wang, S. et al. (2018). High-performance flexible perovskite solar cells with effective interfacial optimization processed at low temperatures. *ChemSusChem* 11 (23): 4131–4138.

112 Wilkes, G.C., Deng, X., Choi, J.J., and Gupta, M.C. (2018). Laser annealing of TiO_2 Electron-transporting layer in perovskite solar cells. *ACS Appl. Mater. Interfaces* 10 (48): 41312–41317.

113 Han, G.S., Lee, S., Duff, M.L. et al. (2018). Highly bendable flexible perovskite solar cells on a nanoscale surface oxide layer of titanium metal plates. *ACS Appl. Mater. Interfaces* 10 (5): 4697–4704.

114 Jiang, J., Wang, S., Jia, X. et al. (2018). Totally room-temperature solution-processing method for fabricating flexible perovskite solar cells using an Nb_2O_5-TiO_2 electron transport layer. *RSC Adv.* 8 (23): 12823–12831.

115 Singh, R., Ryu, I., Yadav, H. et al. (2019). Non-hydrolytic sol-gel route to synthesize TiO_2 nanoparticles under ambient condition for highly efficient and stable perovskite solar cells. *Sol. Energy* 185 (March): 307–314.

116 Cao, B., Yang, L., Jiang, S. et al. (2019). Flexible quintuple cation perovskite solar cells with high efficiency. *J. Mater. Chem. A* 7 (9): 4960–4970.

117 Subbiah, A.S., Mathews, N., Mhaisalkar, S., and Sarkar, S.K. (2018). Novel plasma-assisted low-temperature-processed SnO_2 thin films for efficient flexible perovskite photovoltaics. *ACS Energy Lett.* 3 (7): 1482–1491.

118 Park, S.Y., Baek, M.Y., Ju, Y. et al. (2018). Simultaneous ligand exchange fabrication of flexible perovskite solar cells using newly synthesized uniform tin oxide quantum dots. *J. Phys. Chem. Lett.* 9 (18): 5460–5467.

119 Yang, F., Liu, J., Lim, H.E. et al. (2018). High bending durability of efficient flexible perovskite solar cells using metal oxide electron transport layer. *J. Phys. Chem. C* 122 (30): 17088–17095.

120 Bu, T., Shi, S., Li, J. et al. (2018). Low-temperature presynthesized crystalline tin oxide for efficient flexible perovskite solar cells and modules. *ACS Appl. Mater. Interfaces* 10 (17): 14922–14929.

121 Shi, S., Li, J., Bu, T. et al. (2019). Room-temperature synthesized SnO_2 electron transport layers for efficient perovskite solar cells. *RSC Adv.* 9 (18): 9946–9950.

122 Qiu, X., Yang, B., Chen, H. et al. (2018). Efficient, stable and flexible perovskite solar cells using two-step solution-processed SnO_2 layers as electron-transport-material. *Org. Electron. Phys. Mater. Appl.* 58 (April): 126–132.

123 Bu, T., Li, J., Zheng, F. et al. (2018). Universal passivation strategy to slot-die printed SnO_2 for hysteresis-free efficient flexible perovskite solar module. *Nat. Commun.* 9 (1): 4609.

124 Yang, D., Yang, R., Wang, K. et al. (2018). High efficiency planar-type perovskite solar cells with negligible hysteresis using EDTA-complexed SnO_2. *Nat. Commun.* 9: 3239 (1–11). https://doi.org/10.1038/s41467-018-05760-x.

125 Zhao, X., Tian, L., Liu, T. et al. (2019). Room-temperature-processed fullerene single-crystalline nanoparticles for high-performance flexible perovskite photovoltaics. *J. Mater. Chem. A* 7 (4): 1509–1518.

126 Li, Y., Qi, X., Liu, G. et al. (2019). High performance of low-temperature processed perovskite solar cells based on a polyelectrolyte interfacial layer of PEI. *Org. Electron. Phys. Mater. Appl.* 65 (September 2018): 19–25.

127 Pisoni, S., Carron, R., moser, T. et al. (2018). Tailored lead iodide growth for efficient flexible perovskite solar cells and thin-film tandem devices. *NPG Asia Mater.* 10 (11): 1076–1085.

128 Pisoni, S., Fu, F., Widmer, R. et al. (2018). Impact of interlayer application on band bending for improved electron extraction for efficient flexible perovskite mini-modules. *Nano Energy* 49 (April): 300–307.

129 Li, K., Xiao, J., Yu, X. et al. (2018). An efficient, flexible perovskite solar module exceeding 8% prepared with an ultrafast PbI_2 deposition rate. *Sci. Rep.* 8 (1): 442.

130 Liu, Z., Xie, X., Liu, G., and Lee, E.C. (2019). High-performance metal-oxide-free perovskite solar cells based on organic electron transport layer and cathode. *Org. Electron. Phys. Mater. Appl.* 64 (June 2018): 195–201.

131 Hou, L., Wang, Y., Liu, X. et al. (2018). 18.0% efficiency flexible perovskite solar cells based on double hole transport layers and $CH_3NH_3PbI_{3-x}$ Cl_x with dual additives. *J. Mater. Chem. C* 6 (32): 8770–8777.

132 Lin, M.Y. and Chen, S.L. (2018). Fabrication of imprinted flexible solar PV maximum power point tracking system for smart cities sustainable development. In: *2018 1st International Cognitive Cities Conference (IC3 2018)*, 190–191. https://doi.org/10.1109/IC3.2018.00-29.

133 Jung, J.W., Bae, J.H., Ko, J.H., and Lee, W. (2018). Fully solution-processed indium tin oxide-free textile-based flexible solar cells made of an organic–inorganic perovskite absorber: toward a wearable power source. *J. Power Sources* 402 (September): 327–332.

134 Duan, X., Huang, Z., Liu, C. et al. (2019). A bendable nickel oxide interfacial layer via polydopamine crosslinking for flexible perovskite solar cells. *Chem. Commun.* 55 (25): 3666–3669.

135 Chen, W., Wu, Y., Fan, J. et al. (2018). Understanding the doping effect on NiO: toward high-performance inverted perovskite solar cells. *Adv. Energy Mater.* 8 (19): 1–10.

136 Zhang, F., Mezgeb, M.M., Guo, W. et al. (2018). An efficient and simple dual effect by under-layer abduction design for highly flexible NiO_x-based perovskite solar cells. *J. Power Sources* 399 (May): 246–253.

137 Rao, H., Ye, S., Gu, F. et al. (2018). Morphology controlling of all-inorganic perovskite at low temperature for efficient rigid and flexible solar cells. *Adv. Energy Mater.* 8 (23): 1–8.

138 Jokar, E., huang, Z. Y., Narra, S. et al. (2018). Anomalous charge-extraction behavior for graphene-oxide (GO) and reduced graphene-oxide (rGO) films as efficient p-contact layers for high-performance perovskite solar cells. *Adv. Energy Mater.* 8 (3): 1–10.

139 Lei, T., Dong, H., Xi, J. et al. (2018). Highly-efficient and lowerature perovskite solar cells by employing a Bi-hole transport layer consisting of vanadium oxide and copper phthalocyanine. *Chem. Commun.* 54 (48): 6177–6180.

140 Xu, C., Liu, Z., and Lee, E.C. (2018). High-performance metal oxide-free inverted perovskite solar cells using poly(bis(4-phenyl)(2,4,6-trimethylphenyl)amine) as the hole transport layer. *J. Mater. Chem. C* 6 (26): 6975–6981.

141 Khenkin, M.V., Katz, E., Lira-Cantu, M. et al. (2020). Consensus statement for stability assessment and reporting for perovskite photovoltaics based on ISOS procedures. *Nat. Energy* 5 (1): 35–49.

142 Guerrero, A., García-Belomonte, G, Mora-Sero, I. et al. (2016). Properties of contact and bulk impedances in hybrid Lead halide perovskite solar cells including inductive loop elements. *J. Phys. Chem. C* 120 (15): 8023–8032.

143 Yang, L., Dall'Agnese, Y., Hantanasirisakul, K. et al. (2019). SnO_2 -Ti_3 C_2 MXene electron transport layers for perovskite solar cells. *J. Mater. Chem. A* 7 (10): 5635–5642.

144 Alidaei, M., Izadifard, M., Ghazi, M.E. et al. (2019). Interfacial defect passivation in $CH_3NH_3PbI_3$ perovskite solar cells using modifying of hole transport layer. *J. Mater. Sci. Mater. Electron.* 30: 6936–6946. https://doi.org/10.1007/s10854-019-01009-5.

145 Cai, Y., Zhang, Z., Zhou, Y. et al. (2018). Enhancing the efficiency of low-temperature planar perovskite solar cells by modifying the interface between perovskite and hole transport layer with polymers. *Electrochim. Acta* 261: 445–453.

146 Cerdán-Pasarán, A., Sidhik, S., López-Luke, T., and la Rosa, E. (2019). Improved performance of inverted planar $MAPbI_3$ based perovskite solar cells using bromide post-synthesis treatment. *Sol. Energy*: 538–544.

147 Pang, A., Shen, D., Wei, M., and Chen, Z.-N. (2018). Highly efficient perovskite solar cells based on $Zn_2Ti_3O_8$ nanoparticles as electron transport material. *ChemSusChem* 11 (2): 424–431.

148 Kim, G.-W., Kang, G., Choi, K. et al. (2018). Solution processable inorganic–organic double-layered hole transport layer for highly stable planar perovskite solar cells. *Adv. Energy Mater.* 8, 8 (26): 1801386. (1–7).

149 Kim, H.I., kim, M.-J., Choi, K., Lim, C. et al. (2018). Improving the performance and stability of inverted planar flexible perovskite solar cells employing a novel NDI-based polymer as the electron transport layer. *Adv. Energy Mater.* 8 (16).

150 Li, X., Zhao, X., Hao, F., Yin, X. et al. (2018). Bifacial modified charge transport materials for highly efficient and stable inverted perovskite solar cells. *ACS Appl. Mater. Interfaces* 10 (21): 17861–17870.

151 Lv, Y., Cai, B., Ma, Q. et al. (2018). Highly crystalline Nb-doped TiO_2 nanospindles as superior electron transporting materials for high-performance planar structured perovskite solar cells. *RSC Adv.* 8 (37): 20982–20989.

152 Parida, B., Singh, A., Oh, M. et al. (2019). Effect of compact TiO_2 layer on structural, optical, and performance characteristics of mesoporous perovskite solar cells. *Mater. Today Commun.* 18: 176–183.

153 Shit, A., Chal, P., and Nandi, A.K. (2018). Copolymers of poly(3-thiopheneacetic acid) with poly(3-hexylthiophene) as hole-transporting material for interfacially engineered perovskite solar cell by modulating band positions for higher efficiency. *Phys. Chem. Chem. Phys.* 20 (23): 15890–15900.

154 Wang, D., Elumanai, N.K., Mahmud, M. A. et al. (2018). V_2O_5-PEDOT: PSS bilayer as hole transport layer for highly efficient and stable perovskite solar cells. *Org. Electron. Phys. Mater. Appl.* 53: 66–73.

155 Yang, Y., Xiao, J., Wei., H. et al. (2014). An all-carbon counter electrode for highly efficient hole-conductor-free organo-metal perovskite solar cells. *RSC Adv.* 4 (95): 52825–52830.

156 Xiong, L., Qin, M., Chen, C. et al. (2018). Fully high-temperature-processed SnO_2 as blocking layer and scaffold for efficient, stable, and hysteresis-free mesoporous perovskite solar cells. *Adv. Funct. Mater.* 28 (10).

157 Ulfa, M., Wang, P., Zhang, J. et al. (2018). Charge injection and electrical response in low-temperature SnO_2-based efficient perovskite solar cells. *ACS Appl. Mater. Interfaces* 10 (41): 35118–35128.

158 Yang, L., Dall'Agnese, Y., Hantanasirisakul, K. et al. (2019). SnO_2–Ti_3C_2 MXene electron transport layers for perovskite solar cells. *J. Mater. Chem. A* 7 (10): 5635–5642.

159 Pascoe, A.R., Duffy, N.W., Scully, A.D. et al. (2015). Insights into planar $CH_3NH_3PbI_3$ perovskite solar cells using impedance spectroscopy. *J. Phys. Chem. C* 119 (9): 4444–4453.

160 Bisquert, J., Bertoluzzi, L., Mora-Sero, I., and Garcia-Belmonte, G. (2014). Theory of impedance and capacitance spectroscopy of solar cells with dielectric relaxation, drift-diffusion transport, and recombination. *J. Phys. Chem. C* 118 (33): 18983–18991.

161 Pham, N.D., Tiong, V.T., Yao, D. et al. (2017). Guanidinium thiocyanate selective Ostwald ripening induced large grain for high performance perovskite solar cells. *Nano Energy* 41: 476–487.

162 Ngo, T.T., Barea, E.M., Tena-Zaera, R., and Mora-Sero, I. (2018). Spray-pyrolyzed ZnO as electron selective contact for long-term stable planar $CH_3NH_3PbI_3$ perovskite solar cells. *ACS Appl. Energy Mater.* 1 (8): 4057–4064.

9

Electron Transport Layers in Perovskite Solar Cells

Fatemeh Jafari[1,2,3], *Mehrad Ahmadpour*[1], *Um Kanta Aryal*[1], *Mariam Ahmad*[1], *Michela Prete*[1], *Naeimen Torabi*[3], *Vida Turkovic*[1], *Horst-Günter Rubahn*[1], *Abbas Behjat*[2,3], *and Morten Madsen*[1]

[1] *University of Southern Denmark, SDU NanoSYD, Mads Clausen Institute, Alsion 2, Sønderborg, 6400, Denmark*
[2] *Yazd University, Department of Physics, University Blvd, Safayieh, Yazd, 8915818411, Iran*
[3] *Yazd University, Photonics Research Group, University Blvd, Safayieh, Yazd, 8915818411, Iran*

9.1 Introduction

The large improvement in the performance of perovskite solar cells has attracted much attention across the scientific communities, along with rising commercial activities. Further optimization of the solar cells relies on careful optimization of all involved layers in the cells, which may in fact have different requirements when made on an industrial scale than those typically used on a research scale. In this development, it is important to acknowledge the strong role that interlayers play on both device performance and device stability, gaining more importance now that the technology has proven efficient for many applications. The electron transport layer (ETL) is sandwiched between the perovskite absorber layer and the cathode electrode to support the efficient extraction of electrons from the perovskite upon photoexcitation. In the past years, the introduction of such transport layers in perovskite solar cells benefitted from the vast amount of work done within organic solar cells on this topic, significantly speeding up the integration in perovskite solar cells. The interlayers mainly tune the work function of the electrodes to make them more suitable electron and hole contacts. The ETL also supports carrier selective transport. It blocks holes to reduce charge recombination at the cathode interface [1, 2]. This is similar to the hole transport layer (HTL), being selective for holes alone in an ideal scenario. This book chapter investigates the requirements for ideal ETL in perovskite solar cells, considering the vast amount of research conducted on this topic in the past. We briefly discuss the different types of electron transport materials employed in perovskite solar cells over the years. Moreover, we touch upon the role of the perovskite solar cell architectures used for the various types of ETL. This chapter serves as a mini-review of ETL for perovskite solar cells and discusses its important role in device performance and stability.

Perovskite Solar Cells: Materials, Processes, and Devices, First Edition.
Edited by Shahzada Ahmad, Samrana Kazim, and Michael Grätzel.

9.2 Requirements of Ideal Electron Transport Layers (ETL)

The ideal ETL should in perovskite solar cells fulfill several requirements on electronic properties, morphology, and defect states. The most basic property of an ETL is its optical transparency, which is required when light is illuminated through the ETL side. It is the case in most perovskite solar cells to guarantee maximum light absorption by the perovskite layer [2]. One main requirement for electronic properties is energy level alignment between the ETL and perovskite layer. The ETL should have the lowest unoccupied molecular orbital (LUMO), or conduction band, that aligns well to the perovskite conduction band to facilitate efficient electron extraction and transport [3]. Furthermore, due to the ambipolar nature (i.e. both p- and n-conductive) of perovskite materials [4], a wider bandgap with deeper-lying highest occupied molecular orbital (HOMO) or valence band than that of the perovskite valence band is required to achieve an efficient hole-blocking function [5]. A higher-lying LUMO level in the ETL would result in an extraction barrier and severe s-shape in the device J–V characteristics, and well-known within organic solar cells [6]. A lower-lying LUMO level (compared to the conduction band of the perovskite) for the ETL would result in an injection barrier and a possible loss of open-circuit voltage (V_{oc}) [7]. This makes selecting an ETL with appropriate energy levels necessary for acceptable V_{oc}, and device performance in general.

The energy levels can be measured from photoelectron spectroscopy techniques, such as ultraviolet photoelectron spectroscopy (UPS), which probes the occupied density of states and work function to extract the ionization energy (IE). To get access to the unoccupied density of states and the electron affinity (EA), inverse photoelectron spectroscopy (IPES), or optical band gap measurements need also to be conducted. Such techniques can be utilized for both the interface transport [8–10] and perovskite [11] layers individually. However, to determine complete and correct energy level alignment between an interlayer and the perovskite layer, photoelectron spectroscopy experiments were conducted while depositing one layer on top of the other [11]. This is crucial due to band bending and interface dipoles that may occur at such interfaces, distorting the interface energy levels from their non-perturbed values. Another challenge has been the discrepancies in reported energy levels using photoelectron spectroscopy of perovskite solar cells recently demonstrated by Olthof [11]. Different processing conditions, structures, defect densities, etc. can affect the exact energy level positions. Different data evaluation schemes have been shown to result in different valence band onsets from photoelectron spectroscopy, e.g. perovskite/fullerene ETL interfaces, as demonstrated in Figure 9.1. Regardless, an energy level band alignment presented in Figure 9.1a would be ideal for an ETL/perovskite interface. It presents an almost perfect alignment of the fullerene LUMO with the perovskite conduction band. Simultaneously, it enables hole blocking due to the deep-lying fullerene HOMO level, compared to the perovskite valence band.

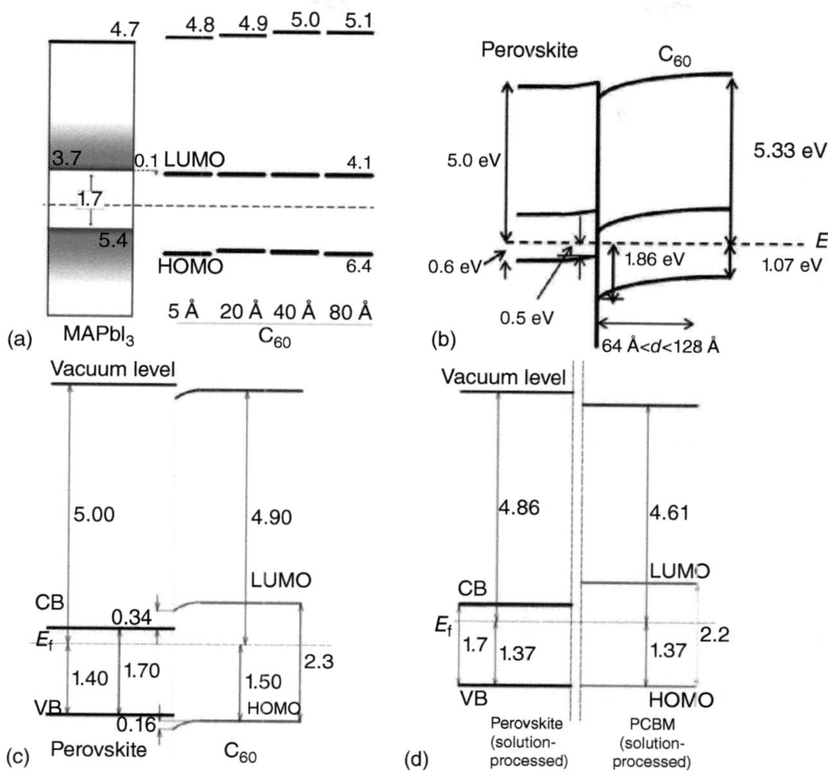

Figure 9.1 Energy level band alignment of fullerene/perovskite interfaces demonstrates differences in the reported values (a–d) and the band alignment schemes. Source: Olthof [11]. Licenced under CC BY-4.0.

The electron mobility of the ETL is another important factor that is decisive for the perovskite solar cell (PSC) performance. The ETL electron mobility should be higher than that of the perovskite active layer. This is to obtain efficient transport and collection of the photogenerated electrons, and avoid any charge accumulation at the interface, increasing the charge recombination in the solar cell [12]. In many metal oxide electron transport materials (ETM), large variations in electron mobilities between different oxide materials are seen, as presented in Section 9.3.

The device behavior may also be significantly affected by the morphology and the structure of the ETL thin films. A pinhole-free compact layer improves the contact between the perovskite layer and the ETL. The existence of pinholes and defects in the ETL may increase charge recombination and reduce shunt resistance. This reduces the fill factor and power conversion efficiency, emphasizing the importance of ETL film morphology [13]. The stability of the devices has been demonstrated to be related to the film morphology and internal structure, where crystalline contact layers have demonstrated to result in prolonged perovskite lifetimes due to improved chemical stability [14]. Similar effects have been studied and shown for transport layers used in organic solar cells and appear to route long-term performance for such interface layers [9]. As the transport properties of the interface layers are also affected highly by the morphology and structure, it makes these parameters even more important for the performance of perovskite solar cells. The ETL morphology can be controlled by different deposition methods, such as sputtering, solution processing, atomic layer deposition, and thermal evaporation, with careful tuning of all deposition parameters for each processing step [15].

The interface between the ETL and the perovskite active layer is critical for high device performance. Trap states on the ETL surface may increase charge recombination and reduce device stability, making interface engineering a well-studied method to reduce interface charge recombination and enhance device stability. They also suppress hysteresis, observed as the difference in J–V characteristics upon reversing the scan directions [16–19]. The performance of planar devices, which is discussed further in Section 9.4 of this chapter, is very sensitive to the ETL/perovskite interface, so interfacial engineering is even more important to achieve high device efficiency [20]. Interface engineering is also investigated to improve the device lifetime, where interfaces play an important role.

In connection, an ideal ETL should have high intrinsic chemical stability to prevent chemical degradation of the interlayer itself and avoid chemical reactions with the nearby layers [21]. Finally, hydrophobicity is also a feature that may be important for ETL in inverted device architecture to avoid moisture penetration into the perovskite active layer, thus prolonging the device lifetime [3, 22]. In the following, a review of the types of electron transport materials used in perovskite solar cells is provided, focusing both on metal oxide and organic transport layers.

9.3 Overview of Electron Transport Materials

In this chapter, different ETM used as ETL in perovskite solar cell devices are reviewed, focusing on metal oxides and organic ETL classes. The energy levels of the different materials are shown in Figure 9.2.

9.3.1 Metal Oxide Electron Transport Materials

Metal oxide materials such as TiO_2, ZnO, and SnO_2 are appropriate choices as electron transport materials for perovskite solar cells due to n-type character with work functions and conduction bands that match well with the conduction band of many well-known hybrid perovskite active layer systems. Furthermore, with their band gaps above 3 eV, the valence bands of these oxides are deep-lying with a large offset to the perovskite valence band, providing a sufficient hole-blocking barrier. Additional advantages of these oxides are low-cost fabrication, relatively long-term stability (although depending a lot on the oxide), and high efficiency in perovskite cells due to band alignment and typical high electron mobilities [30]. TiO_2 is currently the most studied ETM in perovskite solar cells, where it provides a dense and mesoscopic layer that can be formed in anatase, brookite, and rutile crystalline phases. TiO_2 has a large bandgap of around 3.2 eV for the anatase phase

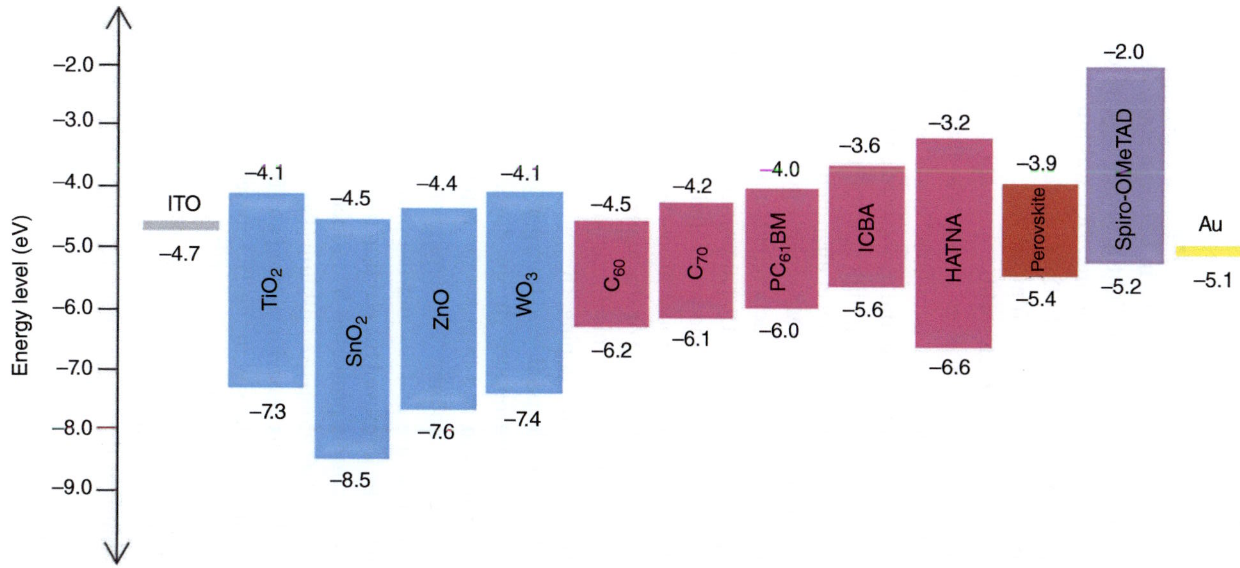

Figure 9.2 Energy level diagram of selected main ETM. Source: Lian et al. [3]; Mahmud et al. [23]; Mahmood et al. [24]; Park et al. [25]; Chen et al. [26]; Patil et al. [27]; Wu et al. [28]; Zhao et al. [29].

[31] and around 3.0 eV for the rutile phase [32]. The conduction band minimum of around 4.1 eV is slightly lower than that of $MAPbI_3$ (3.9 eV), matching well for electron extraction from the perovskite active layer to the ETL conduction band. The valence band maximum is positioned at 7.3 eV, as required for hole-blocking properties [13]. In 1991, the mesoscopic anatase TiO_2 film was for the first time used in dye-sensitized solar cells (DSSCs) by Grätzel and coworkers [33]. Since then, TiO_2 has become the most common ETL in DSSC. In 2009, Miyasaka's group reported the first PSC using a perovskite light absorber instead of a dye in the same structure as DSSC [34]. From the beginning, TiO_2 was introduced as the most important ETL candidate of PSC, as it was in DSSC devices. TiO_2 nanoparticles are commonly used to form mesoscopic ETL film for PSC, for which it is important to synthesize well-dispersed nanoparticles with controlled sizes. Extensive studies have been conducted in TiO_2 particle sizes to improve the PSC efficiency [30]. Furthermore, nanorods, nanotubes, and nanofibers could provide even improved electron transport properties compared to the nanoparticle films due to their directional charge-transport properties, yielding high electron mobilities [35]. Although the advantages, such as favorable energy level alignment, ease of fabrication, and low cost, make TiO_2 the most common ETM, a relatively low electron mobility and high density of trap states below the conduction band minimum may influence the efficiency and stability (under ultraviolet [UV] illumination) of PSC devices [3]. The deep traps reduce the quasi-Fermi level and conductivity, and act as recombination sites for electrons and holes. The electronic properties of TiO_2 can be modified by doping, where the metal (Ti^{4+}cation) or the nonmetal (O^{2-} anion) component can be replaced in order to modify the electronic and transport properties of the materials [36]. Several methods for doping TiO_2 with other materials have been employed to improve the TiO_2 electronic and transport properties and further boost PSC performance. Zhou et al. introduced yttrium doped TiO_2 (Y-TiO_2) to increase the charge-carrier concentration and improve the conductivity of this layer [17]. Kim et al. studied the effect of niobium (Nb) doping of anatase TiO_2 nanoparticles on the electronic band structure of TiO_2 and the photovoltaic properties of PSC. They showed that Nb doping improved the electron transport properties of the TiO_2 layer [37]. To improve electronic properties by reducing electronic trap states and recombination in PSC, Grätzel group used lithium (Li) doped TiO_2 as a mesoscopic ETL. Devices with low hysteresis behavior showed an efficiency above 19%, mainly by reducing the electronic trap density in the TiO_2-based ETL, thus improving the transport properties [38]. Studies on doping of TiO_2 layers in PSC are still ongoing [39, 40]. However, the TiO_2 mesoporous layer requires high-temperature fabrication processes (up to 500 °C), increasing manufacturing costs and limiting printable and flexible fabrication. Solutions that can mitigate that may be in high demand in the future.

ZnO has shown higher electron mobility (200–300 $cm^2\ V^{-1}\ s^{-1}$) than TiO_2 (0.1–4 $cm^2\ V^{-1}\ s^{-1}$) [41], which makes it an ideal choice for an ETL. Its electronic energy levels are quite similar to that of TiO_2, with a conduction band minimum just above 4 eV, which facilitates electron extraction from many perovskite active layers. The valence band maximum energy level of around 7.6 eV is also more than enough to block holes from the valence band of the perovskite layer, similarly to TiO_2 [23, 24]. ZnO can easily be solution-processed at low temperatures with high structural quality. PSC based on ZnO compact layers developed without thermal sintering and mesoporous scaffolds exhibit efficient electron extraction and high photovoltaic device performance [42]. Like TiO_2, the doping method effectively reduces defects and improves the electronic transport properties of ZnO. The use of electron-rich and nitrogen-doped ZnO (N:ZnO) nanorod-based films has been shown to increase the charge-carrier density and electron mobility of ETL [43]. The result of using nanopillar iodine-doped ZnO (ZnO:I) ETL films has been shown to include high optical transparency, reduced work function, and superior electron extraction ability, overall enhancing device performance [44]. Perovskite films deposited on aluminum-doped ZnO (AZO) ETL films have demonstrated improved thermal stability compared to reference ZnO ETL [45]. However, improved device stability was not demonstrated. It was challenging with ZnO ETL due to chemical reactivity between the perovskite layer and the ZnO at elevated temperatures [3, 12], preventing its use in perovskite solar cells.

SnO_2 is a good alternative ETL for PSC due to its stability under solar irradiation and its high electron mobility up to 240 $cm^2\ V^{-1}\ s^{-1}$, which is 100 times that of TiO_2 [46]. Furthermore, SnO_2 has a wide optical bandgap (3.6–4.0 eV) and quite a similar conduction band position and work function compared to TiO_2 and ZnO, providing efficient electron extraction properties [25, 26]. Ke et al. have shown that although the high temperature annealed SnO_2 (HT-SnO_2) exhibits much higher crystallinity compared to low-temperature annealed SnO_2 (LT-SnO_2), the LT-SnO_2 exhibits much better performance than HT-SnO_2 in planar PSC architectures. This is due to better film coverage, higher electron mobility, and wider bandgap of the LT-SnO_2 ETL, providing improved electron transport properties and less interface recombination [47]. That opens up routes for the fabrication of efficient perovskite solar cells on printable and flexible substrates, using SnO_2 as the ETL. Similar to TiO_2 and ZnO, doping as an optimization method can be effective in PSC based on SnO_2 ETL. Sn^{4+} ions can be replaced with suitable dopant elements such as Mg [48], Li [25], Y [49], Sb [50], and Nb [51] to tune the electronic and transport properties. The low-temperature atomic layer deposited SnO_2 in a sandwich with AZO has also led to significantly prolonged device lifetimes for perovskite solar cells [21]. The main reason for this effect is the suppression of out-diffusion of decomposition products, which recently has been explored as a very important property for device interlayers in perovskite solar cells (see Figure 9.3). The reactivity of several oxides with the perovskite layer has also been reviewed recently by Olthof and Riedl [21].

9.3.2 Organic Electron Transport Materials

Organic electron transport materials can be divided into fullerene and derivatives and non-fullerene – small organic molecules or polymer materials. Fullerenes and their derivatives are commonly used as electron transport materials in inverted PSC due to their low fabrication temperature, suitable energy level alignment, relatively high electron mobility (compared to various organic ETL). In some cases,

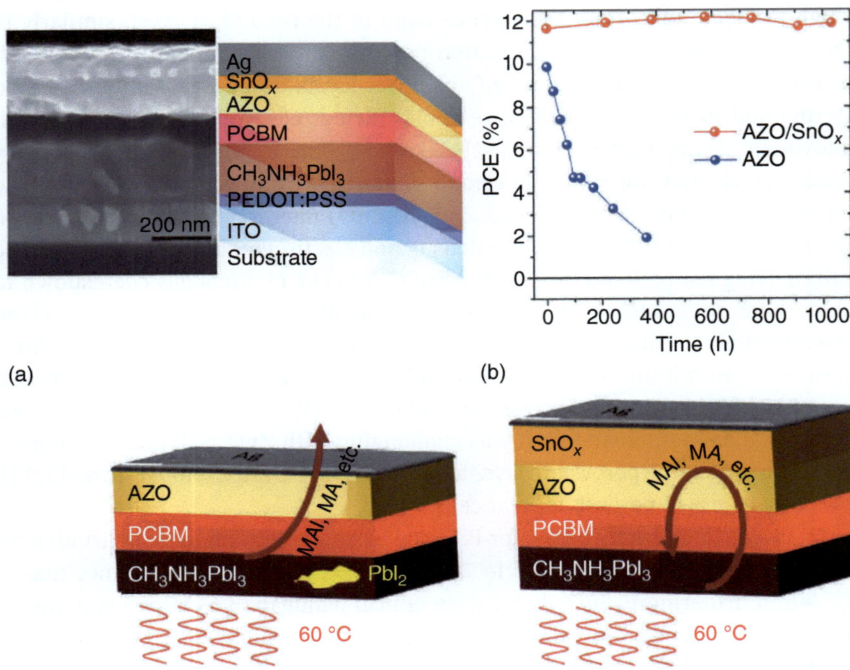

Figure 9.3 AZO (a) and SnO_2/AZO composite (b) interlayers used as efficient and stable ETL in perovskite solar cells. Source: Oltof and Riedl [21].

efficient hole-blocking properties [52–54]. In 2013, various fullerene derivatives (C_{60}, $PC_{61}BM$, and ICBA) were used as ETL in inverted planar PSC for the first time, with the highest efficiency of 3.9% [55]. Within a rather short time, the efficiency of PSC with organic-based ETL has risen from the initial 3.9% to over 20% [56]. $PC_{61}BM$ ([6,6]-phenyl-C_{61}-butyric acid methyl ester) has nowadays been widely used as organic ETL in PSC owing to its advantages such as low-temperature solution-processing on flexible substrates and elimination of photocurrent hysteresis, as well as promoting efficient electron transport [57, 58]. Using $PC_{61}BM$ in inverted PSC devices has been shown to effectively passivate charge trap states at the $CH_3NH_3PbI_3$ perovskite film surfaces and grain boundaries [59], promoting the hysteresis-free behavior. Compared with PCBM, the low-cost C_{60} can be packed more densely to facilitate improved electron transport and has higher electron mobility (1.6 cm^2 V^{-1} s^{-1}) and conductivity (2.3×10^{-3} S cm^{-1}) compared to that of PCBM (6.1×10^{-2} cm^2 V^{-1} s^{-1} and 3.2×10^{-4} S cm^{-1}, respectively) [60, 61]. PSC based on C_{60} shows negligible hysteresis (Figure 9.4), indicating that the hysteresis in PSC may also originate from space charge accumulation at interfaces with poor charge-carrier extraction [62].

C_{70} is another fullerene derivative utilized heavily in organic solar cell development [63–65] and can potentially be utilized as an efficient ETL in PSC. C_{70} has a LUMO energy level of 4.2 eV [27], which provides a suitable ETL for $CH_3NH_3PbI_3$ perovskite absorber-based PSC [55]. Also, the hydrophobic surface of C_{70} compared

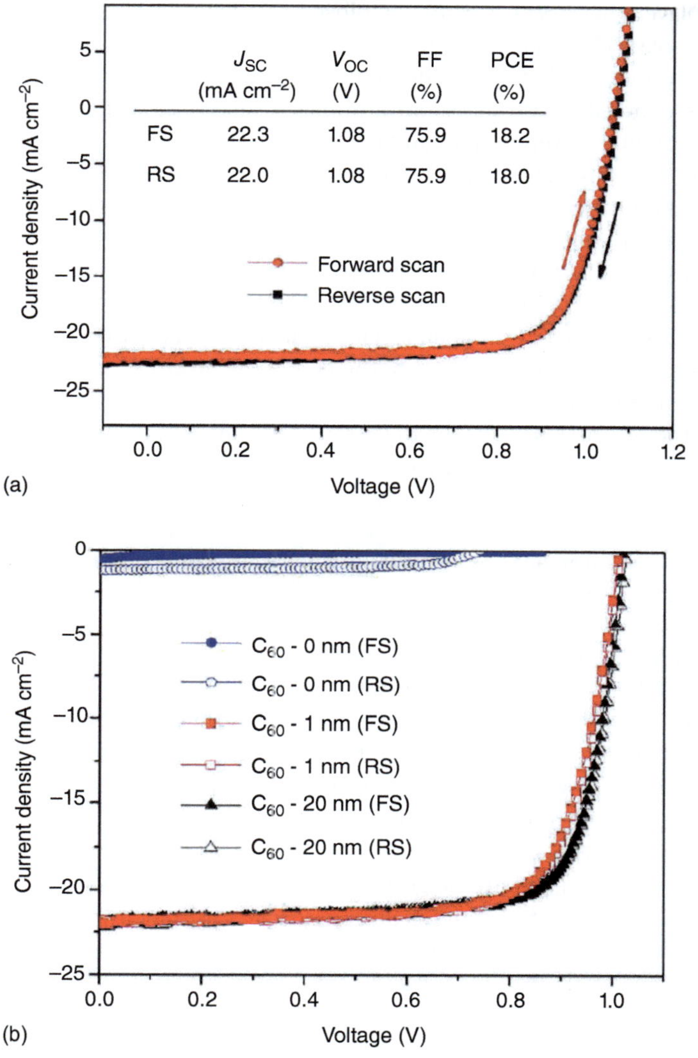

Figure 9.4 (a) $J-V$ curves of the champion device with 1 nm C60 as ETL measured with reverse and forward scanning directions under 1-sun illumination and (b) $J-V$ curves for the devices with various thicknesses of C60 measured under reverse and forward scans at a rate of 50 mV s^{-1}. Source: Liu et al. [62].

to PCBM (water contact angle of 99.6° [66] vs. 86.5° [67]) can result in enhanced stability of PSC in ambient atmosphere. However, pristine fullerenes (C_{60} and C_{70}) have poor solubility in common organic solvents. Therefore, they are usually prepared using vacuum deposition. The vacuum deposition of pristine fullerenes requires high purity samples and relatively high temperature, making a cost-benefit analysis important for up-scaled production [57]. Solution-processed pristine C_{60}

and C_{70} compact layers have been recently employed as efficient ETL in normal (n–i–p) perovskite solar cells as an alternative route [68, 69].

Another C_{60} derivative, e.g. indene-C_{60} bisadduct (ICBA), has also been used as ETL in perovskite solar cells. ICBA has some advantages, such as ease of synthesis from C_{60} and indene and relatively high solubility in many common organic solvents [70]. Compared to PCBM, the LUMO level of ICBA is closer to the perovskite conduction band, facilitating more efficient charge extraction from the perovskite layer [28]. Recently, a new series of fullerene derivatives, C_{60}-DPM-OE, C_{60}-(DPM-OE)$_2$, C_{70}-DPM-OE, and C_{70}-(DPM-OE)$_2$, were used as ETL in an inverted structure of PSC (p–i–n). They were synthesized by phenyl groups of diphenylmethanofullerene (DPM) moiety with oligoether (OE) chains. These chains modify the cathodes' work function and passivate the trap states of the perovskite layer [71]. Although the fullerenes and their derivatives are the most commonly used organic ETL for the planar structure of PSC, due to their relatively poor ambient stability and aggregation upon annealing, it is necessary to design and develop efficient non-fullerene organic small molecules or polymers as alternative ETL for PSC [35].

A non-fullerene small organic molecule, perylene diimide derivative (PDI), showed promising performance as n-type organic semiconductors in the PSC due to their relatively high electron affinity, high electron mobility ($>1.0\times10^{-3}$ cm^2 V^{-1} s^{-1}), and good thermal stability. The small-molecule PDI is also easy to functionalize and purify compared with polymeric molecules, and the properties of PDI can be controlled by molecular design and morphological regulation [72–74]. The fluorinated perylene diimide derivative (FPDI) showed high electron mobility with excellent air stability. However, the best efficiency of the devices obtained was only 7.93% [72]. Zhang et al. used the amino-substituted PDI (N-PDI) as the ETL in the normal structure of perovskite solar cells with a competitive efficiency of 17.66% on glass/FTO rigid substrates and 14.32% on PET/ITO flexible substrates [74]. Although the efficiency of PSC using PDI derivatives as ETL is still outperformed by the fullerenes, the stability of the devices is high. Therefore, a combination of PDI and fullerene, namely PDI-C_{60}, was used to combine the advantages of both materials (see Figure 9.5) [75]. Compared with PCBM, the novel PDI/fullerene hybrid (PDI-C_{60}) as ETL showed higher electron mobility and stronger hydrophobic capacity. As a result, the devices based on PDI-C_{60} achieved an excellent power conversion efficiency (PCE) of 18.6%, which is significantly higher than the control devices of PCBM (16.6%) and PDI (13.8%), combined with improved device stability [75].

Some other small organic molecule materials, such as hexaazatrinaphthylene (HATNA) derivatives, have attracted attention as promising ETL. Although the HATNA derivatives are not common in perovskite solar cells, they may provide a good alternative to fullerenes due to their low cost, high electron mobility, and large bandgap, supporting carrier selective transport. The devices based on HATNA derivatives also exhibited improved device stability due to their hydrophobic nature and surface passivation properties [29]. By employing the HATNA-based small molecular hexaazatrinaphtho[2,3-*c*][1,2,5] thiadiazole as ETL in low temperature and solution-processed planar structure of PSC, a high PCE

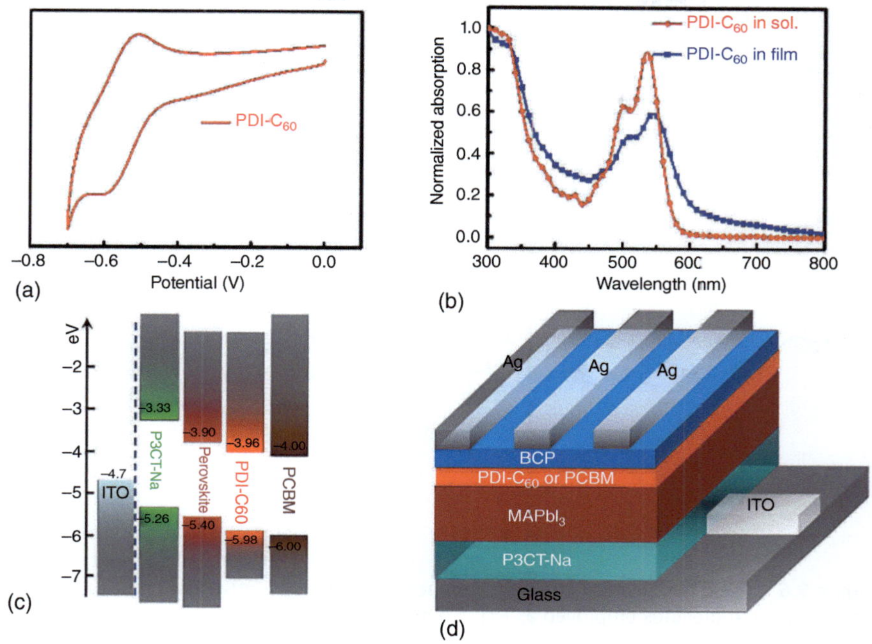

Figure 9.5 (a) Cyclic voltammetry scans for PDI-C_{60}; (b) UV–vis absorption spectra of PDI-C_{60} in $CHCl_3$ solution and in film; (c) Energy level diagram of the materials used in PSCs; (d) Device architecture of the inverted PSCs studied in this work. Source: Luo et al. [75].

of 18.1% was obtained, which is comparable with the standard $PC_{61}BM$-based devices [76]. An example of organic ETL, novel n-type conjugated polymers such as poly(naphthodiperylenetetraimide-vinylene) (NDP-V) [77], P(NDI2DT-TTCN) [78], PN, PN-F25%, PN-F50% [79] were also developed as efficient ETM, leading to relatively high device stability as of the fullerene ETL. The structure of the organic ETM is shown in Figure 9.6.

9.4 The Architectures of Perovskite Solar Cells

Perovskite solar cell architectures can be divided into three types based on the ETL and the device stack used, namely mesoscopic PSC, normal planar PSC, and inverted planar PSC. These are shown in Figure 9.7. In this last part of the mini-review on perovskite ETL, we will briefly discuss the ETL integration into these various device architectures.

9.4.1 Mesoscopic Perovskite Solar Cells

PSC with normal configuration based on a mesoporous scaffold of ETL usually consists of a mesoporous ETL, a perovskite absorber, and an HTL, sandwiched between a transparent conducting oxide (TCO) substrate and a metal contact

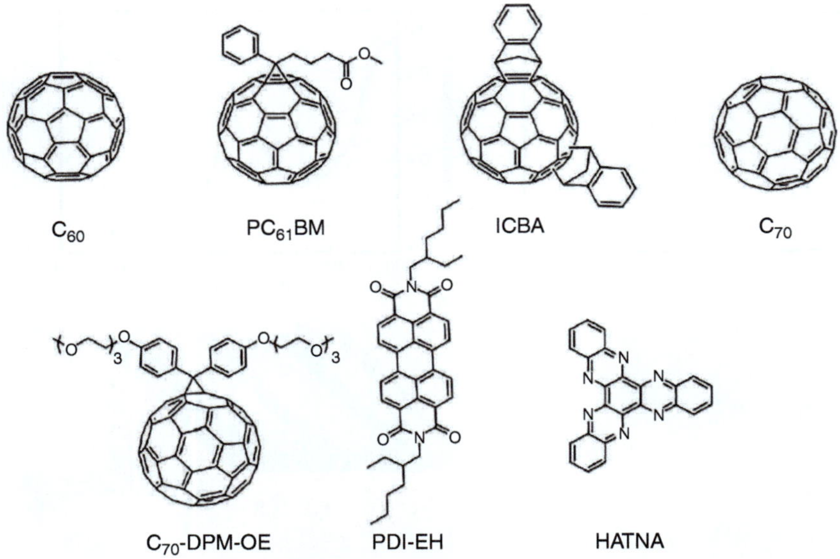

Figure 9.6 The molecular structures of common fullerene derivatives and non-fullerene small organic molecules employed as ETL. Source: Zheng et al. [35].

(Figure 9.7a) [80]. Mesoscopic PSC evolved from dye-sensitized solar cells, and the ground-breaking first report on their use came in 2009 by the Miyasaka group [34]. They used $CH_3NH_3PbI_3$ and $CH_3NH_3PbBr_3$ as sensitizers on mesoporous TiO_2, instead of dye molecules, and a liquid electrolyte as the hole transport material, resulting in power conversion efficiencies of 3–4%. In addition to low device efficiency, these liquid electrolyte-based devices also suffered from poor device stability [34]. In 2012, Kim et al. replaced the liquid electrolyte with spiro-OMeTAD as a new solid hole transporting material in the mesoscopic TiO_2 based PSC, achieving 9.7% power conversion efficiency [83]. Most mesoscopic PSC uses a TiO_2 compact, thin film as a combined ETL and hole-blocking layer, to avoid shunting losses. Then, the mesoporous TiO_2 is a scaffold film filled with the perovskite absorber. Another category of mesoscopic PSC, meso-superstructured solar cells, utilize Al_2O_3 as an insulating scaffold layer instead of TiO_2, showing suppressed voltage losses otherwise arising from the presence of subgap states [84, 85]. Although the mesoscopic PSC showed high efficiency, the question arose if PSC must use mesoporous layers as scaffolds.

Mesoporous layers extend the interface surface area, an advantage for solar cells with light absorbers with a short charge-diffusion length and tightly bound excitons. The increased interface area may facilitate more efficient charge separation and extraction processes. However, this advantage is not useful in PSC due to the long charge-diffusion length [86] and the relatively low exciton binding energy of perovskite materials. Excitons are easily thermally dissociated at ambient temperature in most perovskite materials. The mesoporous ETL structure may create scattering effects and consequently increase light absorption in the solar cells. However, the

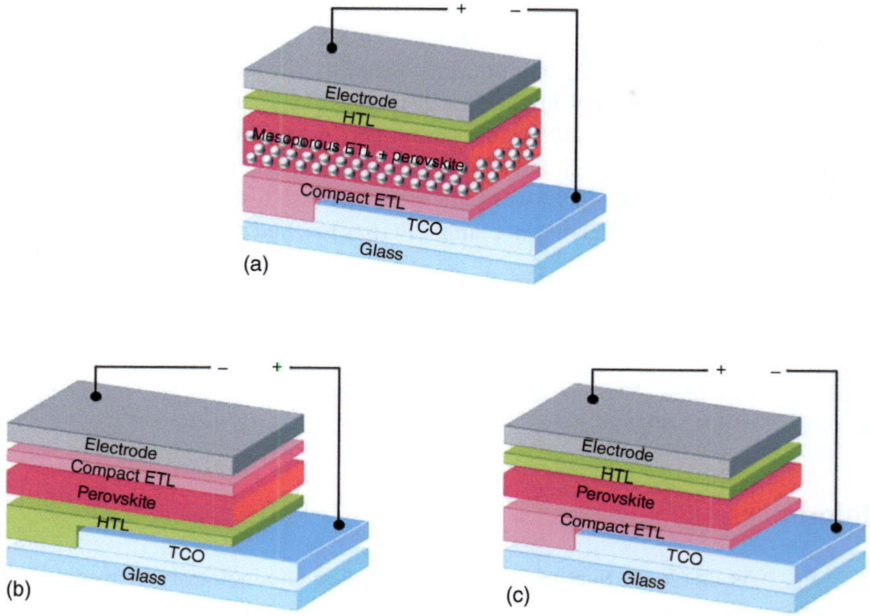

Figure 9.7 Schematic PSC device architectures: (a) mesoscopic PSC. Source: Modified from Asghar et al. [80]. (b) Inverted planar PSC. Source: Liu et al. [81]; Heo et al. [82]. (c) Normal planar PSC.

perovskite materials have an excellent light-harvesting ability with high absorption coefficients (more than $4.5 \times 10^4\,cm^{-1}$ for $MAPbI_3$ in the whole visible absorption range), making this property less relevant for PSC [3, 87]. Also, the most mesoporous ETL require high-temperature treatment (>450 °C), which is not compatible with flexible, low-cost devices [88].

9.4.2 Planar Perovskite Solar Cells

Planar PSC can simplify the fabrication process by eliminating the use of the mesoporous layers as scaffolds. Planar PSC includes normal (n–i–p) and inverted (p–i–n) structures, which were evolved from organic solar cells. The n-type ETL is located between the bottom TCO electrode and the perovskite absorber layer in the normal structure. In general, the normal device architecture of $MAPbI_3$ perovskite solar cells is composed of (i) TCO, such as fluorine-doped tin oxide (FTO) or indium tin oxide (ITO), (ii) n-type electron transporting materials, e.g. compact TiO_2, ZnO, SnO_2 or some organic materials, (iii) $MAPbI_3$, (iv) p-type hole-transporting materials, e.g. spiro-OMeTAD, PTAA (polytriarylamine), P3HT (poly-3-hexylthiophene), CNT (carbon nanotube) composite, etc., and (v) metal electrode, so it forms a TCO/n-ETL/$MAPbI_3$/p-HTL/metal stack (see Figure 9.7c). In PSC with inverted structure, the location of ETL is at the top of the perovskite layer, so the common structure of such devices consists of (i) TCO such as ITO, (ii) HTL e.g. PEDOT:PSS or NiO_x, (iii) $MAPbI_3$, (iv) n-type organic materials, e.g. C_{60}, $PC_{61}BM$, $PC_{71}BM$, ICBA, or some low temperature processed metal oxide, e.g. ZnO, and (v) metal electrode, thus forming a ITO/HTL/$MAPbI_3$/ETL/metal stack (see Figure 9.7b) [81, 82]. Solution-processability, simplicity of device architecture, and low cost are the most important advantages of planar structures. Moreover, the low-temperature processable planar structures are frequently used to make flexible perovskite solar cells [89]. Since Jeng et al. first reported the low-temperature PSC in 2013, where a PCE of 3.9% was achieved [55], much research has been done in this area that today, the efficiency of planar PSC has reached more than 22% [90, 91].

Acknowledgments

M.M. acknowledges 'Danmarks Frie Forskningsfond, DFF FTP' for funding React-PV, No. 8022-00389B.

References

1 Yang, D. et al. (2016). Surface optimization to eliminate hysteresis for record efficiency planar perovskite solar cells. *Energy Environ. Sci.* 9 (10): 3071–3078.

2 Pan, H. et al. (2020). Advances in design engineering and merits of electron transporting layers in perovskite solar cells. *Mater. Horiz.* 7 (9): 2276–2291.

3 Lian, J. et al. (2018). Electron-transport materials in perovskite solar cells. *Small Methods* 2 (10): 1800082.

4 Giorgi, G. and Yamashita, K. (2015). Organic–inorganic halide perovskites: an ambipolar class of materials with enhanced photovoltaic performances. *J. Mater. Chem. A* 3 (17): 8981–8991.

5 Völker, S.F., Collavini, S., and Delgado, J.L. (2015). Organic charge carriers for perovskite solar cells. *ChemSusChem* 8 (18): 3012–3028.

6 Trost, S. et al. (2013). Overcoming the "light-soaking" issue in inverted organic solar cells by the use of Al:ZnO electron extraction layers. *Adv. Energy Mater.* 3 (11): 1437–1444.

7 Shao, Y., Yuan, Y., and Huang, J. (2016). Correlation of energy disorder and open-circuit voltage in hybrid perovskite solar cells. *Nat. Energy* 1 (1): 15001.

8 Mirsafaei, M. et al. (2020). Sputter-deposited titanium oxide layers as efficient electron selective contacts in organic photovoltaic devices. *ACS Appl. Energy Mater.* 3 (1): 253–259.

9 Ahmadpour, M. et al. (2019). Crystalline molybdenum oxide layers as efficient and stable hole contacts in organic photovoltaic devices. *ACS Appl. Energy Mater.* 2 (1): 420–427.

10 Cauduro, F., Cauduro, A.L.F. et al. (2017). Crystalline molybdenum oxide thin-films for application as interfacial layers in optoelectronic devices. *ACS Appl. Mater. Interf.* 9 (8): 7717–7724.

11 Olthof, S. (2016). Research update: the electronic structure of hybrid perovskite layers and their energetic alignment in devices. *APL Mater.* 4 (9): 091502.

12 Yang, G. et al. (2016) Recent progress in electron transport layers for efficient perovskite solar cells. *J. Mater. Chem. A* 4 (11): 3970–3990.

13 Liu, H. et al. (2016). Nano-structured electron transporting materials for perovskite solar cells. *Nanoscale* 8 (12): 6209–6221.

14 Fakharuddin, A. et al. (2015). Role of morphology and crystallinity of nanorod and planar electron transport layers on the performance and long term durability of perovskite solar cells. *J. Power Sources* 283: 61–67.

15 Nimens, W.J. et al. (2018). Morphology and optoelectronic variations underlying the nature of the electron transport layer in perovskite solar cells. *ACS Appl. Energy Mater.* 1 (2): 602–615.

16 Li, Y. et al. (2015). Multifunctional fullerene derivative for interface engineering in perovskite solar cells. *J. Am. Chem. Soc.* 137 (49): 15540–15547.

17 Zhou, H. et al. (2014). Interface engineering of highly efficient perovskite solar cells. *Science* 345 (6196): 542–546.

18 Shahvaranfard, F. et al. (2020). Engineering of the electron transport layer/perovskite interface in solar cells designed on TiO_2 rutile nanorods. *Adv. Funct. Mater.* 30 (10): 1909738.

19 Jeon, N.J. et al. (2014). Solvent engineering for high-performance inorganic–organic hybrid perovskite solar cells. *Nat. Mater.* 13 (9): 897–903.

20 Yang, G. et al. (2017) Interface engineering in planar perovskite solar cells: energy level alignment, perovskite morphology control and high performance achievement. *J. Mater. Chem. A* 5 (4): 1658–1666.

21 Olthof, S. and Riedl, T. (2019). Metal-oxide interface materials for organic and perovskite solar cells. In: *World Scientific Reference of Hybrid Materials, Devices from Hybrid and Organic Materials*, vol. 2, 61–104. World Scientific.

22 Wang, H. et al. (2019). Achieving efficient inverted perovskite solar cells with excellent electron transport and stability by employing a ladder-conjugated perylene diimide dimer. *J. Mater. Chem. A* 7 (42): 24191–24198.

23 Mahmud, M.A. et al. (2017). Low temperature processed ZnO thin film as electron transport layer for efficient perovskite solar cells. *Sol. Energy Mater. Sol. Cells* 159: 251–264.

24 Mahmood, K., Swain, B.S., and Amassian, A. (2014). Double-layered ZnO nanostructures for efficient perovskite solar cells. *Nanoscale* 6 (24): 14674–14678.

25 Park, M. et al. (2016). Low-temperature solution-processed Li-doped SnO_2 as an effective electron transporting layer for high-performance flexible and wearable perovskite solar cells. *Nano Energy* 26: 208–215.

26 Chen, J.-Y. et al. (2017). Low-temperature electrodeposited crystalline SnO_2 as an efficient electron-transporting layer for conventional perovskite solar cells. *Sol. Energy Mater. Sol. Cells* 164: 47–55.

27 Patil, B.R. et al. (2018). Area dependent behavior of bathocuproine (BCP) as cathode interfacial layers in organic photovoltaic cells. *Sci. Rep.* 8 (1): 12608.

28 Wu, F. et al. (2018). Enhanced photovoltaic performance and reduced hysteresis in perovskite-ICBA-based solar cells. *Org. Electron.* 58: 6–11.

29 Zhao, D. et al. (2016). Hexaazatrinaphthylene derivatives: efficient electron-transporting materials with tunable energy levels for inverted perovskite solar cells. *Angew. Chem. Int. Ed.* 55 (31): 8999–9003.

30 Shin, S.S., Lee, S.J., and Seok, S.I. (2019). Metal oxide charge transport layers for efficient and stable perovskite solar cells. *Adv. Funct. Mater.* 29 (47): 1900455.

31 Dette, C. et al. (2014). TiO_2 anatase with a bandgap in the visible region. *Nano Lett.* 14 (11): 6533–6538.

32 Tang, H. et al. (1994). Electrical and optical properties of TiO_2 anatase thin films. *J. Appl. Phys.* 75 (4): 2042–2047.

33 O'Regan, B. and Grätzel, M. (1991). A low-cost, high-efficiency solar cell based on dye-sensitized colloidal TiO_2 films. *Nature* 353 (6346): 737–740.

34 Kojima, A. et al. (2009). Organometal halide perovskites as visible-light sensitizers for photovoltaic cells. *J. Am. Chem. Soc.* 131 (17): 6050–6051.

35 Zheng, S. et al. (2019). Materials and structures for the electron transport layer of efficient and stable perovskite solar cells. *Sci. China Chem.* 62 (7): 800–809.

36 Chen, X. and Mao, S.S. (2007). Titanium dioxide nanomaterials: synthesis, properties, modifications, and applications. *Chem. Rev.* 107 (7): 2891–2959.

37 Kim, D.H. et al. (2015). Niobium doping effects on TiO_2 mesoscopic electron transport layer-based perovskite solar cells. *ChemSusChem* 8 (14): 2392–2398.

38 Giordano, F. et al. (2016). Enhanced electronic properties in mesoporous TiO_2 via lithium doping for high-efficiency perovskite solar cells. *Nat. Commun.* 7 (1): 10379.

39 Wu, T. et al. (2019). Gradient Sn-doped heteroepitaxial film of faceted rutile TiO_2 as an electron selective layer for efficient perovskite solar cells. *ACS Appl. Mater. Interf.* 11 (21): 19638–19646.

40 Sanehira, Y. et al. (2020). Low-temperature synthesized Nb-doped TiO_2 electron transport layer enabling high-efficiency perovskite solar cells by band alignment tuning. *ACS Appl. Mater. Interf.* 12 (13): 15175–15182.

41 Chandiran, A.K. et al. (2014). Analysis of electron transfer properties of ZnO and TiO_2 photoanodes for dye-sensitized solar cells. *ACS Nano* 8 (3): 2261–2268.

42 Liu, D. and Kelly, T.L. (2014). Perovskite solar cells with a planar heterojunction structure prepared using room-temperature solution processing techniques. *Nat. Photonics* 8 (2): 133–138.

43 Mahmood, K., Swain, B.S., and Amassian, A. (2015). 16.1% efficient hysteresis-free mesostructured perovskite solar cells based on synergistically improved ZnO nanorod arrays. *Adv. Energy Mater.* 5 (17): 1500568.

44 Zheng, Y.-Z. et al. (2017). Iodine-doped ZnO nanopillar arrays for perovskite solar cells with high efficiency up to 18.24%. *J. Mater. Chem. A* 5 (24): 12416–12425.

45 Tseng, Z.-L. et al. (2016). Surface engineering of ZnO electron transporting layer via Al doping for high efficiency planar perovskite solar cells. *Nano Energy* 28: 311–318.

46 Snaith, H.J. and Ducati, C. (2010). SnO_2-based dye-sensitized hybrid solar cells exhibiting near unity absorbed photon-to-electron conversion efficiency. *Nano Lett.* 10 (4): 1259–1265.

47 Ke, W. et al. (2015). Effects of annealing temperature of tin oxide electron selective layers on the performance of perovskite solar cells. *J. Mater. Chem. A* 3 (47): 24163–24168.

48 Xiong, L. et al. (2016). Performance enhancement of high temperature SnO_2-based planar perovskite solar cells: electrical characterization and understanding of the mechanism. *J. Mater. Chem. A* 4 (21): 8374–8383.

49 Yang, G. et al. (2017). Reducing hysteresis and enhancing performance of perovskite solar cells using low-temperature processed Y-doped SnO_2 nanosheets as electron selective layers. *Small* 13 (2): 1601769.

50 Bai, Y. et al. (2016). Low temperature solution-processed Sb:SnO_2 nanocrystals for efficient planar perovskite solar cells. *ChemSusChem* 9 (18): 2686–2691.

51 Halvani Anaraki, E. et al. (2018). Low-temperature Nb-doped SnO_2 electron-selective contact yields over 20% efficiency in planar perovskite solar cells. *ACS Energy Lett.* 3 (4): 773–778.

52 Malinkiewicz, O. et al. (2013). Perovskite solar cells employing organic charge-transport layers. *Nat. Photonics* 8: 128.

53 Docampo, P. et al. (2013). Efficient organometal trihalide perovskite planar-heterojunction solar cells on flexible polymer substrates. *Nat. Commun.* 4 (1): 2761.

54 Said, A.A., Xie, J., and Zhang, Q. (2019). Recent progress in organic electron transport materials in inverted perovskite solar cells. *Small* 15 (27): 1900854.

55 Jeng, J.-Y. et al. (2013). $CH_3NH_3PbI_3$ perovskite/fullerene planar-heterojunction hybrid solar cells. *Adv. Mater.* 25 (27): 3727–3732.

56 Zheng, X. et al. (2017). Defect passivation in hybrid perovskite solar cells using quaternary ammonium halide anions and cations. *Nat. Energy* 2 (7): 17102.

57 Dai, S.-M. et al. (2017). Pristine fullerenes mixed by vacuum-free solution process: efficient electron transport layer for planar perovskite solar cells. *J. Power Sources* 339: 27–32.

58 Heo, J.H. et al. (2015). Hysteresis-less inverted $CH_3NH_3PbI_3$ planar perovskite hybrid solar cells with 18.1% power conversion efficiency. *Energy Environ. Sci.* 8 (5): 1602–1608.

59 Shao, Y. et al. (2014). Origin and elimination of photocurrent hysteresis by fullerene passivation in $CH_3NH_3PbI_3$ planar heterojunction solar cells. *Nat. Commun.* 5: 5784.

60 Liu, X. et al. (2018). Novel efficient C_{60}-based inverted perovskite solar cells with negligible hysteresis. *Electrochim. Acta* 288: 115–125.

61 Yoon, H. et al. (2016). Hysteresis-free low-temperature-processed planar perovskite solar cells with 19.1% efficiency. *Energy Environ. Sci.* 9 (7): 2262–2266.

62 Liu, D. et al. (2018). Impact of ultrathin C_{60} on perovskite photovoltaic devices. *ACS Nano* 12 (1): 876–883.

63 Sherafatipour, G. et al. (2019). Degradation pathways in standard and inverted DBP-C_{70} based organic solar cells. *Sci. Rep.* 9 (1): 4024.

64 Jafari, F. et al. (2019). Inverted organic solar cells with non-clustering bathocuproine (BCP) cathode interlayers obtained by fullerene doping. *Sci. Rep.* 9 (1): 10422.

65 Patil, B.R. et al. (2017). 4P-NPD ultra-thin films as efficient exciton blocking layers in DBP/C_{70} based organic solar cells. *J. Phys. D. Appl. Phys.* 50 (38): 385101.

66 Sharma, P., Singhal, R. et al. (2016). Surface and structural studies of fullerene C_{70} under ion irradiation. *Surf. Eng.* 32 (11): 846–852.

67 Lin, X. et al. (2014). Morphological investigation of P3HT/PCBM heterojunction and its effects on the performance of bilayer organic solar cells. *Synth. Met.* 196: 145–150.

68 Wojciechowski, K. et al. (2015). C_{60} as an efficient N-type compact layer in perovskite solar cells. *J. Phys. Chem. Lett.* 6 (12): 2399–2405.

69 Collavini, S. et al. (2016). Efficient regular perovskite solar cells based on pristine [70]fullerene as electron-selective contact. *ChemSusChem* 9 (11): 1263–1270.

70 He, Y. et al. (2010). Indene–C_{60} bisadduct: a new acceptor for high-performance polymer solar cells. *J. Am. Chem. Soc.* 132 (4): 1377–1382.

71 Xing, Y. et al. (2016). New fullerene design enables efficient passivation of surface traps in high performance p–i–n heterojunction perovskite solar cells. *Nano Energy* 26: 7–15.

72 Huang, J. et al. (2016). Morphology control of planar heterojunction perovskite solar cells with fluorinated PDI films as organic electron transport layer. *Sol. Energy* 133: 331–338.

73 Huang, J. et al. (2018). Lead-free $(CH_3NH_3)_3Bi_2I_9$ perovskite solar cells with fluorinated PDI films as organic electron transport layer. *J. Alloys Compd.* 767: 870–876.

74 Zhang, H. et al. (2016). New generation perovskite solar cells with solution-processed amino-substituted perylene diimide derivative as electron-transport layer. *J. Mater. Chem. A* 4 (22): 8724–8733.

75 Luo, Z. et al. (2019). Designing a perylene diimide/fullerene hybrid as effective electron transporting material in inverted perovskite solar cells with enhanced efficiency and stability. *Angew. Chem. Int. Ed.* 58 (25): 8520–8525.

76 Wang, N. et al. (2017). Improving interfacial charge recombination in planar heterojunction perovskite photovoltaics with small molecule as electron transport layer. *Adv. Energy Mater.* 7 (18): 1700522.

77 Jiang, K. et al. (2018). Naphthodiperylenetetraimide-based polymer as electron-transporting material for efficient inverted perovskite solar cells. *ACS Appl. Mater. Interf.* 10 (42): 36549–36555.

78 Kim, H.I. et al. (2018). Improving the performance and stability of inverted planar flexible perovskite solar cells employing a novel NDI-based polymer as the electron transport layer. *Adv. Energy Mater.* 8 (16): 1702872.

79 Tian, L. et al. (2019). Fluoro- and amino-functionalized conjugated polymers as electron transport materials for perovskite solar cells with improved efficiency and stability. *ACS Appl. Mater. Interf.* 11 (5): 5289–5297.

80 Asghar, M.I. et al. (2017). Device stability of perovskite solar cells – a review. *Renew. Sust. Energ. Rev.* 77: 131–146.

81 Liu, M., Johnston, M.B., and Snaith, H.J. (2013). Efficient planar heterojunction perovskite solar cells by vapour deposition. *Nature* 501 (7467): 395–398.

82 Heo, J.H. et al. (2016). Highly efficient low temperature solution processable planar type $CH_3NH_3PbI_3$ perovskite flexible solar cells. *J. Mater. Chem. A* 4 (5): 1572–1578.

83 Kim, H.-S. et al. (2012). Lead iodide perovskite sensitized all-solid-state submicron thin film mesoscopic solar cell with efficiency exceeding 9%. *Sci. Rep.* 2 (1): 591.

84 Lee, M.M. et al. (2012). Efficient hybrid solar cells based on meso-superstructured organometal halide perovskites. *Science* 338 (6107): 643–647.

85 Leijtens, T. et al. (2014). Electronic properties of meso-superstructured and planar organometal halide perovskite films: charge trapping, photodoping, and carrier mobility. *ACS Nano* 8 (7): 7147–7155.

86 Stranks, S.D. et al. (2013). Electron-hole diffusion lengths exceeding 1 micrometer in an organometal trihalide perovskite absorber. *Science* 342 (6156): 341–344.

87 Park, J.-S. et al. (2015). Electronic structure and optical properties of α-$CH_3NH_3PbBr_3$ perovskite single crystal. *J. Phys. Chem. Lett.* 6 (21): 4304–4308.

88 Bao, X. et al. (2015). Efficient planar perovskite solar cells with large fill factor and excellent stability. *J. Power Sources* 297 (Supplement C): 53–58.

89 Zhang, J. et al. (2020). Critical review of recent progress of flexible perovskite solar cells. *Mater. Today* 39: 66–88.

90 Wang, P. et al. (2020). Gradient energy alignment engineering for planar perovskite solar cells with efficiency over 23%. *Adv. Mater.* 32 (6): 1905766.

91 Hu, X. et al. (2020). 22% efficiency inverted perovskite photovoltaic cell using cation-doped brookite TiO_2 top buffer. *Adv. Sci.* 7 (16): 2001285.

10

Dopant-Free Hole-Transporting Materials for Perovskite Solar Cells

Meenakshi Pegu[1], Shahzada Ahmad[1,2], and Samrana Kazim[1,2]

[1] *BCMaterials, Basque Center for Materials, Applications and Nanostructures, UPV/EHU Science Park, Leioa, 48940, Spain*
[2] *IKERBASQUE, Basque Foundation for Science, Plaza Euskadi 5, Bilbao, 48009, Spain*

10.1 Introduction

To meet the societal energy demand and limit the greenhouse emission, harvesting sunlight with cost-effective technology are of paramount importance. In the past, inorganic crystalline semiconductors, such as silicon, cadmium telluride (CdTe), and copper indium gallium selenide/sulfide (CIGS), were utilized as absorbers, for solar cells fabrication. However, it requires high energy input for manufacturing, involving high-temperature and high-vacuum processes [1–3]. Among the potentially transformative technologies, currently perovskite solar cells (PSCs) are seen as the preferred ones, which can be made lightweight and flexible, similar to organic and dye-sensitized solar cells (DSSCs).

The PSCs are strong contenders in the emerging photovoltaic (PV) category and has earned significant interest [4]. Lead halide-based perovskites owing to their unique semiconducting properties are being intensively investigated since its first report by Miyasaka and coworkers [5]. The power conversion efficiency (PCE) is increasing exponentially, and currently 25.5% was reported for lab-sized solar cells [6]. For the first time, $CH_3NH_3PbI_3$ ($MAPbI_3$) and $CH_3NH_3PbBr_3$ ($MAPbBr_3$)were exploited as visible-light sensitizers into DSSCs using liquid electrolyte as redox couple that yielded 3.8% and 3.1% PCE respectively. However, devices were not stable due to dissolution of the perovskite in liquid electrolyte [5].

Later, in 2012, a major breakthrough occurred by the use of solid state PSCs, when liquid electrolyte was replaced with solid state hole transporting material known as 2,2′,7.7′-tetrakis(N,N-di-p-methoxyphenylamine)-9,9′-spirobifluorene (Spiro-OMeTAD). The PCEs of 9.7% and 10.9% was achieved using $MAPbI_3$ and $MAPbI_{3-x}Clx$ respectively as light absorber deposited onto a mesoporous TiO_2 layer and mesostructured insulating Al_2O_3 scaffold [7, 8]. PCE stems from the exceptional opto-electrical properties of perovskites, such as strong optical absorption coefficient, direct band gap, long carrier lifetime, long diffusion length, efficient ambipolar charge transport and high electron/hole mobility in the

Perovskite Solar Cells: Materials, Processes, and Devices, First Edition.
Edited by Shahzada Ahmad, Samrana Kazim, and Michael Grätzel.

crystalline state [9–11]. Hybrid perovskites generate loosely bounded excitons under photo-irradiation, with a small binding energy facilitating the generation of free charge carriers within the perovskite with a low driving force. In spite of the competitive performance shown by PSCs, challenges remain that needs to be addressed to validate their commercial success. Studies are being carried out to improve the operational stability of PSCs, and to uncover environmentally benign and stable perovskite. The intrinsic degradation of $MAPbI_3$ under exposure to moisture, heat, and prolonged illumination under air, heat, and light needs to be addressed [12–14]. We outline the following advances in different architecture of PSCs using dopant-free organic hole transporting materials (HTMs) toward its application.

10.1.1 Device Structure of Perovskite Solar Cells

During the current decade, significant progress has been made in using perovskites as a light harvester. Typically, perovskite is expressed as ABX_3, where A stands for organic or inorganic cation, B is a metal ion, and X is an anion as shown in Figure 10.1a [15]. In addition, Figure 10.1b depicts the rise in PCE with the use of dopant-free hole transport materials (HTMs).

The ambipolar charge transport behaviour of perovskite allow versatility in designing different structures of PSCs. In general, PSC architecture is comprises of five layers such as transparent conducting electrode (TCO), n-type semiconducting electron-transporting layers (ETLs), perovskite absorber as an

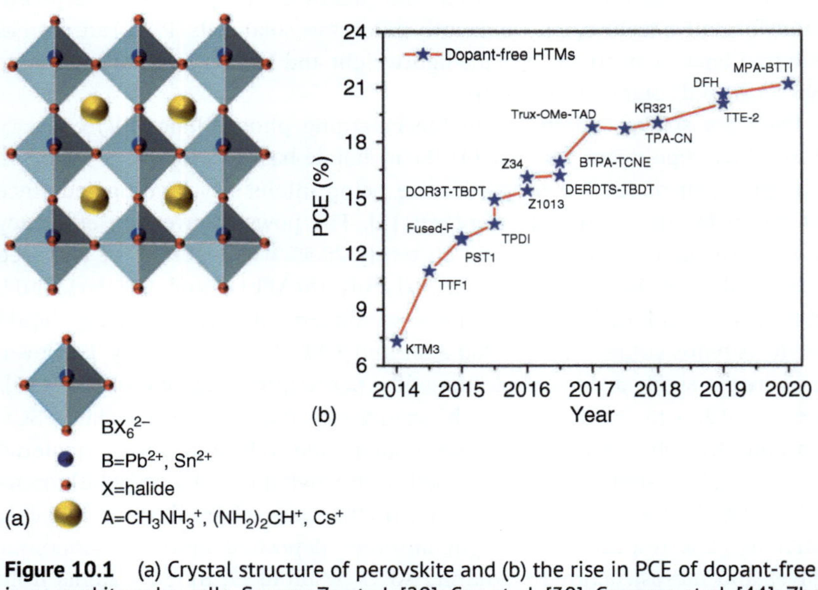

Figure 10.1 (a) Crystal structure of perovskite and (b) the rise in PCE of dopant-free HTMs in perovskite solar cells. Source: Za et al. [29]; Cao et al. [30]; Ganesan et al. [44]; Zhang et al. [53]; Zhang et al. [54]; Qin et al. [55]; Li et al. [59]; Paek et al. [60]; Zhang et al. [66]; Rakstys et al. [68]; Han et al. [73]; Krishnamoorthy et al. [78]; Zheng et al. [79]; Huang et al. [94]; Shen et al. [99]; Wang et al. [100].

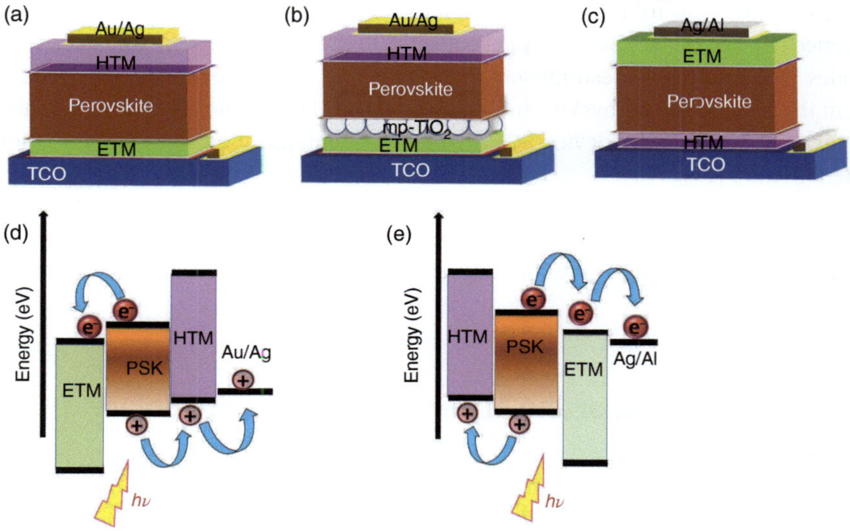

Figure 10.2 Perovskite device structure: (a) planar n–i–p, (b) mesoscopic n–i–p, and (c) inverted planar p–i–n and (d, e) corresponding charge transfer mechanism in PSCs.

intrinsic semiconductor, p-type semiconducting hole-transporting layers (HTLs), and metallic electrode. In a typical PSC, perovskite layer is implemented between an ETL and HTL. Due to ambipolar nature, perovskite can be irradiated through either ETL or HTL side, thus both n–i–p and p–i–n structure can be fabricated [16]. Depending on the morphology of the n-type layer, PSCs can be classified into mesoscopic or planar heterojunction structures as shown in Figure 10.1(a–b).

The three most common PSC architectures are planar (n–i–p) (Figure 10.2a), mesoscopic n–i–p (Figure 10.2b), and inverted p–i–n (Figure 10.2c), where the perovskite is light-absorbing semiconductor, and the (n–i–p) or p–i–n device structure is defined based on the overlaying ETLs and HTLs on the surface of the transparent electrode (Figure 10.2a–c). All three PSCs architecture exhibited competitive PV performance [17–20]. The transparent glass electrode made up of fluorine-doped tin oxide (FTO) or indium doped tin oxide (ITO) is coated with a thin ETL or HTL depending on the architect used. For the conventional mesoscopic n-i-p structure, an additional mesoscopic TiO_2 scaffold is deposited on top of compact layer of ETL, followed by infiltration of perovskite, HTL and metallic electrode (Figure 10.2b). Metallic electrode is usually made of a high work function metal, such as gold, silver, aluminum, or copper. HTLs and ETLs can be either inorganic or organic in nature. PSCs can yield high PCE depending on the charge carrier mobility of both doped and undoped inorganic and organic stacked layers.

10.1.2 Charge Transport in Perovskite Solar Cells and Role of HTM

Figure 10.2d, e depicts the charge transfer mechanism in PSCs. On photoexcitation, the perovskite layer absorbs light and generates electron and hole pairs. The electrons move toward the conduction band of ETL, and holes are transferred to the valence band of HTM, and finally get collected to the respective front and back electrodes [21, 22]. The undesirable charge transfer, i.e. the back transfer of electrons from the ETM to the perovskite and holes from the HTM to the perovskite, can also occur, thus limiting the device PCE by means of enhanced charge recombination phenomenon [23].

Here, we have elaborated the charge transport properties and device performance by employing dopant free small organic molecules as the HTL. p-type dopants used for the conductivity and mobility enhancement of the HTMs are also briefly discussed.

10.2 Hole-Transporting Material for Perovskite Solar Cells

10.2.1 Characteristics of an HTM and Interaction with Perovskite

Organic semiconductors (p-type) as HTM play a vital role in improving the PSC performance [24]. The role of HTM is to extract and transfer the holes from the perovskite layer to the respective electrode. The optical properties of the HTMs affect to the performance of the PSCs [25]. Various transitions such as n–π*, π–π*, or intramolecular charge transfer (ICT) influence the PCE and functions of the PSCs. The most frequently employed HTL in the PSCs is the Spiro-OMeTAD. The well-known UV–vis peak of spiro-OMeTAD, shows two binary peaks at 302 and 398 nm, respectively in undoped form [26]. On p-type doping of HTM, it converts into the oxidized form due to the formation of a radical cation, confirmed by a new absorption band in the visible region between 500 and 800 nm [40] as shown in Figure 10.3a. The oxidized HTMs exhibit improved electrical conductivity and enhance the charge transfer at the perovskite/HTM/Au interface. A strong absorption in the UV range below 400 nm and intense re-emission in the visible and near-infrared range is advantageous as shown in Figure 10.3b [28].

Compare to spiro-OMeTAD, twisted triphenylamine based organic molecules with small core size restricts the conjugation length and thus desired optical property can be obtained [28]. A large stokes shift of organic HTM improve the PV performance by preventing UV induced photo-degradation along with the increased light harvesting efficiency of the perovskite layer in visible and NIR region. Moreover, a large Stokes shift bring about a big change in the molecular orientation or flexibility between ground and excited states. Thus, it can also support infiltration and pore filling of HTM in PSCs by light soaking post-treatment or simple heating [31, 32]. The charge recombination and collection occurs at the perovskite/HTM interface and can be explained by built-in potential (V_{bi}) or interfacial defects or characterized by a lower photoluminescence (PL) intensity of the perovskite. The decrease in higher energetic barrier at the perovskite/HTM interface can increase the electric field strength point toward higher charge extraction from the perovskite layer and minimizes the charge recombination at the interface. If the PL emission

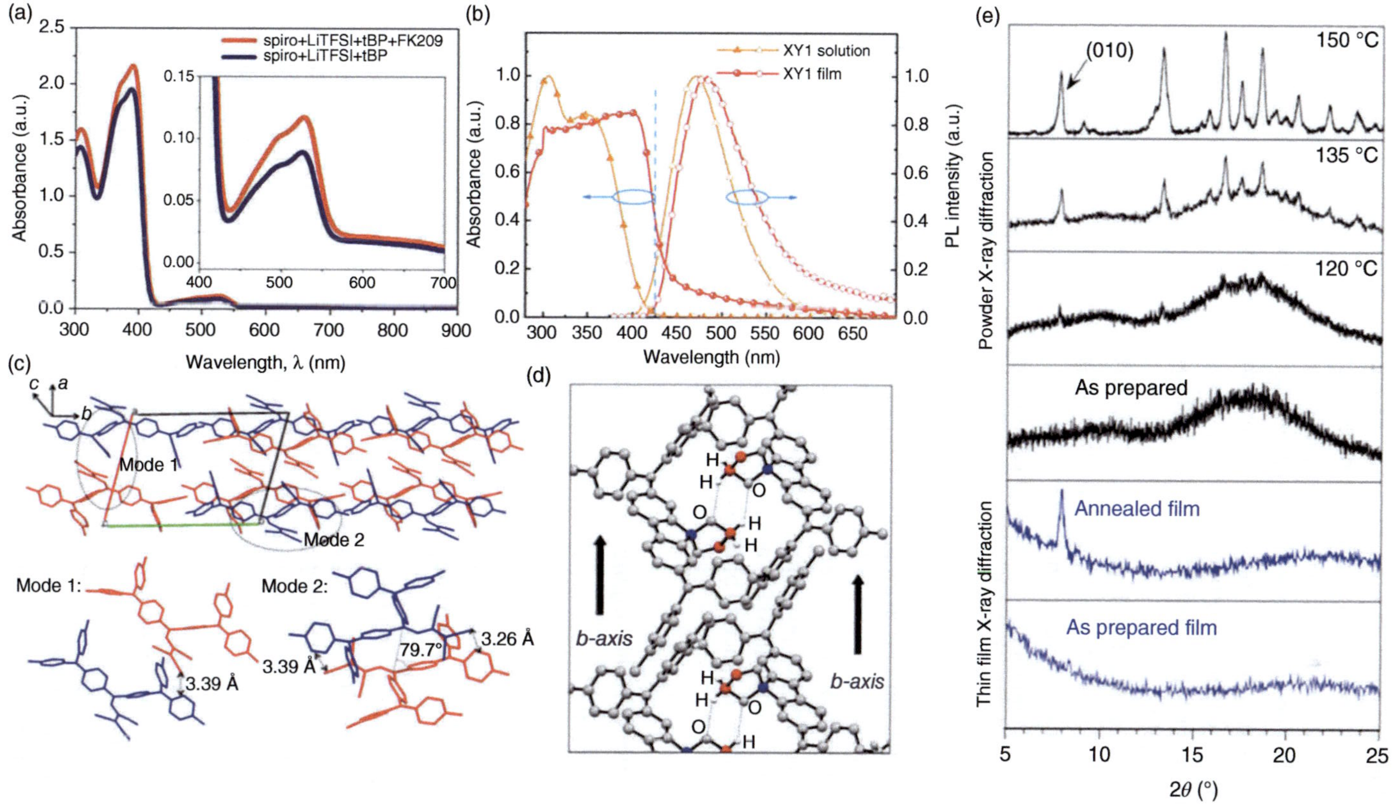

Figure 10.3 (a) UV–vis absorption spectra of spiro-OMeTAD doped with LiTFSi + *t*-BP + FK209 and LiTFSi + *t*-BP. Source: Calio et al. [40]. (b) UV–vis absorption and PL spectra of XY1 in CH_2Cl_2 solution and film. Source: Yang et al. [28]. (c) Two molecular packing modes between different isomer [1, 3] in the crystal structure of BTPA-TCNE viewed along *c*-axis. Source: Li Za et al. [29]. (d) Propensity of CH· · ·O interactions between the DFH molecules derived from the crystallographic data. (e) Powder XRD diffractograms of DFH before and after thermal crystallization (black) and GIXD traces of spin-coated DFH films before and after annealing at 150 °C (blue). Source: Cao et al. [30].

peak of HTM-coated perovskite shifts toward blue wavelength region with respect to the pristine perovskite, it reflects the well-ordered passivation of the trap states at the interface, thereby leading to higher hole injection.

In Figure 10.3c, the intermolecular interaction between neighboring functional groups of HTMs creates a supramolecular geometry with antiparallel molecular packing that helps in charge transport [29]. In some cases, these molecular interactions can also enable amorphous HTM to transform to paracrystalline dimers at a certain temperature that induces the increased hole mobility as shown in Figure 10.3d, e [30].

Thus, an ideal HTM for high-performance PSCs should fulfill basic characteristic such as (i) easy synthetic and purification process; (ii) solubility in orthogonal solvents; (iii) cost-competitiveness; (iv) thermal and photochemical stability; (v) energy level alignment, i.e. the highest occupied molecular orbital (HOMO) slightly higher in energy level than the valence band of the perovskite layer for efficient hole extraction; (vi) lowest unoccupied molecular orbital (LUMO) energy level significantly higher than that of the conduction band of the perovskite absorber; (vii) moisture tolerant, i.e. hydrophobic nature for device stability; (viii) high hole mobility (possibly $> 10^{-3}$ cm^2 V^{-1} s^{-1}) to facilitate hole conduction and prevent charge recombination; (ix) lower tendency towards crystallization to avoid phase transition during device processing and operation, (x) low light absorption in the visible spectrum; and (xi) HTM acting as a barrier between perovskite layer and cathode (top metal contact) to block electrons.

10.2.2 Nature of HTM: Organometallic, Inorganic, and Organic (Small Molecules and Polymers)

Arguably, inorganic materials are expected to be stable in terms of high temperature and moisture tolerance. A variety of inorganic HTMs have been reported for PSCs, however few of them have shown good PV performance, which includes nickel oxide (NiO_x), copper chromium oxide ($CuCrO_2$), copper gallium oxide ($CuGaO_2$), and copper thiocyanate (CuSCN) [23, 24]. The use of organometallic phthalocyanines or porphyrinoids as the HTM was also employed to fabricate efficient and stable PSCs. These metal–organic complexes with 18π electrons equip themselves to act as semiconductor by attaining properties such as high charge mobility and excellent thermal and chemical stabilities. Cu(II)-, Ni(II)-, and Zn(II)-based phthalocyanines have recently attracted considerable attention in PSCs [23, 24] and exhibited good performance. However, most of the phthalocyanines have poor solubility in common organic solvents, and thus thermal evaporation methods are preferred.

Depending on the molecular weight, organic HTMs can be divided into two subclasses: small molecules and polymers. Organic p-type semiconducting polymers used in opto-electrical devices are also suitable as HTM in PSCs. Among them, poly(3-hexylthiophene-2,5-diyl) P3HT, poly(triarylamine) PTAA, and poly(2,3-dihydrothieno-1,4-dioxin)-poly(styrenesulfonate) PEDOT:PSS are more commonly reported. The important advantages of these polymers are in terms of scale-up synthesis, high thermomechanical stability, and high intrinsic hole mobility. However, in some cases, there are mismatch of band alignment with perovskite layer that suppresses the efficient charge extraction and transfer and presence of defects due to long chain, which may lead to significant loss of open-circuit voltage and reproducibility in device. The frequently used HTM employed in PSCs is spiro-OMeTAD as shown in Figure 10.4a. However, it suffers from high cost, originated due to challenging synthesis and purification steps. Moreover, it needs dopants and additives to improve its electrical properties for performance enhancement. Several dopants have been reported for spiro-OMeTAD and other HTM (Figure 10.4b).

10.2.3 Doping of Hole-Transporting Materials in PSCs

To increase the carrier concentration of the organic HTMs, dopant and additives are being used [33], to improve the performance of optoelectronic devices. The basic principle of doping is similar to inorganic semiconducting materials, generating a number of free charge carriers by the addition of electron acceptor or donor molecules and tuning the electronic properties. These external acceptor or electron donor molecules are known as p-type (n-type) dopants will improve the electrical properties by the removal or addition of electron from the matrix of host organic semiconductor. Chemical p-type doping of HTM can be defined as electron transfer from the HOMO of a matrix of host HTM to the LUMO of a dopant molecule, thus leaving behind a mobile hole in the HOMO of the organic semiconductor. Thus, the p-doping in organic semiconductor-based HTM can take place only when the electron affinity of the dopant is equivalent or larger than ionization potential of the host matrix (Figure 10.4c). As a result of p-type doping, electrical conductivity increases due to the generation of free charge carrier concentration, and this shifts the Fermi level toward HOMO of the host HTM and minimizes the energy level mismatch [34] to reduce Ohmic losses [35, 36]. Additionally, doping of organic semiconductors facilitates the charge carrier injection from electrodes to the doped molecules. Arguably, in PSCs, chemical doping of organic HTMs in a controlled amount tunes the energy level alignment at interfaces [31], which facilitates efficient charge carrier extraction and device voltage.

Various p-type dopants for the HTM have been developed and employed in perovskite solar cells. The most commonly used dopant includes 2,3,5,6-tetra-fluoro-7,7, 8,8-tetracyanoquinodimethane (F4TCNQ), and cobalt(III) complexes [26, 37, 38]. Simultaneous addition of cobalt complex-based tris(2-(1*H*-pyrazol-1-yl)-4-*tert*-butylpyridine)cobalt(III) tri[bis(trifluoromethane) sulfonimide] (FK209) and lithium bis(tri-fluoromethanesulfonyl)imide (LiTFSI) with additives of 4-tert-butylpyridine (*t*-BP) to spiro-OMeTAD improves the electrical conductivity significantly and increases the V_{oc} [38]. LiTFSI favors the oxidation, and *t*-BP additive prevents the aggregation of LiTFSI and improves its distribution, thus providing a better interface between the HTM and the perovskite layer. Recently, zinc bis(tri-fluoromethanesulfonyl)imide ($Zn(TFSI)_2$) has been used as a dopant that increases hole mobility and stability of the HTM [39]. The ionic dopants

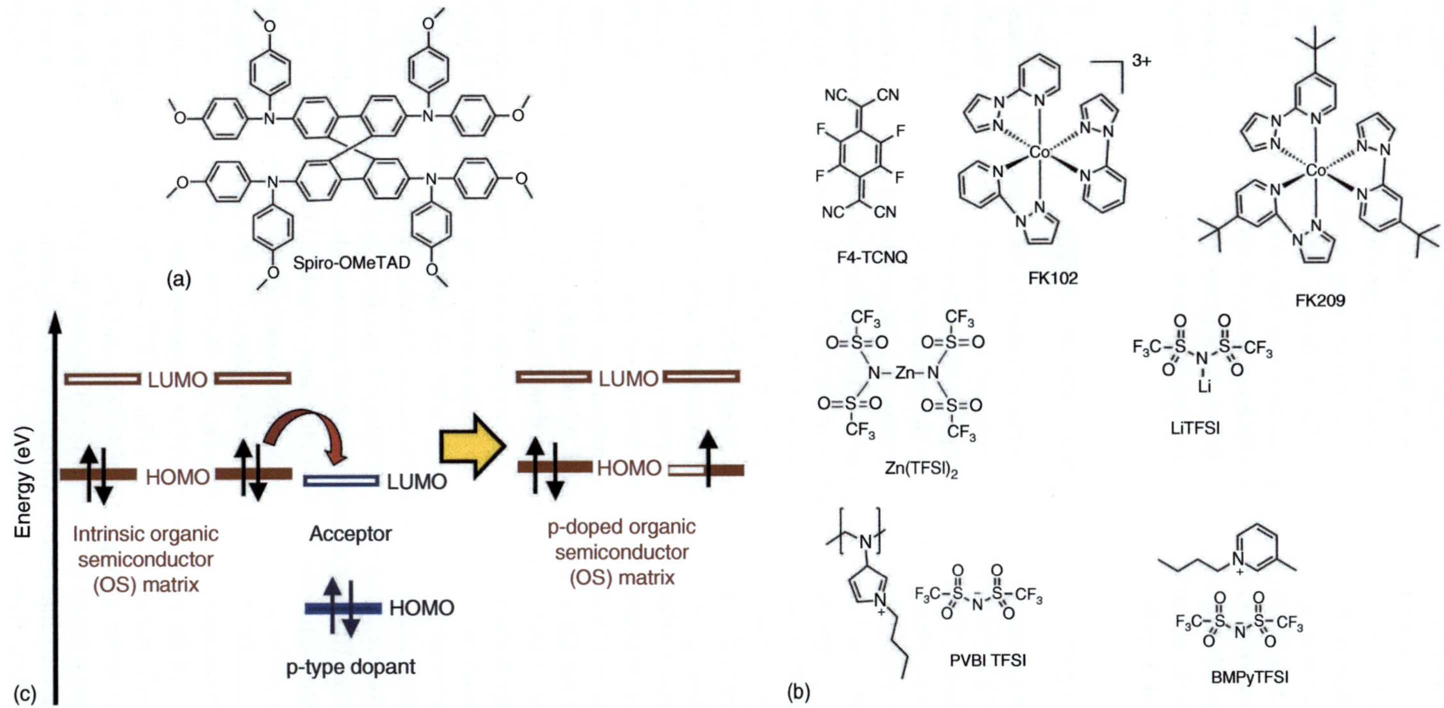

Figure 10.4 Structure of (a) spiro-OMeTAD and (b) p-type dopants in HTMs. (c) Schematic energy level diagram of an organic semiconductor's matrix: p-dopant molecule representing of molecular p-doping in organic semiconductors. Usually, electron transfer from the HOMO of host matrix molecule to the LUMO of a dopant molecule assumes p-type doping. Source: Walzer et al. [34].

are hygroscopic in nature and will induce instability to the devices through moisture absorption and ion migration that can accelerate the degradation of the perovskite layer. Furthermore, if the molecular dopants are not covalently linked to the organic HTMs, they may also diffuse. To overcome this, ionic liquid 1-butyl-4-methylpyridiniumbis(trifluoromethanesulfonyl)imide [BMPyTFSI] was reported as hydrophobic dopants to substitute cobalt complex, LiTFSI, and *t*-BP additive [40]. It increases the conductivity of spiro-OMeTAD and reduces charge recombination at optimized concentration and demonstrated on par PCE, but with extended lifetime. Another ionic liquid PVBI–TFSI was reported to increase spiro-OMeTAD conductivity by four orders of magnitude and showed reduced hysteresis as compared with conventional dopants [41].

However, the development of dopant-free HTMs is paramount to fabricate stable PSCs. Currently, the maximum PCEs over 21% with dopant-free HTM-based device have been reported. By eliminating the use of dopants, device stability can be extended. Several strategies have been used to design efficient dopant-free HTMs based on different core structures to address the issues related to the intrinsic and extrinsic stability of the PSCs. In this direction, planar, rigid, and fully conjugated molecule, with suitable energy levels and larger π-conjugated systems, are being developed.

The advantages and synthetic approaches of small molecules as the dopant-free HTMs will be taken into account. The energetic level of the HTMs with respect to the HOMO level of the perovskite is depicted in Figure 10.5a. It shows the ideal features of the doped (case i) and undoped HTMs (case ii) for an improved charge carrier mobility and device performance [27].

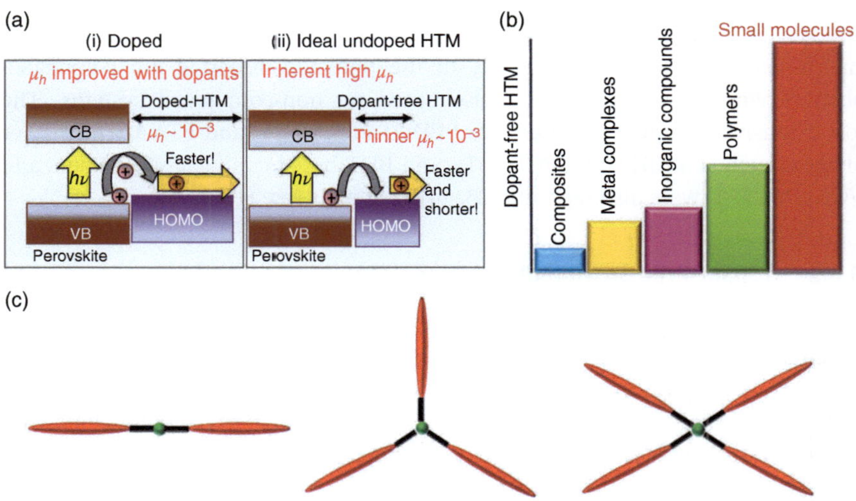

Figure 10.5 (a) Different conditions for the HTM in a perovskite solar cell. Source: Kim et al. [27]. (b) Bar representation of dopants free HTMs developed to date. (c) Different conformations of organic molecular core: linear structure, star-shaped structure, and spiro structure.

10.3 Dopant-Free Organic HTMs for Perovskite Solar Cells

Dopant-free HTMs in some cases demonstrate properties that supersede the use of spiro-OMeTAD. A majority of HTMs reported without the use of dopants are based on small organic molecules or polymers (Figure 10.5b). Arguably, the molecular design of a dopant-free HTMs should be embodied with π-conjugated structures and specific functional groups having electron-donating and electron-accepting properties. These HTMs can be synthesized using different cross-coupling reaction processes such as Buchwald–Hartwig coupling, Suzuki–Miyuara, Stille cross-coupling, Sonogashira coupling, Heck reactions, BuLi treatment, or Schiff base reactions. The small organic molecules according to the structural characteristics of their core group and functional groups can be in linear, star-shaped, or spiro-type structure as shown in Figure 10.5c. If the molecules display a planar and rigid geometry accompanied by a highly π-conjugated molecular structure, exhibiting π–π stacking, deep-lying HOMO energy level, and donor–acceptor conjugated structure, it can facilitate the higher charge transport behavior, thereby improving the hole mobility and conductivity. The alkoxy groups (–OR) in the HTM structure are responsible for adjusting the HOMO level of the material and play a vital role in anchoring the material onto the underlying perovskite layer and improve its solubility. The ease of synthesis and purification and moreover, their ability to form a uniform film followed by reduced pinholes favor their employment.

10.3.1 Dopant-Free Organic Polymer As HTM

Polymers are mainly of two types: homopolymer and donor–acceptor copolymer. As an HTM, these should also possesses the basic requirements as of small-molecule-based HTM. Majority of the polymers reported are conjugated polymers while few studies were also based on non-conjugated systems. The most extensively studied polymers are PEDOT:PSS, P3HT, and PTAA while other polymers such as PCBTDPP-, PCDTBT-, and BDT-based conjugated polymers and PVCz-OMeDAD were also reported [23, 24] as dopant-free HTM.

10.3.2 Dopant-Free Small Molecules as HTM

To provide easy grasp of the properties, we divided these in parts depending on their molecular core and structure. The chemical structures of the representative small molecules as dopant-free HTMs for PSCs with the PCE of devices shown in the parenthesis are depicted in Figures 10.6–10.8 and the summary of Photovoltaics and electrical properties is shown in Tables 10.1 and 10.2.

10.3.2.1 Triarylamine-Based HTM

Triarylamine-based dopant-free HTMs have been extensively studied in different optoelectronic devices due to their efficient hole mobilities and conductivities. The facile synthesis and purification route along with their opto-electrical properties

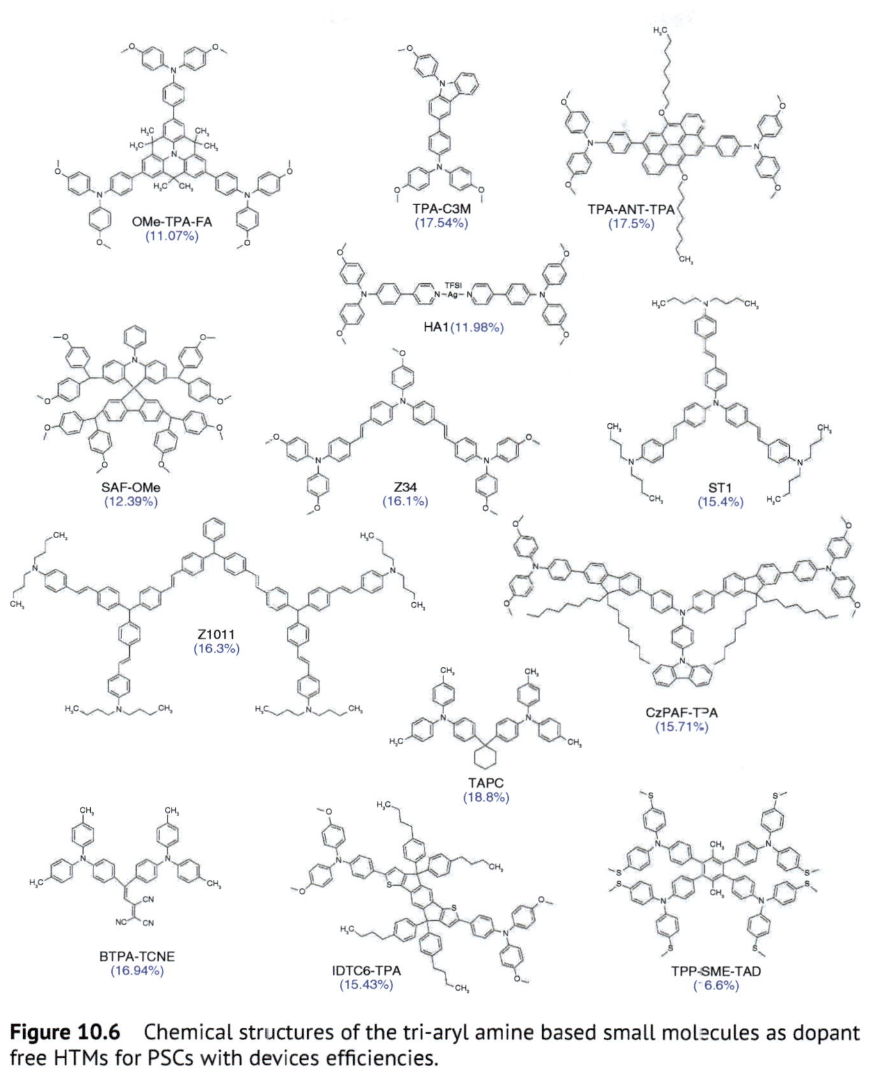

Figure 10.6 Chemical structures of the tri-aryl amine based small molecules as dopant free HTMs for PSCs with devices efficiencies.

validate its application in PSCs. Also, they possess low-lying HOMO level that is appropriately aligned with the perovskite absorbers, along with effective hole extraction. Triarylamine-based compounds as a building blocks or side groups in the HTMs have been reported in the past. spiro-OMeTAD itself has the triphenylamine group and shows ability to transport positive charge. The non-planar geometry of the molecule in the solid state influences their opto-electrical properties.

Franckevičius et al. reported triarylamine-based **HTM1** that showed redshifted absorption in both solution and thin films as compared to spiro-OMeTAD and achieved a PCE of 13.4% [42]. Ma et al. achieved a PCE of 10.4% with an open-circuit voltage (V_{oc}) of 930 mV with spiro-thiophene **(SCPDT-BiT)**-based molecule as

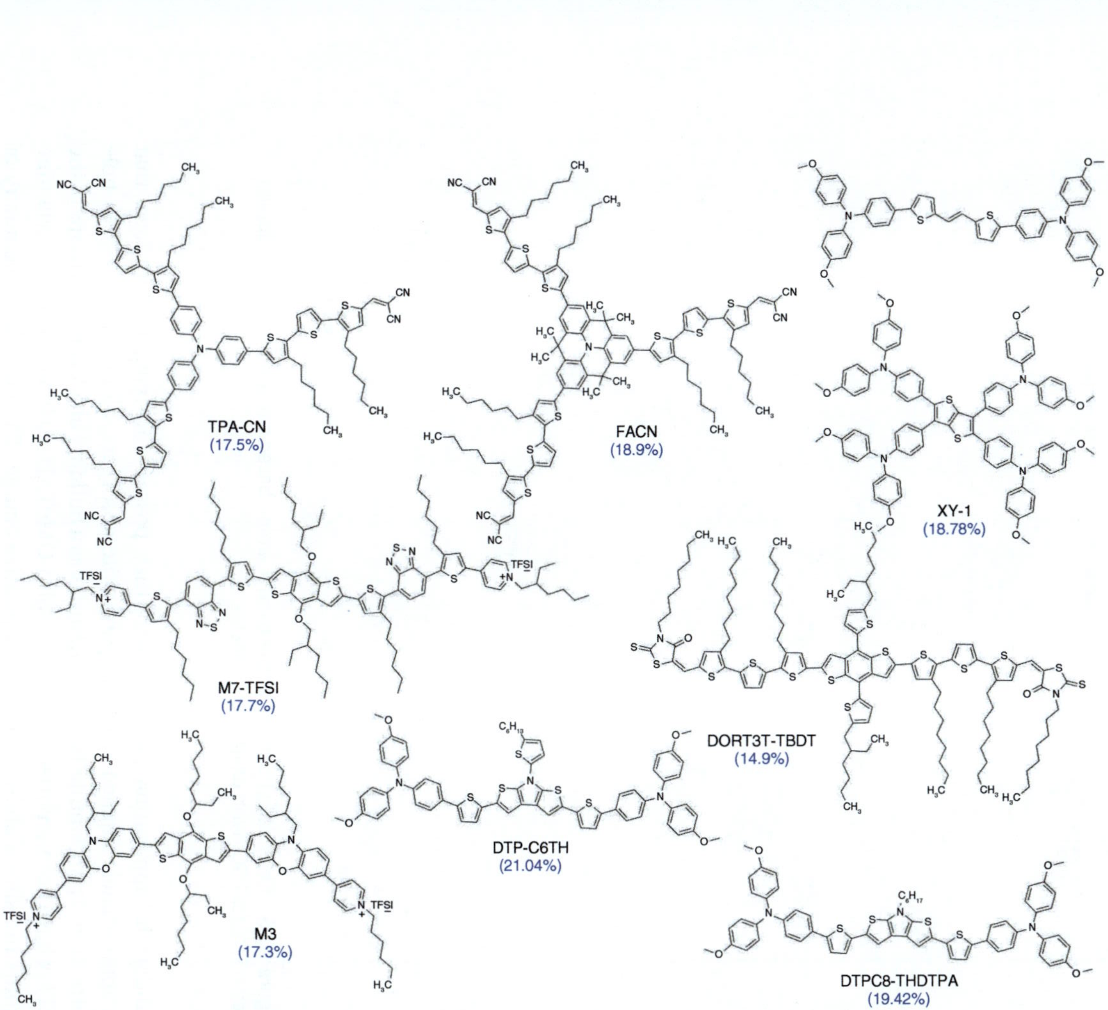

Figure 10.7 Chemical structures of the thiophene based small molecules as dopant free HTMs for PSCs with devices efficiencies.

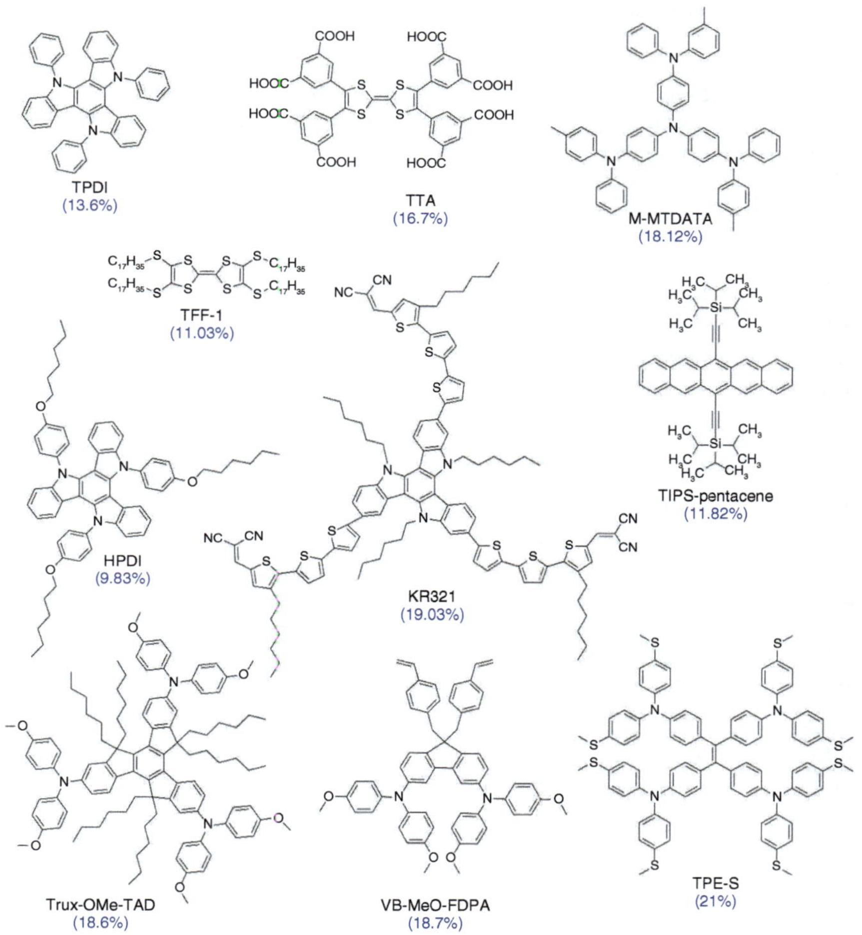

Figure 10.8 Chemical structures of the different core based small molecules as dopant free HTMs for PSCs with devices efficiencies.

dopant-free HTMs [43]. These are found to be promising due to the presence of the sulfur atoms as electron transfer group. Ganesan et al. employed another HTM **PST1** in PSCs that yielded a PCE of 12.74% for PSCs with J_{sc} of 19.21 mA cm^{-2}, V_{oc} of 937 mV, and FF of 69% [44]. They showed multiple CH–π, O–C, and π–π intermolecular interactions in the crystal structure of the molecule that helps in the better charge transport. Choi et al. developed **OMeTPA-FA** that gave an average PCE of 11.07% [45]. Xiao et al. reported easy to synthesize **PNBA** for PSCs with PCE of 11.4% [46]. *N*,*N*′-Di(3-methylphenyl)-*N*,*N*′-diphenyl-4,4′-diaminobiphenyl (**TPD**) as dopant-free HTM in PSCs was reported, but with lower PCE of 3.2%. However, when used as a buffer layer of **HAT-CN**, PCE increases to 7.1% [47]. Lv et al. introduced the **HTMs** [3–6] and achieved a maximum PCE of 12% [48, 49]. **SAF-OMe** based on spiro and TPA is synthesized and a PCE of 12.39%

Table 10.1 A summary of photovoltaics and electrical properties of dopant free HTMs for n–i–p configuration.

HTMs	Device configuration	V_{oc} (V)	J_{sc} (mA cm^{-2})	FF (%)	PCE (%)	Hole mobility (cm^2 V^{-1}s^{-1})	References
HTM1	Mesoporous	0.97	19.3	72	13.4	3.0×10^{-5}	[42]
SCPDT-BiT	Mesoporous	0.94	16.54	67	10.39	4.5×10^{-6}	[43]
PST1	Mesoporous	0.93	19.21	69	12.74	—	[44]
OMeTPA-FA	Mesoporous	0.90	19.42	63	11.07	3.67×10^{-4}	[45]
PNBA	Mesoporous	0.94	17.5	68	11.4	4.92×10^{-4}	[46]
TPD	Mesoporous	0.92	13.1	59	7.1	3.5×10^{-5}	[47]
HTM3	Mesoporous	0.92	18.1	68	11.34	2.98×10^{-3}	[48]
HTM4	Mesoporous	0.94	17.9	69	11.63	1.27×10^{-3}	[48]
HTM5	Mesoporous	0.93	15.4	63	9.0	5.66×10^{-4}	[49]
HTM6	Mesoporous	0.93	18.4	70	12.0	7.23×10^{-4}	[49]
TPA-MeOPh	Mesoporous	0.94	15.01	52	7.45	1.06×10^{-4}	[50]
FA-OMePh	Mesoporous	0.90	16.97	59	9.18	4.92×10^{-4}	[50]
ST1	Mesoporous	1.05	21.07	66	15.4	4.57×10^{-4}	[51]
HA1	Mesoporous	0.92	17.28	76	11.98	6.49×10^{-4}	[52]
HA2	Mesoporous	0.91	15.12	78	10.79	8.38×10^{-4}	[52]
Z1011	Mesoporous	1.09	20.52	70	16.3	8.49×10^{-4}	[53]
Z1013	Mesoporous	1.02	21.33	70.2	15.4	6.67×10^{-4}	[53]
Z34	Mesoporous	1.05	21.24	70	16.1	7.46×10^{-4}	[54]
Fused-F	Mesoporous	1.03	17.9	68	12.83	$(3–6)\times10^{-5}$	[55]
SGT-405 (2,7)	Mesoporous	1.05	22.32	76.32	18.00	1.05×10^{-4}	[56]
SGT-405 (3,6)	Mesoporous	1.04	22.92	78.65	18.87	1.80×10^{-4}	[56]
SGT-410 (3,6)	Mesoporous	1.04	20.57	75.25	16.20	5.27×10^{-5}	[56]
SGT-411 (3,6)	Mesoporous	1.03	21.19	75.22	16.50	8.74×10^{-5}	[56]
CMT	Mesoporous	1.03	21.82	58.23	13.05	2.6×10^{-5}	[57]
H101	Mesoporous	0.97	18.9	57	10.6	—	[58]
KTM3	Mesoporous	0.99	10.3	70.9	7.3	—	[59]
FA-CN	Mesoporous	1.13	20.85	77	18.9	1.20×10^{-4}	[60]
TPA-CN	Mesoporous	1.09	20.85	77	17.5	1.1×10^{-4}	[60]
DR3TBDTT	Mesoporous	0.95	15.3	60	8.8	—	[61]
M7-TFSI	Mesoporous	1.09	22.84	70.6	17.7	3.4×10^{-4}	[62]
TPA-ANT-TPA	Mesoporous	1.03	21.05	79.6	17.5	2.6×10^{-4}	[63]
M3	Mesoporous	1.08	20.7	69.7	14.8	2.67×10^{-4}	[64]
M4	Mesoporous	0.98	19.7	68.6	12.4	1.29×10^{-4}	[64]
HMDI	Mesoporous	0.93	14.43	63.7	8.62	—	[65]
HPDI	Mesoporous	0.89	16.51	66	9.83	—	[65]

(Continued)

Table 10.1 (Continued)

HTMs	Device configuration	V_{oc} (V)	J_{sc} (mA cm^{-2})	FF (%)	PCE (%)	Hole mobility (cm^2 V^{-1}s^{-1})	References
TPDI	Mesoporous	1.03	18.7	70.8	13.6	3.5×10^{-3}	[66]
TBDI	Mesoporous	1.07	16.2	42	7.3	5.95×10^{-3}	[67]
KR321	Mesoporous	1.13	21.70	78	19.03	2.6×10^{-4}	[68]
TIPS-Pentacene	Mesoporous	0.91	20.84	11.51	11.07	—	[69]
HTM-7	Mesoporous	0.99	16.4	65	10.5	7.3×10^{-5}	[70]
HTM-8	Mesoporous	0.90	15.2	68	9.5	6.6×10^{-5}	[70]
HTM-9	Mesoporous	0.95	16.6	72	11.4	4.2×10^{-5}	[71]
HTM10	Mesoporous	0.89	16.3	70	10.03	3.5×10^{-5}	[71]
HTM11	Mesoporous	1.02	16.1	44	7.3	—	[72]
TTF-1	Mesoporous	0.86	19.9	64	11.03	—	[73]

was measured [75]. In another work, Choi et al. introduced **TPA-MeOPh and FA-OMePh** without the use of any additive, achieving PCEs of 7.45% and 9.18%, respectively [50]. Zhao et al. reported a one-step synthesis for a star-shaped **HTM ST1** employed in PSCs without using any dopants and measured a PCE of 15.4% along with long-term air stability [51]. A metal–organic-based HTM (**HA1 and HA2)** was developed by Hua and group that led to high conductivities, and PCE of 11.98% and 10.79% with an extraordinary FF of 76% and 78%, respectively, was noted [52]. Star-shaped HTMs **Z1011 and Z1013** were synthesized by Zhang et al., and its performance in PSCs with an average PCE of 12.4%, 15.4%, and 16.3% were reported [53, 54]. These star-shaped molecules have long branched arms and longer π-conjugated bonds. Qin et al. have designed a low band-gap HTM with triarylamine core named **fused-F** for PSC and reported a PCE of 12.8% [55]. **DFTAB and TAE** introduced by Labban et al. were promising due to easy synthetic route and low cost and were also found to have positive effect on the long-term stability of the PSCs [91]. Triarylamine-based small molecules **NTPA and BTPA** were employed as HTMs in PSC with hole mobility of 10^{-3} cm^2 V^{-1} s^{-1} [92]. Li et al. introduced a donor–acceptor-type triphenylamine-based **BTPA-TCNE** that yielded 17% PCE in n–i–p type PSCs [29]. Furthermore, they reported another **IDTT-TPA** with PCE of as high as 15.7% [76]. The group developed two new IDT-based compounds, **IDTC4-TPA** and **IDTC6- TPA** [77], and the π–π interactions in the molecular arrangement of **IDTC6** led to higher hole mobility of 10^{-4} cm^2 V^{-1} s^{-1} and gave a PCE of 15.43%. Chen et al. developed two HTMs, **TPP-OMeTAD** and **TPP-SMeTAD**, and the best p–i–n-type PSCs measured a PCE of 14.6% and 16.6%, respectively. The lower HOMO level of these HTM were favorable and aligned to the perovskite, which helped to extract hole efficiently [81]. Paek et al. developed **FA-CN and TPA-CN** using star-shaped D–π–A structure as stable (remarkable stability up to 500 hours) PSCs with a PCE of 18.9% and 17.55%, respectively [60]. Three other

Table 10.2 A summary of photovoltaics and electrical properties of dopant free HTMs for p–i–n configuration.

HTMs	Device configuration	V_{oc} (V)	J_{sc} (mA cm^{-2})	FF (%)	PCE (%)	Hole mobility (cm^2 V^{-1} s^{-1})	References
M1	n–i–p	1.01	19.1	67.8	13.2	2.71×10^{-4}	[74]
SAF-OMe	n–i–p	0.97	20.59	62	12.39	9.9×10^{-4}	[75]
BTPA-TCNE	n–i–p	1.04	20.84	78.2	16.94	3.14×10^{-5}	[29]
IDTT-TPA	n–i–p	1.03	21.55	67.8	15.1	6.46×10^{-4}	[76]
IDTC4-TPA	n–i–p	0.98	21.05	70.01	14.45	1.05×10^{-4}	[77]
IDTC6-TPA	n–i–p	1.00	21.36	72.45	15.43	4.26×10^{-4}	[77]
DERDTS-TBDT	n–i–p	1.02	21.2	72.8	16.2	1.0×10^{-4}	[78]
DOR3T-TBDT	n–i–p	0.97	20.7	74	14.9	—	[79]
BDT-C1	n–i–p	0.98	19.6	71.2	13.4	3.8×10^{-4}	[80]
TPP-OMeTAD	n–i–p	1.00	18.56	79	14.6	—	[81]
TPP-SMeTAD	n–i–p	1.07	20.15	77	16.6	—	[81]
DTP-C6Th	n–i–p	1.15	22.76	79.9	21.04	4.18×10^{-4}	[82]
TTA	n–i–p	1.04	20.4	78	16.7	9.68×10^{-4}	[83]
TPACOM	p–i–n	0.91	20.73	74	13.92	1.2×10^{-5}	[84]
TPAC2M	p–i–n	0.99	22.58	71	15.77	1.0×10^{-5}	[84]
TPAC3M	p–i–n	1.00	22.79	78	17.54	1.1×10^{-5}	[84]
TPA-TVT-TPA	p–i–n	1.07	21.49	71	16.32	2.1×10^{-5}	[85]
TPA-NAP-TPA	p–i–n	0.96	21.55	71	14.63	5.1×10^{-5}	[85]
CzPAF-TPA	p–i–n	1.05	20.00	74.58	15.71	3.3×10^{-4}	[86]
TAPC	p–i–n	1.04	22.32	81.15	18.8	4.99×10^{-4}	[87]
DFH	p–i–n	1.10	22.10	82.9	20.2	1.0×10^{-3}	[30]
TPE-S	p–i–n	1.26	15.6	78.5	15.4	1.25×10^{-5}	[88]
VB-MeO-FDPA	p–i–n	1.15	20.89	77.8	18.70	3.0×10^{-4}	[89]
D106	p–i–n	1.05	22.32	77.8	18.24	1.65×10^{-4}	[90]
DFTAB	p–i–n	1.07	17.5	69	12.4	2.4×10^{-5}	[91]
TAE	p–i–n	1.0	16.7	66	11.5	1.6×10^{-5}	[91]
NTPA	p–i–n	1.02	17.4	73.3	13	1.1×10^{-3}	[92]
BTPA	p–i–n	0.98	17.4	70.8	12.1	4.7×10^{-4}	[92]
Trux-1	p–i–n	0.98	17.7	63	10.2	—	[93]
Trux-2	p–i–n	0.98	19.7	69	13.4	—	[93]
Trux-OMeTAD	p–i–n	1.02	23.2	79	18.6	2.3×10^{-3}	[94]

HTMs called **TPAC0M, TPAC2M, and TPAC3M** with varying number of methoxy groups in the structures were also reported [84]. Triarylamine-based HTMs termed as **TPA-TVT-TPA and TPA-NAP-TPA** were developed by Pham et al. and when employed in inverted PSCs yielded a PCE of 16.32% for **TPA-TVT-TPA** and 14.63% for **TPA-NAP-TPA** [85]. Furthermore, the authors reported humidity-controlled PSCs based on **TPA-ANT-TPA** in a mesoscopic PSCs with a PCE of 17.5% [63]. Reddy et al. employed **CzPAF-TPA** as HTM and measured a PCE of 15.71% [86]. Yang et al. introduced another π-conjugated molecule **TAPC** and achieved a PCE of 18.8% [87]. They explained the increase in PCE is due to thermal annealing that could improve the surface crystallinity, thereby enhancing the hole conductivity. The use of **DFH** was reported, and remarkable PCE of 20.6% in an inverted (p–i–n) PSCs was measured (Figure 10.9a, b) [30]. The authors noted that annealing of the HTM at a certain temperature induced a preferential molecular organization normal to the substrate and improved optical properties at 135 °C (Figure 10.9c, d). It was also explained that DFH could trigger the intermolecular C–H···O and

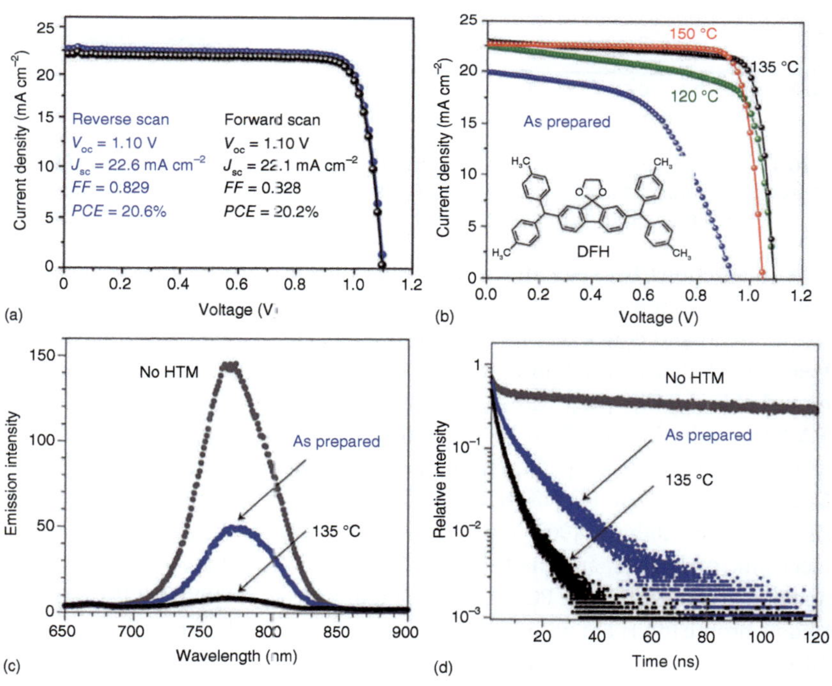

Figure 10.9 (a) J–V curves of PSCs based on DFH as hole transport layer, (b) J–V curves of PSCs based on DFH as hole transport layer at variable temperature, (c) steady-state fluorescence quenching of perovskite thin films by the underlying DFH layers, and (d) fluorescence decay kinetics of PMMA-encapsulated perovskite thin films on different DFH thin films illuminated by a 553 nm laser. Source: Cao et al. [30].

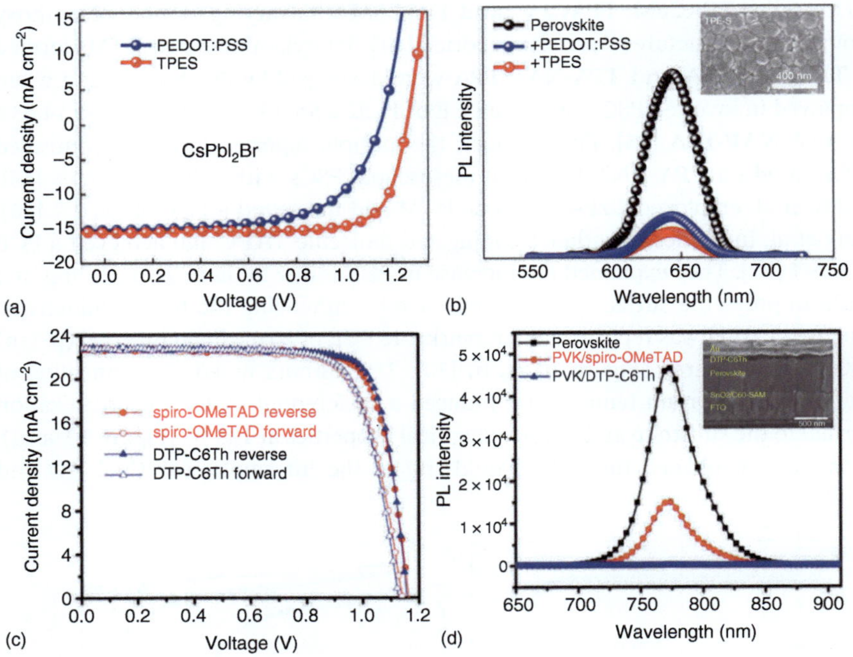

Figure 10.10 *J–V* curves and steady-state PL graphs of PSCs and thin films. (a, b) **TPE-S** as dopant-free hole transport layers. (c, d) **DTP-C6Th** as dopant free hole transport layers. Source: (a, b) Jiang et al. [88], (c, d) Yin et al. [57].

C–H···π interactions between the neighboring dioxane groups that help in the transformation of amorphous DFH to polycrystalline dimers when heated above T_g, 120 °C. Shao and coworkers reported a fluorene core based cross-linked HTM **VB-MeO-FDPA** that showed an excellent PCE of 18.7% with negligible hysteresis and remarkable stability [89]. Wang et al. developed a long alkyl chain-based HTM **CZTPA-2** that exhibited a PCE of 11.79% with good stability [95].

Xue and coworkers developed three pyridine core-based HTM, and among them **D106** showed an excellent PCE of 18.24% while the hydrophobic nature of the HTMs contributed to the stability [90]. Yip et al. reported **TPE-S** that showed excellent optoelectronic properties, and the HTM in all inorganic-based perovskite $CsPbI_2Br$ gave long-term stability with a PCE of 15.4% (Figure 10.10a, b) [88]. **DTP-C6Th** was reported as promising HTM with an improved electro-optical merits, hole mobility of 4.18×10^{-4} $cm^2V^{-1}s^{-1}$ and a champion PCE of 21.04% with long term stability (Figure 10.10c,d) [57].

10.3.2.2 Carbazole-Based HTMs

(2-Ethylhexyl)-9*H*-carbazole is used as core, and *N,N′*-di-*p*-methylthiophenylamine **CMT** as end groups with PCE of 13.05% (V_{oc} of 1030 mV and J_{sc} of 21.82%) [56]. They introduced methyl thiol groups in place of methoxy groups at the para position of the diphenylamine to lower the HOMO level of the molecule. Lu et al. reported three carbazole-based additive-free HTMs, namely, **SGT-405**, **SGT-410**, and **SGT-411** [96]. The substitution position of the carbazole moiety in three HTMs are altered accordingly from 2,7 to 3,6 positions, and the enhancement in the hole mobility was noted in the substituted molecule as compared with the parent SGT-405 and spiro-OMeTAD. **LD29** showed competitive performance with PCE of 14.29%, together with an V_{oc} of 1020 mV, J_{sc} of 19.60 mA cm^{-2}, and fill factor of 0.71 [97]. Carbazole-based HTMs **PhCz-4MeOTPA** and **BDT-4MeOTPA** with suitable energy levels, high glass transition temperatures, and high hole mobilities, having PCEs of 8.78% and 16.04% (V_{oc} of 1060 and 1080 mV; J_{sc} of 20.16 and 21.52 mA cm^{-2}, respectively), were reported [58].

10.3.2.3 Thiophene-Based HTMs

Thiophene or polythiophene family are widely used as HTMs and are being investigated in PSCs. Li et al. reported an electron-rich small molecule **H101** for PSCs [59]. They obtained an average PCE of 10.6% without the use of any dopant under AM1.5G solar simulation. Another molecule developed was **KTM3**, a novel 3,3′-thiophene, which gave a PCE of 7.3% [78]. **DERDTS-TBDT** as dopant-free HTM has been developed with a competitive PCE of 16.2% in a conventional planar structure [61]. Zheng et al. developed a highly hydrophobic oligothiophene derivative named **DR3TBDTT** with the donor–acceptor–donor-type molecule to achieve a PCE of 8% [79]. However, the authors used poly(dimethylsiloxane) (PDMS) to achieve the desired hydrophobicity and the PCE remains unchanged after storing in 20% humidity. **DOR3T-TBDT**, another linear donor–acceptor-conjugated small molecule, yielded a PCE of 14.9% in a planar PSCs [80]. Chen et al. introduced a novel small-molecule HTMs with benzodithiophene as a core unit, and the PSCs fabricated with **BDT-C1** showed an average PCE of 13.4% [62]. Various HTMs representing the acceptor–donor–acceptor-conjugated molecule such as **M7-TFSI** and **M1** were reported [64, 74]. Such molecules shared benzodithiophene as the donor in their respective cores, and average PCEs of 17.4% and 13.2%, respectively, were measured. Another acceptor–donor–acceptor-conjugated small molecules M3 and M4 have been designed and employed in PSCs. The PSCs based on **M3** and **M4** with the structure FTO/compact TiO_2/mesoporous $TiO_2/(FAPbI_3)_{0.85}(MAPbBr_3)_{0.15}$/HTM/Au exhibited PCE of 14.8% and 12.3%, respectively [98]. However, the incorporation of $PC_{70}BM$ with **M3 or M4** to form a bulk heterojunction (BHJ) has facilitated exciton dissociation, which allowed to improve PCE of 16.2% and 15.0% without any additives. Liu et al. developed a π-conjugated thiophene-based molecule **IDTT** extended with strong electron-donating TPA groups that unveiled the strong π–π interactions yielding a hole mobility of 10^{-4} cm^2 V^{-1} s^{-1} and PCE of 15.7%. Yun et al. reported another series of thiophene-based molecule with different fluorine substitution that achieved the maximum PCE of 14.45%. A PCE of 17.4% was reported with acceptor–donor–acceptor (A–D–A)-conjugated ionic molecule **M7-TFSI** system in n–i–p-type PSCs. In both the cases, HTMs showed excellent higher hole mobility and conductivity without any additives, and high PCE was attributed to a deep HOMO energy level of the molecule. Star-shaped small molecules represent another notable building block that was used in opto-electrical devices. They have been widely used in organic light-emitting diodes (OLEDs) as they possess the face-on stacking on the surface, flat structure, and extended delocalized π-system that allows further substitution to tune their opto-electrical properties. In a recent report, Zhu et al. reported two dopant-free **TTE 1** and **TTE 2** PSCs exhibiting a significant PCE of over 20% along with long-term stability, achieving to date the best performances in a planar n–i–p-type PSCs (Figure 10.11a–g) [99]. Recently, two donor–acceptor–donor-based additive-free HTMs **MPA-BTI** and **MPA-BTTI** in an inverted planar system measured the highest PCE of 21.17% with reproducibility and long-term stability (Figure 10.12a–d) [100]. Gao and coworkers reported benzodithiophene-based copolymer as HTM **PBDT[2F]T** yielding a PCE of 17.52% [82]. **DTP-C6Th** was reported as promising HTM with an improved electro-optical properties, a hole mobility of 4.18×10^{-4} cm^2 V^{-1} s^{-1}, and an excellent PCE of 21.04% with long-term stability (Figure 10.10c, d) [57].

10.3.2.4 Acene-Based HTMs

Kazim et al. were the first to report the well-known π-conjugated **TIPS-pentacene** molecule as HTMs, which at that point of time gave an PCE of 11.8% in its pristine form with $MAPbI_3$. This study put forward new direction of research for the dopant-free HTM by highlighting its advantages [69]. Various planar and star-shaped flat molecules (two-dimensional) have been introduced in the molecular structure core that enhances the intermolecular π–π packing interactions and face-on stacking on the surface and increases the lifetime of the charge separation, leading to an efficient HTM. Qin et al. reported the star-shaped molecules **HTMs 7-8** in PSCs [70] and they developed molecules with π-conjugated oligothiophenes based on *S,N*-heteropentacene, gave a PCE of 9.5–10.5%. Steck et al. reported another class of low band-gap **HTMs 9-10** composed of *S,N*-heteropentacene units to yield PCEs of 10.3–11.4%, respectively [71]. In 2016, another *S,N*-based heteropentacene molecule **HTM11** with mixed perovskite having π-bridge connecting to the donor and acceptor group was reported that gave a PCE of 16.9% [72].

10.3.2.5 Triazatruxene-Based HTMs

Triazatruxene (TAT) has been widely used as HTMs in PSCs, owing to its easy synthetic approach and delocalized π-system with reactive sites on the indole *N* atoms and benzene rings that enable the tuning of its opto-electrical and morphological properties. TAT also has an electron-rich rigid planar molecular structure, and the reactive sites allow more substitution to improve the opto-electrical properties of materials, as one benzene ring (delocalized electron) is connected to three indole rings. Ahmad and coworkers were the first to report TAT-based HTMs **HMDI** and **HPDI**, and its suitability were shown in PSCs [65]. The average PCEs were measured to be 8.62% and 9.83%, respectively. However, when doped with the p-type dopants, the PCE of **HPDI** increased to 10.82%. Further modification of this molecule by placing rational side arms, star-shaped planar triazatruxene-based core, with thiophene-based side arms, **KR321**, has been reported and gave unparalleled

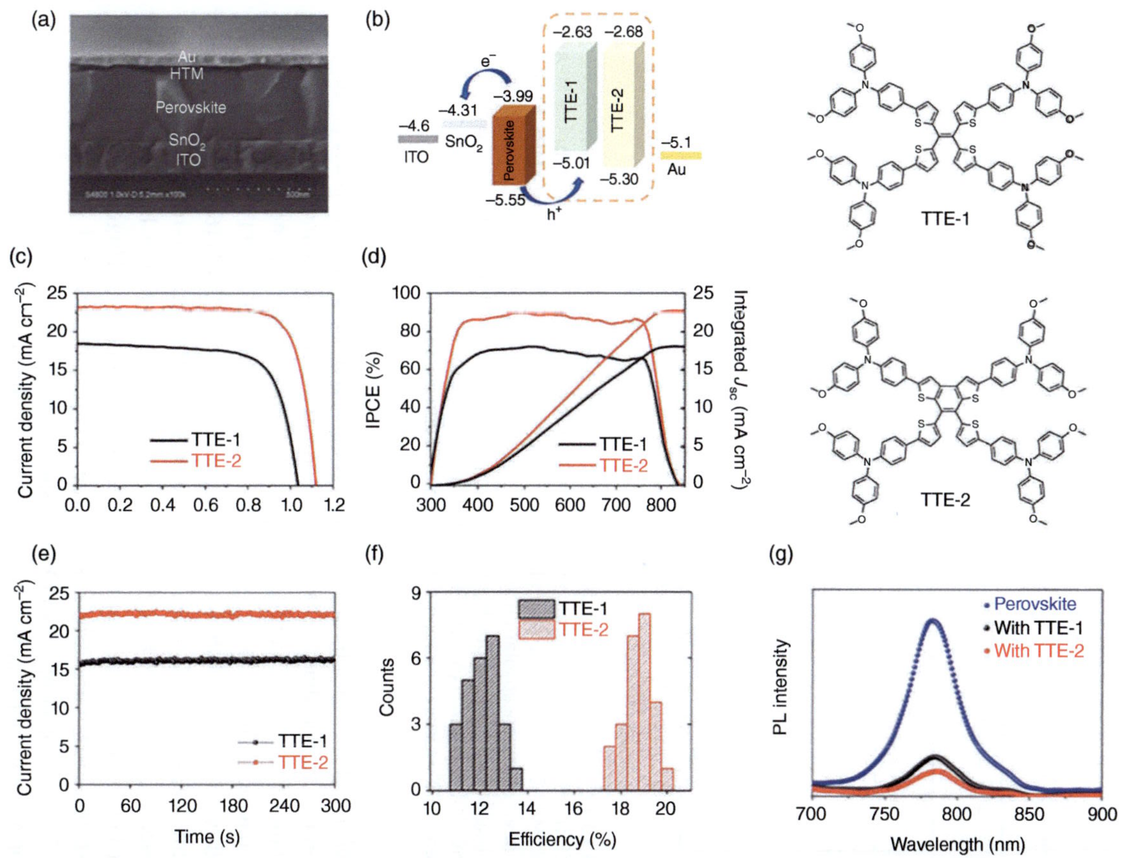

Figure 10.11 (a) Cross-sectional SEM image of the TTE-2-based planar n–i–p device, (b) energy level alignment of component in PSC with different HTMs, (c) *J–V* curves of the best performing TTE-1- and TTE-2-based PSCS, (d) IPCE curves along with the integrated short-circuit current density for the best performing TTE-1- and TTE-2-based PSCs, (e) stabilized PCE measurement for the best device for TTE-1 and TTE-2, (f) histogram of PCE from 25 individual devices based on TTE-1 and TTE-2, and (g) steady-state PL spectra of the perovskite with and without different HTMs. Source: Shen et al. [99].

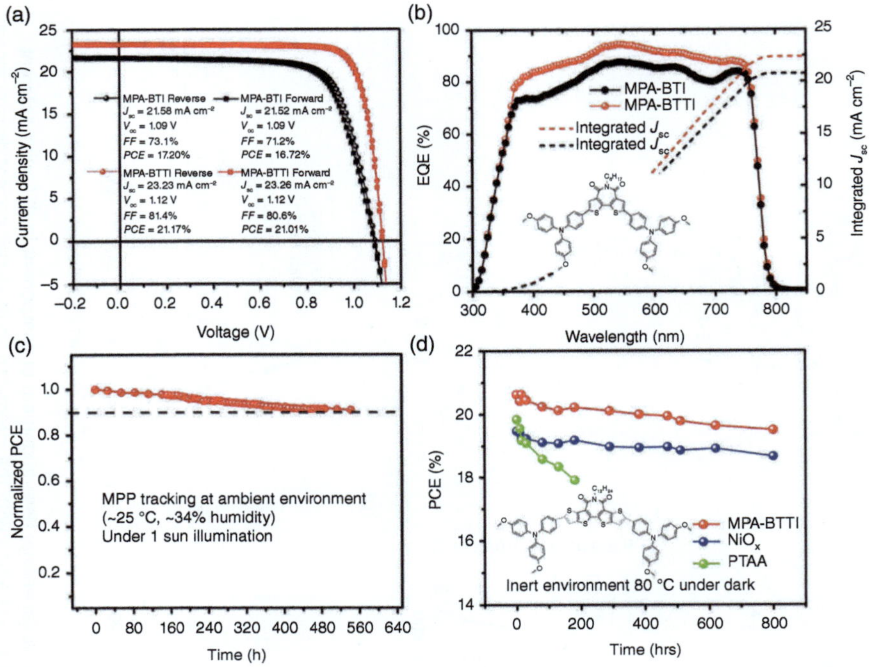

Figure 10.12 (a) *J*–*V* curves of the best performing MPA-BTI- and MPA-BTTI-based PSCs, (b) IPCE curves with the integrated short-circuit current density (dashed lines) for the MPA-BTI and MPA-BTTI, (c) device stability with the normalized PCE obtained from MPA-BTTI-based device with encapsulation at ambient environment under maximum power point (MPP) tracking and continuous light irradiation tracking and continuous light irradiation (AM1.5G, 100 mW cm^{-2}) (white LED), and (d) long-term stability of MPA-BTTI-based device stressed at 80° in inert environment in the dark. Source: Wang et al. [100].

PCE of over 19% with mixed perovskite composition [68]. The molecule possesses face-on orientation that facilitates the charge transfer, thereby increasing the PCE and stability of PSCs. Zhang et al. reported the use of pristine and doped **TPDI** as HTM having PCEs of 15.5% and 13.6% measured in a PSCs. **TPDI** showed high solubility, thermal stability, and efficient hole mobility in PSCs [66]. Furthermore, vacuum-deposited TAT-based HTM named as **TBDI** in p–i–n-structured PSC was reported with a PCE of 7.3%. However, by placing a thin layer of MoO_3 as buffer layer, the PCE increases significantly to 14.83% [67]. Huang et al. reported **Trux-OMe-TAD** with C3h truxene core and planar and rigid geometry with a HOMO energy level of −5.28 eV that showed well-matched energetic levels over that of $MAPbI_3$ perovskite (−5.43 eV) [94]. It showed a high PCE of 18.6% and hole mobility of 10^{-3} cm^2 V^{-1} s^{-1} in a planar p–i–n-type PSCs. Recently, Grisorio et al. designed two new truxene-based HTMs **Trux1** and **Trux2** in both typical and inverted structured PSCs [93]. The inverted structured PSCs showed PCEs of 13.4%, while typical structured was 10.2%. Kil et al. reported a TAT-based HTM with an end-capped acceptor **SGT-462** yielding a PCE of 11.7% [101].

10.3.2.6 Tetrathiafulvalene-Based HTM

The first dopant-free molecular HTM from tetrathiafulvalene derivatives for PSCs was **TTF-1** [73], achieving a competitive PCE of 11.03%. **TTF-1**-based devices showed better hole conductivity and stability in the ambient air at a relative humidity of 40%. However, the molecule showed poor solubility in the common organic solvents that directed its use toward vacuum deposition process. Recently, Chen et al. introduced another tetrathiafulvalene derivative **TTA** in PSCs [83]. **TTA** showed improved solubility as compared with **TTF-1**, as well as easy solution processability. The inverted planar PSCs based on **TTA** measured a PCE of 16.7% owing to its well-aligned energy level and high hole conductivity.

10.3.2.7 Organometallic Compounds and Other Molecules as HTM

The organometallic compounds (Figure 10.13) with phthalocyanine or porphyrin cores are being developed as dopant-free HTMs [102]. The porphyrin cores are generally complexed with divalent metal cations such as Zn^{2+}, Cu^{2+}, Ni^{2+}, and Co^{2+}. Their optoelectronic properties can be tuned by modifying the metal center and substituents on its periphery that could improve PV device performances. The fully vacuum-processed planar PSCs with copper phthalocyanine (**CuPc**) as the HTM were reported. Excellent device showed an overall PCE of 15.42%, and in addition the encapsulated device maintained the stability when stored in the dark under 20% humidity for 1000 hours [103]. Jiang et al. developed phthalocyanines (**CuPc-TIPS**, **CuPc-Bu**, **CuPc-OBu**) derivatives by modifying the non-peripheral substituents in the structure that delivered notable PCE and device stability [104]. A face-on oriented **CuPrPc** exhibits a PCE of 17.85% and retained over 94% of its initial PCE after 800 hours of storage [105]. Solution-processable and cost-effective **OMe-TPA-CuPc** was reported by Feng et al. showing an excellent PCE of 19.67% [106]. A nickel-based phthalocyanine derivative **NiPc-(OBu)$_8$** with V_2O_5 as the buffer layer was developed and delivered a PCE of 18.3% [107]. A Zn(II)-based **ZnP** as HTM in mesoscopic PSCs was reported with a PCE of 17.78% [108]. Gao et al. developed a near-infrared (NIR)-absorbing phthalocyanine (**DPPEZnPTEH**) and blended it with PCBM to be used in planar PSCs, exhibiting a high PCE of 19% [109]. Furthermore, **Zn(II)** porphyrin as HTM in PSCs offered a PCE of 14.5% with long-term stability [110]. **Co(II)/Co(III)**-based mixed HTMs were introduced and afforded a PCE of 20.4% and high thermal stability at 85 °C for 1000 hours (Figure 10.14a–c) [111].

Tetra-arylethene-based HTM named as **TAE-1** for PSCs exhibited a PCE of 11.0% [112]. Furthermore, Zhu et al. reported **TTBCPE** having electron-rich moieties with conductivity of 10^{-5} S cm^{-1} and hole mobility of 10^{-5} cm^2 V^{-1} s^{-1} that is higher than undoped spiro-OMeTAD and gave a PCE of 12.65% [113]. Two molecules, **BPZTPA and MeO-BPZTPA**, were developed and exploited as HTMs in a mesoscopic $TiO_2/CH_3NH_3PbI_3$/HTM/Au architect and yielded PCE of 8.2% and 10.36% [114]. Jeon et al. introduced a novel conjugated materials **DPIO and DPIE** as HTMs in PSCs [115], which have similar molecular configuration and consist of electron-donating and electron-accepting groups. Phenol-based HTMs **TCP-OH** and **TCP-OC$_8$** with hole mobilities of 10^{-7} cm^2 V^{-1} s^{-1} and 10^{-3} cm^2 V^{-1} s^{-1}, respectively, was reported, exhibiting PCEs of 16.97% and 15.28% [116]. An

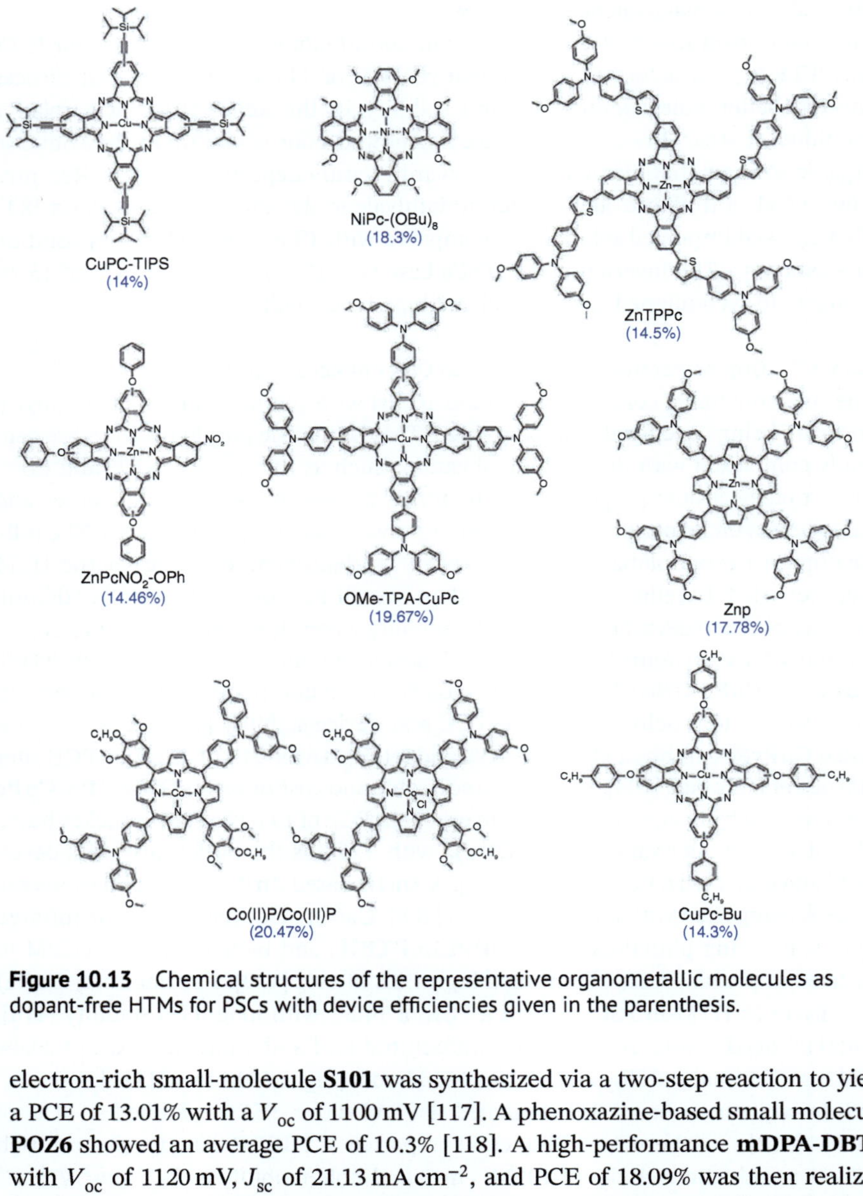

Figure 10.13 Chemical structures of the representative organometallic molecules as dopant-free HTMs for PSCs with device efficiencies given in the parenthesis.

electron-rich small-molecule **S101** was synthesized via a two-step reaction to yield a PCE of 13.01% with a V_{oc} of 1100 mV [117]. A phenoxazine-based small molecule **POZ6** showed an average PCE of 10.3% [118]. A high-performance **mDPA-DBTP** with V_{oc} of 1120 mV, J_{sc} of 21.13 mA cm^{-2}, and PCE of 18.09% was then realized [119], showing a negligible hysteresis with performance superseding of doped spiro-OMeTAD. A series of humidity-stable HTMs termed as **ACE-QA-ACE**, **TPA-QA-TPE**, and **DPA-QA-DPA** based on quinacridone dyes were reported, and PCEs of 18.2%, 16.6%, and 15.5% were measured [120]. The results were superior to the doped spiro-OMeTAD under identical conditions. As calculated by space charge-limited current (SCLC) method, all three HTMs possessed a hole mobility of 10^{-4} cm^2 V^{-1} s^{-1}. spiro-type HTM **Y1** achieving a PCE of 16.29% was put forward [121]. **BTF-4** is employed in both n–i–p and p–i–n configuration showing

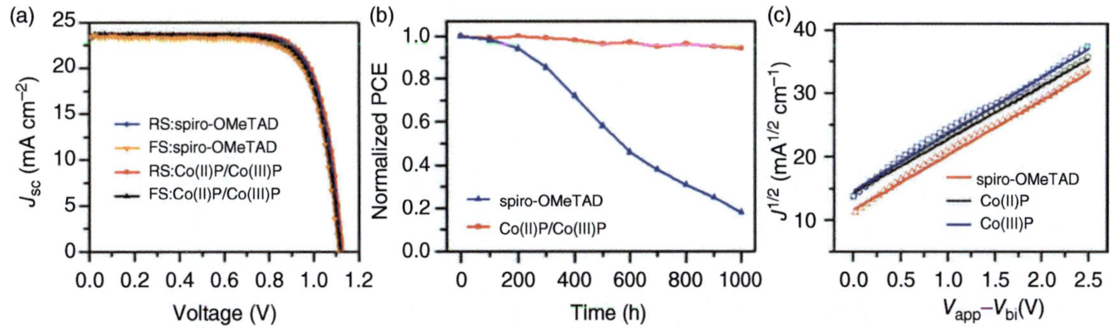

Figure 10.14 (a) *J–V* curves of PSCs based on dopant-free **Co(II)P/Co(III)P** HTM, (b) stability measured at 85 °C, and (c) electron mobility of spiro-OMeTAD and Co(II)Co(III) porphyrins. Source: Cao et al. [111].

remarkable PCEs of 18.03% and 17.01%, respectively [122], while the same group also reported HTMs **P2** and **P3** exhibiting stabilized PCEs of 19.6% and 20.3%, respectively [123].

10.4 Conclusion and Outlook

One of the vital components of PSCs is the development of HTMs. Among the organic molecules, the dopant-free small molecules is being optimized to surpass the performance in terms of stability and efficiency. We summarized the electro-optical properties of dopant-free HTMs and its applications in the PSCs. The main features of HTMs that need to be taken into account are high hole conductivity, low-lying HOMO level located above the valence band of perovskite, wide optical band gap (E_g), LUMO with high lying energy level that blocks electron transfer at the interface of perovskite and HTM, film-forming ability, effective molecular weight, solubility in orthogonal solvents, thermal and chemical stability, ease of functionalization and π-conjugated bonds, and low production cost. In some cases, the small molecules demand multistep synthesis and produce low yield that make them cost uneffective. It is beneficial to design planar molecules with a face-on orientation that helps in better charge and exciton transport and allows larger absorption strength when the transition dipole moment of the planar molecule (typically along its long axis) is aligned with the electric field of the incident light. Molecule that adopts face-on orientations enhances the substrate–molecule interactions and dominates molecule–molecule interactions. **MPA-BTTI** is the one such dopant-free HTM, gave competitive performances in inverted planar p–i–n-structured PSCs. This orientation facilitates the vertical charge transfer, which in turn yields high efficiencies. Typically doped HTMs give higher performance than their pristine counterparts, however it is trade off with device stability and make PSCs prone to degradation.

Acknowledgments

This work received funding from the European Union H2020 Programme under a European Research Council Consolidator grant [MOLEMAT, 726360] and PARASOL (RTI2018-102292-B-I00) from Spanish Ministry of Science and Innovation.

List of Abbreviations

ACE-QA-ACE	2,9-bis(1,2-dihydroacenaphthylen-5-yl)-5,12-dioctyl-5,12-dihydroquinolino[2,3-*b*]acridine-7,14-dione
BPZTPA	5,12-bis{*N*-[4,4′-bis-(phenyl)-aminophen-4″-yl]}-phenonaphthazine
BTF-4	3,8-bis(bis(4-methoxyphenyl)amino)fluoranthene-4,5-dicarbonitrile

BTPA	*N*4,*N*4,*N*4‴,*N*4‴-tetrakis(4-methoxyphenyl)-[1,1′:4′,1″:4″,1‴-quaterphenyl]-4,4‴-diamine
CMT	(2-ethylhexyl)-9*H*-carbazole substituted by *N*,*N*-di-*para*-methylthiophenylamine
CuPc	copper phthalocyanine
CzPAF-TPA	(*N*-(4-(9*H*-carbazol-9-yl)phenyl)-7-(4-(bis(4-methoxyphenyl)amino)phenyl)-*N*-(7-(4-(bis(4-methoxyphenyl)amino)phenyl)-9,9-dioctyl-9*H*-fluoren-2-yl)-9,9-dioctyl-9*H*-fluoren-2-amine
CZTPA	4,4′-(9-methyl-9*H*-carbazole-3,6-diyl)bis (*N*,*N*-bis(4-methoxyphenyl)aniline)
DERDTS-TBDT	2,8-bis(4-(bis(4-methoxyphenyl)amino)phenyl)-5-octyl-4*H*-dithieno[3,2-*c*:2′,3′-*e*]azepine-4,6(5*H*)-dione
DFH	N^2,N^2,N^7,N^7-tetra-ptolylspiro[fluorene-9,20-[1,3]dioxolane]-2,7-diamine
DFTAB	6-difluoro-N^1,N^1,N^2,N^2,N^4,N^4,N^5,N^5-octakis(4-methoxyphenyl) benzene-1,2,4,5
DPA-QA-DPA	2,9-bis(bis(4-methoxyphenyl)amino)-5,12-dioctyl-5,12-dihydroquinolino[2,3-*b*]acridine-7,14-dione
DPPEZnPTEH	5,15-bis(2,5-bis-(2-ethyl-hexyl)-3,6-di-thienyl-2-yl-2,5-dihydro-pyrrolo[3,4-*c*] pyrrole-1,4-dione-5′-yl-ethynyl)-10,20-bis(5-(2-ethylhexyl)-thienyl)-porphyrin zinc(II)
ETM	electron transport material
FA-CN	Benzo[1,9] quinolizino[3,4,5,6,7-defg]acridine
FA-OMePh	tris{bis(4-methoxyphenylethenyl)-*N*-phenyl}amine quinolizino acridine
FF	fill factor
GIXD	Grazing-incidence X-ray diffraction
H101	2,5-bis(4,4’-bis(methoxyphenyl)aminophen-4″-yl)-3,4-ethylenedioxythiophene
HAT-CN	1,4,5,8,9,11-hexaazatriphenylenehexacarbonitrile
HMDI	5,10,15-trihexyl-3,8,13-trimethoxy10,15-dihydro-5*H*-diindolo [3,2-*a*:30,20-]carbazole
HPDI	5,10,15-tris(4-(hexyloxy)phenyl)-10,15-dihydro-5*H*-diindolo[3,2-*a* 30, 20-*c*]carbazole
HTM	hole transport material
IDTT	indacenodithienothiophene
IPCE	incidence photon to electron conversion efficiency
KTM3	2,2′,5,5′-Tetrakis [*N*,*N*-di(4-methoxyphenyl)amino]-3,3′-bithiophene
MeO-BPZTPA	5,12-bis{*N*-[4,4′-bis(methoxy-phenyl)aminophen-4″-yl]}-phenonaphthazine
MPABTI	2,8-bis(4-(bis(4-methoxyphenyl)amino)phenyl)-5-octyl-4*H*-dithieno[3,2-*c*:2′,3′-*e*]azepine-4,6(5*H*)-dione

MPA-BTTI	2,8-bis(4-(bis(4-methoxyphenyl)amino)phenyl)-5-dodecyl-4*H*-thieno[2′,3′:4,5]thieno[3,2-*c*]thieno[2′,3′:4,5]thieno[2,3-*e*] azepine-4,6(5*H*)-dione
MPABTTI	2,8-bis(4-(bis(4-methoxyphenyl)amino)phenyl)-5-dodecyl-4*H*-thieno[2′,3′:4,5] thieno[3,2-c]thieno[2′,3′:4,5]thieno[2,3-*e*] azepine-4,6(5*H*)-dione
NTPA	4,4′-(Naphthalene-2,6-diyl)bis(*N*,*N*-bis(4-methoxyphenyl) aniline)
OMeTPA-FA	tris{*N*,*N*-bis(4-methoxyphenyl)-*N*-phenyl}amine quinolizino acridine
OMeTPA-FA	tris{*N*,*N*-bis(4-methoxyphenyl)-*N*-phenyl}amine quinolizino acridine
PBDT	poly (benzo[1,2-*b*:4,5-*b*′]dithiophene)
PCBM	[6,6]-Phenyl C61 butyric acid methyl ester
PCBTDPP	poly[*N*-9′-heptadecanyl-2,7carbazole-alt-3,6-bis(thiophen-5-yl)-2,5-dioctyl-2,5-dihydropyrrolo[3,4]pyrrole-1,4-dione]
PCDTBT	poly[*N*-9′-heptadecanyl-2,7-carbazole-alt-5,5-(4′,7′-di-2-thienyl-2′,1′,3′-benzothiadiazole)]
PMMA	poly (methyl methacrylate)
PNBA	4-(4-phenyl-4-*α*-naphthyl butadienyl)-*N*,*N*-bis(4-benzyl)-aniline
poly(DTSTPD-r-BThTPD)	poly-[(dithieno[3,2-*b*:2′,3′-*d*]silolethieno[3,4-*c*]pyrrole-4,6-dione)-random-(2,2′-bithiophenethieno[3,4-*c*]pyrrole-4,6-dione)]
PSCS	perovskite solar cells
PST1	4,4′,4″,4‴-(2*H*,2′*H*,4*H*,4′*H*,3,3′-spiro-bi[thieno[3,4-*b*][1,4] dioxepine]-6,6′,8,8′-tetrayl)tetrakis(*N*,*N*-bis(4-methoxyphenyl) aniline)
S101	4,4′-(5,5-dihexyl-5*H*-dibenzo[*b*,*d*]silole-3,7-diyl)bis(*N*,*N*-bis(4-methoxyphenyl)aniline)
SAF-OMe	N2,N2,N2′,N2′,N7,N7,N7′,N7′-octakis(4-methoxyphenyl)-10-phenyl-10H-spiro[acridine-9,9′-fluorene]-2,2′,7,7′-tetraamine
SCPDT	4,4′-spirobi[cyclopenta[2,1-*b*;3,4-*b*′]dithiophene]
SEM	scanning electron microscope
Spiro-OMeTAD	2,2′,7.7′-tetrakis(*N*,*N*-di-*p*-methoxyphenylamine)-9,9′-spirpbifluorene
ST1	4-(4-(bis(4-(4-(dibutylamino)styryl) phenyl)-amino)styryl)-*N*,*N*-dibutylaniline
TAE	4,4′,4″,4‴-(ethene1,1,2,2-tetrayl)tetrakis(*N*,*N*-bis(4-methoxyphenyl)aniline)
TAPC	4,4′-cyclohexylidenebis[*N*,*N*-bis(4-methylphenyl) benzenamine]
TBDI	5,10,15-tribenzyl-5*H*-diindolo[3,2-*a*:3′,2′-*c*]-carbazole
TCNE	tricyanovinylene
TCP-OC_8	3,3′,3″-(2-(octyloxy)-benzene-1,3,5-triyl)tris(9-phenyl-9H-carbazole)

TCP-OH	2,4,6-tris(9-phenyl-9*H*-carbazol-3-yl)phenol
TIPS	6,13-bis(triisopropylsilylethinyl)
TPA	methoxy tripheny lamine
TPA-ANT-TPA	4,4′-(6,12-bis(octyloxy)-6,12-dihydronaphtho[7,8,1,2,3-nopqr]-tetraphene-4,10-diyl)bis(*N*,*N*-bis(4-methoxyphenyl)aniline)
TPA-MeOPh	tris{bis(4-methoxyphenylethenyl)-*N*-phenyl}amine
TPA-NAP-TPA	4,4′-(naphthalene-2,6-diyl)bis(*N*,*N*-bis(4-methoxyphenyl) aniline)
TPA-QA-TPA	2,9-bis(4-(bis(4-methoxyphenyl)amino)phenyl)-5,12-dioctyl-5, 12-dihydroquinolino[2,3-*b*]acridine-7,14-dione
TPA-TVT-TPA	(*E*)-4,4′-(ethene-1,2-diylbis (thiophene-5,2-diyl))bis(*N*,*N*-bis (4-methoxyphenyl)aniline
TPD	*N*,*N*′-di(3-methylphenyl)-*N*,*N*′ diphenyl-4,4’-di amino biphenyl
TPDI	5,10,15-triphenyl-5*H*-diindolo[3,2-*a*:30, 20-*c*]carbazole
TPE-S	4,4′,4″,4‴-(ethene-1,1,2,2-tetrayl)tetrakis(*N*,*N*-bis(4-(methylthio)phenyl)aniline)
TTA	5,5′, 5″,5‴-(tetrathiafulvalene-2,3,6,7-tetrayl)tetraisophthalic acid
TTBCPE	1,1,2,2-tetrakis[4-(3,6-di-tertbutyl-9H-carbazol-9-yl)phenyl] ethane
TTE	tetrathienylethene
VB-MeO-FDPA	2,7-di-(N,N′-di-4-methoxyphenylamino)-9,9-bis-(4-vinyl-benzyl)-9*H*-fluorene
XRD	X-ray diffraction
ZnTPPc	zinc(II) 2,9,16,23-tetra-thienylmethoxytriphenylamine-29*H*,31*H*-phthalocyanine

References

1 Alivisatos, A.P. (1996). Semiconductor clusters, nanocrystals, and quantum dots. *Science* 271 (5251): 933–937.

2 Eberspacher, C., Ermer, J., Fredric, C. et al. (eds.) (1991). Recent advances in single-junction and tandem thin film modules based on $CuInSe_2$. In: *Tenth EC Photovoltaic Solar Energy Conference*. Springer.

3 Repins, I., Contreras, M.A., Egaas, B. et al. (2008). 19.9%-efficient ZnO/CdS/$CuInGaSe_2$ solar cell with 81.2% fill factor. *Prog. Photovolt.* 16 (3): 235–239.

4 Green, M., Emery, K., Hishikawa, Y. et al. (2012). Solar cell efficiency tables (version 40). *Prog. Photovolt.* 20: 606–614.

5 Kojima, A., Teshima, K., Shirai, Y., and Miyasaka, T. (2009). Organometal halide perovskites as visible-light sensitizers for photovoltaic cells. *J. Am. Chem. Soc.* 131 (17): 6050–6051.

6 NREL, *Best Research-Cell Efficiency Chart*. 2019. https://www.nrel.gov/pv/assets/pdfs/best-research-cell-efficiencies.20200104.pdf.

7 Kim, H.-S., Lee, C.-R., Im, J.-H. et al. (2012). Lead iodide perovskite sensitized all-solid-state submicron thin film mesoscopic solar cell with efficiency exceeding 9%. *Sci. Rep.* 2: 591.

8 Lee, M.M., Teuscher, J., Miyasaka, T. et al. (2012). Efficient hybrid solar cells based on meso-superstructured organometal halide perovskites. *Science* 338 (6107): 643–647.

9 Zhao, T., Shi, W., Xi, J. et al. (2016). Intrinsic and extrinsic charge transport in $CH_3NH_3PbI_3$ perovskites predicted from first-principles. *Sci. Rep.* 6 (1): 1–9.

10 Dong, Q., Fang, Y., Shao, Y. et al. (2015). Electron-hole diffusion lengths> 175 μm in solution-grown $CH_3NH_3PbI_3$ single crystals. *Science* 347 (6225): 967–970.

11 Savenije, T.J., Ponseca, C.S. Jr., Kunneman, L. et al. (2014). Thermally activated exciton dissociation and recombination control the carrier dynamics in organometal halide perovskite. *J. Phys. Chem. Lett.* 5 (13): 2189–2194.

12 Niu, G., Li, W., Meng, F. et al. (2014). Study on the stability of CH_3 NH_3 PbI_3 films and the effect of post-modification by aluminum oxide in all-solid-state hybrid solar cells. *J. Mater. Chem. A* 2 (3): 705–710.

13 Divitini, G., Cacovich, S., Matteocci, F. et al. (2016). In situ observation of heat-induced degradation of perovskite solar cells. *Nat. Energy* 1 (2): 1–6.

14 Leijtens, T., Eperon, G.E., Pathak, S. et al. (2013). Overcoming ultraviolet light instability of sensitized TiO_2 with meso-superstructured organometal tri-halide perovskite solar cells. *Nat. Commun.* 4 (1): 1–8.

15 Park, N.-G. (2015). Perovskite solar cells: an emerging photovoltaic technology. *Mater. Today* 18 (2): 65–72.

16 Xing, G., Mathews, N., Sun, S. et al. (2013). Long-range balanced electron-and hole-transport lengths in organic-inorganic $CH_3NH_3PbI_3$. *Science* 342 (6156): 344–347.

17 Ball, J.M., Lee, M.M., Hey, A., and Snaith, H.J. (2013). Low-temperature processed meso-superstructured to thin-film perovskite solar cells. *Energy Environ. Sci.* 6 (6): 1739–1743.

18 Cha, J.-M., Lee, J.-W., Son, D.-Y. et al. (2016). Mesoscopic perovskite solar cells with an admixture of nanocrystalline TiO_2 and Al_2O_3: role of interconnectivity of TiO_2 in charge collection. *Nanoscale* 8 (12): 6341–6351.

19 Chen, Q., Zhou, H., Hong, Z. et al. (2014). Planar heterojunction perovskite solar cells via vapor-assisted solution process. *J. Am. Chem. Soc.* 136 (2): 622–625.

20 Kim, H., Lim, K.-G., and Lee, T.-W. (2016). Planar heterojunction organometal halide perovskite solar cells: roles of interfacial layers. *Energy Environ. Sci.* 9 (1): 12–30.

21 Marchioro, A., Teuscher, J., Friedrich, D. et al. (2014). Unravelling the mechanism of photoinduced charge transfer processes in lead iodide perovskite solar cells. *Nat. Photonics* 8 (3): 250.

22 Guerrero, A., Garcia-Belmonte, G., Mora-Sero, I. et al. (2016). Properties of contact and bulk impedances in hybrid lead halide perovskite solar cells including inductive loop elements. *J. Phys. Chem. C* 120 (15): 8023–8032.

23 Rakstys, K., Igci, C., Nazeeruddin, M.K. (2019). Efficiency vs. stability: dopant-free hole transporting materials towards stabilized perovskite solar cells. *Chem Sci.* 10 (28): 6748–6769.

24 Calió, L., Kazim, S., Grätzel, M., and Ahmad, S. (2016). Hole-transport materials for perovskite solar cells. *Angew. Chem. Int. Ed.* 55 (47): 14522–14545.

25 Liu, Y., Zhang, H., Zhang, Y. et al. (2018). Influence of hole transport layers on internal absorption, charge recombination and collection in HC $(NH_2)_2$ PbI_3 perovskite solar cells. *J. Mater. Chem. A* 6 (17): 7922–7932.

26 Burschka, J., Dualeh, A., Kessler, F. et al. (2011). Tris(2-(1 H-pyrazol-1-yl)pyridine)cobalt(III) as p-type dopant for organic semiconductors and its application in highly efficient solid-state dye-sensitized solar cells. *J. Am. Chem. Soc.* 133 (45): 18042–18045.

27 Kim, G.-W., Kang, G., Kim, J. et al. (2016). Dopant-free polymeric hole transport materials for highly efficient and stable perovskite solar cells. *Energy Environ. Sci.* 9 (7): 2326–2333.

28 Yang, X., Xi, J., Sun, Y. et al. (2019). A dopant-free twisted organic small-molecule hole transport material for inverted planar perovskite solar cells with enhanced efficiency and operational stability. *Nano Energy* 64: 103946.

29 Li Za, Z.Z., Chueh, C.-C., Jo, S.B. et al. (2016). Rational design of dipolar chromophore as an efficient dopant-free hole-transporting material for perovskite solar cells. *J. Am. Chem. Soc.* 138 (36): 11833–11839.

30 Cao, Y., Li, Y., Morrissey, T. et al. (2019). Dopant-free molecular hole transport material that mediates a 20% power conversion efficiency in a perovskite solar cell. *Energy Environ. Sci.* 12 (12): 3502–3507.

31 Liu, X. and Liu, X. (2019). Optimizing electron-rich arylamine derivatives in thiophene-fused derivatives as π bridge-based hole transporting materials for perovskite solar cells. *RSC Adv.* 9 (43): 24733–24741.

32 Li, H., Fu, K., Hagfeldt, A. et al. (2014). A simple 3,4-ethylenedioxythiophene based hole-transporting material for perovskite solar cells. *Angew. Chem. Int. Ed.* 126 (16): 4169–4172.

33 Salzmann, I., Heimel, G., Oehzelt, M. et al. (2016). Molecular electrical doping of organic semiconductors: fundamental mechanisms and emerging dopant design rules. *Acc. Chem. Res.* 49 (3): 370–378.

34 Walzer, K., Maennig, B., Pfeiffer, M., and Leo, K. (2007). Highly efficient organic devices based on electrically doped transport layers. *Chem. Rev.* 107 (4): 1233–1271.

35 Ono, L.K., Schulz, P., Endres, J.J. et al. (2014). Air-exposure-induced gas-molecule incorporation into spiro-MeOTAD films. *J. Phys. Chem. Lett.* 5 (8): 1374–1379.

36 Abate, A., Leijtens, T., Pathak, S. et al. (2013). Lithium salts as "redox active" p-type dopants for organic semiconductors and their impact in solid-state dye-sensitized solar cells. *Phys. Chem. Chem. Phys.* 15 (7): 2572–2579.

37 Gong, G., Zhao, N., Li, J. et al. (2016). F4TCNQ-doped DEPT-SC as hole transporting material for stable perovskite solar cells. *Org. Electron.* 35: 171–175.

38 Noh, J.H., Jeon, N.J., Choi, Y.C. et al. (2013). Nanostructured $TiO_2/CH_3NH_3PbI_3$ heterojunction solar cells employing spiro-OMeTAD/Co-complex as hole-transporting material. *J. Mater. Chem. A.* 1 (38): 11842–11847.

39 Seo, J.-Y., Kim, H.-S., Akin, S. et al. (2018). Novel p-dopant toward highly efficient and stable perovskite solar cells. *Energy Environ. Sci.* 11 (10): 2985–2992.

40 Caliò, L., Salado, M., Kazim, S., and Ahmad, S. (2018). A generic route of hydrophobic doping in hole transporting material to increase longevity of perovskite solar cells. *Joule* 2 (9): 1800–1815.

41 Geffroy, C., Grana, E., Bessho, T. et al. (2020). p-Doping of a hole transport material via a poly (ionic liquid) for over 20% efficiency and hysteresis-free perovskite solar cells. *ACS Appl. Energy Mater.* 3 (2): 1393–1401.

42 Franckevičius, M., Mishra, A., Kreuzer, F. et al. (2015). A dopant-free spirobi [cyclopenta [2,1-b: 3,4-b′] dithiophene] based hole-transport material for efficient perovskite solar cells. *Mater. Horiz.* 2 (6): 613–618.

43 Ma, S., Zhang, H., Zhao, N. et al. (2015). Spiro-thiophene derivatives as hole-transport materials for perovskite solar cells. *J. Mater. Chem. A* 3 (23): 12139–12144.

44 Ganesan, P., Fu, K., Gao, P. et al. (2015). A simple spiro-type hole transporting material for efficient perovskite solar cells. *Energy Environ. Sci.* 8 (7): 1986–1991.

45 Choi, H., Paek, S., Lim, N. et al. (2014). Efficient perovskite solar cells with 13.63% efficiency based on planar triphenylamine hole conductors. *Chem. Eur. J.* 20 (35): 10894–10899.

46 Xiao, J., Han, L., Zhu, L. et al. (2014). A thin pristine non-triarylamine hole-transporting material layer for efficient $CH_3NH_3PbI_3$ perovskite solar cells. *RSC Adv.* 4 (62): 32918–32923.

47 Ma, Y., Chung, Y.-H., Zheng, L. et al. (2015). Improved hole-transporting property via HAT-CN for perovskite solar cells without lithium salts. *ACS Appl. Mater. Interfaces* 7 (12): 6406–6411.

48 Lv, S., Han, L., Xiao, J. et al. (2014). Mesoscopic $TiO_2/CH_3NH_3PbI_3$ perovskite solar cells with new hole-transporting materials containing butadiene derivatives. *ChemComm* 50 (52): 6931–6934.

49 Lv, S., Song, Y., Xiao, J. et al. (2015). Simple triphenylamine-based hole-transporting materials for perovskite solar cells. *Electrochim. Acta* 182: 733–741.

50 Choi, H., Park, S., Paek, S. et al. (2014). Efficient star-shaped hole transporting materials with diphenylethenyl side arms for an efficient perovskite solar cell. *J. Mater. Chem. A* 2 (45): 19136–19140.

51 Zhao, X., Zhang, F., Yi, C. et al. (2016). A novel one-step synthesized and dopant-free hole transport material for efficient and stable perovskite solar cells. *J. Mater. Chem. A* 4 (42): 16330–16334.

52 Hua, Y., Xu, B., Liu, P. et al. (2016). High conductivity Ag-based metal organic complexes as dopant-free hole-transport materials for perovskite solar cells with high fill factors. *Chem. Sci.* 7 (4): 2633–2638.

53 Zhang, F., Yi, C., Wei, P. et al. (2016). A novel dopant-free triphenylamine based molecular "Butterfly" hole-transport material for highly efficient and stable perovskite solar cells. *Adv. Energy Mater.* 6 (14): 1600401.

54 Zhang, F., Zhao, X., Yi, C. et al. (2017). Dopant-free star-shaped hole-transport materials for efficient and stable perovskite solar cells. *Dyes Pigm.* 136: 273–277.

55 Qin, P., Paek, S., Dar, M.I. et al. (2014). Perovskite solar cells with 12.8% efficiency by using conjugated quinolizino acridine based hole transporting material. *J. Am. Chem. Soc.* 136 (24): 8516–8519.

56 Xu, L., Huang, P., Zhang, J. et al. (2017). *N,N*-Di-para-methylthiophenylamine-substituted (2-ethylhexyl)-9*H*-carbazole: a simple, dopant-free hole-transporting material for planar perovskite solar cells. *J. Phys. Chem. C* 121 (40): 21821–21826.

57 Yin, X., Zhou, J., Song, Z. et al. (2019). Dithieno [3,2-b: 2′,3′-d] pyrrol-cored hole transport material enabling over 21% efficiency dopant-free perovskite solar cells. *Adv. Funct. Mater.* 29 (38): 1904300.

58 Qiu, J., Liu, H., Li, X. et al. (2019). Impact of 9-(4-methoxyphenyl) carbazole and benzodithiophene cores on performance and stability for perovskite solar cells based on dopant-free hole-transporting materials. *Sol. RRL* 3 (10): 1900202.

59 Li, H., Fu, K., Hagfeldt, A. et al. (2014). A simple 3,4-ethylenedioxythiophene based hole-transporting material for perovskite solar cells. *Angew. Chem. Int. Ed.* 53 (16): 4085–4088.

60 Paek, S., Qin, P., Lee, Y. et al. (2017). Dopant-free hole-transporting materials for stable and efficient perovskite solar cells. *Adv. Mater.* 29 (35): 1606555.

61 Liu, Y., Hong, Z., Chen, Q. et al. (2016). Perovskite solar cells employing dopant-free organic hole transport materials with tunable energy levels. *Adv. Mater.* 28 (3): 440–446.

62 Chen, C., Cheng, M., Liu, P. et al. (2016). Application of benzodithiophene based A–D–A structured materials in efficient perovskite solar cells and organic solar cells. *Nano Energy* 23: 40–49.

63 Pham, H.D., Do, T.T., Kim, J. et al. (2018). Molecular engineering using an anthanthrone dye for low-cost hole transport materials: a strategy for dopant-free, high-efficiency, and stable perovskite solar cells. *Adv. Energy Mater.* 8 (16): 1703007.

64 Cheng, M., Xu, B., Chen, C. et al. (2015). Phenoxazine-based small molecule material for efficient perovskite solar cells and bulk heterojunction organic solar cells. *Adv. Energy Mater.* 5 (8): 1401720.

65 Ramos, F.J., Rakstys, K., Kazim, S. et al. (2015). Rational design of triazatruxene-based hole conductors for perovskite solar cells. *RSC Adv.* 5 (66): 53426–53432.

66 Zhang, F., Yang, X., Cheng, M. et al. (2015). Engineering of hole-selective contact for low temperature-processed carbon counter electrode-based perovskite solar cells. *J. Mater. Chem. A* 3 (48): 24272–24280.

67 Calió, L., Momblona, C., Gil-Escrig, L. et al. (2017). Vacuum deposited perovskite solar cells employing dopant-free triazatruxene as the hole transport material. *Sol. Energy Mater. Sol. Cells* 163: 237–241.

68 Rakstys, K., Paek, S., Gao, P. et al. (2017). Molecular engineering of face-on oriented dopant-free hole transporting material for perovskite solar cells with 19% PCE. *J. Mater. Chem. A* 5 (17): 7811–7815.

69 Kazim, S., Ramos, F.J., Gao, P. et al. (2015). A dopant free linear acene derivative as a hole transport material for perovskite pigmented solar cells. *Energy Environ. Sci.* 8 (6): 1816–1823.

70 Qin, P., Kast, H., Nazeeruddin, M.K. et al. (2014). Low band gap *S*, *N*-heteroacene-based oligothiophenes as hole-transporting and light absorbing materials for efficient perovskite-based solar cells. *Energy Environ. Sci.* 7 (9): 2981–2985.

71 Steck, C., Franckevičius, M., Zakeeruddin, S.M. et al. (2015). A–D–A-type S,*N*-heteropentacene-based hole transport materials for dopant-free perovskite solar cells. *J. Mater. Chem. A* 3 (34): 17738–17746.

72 Bi, D., Mishra, A., Gao, P. et al. (2016). High-efficiency perovskite solar cells employing a S,*N*-Heteropentacene-based D–A hole-transport material. *ChemSusChem* 9 (5): 433–438.

73 Liu, J., Wu, Y., Qin, C. et al. (2014). A dopant-free hole-transporting material for efficient and stable perovskite solar cells. *Energy Environ. Sci.* 7 (9): 2963–2967.

74 Cheng, M., Aitola, K., Chen, C. et al. (2016). Acceptor–donor–acceptor type ionic molecule materials for efficient perovskite solar cells and organic solar cells. *Nano Energy* 30: 387–397.

75 Wang, Y.K., Yuan, Z.C., Shi, G.Z. et al. (2016). Dopant-free spiro-triphenylamine/fluorene as hole-transporting material for perovskite solar cells with enhanced efficiency and stability. *Adv. Funct. Mater.* 26 (9): 1375–1381.

76 Liu, X., Zheng, X., Wang, Y. et al. (2017). Dopant-free hole-transport materials based on methoxytriphenylamine-substituted indacenodithienothiophene for solution-processed perovskite solar cells. *ChemSusChem* 10 (13): 2833–2838.

77 Liu, X., Rezaee, E., Shan, H. et al. (2018). Dopant-free hole transport materials based on alkyl-substituted indacenodithiophene for planar perovskite solar cells. *J. Mater. Chem. C* 6 (17): 4706–4713.

78 Krishnamoorthy, T., Kunwu, F., Boix, P.P. et al. (2014). A swivel-cruciform thiophene based hole-transporting material for efficient perovskite solar cells. *J. Mater. Chem. A* 2 (18): 6305–6309.

79 Zheng, L., Chung, Y.-H., Ma, Y. et al. (2014). A hydrophobic hole transporting oligothiophene for planar perovskite solar cells with improved stability. *ChemComm* 50 (76): 11196–11199.

80 Liu, Y., Chen, Q., Duan, H.-S. et al. (2015). A dopant-free organic hole transport material for efficient planar heterojunction perovskite solar cells. *J, Mater. Chem. A* 3 (22): 11940–11947.

81 Chen, H., Fu, W., Huang, C. et al. (2017). Molecular engineered hole-extraction materials to enable dopant-free, efficient p-i-n perovskite solar cells. *Adv. Energy Mater.* 7 (18): 1700012.

82 Kong, X., Jiang, Y., Wu, X. et al. (2020). Dopant-free F-substituted benzodithiophene copolymer hole-transporting material for efficient and stable perovskite solar cells. *J. Mater. Chem. A*.

83 Chen, Q., Li, X., Jiu, T. et al. (2017). Tetrathiafulvalene derivative as a new hole-transporting material for highly efficient perovskite solar cell. *Dyes Pigm.* 147: 113–119.

84 Park, S.J., Jeon, S., Lee, I.K. et al. (2017). Inverted planar perovskite solar cells with dopant free hole transporting material: Lewis base-assisted passivation and reduced charge recombination. *J. Mater. Chem. A* 5 (25): 13220–13227.

85 Pham, H.D., Hu, H., Feron, K. et al. (2017). Thienylvinylenethienyl and naphthalene core substituted with triphenylamines – highly efficient hole transporting materials and their comparative study for inverted perovskite solar cells. *Sol. RRL* 1 (8): 1700105.

86 Reddy, S.S., Shin, S., Aryal, U.K. et al. (2017). Highly efficient air-stable/hysteresis-free flexible inverted-type planar perovskite and organic solar cells employing a small molecular organic hole transporting material. *Nano Energy* 41: 10–17.

87 Yang, L., Cai, F., Yan, Y. et al. (2017). Conjugated small molecule for efficient hole transport in high-performance p–i–n type perovskite solar cells. *Adv. Funct. Mater.* 27 (31): 1702613.

88 Jiang, K., Wang, J., Wu, F. et al. (2020). Dopant-free organic hole-transporting material for efficient and stable inverted all-inorganic and hybrid perovskite solar cells. *Adv. Mater.* 32 (16): 1908011.

89 Zhang, Y., Kou, C., Zhang, J. et al. (2019). Crosslinked and dopant free hole transport materials for efficient and stable planar perovskite solar cells. *J, Mater. Chem. A* 7 (10): 5522–5529.

90 Duan, L., Chen, Y., Jia, J. et al. (2020). Dopant-free hole-transport materials based on 2,4,6-triarylpyridine for inverted planar perovskite solar cells. *ACS Appl. Energy Mater.* 3 (2): 1672–1683.

91 Labban, A.E., Chen, H., Kirkus, M. et al. (2016). Improved efficiency in inverted perovskite solar cells employing a novel diarylamino-substituted molecule as PEDOT: PSS replacement. *Adv. Energy Mater.* 6 (11): 1502101.

92 Neophytou, M., Griffiths, J., Fraser, J. et al. (2017). High mobility, hole transport materials for highly efficient PEDOT: PSS replacement in inverted perovskite solar cells. *J. Mater. Chem. C* 5 (20): 4940–4945.

93 Grisorio, R., Iacobellis, R., Listorti, A. et al. (2017). Rational design of molecular hole-transporting materials for perovskite solar cells: direct versus inverted device configurations. *ACS Appl. Mater. Interfaces* 9 (29): 24778–24787.

94 Huang, C., Fu, W., Li, C.-Z. et al. (2016). Dopant-free hole-transporting material with a C3h symmetrical truxene core for highly efficient perovskite solar cells. *J. Am. Chem. Soc.* 138 (8): 2528–2531.

95 Wang, K., Chen, H., Niu, T. et al. (2019). Dopant-free hole transport materials with a long alkyl chain for stable perovskite solar cells. *Nanomaterials* 9 (7): 935.

96 Lu, C., Choi, I.T., Kim, J., and Kim, H.K. (2017). Simple synthesis and molecular engineering of low-cost and star-shaped carbazole-based hole transporting materials for highly efficient perovskite solar cells. *J. Mater. Chem. A* 5 (38): 20263–20276.

97 Liu, X., Ding, X., Ren, Y. et al. (2018). A star-shaped carbazole-based hole-transporting material with triphenylamine side arms for perovskite solar cells. *J. Mater. Chem. C* 6 (47): 12912–12918.

98 Cheng, M., Chen, C., Aitola, K. et al. (2016). Highly efficient integrated perovskite solar cells containing a small molecule-PC70BM bulk heterojunction layer with an extended photovoltaic response up to 900 nm. *Chem. Mater.* 28 (23): 8631–8639.

99 Shen, C., Wu, Y., Zhang, H. et al. (2019). Semi-locked tetrathienylethene as a building block for hole-transporting materials: toward efficient and stable perovskite solar cells. *Angew. Chem. Int. Ed.* 58 (12): 3784–3789.

100 Wang, Y., Chen, W., Wang, L. et al. (2019). Dopant-free small-molecule hole-transporting material for inverted perovskite solar cells with efficiency exceeding 21%. *Adv. Mater.* 31 (35): 1902781.

101 Kil, D.R., Lu, C., Ji, J.-M. et al. (2018). Dopant-free triazatruxene-based hole transporting materials with asymmetrical π-bridges for stable and efficient perovskite solar cells. *Nanomaterials* 43 (2): 2963–2967.

102 Duong, T., Peng, J., Walter, D. et al. (2018). Perovskite solar cells employing copper phthalocyanine hole-transport material with an efficiency over 20% and excellent thermal stability. *ACS Energy Lett.* 3 (10): 2441–2448.

103 Ke, W., Zhao, D., Grice, C.R. et al. (2015). Efficient fully-vacuum-processed perovskite solar cells using copper phthalocyanine as hole selective layers. *J, Mater. Chem. A* 3 (47): 23888–23894.

104 Jiang, X., Wang, D., Yu, Z. et al. (2019). Molecular engineering of copper phthalocyanines: a strategy in developing dopant-free hole-transporting materials for efficient and ambient-stable perovskite solar cells. *Adv. Energy Mater.* 9 (4): 1803287.

105 Liu, X., Wang, Y., Rezaee, E. et al. (2018). Tetra-propyl-substituted copper(II) phthalocyanine as dopant-free hole transporting material for planar perovskite solar cells. *Sol. RRL* 2 (7): 1800050.

106 Feng, Y., Hu, Q., Rezaee, E. et al. (2019). High-performance and stable perovskite solar cells based on dopant-free arylamine-substituted copper(II) phthalocyanine hole-transporting materials. *Adv. Energy Mater.* 9 (26): 1901019.

107 Cheng, M., Li, Y., Safdari, M. et al. (2017). Efficient perovskite solar cells based on a solution processable nickel(II) phthalocyanine and vanadium oxide integrated hole transport layer. *Adv. Energy Mater.* 7 (14): 1602556.

108 Chen, S., Liu, P., Hua, Y. et al. (2017). Study of arylamine-substituted porphyrins as hole-transporting materials in high-performance perovskite solar cells. *ACS Appl. Mater. Interfaces* 9 (15): 13231–13239.

109 Gao, K., Zhu, Z., Xu, B. et al. (2017). Highly efficient porphyrin-based OPV/perovskite hybrid solar cells with extended photoresponse and high fill factor. *Adv. Mater.* 29 (47): 1703980.

110 Huang, P., Hernández, A., Kazim, S. et al. (2020). Molecularly engineered thienyl-triphenylamine substituted zinc phthalocyanine as dopant free hole transporting materials in perovskite solar cells. *Sustain. Energy Fuels* 4 (12): 6188–6195.

111 Cao, J., Lv, X., Zhang, P. et al. (2018). Plant sunscreen and Co(II)/(III) porphyrins for UV-resistant and thermally stable perovskite solar cells: from natural to artificial. *Adv. Mater.* 30 (27): 1800568.

112 Rezaee, E., Liu, X., Hu, Q. et al. (2018). Dopant-free hole transporting materials for perovskite solar cells. *Sol. RRL* 2 (11): 1800200.

113 Zhu, H., Zhang, F., Liu, X. et al. (2017). Dopant-free hole-transport material with a tetraphenylethene core for efficient perovskite solar cells. *Energy Technol.* 5 (8): 1257–1264.

114 Liu, X., Zhang, F., Liu, X. et al. (2016). Application of phenonaphthazine derivatives as hole-transporting materials for perovskite solar cells. *J. Energy Chem.* 25 (4): 702–708.

115 Jeon, S., Thakur, U.K., Lee, D. et al. (2016). *N*-phenylindole-diketopyrrolopyrrole-containing narrow band-gap materials for dopant-free hole transporting layer of perovskite solar cell. *Org. Electron.* 37: 134–140.

116 Xue, Y., Wu, Y., and Li, Y. (2017). Readily synthesized dopant-free hole transport materials with phenol core for stabilized mixed perovskite solar cells. *J. Power Sources* 344: 160–169.

117 Krishna, A., Sabba, D., Yin, J. et al. (2016). Facile synthesis of a hole transporting material with a silafluorene core for efficient mesoscopic $CH_3NH_3PbI_3$ perovskite solar cells. *J, Mater. Chem. A* 4 (22): 8750–8754.

118 Cheng, M., Chen, C., Xu, B. et al. (2015). A novel phenoxazine-based hole transport material for efficient perovskite solar cell. *J. Energy Chem.* 24 (6): 698–706.

119 Pham, H.D., Yang, T.C.J., Jain, S.M. et al. (2020). Development of dopant-free organic hole transporting materials for perovskite solar cells. *Adv. Energy Mater.* 10 (13): 1903326.

120 Pham, H.D., Jain, S.M., Li, M. et al. (2019). Dopant-free novel hole-transporting materials based on quinacridone dye for high-performance and humidity-stable mesoporous perovskite solar cells. *J, Mater. Chem. A* 7 (10): 5315–5323.

121 Zhang, Y., Wu, F., Chen, L. et al. (2020). Molecular engineering of highly efficient dopant-free spiro-type hole transporting materials for perovskite solar cells. *Sol. Energy Mater. Sol. Cells* 212: 110534.

122 Sun, X., Xue, Q., Zhu, Z. et al. (2018). Fluoranthene-based dopant-free hole transporting materials for efficient perovskite solar cells. *Chem. Sci.* 9 (10): 2698–2704.

123 Zhang, F., Yao, Z., Guo, Y. et al. (2019). Polymeric, cost-effective, dopant-free hole transport materials for efficient and stable perovskite solar cells. *J. Am. Chem. Soc.* 141 (50): 19700–19707.

11

Impact of Monovalent Metal Halides on the Structural and Photophysical Properties of Halide Perovskite

Mojtaba Abdi-Jalebi[1] and M. Ibrahim Dar[2]

[1] *University College London, Institute for Material Discovery, Malet Place, London WC1E 7JE, UK*
[2] *University of Cambridge, Cavendish Laboratory, Department of Physics, Cambridge CB3 0HE, UK*

11.1 Introduction

Additive engineering has been a key enabler to tune and optimize material and optoelectronic properties of metal halide perovskites, enhancing their performance and stability in various applications such as solar cells [1, 2] and light-emitting diodes [3, 4] (LEDs), field-effect transistors [5] and X-ray detectors [6]. These properties including morphology, crystallinity, trap-state density, and luminescence efficiency, directly impact the performance of perovskite devices [7]. Halide perovskites are remarkably tolerant to additives due to their unique crystal and band structure, providing a wide range of creative space for additive engineering [8]. The main two categories of additives include inorganic adjunct (e.g. metal halides [9, 10], nanoparticles [11], and inorganic acids [12]) and organic additives (e.g. small molecules [13], ionic liquids [14], polymers [15], and fullerenes [16]). The latter is less prominent since its limited impact on surface states, whereas the density of traps in bulk is prominent. Furthermore, the organic elements in halide perovskites are shown to be the source of instability. Adding organic adjunct could potentially impact negatively on the long-term stability of perovskite devices [17]. In particular metal cations, the inorganic additives have been introduced as the key influencers on both surfaces and the bulk of the halide perovskites to enhance their structural and optoelectronic properties.

11.2 Metal Halides

Upon adding metal halides into the perovskite system with the ABX_3 formula, the metal cation could directly impact the A- or B- sites based on the ionic radius and its valence. The halide will be incorporated in the X site. Thus far, various metal cations with a wide range of ionic radius have been added to the perovskite systems (Figure 11.1a). Notably, the metal cations could be incorporated into the

Perovskite Solar Cells: Materials, Processes, and Devices, First Edition.
Edited by Shahzada Ahmad, Samrana Kazim, and Michael Grätzel.

perovskite structure or located at the crystalline surfaces such as grain boundaries and interfaces. The incorporation of the metal cations could be in the form of doping [21, 22] (e.g. interstitial or substitutional) or the formation of alloyed structures [23, 24]. Notably, most metal cations are mismatched to sustain a photoactive ABX_3 perovskite with an appropriate Goldschmidt tolerance factor between 0.8 and 1.0. Therefore, it does not form the alloyed structures (Figure 11.1b) [19]. The metal halide additives are usually categorized into monovalent [9], divalent [25, 26], and trivalent [22] cation halides based on the nature of the metal cation. Monovalent metal halides are the most common type of metal halide additives. We will discuss their impact on the structural and optoelectronic properties of perovskite materials and photovoltaic devices in Sections (11.4 and 11.5).

11.3 Monovalent Metal Halides

This category of metal halide additives includes alkali metal cations [27] and several transition metals such as silver [28] and copper [9, 29]. Among various monovalent cations, the ionic radius of cesium and rubidium are closer to the A- site. The ionic radius of sodium and silver are the closest match to the B- cite (e.g. Pb). Although the other monovalent metal cations' ionic radius (e.g. Li, Cu) would not match neither A- site nor B- site, they have been used as the surface modifiers or located at the interstitial site within the perovskite crystal structures [10].

Cesium is the most common monovalent cation that could fit within the A-site in ABX_3 perovskite structure and form alloyed compositions with the most common organic A- site cation such as MA = methylammonium, $CH_3NH_3^+$, and FA = formamidinium, $CH_3(NH_2)_2^+$. These cations are placed within the appropriate range of the Goldschmidt tolerance factor (i.e. 0.8 and 1.0) and form black α-phase perovskite [23]. Cs was first introduced in the perovskite system, inhibiting the undesired phase transition and enhancing the photo- and moisture stability of the FA-based perovskites. It led to the shrinkage of cubo-octahedral volume and stronger chemical interaction between FA and iodine, leading to a reduction in trap density and ionic migration [30]. Notably, a combination of Cs with FA and MA where FA is the dominant cation known as alloyed triple-action perovskite (e.g. $(Cs_{0.06}FA_{0.79}MA_{0.15})Pb(I_{0.85}Br_{0.15})_3$). It became the state-of-the-art composition in the field since it is thermally more stable, contains fewer phase impurities, and is less sensitive to processing conditions [23, 31].

The implementation of monovalent metal halide started in 2016 by Abdi-Jalebi et al. via the addition of Cu, Na, and Ag with similar ionic radii to Pb into the $MAPbI_3$ to explore the possibility of doping in this system [9]. These monovalent cations remarkably impact the structural and optoelectronic properties and alter the electronic band structure of the perovskites (Figure 11.1c) [20]. In these studies followed by other works, the addition of Rb in the complex composition of triple cation perovskites to enhance power conversion efficiency (PCE), thermal, and photostability is attributed to the enlarged grain size with the reduced non-radiative recombination and entropic gains through a unique defect passivation mechanism [19]. The

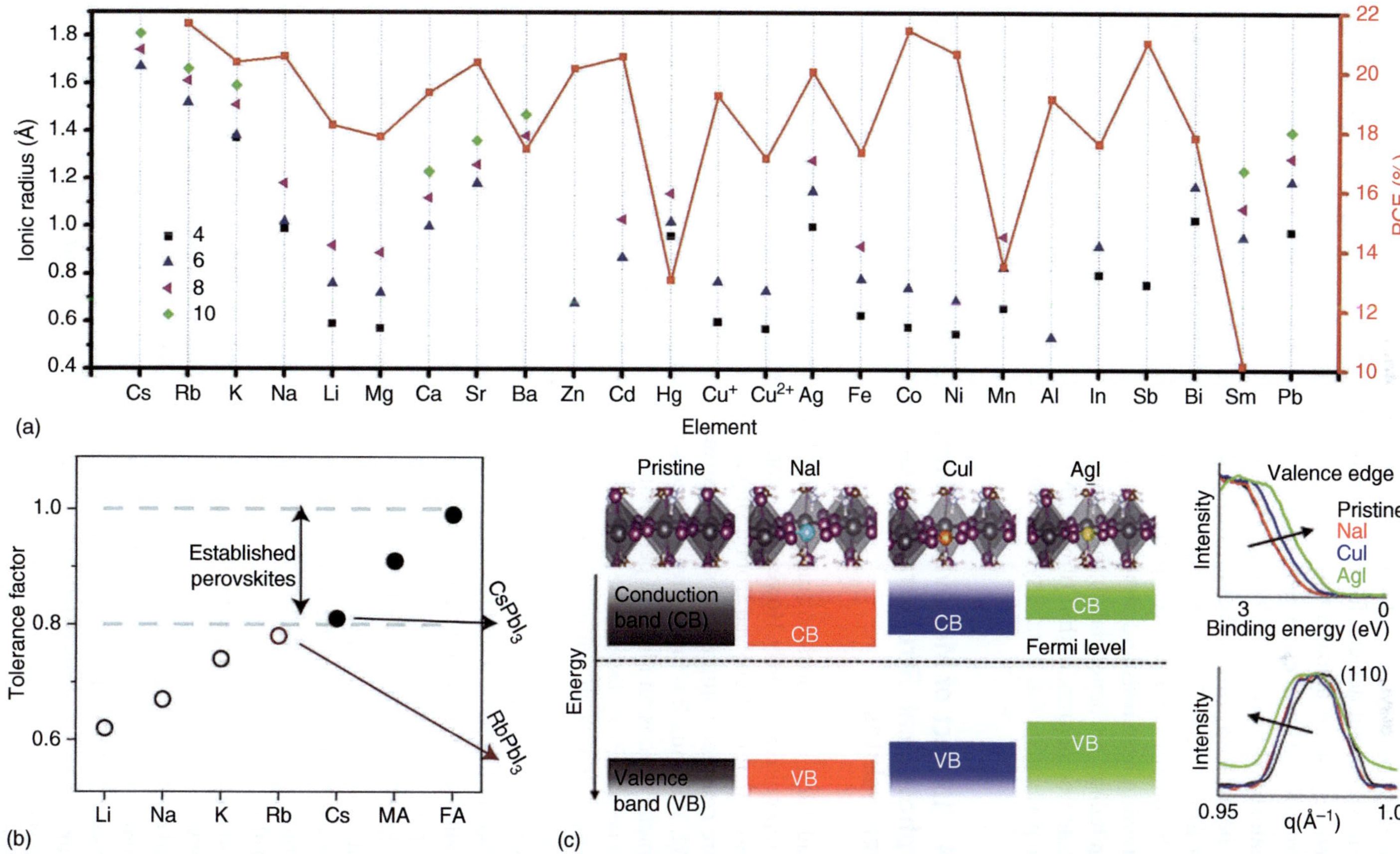

Figure 11.1 Incorporation of metal halide additives in perovskites. (a) The ionic radius for a range of cations with different coordination numbers vs. the highest achieved PCE of the perovskite solar cells with metal cation additives. Source: Wang et al. [18]. (b) Tolerance factor of $APbI_3$ perovskite with the oxidation-stable A (Li, Na, K, Rb, or Cs) and MA or FA. The dashed lines (tolerance factor between 0.8 and 1.0) represent the range for the formation of perovskite with photoactive black phase (solid circles) compared to nonphotoactive phases (open circles). Source: Saliba et al. [19]. (c) Schematic representation of the valence and conduction band edges position vs. the Fermi level (black dot line) induced by incorporating monovalent metal halides. The two plots on the right show the position of valence band edge and the main perovskite diffraction peak (110) of pristine and perovskite thin films with monovalent metal halides. Source: Abdi-Jalebi et al. [20].

next breakthrough in the field happened by adding potassium into the alloyed triple cation perovskite system. This leads to the inhibition of ionic migration and formation of Frenkel defects, forming significant bandgap stability in mixed halide perovskite systems across a wide range of bandgap (e.g. 1.7–1.9 eV) and the complete removal of hysteresis in perovskite solar cells (PSCs) [10, 32]. The enhancement mechanism with potassium halide additives includes decoration of the surface and grain boundaries with potassium while filling the halide vacancy with the excess halide from KI [33]. Due to the sequestration of K, excess halide is immobilized at the grain boundaries and surfaces, thereby inhibiting halide migration in parallel with suppressing hysteresis and photo-induced ion segregation [33]. Monovalent metal cation halides are beneficial for the grain growth [34] and the formation of uniform pin-hole free perovskite thin films with larger grains, leading to higher optoelectronic performance and stability in the subsequent devices [18].

11.4 Impact of Monovalent Metal Halides on the Morphological, Structural and Optoelectronic Properties of Perovskites

The addition of monovalent metal cation halides to the perovskite materials has been reported extensively. However, the composition of the perovskites in these studies was different. It evolved from a single cation (e.g. $MAPbI_3$, $FAPbI_3$) [9, 35] to a more complex composition such as mixed cation lead mixed halide perovskite [19, 32, 36]. Thus, it is imperative to discuss and compare the influences of monovalent metal cations on identical perovskites to obtain an optimal matching. In most of the perovskite systems, these additives alter the grain growth and morphology of the perovskites, as shown in Figure 11.2. Although there is no general trend for all of these cations, the increase in the average size of the grains, higher uniformity, and smoothness with better surface coverage and removal of the pin-holes are the common enhancement observations in the morphology of the perovskite thin films after addition of these additives (Figure 11.2). These enhancements in the film quality of perovskites upon adding the additives are mostly attributed to the controlled crystal growth of perovskite thin films [9, 37]. The formation of larger grains in perovskite thin films is favorable for the optoelectronic performance of the subsequent devices as the density of the grain boundaries that are known as the main source of non-radiative recombination centers will significantly diminish upon increasing the grain size [38].

Monovalent metal halide additives play effective roles in defining structural properties such as the average crystalline size, crystalline phase, lattice constant, and crystal orientation of the perovskite materials (Figure 11.1c and Figure 11.3) [35]. In particular, one of the key methods to ensure the complete conversion of PbI_2 into perovskite in the conventional sequential two-step deposition technique is the disappearance of the corresponding PbI_2 feature in the X-ray diffraction spectra [40]. For instance, the addition of NaI leads to complete conversion of lead iodide into $MAPbI_3$ perovskite, evidenced by the vanished diffraction peak of PbI_2, as shown in

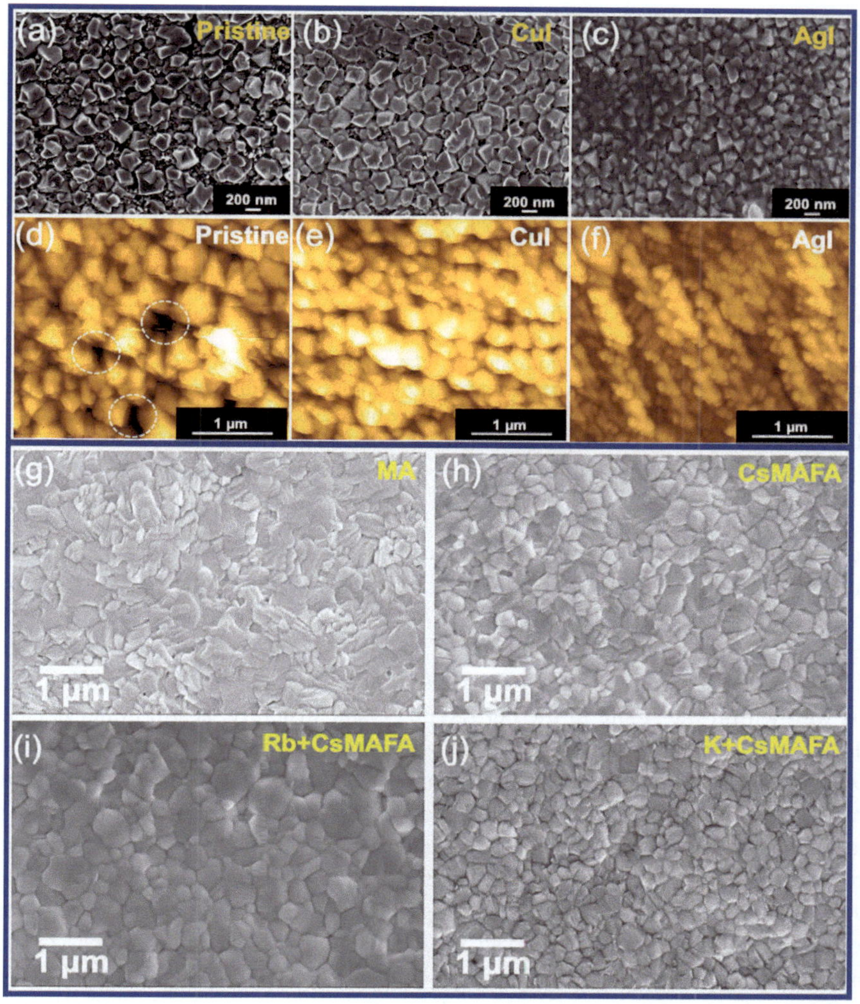

Figure 11.2 Morphological modifications of perovskites with monovalent metal halides. Top-view scanning electron microscopy (SEM) and atomic force microscopy (AFM) images of (a, d) pristine, (b, e) CuI-, (c, f) AgI-based $MAPbI_3$ deposited on a mesoporous TiO_2-coated fluorine-doped tin oxide (FTO) glass by two-step sequential spin-coating [9]. SEM images of (g) MA-, (h) CsMAFA-, (i)Rb – CsMAFA-, and (j) K+CsMAFA-based lead mixed halides ($APb(I_{0.85}Br_{0.15})_3$) deposited on FTO coated glass via single step anti solvent spin-coating. Source: (g–j) Feldmann et al. [31].

Figure 11.3a. In the complex composition of mixed cation lead mixed halide with excess lead iodide, the addition of potassium reduces the amount of PbI_2 residue where at higher concentration of potassium (e.g. $x = 0.40$), the PbI_2 peak has completely vanished. A new phase rich in potassium appears and becomes stronger at higher potassium concentrations (Figure 11.4). It is shown that the addition of monovalent metal cations such as Li, Na, and K increases the lattice constant of the

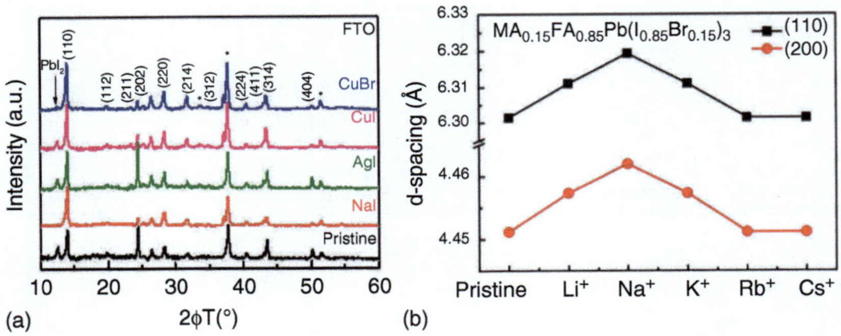

Figure 11.3 Impact of monovalent metal halides additives on the structural properties of perovskites. (a) X-ray diffraction spectra of pristine and $CH_3NH_3PbI_3$ perovskite with metal halide additives grown on mesoporous TiO_2 film, which is deposited on the FTO-coated glass. Source: Abdi-Jalebi et al. [9]. (b) The d-spacing of alkali cations incorporated-$MA_{0.15}FA_{0.85}Pb(I_{0.85}Br_{0.15})_3$. Source: Tang et al. 2018 [39].

perovskite. However, the larger cations such as Rb and Cs have no impact on the lattice parameter (Figure 11.3b). Considering the ionic radius and tolerance factor, the Rb and Cs can be settled down on the A-site while the Li, Na, and K are difficult to occupy the A-site.

Notably, the location of these monovalent metal cations, excluding cesium that fits well within the A-site in the perovskite crystal structure, is still under debate. However, interstitials sites [41], grain boundaries [32], and crystalline surfaces [9] are plausible locations. On this front, the scanning transmission electron microscopy energy dispersive X-ray spectroscopy (STEM-EDX) together with a non-negative matrix factorization (NMF) algorithm [42] and hard-X-ray photoelectron spectroscopy (HAXPES) [20, 43] were the two key advanced techniques that were used to locate these cations. Utilizing STEM-EDX with NMF analysis, various compositional phases were observed for the potassium- and rubidium-passivated mixed cation lead mixed halides systems that include a monovalent cation halide additive-rich phase and a perovskite phase (Figure 11.5a–f). In the alloyed perovskite systems with potassium addition, the K-(additive)-rich phase is composed of K and Br situated at the grain boundaries and interfaces of the perovskite film [32]. However, the addition of rubidium in the same system with identical concentration forms large, micron-sized crystals rich in Rb and iodide (Figure 11.5d–f) [43]. Soft X-ray photoelectron spectroscopy (XPS, 1486.6 eV photon energy) and synchrotron HAXPES (4000 eV photon energy) were utilized [20] for deeper insight into the chemical composition, distribution, and electronic structure of the perovskite films with monovalent metal cation halide additives at the surface and the bulk. These measurements in the alloyed perovskite systems with potassium and rubidium halide additives reveal that Rb is uniformly distributed throughout the film, with a negligible change when probing the surface (XPS) and probing deeper into the bulk (via the use of hard X-rays, HAXPES). On the other hand, the K content is higher on the surface than in the bulk (Figure 11.5g–j).

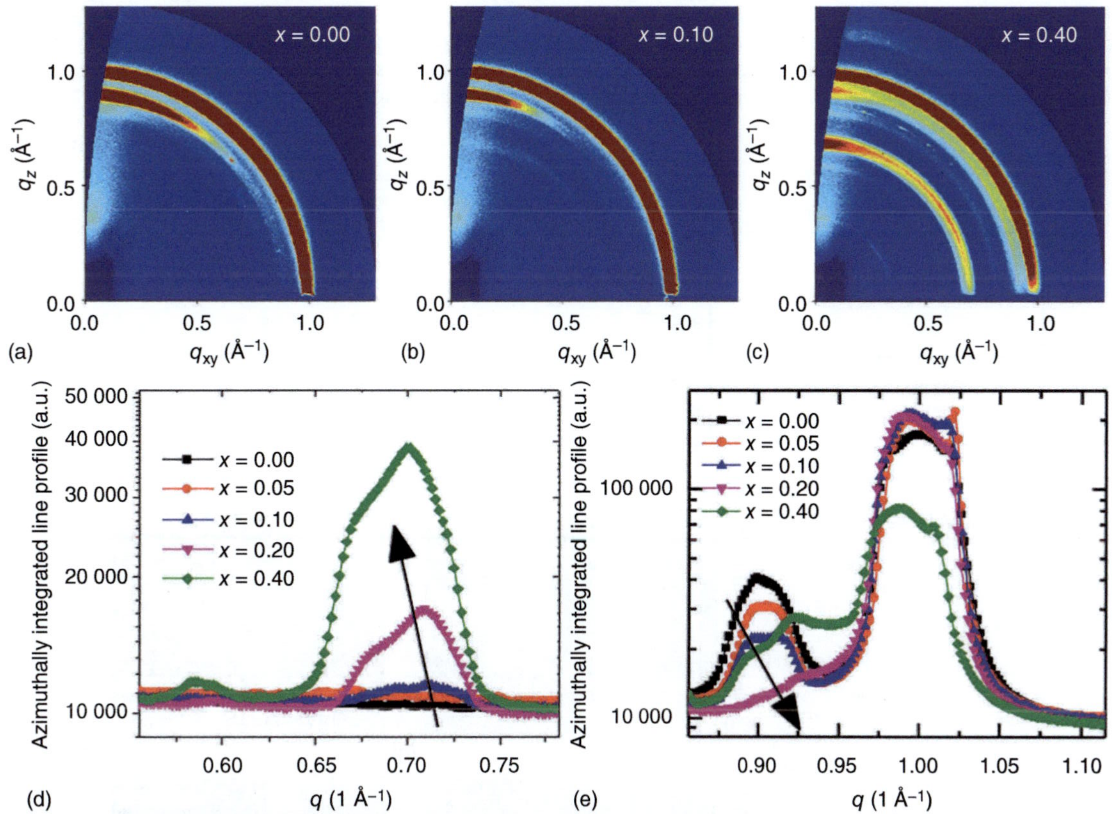

Figure 11.4 Impact of potassium iodide addition on the crystal structure of triple cation lead mixed halides ((Cs,FA,MA)Pb($I_{0.85}Br_{0.15})_3$). The diffraction patterns of thin (Cs, FA, MA)Pb($I_{0.85}Br_{0.15})_3$ films collected at low angle using GIWAXS for (a) $x = 0.00$, (b) $x = 0.10$, and (c) $x = 0.40$. (d, e) The high-resolution line profiles azimuthally integrated over the entire GIWAX profile for different fractions of potassium at $0.5 \leq q \leq 0.8$ (d) and $0.8 \leq q \leq 1.1$ (e). Source: Abdi-Jalebi et al. [32].

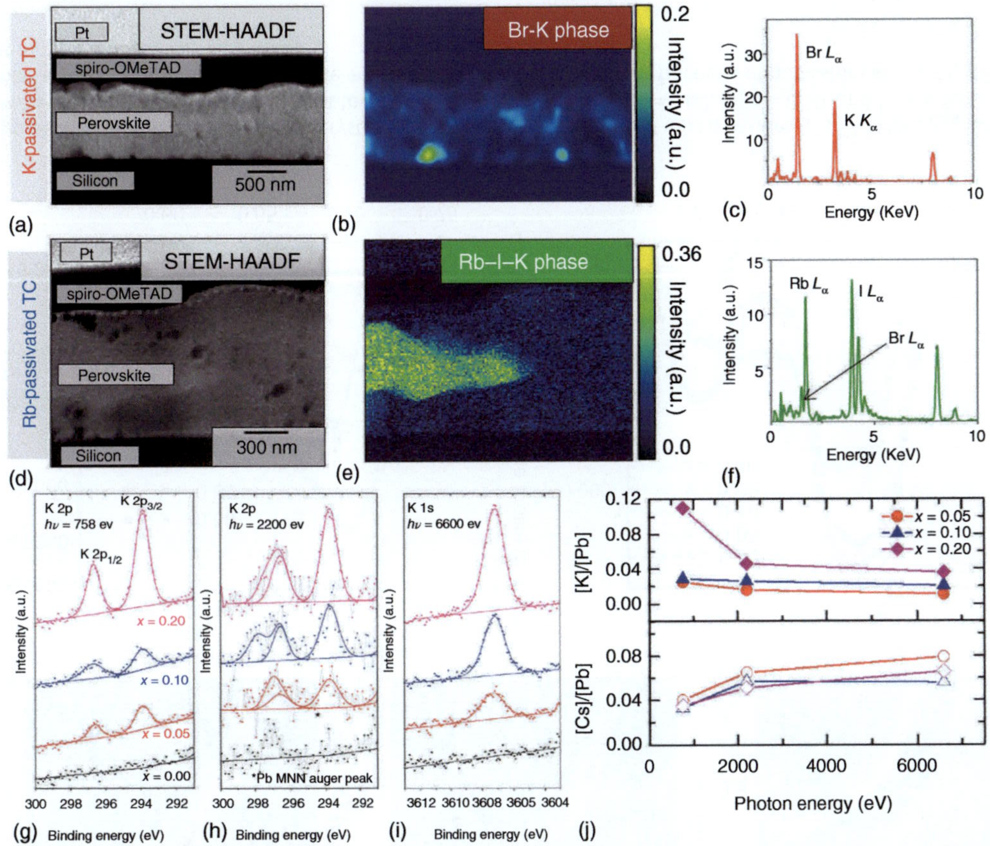

Figure 11.5 Locating the monovalent metal halide additives in perovskite thin films with STEM-EDX and HAXPES. (a) HAADF STEM cross-sectional image of a triple cation alloyed perovskite thin film potassium addition ($x = 0.10$). (b) NMF decomposition of this potassium-based perovskite showing the KBr phase and (c) the corresponding EDX spectra. (d) HAADF STEM cross-sectional image of a Rb-based perovskite ($x = 0.10$). (e) NMF decomposition of the Rb-based perovskite showing the Rb–I–Br phase and (f) the corresponding EDX spectra. Source: (a–e) Abdi-Jalebi et al. [43], (f) Abdi-Jalebi et al. [43]. Licenced under CC BY-4.0. The HAXPES spectra of the K-based perovskite films for 2P and 1S core levels recorded with a photon energy of (g) 1486.6 eV, (h) 2200 eV, and (i) 6600 eV. (j) Intensity ratio between core levels ([Cs]/[Pb] and [K]/[Pb]) calculated from the experimental results as a function of photon energy (measurements at 758, 2200 and 6600 eV). Source: (j) Abdi-Jalebi et al. [32].

In addition to the growth mechanism, morphology, and structure of halide perovskites, their optoelectronic properties are remarkably enhanced via an optimized amount of monovalent metal cation halide into the system. Although the bandgap and emission peak of the perovskite from the absorption and photoluminescence spectra have not been affected by these additives, the charge career lifetime and photoluminescence quantum yield (PLQE) enhanced significantly. For instance, Na, Cu, and Ag in the $MAPbI_3$ perovskites increased the PLQE (from less than 5% to about 20% for Na addition) and diminished the Urbach energy (i.e. sharper band edge) that is representative of the energetic disorder in the semiconductor (Figure 11.6). The enhanced PLQE and PL lifetime collectively suggest that these monovalent metal cation additives can eliminate non-radiative decay channels in the perovskite thin films and achieve a long carrier lifetime (Figure 11.6).

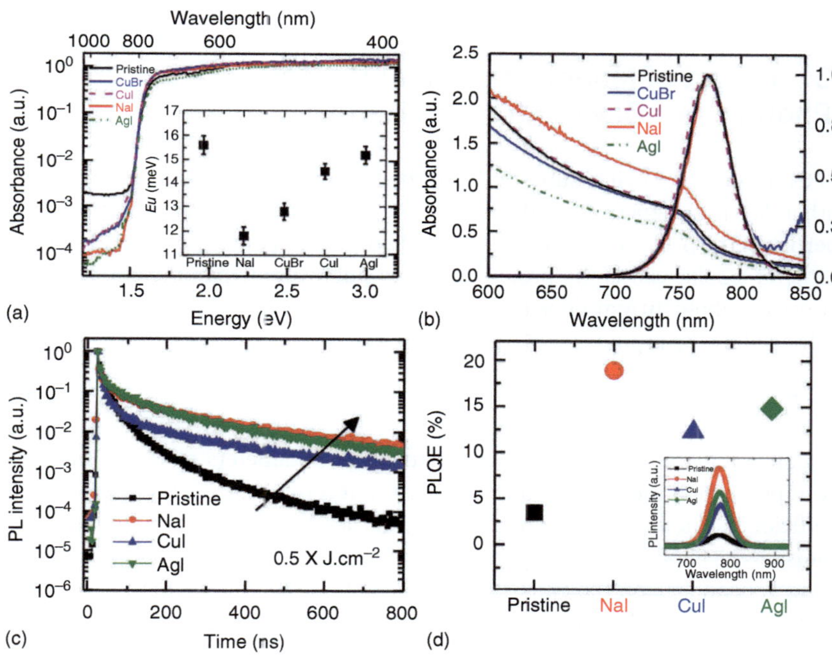

Figure 11.6 Enhanced optoelectronic quality of perovskite thin film via addition of monovalent cation halides (a) The photothermal deflection spectroscopy absorption spectra of perovskite films derived from pristine and monovalent metal halide additive based perovskites measured. The inset shows the corresponding Urbach energies for all samples. (b) Steady-state absorption and photoluminescence spectra for pristine and additives-based perovskite films. Source: Abdi-Jalebi et al. [9]. (c) Time-resolved PL decays of the doped perovskite films, with excitation at 400 nm and a pulse fluence of 0.5 μJ cm^{-2}. (d) PLQE of the doped perovskite films measured under illumination with a 532 nm laser. The inset shows the PL spectrum of the pristine and doped perovskite films. Source: Abdi-Jalebi et al. [20].

It is notable that for a solar cell or light-emitting diode to approach its efficiency limit, all recombination should be radiative, and the luminescence should be maximized [32]. The addition of monovalent metal cation halides, particularly potassium iodide, has been the most effective approach to suppress the substantial non-radiative losses. It originates from charge-carrier trap states present in the state-of-the-art perovskite films [44–46]. The alloyed perovskite film with potassium iodide additives reached a very high stable PLQE of 66% over time (Figure 11.7a). At different excitation power (Figure 11.7b), it confirms the substantial removal of the non-radiative losses within the perovskite [32]. This finding is also observed in micro-photoluminescence maps of the perovskite films acquired by confocal photoluminescence microscopy that shows the continuous enactment of emission intensity at higher concentration of potassium (Figure 11.7d–f) [32]. The remarkable increases in luminescence efficiency of alloyed perovskite film with potassium compare to other monovalent metal cations, such as Rb, attributed to the tolerance of these perovskites for higher loadings of K than Rb (Figure 11.7c) [43].

11.5 Impact of Monovalent Metal Halides on Photovoltaic Device Characterizations

The PCE of the solar cell is subservient to the photophysical properties of the light absorber layer. It depends on the compositional, structural, and morphological properties of the perovskite layer, determining the output power (P_{out}) and stability. The P_{out} of a solar cell is the product of three photovoltaic parameters, including short-circuit photocurrent density (J_{SC}), open-circuit voltage (V_{OC}), and fill factor (FF) divided by the input power (P_{in}) (Eq. (11.1)).

$$P_{out} = \frac{J_{SC} \cdot V_{oc} \cdot \mathrm{FF}}{\mathrm{P_{in}}} \tag{11.1}$$

To realize maximum P_{out}, all three photovoltaic parameters need to simultaneously improve their thermodynamic limits without compromising the charge collection efficiency. The J_{SC} could be improved by harnessing more solar photons, which depends on the bandgap of the light absorber layer. On the other hand, V_{OC} exhibits a strong dependence on the emission characteristics, including emission quantum yield and charge-carrier recombination dynamics. FF is primarily influenced by the series resistance and the ideality factor.

In particular, high absorption coefficient [47], high emission quantum yield [48], and long charge-carrier lifetimes [49, 50] are essential to maximizing the performance, achieved primarily as a result of device architecture modification [51, 52], compositional [53], and solvent engineering [54] in different PSC architectures. As illustrated previously, the optoelectronic properties could be tailored by introducing an optimized amount of monovalent metal halide additives into the perovskite precursor solutions. For example, the conversion of PbI_2 to $MAPbI_3$ can be improved by adding MI (M = Na, Ag, and Cu) additives. The devices were fabricated using FTO/compact-TiO_2/mesoporous-TiO_2/$MAPbI_3$/spiro-MeOTAD/Au architecture

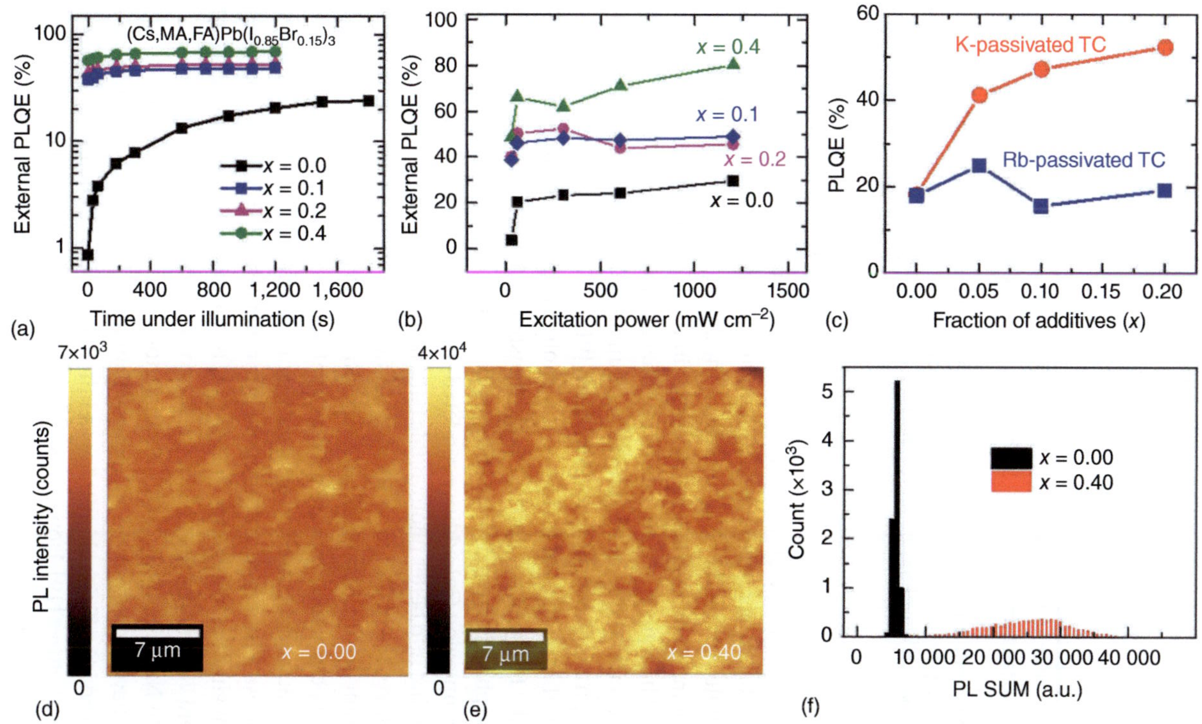

Figure 11.7 Increased radiative efficiency of perovskite thin films via addition of monovalent metal halides. (a) PLQE time course for (Cs, FA, MA)$Pb(I_{0.85}Br_{0.15})_3$ films with different fractions of potassium illuminated with a 532-nm laser at an excitation intensity equivalent to approximately 1 sun (60 mW cm^{-2}) in an ambient atmosphere. (b) PLQE as a function of excitation power measured by a 532-nm continuous-wave laser for potassium-based perovskite thin films in an ambient atmosphere. Source: (a, b, d, e) Abdi-Jalebi et al. [32]. (c) PLQE of K- and Rb-based perovskite thin films measured under illumination with a 532 nm laser at an excitation intensity equivalent to ~1 sun (~60 mW cm^{-2}). Source: (c) Abdi-Jalebi et al. [43]. Licenced under CC BY-4.0. Confocal photoluminescence intensity maps with 405-nm excitation measured in ambient atmosphere for (Cs,FA,MA)$Pb(I_{0.85}Br_{0.15})_3$ perovskite thin films with (d) $x = 0.0$ and (e) $x = 0.40$ fraction of potassium. (f) Histograms of the absolute photoluminescence intensities extracted from the respective maps for $x = 0$ and $x = 0.40$.

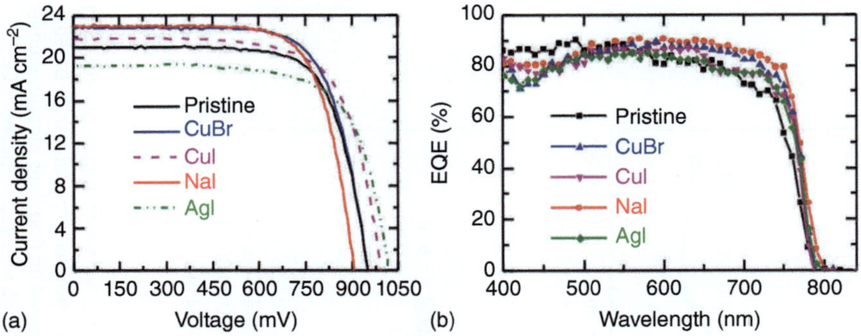

Figure 11.8 Photovoltaic performance of solar cells containing NaI, AgI, CuI, and CuBr based $MAPbI_3$ films. (a) *J–V* data recorded under standard illumination and (b) EQE as a function of illumination wavelength. Source: Abdi-Jalebi et al. [9].

and were examined under simulated air mass 1.5 global (AM1.5G) solar irradiation. The current density–voltage (*J–V*) characteristics presented in Figure 11.8a of the pristine- and MI-based perovskite devices demonstrated improvement in the PCE. A relatively smaller cation, i.e. Na^+ increased the J_{SC} by 2 mA cm^{-2}, and FF from 70% to 73%, leading to an overall enhancement in the PCE to 15% (PCE of control devices = 14%). This gain in output power is realized at the expense of a 50 mV drop in V_{OC}, which occurred due to the presence of undesired contacts associated with the relatively high roughness of perovskite films [9].

In contrast, AgI introduction increased the photovoltage significantly by 100 mV and improved the FF by balancing the charge transport behavior. However, a significant drop in the photocurrent density marginally improved the overall performance of the device. Employing CuI additive rendered the $MAPbI_3$ films smooth and contiguous, translating into better photovoltaic parameters, including J_{SC}, V_{OC}, and FF. By changing the source of Cu^+ to CuBr, P_{out} of 15.6% was obtained, primarily due to high photocurrent density approaching 23 mA cm^{-2}. Furthermore, when NaI-, CuI-, and CuBr-based perovskite films are used as light absorbers, the photocurrent densities are enhanced as supported by external quantum efficiency (EQE) data shown in Figure 11.8b. As a function of illumination wavelength, EQE realized that these additives do not alter the bandgap of the $MAPbI_3$ as the generation of photocurrents begins at around 780 nm irrespective of the monovalent metal cation [9].

In addition to Na^+, relatively larger alkali cations, including rubidium (Rb) and potassium (K), have been explored as a monovalent metal halide additive to improve the power conversion efficiencies. Rb^+ was initially reported to be incorporated in the perovskite structures [19]. However, structural studies established that Rb^+ is too small to occupy the A-site in the 3D $MPbX_3$ lattice [55, 56]. The influence of Rb^+ as a monovalent additive cation on PSCs' performance was further explored, as shown in Figure 11.9a. This study showed that RbI incorporation into the precursor solution increased the device's open-circuit voltage by 100 mV, leading to the realization of V_{OC} as high as 1.23 V under standard illumination conditions Figure 11.9b.

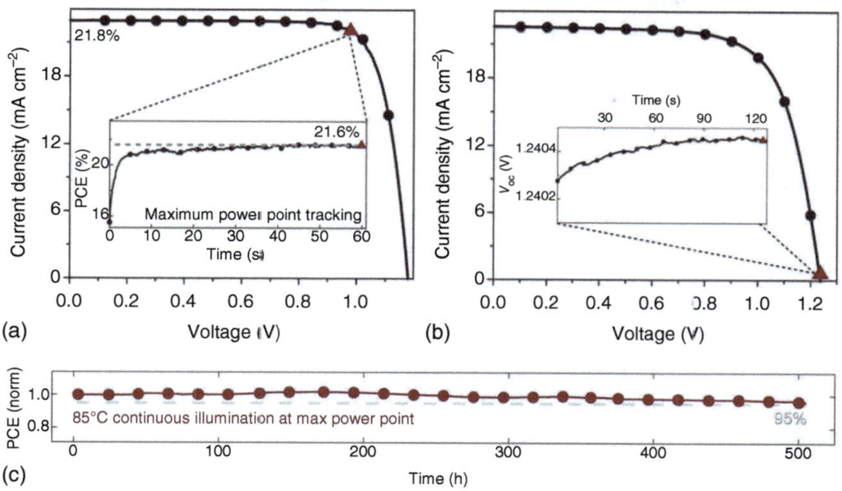

Figure 11.9 Photovoltaic performance of solar cells containing RbI based perovskite films. (a) Current density–voltage (J–V) curve, taken at 10 mV s^{-1} scan rate, of the solar cell with 21.8% efficiency (V_{oc} = 1180 mV, J_{sc} = 22.8 mA cm^{-2}, and FF = 81%). The inset shows the scan rate-independent MPP tracking for 60 s, resulting in a stabilized efficiency of 21.6% at 977 mV and 22.1 mA cm^{-2} (displayed as triangles in the J–V and MPP scans). (b) J–V curve of the highest-V_{oc} device. The inset shows the V_{oc} over 120 s, resulting in 1240 mV (displayed as the red triangles in the J–V and V_{oc} scans). (c) Operational stability under full-sun illumination at 85 °C recorded for the perovskite solar cell containing RbI additive.

This highlights a powerful design strategy for high V_{OC} and more effective charge extraction in PSCs through energy level engineering of the interfaces. A general comparison establishes that remarkable V_{OC} is a key photovoltaic metric that distinguishes PSCs from other photovoltaic technologies. Therefore, a high degree of interest focusing on V_{OC} has emerged in PSCs involving applying various additive-based light absorbers. Impressive progress has been made, especially in V_{OC}, which has reached an unprecedented mark of 1.26 V. Besides high V_{OC}, the RbI-based device yields a high FF value. Electrochemical impedance and other electrical characterization techniques revealed the improvement of charge transport across the interfaces on either side of the RbI-based perovskite absorber layer [19]. In summary, this combination of small perturbation techniques, including intensity-modulated photocurrent spectroscopy (IMPS) and intensity-modulated photovoltage spectroscopy (IMVS), revealed that Rb incorporated PSCs show faster photocurrent transient response, enhanced charge transport, slower charge-carrier recombination, and less capacitance [19, 56].

To understand the role of Rb^+, a controlled experiment involving mixing of RbI and $MAPbI_3$ was undertaken, and the photocurrent density–voltage (J–V) (Figure 11.10a) measurements were carried out on the devices involving $MAPbI_3$ and $Rb_xMA_{1-x}PbI_3$ perovskite films. The performance decreased from 19.0% to 13.4%, and V_{OC} decreased from 1086 mV for $MAPbI_3$ to 1010 mV by introducing 15% RbI into the solution, i.e when $Rb_{0.15}MA_{1-0.15}PbI_3$ films are used as the light absorbers (Figure 11.10b). The undesired formation of pin-holes and the occurrence

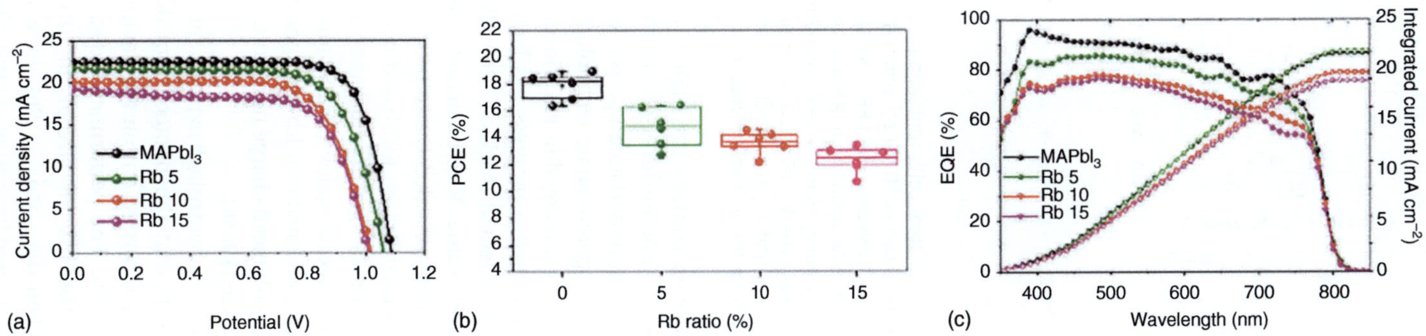

Figure 11.10 Photovoltaic performance of solar cells containing RbI based $MAPbI_3$ films. (a) *J–V* data recorded under standard illumination, (b) the extracted PCE values showing the dependence of efficiency on RbI concentration, and (c) EQE as a function of illumination wavelength. Source: Uchida et al. [57].

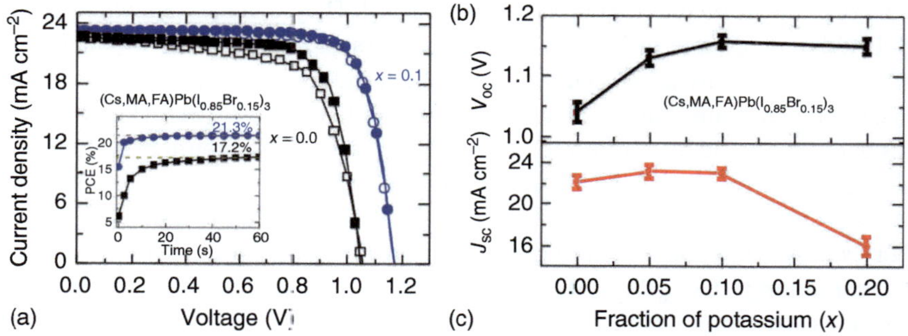

Figure 11.11 Photovoltaic performance of solar cells containing KI-based perovskite films. (a) Forward (open symbols) and reverse (closed symbols) *J–V* curves of the best-performing solar cells with $(Cs_{0.06}FA_{0.79}MA_{0.15})Pb(I_{0.85}Br_{0.15})_3$ absorbers without ($x = 0$) and with ($x = 0.1$) KI, measured under AM1.5, 100 mW cm^{-2}. Inset stabilized power output under the same conditions. (b) Open-circuit voltage (V_{OC}) and (c) short-circuit current density (J_{SC}) as functions of potassium fraction on *x*, with error bars representing the standard deviation across 10 devices for each composition. Source: Abdi-Jalebi et al. [32].

of increased parasitic non-radiative recombinations in the $Rb_xMA_{1-x}PbI_3$ films could explain the drop in the photovoltaic performance of the devices. The $MAPbI_3$ perovskite band edge remains unchanged with increasing Rb^+ content from 5% to 15%. As the EQE) (Figure 11.10c) recorded by monitoring the short-circuit current at different incident wavelengths, the photocurrent generation begins at around the same incident wavelength irrespective of the composition [57].

Given the remarkable photovoltaic performances obtained from judiciously optimized perovskite composition containing RbI, rational studies were carried by choosing mixed-halide ($MAPb(I_{0.85}Br_{0.15})_3$) and mixed-cation ($FA_{0.85}MA_{0.15}PbI_3$) systems. The devices with normal architecture fabricated using $Rb_xMA_{1-x}Pb(I_{0.85}Br_{0.15})_3$ and $Rb_x(FA_{0.85}MA_{0.15})_{1-x}PbI_3$ films amply established that the presence of Rb^+ has a detrimental effect on the PCE of the solar cell. The EQE spectra recorded for the devices involving Rb-based absorbers, i.e. $Rb_xMA_{1-x}Pb(I_{0.85}Br_{0.15})_3$ and $Rb_x(FA_{0.85}MA_{0.15})_{1-x}PbI_3$ rule out the incorporation of Rb^+ in these 3D perovskite lattices as the current generation as a function of illumination wavelength starts at the same wavelength [56].

In a similar direction, careful studies revealed that monovalent halides based on K^+ could influence various photovoltaic parameters. For example, the introduction of KI into $(Cs_{0.06}FA_{0.79}MA_{0.15})Pb(I_{0.85}Br_{0.15})_3$, a state-of-the-art triple-cation perovskite composition, greatly improved the external PLQEs, with internal yields exceeding 95%. This consequently improved the open-circuit voltage from 1.05 to 1.17 V and FF from 73% to 79%, translating into PCEs greater than 21% (Figure 11.11) [32]. As compared to Rb^- incorporation, K^+ seems to exhibit a profound effect on the luminescence due to its relatively higher tolerance limit for the triple-cation perovskites. The structural and elemental analyses revealed that Rb^+ exhibits a selective affinity toward iodide, whereas K^+ binds to bromide [43].

It is worth emphasizing that a particular challenge encountered while exploring the full potential of perovskite devices is to understand the origin of degradation and nature of parasitic recombination processes [58, 59]. They adversely affect the device stability, dynamics, and diffusion of charge carriers and increase the V_{OC} deficit in PSCs [60]. The nature and origin of traps, which induce non-radiative recombinations, seem predominantly associated with the halide vacancies [61]. The density of traps or defects could be minimized by introducing an excess amount of monovalent cation-based halides such as KI and RbI [33]. In addition to PCE, the monovalent metal cations were also probed for their impact on the stability of PSCs.

For example, RbI was found to greatly improve the operational stability of solar cells containing $(Cs_{0.06}FA_{0.79}MA_{0.15})Pb(I_{0.85}Br_{0.15})_3$ based light absorber layer. These measurements were carried out at the maximum output power under continuous illumination at 85 °C (Figure 11.9c). Various investigations revealed that KI-treatment greatly mitigates the photoinduced phase segregation in mixed-halide perovskites, which consequently improved the device stability under different stresses [10, 32]. Fundamentally, it is impossible to have water or moisture-resistant perovskite film or a fully assembled device, primarily because of the ionic nature of metal halide perovskites. By decorating these perovskite structures or films with hydrophobic or less reactive species without compromising the charge transport properties, they could render these perovskite systems immune to moisture. Given that these PSCs will be sealed for large-scale deployment, obtaining operationally stable devices under inert conditions in the laboratory could be sufficient to provide proof of the concept. However, photo- and thermal-induced degradation occurring in the sealed perovskite devices arising from the ion migration within the bulk of the perovskite layer, chemical instability of interfaces, and charge selective contacts. These monovalent metal halides greatly minimize these detrimental processes and consequently improve the overall stability of the devices.

References

1 Bai, S., Da, P., Li, C. et al. (2019). Planar perovskite solar cells with long-term stability using ionic liquid additives. *Nature* 571 (7764): 245–250. https://doi.org/10.1038/s41586-019-1357-2.

2 Kim, D., Jung, H.J., Park, I.J. et al. (2020). Efficient, stable silicon tandem cells enabled by anion-engineered wide-bandgap perovskites. *Science* 368 (6487): 155–160. https://doi.org/10.1126/science.aba3433.

3 Ban, M., Zou, Y., Rivett, J.P.H. et al. (2018). Solution-processed perovskite light emitting diodes with efficiency exceeding 15% through additive-controlled nanostructure tailoring. *Nat. Commun.* 9 (1): 3892. https://doi.org/10.1038/s41467-018-06425-5.

4 Lin, K., Xing, J., Quan, L.N. et al. (2018). Perovskite light-emitting diodes with external quantum efficiency exceeding 20 percent. *Nature* 562 (7726): 245–248. https://doi.org/10.1038/s41586-018-0575-3.

5 Senanayak, S.P., Abdi-Jalebi, M., Kamboj, V.S. et al. (2020). A general approach for hysteresis-free, operationally stable metal halide perovskite field-effect transistors. *Sci. Adv.* 6 (15): eaaz4948. https://doi.org/10.1126/sciadv.aaz4948.

6 Basiricò, L., Senanayak, S.P., Ciavatti, A. et al. (2019). Detection of X-rays by solution-processed cesium-containing mixed triple cation perovskite thin films. *Adv. Funct. Mater.* 29 (34): 1902346. https://doi.org/10.1002/adfm.201902346.

7 Li, Z., Klein, T.R., Kim, D.H. et al. (2018). Scalable fabrication of perovskite solar cells. *Nat. Rev. Mater.* 3 (4): 1–20. https://doi.org/10.1038/natrevmats.2018.17.

8 Tennyson, E.M., Doherty, T.A.S., and Stranks, S.D. (2019). Heterogeneity at multiple length scales in halide perovskite semiconductors. *Nat. Rev. Mater.* 4 (9): 573–587. https://doi.org/10.1038/s41578-019-0125-0.

9 Abdi-Jalebi, M., Dar, M.I., Sadhanala, A. et al. (2016). Impact of monovalent cation halide additives on the structural and optoelectronic properties of $CH_3NH_3PbI_3$ perovskite. *Adv. Energy Mater.* 6: 1502472.

10 Son, D.-Y., Kim, S.G., Seo, J.Y. et al. (2018). Universal approach toward hysteresis – free perovskite solar cell via defect engineering. *J. Am. Chem. Soc.* 140 (4): 1358–1364. https://doi.org/10.1021/jacs.7b10430.

11 Han, J., Luo, S., Yin, X. et al. (2018). Hybrid PbS quantum-dot-in-perovskite for high-efficiency perovskite solar cell. *Small* 14 (31): 1801016. https://doi.org/10.1002/smll.201801016.

12 McMeekin, D.P., Sadoughi, G., Rehman, W. et al. (2016). A mixed-cation lead mixed-halide perovskite absorber for tandem solar cells. *Science* 351: 151–155.

13 Azam, M., Liu, K., Yue, S. et al. (2019). The positive function of incorporation of small molecules into perovskite materials for high-efficient stable solar cells. *Sol. RRL* 3 (3): 1800327. https://doi.org/10.1002/solr.201800327.

14 Lin, Y.H., Sakai, N., Da, P. et al. (2020). A piperidinium salt stabilizes efficient metal-halide perovskite solar cells. *Science* 369: 96–102.

15 Han, T.H., Lee, J.W., Choi, C. et al. (2019). Perovskite-polymer composite cross-linker approach for highly-stable and efficient perovskite solar cells. *Nat. Commun.* 10 (1): 1–10. https://doi.org/10.1038/s41467-019-08455-z.

16 Xu, J., Buin, A., Ip, A.H. et al. (2015). Perovskite-fullerene hybrid materials suppress hysteresis in planar diodes. *Nat. Commun.* 6 (1): 1–8. https://doi.org/10.1038/ncomms8081.

17 Turren-Cruz, S.H., Hagfeldt, A., and Saliba, M. (2018). Methylammonium-free, high-performance, and stable perovskite solar cells on a planar architecture. *Science* https://doi.org/10.1126/science.aat3583.

18 Wang, K., Subhani, W.S., Wang, Y. et al. (2019). Metal cations in efficient perovskite solar cells: progress and perspective. *Adv. Mater.* 31 (50): 1902037. https://doi.org/10.1002/adma.201902037.

19 Saliba, M., Matsui, T., Domanski, K. et al. (2016). Incorporation of rubidium cations into perovskite solar cells improves photovoltaic performance. *Science* 354: 206–209.

20 Abdi-Jalebi, M., Pazoki, M., Philippe, B. et al. (2018). Dedoping of lead halide perovskites incorporating monovalent cations. *ACS Nano* 12: 7301–7311.

21 Abdi-Jalebi, M., Dar, M.I., Sadhanala, A. et al. (2017). Monovalent cation doping of $CH_3NH_3PbI_3$ for efficient perovskite solar cells. *J. Vis. Exp.* (121): e55307, https://doi.org/10.3791/55307.

22 Mir, W.J., Sheikh, T., Arfin, H. et al. (2020). Lanthanide doping in metal halide perovskite nanocrystals: spectral shifting, quantum cutting and optoelectronic applications. *NPG Asia Mater.* https://doi.org/10.1038/s41427-019-0192-0.

23 Saliba, M., Matsui, T., Seo, J.Y. et al. (2016). Cesium-containing triple cation perovskite solar cells: improved stability, reproducibility and high efficiency. *Energy Environ. Sci.* 9: 1989–1997.

24 Andaji-Garmaroudi, Z., Abdi-Jalebi, M., Guo, D. et al. (2019). A highly emissive surface layer in mixed-halide multication perovskites. *Adv. Mater.* 31 (42): 1902374. https://doi.org/10.1002/adma.201902374.

25 Bowman, A.R., Klug, M.T., Doherty, T.A. et al. (2019). Microsecond carrier lifetimes, controlled p-doping, and enhanced air stability in low-bandgap metal halide perovskites. *ACS Energy Lett.* 4 (9): 2301–2307. https://doi.org/10.1021/acsenergylett.9b01446.

26 Gong, J., Yang, M., Rebollar, D. et al. (2018). Divalent anionic doping in perovskite solar cells for enhanced chemical stability. *Adv. Mater.* 30 (34): 1800973. https://doi.org/10.1002/adma.201800973.

27 Zhao, W., Yao, Z., Yu, F. et al. (2018). Alkali metal doping for improved $CH_3NH_3PbI_3$ perovskite solar cells. *Adv. Sci.* https://doi.org/10.1002/advs.201700131.

28 Shahbazi, S., Tsai, C.M., Narra, S. et al. (2017). Ag doping of organometal lead halide perovskites: morphology modification and p-type character. *J. Phys. Chem. C* 121: 3673–3679.

29 Bi, C., Wang, S., Li, Q. et al. (2019). Thermally stable copper(II)-doped cesium lead halide perovskite quantum dots with strong blue emission. *J. Phys. Chem. Lett.* 10: 943–952.

30 Lee, J.W., Kim, D.H., Kim, H.S. et al. (2015). Formamidinium and cesium hybridization for photo- and moisture-stable perovskite solar cell. *Adv. Energy Mater.* 5 (20): 1501310. https://doi.org/10.1002/aenm.201501310.

31 Feldmann, S., Macpherson, S., Senanayak, S.P. et al. (2020). Photodoping through local charge carrier accumulation in alloyed hybrid perovskites for highly efficient luminescence. *Nat. Photonics* 14 (2): 123–128. https://doi.org/10.1038/s41566-019-0546-8.

32 Abdi-Jalebi, M., Andaji-Garmaroudi, Z., Cacovich, S. et al. (2018). Maximizing and stabilizing luminescence from halide perovskites with potassium passivation. *Nature* 555: 497–501.

33 Jalebi, M. A. (2018) Chemical modifications and passivation approaches in metal halide perovskite solar cells. https://doi.org/10.17863/CAM.30584.

34 Tailor, N.K., Abdi-Jalebi, M., Gupta, V. et al. (2020). Recent progress in morphology optimization in perovskite solar cell. *J. Mater. Chem. A* 8 (41): 21356–21386. https://doi.org/10.1039/D0TA00143K.

35 Chen, Y., Li, N., Wang, L. et al. (2019). Impacts of alkaline on the defects property and crystallization kinetics in perovskite solar cells. *Nat. Commun.* 10 (1): 1–10. https://doi.org/10.1038/s41467-019-09093-1.

36 Guo, D., Andaji Garmaroudi, Z., Abdi-Jalebi, M. et al. (2019). Reversible removal of intermixed shallow states by light soaking in multication mixed halide perovskite films. *ACS Energy Lett.* https://doi.org/10.1021/acsenergylett.9b01726.

37 Liu, C., Cheng, Y.B., and Ge, Z. (2020). Understanding of perovskite crystal growth and film formation in scalable deposition processes. *Chem. Soc. Rev.* https://doi.org/10.1039/c9cs00711c.

38 Nie, W., Tsai, H., Asadpour, R. et al. (2015). High-efficiency solution-processed perovskite solar cells with millimeter-scale grains. *Science* 347: 522–525.

39 Tang, Z., Uchida, S., Bessho, T. et al. (2018). Modulations of various alkali metal cations on organometal halide perovskites and their influence on photovoltaic performance. *Nano Energy* 45: 184–192. https://doi.org/10.1016/j.nanoen.2017.12.047.

40 Dar, M.I., Abdi-Jalebi, M., Arora, N. et al. (2016). Growth engineering of $CH_3NH_3PbI_3$ structures for high-efficiency solar cells. *Adv. Energy Mater.* 6: 1501358.

41 Zheng, F., Chen, W., Bu, T. et al. (2019). Triggering the passivation effect of potassium doping in mixed-cation mixed-halide perovskite by light illumination. *Adv. Energy Mater.* 9 (24): 1901016. https://doi.org/10.1002/aenm.201901016.

42 Cacovich, S., Matteocci, F., Abdi-Jalebi, M. et al. (2018). Unveiling the chemical composition of halide perovskite films using multivariate statistical analyses. *ACS Appl. Energy Mater.* 1 (12): 7174–7181. https://doi.org/10.1021/acsaem.8b01622.

43 Abdi-Jalebi, M., Andaji-Garmaroudi, Z., Pearson, A.J. et al. (2018). Potassium-and rubidium-passivated alloyed perovskite films: optoelectronic properties and moisture stability. *ACS Energy Lett.* 3 (11): 2671–2678. https://doi.org/10.1021/acsenergylett.8b01504.

44 Zhang, R., Li, M., Huan, Y. et al. (2019). A potassium thiocyanate additive for hysteresis elimination in highly efficient perovskite solar cells. *Inorg. Chem. Front.* 6 (2): 434–442. https://doi.org/10.1039/c8qi01020j.

45 Nam, J.K., Chai, S.U., Cha, W. et al. (2017). Potassium incorporation for enhanced performance and stability of fully inorganic cesium lead halide perovskite solar cells. *Nano Lett.* 17: 2028–2033.

46 Kim, S.G., Li, C., Guerrero, A. et al. (2019). Potassium ions as a kinetic controller in ionic double layers for hysteresis-free perovskite solar cells. *J. Mater. Chem. A* 7 (32): 18807–18815. https://doi.org/10.1039/c9ta07595j.

47 Abdi-Jalebi, M., Ibrahim Dar, M., Sadhanala, A. et al. (2020). Optical absorption and photoluminescence spectroscopy. In: *Characterization Techniques for Perovskite Solar Cell Materials*, 49–79. Elsevier https://doi.org/10.1016/B978-0-12-814727-6.00003-7.

48 Richter, J.M., Abdi-Jalebi, M., Sadhanala, A. et al. (2016). Enhancing photoluminescence yields in lead halide perovskites by photon recycling and light out-coupling. *Nat. Commun.* 7: 13941.

49 Longo, G., Mahesh, S., Buizza, L.R. et al. (2020). Understanding the performance-limiting factors of $Cs_2AgBiBr_6$ double-perovskite solar cells. *ACS Energy Lett.* 5 (7): 2200–2207. https://doi.org/10.1021/acsenergylett.0c01020.

50 Doherty, T.A.S., Winchester, A.J., Macpherson, S. et al. (2020). Performance-limiting nanoscale trap clusters at grain junctions in halide perovskites. *Nature* 580 (7803): 360–366. https://doi.org/10.1038/s41586-020-2184-1.

51 Pazos-Outon, L.M., Szumilo, M., Lamboll, R. et al. (2016). Photon recycling in lead iodide perovskite solar cells. *Science* 351: 1430–1433.

52 Friend, R.H., Deschler, F., Pazos-Outón, L.M. et al. (2019). Back-contact perovskite solar cells. *Sci. Video Protoc.* 1: 1–10.

53 Merdasa, A., Kiligaridis, A., Rehermann, C. et al. (2019). Impact of excess lead iodide on the recombination kinetics in metal halide perovskites. *ACS Energy Lett.* 4 (6): 1370–1137. https://doi.org/10.1021/acsenergylett.9b00774.

54 Jeon, N.J., Noh, J.H., Kim, Y.C. et al. (2014). Solvent engineering for high-performance inorganic? Organic hybrid perovskite solar cells. *Nat. Mater.* 13: 897–903.

55 Kubicki, D.J., Prochowicz, D., Hofstetter, A. et al. (2017). Phase segregation in Cs-, Rb- and K-doped mixed-Cation $(MA)_x(FA)_{1-x}PbI_3$ hybrid perovskites from solid-state NMR. *J. Am. Chem. Soc.* 139: 14173–14180.

56 Yadav, P., Dar, M.I., Arora, N. et al. (2017). The role of rubidium in multiple-cation-based high-efficiency perovskite solar cells. *Adv. Mater.* 29: 1701077.

57 Uchida, R., Binet, S., Arora, N. et al. (2018). Insights about the absence of Rb cation from the 3D perovskite lattice: effect on the structural, morphological, and photophysical properties and photovoltaic performance. *Small* 14 (36): 1802033. https://doi.org/10.1002/smll.201802033.

58 Abdi-Jalebi, M., Dar, M.I., Senanayak, S.P. et al. (2019). Charge extraction via graded doping of hole transport layers gives highly luminescent and stable metal halide perovskite devices. *Sci. Adv.* 5 (2): eaav2012. https://doi.org/10.1126/sciadv.aav2012.

59 Stavrakas, C., Zhumekenov, A.A., Brenes, R. et al. (2018). Probing buried recombination pathways in perovskite structures using 3D photoluminescence tomography. *Energy Environ. Sci.* 11 (10): 2846–2852. https://doi.org/10.1039/c8ee00928g.

60 Zhao, B., Abdi-Jalebi, M., Tabachnyk, M. et al. (2017). High open-circuit voltages in tin-rich low-bandgap perovskite-based planar heterojunction photovoltaics. *Adv. Mater.* 29: 1604744.

61 Jones, T.W., Osherov, A., Alsari, M. et al. (2019). Lattice strain causes non-radiative losses in halide perovskites. *Energy Environ. Sci.* 12 (2): 596–606. https://doi.org/10.1039/c8ee02751j.

12

Charge Carrier Dynamics in Perovskite Solar Cells

Mohd T. Khan[1], Abdullah Almohammedi[1], Samrana Kazim[2,3], and Shahzada Ahmad[2,3]

[1] *Islamic University of Madinah, Department of Physics, Faculty of Science, Prince Naifbin Abdulaziz, Al Jamiah, Madinah 42351, Kingdom of Saudi Arabia*
[2] *BCMaterials, Basque Center for Materials, Applications and Nanostructures, UPV/EHU Science Park, 48940 Leioa, Spain*
[3] *IKERBASQUE, Basque Foundation for Science, Bilbao, 48013, Spain*

12.1 Introduction

Organo-lead halide perovskites possess superior optoelectrical properties such as high absorption coefficient of ~10^5 cm^{-1} [1], ambipolar charge transport behavior [2], large charge diffusion length of up to several micrometers [3–5], high charge carrier mobility [6, 7], low exciton binding energy [8, 9], and tunable energy band gap [10–13]. These properties leads to unparalleled success in the field of perovskite solar cells (PSCs) in a short period with an electrifying power conversion efficiencies (PCE) in excess of 25% [14]. It is expected that the PCE of PSCs in the near future will surpass that of crystalline silicon solar cells. Despite the rapid progress in the performance of PSCs, there are several key issues such as non-radiative recombination losses, electrical instability charge accumulation, giant low-frequency capacitance, and others that are not observed in conventional solar cells. These factors hinder the efficient charge transport in perovskite layers and thus prevent the PCE to reach the Shockley–Queisser limit [15, 16]. To overcome such challenges, numerous potential origins have been proposed: trapping of charge carriers [17, 18], ionic migration in the perovskite film [19], chemical or structural changes in the material [20, 21], and redistribution of slowly moving ions [22, 23]. The primary cause of this phenomenon is the presence of defect states in perovskite layers. The defect that arises in perovskite layer is due to the presence of impurities, or lattice disorder. Moreover, hybrid perovskites exhibit both ionic and electronic conductivities. Ionic conductivity includes the transport of anions (I^-, Cl^-, Br^-) and cation (MA^+, Pb^{2+}). The migration of ionic species creates ionic vacancy/lattice defects in perovskite structure [24]. Another cause of the defect in the perovskite-based devices is electrode polarization. The charge carriers are captured in these trap states (Figure 12.1) and influence the properties of free

Perovskite Solar Cells: Materials, Processes, and Devices, First Edition.
Edited by Shahzada Ahmad, Samrana Kazim, and Michael Grätzel.

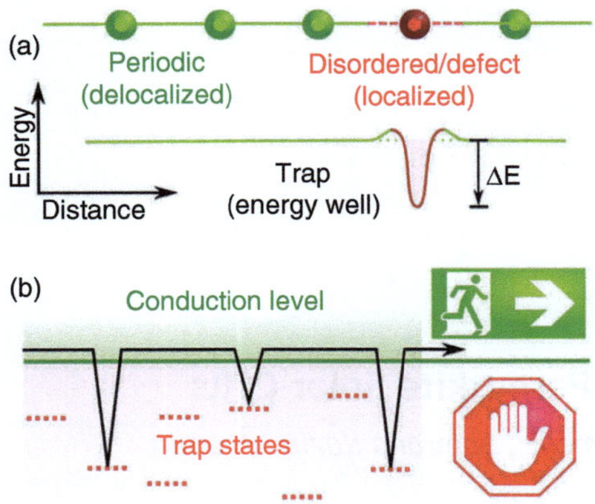

Figure 12.1 (a) Scheme of trap levels (bottom) caused by non-periodic disorder/defect in a semiconductor lattice (top) and (b) charge trapping kinetics, where trap slows down charge carrier transport through trapping and detrapping events. Source: Jin et al. [25].

carriers moving through these states. These traps cause charge scattering [26], *J–V* hysteresis [27], and recombination of charge carriers causing radiative or non-radiative losses [28]. In this context, the understanding of the charge carrier dynamics in perovskites layer is of importance for the further development of these materials and device optimization, revealing the potential they hold.

The effect of traps on the charge carrier dynamics has been systematically elucidated via various electro-optical techniques such as transient optical spectroscopy, immittance spectroscopy, and space charge-limited current (SCLC) techniques [29–32]. This chapter focuses on the understanding of the fundamental photophysics, exciton/charge carrier dynamics, and interfacial energy alignment processes in PSCs and, for clarity purposes, has been divided into three sections. In Section 12.2, the estimation of important electrical parameters such as charge carrier mobility, electrical conductivity, carrier concentration, trap distribution, and density by SCLC method has been discussed. In Section 12.3, numerous immittance spectroscopy techniques are illustrated such as impedance spectroscopy, capacitance spectroscopy including MS analysis, and thermal admittance spectroscopy. Finally, in Section 12.4, we discussed the transient spectroscopic techniques such as transient absorption spectroscopy (TAS), time-resolved photoluminescence (TRPL), and time-resolved microwave conductivity (TRMC).

12.2 Space Charge-Limited Conduction

SCLC is a basic characterization technique to unravel the charge transport properties in a semiconductor thin film. This techniques have been extensively employed to estimate the electrical parameters of perovskites that affect the performance of solar cells [33–37]. Also, this is a steady-state technique where dark current density (J) as a function of applied bias (V) is measured for a semiconductor material sandwiched between two metal electrodes in either electron-only or hole-only device configuration (Figure 12.2a,b, respectively). Dark *J–V* characteristics of perovskites generally show three different regions, namely, (i) ohmic region, (ii) trap-filling SCLC region,

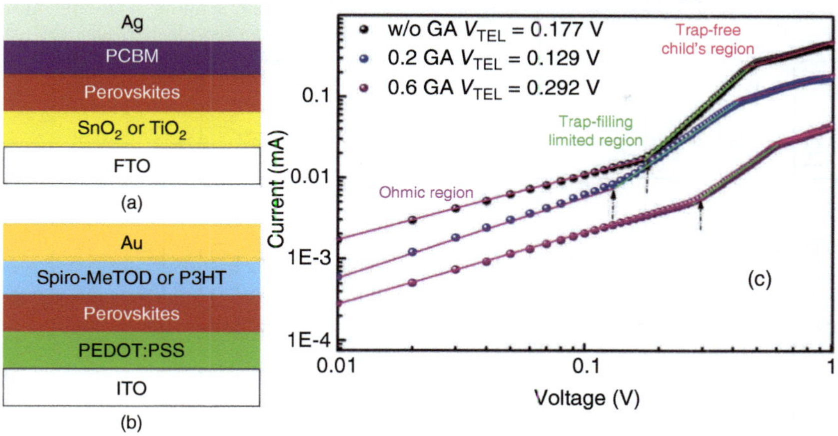

Figure 12.2 (a) Electron-only (b) and hole-only device configurations to measure the SCLC characteristics of perovskite layer and (c) dark J–V characteristics of quadruple-cation perovskite in electron-only device configuration, viz. FTO/SnO_2/$Cs_{0.05}(FA_{0.83}(MA_{1-x}GA_x)_{0.17})_{0.95}Pb(I_{0.83}Br_{0.17})_3$/PCBM/Ag with different weight content of Guanidinium (GA). The plot visualized three different transport regions: (i) ohmic region ($J \propto V$), (ii) trap-filling region ($J \propto V^{n>2}$), and (iii) trap-free Child's region ($J \propto V^2$). Source: Zhang et al. [38].

and (iii) trap-free Child's region as shown in Figure 12.2c. From the measured dark J–V characteristics, electrical transport parameters such as charge carrier mobility (μ), electrical conductivity (σ), charge carrier concentrations (n_c), trap density (n_{trap}), and activation energy (E_A) of semiconductor thin films can be extracted. The electron or hole mobility can be individually evaluated from the quadratic part of the J–V curve through the Mott–Gurney relation [39–41]

$$J = \frac{9}{8}\varepsilon_0 \varepsilon \mu \frac{V^2}{L^3} \tag{12.1}$$

where ε_0, ε, and L are the permittivity, dielectric constant, and thickness of the perovskite layer, respectively. The conductivity (σ) of thin film can be calculated from the linear part of J–V characteristics via the relation $J = \sigma E = \sigma V/L$. The free carrier concentration in perovskite thin films can be determined from estimated conductivity and mobility values through the equation

$$n_c = \frac{\sigma}{e\mu} \tag{12.2}$$

Table 12.1 demonstrates the evaluated electrical parameters of various perovskite materials using SCLC method, where it can be noted that the electrical parameters of the perovskites depend on the growth conditions, organic cation, halide materials, etc. Moreover, the carrier mobility strongly depends on the structure of perovskites. The mobility of single crystal perovskites was found to be higher (up to 164 $cm^2\ V^{-1}\ s^{-1}$ for $MAPbI_3$ [45], 60 $cm^2\ V^{-1}\ s^{-1}$ for $MAPbBr_3$ [2]) SC than of polycrystalline thin films (up to 2×10^{-2} $cm^2\ V^{-1}\ s^{-1}$ for $MAPbI_3$ [49], 0.26 $cm^2\ V^{-1}\ s^{-1}$ for $MAPbBr_3$ [2]). From the evaluated carrier mobility, the electron

Table 12.1 An overview of electrical parameters of organo-lead halide perovskites evaluated through from SCLC measurement.

Active layer	Remark	σ (Ω cm)$^{-1}$	μ (cm^2 V^{-1} s^{-1})	n_C (cm^{-3})	n_{trap} (cm^{-3})	References
$MAPbI_3$ (single crystal)	AVC	1×10^{-8}	2.5	2×0^{10}	3.3×10^{10}	[42]
	ITC		67.2		1.4×10^{10}	[43]
	Film				3.6×10^{10}	[42]
	Hole	3×10^{-8}	65	4×10^9	3×10^{10}	[44]
	Electron		53		3×10^{11}	
	Hole		$\mu_h=164\pm25$		2.0×10^{15}	[45]
	Electron		$\mu_e=24.8\pm4.1$		4.5×10^{10}	
$MAPbI_3$ (poly crystalline)	Pure	10^{-9}	$\mu_e=0.02$			
			$\mu_h=0.008$			
	Doped with NaI	5×10^{-8}	$\mu_e=0.04$			
			$\mu_h=0.07$			
$MAPbBr_3$ (single crystal)	AVC		38		5.8×10^9	[42]
	ITC		24		3.0×10^{10}	[43]
	Film		60			[2]
$MAPbBr_3$ (poly crystalline)	Film		0.26			
$FAPbI_3$ (single crystal)	α-Phase	1.1×10^{-7}	4.4	1.5×10^{11}	6.2×10^{11}	[46]
	δ-Phase	8.9×10^{-9}	0.179	3.1×10^{11}	2.6×10^{12}	
		2.2×10^{-8}	35	3.9×10^9	1.13×10^{10}	[47]
$FAPbBr_3$ (single crystal)		1.5×10^{-8}	62	1.5×10^9	9.6×10^9	
$MA_{0.60}FA_{0.40}PbI_3$	Hole		19.96			[48]
	Electron		18.38			

AVC, antisolvent vapor-assisted crystallization; ITC, inverse temperature crystallization; PC, polycrystalline.

and hole diffusion length L_D can be estimated from

$$L_D = \sqrt{\frac{\mu\tau k_B T}{e}} \tag{12.3}$$

where k_B is the Boltzmann's constant, T is the absolute temperature, and τ is the carrier lifetime. The carrier diffusions of $MAPbI_3$ and $MAPbBr_3$ single crystals grown by inverse temperature crystallization (ITC) process were found to be ~10.0 and 4.3 μm, respectively [43]. The $MAPbI_3$ single crystal grown via low-temperature solution approach shows exceptional high carrier diffusion length of >175 μm [42]. The carrier diffusion lengths of $MAPbBr_3$-integrated single crystal and polycrystalline films were calculated to be 5 and 0.3 μm, respectively, which are smaller than the freestanding $MAPbBr_3$ single crystal [2]. Moreover, the carrier diffusion lengths of $FAPbI_3$ and $FAPbBr_3$ crystals were estimated to be 6.6 and 19.0 μm, respectively [47], whereas the shorter diffusion length was calculated for mixed-cation perovskite $MA_{0.60}FA_{0.40}PbI_3$ with a value of 627 ± 9 nm for electron and 504 ± 8 nm for hole [48].

The density of trap states in the band gap of perovskites can be calculated using equation (12.4) [50]:

$$n_{trap} = \frac{2\varepsilon_0 \varepsilon V_{TFL}}{eL^2} \tag{12.4}$$

The transition voltage V_{TFL} is determined from the kink of J–V curve between ohmic and trap-filling region depicted in Figure 12.2c. The values of traps in different perovskites crystals are shown in Table 12.1. The trap density in polycrystalline perovskites thin films was found to be in the range of 10^{15}–10^{17} cm^{-3}, whereas perovskite single crystals have lower trap density of the order of 10^{9}–10^{11} cm^{-3}. The trap density in organo-lead halide perovskites is comparable with the trap densities in a wide range of traditional semiconductor polycrystalline Si (10^{13}–10^{14} cm^{-3}) [51], single crystal Si (~10^{8} cm^{-3}) [52], GaAs (10^{13}–10^{15} cm^{-3}) [53], CdTe (10^{13}–10^{16} cm^{-3}) [54, 55], GaN (~5×10^{15} cm^{-3}) [56], and CIGS (10^{11}–10^{15} cm^{-3}) [57, 58].

In addition to quantitative evaluation of traps, the techniques can also be used to investigate the qualitative nature of traps in semiconducting thin films. The trap states that are located above the conduction band (CB) edge or below the valence band (VB) edge does not impact the charge transport properties of free carriers (Figure 12.3). The charge carriers gets trapped only in the defect states that reside in the band gap of semiconductor. Various trap states located in the band gap of semiconductor are depicted in Figure 12.3 [25]. The shallow traps are located near the edge of CB or VB with energy depth $\Delta E < k_B T$. The charge carriers trapped can detrapped by means of thermal excitations. Hence, temperature-dependent SCLC measurements can be used to investigate the trap states near the band edge of CB and VB of perovskite layer. The electron-only device configuration can be used to probe the trap states near the CB edge, whereas the traps near VB can be investigated by using hole-only device configuration. Sargent and coworkers [44] have measured the temperature-dependent dark J–V characteristics to evaluate the trap density in $MAPbI_3$ single crystal by using both electron-only and hole-only

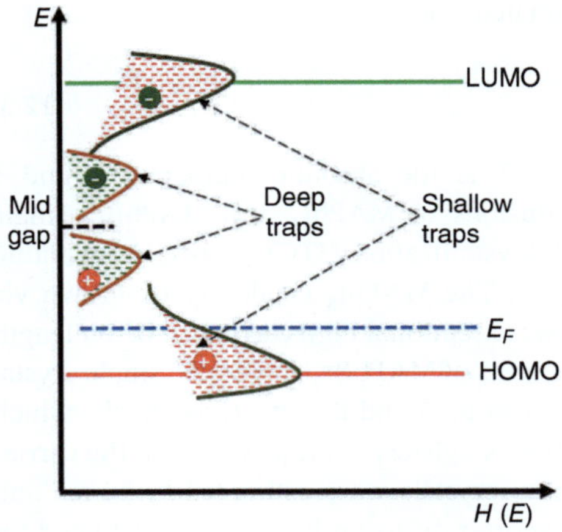

Figure 12.3 Visualization of density of shallow and deep traps for electrons and holes.

device configurations. The total density of charge carriers n_s (trapped and free) can be expressed through the expression [44, 59]

$$\frac{dn_s}{dE_{Fn}} = \frac{1}{k_B T} \frac{\varepsilon_0 \varepsilon}{eL^2} \frac{(2l-1)}{l^2}(1+C) \tag{12.5}$$

where l is the slope of log$J$ vs. log$V$ curve and the parameter C is defined as

$$C = \frac{B(2l-1) + B^2(3m-2) + d[\ln(1 + B/d(\ln V))]}{1 + B(l-1)}$$

where

$$B = \frac{-dl/d(\ln V)}{[m(m-1)(m-2)]}$$

Using Eq. (12.5), the trap density of states (DOS) at the edge of CB $n_{tCB} = 3 \times 10^{11}$ cm^{-3} was found to be exactly 1 order higher than the trap density close to the VB ($n_{tVB} = 3 \times 10^{10}$ cm^{-3}) [44]. Additionally, the traps close to the VB (peak around = 0.1 eV) are found to be shallower than the traps near to the CB (peak around = 0.2 eV).

The charge carriers captured by the traps near the band edge can be thermal activated. The charge carriers captured in these defect states can be released by thermal energy. The energy required to detrap the charge carrier from defect state is called activation energy E_A, which can be calculated from the temperature-dependent J–V measurements from the slope of log(J) and ($1/T$) plot at each voltage point through the equation

$$E_A = -\frac{d(\ln J)}{d(k_B T)^{-1}} \tag{12.6}$$

The effect of the monovalent cation halide NaI doping on the electrical properties of $MAPbI_3$ was probed through SCLC method [39]. It was found that the doping of halide cation reduces the activation energy for hole transport from 198 to 137 meV and electron transport from 135 to 77 meV due to the enhancement of charge carrier mobility (Table 12.1). The decrease in activation energy is attributed to enhancement of carrier concentration in $MAPbI_3$ thin films due to NaI doping, which resulted in filling up the traps at the transport level and decreases the barrier for transport. Moreover, when tetracene layer was inserted in between the perovskite layer and HTL, a significant drop in the activation energy was detected [60].

12.3 Immitance Spectroscopy

The immittance spectroscopy includes the techniques where electrical response of materials to an alternating voltage is measured [61]. This comprises a range of electrical methods such as impedance, admittance, capacitance, and dielectric spectroscopies. These methods provides the insight of the electrical properties such as electrical and ionic transport, charge accumulation, charge relaxation process, trap density of states, interfacial recombination, lattice distortions, and electrode polarization in PSCs [62, 63].

12.3.1 Impedance Spectroscopy

Impedance spectroscopy is profound technique to investigate the various physical processes in PSCs by providing full device investigation through single and simple experimental setup. Several groups have investigated the interfacial recombination processes at the charge extraction layer (ETL or HTL)/perovskite interface in PSCs. The electrical response of the PSCs in the frequency domain is generally analyzed by fitting to an equivalent circuit, and important electrical parameters, among others, such as resistance, capacitance, and time constant can be extracted. The impedance *C–f* measurement is performed through impedance spectroscopy technique, where an AC voltage $V(t) = V_0e^{i\omega t}$ is applied at a given frequency (ω) and then response of AC current $I(t) = I_0e^{i(\omega t + \phi)}$ flowing through the sample is probed. Where V_0 and I_0 are the peak value of the AC voltage and current, respectively, ϕ is the phase difference between $V(t)$ and $I(t)$, and t is time. The impedance spectra of sample are obtained through the relation

$$Z(\omega) = \frac{V(t)}{I(t)} = Z_0e^{-i\phi} \tag{12.7}$$

with $Z_0 = {}^{V_0}/_{I_0}$. In a complex plane the impedance is expressed as

$$Z = Z' + iZ'' \tag{12.8}$$

where real part of impedance Z' represents the resistance R of the sample and imaginary part Z'' ascribed as reactance. The reactance Z'' may be positive or negative. If Z″ is negative, the impedance is dominated by capacitance, whereas if Z'' is positive, the reactance is inductive. The basic impedance spectroscopy elements are defined in Table 12.2, and the corresponding Nyquist plot (Z' vs. Z'') are shown in Figure 12.4.

The combination of two or more passive elements modifies the Nyquist plot shown in Figure 12.5. The impedance of series RC circuit shown in Figure 12.5a,b

Table 12.2 Basic elements of impedance spectroscopy.

Element	Symbol	Impedance
Resistance	R	$Z = R$
Capacitance	C	$Z = 1/i2\pi fC$
Inductance	L	$Z = i2\pi f L$
Gerischer	G	$Z = R_{G1}/\sqrt{1 + i2\pi\tau_{G1}f}$
Warburg	W1	$Z = \sqrt{2}\sigma/\sqrt{i2\pi f}$
Constant phase element (CPE)	Q1	$Z = 1/Q_1(i2\pi f)^{\alpha_1}$

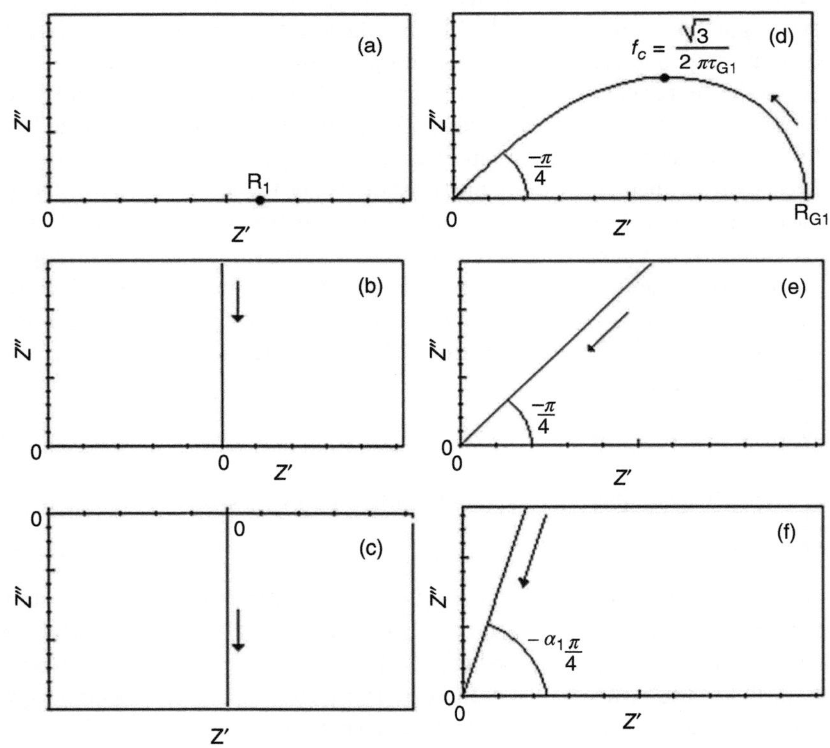

Figure 12.4 Nyquist plot (Z' vs. Z'') of (a) resistance, (b) capacitance, (c) inductor, (d) Gerischer, (e) Warburg, and (f) constant phase element.

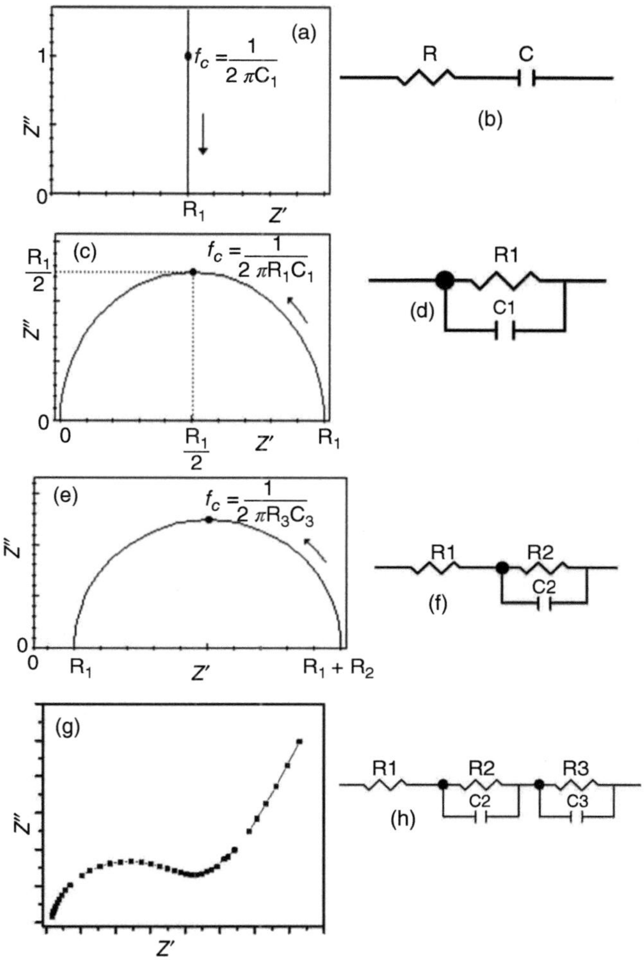

Figure 12.5 Nyquist plots and corresponding equivalent circuits for (a, b) series R–C circuit, (c, d) a parallel R–C circuit, (e, f) a parallel R–C circuit including contact resistance R_1 (g, h) two component parallel R–C circuit.

is represented by

$$Z = R_1 + \frac{1}{i2\pi fC_1} \tag{12.9}$$

Impedance of parallel RC circuit shown in Figure 12.5c,d is equal to

$$Z = \frac{R_1}{1 + i2\pi fR_1C_1} \tag{12.10}$$

The commonly used impedance spectra of a PSC are shown in Figure 12.5e [64, 65]. Also, the equivalent circuit of the corresponding impedance spectra is shown in Figure 12.5f, which contain a series resistance R_1 ascribed to contact resistance and an RC element associated with the ionic diffusion, a recombination resistance R_2, and a capacitance C_2 related to the perovskite dielectric relaxation capacitance featured at low frequency. The net impedance of this circuit is given by

$$Z = R_1 + \frac{R_2}{1 + i2\pi f R_2 C_2} \tag{12.11}$$

To improve the electron transfer and collection efficiency, electron transport layer (ETL) is inserted between anode and perovskite layer. Similarly hole transport layer (HTL) is inserted between perovskite layer and cathode. These charge extraction layer provide additional RC element in the equivalent circuit (Figure 12.5h). The resistance R_3 ascribed to the charge transport resistance of CEL and their interfaces with the perovskite absorber [62, 63], whereas the capacitance C_3 attributed to the interfacial charge accumulation [66]. The interfacial feature appears at high frequency (HF), whereas the bulk phenomenon occurs at intermediate frequency (IF) in the Nyquist plot [67, 68]. To compensate the inhomogeneous layer or interface due to roughness or variations in the thickness, normally a constant phase elements (CPE) is used in equivalent circuit instead of an ideal capacitor [69, 70].

Apart from RC circuit usage, transmission line (TL) and Gerischer (G) model for PSCs were also reported. TL pattern is typically noted in devices where the charge transport resistance R_{Ct} is lesser than the recombination resistance R_{rec}. However, if the recombination resistance is reduced instead of transport resistance, a Gerischer (G) pattern was noted [71, 72]. TL is a typical feature observed in dye-sensitized solar cells (DSSCs) and mesostructured PSCs, in which transport is coupled with recombination [71].

By employing TL model, Bisquert et al. [64] analyze the IS of $MAPbI_3$-based PSCs with different ETLs, i.e. either nanostructured TiO_2 or ZrO_2. TL feature was noted in the case of the coating thickness of 2–3 μm, while for very thin electrodes, TL feature was absent. Graetzel group have used the TL model (Figure 12.6a) similar to the report in DSSCs (Figure 12.6b) to investigate the mesoporous TiO_2/$MAPbI_3$/*spiro*-OMeTAD-based solar cells [68]. In this model, they introduced additional RC element describing the c-TiO_2 in contact with the perovskite layer instead of the TL branch for the m-TiO_2. Moreover, an additional RC element (R_{HTM} and C_{HTM}) was also implemented instead of the Warburg impedance to represent the charge transport inside the HTM to compensate the low hole mobility (Figure 12.6c). Garcia-Belmonte and coworkers [75] have presented a simplified TL model to investigate the ambipolar carrier transport and dielectric polarization in perovskite layer. In modified TL model, they observed three arcs: low frequency, intermediate frequency, and high frequency that are associated with the dielectric relaxation process, drift–diffusion/recombination, and geometric capacitance in parallel with transport resistances, respectively. The impedance response of planar PSC is much simpler than that of mesostructured PSCs. TL behavior was not observed in planar PSCs due to rapid transport of carriers through the perovskite films and negligible low-frequency effect [76]. Recently, this behavior was also reported for FTO/TiO_2/$MAPbI_3$/*spiro*-OMeTAD [77]. However, this model was not found to be very useful to interpret the IS signal of PSCs as of DSSCs.

The impedance spectra also provide insight on the polarization of the perovskite. For instance, three distinguished arc appeared in the impedance plot of PSCs based

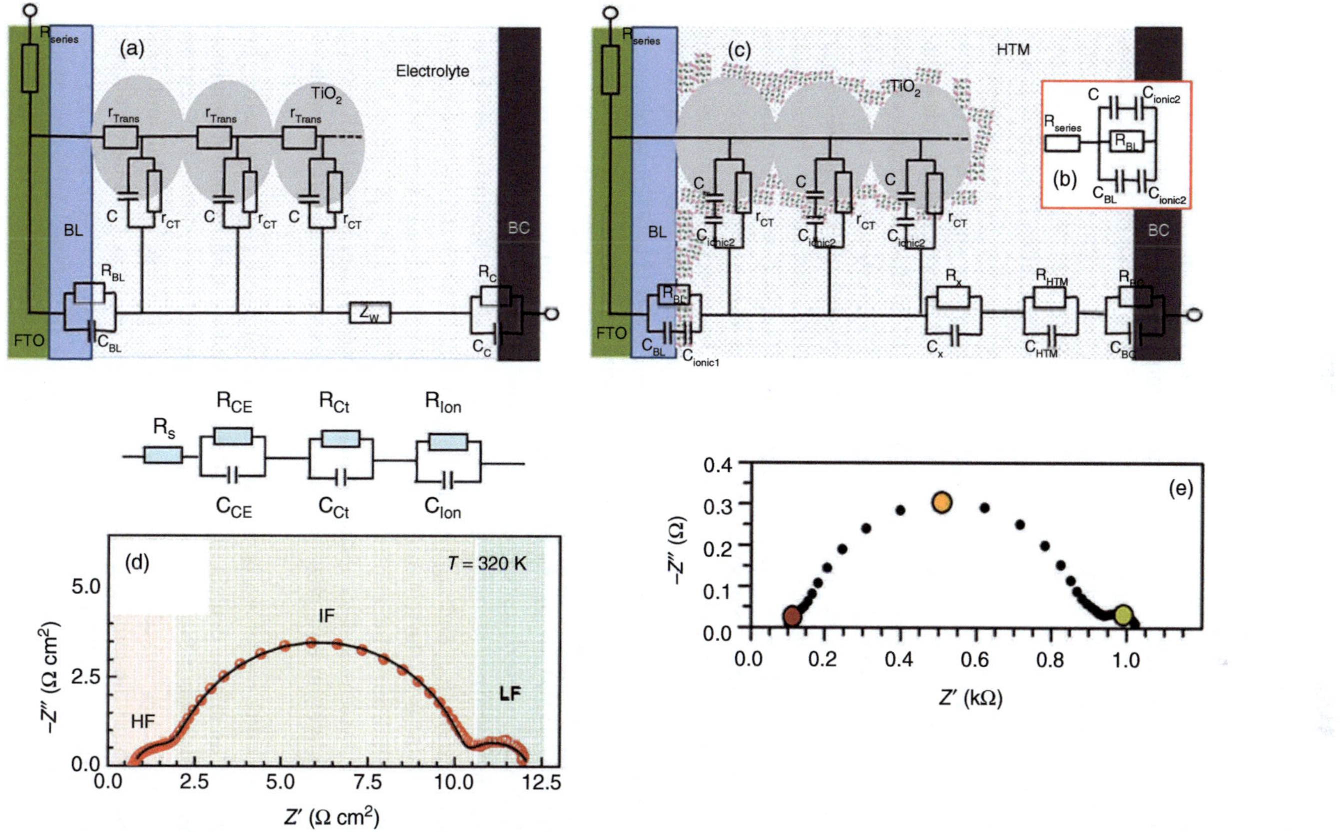

Figure 12.6 (a) Equivalent circuit of a DSSC. (b, c) Full model of a solid-state device with perovskite. Source: Modified from Zhang et al. [73]. (d) Top: Equivalent circuit for fitting of electronic IS and bottom: IS of c-TiO_2/$MAPbI_3$/*spiro*-OMeTAD PSC at 320 K. Source: Modified from Sanchez et al. [74]. (e) IS of perovskite solar cell based on $MAPbI_{3-x}Cl_x$ under dark conditions at 0.9 V applied DC bias. Source: Modified from Bisquert et al. [75].

on $MAPbI_{3-x}Cl_x$, shown in Figure 12.6e [74]. The arcs at low, intermediate, and high frequencies are with the polarization, chemical capacitance, and selective contacts, respectively. Pauporté and coworkers [78] have used *i*-ZnO as ETL and observed a semicircle at high frequency assigned to the interfacial charge transfer and recombination while the origin of the arc at LF was not clear to them. Later, they ascribed the LF feature to the dielectric relaxation and recombination process in perovskite layer using two-component series RC circuit [79, 80]. The role of 3D mesoporous and 1D nanocolumn TiO_2 structure was investigated on the charge transport properties of c-TiO_2/$MAPbI_3$/*spiro*-OMeTAD-based PSCs [81]. We have employed two-component series RC circuit to extract the HF recombination resistance. It was found that different TiO_2 structure leads to different ideality factors, suggesting that electrical properties of perovskite layer depends on the structure of the ETL on which it is deposited. Mora-Sero and coworkers [66] investigated the role of the different ETL (TiO_2, ZnO, and CdS) in PSC by using IS. In their investigation, they found that the transport resistance is dominated by HTL instead of ETL and main recombination pathway is interfacial rather than bulk at low applied bias. Moreover, selective contacts are responsible for the (i) transport resistance at the selector layer, (ii) charge transfer rate at the interface, and (iii) surface recombination at the selective contact interface. IS of solid-state DSSC based on nanocrystal $CH_3NH_3PbI_3$ absorber layer on top of rutile TiO_2 showed a low-frequency feature in addition to the two arcs that was attributed to slow charge transport and does not have direct relevance to the device physics [82]. Wang and coworkers [73] have observed three arc in their impedance measurement and thus used three-component series RC-circuit (Figure 12.6d) to fit the measured impedance signal. They describe the HF (100–600 kHz) arc to the charge exchange (R_{CE} and C_{CE}) process at the *spiro*-OMeTAD/Au interface, the semicircle in the IF region (50 Hz to 100 kHz) to interfacial charge carrier recombination (R_{Ct} and C_{Ct}), and the feature at very LF to the ionic migration (R_{ion} and C_{ion}) inside the perovskite. The charge recombination lifetime evaluated from the relation $\tau = R_{ct}C$ shows that it decreases with increasing temperature, indicating incomplete dissociation of excitons at low temperature. Moreover, *J*–*V* hysteresis effect of such devices becomes weaker below 180 K, owing to the ionic migration and phase transition from tetragonal to orthorhombic of perovskite.

Linear series RC model explains the dominant physical processes inside the PSC but was unable to explain the hysteresis resulting from trapping behavior that requires nonlinear RC components [83]. Moreover, the simplified RC circuit cannot provide the mixed electronic/ionic nature of these material and was unable to separate the ionic and electronic transport features. Venkataraman and coworkers [84] have used an equivalent circuit based on mixed conductor systems shown in Figure 12.7b for IS displayed in Figure 12.7a. In this model, low-frequency component consists of a charge transfer resistance (R_{CT}) coupled with Warburg element W_S in parallel to interfacial double-layer capacitance (C_{dl}). The capacitance C_{dl} provides the information about charge accumulation at the interface, and W_S is related to the ion diffusion toward the interface. The chemical capacitance (C_μ) represents the total charge stored in the perovskite layer, while R_{electr} ascribed

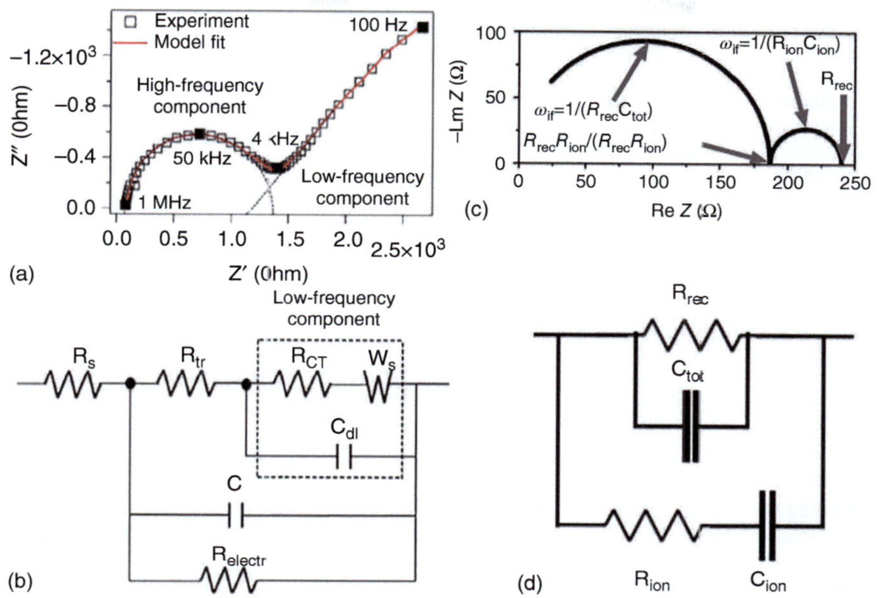

Figure 12.7 (a) EIS plot of $MAPbI_3$ at 45 °C at 100 mW cm^{-2} light intensity and 0 V applied bias. (b) Equivalent circuit diagram of PSC showing combined charge and ion transport impedance. Source: Bag et al. [84]. (c) Equivalent circuit for perovskite thin films, including the effect of ionic migration. (d) Calculated impedance response of the equivalent circuit (c) using C_{tot} = 14 nF and the experimentally determined parameters R_{rec} = 240 Ω, R_{ion} = 840 Ω, and C_{ion} = 0 : 28 mF. Source: (d) Modified from Chen et al. [85].

to free carrier recombination/transport (electrons and holes or both). Recently, Miyano and coworkers [85] have introduce a similar equivalent circuit to model the ionic migration in $MAPbI_3$ perovskites where they merge the two resistances R_{tr} and R_{Ct} and remove the C_{dl}. Grätzel and coworkers [86] have investigated the temperature-dependence electrical properties of $MAPbI_3$ pellets by means of impedance spectroscopy and found that the $MAPbI_3$ perovskite does not exhibit ferroelectricity while clearly reveal the mixed conducting (i.e. ionic and electronic) behavior. To explain the mixed conducting behavior and DC polarization in perovskite, they have added ionic resistance R_{ion} in addition to electronic resistance R_e. However, they found this model unable to describe the $\sqrt{t}$ law for a short time. Radaelli and coworkers [87] also included the effects of ionic migration in their model (Figure 12.7d) that produces an additional arc at low frequencies (see Figure 12.7c). They concluded that the low-frequency feature is purely related to conductance rather than of dielectric effects, as reported earlier [74]. They found that non-ferroelectric contribution is dominant and ruled out the ferroelectricity as the primary source of *J–V* hysteresis observed in PSC. The estimated ionic mobility, ionic conductivities, and carrier density in $MAPbI_3$ perovskite were of the order of 10^{-10} cm^2 V^{-1} s^{-1}, 10^{-7} $Ω^{-1}$ cm^{-1}, and 10^{20} cm^{-3}, respectively.

Fan and coworkers [88] also found the non-ferroelectric behavior of $MAPbI_3$ in their study by using similar model as of Grätzel and coworkers [86] and Radaelli and

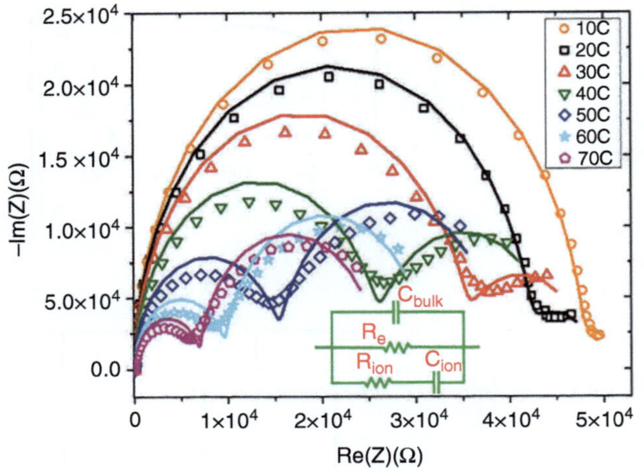

Figure 12.8 The fittings to the measured IS using the associated model, from which electronic resistance R_e, ionic resistance R_{ion}, and geometric and chemical capacitances C_{bulk} and C_{ion} for contact blocking of mobile ions were extracted. The inset shows the equivalent circuit to model the impedance spectrum. Source: Modified from Hoque et al. [88].

coworkers [87] to study the ionic transport. Temperature-dependence impedance spectroscopy (Figure 12.8) shows that the $MAPbI_3$ exhibits different electrical behaviors below 40 °C and above 50 °C, due to the phase transition from tetragonal to cubic structure across this temperature range. The ionic diffusion component also required to fit the impedance spectroscopy of spin-coated or sprayed $MAPbI_3$, $FA_{0.85}Cs_{0.15}PbI_3$, and $MAPbBr_3$ perovskite films in contact with non-aqueous electrolyte composed of 0.1 M $TBAPF_6$ in CH_2Cl_2 [89].

12.3.2 Capacitance Spectroscopy

Capacitance spectroscopy provides the information of built-in potential, trap density, distribution profile of trap density, doping density, and polarization processes. Moreover, this constitutes information about the mechanism of undesired *J–V* hysteresis in PSCs [90]. In addition, there are different mechanisms contributing to the capacitance, which may be chemical or dielectric, including contact and bulk capacitance and possible ferroelectric effects [76].

The basic capacitive response of PSCs is the geometrical capacitance (C_g) that is equal to the sum of the capacitance of individual layer between two contact electrodes. The geometrical capacitance per unit area is expressed through the relation [91]

$$\frac{1}{C_g} = \sum_n \left[\frac{\varepsilon_0 \varepsilon_{\infty,n}}{L_n}\right]^{-1} \tag{12.12}$$

where L_n represents the thickness of the nth layer with high-frequency dielectric constant of $\varepsilon_{\infty,n}$. It is obvious from above equation that the geometrical capacitance of device is ruled by the layer with least dielectric constant. In general, the geometrical capacitance is the smallest capacitance of a device and observed at very high frequencies [75].

Due to charge accumulation at the interface of perovskite/charge extraction layer, the solar cells exhibit the second capacitive response, namely, depletion-layer capacitance C_{dl} that can be expressed as

$$C_{dl} = \frac{\varepsilon_0 \varepsilon_r}{w} = \sqrt{\frac{q\varepsilon_0 \varepsilon_r N_{eff}}{2(V_{bi} - V)}} \tag{12.13}$$

where N_{eff} is the effective concentration of fixed ionized species (N_A or N_D) at depletion zone, W is the width of depletion layer, and V_{bi} is the built-in voltage, also known as flat-band potential. As depletion width is small as compared with film thickness, i.e. $W \leq L$, therefore, in general $C_{dl} \geq C_g$ at IF range (1 kHz to 1 MHz).

Another capacitance dominating above V_{bi} under forward bias is called chemical capacitance C_μ. This is observed due to the shift of the Fermi level with respect to the band edge and pronounces the accumulation of charge carriers in the absorber layer and charging and discharging processes during continuous current at a given voltage [92]. In contrast to the reciprocal dependency on film thickness of geometrical capacitance in the above equation, the chemical capacitance increases with film thickness. The chemical capacitance per unit area is given by [91]

$$C_\mu = q^2 \frac{LN_\mu}{k_B T} \exp\left[\frac{qV}{\beta k_B T}\right] \tag{12.14}$$

where N_μ is the effective total equilibrium charge density that contributes to C_μ and β is a statistical constant. The equation above is true for the systems following Boltzmann statistics; however, for other system, this transforms to [92]

$$C_\mu = q^2 \frac{dN}{dE_{Fn}} = \frac{q^2 n}{k_B T} \tag{12.15}$$

12.3.2.1 Capacitance vs. Frequency (*C*–*f*) Measurements

The capacitance of PSCs can be derived from impedance data through the relation $C = 1/i\omega Z$. The C vs. f plot of $(FAPbI_3)_{0.85}(MAPbBr_3)_{0.15}$-based PSCs is shown in Figure 12.9 [62]. Three distinct regions can be deducted in the C–f spectra, which are associated with different mechanisms. Generally, a high capacitance is observed at LF that drops rapidly with frequency. The LF region is attributed to the polarization of interfaces due to the charge accumulation at the perovskite/charge-selective contact interface [95–98]. A stable plateau in the mid-frequency region corresponds to dielectric relaxation in the perovskite layer and is determined by the geometrical capacitance per unit area given by Eq. (12.12). The capacitance in the HF region further decreases that ascribed to the effect of the series resistance caused by contact electrode. The value of capacitance further amplified under illumination [93, 98] or by charge injection at applied bias or increase of temperature [62] as shown in Figure 12.9a,b. Photoinduced carriers modify the local unit cell equilibrium and change the polarizability, assisted by the freedom of rotation of MA, which results

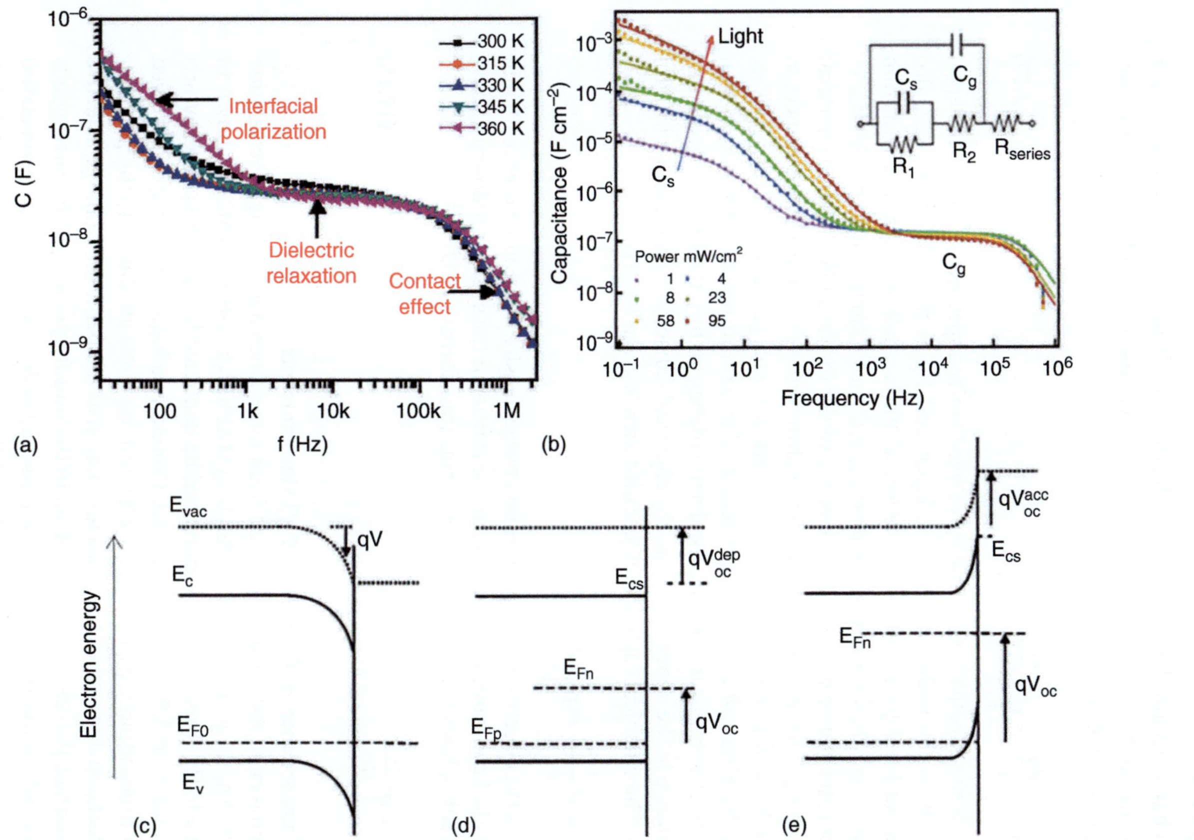

Figure 12.9 (a) C–f spectra of mixed perovskites $(FAPbI_3)_{0.85}(MAPbBr_3)_{0.15}$-based device at different temperatures, as indicated. Source: Khan et al. [62]. (b) Capacitance spectra measured under short-circuit conditions at varying irradiation intensities corresponding to a device of planar structure FTO/TiO_2/$CH_3NH_3PbI_3$/spiro-OMeTAD/Au. Source: Almora et al. [93]. (c–e) Energy diagram for ETL to the perovskite layer at open-circuit conditions. (c) Depletion layer at equilibrium. Generation of minority carriers produces flat-band condition (d) and further accumulation (e) of majority carrier. Source: Zarazua et al. [94].

in giant dielectric constant. The ions migrate under applied bias or photovoltage generates the charges at the perovskite/ETL and perovskite/HTL interfaces that participate in the capacitive response. Moreover, due to the inefficient extraction of photoexcited charges under illumination condition, there will be a large interfacial electronic dipole polarization induced by accumulated electrons and holes at the perovskite/electrode interface [99, 100].

It was found that the *J–V* hysteresis of PSCs was almost eliminated when the capacitance of the device was reduced [17]. Nazeeruddin and coworkers have describe the evolution of low-frequency capacitance by the Guoy–Chapman model, whose voltage dependence for voltages above $k_B T/q$ is given by [96]

$$C = \sqrt{\frac{z^2 q^2 n_i^0 \varepsilon_r \varepsilon_0}{2k_B T}} \exp\left(\frac{zqV}{m_c k_B T}\right) \tag{12.16}$$

where z is the charge number of the ions, n_i^0 the ionic concentration in the bulk, and m_c is a capacitive exponent with the value 2. They observed that the devices with m-TiO_2 exhibit a sharp increase in the capacitance of above 0.6 V, whereas the device with planar c-TiO_2 anticipated a negative capacitance region of around 100 Hz.

The device without any ETL also exhibits exponential evolution of the low-frequency capacitance but without negative capacitance trend. The negative capacitance arises due to the charging and discharging of charge carriers at the perovskite/c-TiO_2 interface [101]. The $FAPbI_{3-x}Cl_x$-based PSCs also show an enhancement of low-frequency capacitance under illumination [94], which was ascribed to electronic accumulation zone at the perovskite/TiO_2 interface. The interfacial capacitance is given through the similar expression as Eq. (12.16) [94]:

$$C_S = \frac{dQ}{dt} = \frac{qp_b^{1/2}}{\sqrt{2}} \sqrt{\left(\frac{\varepsilon_r \varepsilon_0}{k_B T}\right)} e^{qV/2k_B T} \tag{12.17}$$

where p_b is the bulk carrier density corresponding to dopant density and qV corresponds to the local separation of vacuum level from bulk value.

The origin of anomalous *J–V* hysteresis in a PSC was probed by means of temperature, frequency, and film thickness-dependent capacitance spectroscopy [3]. Low-frequency capacitance of the device was found to be independent of layer thickness. Arguably, it can be concluded that the electrode polarization due to ion accumulation in the vicinity of the outer interfaces might be the cause of dark current hysteresis. In the capacitance spectra, two distinct processes was observed. The dielectric constant measured at 1 kHz shows a sharp rise at $T = 160$ K, followed by a slight decrease at higher temperatures (Figure 12.10). The enhancement in the dielectric value is associated with the phase transitions of $MAPbI_3$ from orthorhombic structure (γ-phase) to tetragonal structure (β-phase) where the super cell symmetry is enhanced [102, 103]. Additionally, it can be noted that at high temperature, the capacitance first decreases with temperature and subsequently starts increasing further with the increase in temperature beyond 330 K. Such kind of capacitance behavior indicates ferroelectric-to-paraelectric transition, which originates due to the phase transition of perovskites from tetragonal to cubic structure in the temperature range of 315–330 K [62, 104].

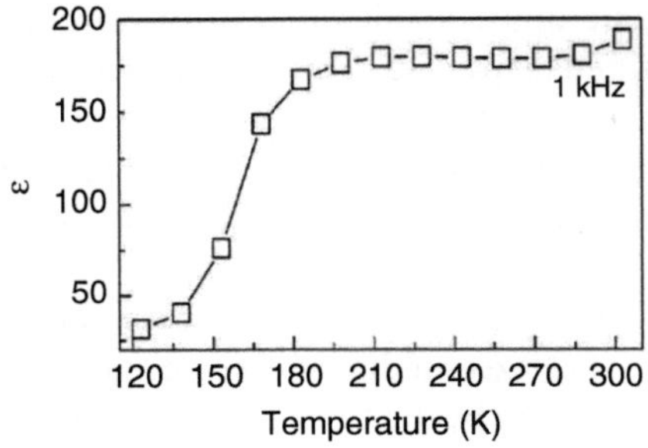

Figure 12.10 Temperature-dependent dielectric constant of $MAPbI_{3-x}Cl_x$-based PSCs measured at 1 kHz. Source: Almora et al. [91].

12.3.2.2 Capacitance vs. Voltage ($C-V$) Measurements and Mott–Schottky Analysis

As discussed above, in the IF range the depletion layer capacitance dominates the dielectric polarization capacitance. However, at HF the resistive and inductive losses can be neglected, and C_g is shunted by C_{dl}. In reverse bias, depilation layer width increases that leads to the decrease of C_{dl}, following an inverse square root law $C_{dl} \propto 1/\sqrt{V}$, whereas, when the applied bias approaches toward V_{bi}, depletion width shrink and C_{dl} increases. MS plot is obtained by plotting the graph between C_{dl}^{-2} and applied bias V. The expression (12.13) can be rewritten as

$$C_{dl}^{-2} = \frac{2(V_{bi} - V)}{q\varepsilon_0\varepsilon_r N_{eff}} \tag{12.18}$$

MS analysis provides the dopant density (N_{eff}), depletion layer width (W), and built-in potential (V_{bi}) at a PN junction. At the perovskite/charge-selective contact interface, the depletion zone width extends to a conspicuous portion of the perovskite layer thickness, where the defect density N entails a homogeneous charge distribution $\rho = qN$ as shown in the inset of Figure 12.11a [105]. The doping density N_{eff} can be evaluated through the slope of straight line in the MS plot. The intersection of the straight line on the voltage axis determines V_{bi} as shown in Figure 12.11a. The width of depletion layer corresponding to the zero applied bias near perovskite/charge-selective contact interface can be evaluated through

$$W = \sqrt{\frac{2\varepsilon\varepsilon_0 V_{bi}}{qN_{eff}}} \tag{12.19}$$

It is important to mention here that the linear region related to the C_{dl} is present only for limited range as can be seen in Figure 12.11a. In reverse bias, the capacitance deviates from reverse bias giving a saturated capacitance ascribed to C_g. As the applied bias approaches toward the built-in voltage V_{bi} ($V > 0.6$ V), the linear region further deviates and shows an exponential dependence of voltage given by Eq. (12.16) as evident in Figure 12.11a,b. The transition from C_g to C_{dl} is not always visible (Figure 12.11b), where C_{dl} region is absent for $MAPbI_3$-based PSCs. On the other hand, when voltage was sweep from positive to negative direction, the influence of interfacial accumulation capacitance C_s was found to be absent (Figure 12.11c). It is important to note that the MS plot shown in Figure 12.11a,c is from same device. The only difference between these plots is the

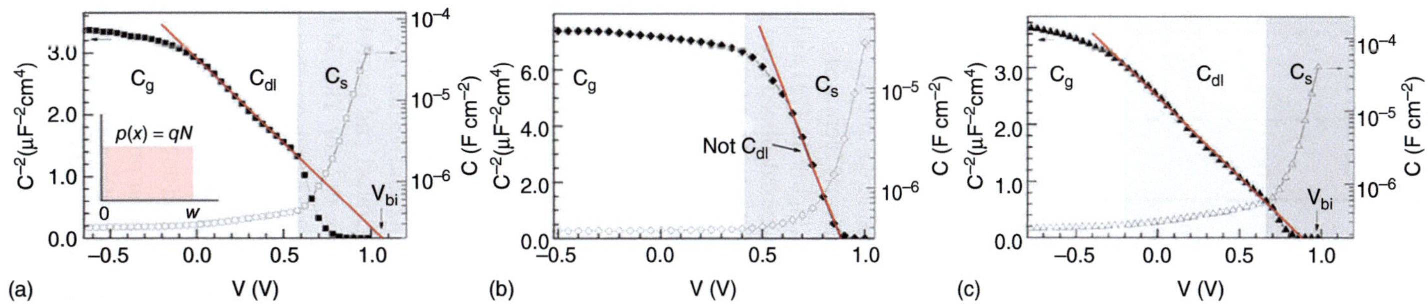

Figure 12.11 Mott–Schottky plots (black dots) and respective *C*–*V* curves (gray dots) for (a) mix halide perovskite $MAPbI_{3-x}Cl_x$ solar cells with voltage sweep from negative to positive. (b) $MAPbI_3$ solar cells and (c) $MAPbI_{3-x}Cl_x$ solar cells with voltage sweep from positive to negative. Source: Almora et al. [105].

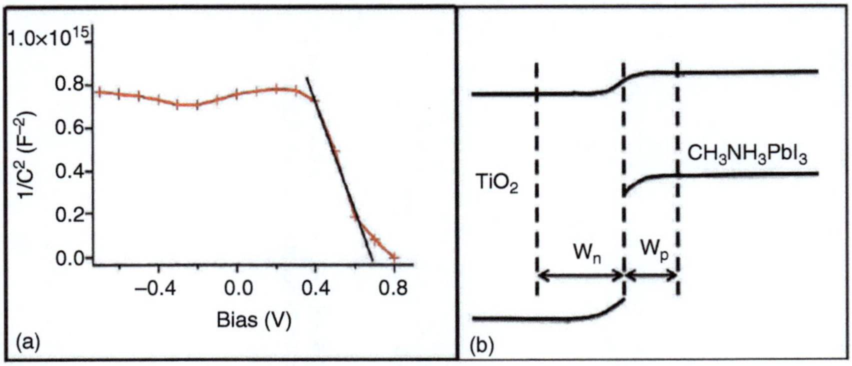

Figure 12.12 (a) MS analysis for the $MAPbI_3/TiO_2$ interface. (b) Schematic band-bending diagram: the depletion width on the n-side (W_n) and on the p-side (W_p). Source: Laban and Etgar [106].

earlier voltage sweep from negative to positive and later voltage sweep from positive to negative. Therefore, during the measurement and analysis of MS plot, all the aspects (Figure 12.11a–c) should be considered, and MS analysis is only significant when the depletion capacitance can be clearly identified.

MS plot for $MAPbI_3/TiO_2$ interface (Figure 12.12a) and the slope of the linear region and the doping density [106] can be calculated from Eq. (12.18). The doping density was found to be 2.14×10^{17} cm^{-3} in the $MAPbI_3$ thin film. Built-in potential V_{bi} was evaluated through the intersection of liner region on voltage axis and found to be 0.67 V [106]. Additionally, the depletion layer width was calculated by relation

$$W_{p,n} = \frac{1}{N_{a,d}}\sqrt{\frac{2\varepsilon\varepsilon_0 V_{bi}}{q\left(\frac{1}{N_a+N_d}\right)}} \tag{12.20}$$

where N_a and N_d represent the doping density of the acceptor and the donor in ETL and perovskite, respectively.

The doping density in TiO_2 layer found to be of the order of $N_a = 1 \times 10^{16}$ cm^{-3} [107–109]. By using the above equation, the depletion widths on the n-side and on the p-side were calculated to be 162 and 75 nm, respectively. Figure 12.12b shows the band bending at interface. Grätzel and coworkers have estimated V_{bi} of ~1.0 V and defect density of 10^{17} cm^{-3} for $TiO_2/MAPbI_3$ perovskite interface [110]. MS analyses of device were based on the triple-cation perovskite $(Cs_{0.05}FA_{0.54}MA_{0.41})Pb(I_{0.98}Br_{0.02})_3$ (CsMAFA) and CsMAFA treated with sodium fluoride (CsMAFA-F) of the data measured at a frequency of 1 kHz [111]. Built-in potentials of the CsFAMA and CsFAMA-F samples are 0.80 and 0.90 V, and the corresponding depletion widths are 133 and 207 nm, respectively.

Fang et al. have investigated the junction properties of $MAPbI_3/SnO_2$ interface before and after the modification of SnO_2 with 3-aminopropyltriethoxysilane (APTES) self-assembled monolayer by using MS analysis [112]. It was noticed that

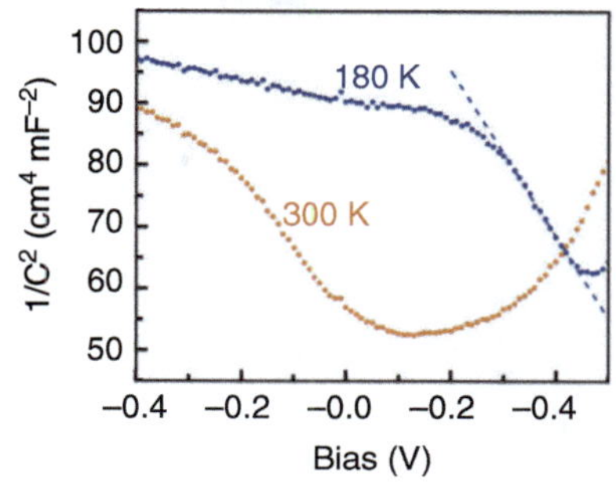

Figure 12.13 Mott–Schottky plot of an inverted perovskite solar cell $NiO_x/MAPbI_3/C_{60}$ measured at 300 and 180 K in the dark with an AC perturbation of 10 mV at 10 kHz. Source: Futscher et al. [113].

the V_{bi} of APTES-treated $SnO_2/MAPbI_3$ perovskite was about 0.1 V higher than that of untreated $SnO_2/MAPbI_3$ device. If the contribution of ionic capacitance dominates the depletion capacitance at room temperatures, then the measurement should be carried out at low temperature as shown in Figure 12.13 [113].

From the MS plot of the inverted PSCs based on ($NiO_x/MAPbI_3/C_{60}$), V_{bi} was found to be ~0.92 V, and doping density N of ~7.0×10^{16} cm^{-3}. In similar device configuration (i.e. ITO/NiO_x/0–1.0% Ag-incorporated $MAPbI_3/C_{60}$/Ag), N was found to be decreasing as the Ag concentration increases from 0% (5.07×10^{16} cm^{-3}) to 0.5% (3.28×10^{16} cm^{-3}) and 1.0% (1.58×10^{16} cm^{-3}), reflecting a reduction in electron concentration [114], whereas V_{bi} remains almost same. It was also found that the doping concentration decreases when device was annealed up to 130 °C [115].

12.3.2.3 Thermal Admittance Spectroscopy

Temperature-dependent C–f measurement often known as thermal admittance spectroscopy is a powerful techniques to analyze defect states in the band gap of semiconductor device. By using thermal admittance spectroscopy, the profile of the density of trap states (N_T), the emission rate of charge carriers from the trap states, and trap activation energy (E_A) can be evaluated. As discussed in Section 12.3.2.1, the excess LF capacitance is originated by interfacial charge accumulation. Besides the charge accumulation at the interface, the trapping and detrapping of a trap states also contribute to the LF capacitance [116]. Therefore, trapping/detrapping of defect states with changes in the AC frequency of applied bias influences the capacitance of device. The charging/discharging of defects at a location where the defect energy level crosses the Fermi level is shown in Figure 12.14a. The electron trap states above the Fermi level (E_F) are considered vacant, whereas the trap states below E_F are filled. When the frequency of AC signal corresponds to the emission frequency of a trap state, the capacitance of the device significantly changes. Additionally, the trapped charge could be thermally activated as a result of change in capacitance. Therefore, the junction capacitance of a device is related to the applied frequency of AC bias and measuring temperature of device [18, 118]. The deep trap states have low emission rate and can contribute to the capacitance at LF and at high temperature, whereas the shallow traps follow both HF and LF as shown schematically in Figure 12.14b.

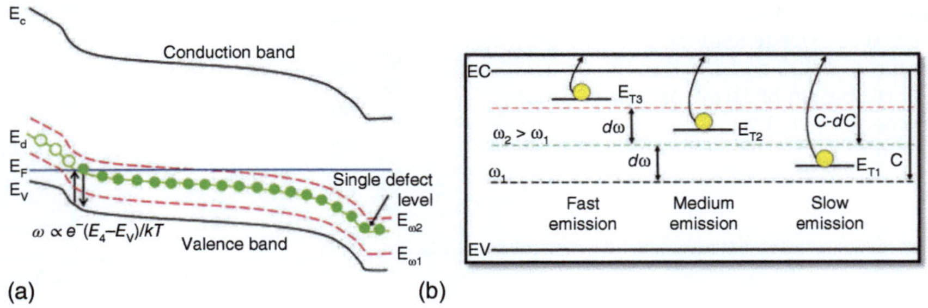

Figure 12.14 (a) Schematic band diagram of p-type semiconductor junction with a single trap level E_d and two measurement frequencies ω1 and ω2. Source: Duan et al. [18]. (b) Schematic representation of shallow and deep traps. Shallow traps have very fast emission rate and therefore contribute to the measured capacitance at HF and LF. Deep trap has low emission rate and responds at LF. Source: Modified from Yin et al. [117].

The rate of release an electrons from a trap state located below E_T to the conduction band edge E_C is given by [119, 120]

$$e_n = N_C v_{\text{th}} \sigma_n \exp\left(-\frac{E_C - E_T}{k_B T}\right) \tag{12.21}$$

where N_C, v_{th}, and σ_n represent the effective density, root mean square velocity of electrons, and trap capture cross section, respectively. The quantity $N_C v_{\text{th}} \sigma_n$ is defined as an attempt to escape frequency (v_0). The effective density of state is directly proportional to the temperature to the power of 3/2, i.e. $N_C \propto T^{3/2}$, and rms speed, i.e. $v_{\text{th}} \propto T^{1/2}$, subsequently $v_0 \propto T^2$. Then Eq. (12.21) transformed to

$$e_n = \xi T^2 \exp\left(-\frac{E_A}{k_B T}\right)$$

or

$$\ln\left(\frac{e_n}{T^2}\right) = \ln \xi - \frac{E_A}{k_B T} \tag{12.22}$$

where ξ is the proportionality constant and $E_A = E_C - E_T$ is the activation energy of the traps.

Yang and coworkers [18] have investigated the distribution of traps in $MAPbI_3$ perovskite films by measuring temperature-dependence *C-f* spectra shown in Figure 12.15a. The activation energy E_A could be deduced from the slope of Arrhenius plot, $\ln\left(\frac{f_{\text{peak}}}{T^2}\right)$ vs. $\frac{1}{T}$ plot. The attempt to escape frequency can be extracted by the intercept of the line in *y*-axis. The characteristic frequency f_{peak} is the frequency where peak of the $[-f\,dC/df]$ vs. f plot was observed as shown in Figure 12.15b. The distribution of the density of trap states is given by [62, 111]

$$N_T = -\frac{V_{\text{bi}}}{qWk_B T}\left(f\frac{dC}{df}\right) \tag{12.23}$$

In above equations, V_{bi} and W can be extracted from MS analysis as discussed in Section 12.3.2.2. The demarcation energy E_ω explores the band-gap density of states contributing to the capacitance, which is the energy difference between defect site

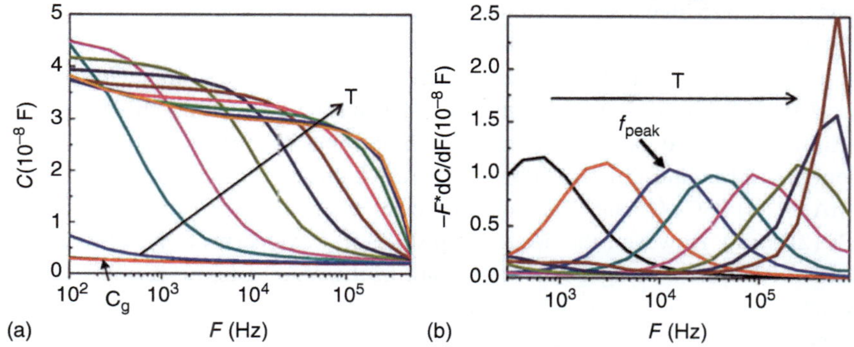

Figure 12.15 (a) C–f spectra at different temperatures and (b) the derivative of the capacitance spectra showing the characteristic frequency at each temperature. Source: Duan et al. [18].

and band edge. The demarcation energy separates the trap states that respond to the voltage modulation at frequency ω_{trap} and those that do not. The demarcation energy is defined through [121]

$$\Delta E = E_{\omega} = k_B T \ln\left(\frac{\omega_0}{\omega_{trap}}\right) \tag{12.24}$$

The characteristic timescale τ_{trap} for thermal emission of a carrier from a defect is given by [122, 123]

$$\tau_{trap} = \nu_0^{-1} \exp\left(\frac{\Delta E}{k_B T}\right) \tag{12.25}$$

The defect activation energy and distribution of trap density of CsFAMA- and CsFAMA-F-based perovskite solar cells can be calculated through Arrhenius plot [111]. The defect activation energies in the CsFAMA devices were found to be 0.275 eV, and peak position of trap density was observed at ~0.26 eV with a peak value of trap density of states of ~2.24×10^{14} cm^{-3} eV^{-1}. The incorporation of fluoride in CsFAMA perovskites effectively reduces the activation energy to 0.201 eV, peak position of trap density to 0.18 eV, and density of trap states to 1.49×10^{15} cm^{-3} eV^{-1}. Recently Park and coworkers have investigated the effect of annealing of different forms of MA perovskite solar cells [i.e. $MAPbI_3$, $(FA_{0.83}MA_{0.17})Pb(I_{0.83}Br_{0.17})_3$, and $(Cs_{0.1}FA_{0.75}MA_{0.15})Pb(I_{0.85}Br_{0.15})_3$] on trap density of states [116]. The trap density in $MAPbI_3$-based device was found to be decreasing from 9.24×10^{17} cm^{-3} to 8.79×10^{17} cm^{-3} when device was annealed at 100 °C for 60 minutes. The reduction of trap density of annealed $MAPbI_3$ device ascribed to the enlargement of grain boundaries. This finding suggests that grain boundaries are primary cause of electronic traps in $MAPbI_3$, contradicting the theoretical calculations that grain boundaries do not generate deep trap states in the band gap [117, 124]. The deep level traps in $Cs_{0.1}FA_{0.75}MA_{0.15})Pb(I_{0.85}Br_{0.15})_3$ were found to be decreasing as compared with both $MAPbI_3$ and $(FA_{0.83}MA_{0.17})Pb(I_{0.83}Br_{0.17})_3$. Moreover, the deep trap level depth (~0.6 eV) was found to be shallower in $Cs_{0.1}FA_{0.75}MA_{0.15})Pb(I_{0.85}Br_{0.15})_3$ than in other two perovskites (~0.7 eV).

The peak trap densities of states in $MAPbI_3$ and $(FAPbI_3)_{0.85}(MAPbBr_3)_{0.15}$ PSCs were found to be at 0.42 and 0.70 eV with peak values of 1.37×10^{17} and 8.84×10^{16} eV^{-1} cm^{-3} and activation energies of 83 and 123 meV [62]. Grätzel and coworkers [125] have compared the defect density of states in $MAPbI_3$ thin films synthesized via (i) antisolvent deposition method [$MAPbI_3$(m)] and (ii) equimolar amounts of MAI and PbI_2 [$MAPbI_3$(s)]. It was found that the activation energies of $MAPbI_3$(s) PSCs (17 meV) are slightly lower than the device with $MAPbI_3$(m) PSCs (21 meV). However, the deeper trap states in $MAPbI_3$(m) (5.57×10^{17} cm^{-3}) was about 1 order of magnitude lower than in the $MAPbI_3$(s) (2.76×10^{18} cm^{-3}). The admittance spectra of mix halide perovskite $MAPbI_xCl_{1-x}$ (Figure 12.16a) exhibit a secondary peak at high frequency (inset) in addition to main peak at low frequency [126]. The additional peak indicates that two defect energy levels of around ~0.24 and 0.66 eV located above the valance band edge are evaluated from the slope of Arrhenius plot (Figure 12.16b). The attempt-to-escape frequency υ_0 was found to be 2×10^{11} s^{-1}. Furthermore, the peak of trap density of states was located around 0.65 eV below conduction band deduced to be 3×10^{16} eV^{-1} cm^{-3}.

$MAPbI_3$ films fabricated via interdiffusion of double layers of PbI_2 and MAI in p–i–n configuration show relatively lower trap density of states of ~10^{12} cm^{-3} eV^{-1} with defect energy of ~0.35 eV. The trap density was found to be decreasing when device was annealed at 105 °C for 15–45 minutes [17]. Additionally, the device without PCBM gave slightly higher value of trap density of states of ~10^{13} cm^{-3} eV^{-1}. These finding suggest that PCBM effectively passivates the trap states in perovskite, and when device was thermally annealed, PCBM diffused into perovskite layers along the grain boundaries, resulting in the decrease of trap density of states. S. H. Im et al. have fabricated $MAPbI_3$ PSCs by using either 40 wt% $MAPbI_3$/DMF with 100 μl HI or 40 wt% $MAPbI_3$/DMSO in device configuration, viz. FTO/c-TiO_2 (≈50 nm)/$MAPbI_3$ (≈300 nm)/PTAA (≈40 nm)/Au [127]. The device fabricated with DMSO shows shallower trap energy level of 0.27 eV and trap energy density of ≈10^{17} cm^{-3}, whereas the device fabricated via DMF and HI exhibits relatively lower trap density of states of ~10^{17} cm^{-3} with trap energy level of 0.28 eV below conduction band edge.

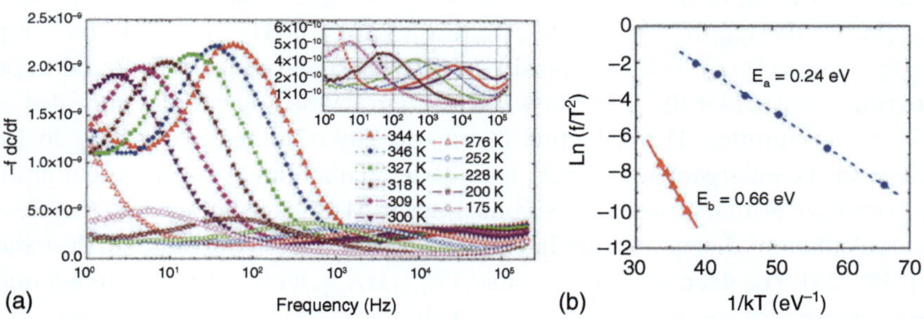

Figure 12.16 (a) Thermal admittance spectra of mix halide perovskite $MAPbI_xCl_{1-x}$-based solar cells. (b) Arrhenius plot of $MAPbI_xCl_{1-x}$-based solar cells. Source: Samiee et al. [126].

Although, thermal admittance spectroscopy is an effective technique to qualitatively analyze the defect density in PSCs. However, the technique has some limitations in principle [128] and is difficult to differentiate between VB and CB trap states. If both VB and CB have similar trap density and shape, the investigation of the measured data would lead to an overestimated defect density [122]. Then, the trap states below the energy demarcation can be trapped/detrapped, whereas trap states does not contribute to the capacitance signal, which leads to incomplete estimation of the defect density. Finally, a trap state with an extended emission time τ_{trap} cannot contribute to the capacitance signal, which also reduces the accuracy of the measured defect density.

12.4 Transient Spectroscopy

Transient spectroscopy is a dominant method that gives direct insight into the dynamics of photogenerated charges carriers in optoelectronic devices.

12.4.1 Time-Resolved Microwave Conductivity Measurements

TRMC is an electrodeless techniques to reveal the charge carrier dynamics in PSCs under solar radiation [129–132]. This method allow us to determine the charge carrier mobilities and lifetimes from the time-dependent change in conductance (ΔG) upon photoexcitation [133]. Additionally, TRMC measurements also allow us to investigate the nature of trap states, which are not possible with other techniques such as photoluminescence and SCLC.

The charge transport properties in perovskite thin films are related to the carrier density of charge carriers and strongly depend upon the generation process. Upon optical excitation of perovskite, the charge carrier density of perovskite layer is significantly enhanced and results in the change of the electrical conductivity of perovskites. The photoconductance as a function of time is expressed by [134]

$$\Delta G(t) = e[n_e\mu_e(t) + n_h\mu_h(t))]\beta L \tag{12.26}$$

where n_e and n_h are the density of photogenerated charge carriers, β is the ratio of the inner dimensions of the microwave cell, and $\mu_e(t)$ and $\mu_h(t)$ are the electron and hole mobility, respectively. To quantitative analyze the charge carrier dynamics in the perovskite thin films, Savenije and coworkers have proposed a kinetic model [7]:

$$\begin{aligned}
\frac{dn_e}{dt} &= G_C - k_2 n_e(n_h + p_0) - k_T n_e(N_T - n_t)\\
\frac{dn_h}{dt} &= G_C - k_2 n_e(n_h + p_0) - k_R n_t(n_h + p_0)\\
\frac{dn_e}{dt} &= k_T n_e(N_T - n_t) - k_R n_t(n_h + p_0)
\end{aligned} \tag{12.27}$$

where G_C represents the photoexcitation of electron from VB to CB; k_2, k_T, and k_R represent the rate constants for radiative electron–hole recombination, trap filling,

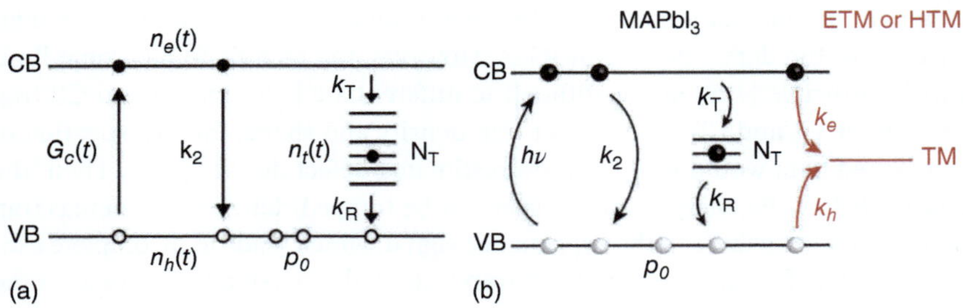

Figure 12.17 (a) Kinetic model of processes occurring upon photoexcitation of a perovskite material. Source: Hutter et al. [7]. (b) Kinetic model of processes occurring in a bilayer of $MAPbI_3$ perovskite and charge extraction layer. Source: Hutter et al. [135].

and detrapping, respectively (Figure 12.17a); N_T is the intra-band trap density; and p_0 is the non-intrinsic charge carrier density in perovskites.

Using the above model, Savenije and coworkers [7] have investigated the dynamic of photogenerated charge carrier in planer configuration of $MAPbI_3$ and $MAPbI_{3-x}Cl_x$ (Figure 12.18a,b), and the results were compared with meso-Al_2O_3 scaffold (Figure 12.18c,d). In Figure 12.18, the fast rising signal as a function of time is associated with the formation of mobile charge carriers, whereas the decay of the TRMC signals after excitation is attributed to a decreasing concentration of mobile carriers. The peak value of TRMC signal gives the value of effective charge carrier mobility. From Figure 12.18a,b, we noted that the mobility of charge carriers in both planer thin films is $\sim 30\,cm^2\,V^{-1}\,s^{-1}$ for the incident light intensities less than 10^{10} photons per cm^2. In contrast, for the meso-structured films, the effective mobilities are less than $6\,cm^2\,V^{-1}\,s^{-1}$. The decrease in the mobility in *meso*-structured device is related to the small size of the crystal domains, which reduces the effective mobilities observed at microwave frequencies. The kinetic model given by Eq. (12.27) was used to evaluate the rate constants, carrier concentrations, and density of trap states. The extracted value of dark carrier concentration was found to be about $10^{15}\,cm^{-3}$ for $MAPbI_{3-x}Cl_x$. The trap density in $MAPbI_{3-x}Cl_x/Al_2O_3$ and $MAPbI_3/Al_2O_3$ was of the order of $10^{15}\,cm^{-3}$ and in $MAPbI_3$ films without any Al_2O_3 was about $6\times10^{16}\,cm^{-3}$.

Savenije and coworkers [136] have investigated the charge carrier dynamics in bilayer $MAPbI_3$/spiro-OMeTAD and $MAPbI_3$/PCBM by using the above kinetic model (Figure 12.19). Both $MAPbI_3$ and $MAPbI_3$/spiro-OMeTAD shows similar TRMC traces because $MAPbI_3$ already has dark concentration of holes (p_0). The optical excitation leads to the generation of electrons in CB and holes in VB. There are two possibilities of photogenerated electrons: either it can recombine with hole in the VB of perovskites, or it can transfer to the spiro-OMeTAD. As the concentration of photogenerated electrons is generally less than the total concentration of holes, therefore recombination of electron and hole either in $MAPbI_3$ or in spiro-OMeTAD does not affect the recombination kinetic. $MAPbI_3$/PCBM trace shows faster decay than $MAPbI_3$ and $MAPbI_3$/spiro-OMeTAD. The faster decay of

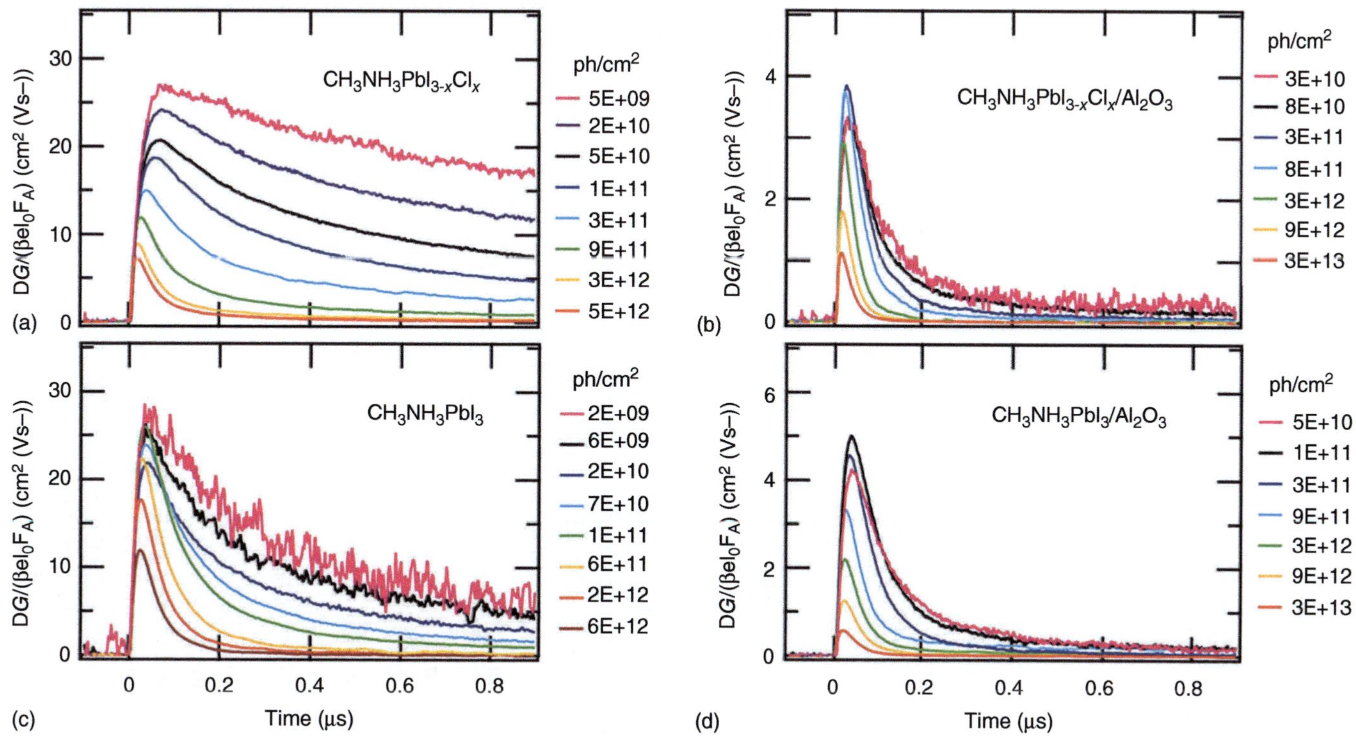

Figure 12.18 TRMC graphs for (a) $MAPbI_{3-x}Cl_x$, (b) $MAPbI_3$, (c) $MAPbI_{3-x}Cl_x/Al_2O_3$, and (d) $MAPbI_3/Al_2O_3$. Here ΔG is the photoconductance normalized for the absorbed number of photons. Source: Hutter et al. [7].

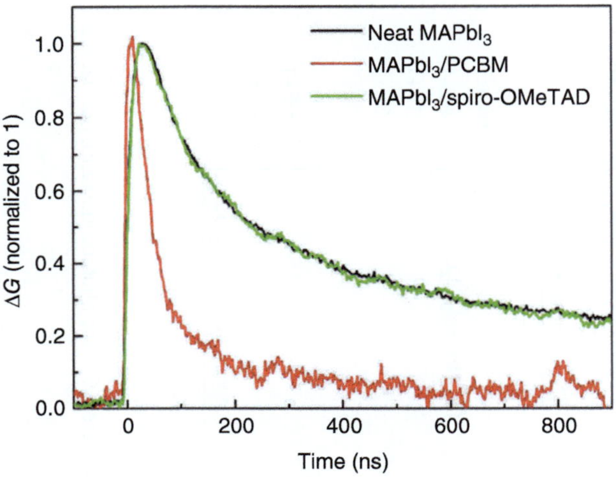

Figure 12.19 TRMC traces for $MAPbI_3$, $MAPbI_3$/spiro-OMeTAD, and $MAPbI_3$/PCBM recorded at the intensity of 1.0×10^{11} photons per cm^2 per pulse $\lambda_{pump} = 600$ nm. Source: Ponseca et al. [136].

photoconductance suggests that the electron–hole recombination process is faster than the electron injection process at perovskite/PCBM interface.

The model represented by Eq. (12.27) did not account the additional pathways introduced by the electron or hole transport material at the perovskite/ETL or HTL interface (Figure 12.17b). Therefore, Sundstrom and coworkers [6] have updated the model by introducing interfacial recombination terms $k_e n_e$ and $k_h n_h$ in the above model. For perovskite/ETL or HTL bilayer system, the kinetic model above changed to

$$\begin{aligned}
\frac{dn_{\mathrm{CB}}}{dt} &= \frac{dn_e}{dt} = G_C - k_2 n_e(n_h + p_0) - k_T n_e(N_T - n_t) - k_e n_e \\
\frac{dn_{\mathrm{VB}}}{dt} &= \frac{dn_h}{dt} = G_C - k_2 n_e(n_h + p_0) - k_R n_t(n_h + p_0) + k_h n_h \\
\frac{dn_e}{dt} &= k_T n_e(N_T - n_t) - k_R n_t(n_h + p_0) \\
\frac{dn_{\mathrm{TM}}}{dt} &= k_e n_e - k_h n_h
\end{aligned} \tag{12.28}$$

This model was used to examine the TRMC traces for bilayer system, i.e. $MAPbI_3$/HTLs and $MAPbI_3$/ETL with different HTLs and ETLs [135]. For $MAPbI_3$ thin films prepared via sequential physical vapor deposition, the dark hole concentration (p_0) was found to be 4×10^{14} cm^{-3}. The electron–hole recombination rate of $k_2 = 2.8 \times 10^{-9}$ cm^3 s^{-1}, trap filling rate of $k_T = 6 \times 10^{-9}$ cm^3 s^{-1}, and exceptionally low trap density of 3×10^{14} cm^{-3} were estimated for $MAPbI_3$ thin films. The rate of hole injection from $MAPbI_3$ to spiro-OMeTAD was estimated to be $k_h = 8 \times 10^8$ s^{-1}. It was found that the charge injection into PCBM and C_{60} was significantly more efficient than into ICBA.

Seki et al. [137] have investigated the charge carrier dynamics in $MAPbI_3$, $MAPbI_3/TiO_2$, and $MAPbI_3/Al_2O_3$ by using TRMC. $MAPbI_3$ layer were deposited on mesoporous TiO_2 and Al_2O_3 by one- or two-step sequential deposition. $MAPbI_3$ thin films deposited on TiO_2 exhibited a charge carrier mobility of $20\,cm^2\,V^{-1}\,s^{-1}$, whereas the perovskite layer deposited on Al_2O_3 exhibits about 50% lower mobility.

12.4.2 Transient Absorption Spectroscopy

TAS is a technique where a fraction (0.1–10%) of molecules are excited by means of a pump pulse. A probe pulse with a delay Δt is sent through the sample, and the difference of the two absorption spectra (ΔA) is then calculated [138, 139]. The absorption of the probe through the sample is measured at various delays both with and without the pump pulse to generate a transient absorption spectrum (Figure 12.20). TAS technique was able to detect most of the photogenerated species (e.g. polarons, excitons, free charges, etc.) in PSCs and can also reveal the photodynamic phenomenon occurring in a very wide range of timescales (10^{-13} to 10^{-4} seconds) [141]. This technique is reported in one of two conventions: differential absorption (Figure 12.20) or differential transmission spectrum (Figure 12.21). For a differential transmission spectrum, the transmission measured without the pump pulse is subtracted from the transmission measured with the pump pulse. In this convention, positive features correspond to photoinduced transmission due to the decreased population of the ground state (ground-state bleach) or stimulated emission, and negative features correspond to photoinduced absorption due to the increased population of photoexcited states. Similarly, a differential absorption spectrum also involves subtracting the absorption spectrum collected without the pump pulse from the absorption spectrum collected with the pump pulse, i.e. the absorption spectrum of the excited sample minus the absorption spectrum of the sample in the ground state (ΔA). Equation (12.29) defines absorption (A) in terms of transmission (T) where I represents the transmitted intensity and I_0 incident intensity:

$$A = -\log(T) = -\log\left(\frac{I}{I_0}\right) \tag{12.29}$$

The charge separation mechanism in $MAPbI_3$-based PSCs was elucidated via femtosecond TAS using Al_2O_3 and TiO_2 as ETL [140]. Both $MAPbI_3/Al_2O_3$ and $MAPbI_3/TiO_2$ devices exhibit similar TAS spectra with a negative signal peaking at 483 nm and a positive peak at 683 nm (Figure 12.20). The negative peak is attributed to the bleaching of the absorption of the perovskite. An additional negative feature above 700 nm can also be seen for both samples that is attributed to the stimulated emission of excited $MAPbI_3$.

The similar TAS suggest that spectral features in both devices are associated with the same excited state of the $MAPbI_3$ with half-lifetime of 50 ps. The devices with HTL, i.e. spiro/$MAPbI_3/Al_2O_3$ (Figure 12.20c), show the reduced amplitude of the bleached signal at 483 nm as compared with the devices without HTL. The positive absorption signal in spiro/$MAPbI_3/Al_2O_3$ (Figure 12.20c) completely disappeared, alongside with a strong quenching of the stimulated emission above 700 nm. TAS of

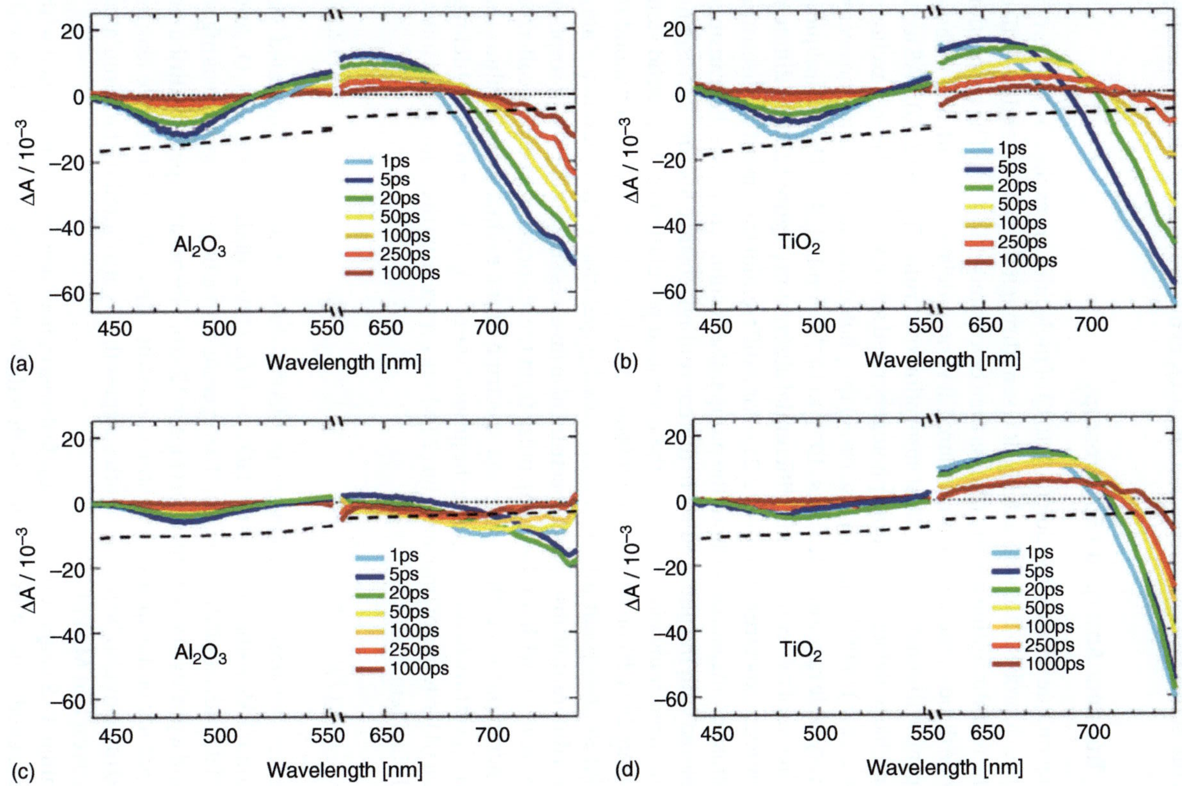

Figure 12.20 Femtosecond TAS with $l_{exc} = 580$ nm and white light continuum probe of (a) $MAPbI_3/Al_2O_3$, (b) $MAPbI_3/TiO_2$, (c) spiro/$MAPbI_3/Al_2O_3$, and (d) spiro/$MAPbI_3/TiO_2$ recorded at various time delays after excitation. Black dashed lines represent the absorbance spectrum of the sample. Source: Kim et al. [140].

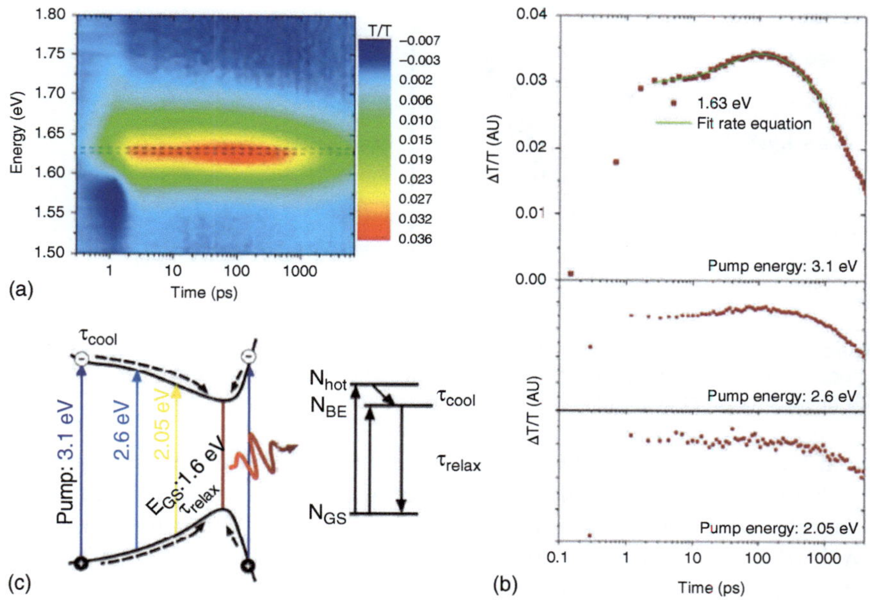

Figure 12.21 (a) 2D contour map of TAS of $MAPbI_3$(Cl) films after excitation with a ~150 fs pulse at 3.1 eV. (b) The kinetics profile recorded for the pump energies of 3.1, 2.6, and 2.05 eV. (c) Schematic representation of understanding the excitation and relaxation process in perovskites. The populations of the three energy levels, such as (i) ground state (GS), (ii) band edges (BE), and (iii) hot states, are denoted as N_{GS}, N_{BE}, and N_{hot}, with respective lifetimes τ_{relax} and τ_{cool}. Source: Bretschneider et al. [142].

spiro/$MAPbI_3$/TiO_2 (Figure 12.20d) also exhibits quenched bleaching peak and less pronounced stimulated emission peak.

The following equation is used to transform the differential absorption spectrum to differential transmittance spectra:

$$\Delta A = -\log\left(\frac{\Delta T}{T} + 1\right) \tag{12.30}$$

Notably, ΔA and $\Delta T/T$ have opposite signs. In other words, a positive $\Delta T/T$ signal indicating photoinduced transmission corresponds to a negative ΔA signal and vice versa.

The false color 2D map of TAS recorded for excitation at 3.1 eV and fluence $2.5 \pm 0.25\,\mu J\,cm^{-2}$ are shown in Figure 12.21a [142]. The broad feature of around 1.63 eV is attributed to the transition between the top of VB to the bottom of CB [143, 144]. The derivative-like feature between 1.5 and 1.6 eV at 1 ps is related to the change in photoinduced reflectivity [145]. The tail of above 1.7 eV with sub-ps lifetime is due to the population of hot carriers [145, 146]. The kinetics profiles recorded for the pump energies of 3.1, 2.6, and 2.05 eV are also shown in Figure 12.21b. It can be noted that the signal probed at 1.63 eV rises to an amplitude of 0.034 within 100 ± 5 ps. The dynamics profile recorded for the excitation energy of 2.6 eV shows a similar trend. The similar results for these excitation energies

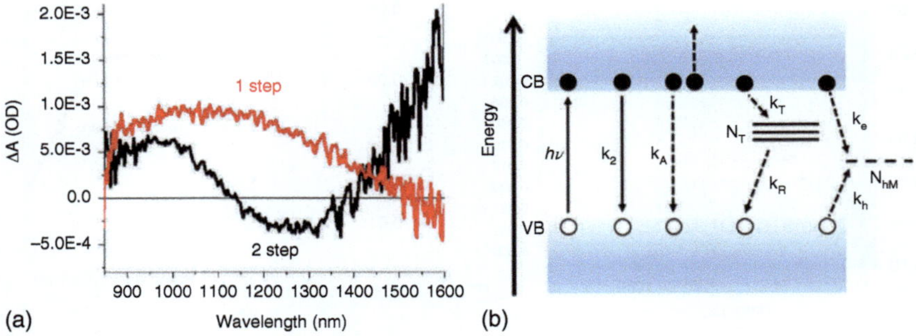

Figure 12.22 (a) TAS $MAPbI_3$ thin films fabricated via 1-step and 2-step processes and (b) schematic representation of recombination kinetics of electrons in CB and holes in VB of $MAPbI_3$. Source: Li et al. [149].

reveal that for the pump energies at 3.1 and 2.6 eV, the electrons and holes are populating far from VB maximum and CB minimum, as illustrated in Figure 12.21c.

Sum et al. have measured the differential transmission ($\Delta T/T$) spectra for $MAPbI_3$, $MAPbI_3$/PCBM, and $MAPbI_3$/spiro-OMeTAD [4]. Two prominent photobleaching (PB) bands were observed at 480 and 760 nm. The band at 760 nm was attributed to state-filling effects. There was no additional peak observed when the ETL or HTL was deposited on $MAPbI_3$. Moreover, it was found that the $MAPbI_{3-x}Cl_x$ film possess about 1 order of magnitude lower traps than $MAPbI_3$ [147]. Recently Schlenker et al. have fabricated the $MAPbI_3$ films by using 1-step and 2-step methods and investigated the TAS of these samples [148]. They observed that the nature and density of defect strongly depend on fabrication process. The film fabricated via 1-step method shows fewer traps (~1.5×10^{17} cm^{-3}) than the films fabricated via 2-step method (~4.6×10^{17} cm^{-3}). They also examine the excited-state dynamics in $MAPbI_3$ thin films using TAS shown in Figure 12.22a [148]. To evaluate the trap density, they have employed similar model used in TRMC analyses (Figure 12.17b and Eq. (12.27)) with an additional term for Auger recombination (k_A) shown in Figure 12.22b. The charge carriers recombine non-radiatively by transferring energy to a third carrier given by the set of differential equations:

$$\frac{dn_{CB}}{dt} = -k_T n_e (N_T - n_t) - k_2 n_e n_h - k_A (n_e^2 n_h + n_e n_h^2) - k_e n_e$$
$$\frac{dn_{VB}}{dt} = -k_R n_t n_h - k_2 n_e n_h + k_A (n_e^2 n_h + n_e n_h^2) + k_h n_h$$
$$\frac{dn_t}{dt} = k_T n_e (N_T - n_t) - k_R n_t n_h$$
$$\frac{dn_{TM}}{dt} = k_e n_e - k_h n_h \qquad (12.31)$$

12.4.3 Time-Resolved Photoluminescence

TRPL is measured by exciting a sample with a pulsed light source and recording the intensity of photoluminescence as a function of time from the excitation. The photoluminescence is directly proportional to the emission of the photoexcited charge carriers, which is useful to elucidate the dynamics of photogenerated charges in a semiconducting materials. $MAPbI_3$ and $MAPbBr_3$ both exhibit dual peak in photoluminescence emission spectra centered at ~2.32 and 2.24 eV, originated from free carrier and bound exciton, respectively [142–148, 150–157]. Thus the photoluminescence of perovskites shows the emission of free carriers and bound excitons. Photoluminescence also provides the information about the charge and energy transfer process between donor/acceptor species in photovoltaics. Moreover, photoluminescence quenching can also be used to determine the exciton diffusion length, diffusion rate, and lifetime [4]. The photoluminescence-quenching experiment can be simply carried out by fabricating perovskite thin films with and without exciton-quenching layer (Figure 12.23) and modeling the photoluminescence decay to a diffusion equation:

$$\frac{\partial n(x,t)}{\partial t} = D\frac{\partial^2 n(x,t)}{\partial x^2} - k(t)n(x,t) \tag{12.32}$$

where D is the diffusion coefficient and $k(t)$ is the photoluminescence decay rate in the absence of any quencher. A generic kinetic model accounting for exciton formation, dissociation into free charges, and trapping of free electrons is given by the rate equations [149]

$$\begin{aligned}
\frac{dn_e}{dt} &= \frac{I}{d} + R_d n_x - R_f n_e n_h - R_{eh} n_e n_h - R_{pop} n_e (N_T - n_T)\\
\frac{dn_x}{dt} &= R_f n_e n_h - R_d n_x - R_x n_x\\
\frac{dn_T}{dt} &= R_{pop} n_e (N_T - n_T) - R_{dep}(n_T^2 + n_T n_e)
\end{aligned} \tag{12.33}$$

where parameters R_f, R_d, and R_x are the rates of exciton formation, dissociation, and decay, respectively and R_{eh}, R_{pop}, and R_{dep} determine the recombination rate of free electrons and holes, trap population, and depopulation by electrons, respectively. The total decay rate k can be determined by fitting the exponential decay to the photoluminescence data measured from perovskite layers with quencher. The average diffusion length L_D of the charge carriers is given by $L_D = \sqrt{D\tau}$,

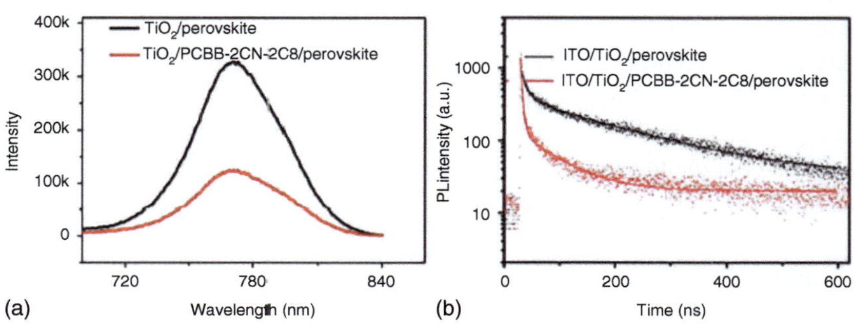

Figure 12.23 (a) Steady-state PL and (b) TRPL spectra of ITO/TiO_2/perovskite and ITO/TiO_2/PCBB-2CN-$2C_8$/perovskite. Source: Li et al. [149].

where τ is recombination time in the absence of quencher. Using the above model, the diffusion length for $MAPbI_{3-x}Cl_x$ was evaluated to be greater than 1 μm, whereas for $MAPbI_3$ it was only 100 nm [5]. The photoluminescence decay of the $MAPbI_{3-x}Cl_x$ film exhibits a time constant of $\tau = 273 \pm 7$ ns that found to be reduced to $\tau = 6.1 \pm 0.1$ ns and $\tau = 5.1 \pm 0.1$ ns for the samples with quenching layer of PCBM and spiro-OMeTAD, respectively. The lifetime of $MAPbI_3$ film was discovered to be only $\tau = 9.6 \pm 0.3$ ns and further reduces to $\tau = 3.17 \pm 0.03$ ns and $\tau = 4.2 \pm 0.1$ ns for $MAPbI_3$ with quencher PCBM and spiro-OMeTAD, respectively.

It has been reported that when a passivation layer of fullerene derivative, namely, PCBB-2CN-2C8, was sandwiched in between the TiO_2 and perovskites layer, the photoluminescence of the device gets quenched as compared with the device without passivation layer (Figure 12.23a) [158]. PL quenching was ascribed to the interfacial charge transfer between perovskites and PCBB-2CN-2C8. TRPL data (Figure 12.2b) of the same device shows a bi-exponential decay with time constants of $\tau_1 = 9.7$ ns (96.82%) and $\tau_2 = 172$ ns (3.18%). The short timescale τ_1 is associated with the bulk recombination in perovskite bulk film, and the delayed time constant τ_2 is attributed to the recombination of trapped charges [158, 159]. For device with PCBB-2CN-2C8 passivation layer (red line), both time constants were reduced ($\tau_1 = 3.6$ ns, $\tau_2 = 63.2$ ns), and excitingly τ_2 was almost overwhelmed by τ_1 due to the enhanced charge transfer from perovskite film and reduction of trap density.

$FAPbI_3$ single crystal possesses two decay processes, $\tau_1 = 32$ ns and $\tau_2 = 484$ ns extracted from the bi-exponential fitting of TRPL spectra [46]. The faster decay ascribes to the high trap density related to the crystal surface conditions, and the slow decay response is associated with the carrier transportation in bulk crystal having fewer defects [160]. Due to the high trap density, thin films show a faster decay than single crystal with lifetimes of $\tau_1 = 29$ ns and $\tau_2 = 227$ ns. Using these values, Bakr et al. have evaluated the diffusion length in $FAPbI_3$ and $FAPbBr_3$ single crystal through the relation $L_D = \sqrt{\mu\tau k_B T/e}$ [43]. The longest diffusion length was calculated to be 19 μm for $MAPbBr_3$ and ~6.6 μm for $MAPbI_3$. The band gap of halide perovskites can be tuned by varying the halide ratio, and so the photoluminescence emission peak energy was found to be tuned continuously from 2.26 eV for $FAPbBr_3$ to 1.50 eV for $FAPbI_3$ with respective diffusion lengths of 1.3 and 3.1 μm [161]. Photoluminescence emission peak of single crystal α-phase $FAPbI_3$ was observed at 820 nm, whereas a blueshift of 15 nm was recorded for 300 nm-thick $FAPbI_3$ film, implying that thin film has higher band gap than the single crystal structure [46].

The density of photogenerated charge carrier density (n_c) is defined through the equations [162, 163]

$$\frac{dn_c(t)}{dt} = -\sum_i a_i n_c(t) n_T(t) - \frac{n_c(t)}{\tau_0}$$
$$\frac{dn_T(t)}{dt} = -a_i n_c(t) n_T(t) \tag{12.34}$$

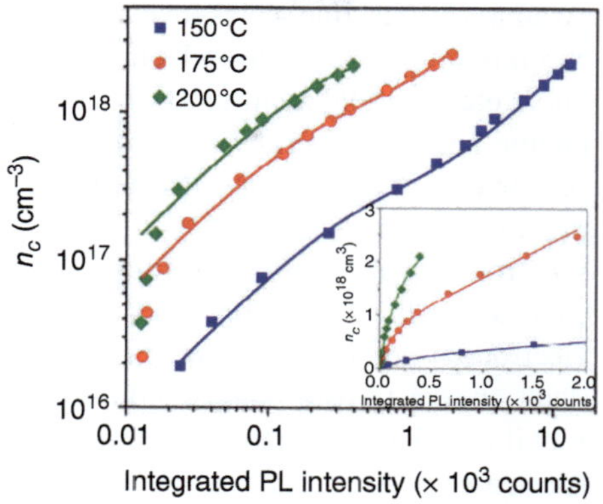

Figure 12.24 Fitting of photoexcited carrier density n_c vs. the I_{PL} for perovskite film $MAPbI_3$. Symbol represents experimental data, and solid lines are numerical fits using Eq. (12.34). Source: Peng et al. [164].

where $n_T(t)$is the trap density and a_i is constant. The photogenerated charge carrier density at $t = 0$ is given by

$$n_c(0) = \sum_i n_T(0)(1 - e^{-a\tau_0 I_{PL}/k}) + I_{PL}/k \tag{12.35}$$

with

$$I_{PL} = k\int_0^\infty [n_c(t)/\tau_0]dt,$$

where k is a constant. The bulk and surface trap density in $MAPbI_3$ thin film was evaluated to be $\sim 5 \times 10^{16}$ cm^{-3} and 1.6×10^{17} cm^{-3}, respectively.

Chabal et al. have used the above model to elucidate the effect of growth temperature on the trap density in $MAPbI_3$ thin films as depicted in Figure 12.24 [164]. It was found that the trap density increased by one order when growth temperature increases from 150 (1.6×10^{17} cm^{-3}) to 200 °C (1.6×10^{18} cm^{-3}). Similar value of trap density ($\sim 2.5 \times 10^{17}$ cm^{-3}) was also evaluated via face emission spectra [165]. These values are higher than that calculated from SCLC method. In contrast to $MAPbI_3$, $FAPbI_3$ thin film possesses slightly lower trap density of $\sim 1.0 \times 10^{17}$ cm^{-3} [165].

12.5 Conclusion

In summary, this chapter provides an insight into charge carrier kinetics and evaluated the important transport parameters in perovskite solar cells using SCLC, immittance spectroscopy, and transient spectroscopy. The SCLC method is useful when investigating the intrinsic electrical properties of the perovskite layer sandwiched between two electrodes of similar work function to facilitate the charge transfer of either type of carrier (electrons or holes) and block another type. This technique enhances the intrinsic electrical parameters of perovskites layers such as conductivity, mobility, trap density, etc., but does not account for charge transport in actual device configuration. To study the charge carrier kinetics in actual device configuration, immittance spectroscopy is an effective and facile technique that includes impedance spectroscopy, thermal admittance spectroscopy, and Mott–Schottky analyses. This technique is commonly used to distinguish the physical phenomena that occur at various time scales such as charge transfer at the interface or in bulk, charge recombination, electrical and ionic transport, electrode polarization, dielectric relaxation, etc. Thermal admittance spectroscopy provides information about interfacial charge accumulation, built-in voltage and depletion width at perovskites/charge selective contact interface, distribution of trap density in the bandgap. The ultrafast phenomena (ns – fs) occur in perovskites solar cells are analyzed by transient spectroscopy. These methods illustrate various phenomena that occur in a working solar cell from photoexcitation of charge carriers to their collection in the respective electrodes including recombination losses, decay paths, formation of intermediate charge transfer states. Additionally, charge carrier lifetime, charge carrier mobilities, and the nature of trap states should be investigated using transient spectroscopy.

Acknowledgments

Authors are also thankful to the deanship of scientific research at Islamic University of Madinah, Saudi Arabia for the financial support to this work through the Tamayyuz programme of the academic year 1441/1442 H (project No: 497). This study was also supported by the Europe excellence project ERC-CoG, MOLEMAT (726360), PARASOL (RTI2018-102292-B-I00), and Spanish ministry of science and technology.

References

1 Chen, Y., Sun, Y., Peng, J. et al. (2017). *Adv. Energy Mater.* 7: 1700162.
2 Saidaminov, M.I., Adinolfi, V., Comin, R. et al. (2015). *Nat. Comm.* 6: 8724.
3 Pascoe, A.R., Duffy, N.W., Scully, A.D. et al. (2015). *J. Phys. Chem. C* 119: 4444.
4 Xing, G., Mathews, N., Sun, S. et al. (2013). *Science* 342 (6156): 344–347.
5 Stranks, S.D., Eperon, G.E., Grancini, G. et al. (2013). *Science* 342 (6156): 341–344.
6 Ponseca, C.S., Savenije, T.J., Abdellah, M. et al. (2014). *J. Am. Chem. Soc.* 136: 5189–5192.
7 Hutter, E.M., Eperon, G.E., Stranks, S.D., and Savenije, T.J. (2015). *J. Phys.Chem. Lett.* 6: 3082–3090.
8 Haynes, J.R. and Hornbeck, J.A. (1955). *Phys. Rev.* 100: 606–615.
9 Hornbeck, J.A. and Haynes, J.R. (1955). *Phys. Rev.* 97: 311–321.
10 Koh, T.M., Fu, K., Fang, Y. et al. (2014). *J. Phys. Chem. C* 118: 16458–16462.
11 Eperon, G.E., Stranks, S.D., Menelaou, C. et al. (2014). *Energy Environ. Sci.* 7: 982–988.
12 Pang, S., Hu, H., Zhang, J. et al. (2014). *Chem. Mater.* 26: 1485–1491.
13 Noh, J.H., Im, S.H., Heo, J.H. et al. (2013). *Nano Lett.* 13: 1764–1769.
14 Heong, J. et al. (2021). *Nature* 592: 381–385. https://doi.org/10.1038/s41586-021-03406-5.

15 Agiorgousis, M.L., Sun, Y.-Y., Zeng, H., and Zhang, S. (2014). *J. Am. Chem. Soc.* 136: 14570–14575.

16 Kim, J., Lee, S.-H., Lee, J.H., and Hong, K.-H. (2014). *Phys. Chem. Lett.* 5: 1312–1317.

17 Shao, Y., Xiao, Z., Bi, C. et al. (2014). *Nat. Commun.* 5: 5784.

18 Duan, H.-S., Zhou, H., Chen, Q. et al. (2015). *Phys. Chem. Chem. Phys.* 17: 112.

19 Snaith, H.J., Abate, A., Ball, J.M. et al. (2014). *J. Phys. Chem. Lett.* 5: 1511.

20 Tress, W., Marinova, N., Moehl, T. et al. (2015). *Energy Environ. Sci.* 8: 995.

21 Contreras, L., Idigoras, J., Todinova, A. et al. (2016). *Phys. Chem. Chem. Phys.* 18: 31033.

22 Eames, C., Frost, J.M., Barnes, P.R.F. et al. (2015). *Nat. Commun* 6: 7497.

23 Unger, E.L., Hoke, E.T., Bailie, C.D. et al. (2014). *Energy Environ. Sci.* 7: 3690.

24 Yuan, Y. and Huang, J. (2016). *Acc. Chem. Res.* 49: 286–293.

25 Jin, H., Debroye, E., Keshavarz, M. et al. (2020). *Mater. Horiz.* 7: 397–410.

26 Conwell, E. and Weisskopf, V.F. (1950). *Phys. Rev.* 77: 388–390.

27 Xiao, Z., Yuan, Y., Shao, Y. et al. (2015). *Nat. Mater.* 14: 193–198.

28 Wetzelaer, G.J., Scheepers, M., Sempere, A.M. et al. (2015). *Adv. Mater.* 27: 1837–1841.

29 Suarez, B., Gonzalez-Pedro, V., Ripolles, T.S. et al. (2014). *J. Phys. Chem. Lett.* 5: 1628–1635.

30 Guillén, E., Ramos, F.J., Anta, J.A., and Ahmad, S. (2014). *J. Phys. Chem. C* 118 (7): 22913–22922.

31 Ke, W., Fang, G., Wang, J. et al. (2014). *ACS Appl. Mater. Interfaces* 6: 15959–15965.

32 Salado, M., Andresini, M., Huang, P. et al. (2020). *Adv. Funct. Mater.* 30: 1910561.

33 Yang, D., Yang, Z., Qin, W. et al. (2015). *J. Mater. Chem. A* 3: 9401.

34 Mahmood, K., Swain, B.S., and Jung, H.S. (2014). *Nanoscale* 6: 9127.

35 Ke, W., Fang, G., Liu, Q. et al. (2015). *J. Am. Chem. Soc.* 137: 6730.

36 Mei, A., Li, X., Liu, L. et al. (2014). *Science* 345: 295.

37 Yang, D., Yang, R., Xuejie, X.R. et al. (2016). *Adv. Mater.* 28: 5206–5213.

38 Zhang, W., Xiong, J., Li, J., and Daoud, W.A. (2019). *J. Mater. Chem. A* 7: 9486.

39 Khan, M.T., Kaur, A., Dhawan, S.K., and Chand, S. (2011). *J. App. Phys.* 109: 114509.

40 Khan, M.T. and Almohammedi, A. (2017). *J. App. Phy.* 122: 075502.

41 Khan, M.T., Agrawal, V., Almohammedi, A., and Gupta, V. (2018). *Solid-State Electronics* 145: 49–53.

42 Shi, D., Adinolfi, V., Comin, R. et al. (2015). *Science* 347: 519.

43 Saidaminov, M.I., Abdelhady, A.L., Murali, B. et al. (2015). *Nat. Comm.* 6: 7586.

44 Adinolfi, V., Yuan, M., Comin, R. et al. (2016). *Adv. Mater.* 28: 3406–3410.

45 Dong, Q., Fang, Y., Shao, Y. et al. (2015). *Science* 347 (6225): 967–970.

46 Han, Q., Bae, S.-H., Sun, P. et al. (2016). *Adv. Mater.* 28: 2253–2258.

47 Zhumekenov, A.A., Saidaminov, M.I., Haque, M.A. et al. (2016). *ACS Energy Letters* 1 (1): 32–37.

48 Luo, D., Zhao, L., Wu, J. et al. (2017). *Adv. Mater.* 29: 1604758.

49 Abdi-Jalebi, M., Dar, M.I., Sadhanala, A. et al. (2016). *Adv. Energy Mater.* 6: 1502472.

50 Kao, K.C. and Hwang, W. (1981). *Electrical Transport in Solids*. Oxford: Pergamon.

51 Ayres, J. (1993). *J. Appl. Phys.* 74: 1787–1792.

52 Ueki, T., Itsumi, M., and Takeda, T. (1997). *Appl. Phys. Lett.* 70: 1248–1250.

53 Schricker, A.D., Davidson, F.M. III, Wiacek, R.J., and Korgel, B.A. (2006). *Nanotechnology* 17: 2681.

54 Balcioglu, A., Ahrenkiel, R., and Hasoon, F. (2000). *J. Appl. Phys.* 88: 7175–7178.

55 V. Kaydanov and T. Ohno (2007) NREL Subcontract Report, NREL/SR-520-41129. https://www.nrel.gov/docs/fy07osti/41129.pdf.

56 Shen, X., Zhao, D., Liu, Z. et al. (2005). *Solid State Electron.* 49: 847–852.

57 Kerr, L., Li, S.S., Johnston, S. et al. (2004). *Solid-State Electron.* 48: 1579–1586.

58 Rajan, G., Aryal, K., Karki, S. et al. (2018). *J. Spectrosc.* 2018: 852749.

59 Queisser, H.J. (1972). *J. Appl. Phys.* 43: 3892.

60 Abdi-Jalebi, M., Dar, M.I., Senanayak, S.P. et al. (2019). *Sci. Adv.* 5: eaav2012.

61 Hauff, E. (2019). *J. Phys. Chem. C* 123: 11329–11346.

62 Khan, M.T., Salado, M., Almohammedi, A. et al. (2019). *Adv. Mater. Interfaces* 6 (21): 1901193.

63 Khan, M.T., Almohammedi, A., Kazim, S., and Ahmad, S. (2019). *Chem. Rec.* 20 (5): 452–465.

64 Kim, H.-S., Mora-Sero, I., Gonzalez-Pedro, V. et al. (2013). *Nature Commun.* 4: 2242.

65 Todinova, A., Contreras-Bernal, L., Salado, M. et al. (2017). *ChemElectroChem* 4: 2891–2901.

66 Juarez-Perez, E.J., Wußler, M., Fabregat-Santiago, F. et al. (2014). *J. Phys. Chem. Lett.* 5: 680–685.

67 Gonzalez-Pedro, E., Juarez-Perez, J., Arsyad, W.S. et al. (2014). *Nano Lett.* 14: 888–893.

68 Dualeh, A., Moehl, T., Tétreault, N. et al. (2014). *ACS Nano* 8: 362–373.

69 Mulder, W.H., Sluyters, J.H., Pajkossy, T., and Nyikos, L. (1990). *J. Electroanal. Chem.* 285: 103–115.

70 Vogelsang, J. and Strunz, W. (2001). *Electrochim. Acta* 46: 3619–3625.

71 Bisquert, J. (2002). *J. Phys. Chem. B* 106: 325–333.

72 Bisquert, J., Mora-Sero, I., and Fabregat-Santiago, F. (2013). *Chem. Electro. Chem.* 1 (1): 289–296.

73 Zhang, H., Qiao, X., Shen, Y. et al. (2015). *J. Mater. Chem. A* 3: 11762–11767.

74 Sanchez, R.S., Gonzalez-Pedro, V., Lee, J.-W. et al. (2014). *J. Phys. Chem. Lett.* 5: 2357.

75 Bisquert, J., Bertoluzzi, L., Mora-Sero, I., and Garcia-Belmonte, G. (2014). *J. Phys. Chem. C* 118 (33): 18983–18991.

76 Pockett, A., Eperon, G.E., Peltola, T. et al. (2015). *J. Phys. Chem. C* 119: 3456–3465.

77 Lei, L., Zhang, S., Yang, S. et al. (2018). *Nanotechnology* 29: 255201.

78 Zhang, J., Barboux, P., and Pauporté, T. (2014). *Adv. Ener. Mater.* 4 (18): 1400932.

79 Zhang, J., José Juárez-Pérez, E., Mora-Seró, I. et al. (2015). *Journal of Materials Chemistry A* 3: 4909–4915.

80 Zhang, J. and Pauporté, T. (2015). *The Journal of Physical Chemistry C* 119 (27): 14919–14928.

81 Salado, M., Calió, L., Bernal, L.C. et al. (2018). *Materials (Basel)* 11 (7): 1073.

82 Kim, H.-S., Lee, J.-W., Yantara, N. et al. (2013). *Nano Lett* 13: 2412–2417.

83 Rep, D.B.A. and Prins, M.W.J. (1999). *J. Appl. Phys.* 85: 7923.

84 Bag, M., Renna, L.A., Adhikari, R. et al. (2015). *J. Am. Chem. Soc.* 137 (40): 13130–13137.

85 Chen, X., Shirai, Y., Yanagida, M., and Miyano, K. (2019). *J. Phys. Chem. C* 123 (7): 3968–3978.

86 Yang, T.Y., Gregori, G., Pellet, N. et al. (2015). *J. Angew. Chem.* 127: 8016–8021.

87 Beilsten-Edmands, J., Eperon, G., Johnson, R. et al. (2015). *Appl. Phys. Lett.* 106: 173502.

88 Hoque, M.N.F., Yang, M., Li, Z. et al. (2016). *ACS Energy Lett.* 1: 142–149.

89 Li, Z., Mercado, C.C., Yang, M. et al. (2017). *Chem. Commun.* 53: 2467–2470.

90 Almora, O., Zarazua, I., Mas-Marza, E. et al. (2015). *J. Phys. Chem. Lett.* 6: 1645–1652.

91 Lopez-Varo, P., Tejada, J.A.J., Rosell, M.G. et al. (2018). *Adv. Energy Mater.*: 1702772.

92 Bisquert, J. (2003). *Phys. Chem. Chem. Phys.* 5: 5360–5364.

93 Almora, O., Aranda, C., and Belmonte, G.G. (2018). *J. Phys. Chem. C* 122 (25): 13450–13454.

94 Zarazua, I., Bisquert, J., and Belmonte, G.G. (2016). *J. Phys. Chem. Lett.* 7: 525–528.

95 Almora, O., Guerrero, A., and Garcia-Belmonte, G. (2016). *Appl. Phys. Lett.* 108: 043903.

96 Ravishankar, S., Gharibzadeh, S., Roldán-Carmona, C. et al. (2018). *Joule* 2: 788–798.

97 Garcia-Belmonte, G. and Bisquert, J. (2016). *ACS Energy Lett.* 1: 683–688.

98 Juarez-Perez, E.J., Sanchez, R.S., Badia, L. et al. (2014). *J. Phys. Chem. Lett.* 5: 2390–2394.

99 Chen, B., Yang, M., Priya, S., and Zhu, K. (2016). *J. Phys. Chem. Lett.* 7: 5, 905–917.

100 Kim, H.-S., Jang, I.-H., Ahn, N. et al. (2015). *J. Phys. Chem. Lett.* 6: 4633–4639.

101 Ghahremanirad, E., Bou, A., Olyaee, S., and Bisquert, J. (2017). *J. Phys. Chem. Lett.* 8 (7): 1402–1406.

102 Yamamuro, N.O., Matsuo, T., and Suga, H. (1992). *J. Phys. Chem. Solids* 53: 935–939.

103 Stoumpos, C.C., Malliakas, C.D., and Kanatzidis, M.G. (2013). *Inorg. Chem.* 52: 9019–9038.

104 Rakita, Y., Bar-Elli, O., Meirzadeh, E. et al. (2017). *Proc. Nat. Acad. Sci.* 114: E5504–E5512.

105 Almora, O., Aranda, C., Marzá, E.M., and Belmontea, G.G. (2016). *Appl. Phys. Lett.* 109: 173903.

106 Laban, W.A. and Etgar, L. (2013). *Energy Environ. Sci.* 6: 3249–3253.

107 O'Regan, B., Moser, J., Anderson, M., and Graetzel, M. (1990). *J. Phys. Chem.* 94: 8720–8726.

108 Nakade, S., Saito, Y., Kubo, W. et al. (2003). *Electrochem. Commun.* 5: 804–808.

109 Zaban, A., Meier, A., and Gregg, B.A. (1997). *J. Phys. Chem. B* 101: 7985–7990.

110 Yadav, P., Prochowicz, D., Saliba, M. et al. (2017). *Crystals* 7 (8): 252.

111 Li, N., Tao, S., Chen, Y. et al. (2019). *Nature Energy* 4: 408–415.

112 Yang, G., Wang, C., Lei, H. et al. (2017). *J. Mater. Chem. A* 5: 1658–1666.

113 Futscher, M.H., Lee, J.M., McGovern, L. et al. (2019). *Mater. Horiz.* 6: 1497–1503.

114 Chen, Q., Chen, L., Ye, F. et al. (2017). *Nano Lett.* 17 (5): 3231–3237.

115 Song, D., Cui, P., Wang, T. et al. (2015). *J. Phys. Chem. C* 119: 22812–22819.

116 Hwang, T., Yun, A.J., Kim, J. et al. (2019). *ACS Appl. Mater. Interfaces* 11: 6907–6917.

117 Yin, W.-J., Shi, T., and Yan, Y. (2014). *Adv. Mater.* 26: 4653–4658.

118 Almora, O., García-Batlle, M., and Belmonte, G.G. (2019). *J. Phys. Chem. Lett.* 10 (13): 3661–3669.

119 Schroder, D.K. (2005). *Semiconductor Material and Device Characterization*, 3e. Wiley-IEEE Press.

120 Carr, J.A., Elshobaki, M., and Chaudhary, S. (2015). *Appl. Phys. Lett.* 107: 203302.

121 Hegedus, S.S. and Fagen, E.A. (1992). *J. Appl. Phys.* 71: 5941.

122 Walter, T., Herberholz, R., Müller, C., and Schock, H.W. (1996). *J. Appl. Phys.* 80: 4411–4420.

123 Xu, L., Wang, J., and Hsu, J.W.P. (2016). *Phys. Rev. Appl.* 6: 064020.

124 Esfahani, S. (2015). Device physics of organic and perovskite solar cells. Graduate Theses and Dissertations. 14961. https://lib.dr.iastate.edu/etd/14961.

125 Prochowicz, D., Yadav, P., Saliba, M. et al. (2017). *ACS Appl. Mater. Interfaces* 9: 28418–28425.

126 Samiee, M., Konduri, S., Ganapathy, B. et al. (2014). *Appl. Phys. Lett.* 105: 153502.

127 Heo, J.H., Song, D.H., Han, H.J. et al. (2015). *Adv Mater.* 27: 3424–3430.

128 Ran, C., Xu, J., Gao, W. et al. (2018). *Chem. Soc. Rev.* 47: 4581–4610.

129 Valverde-Chávez, D.A., Ponseca, C.S., Stoumpos, C.C. et al. (2015). *Energy Environ. Sci.* 8: 3700–3707.

130 La-o-vorakiat, C., Salim, T., Kadro, J. et al. (2015). *Nat. Commun.* 6: 7903.

131 Milot, R.L., Eperon, G.E., Snaith, H.J. et al. (2015). *Adv. Funct. Mater.* 25: 6218–6227.

132 Wehrenfennig, C., Eperon, G.E., Johnston, M.B. et al. (2014). *Adv. Mater.* 26: 1584–1589.

133 Savenije, T.J., Ferguson, A.J., Kopidakis, N., and Rumbles, G. (2013). *J. Phys. Chem. C* 117: 24085–24103.

134 Ponseca, C.S. Jr., Tian, Y., Sundström, V., and Scheblykin, I.G. (2016). *Nanotechnology* 27: 082001.

135 Hutter, E.M., Hofman, J.-J., Moes, M.P.M. et al. (2017). *Adv. Energy Mater.* 7 (13): 1602349.

136 Ponseca, C.S., Hutter, E.M., Piatkowski, P. et al. (2015). *J. Am. Chem. Soc.* 137 (51): 16043–16048.

137 Oga, H., Saeki, A., Ogomi, Y. et al. (2014). *J. Am. Chem. Soc.* 136 (39): 13818–13825.

138 Polli, D., Rivalta, I., Nenov, A. et al. (2015). *Photochem, Photobiol. Sci.* 14: 213–228.

139 Hsiao, Y.-C., Wu, T., Li, M. et al. (2015). *J. Mater. Chem. A* 3: 15372–15385.

140 Kim, H.-S., Lee, C.-R., Im, J.-H. et al. (2012). *Sci. Rep.* 2: 591.

141 Peng, J., Chen, Y., Zheng, K. et al. (2017). *Chem. Soc. Rev.* 46: 5714–5729.

142 Bretschneider, S.A., Laquai, F., and Bonn, M. (2017). *The Journal of Physical Chemistry C* 121: 11201–11206.

143 Price, M.B., Butkus, J., Jellicoe, T.C. et al. (2015). *Nat. Commun.* 6: 8420.

144 Yang, Y., Ostrowski, D.P., France, R.M. et al. (2015). *Nat. Photonics* 10: 53–59.

145 Herz, L.M. (2016). *Annu. Rev. Phys. Chem.* 67: 65–89.

146 Gao, W., Gao, X., Abtew, T.A. et al. (2016). *Phys. Rev. B* 93: 085202.

147 Wu, X., Trinh, M.T. Niesner, D. et al. (2015). *J. Am. Chem. Soc.* 137: 2089–2096.

148 Pollock, T.P. and Schlenker, C.W. (2019). *J. Phys. Chem. C* 123: 18160–18170.

149 Li, Y., Zhao, Y., Chen, Q. et al. (2015). *J. Am. Chem. Soc.* 137 (49): 15540–15547.

150 Fang, Y.J., Wei, H.T., Dong, Q.F., and Huang, J.S. (2017). *Nat. Commun.* 8: 14417.

151 Priante, D., Dursun, I., Alias, M.S. et al. (2015). *Appl. Phys. Lett.* 106: 081902.

152 Wu, B., Nguyen, H.T., Ku, Z. et al. (2016). *Adv. Energy Mater.* 6: 1600551.

153 Dar, M.I., Jacopin, G., Meloni, S. et al. (2016). *Sci. Adv.* 2: e1601156.

154 Wenger, B., Nayak, P.K., Wen, X. et al. (2017). *Nat. Commun.* 8: 590.

155 Wang, T., Daiber, B., Frost, J.M. et al. (2017). *Energy Environ. Sci.* 10: 509–515.

156 Fu, J., Janaludin, N.F., Wu, B. et al. (2019). *Adv. Energy Mater.* 9: 1803119.

157 Wu, B., Yuan, H., Xu, Q. et al. (2019). *Nat. Commun.* 10: 484.

158 Liang, P.-W., Liao, C.-Y., Chueh, C.-C. et al. (2014). *Adv. Mater.* 26: 3748.

159 Zuo, L., Gu, Z., Ye, T. et al. (2015). *Am. Chem. Soc.* 137: 2674.

160 Sveinbjrnsson, K., Aitola, K., Zhang, X. et al. (2015). *J. Phys. Chem. Lett.* 6: 4259.

161 Rehman, W., Milot, R.L., Eperon, G.E. et al. (2015). *Adv. Mater.* 27: 7938–7944.

162 Stranks, S.D., Burlakov, V.M., Leijtens, T. et al. (2014). *Phys. Rev. App.* 2: 034007.

163 Xing, G., Mathews, N., Lim, S. et al. (2014). *Nature Mater.* 13: 476–480.

164 Peng, W., Anand, B., Liu, L. et al. (2016). *Nanoscale* 8: 1627–1634.

165 Yuan, F., Wu, Z., Dong, H. et al. (2017). *J. Phys. Chem. C* 121: 15318–15325.

13

Printable Mesoscopic Perovskite Solar Cells

Daiyu Li, Yaoguang Rong, Yue Hu, Anyi Mei, and Hongwei Han

Huazhong University of Science and Technology, Wuhan National Laboratory for Optoelectronics, 1037 Luoyu Road, Wuhan 430074, Hubei, China

13.1 Introduction

As a promising photovoltaic (PV) technology candidate with a certified power conversion efficiency (PCE) of 25.5% [1], commercializing organic–inorganic halide perovskite solar cells (PSCs) is the ultimate goal for those studies on them since 2009 [2–4]. The golden triangle of high efficiency, high stability, and low cost is generally pursued to commercialize solar cells when they are up-scaled as modules. Although the efficiency of PSCs with lab size below 1.0 cm^2 is extremely high, up-scaling brings significant efficiency loss. The current reported highest certified efficiency of PSC modules is just 17.25% on a limited area of 19.28 cm^2 [5]. The causes for such loss are because those technologies (such as spin coating) developed in the lab for high efficiency are not suitable for the industrial production of PSC modules with large areas. Those technologies (such as screen printing and slot-die coating) suitable for upscaling are not beneficial for the high efficiency of PSC modules [6]. Also, conventional-structured PSCs suffer from serious stability issues. The stability of PSCs is influenced a lot by the charge transport materials and electrode materials in the devices together with the device structures. Many studies have demonstrated that organic hole transport materials (HTM) such as spiro-OMeTAD, which is generally applied in PSCs for high efficiency and degrades easily under working conditions, restrict the long-term stability of PSCs [7]. Metal electrodes in PSCs also lead to stability problems [8]. The commonly used silver electrode would corrode halide perovskites, and the gold electrode would diffuse in the device to cause shunt concern. The HTM and gold electrode lead to stability and cost concerns since they are not cost-effective. Besides, since the thermal expansion coefficient of halide perovskites differs from that of the selective contacts, the device structure should also be designed to avoid peeling off concerns.

This chapter focuses on the printable mesoscopic PSCs, which avoid the aforementioned concerns by special structure design and fabricating methods choosing. Printable mesoscopic PSCs have shown great advantages on stability and upscaling.

Perovskite Solar Cells: Materials, Processes, and Devices, First Edition.
Edited by Shahzada Ahmad, Samrana Kazim, and Michael Grätzel.

Studied by various groups worldwide, the HTM-free and carbon-electrode-based triple-mesoscopic PSCs have passed many stability tests. Namely, the continuous 1-sun illumination test for over 10 000 hours, the outdoor test in the hot desert (Jeddah, Saudi Arabia) for over a week, the outdoor test in a high-humidity environment (Wuhan, China) for over several months, the thermal stability test at 100 °C for over 1500 hours and the shelf-life stability test for over one year. Meanwhile, printable mesoscopic PSC modules with an area of 3600 cm^2 have also been developed. The device structure, performance optimization strategies, and upscaling progress of printable mesoscopic PSCs are overviewed.

13.2 Device Structures and Working Principles

As shown in Figure 13.1a, printable mesoscopic PSCs consist of a porous TiO_2/ZrO_2/Carbon triple-layer with halide perovskites filled inside the pores on transparent conductive glass (TCO), typically fluorine-doped tin dioxide (FTO) or indium-doped tin oxide (ITO) [9, 10]. The TCO serves as the window electrode. The mesoscopic TiO_2 layer serves as the electron selective layer. It can be replaced by other n-type semiconductors such as SnO_2 and $BaSnO_3$. The porous carbon layer serves as the back electrode and collects holes. The mesoscopic ZrO_2 layer serves as the insulating scaffold to separate the electron selective layer and the back electrode to prevent shunting. It can be replaced by other insulating materials such as Al_2O_3 or SiO_2. Halide perovskites serve as the light absorber and assist hole transport. The energy level diagram of printable mesoscopic PSCs is illustrated in Figure 13.1b. Halide perovskites absorb light and generate electrons and holes. The electrons are collected by the electron selective layer of TiO_2. Holes transport in halide perovskites due to perovskites' excellent hole-transporting properties. Then they are collected by the back electrode. The ZrO_2 scaffold is not involved in the charge generation and transport process. Han group pioneered in this structure and explored both the mesoscopic scaffold and the perovskite absorber that fits the structure.

The configuration of printable mesoscopic PSCs is unique and different from that of those conventional PSCs. On the one hand, no additional organic HMTs are used in printable mesoscopic PSCs. The omitting of organic HMTs releases the stability and cost concern induced by them. On the other hand, printable mesoscopic PSCs are not typically layer-stacked structures as their conventional analogs. A bulk heterojunction is formed between halide perovskite forms and the porous triple layer in printable mesoscopic PSCs. Such structure is beneficial for avoiding the adverse effect of thermal expansion coefficient mismatch. Besides, there are indeed no individual perovskite films in printable mesoscopic PSCs. The morphology of halide perovskites is largely determined by the prepared triple-layer. The triple-layer can be easily upscaled uniformly by wet chemistry techniques such as screen printing or slot-die coating. The mesopores are created by annealing the scaffold at a temperature ≥400 °C or even low temperature lower than 200 °C. Therefore, upscaling printable mesoscopic PSCs is not very challenging.

Besides the special structure design, choosing graphite and carbon black as the back-electrode materials to collect holes is also very important since those carbon

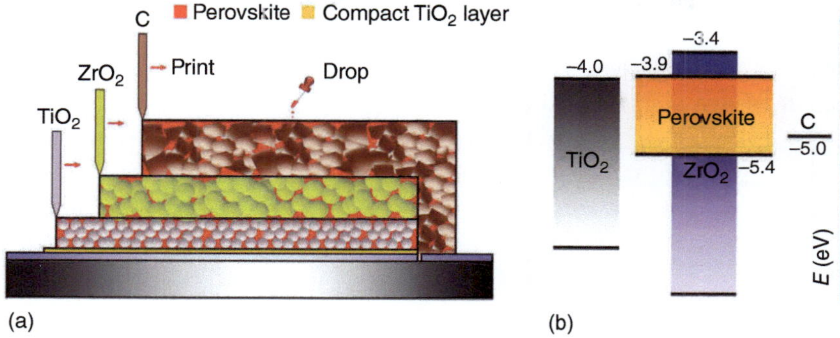

Figure 13.1 (a) Schematic drawing of the printable mesoscopic PSCs. (b) Energy band diagram of the printable mesoscopic PSCs. Source: Mei et al. [9].

materials are cheap and stable. Han group firstly demonstrated the potential application of carbon electrodes in PSCs in 2013 and further obtained a certified efficiency of 12.84% in 2014. Now, various carbon materials such as carbon nanotubes have been developed to serve as back electrodes for PSCs with different configurations, including replacing gold electrodes in conventional PSCs.

13.3 Progress of Efficiency and Stability

Printable mesoscopic PSCs based on carbon electrodes have advantages in fabrication cost and demonstrate promising efficiency and stability. Particularly, an efficiency approaching 19% has been achieved recently. Printable mesoscopic PSCs have survived in the 10 000 hours stability test under continuous illumination. Although the efficiency is relatively lower than that obtained by conventional structured PSCs, printable mesoscopic PSCs have been regarded as the most promising technical route for commercializing this emerging PV technology.

HTM-free and carbon-electrode-based printable mesoscopic PSCs were firstly reported by Han's group in 2013. PSCs based on the structure of TiO_2/ZrO_2/Carbon/Perovskite were also named as "triple-mesoscopic" PSCs [3]. By optimizing the perovskite materials, modifying the interfaces, tuning the structures, and performing theoretical analysis, the efficiency was steadily improved to over 16%, and recent to 18.7% (unpublished). Figure 13.2 presents the evolution of the efficiency improvement of triple-mesoscopic PSCs. The initial efficiency of triple-mesoscopic PSCs was around 6% in 2013 [10]. Then, Han and coworkers employed 5-ammoniumvaleric acid (5-AVA) iodide into the typical perovskite precursor solution and developed a mixed-cation perovskite $(5\text{-AVA})_x(MA)_{1-x}PbI_3$ [9]. Besides lower defect concentration, the mixed-cation perovskite achieved more complete nanopore filling in the triple-layer scaffold of TiO_2/ZrO_2/Carbon. The enhanced quality of the perovskite absorber benefited the charge separation and transport. Correspondingly, the efficiency of the device was improved to 12.84%, which was certified by Newport. PSCs based on TiO_2/Perovskite/spiro-OMeTAD/Au

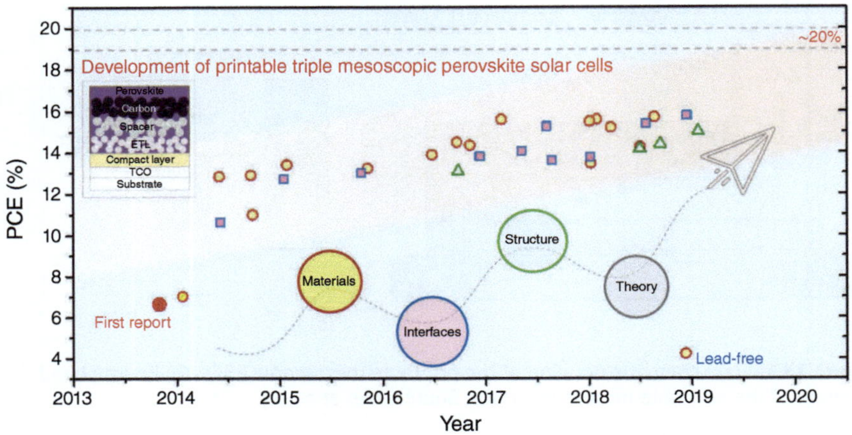

Figure 13.2 The progress of the PCE of printable mesoscopic PSCs. The strategies for improving efficiency are mainly divided into material optimization, interface engineering, structure engineering, and theoretical analysis.

structure have suffered severe stability issues during that time. On the contrary, the PSC devices based on the structure of TiO_2/ZrO_2/Carbon/$(5\text{-AVA})_x(MA)_{1-x}PbI_3$ have demonstrated quite stable performance under continuous illumination, as shown in Figure 13.3b. After 1000 hours of illumination, no loss was observed for such triple-mesoscopic PSCs-based mixed-cation perovskite.

Intensive research attention has been paid to this PSCs structure, and abundant stability results have been reported. Among PSCs based on different device structures, triple-mesoscopic PSCs have demonstrated the most promising stability results [11]. Stability tests under various conditions were also conducted for triple-mesoscopic PSCs. Graetzel et al. tracked the device performance under an outdoor condition in Jeddah, Saudi Arabia, for a week. Then, the devices were exposed to indoor heat stress (80–85 °C) for three months and light soaking for 44 days [12]. No degradation was observed, confirming that triple-mesoscopic PSCs provide a promising path toward realizing efficient and stable perovskite photovoltaics. Ito and coworkers developed an encapsulation method based on UV-curable sealant for triple-mesoscopic PSCs. They investigated the thermal stability of triple-mesoscopic PSCs at 100 °C for over 1500 hours (Figure 13.4a) [13]. Large-area (substrate area $10 \times 10\,cm^2$) triple-mesoscopic PSC mini-modules with efficiency exceeding 10% were also fabricated, exhibiting outdoor stability of one month and shelf-life stability over one year [14]. Nazeeruddin and coworkers reported 10 000 hours stability of triple-mesoscopic PSCs based on $(5\text{-AVA})_x(MA)_{1-x}PbI_3$ under continuous illumination (Figure 13.4b) [15]. The total irradiation equals 10 years outdoor solar irradiation in most areas in Europe, which offers the possibility for being stable at the actual working condition for over 10 years.

For the performance improvement, the strategies mainly focus on optimizing the perovskite absorber, modifying the interfaces, tuning the device structures, and analyzing the theoretical issues. Rong et al. employed guanidinium chloride

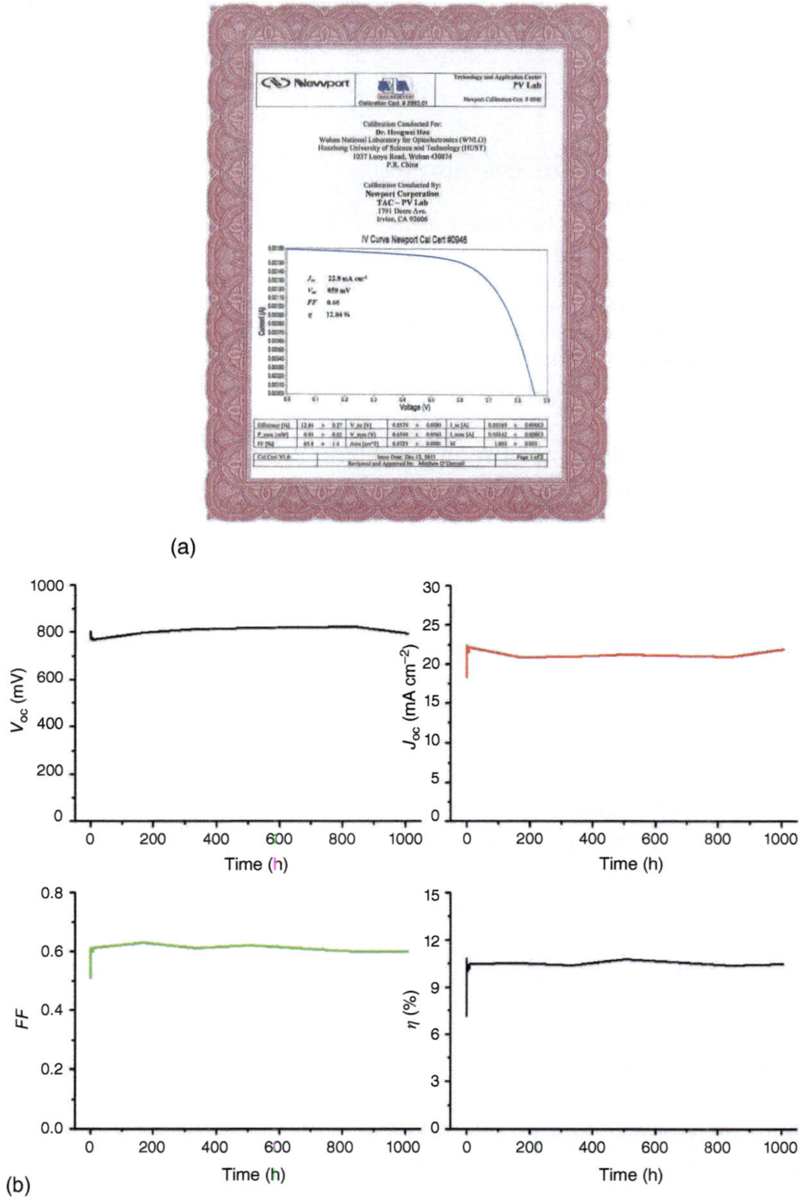

Figure 13.3 (a) The certification of efficiency of 12.84% for triple-mesoscopic PSCs. (b) The stability measurements of triple-mesoscopic PSCs under continuous illumination. Source: Mei et al. [9].

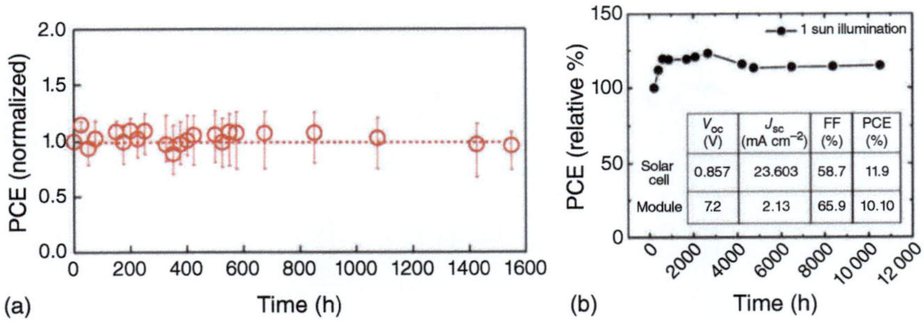

Figure 13.4 (a) Variation of the average normalized PCE of three triple-mesoscopic PSCs fabricated using the side-sealed method during thermal stability testing at 100 °C in the dark. The photovoltaic measurements were performed under AM1.5G simulated sunlight. Each data point represents an average between the forward and reverse scans. Source: Baranwal et al. [13]. (b) Stability test under 1 sun AM1.5G conditions at a stabilized temperature of 55 °C and short circuit conditions. Source: Grancini et al. [14].

(GuaCl) as a multifunctional additive in the typical perovskite precursor. GuaCl acts as a cross-linker among the neighboring grains and improves the quality of the $CH_3NH_3PbI_3$ perovskite absorber. The device fabricated with $CH_3NH_3PbI_3$·GuaCl delivered an efficiency of 14.27% with an enhanced V_{OC} of 1.0 V [16]. Han and coworkers investigated the crystallization of $CH_3NH_3PbI_3$ perovskite in ambient air [17]. By employing NH_4Cl to form an intermediate and controlling the relative humidity (RH), the crystallization of the precursor was monitored. As shown in Figure 13.5, when the as-annealed devices were exposed to ambient air at different RH, the device performance increased dramatically in the first 10–48 hours and then remained stable or decreased. During the ambient air exposure process, the color of the devices (from the glass side) also became darker, while no similar phenomenon was observed in an inert atmosphere. This work investigates the crystallization of the perovskite absorber in ambient air and reveals the importance of the intermediate formation and transition.

For optimizing the compositions of the perovskite absorbers, Chen et al. developed a mixed-anion perovskite of $CH_3NH_3PbI_{(3-x)}(BF_4)_x$ with high charge carrier concentration and conductivity [18]. Correspondingly, the FF of the triple-mesoscopic PSCs was significantly enhanced to 0.76 due to more efficient charge transport and suppressed recombination. Furthermore, Sheng et al. doped the typical perovskite of $CH_3NH_3PbI_3$ with alkali metal salt of LiCl. It was found that LiCl can significantly improve the electronic properties of the perovskite absorber, particularly the elevated conductivity, and at the same time improve the crystallinity and morphology of $CH_3NH_3PbI_3$. Correspondingly, the LiCl·$CH_3NH_3PbI_3$ based triple-mesoscopic PSCs yielded an enhanced PCE of 14.5% with a high FF of 0.77 [19].

For tuning the device structure of triple-mesoscopic PSCs, the research mainly focuses on the construction of the triple-layer mesoporous scaffold of TiO_2/ZrO_2/Carbon, especially the carbon layer. Ku et al. firstly reported the configuration and compared spheroidal graphite and flaky graphite as the main components of the carbon electrodes [20]. Then, Xu et al. prepared a highly ordered

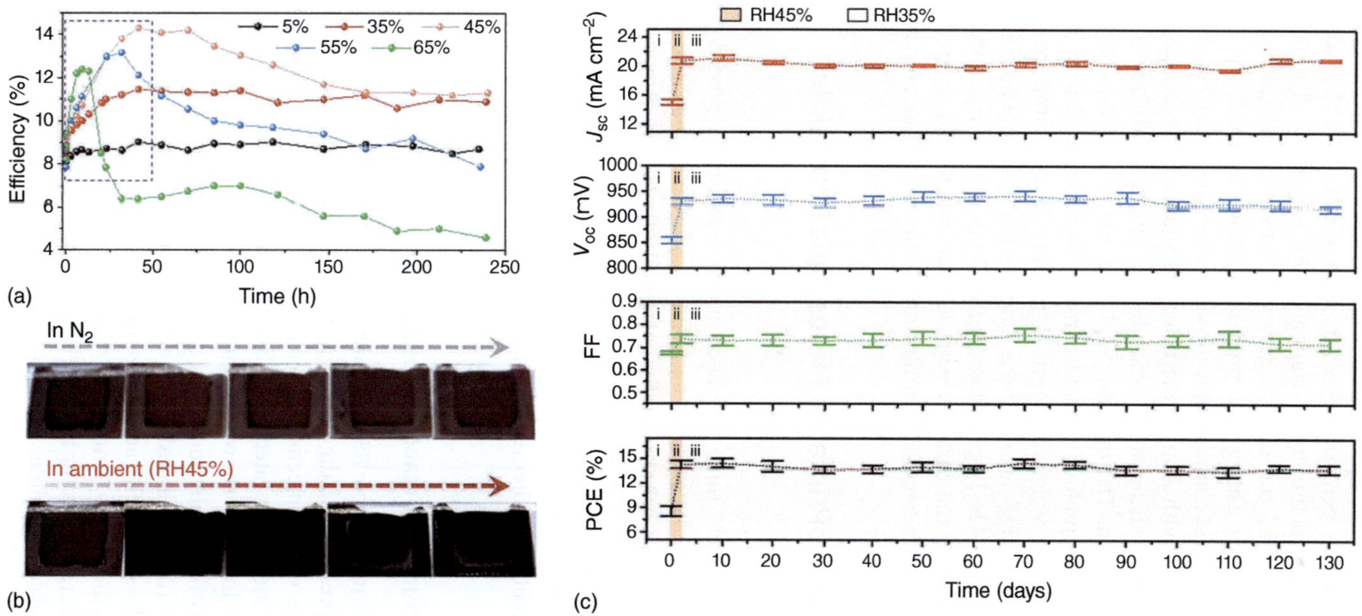

Figure 13.5 (a) PCE evolutions of the as-annealed devices stored under ambient conditions with RH5–65%. The significant improvement in efficiency is marked with a purple rectangle. (b) Digital images (from the glass side) of the as-annealed devices stored in an inert (N_2) and ambient atmosphere (RH45%). (c) Variations of J_{SC}, V_{OC}, FF, and PCE of devices upon fabricating and aging under ambient condition without encapsulation ((i) device fabrication except for the ambient exposure, RH35%; (ii) ambient exposure, RH45%; (iii) long-term storage, RH35%). Source: Rong et al. [17].

mesoporous carbon electrode with well-connected frameworks for constructing the triple-layer scaffold [21]. It was found that the microstructure of the carbon layer mainly affected the perovskite precursor penetration in the scaffold. It correspondingly influenced the crystallization of the perovskite absorber in the scaffold. At the same time, the conductivity will also systematically affect the performance of the devices. Zhang et al. prepared carbon electrodes with graphite of different particle sizes and compared the device performance [22]. For the carbon electrodes using graphite with a larger diameter, the square resistance was higher. For the carbon electrodes using small graphite, the infiltration of the perovskite precursor might be limited. The optimal graphite diameter was around 8 μm for obtaining both sufficient conductivity and porosity. Duan et al. further employed ultrathin graphite (UG) to replace the normal graphite. It was found that UG can effectively increase the surface area of the carbon electrode and obtain comparable conductivity (Figure 13.6). The large surface area facilitated the charge transfer from the perovskite absorber to the carbon electrodes. Meanwhile, the interconnected channels formed by UG can benefit the infiltration of the perovskite precursor [23].

13.4 Scaling-up of Printable Mesoscopic Perovskite Solar Cells

During the past few years, there has been much remarkable progress toward large-scale PSC manufacturing, as shown in Figure 13.7, which shows the evolution of PSCs and PSC modules based on different architectures [24]. Among all PSC architectures, the printable triple mesoscopic PSC module yielded the largest device area of 3600 cm^2 thanks to the easy scale-up screen printing technique. Besides, the replacement of noble metals as back contacts by low-cost carbon materials of graphite and carbon black further reduces the material cost of printable mesoscopic PSCs [14, 25].

13.4.1 The Structure of Printable Mesoscopic PSC Modules

Large-area printable mesoscopic PSC modules were firstly reported by Han's group [14]. Figure 13.8a presents the digital image of the 100 cm^2 printable mesoscopic PSC module in series interconnection with a monolithic architecture. Modules normally use series-connected architecture to divide a large cell into smaller sub-cells to reduce resistance losses, as shown in Figure 13.8b. After comparing 5, 8, and 10 sub-cell designs, 10 cells serially connected modules achieved the best performance (10.4% PCE). In the 10 sub-cell design, the active area was 49 cm^2. Specifically, the length and width of a single cell's active area were 94 and 5.3 mm, respectively. The gap between each sub-cell was 5 mm. Both the PCEs and the active areas are much enhanced in their later work after in-depth research and continuous optimization. For the fabrication of such modules, FTO glass substrates were firstly etched by laser processing. It was then ultrasonically cleaned with detergent, deionized water, and ethanol in turn. Next, a compact-TiO_2 layer was deposited on the FTO substrates

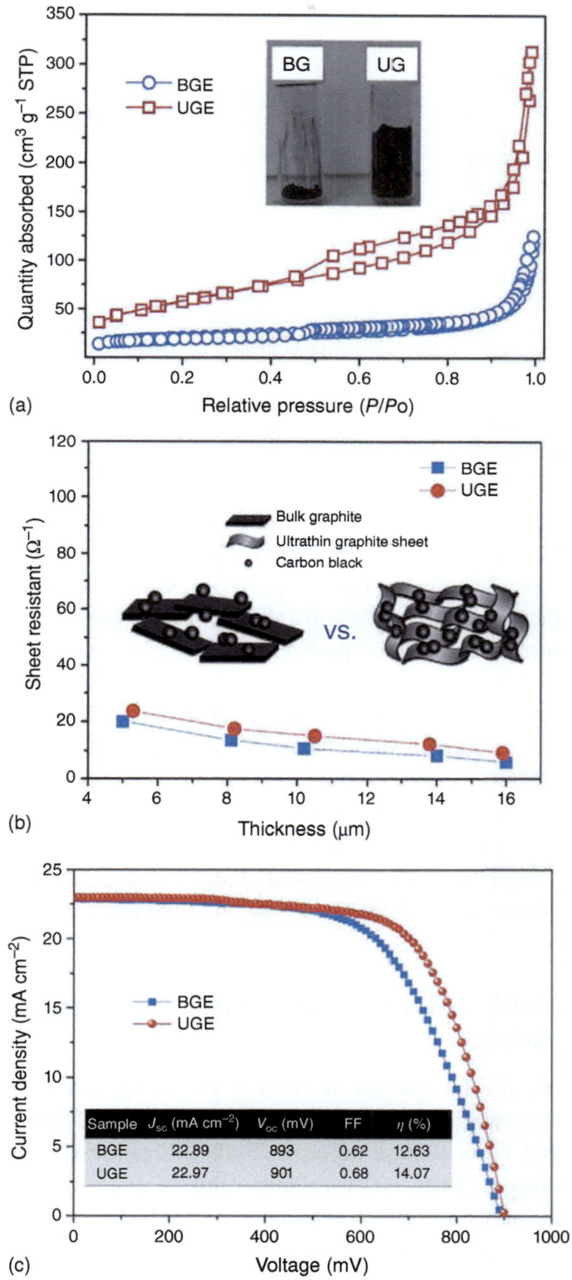

Figure 13.6 (a) Nitrogen gas adsorption and desorption isotherms of bulk graphite/carbon black electrode (BGE) and ultrathin graphite/carbon black electrode (UGE). Digital photographs of 0.05 g BG and 0.05 g UG powders shown in the inset of the figure. (b) Square resistance of BGE and UGE films at different thickness and schematic model of BG and UG conductive network (inset). (c) *J*–*V* curves of the device based on BGE and UGE. Source: Duan et al. [23].

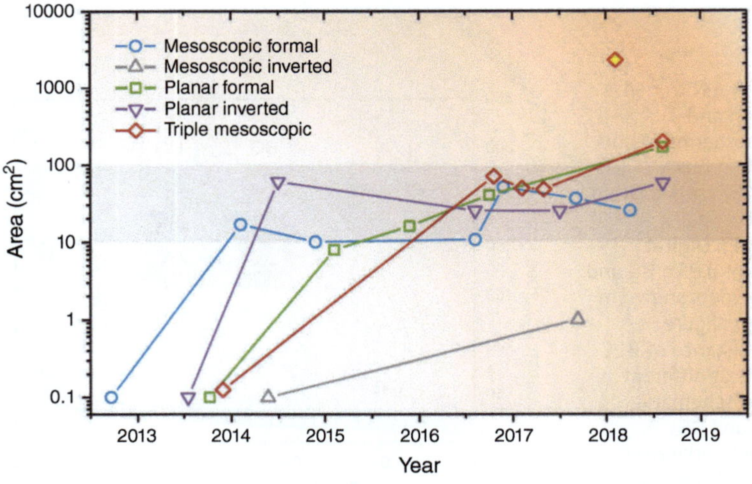

Figure 13.7 The area development of perovskite solar mini-modules, sub-modules, and modules based on different device architectures (mesoscopic formal, mesoscopic inverted, planar formal, planar inverted, triple mesoscopic). Source: Hu et al. [24].

by spray pyrolysis deposition. The mesoporous TiO_2, ZrO_2, and carbon layer were screen printed on the compact-TiO_2 covered FTO substrates. The thickness of each mesoporous layer is controlled by modifying the mesh of the screen mask and the composition of the pastes. After sintering at 400–500 °C for 30–40 minutes, a perovskite precursor solution was dropped on the top of the carbon layer, infiltrating into the mesoscopic structure in ambient air. After drying at 50–70 °C for two to three hours, the fabrication of a printable mesoscopic PSC module is finished.

Figure 13.8c,d show the energy band alignment and charge separation mechanism of the module. The photogenerated electrons and holes are extracted from the perovskite layer to TiO_2 and carbon and then to the front and back contact upon illumination. The separated charge carriers can transport through the serially connected sub-cells and be extracted to the external circuit.

13.4.2 Solution Deposition Methods of Printable Mesoscopic PSC Modules

For the fabrication of PSCs at lab scales (area of $\leq$1.0 cm^2), spin coating is the most popular method. However, it is generally just suitable to a scale under 10 cm and costs a large amount (>90%) of perovskite precursor solution waste in the process. Thus, many scalable solution deposition methods are invented, including but not limited to screen printing [14], slot-die coating, doctor blading [26], spray coating [27], electrodeposition [28], vapor-assisted deposition [29], and roll-to-roll (R2R)-compatible method [30].

Most deposition methods aim at controlling morphology and obtaining uniform films, especially perovskite thin films. Because the thickness of the perovskite layer

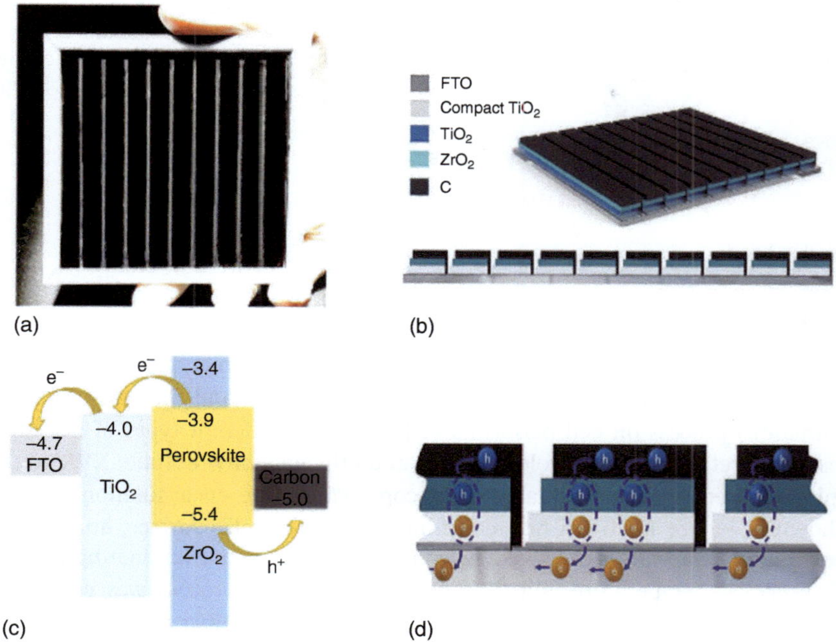

Figure 13.8 (a) Digital image of a large-area printable mesoscopic PSC module. (b) The scheme of the module, which consists of 10 sub-cells in series interconnection. (c) Energy level diagrams of the module. (d) Schematic illustration of the electron and hole separation in a serially connected printable mesoscopic PSC module. Source: Hu et al. [14].

is only a few hundred nanometers, and the film quality directly dominates cell performance. Pinholes in perovskite films lead to direct contact between the electron transport layer (ETL) and hole transport layer (HTL) and cause current leakage and voltage loss for the modules.

Unlike other structures, printable mesoscopic PSCs mainly use drop-casting at lab scales and slot-die coating methods at large scales. The drop-casting method can be easily achieved by pipettes in the lab. As shown in Figure 13.9, the slot-die coating method uses an ink reservoir with a fixed slot gap to deliver solution onto the substrate. The solution is filling the gap between the die and the substrate and forming the coating bead [6]. As the substrate or the die moves, a solution-based wet film is transferred onto the substrate. The slot-die coating method has a larger coating area than the drop-casting method.

In a printable mesoscopic structure, the solution deposition does not need complex installations and harsh conditions. The solution needs to be deposited evenly on the carbon layer. Then it will percolate throughout the triple-mesoscopic layer, as shown in Figure 13.1a. Thus, drop-casting and slot-die coating are the most suitable methods for printable mesoscopic structures. They might be the simplest, most versatile, and reproducible way for high volume factory manufacturing. Certainly, further optimizations on the compositions of the precursors and crystallization process are required.

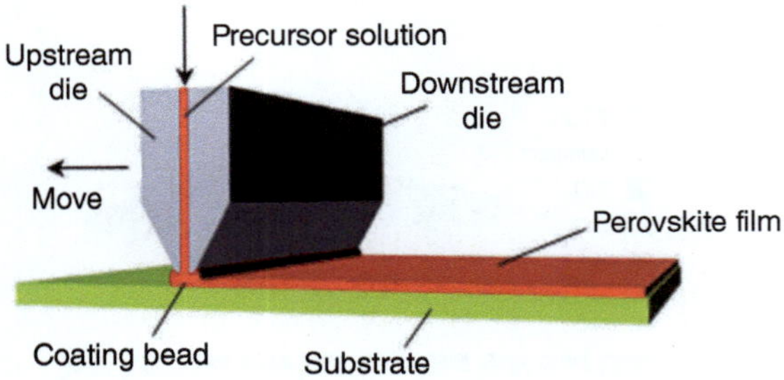

Figure 13.9 Drop-casting method in printable mesoscopic structure. Slot-die coating method. Source: Rong et al. [6].

13.4.3 Encapsulation of Printable Mesoscopic PSCs

The stability issue is still a threshold to be solved in the commercialization of PSCs [31–33]. Han group explored the effects of thermal stress and the encapsulation (lamination) process on cell performance [34]. Three different types of hot melt films (polyurethane, PU; polyolefin, POE; and ethylene vinyl acetate, EVA) were employed to encapsulate printable mesoscopic PSCs. The encapsulation process is shown in Figure 13.10a. A hot melt film covered the cell's surface, and a glass sheet was used as the back cover. After using a commercial laminating system with a critical temperature and time, the encapsulation process was complete. As for the thermal stability of unencapsulated cells, no distinct cell performance degradation was observed among 60–80 °C. However, the efficiency of the cells decreased steadily when the cells heated up to 100–160 °C. Therefore, although POE and EVA are used as encapsulants for silicon and thin film solar cells, they are not suitable for PSCs because of their high laminating temperature (>130 °C) or the corrosion of the perovskite absorber. By contrast, the laminating temperature of PU can be 80 °C. This PU encapsulation can significantly enhance the thermal stability of the cells and modules. Figure 13.10b shows an encapsulated large-area (substrate area 100 cm^2) printable mesoscopic PSC mini-module. Moreover, Figure 13.10c shows the performance of the mini-modules was tracked for 2136 hours under outdoor conditions (location: 39°19′48″N 114°37′26″E). It is found that the mini-modules could maintain 97.52% of the initial efficiency. This encapsulation process is a promising method for printable mesoscopic PSCs and will promote future commercialization of PSCs.

13.4.4 The Recycling of Printable Mesoscopic PSCs

In addition to the stability and mass production, the environmental impact should also be valued. The lead in PSCs is poisonous and can move readily through the biogeochemical cycle due to its water-soluble feature. Thus, the recycling of perovskite modules is required in lead-containing PSCs.

For printable mesoscopic structure, the device scaffold can be reused by washing and reloading perovskite [35]. Figure 13.11a shows that perovskite in device A without encapsulation is partially degraded because of H_2O and O_2 in humid air. The air-exposed devices had 50% performance decrease. By washing away the degraded perovskite in device A with *N*,*N*-dimethylformamide (DMF) and refilling

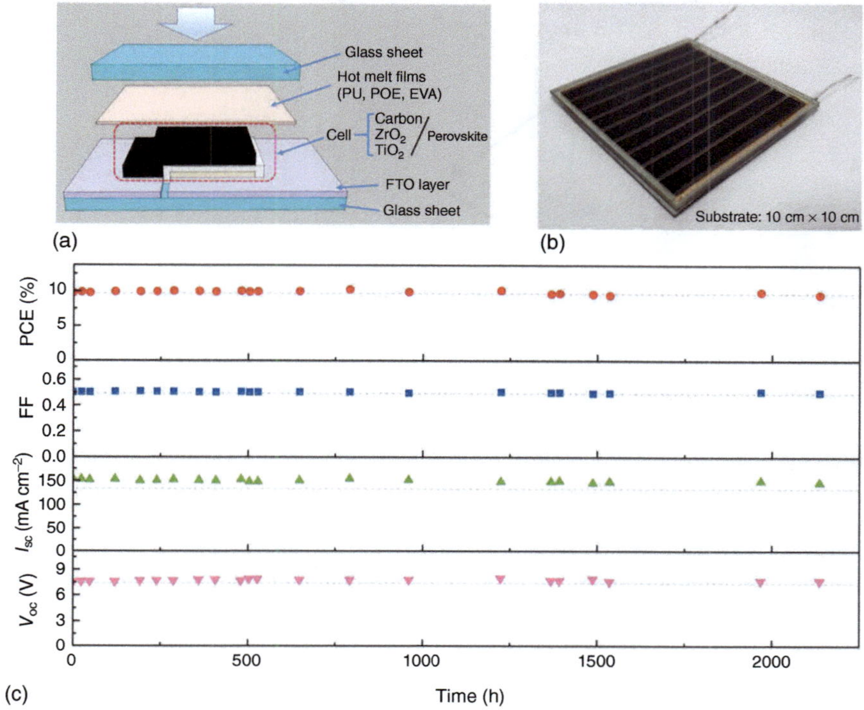

Figure 13.10 (a) The scheme of the encapsulation of printable mesoscopic PSCs based on hot melt films and glass sheets. (b) The digital image of a printable mesoscopic PSC mini-module (substrate area: 100 cm^2; active area: 60.09 cm^2) encapsulated with a PU film as the encapsulant and a glass sheet as the back cover. (c) Evolution of the photovoltaic parameters of printable mesoscopic PSC mini-module under outdoor conditions. Source: Fu et al. [34].

fresh perovskite, the reused PSC device A^+ was obtained. This reuse process could recover almost 90% of their initial PCE. Recently, an improved in situ recycling of printable mesoscopic PSCs through a methylamine gas post-treatment has also been reported [36]. The recycling process is shown in Figure 13.11b. The PSC device was covered with glass for encapsulation, and two holes were left for methylamine gas post-treatment. The holes were sealed with a Kapton tap when the recycling process was not applied. Then the PSC was stored in the aging oven (1 sun illumination, 50 °C) to accelerate aging. During the aging process, $CH_3NH_3PbI_3$ was degraded, and an undesired missing of iodine occurred due to the volatilization of hydrogen iodine. The performance of degraded PSC was below 40% of its initial level. Then, the methylamine gas was introduced to reconstruct the $CH_3NH_3PbI_3$ films, leading to a recovery of the photoelectric properties of the PSCs. After two photodegradation-recovery cycles, the PCE recovered to 91% of the initial value.

It should be noted that it is challenging to fully recover the PCE because some damage would happen during the recycling processes. However, these recycling methods

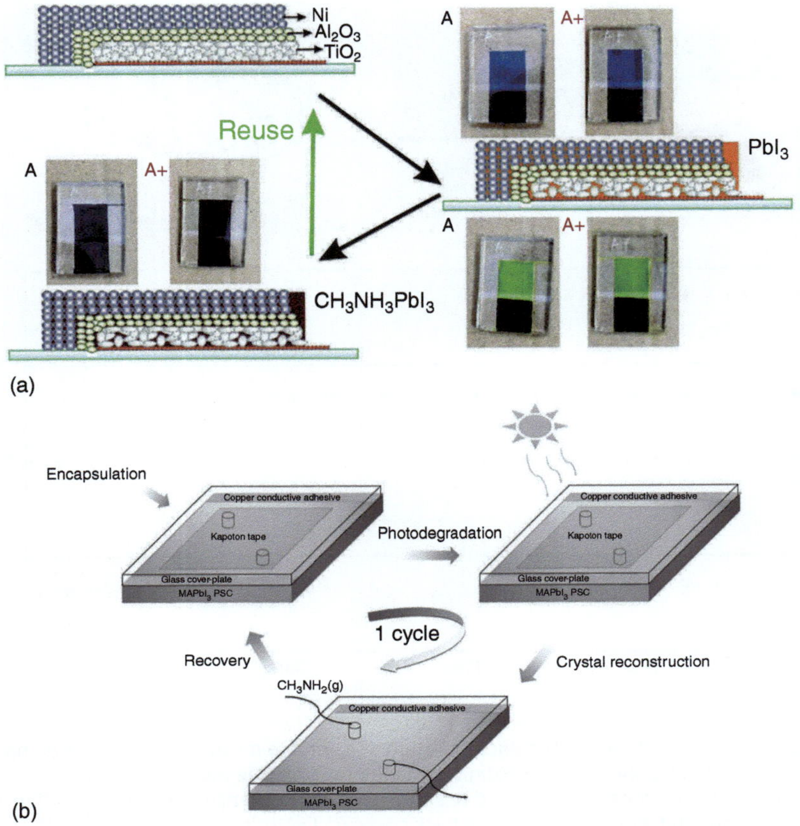

Figure 13.11 (a) The reuse process of the mesoporous Ni-cathode PSC device. Source: Ku et al. [35]. (b) Illustration of one cycle of the photodegradation-recovery of the encapsulated $CH_3NH_3PbI_3$ PSCs. Source: Hong et al [36].

provide a way to extend service life to reduce resource loss and solve environmental issues without substantial performance loss.

13.4.5 Mass-Production of Printable Mesoscopic PSC Modules

Using printable mesoscopic structures, WonderSolar (China) reported the largest PSC module of 3600 cm^2 in September 2017 and established a 110 m^2 perovskite PV system in January 2018, as shown in Figure 13.12 [3]. Considering the time, energy, material, and labor costs, manufacturing modules for industrial production are different from the laboratory.

As mentioned before, series interconnection can reduce resistance losses for modules, but the interconnections occupy module areas inactive for power generation. Scribing process can effectively decrease the interconnection areas. Thus it is normally used to improve geometric fill fact (GFF) to minimize power loss. The P1 scribe separates the bottom FTO substrate. The P2 scribe removes the mesoporous

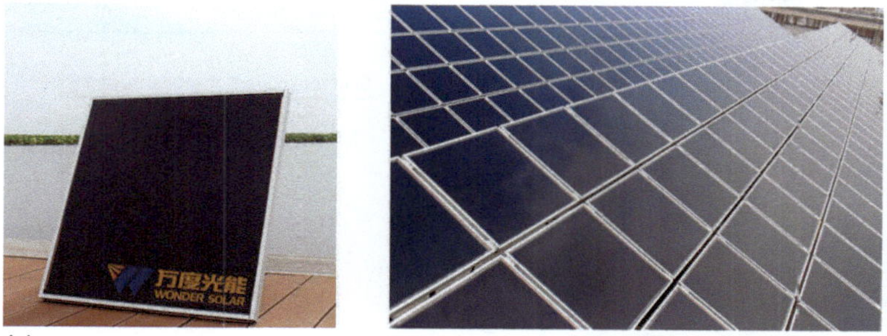

(a) (b)

Figure 13.12 (a) The digital image of a 3600 cm^2 printable mesoscopic PSC module; (b) the digital image of the 110 m^2 perovskite PV system established by WonderSolar (China). Source: Courtesy of Hubei wonder Solar Limited Liability Company.

TiO_2 and ZrO_2 layer and provides a channel to connect the top contact of one cell to the bottom contact of the next cell to form the interconnection. And the P3 scribe isolates the top carbon contact between neighboring cells. By using laser, the three scribes could be narrow and placed as close together as possible. Thus, this interconnection structure and method make modules generally have relatively narrow dead areas, achieving a high GFF [24].

Besides structure, the sintering method changes to adapt to industrial manufacturing. A mesh belt furnace is used to sinter each layer instead of a hot stage. Mesh belt furnace has a high degree of automation, can be used to achieve assembly line production, and can realize gradient sintering for different layers through different temperature areas as the mesh belt runs. It is worth mentioning that every temperature area is independent and stable. Thus, a mesh belt furnace can save time for each annealing step and reduce labor and energy costs. Except for the structure and sintering method, the solution deposition method and encapsulation of printable mesoscopic PSC modules are the same as we mentioned before.

However, there are still many processes that require manual operation, which results in low production efficiency. Future manufacturing needs an auto production line enabling a continuous and controllable fabrication process.

13.4.6 Standardizing the Evaluation of PSC Modules

Establishing an international standard PCE evaluation of solar cells is essential to compare the performance among different laboratories and established technologies. Most laboratories use a steady-state simulator, such as *J–V* measurement, to measure PCEs of PSCs. However, hysteretic behaviors appear in the *J–V* curves, yielding unreliable or even irreproducible PCE results, as shown in Figure 13.13a [37]. To propose a reliable and transferable methodology to accurately reflect the true performance of the device, various measurement protocols have been proposed and updated in recent years. Snaith et al. have introduced the concept of "stabilized power output" by measuring the current continuously at a fixed bias until stabilization occurs (Figure 13.13b) [38]. Until now, many laboratories use this method to

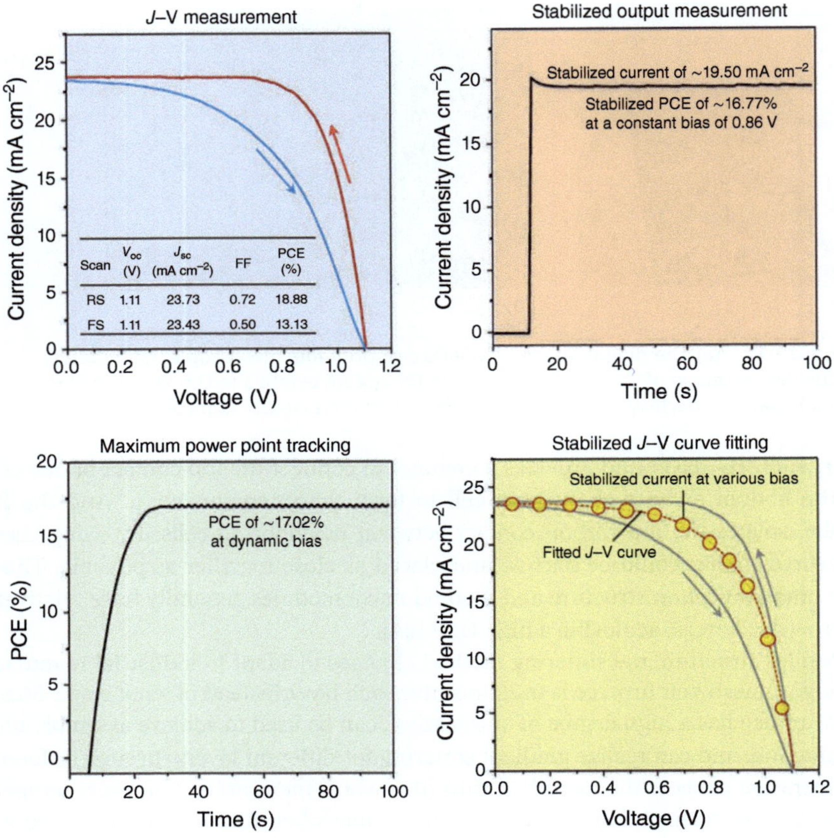

Figure 13.13 (a) *J–V* measurement of PSCs. (b) Stabilized output measurement of PSCs. (c) Maximum power point tracking measurement of PSCs. (d) Stabilized *J–V* curve fitting of PSCs. Source: Hu et al. [24].

confirm the stabilized PCE of PSCs. However, it is doubtful whether the measured value should always be below the true value of PCE. The bias is selected by assuming the maximum power point (MPP) based on "unstabilized" *J–V* curves. Based on the MPP method, the MPP tracking method is proposed, which provided a dynamic bias and consistent approach to the MPP of PSCs (Figure 13.13c) [39]. However, this method still needs a *J–V* curve containing V_{OC}, J_{SC}, and FF information. Recently, a measurement technique called "stabilized *J–V*" or "steady-state *J–V*" has been used in some accredited testing laboratories and an increasing number of research groups [40]. As shown in Figure 13.13d, forward and reverse *J–V* scans are first performed at a rapid scan rate. According to the results, each bias voltage is applied and held until the measured current has stabilized. Finally, a stabilized *J–V* is obtained by fitting the stabilized current values.

The device can be sent to an accredited laboratory that meets the ISO/IEC standard 17025:2017 requirements for certification to receive an internationally recognized device performance. The accredited testing laboratories include National Renewable

Energy Laboratory (NREL), the Commonwealth Scientific and Industrial Research Organization (CSIRO), National Institute of Advanced Industrial Science and Technology (AIST), the Fraunhofer Institute for Solar Energy Systems (ISE), and Sandia National Laboratories (Sandi) [41]. Typically, it is necessary to provide the test environment such as temperature, atmosphere, and humidity. Both forward and reverse *J–V* plots be supplied. Detailed information about the scan conditions like the scan speed, dwell times, or any pre-conditioning protocols should also be included. These laboratories increasingly value the stability of the PSCs and update their test items constantly. For example, at present, they require that the MPP tracking value of PCE be stable for at least 300 seconds. Thus, this evolution of requirements aims to accurately determine the PCE of PSCs and shows concerns to the stability of the PSCs.

13.4.7 Standardizing the Aging Measurements of PSC Modules

Since the lifetime is another critical parameter for all types of solar cells, determining a series of standard testing protocols to compare stability among PSCs is also necessary. For mature PV technologies, international standards such as "IEC 61215 Terrestrial PV modules – design qualification and type approval" have been established and released and become quite complete. The standards call for severe environmental (e.g. temperature, humidity, and irradiation), electrical, and mechanical stress tests to ensure that products meet the standards before entering the market. However, PSCs products are still in the laboratory stage. International standards for PSCs have not been established, so researchers are exploring the stability behavior of PSCs per the methods developed in IEC 61215. As shown in Figure 13.14, great indoor stability results under various conditions have been obtained on laboratory-scale PSCs, such as continuous illumination, heat stress, high humidity, and thermal cycles [13, 42]. Particularly, Nazeeruddin and coworkers has reported that triple mesoscopic printable PSCs enabled stable performance under continuous illumination for ~10 000 hours [15]. As mentioned previously, PU encapsulation can significantly enhance the thermal stability of printable PSCs. The cells did not show any performance degradation at 85 °C after over 325 hours [34].

Simultaneously, the aging test should not only focus on laboratory cells but also transfer to mini-modules and modules, which at least satisfy the damp-heat aging (85 °C and 85%, 1000 hours), thermal cycling (−40 °C to 85 °C, 200 cycles), light-soaking aging (MPP at ~1 sun and 50 °C ± 10 °C), and UV pre-conditioning (15 kWh m^{-2} at 60 °C ± 5 °C) tests. To make the large-scale PSC products truly market-oriented, stricter standards such as higher temperature should be encouraged. This includes but is not limited to, Light-induced degradation (LID)[43], potential-induced degradation (PID) [43], partial shade stress, and mechanical shock tests [44]. Notably, the ion drift between the adjacent series-connected unit cells might exist and should be investigated.

Finally, we would like to point out that it is still too early to talk about the lifetime of PSCs. For PSCs, we need to establish a basic standard for PSCs to survive under general practical conditions. The comparison and analysis of aging behavior will play a guiding role in future commercialization.

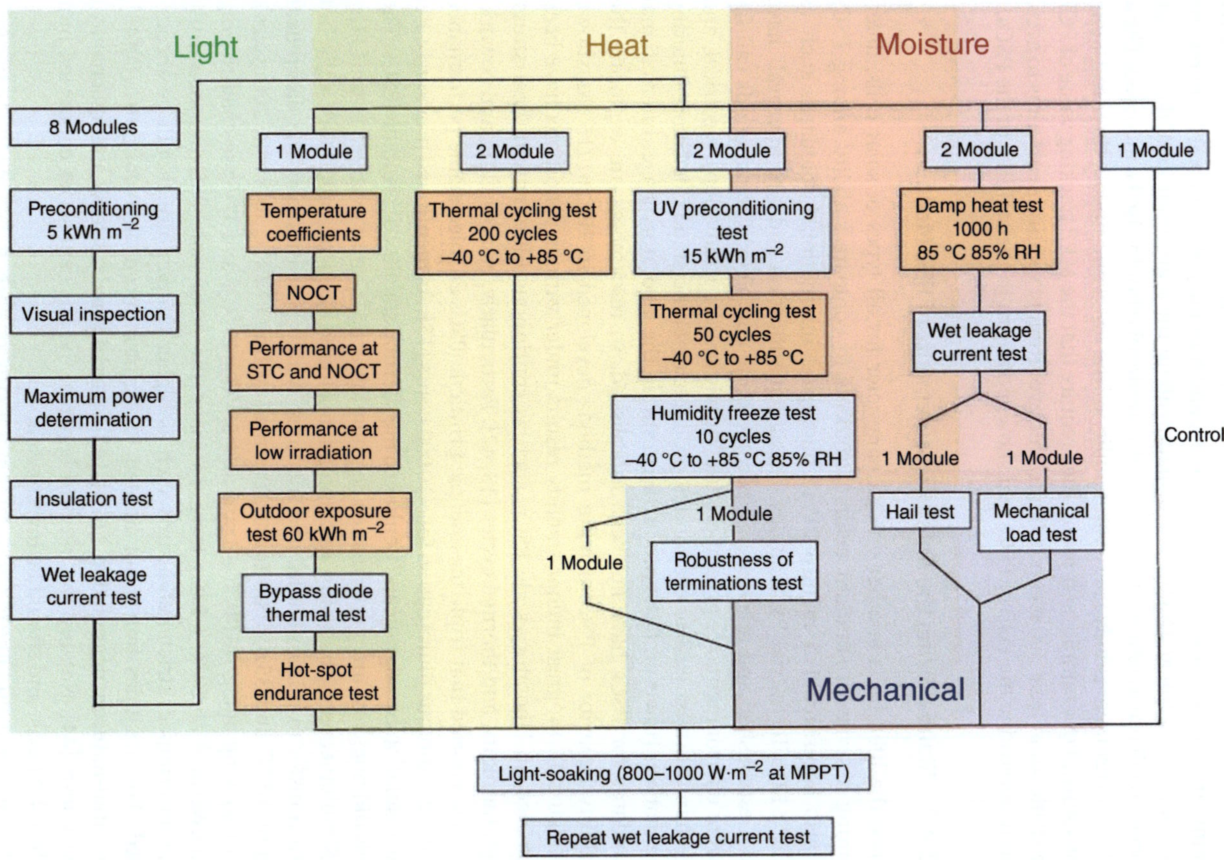

Figure 13.14 The full test flow for design qualification and type approval of PV modules. The laboratory-scale PSCs already passed tests are highlighted in orange. The aging testing should also be focused on the module. Source: Hu et al. [24].

13.5 Conclusions

PSCs have undergone several changes since their initial implementation eight years ago. Among all PSCs, carbon-based PSCs have been paid much attention to from both scientists and industries. The advantages of low cost, easy fabrication, and high stability make carbon-based PSC a fantastic candidate for large-scale applications and commercialization of PSCs. It is important to recognize that the general photovoltaics context has also radically altered and progressed since the first PSCs were developed. The cost of silicon photovoltaics has plummeted. Other types of photovoltaics are also searching for their purpose and opportunities, such as indoor technology for powering the internet of things, building-integrated photovoltaics, etc. The priority and great emphasis put on carbon-based PSCs remains improving the PCE under the condition of low-cost. By exploiting innovative strategies from different aspects, such as materials, interfaces, device structure, and stability measurement consistency, we believe that printable mesoscopic PSCs could still be further promoted. They will pave the way for large-scale commercial deployment of a very promising photovoltaic technology.

References

1 NREL (2021, January 2020). Best Research-Cell Efficiency Chart. Retrieved from https://www.nrel.gov/pv/cell-efficiency.html.

2 Green, M.A., Ho-Baillie, A., and Snaith, H.J. (2014). The emergence of perovskite solar cells. *Nat. Photonics* 8 (7): 506–514.

3 Rong, Y., Hu, Y., Mei, A. et al. (2018). Challenges for commercializing perovskite solar cells. *Science* 361 (6408): eaata8235.

4 Correa-Baena, J.-P., Saliba, M., Buonassisi, T. et al. (2017). Promises and challenges of perovskite solar cells. *Science* 358 (6364): 739.

5 Green, M.A., Dunlop, E.D., Hohl-Ebinger, J. et al. (2020). Solar cell efficiency tables (Version 55). *Prog. Photovoltaics* 28 (1): 3–15.

6 Rong, Y., Ming, Y., Ji, W. et al. (2018). Toward industrial-scale production of perovskite solar cells: screen printing, slot-die coating, and emerging techniques. *J. Phys. Chem. Lett.* 9 (10): 2707–2713.

7 Li, W., Dong, H., Wang, L. et al. (2014). Montmorillonite as bifunctional buffer layer material for hybrid perovskite solar cells with protection from corrosion and retarding recombination. *J. Mater. Chem. A* 2 (33): 13587–13592.

8 Kato, Y., Ono, L.K., Lee, M.V. et al. (2015). Silver iodide formation in methyl ammonium lead iodide perovskite solar cells with silver top electrodes. *Adv. Mater. Interfaces* 2 (13): 1500195.

9 Mei, A., Li, X., Liu, L. et al. (2014). A hole-conductor–free, fully printable mesoscopic perovskite solar cell with high stability. *Science* 345 (6194): 295–298.

10 Ku, Z., Rong, Y., Xu, M. et al. (2013). Full printable processed mesoscopic $CH_3NH_3PbI_3/TiO_2$ heterojunction solar cells with carbon counter electrode. *Sci. Rep.* 3: 3132.

11 Stranks, S.D. and Snaith, H.J. (2015). Metal-halide perovskites for photovoltaic and light-emitting devices. *Nat. Nanotechnol.* 10 (5): 391–402.

12 Li, X., Tschumi, M., Han, H. et al. (2015). Outdoor performance and stability under elevated temperatures and long-term light soaking of triple-layer mesoporous perovskite photovoltaics. *Energy Tech.* 3 (3): 551–555.

13 Baranwal, A.K., Kanaya, S., Peiris, T. et al. (2016). 100° C thermal stability of printable perovskite solar cells using porous carbon counter electrodes. *ChemSusChem* 9 (18): 2604–2608.

14 Hu, Y., Si, S., Mei, A. et al. (2017). Stable large-area ($10 \times 10\,cm^2$) printable mesoscopic perovskite module exceeding 10% efficiency. *Solar RRL* 1 (2): 1600019.

15 Grancini, G., Roldán-Carmona, C., Zimmermann, I. et al. (2017). One-year stable perovskite solar cells by 2D/3D interface engineering. *Nat. Commun.* 8: 15684.

16 Hou, X., Hu, Y., Liu, H. et al. (2016). Effect of guanidinium on mesoscopic perovskite solar cells. *J. Mater. Chem. A*: 5.

17 Rong, Y., Hou, X., Hu, Y. et al. (2017). Synergy of ammonium chloride and moisture on perovskite crystallization for efficient printable mesoscopic solar cells. *Nat. Commun.* 8: 14555.

18 Chen, J., Rong, Y., Mei, A. et al. (2016). Hole-conductor-free fully printable mesoscopic solar cell with mixed-anion perovskite $CH_3NH_3PbI_{(3-x)}(BF_4)_x$. *Adv. Energy Mater.* 6 (5).

19 Sheng, Y., Hu, Y., Mei, A. et al. (2016). Enhanced electronic properties in $CH_3NH_3PbI_3$ via LiCl mixing for hole-conductor-free printable perovskite solar cells. *J. Mater. Chem. A* 4 (42): 16731–16736.

20 Ku, Z., Rong, Y., Xu, M. et al. (2013). Full printable processed mesoscopic $CH_3NH_3PbI_3/TiO_3$ heterojunction solar cells with carbon counter electrode. *Sci. Rep.* 3 (11): 3132.

21 Xu, M., Rong, Y., Ku, Z. et al. (2014). Highly ordered mesoporous carbon for mesoscopic $CH_3NH_3PbI_3/TiO_2$ heterojunction solar cell. *J. Mater. Chem. A* 2 (23): 8607–8611.

22 Zhang, L., Liu, T., Liu, L. et al. (2015). The effect of carbon counter electrodes on fully printable mesoscopic perovskite solar cells. *J. Mater. Chem. A* 3 (17): 9165–9170.

23 Duan, M., Rong, Y., Mei, A. et al. (2017). Efficient hole-conductor-free, fully printable mesoscopic perovskite solar cells with carbon electrode based on ultra-thin graphite. *Carbon* 120: 71–76.

24 Hu, Y., Chu, Y., Wang, Q. et al. (2019). Standardizing perovskite solar modules beyond cells. *Joule* 3: 1–10.

25 Priyadarshi, A., Haur, L.J., Murray, P. et al. (2016). A large area ($70\,cm^2$) monolithic perovskite solar module with a high efficiency and stability. *Energy Environ. Sci.* 9 (12): 3687–3692.

26 Deng, Y., Zheng, X., Bai, Y. et al. (2018). Surfactant-controlled ink drying enables high-speed deposition of perovskite films for efficient photovoltaic modules. *Nat. Energy* 3 (7): 560–566.

27 Heo, J.H., Lee, M.H., Jang, M.H., and Im, S.H. (2016). Highly efficient $CH_3NH_3PbI_{3-x}Cl_x$ mixed halide perovskite solar cells prepared by re-dissolution and crystal grain growth via spray coating. *J. Mater. Chem. A* 4 (45): 17636–17642.

28 Chen, H., Wei, Z., Zheng, X., and Yang, S. (2015). A scalable electrodeposition route to the low-cost, versatile and controllable fabrication of perovskite solar cells. *Nano Energy* 15: 216–226.

29 Chen, X., Cao, H., Yu, H. et al. (2016). Large-area, high-quality organic–inorganic hybrid perovskite thin films via a controlled vapor–solid reaction. *J. Mater. Chem. A* 4 (23): 9124–9132.

30 Chang, N.L., Ho-Baillie, A.W.Y., Vak, D. et al. (2018). Manufacturing cost and market potential analysis of demonstrated roll-to-roll perovskite photovoltaic cell processes. *Sol. Energy Mater. Sol. Cells* 174: 314–324.

31 Rong, Y., Liu, L., Mei, A. et al. (2015). Beyond efficiency: the challenge of stability in mesoscopic perovskite solar cells. *Adv. Energy Mater.* 5 (20): 1501066.

32 Domanski, K., Alharbi, E.A., Hagfeldt, A. et al. (2018). Systematic investigation of the impact of operation conditions on the degradation behaviour of perovskite solar cells. *Nat. Energy* 3 (1): 61–67.

33 Leijtens, T., Eperon, G.E., Noel, N.K. et al. (2015). Stability of metal halide Perovskite solar cells. *Adv. Energy Mater.* 5 (20): 1500963.

34 Fu, Z.Y., Xu, M., Sheng, Y.S. et al. (2019). Encapsulation of printable mesoscopic perovskite solar cells enables high temperature and long-term outdoor stability. *Adv. Funct. Mater.* 29 (16): 1809129.

35 Ku, Z., Xia, X., Shen, H. et al. (2015). A mesoporous nickel counter electrode for printable and reusable perovskite solar cells. *Nanoscale* 7 (32): 13363–13368.

36 Hong, L., Hu, Y., Mei, A. et al. (2017). Improvement and regeneration of perovskite solar cells via methylamine gas post-treatment. *Adv. Funct. Mater.* 27 (43): 1703060.

37 Dunbar, R.B., Duck, B.C., Moriarty, T. et al. (2017). How reliable are efficiency measurements of perovskite solar cells? The first inter-comparison, between two accredited and eight non-accredited laboratories. *J. Mater. Chem. A* 5 (43): 22542–22558.

38 Snaith, H.J., Abate, A., Ball, J.M. et al. (2014). Anomalous hysteresis in perovskite solar cells. *J. Phys. Chem. Lett.* 5 (9): 1511–1515.

39 Rakocevic, L., Ernst, F., Yimga, N.T. et al. (2019). Reliable performance comparison of perovskite solar cells using optimized maximum power point tracking. *Solar RRL* 3 (2): 1800287.

40 Zimmermann, E., Wong, K.K., Müller, M. et al. (2016). Characterization of perovskite solar cells: towards a reliable measurement protocol. *APL Mater.* 4 (9): 091901.

41 Luo, D., Yang, W., Wang, Z. et al. (2018). Enhanced photovoltage for inverted planar heterojunction perovskite solar cells. *Science* 360 (6396): 1442.

42 Kojima, A., Teshima, K., Shirai, Y., and Miyasaka, T. (2009). Organometal halide perovskites as visible-light sensitizers for photovoltaic cells. *J. Am. Chem. Soc.* 131 (17): 6050–6051.

43 Luo, W., Khoo, Y.S., Hacke, P. et al. (2017). Potential-induced degradation in photovoltaic modules: a critical review. *Energy Environ. Sci.* 10 (1): 43–68.

44 Timothy, J. S., Michael, G. D., Chris, D., & Sarah, K. (2015). Partial shade stress test for thin-film photovoltaic modules. Paper presented at the Proc. SPIE, San Diego. Vol. 9563.

45 Nakayashiki, K., Hofstetter, J., Morishige, A.E. et al. (2016). Engineering solutions and root-cause analysis for light-induced degradation in p-type multicrystalline silicon PERC modules. *IEEE J. Photovolt.* 6 (4): 860–868.

14

Upscaling of Perovskite Photovoltaics

Dongju Jang[1,2], *Fu Yang*[1,2,4], *Lirong Dong*[1,2], *Christoph J. Brabec*[1,2,3], *and Hans-Joachim Egelhaaf*[1,2]

[1] *Institute of Materials for Electronics and Energy Technology (i-MEET), Friedrich-Alexander University Erlangen-Nürnberg, Martensstraße 7, 91058 Erlangen, Germany*
[2] *Solar Factory of the Future, Bavarian Center for Applied Energy Research (ZAE Bayern), Fürther Straße 250, 90429 Nürnberg, Germany*
[3] *Helmholtz Institute Erlangen-Nürnberg for Renewable Energy (HIERN), Immerwahrstr. 2, 91058 Erlangen, Germany*
[4] *Laboratory of Advanced Optoelectronic Materials, Suzhou Key Laboratory of Novel Semiconductor Materials and Devices, College of Chemistry, Chemical Engineering and Materials Science, Soochow University, No. 199 Ren'ai Road, Industrial Park, 215123, Suzhou, China*

14.1 Introduction

To reach climate targets and limit global warming, a transition to renewable energy generation is indispensable, with photovoltaics (PVs) being a major pillar of the future energy mix. In the past decade (Figure 14.1a), hybrid organic–inorganic perovskite PVs have seen a tremendous rise in power conversion efficiency (PCE), thus enhancing the chances of providing cheap renewable energy production by PV [1–4].

In 2009, the Miyasaka group first introduced $MAPbI_3$ perovskite as an active layer into dye-sensitized solar cell and achieved 3.8% efficiency, however with low stability due to the corrosive electrolyte [5]. By the Park group, the PCE was increased to 9.7% through replacing the liquid electrolyte with spiro-OMeTAD (2,2′,7,7′-tetrakis[*N*,*N*-di(4-methoxyphenyl)amino]-9,9′-spirobifluorene) as a hole transport layer [6]. The Snaith group enhanced the PCE to 10.9% using $MAPbI_{3-x}Cl_x$ perovskite on compact titanium oxide (TiO_2) and mesoporous aluminum oxide (Al_2O_3) [7]. Seok's group improved the PCE to 12.3% by suggesting $MAPbI_{3-x}Br_x$ perovskites on mesoporous TiO_2 [8]. In 2013, an efficiency of 14.1% was reported by the Grätzel group [9]. A planar perovskite device having 15.4% PCE was achieved via thermal evaporation by the Snaith Group [10]. In 2013, 2014, and 2015, the Seok group pushed the efficiency to 16.2% with $MAPbI_{3-x}Br_x$ [11], to 17.9% with $(FAPbI_3)_{1-x}(MAPbBr_3)_x$ [12] and to 20.1% with $FAPbI_3$ [13]. In 2016, the Grätzel group increased the PCE to 21.02% with a poly(methyl methacrylate) template to control nucleation and crystal growth by using multicomponent perovskite $(FAI)_{0.85}(PbI_2)_{0.85}(MAPbBr_3)_{0.15}$ [14]. In 2017, the Seok group again

Perovskite Solar Cells: Materials, Processes, and Devices, First Edition.
Edited by Shahzada Ahmad, Samrana Kazim, and Michael Grätzel.

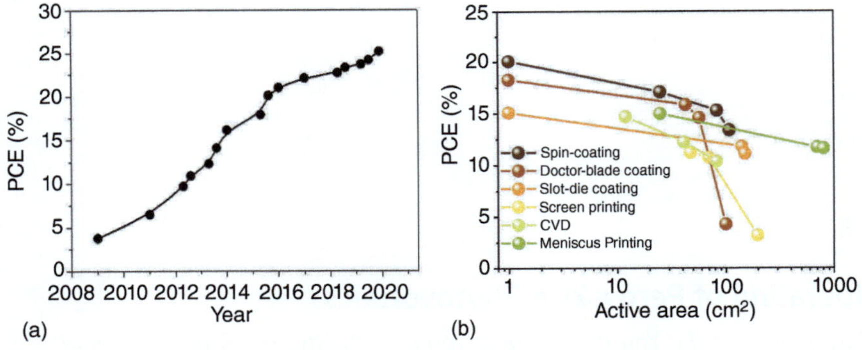

Figure 14.1 (a) Progress in perovskite solar cell efficiency in a small area during the last decade. (b) Efficiency progress of PSCs prepared by different coating techniques related to active area (for active areas above 1 cm^2).

reported a certified record efficiency of 22.1% [15]. In 2019, the Seo group increased the certified PCE value to 22.7% by inserting a wide-bandgap halide between $(FAPbI_3)_{0.95}(MAPbBr_3)_{0.05}$ perovskite layer and P3HT layer [16]. The You group introduced phenethyl ammonium iodide passivation on $(FAMA)PbI_3$ perovskite and increased the certified PCE to 23.3% [17]. With $FAPbI_3$ structure, the Kim group reported a certified PCE of 23.48% by inducing an intermediate phase with MACl [18]. Seok and coworkers achieved a certified PCE of 23.6% by doping with methylene diammonium dichloride to enhance phase stability [19].

This rise of efficiency from less than 5% to more than 25% (Figure 14.1) in just one decade is impressive [20]. However, to deploy up to 10 TW of PVs by 2030, as proposed by Needleman et al. to provide around half of the world's electricity consumption [20, 21], high efficiencies are not sufficient. Two conditions must be met: (i) the generated energy must be cost-competitive with incumbent technologies, i.e. energy generation from fossil fuels; (ii) the PV manufacturing output must be ramped up sustainably [22].

Cost competitiveness is captured by the levelized cost of energy (LCOE), the price per unit of energy [22, 23]. With existing silicon and cadmium telluride (CdTe) technology, energy prices for utility-scale PVs have fallen below grid parity under reasonably good insolation conditions in most countries. Several cost model calculations have shown that perovskite LCOE will keep up with silicon and CdTe if the magic triangle of PV requirements, i.e. high efficiency, low cost, and long lifetime are met simultaneously (Figures 14.2 and 14.3a) [1–4, 24, 26–32].

As for lifetime, approximately 20 years under operational conditions are typically assumed for LCOE calculations [28]. More than 10 000 hours of T80 lifetime (i.e. the time at which efficiency has decayed to 80% of the initial value) have been demonstrated for perovskite solar cells under controlled conditions [4]. However, the error bar is rather large for this variable, considering the lack of long-term experience with outdoor operation of perovskite PV modules. On the efficiency side, PCE values of at least 16% should be achieved for large-area modules to reach competitive LCOE values [22, 23]. Concerning material costs, the general message of the cost studies mentioned above is very clear, albeit the numerical results may

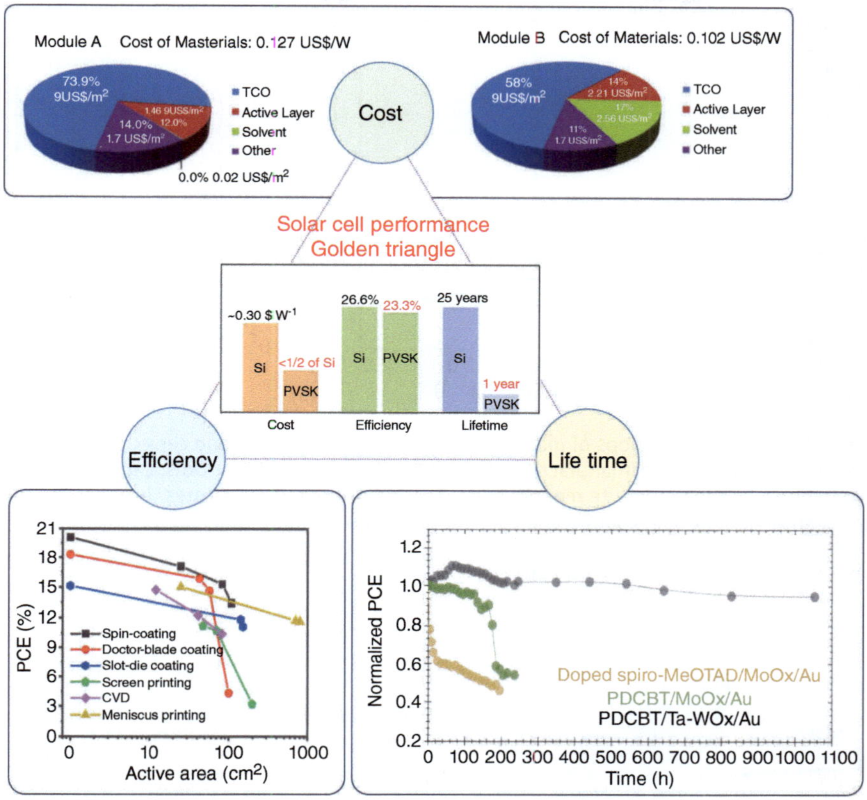

Figure 14.2 The golden triangle of PV. Center: comparison of perovskite and c-Si technologies for the three vertices of the golden triangle today, the graphs on the vertices represent the challenges on the way to commercialization. Cost: Cost of materials distribution of two different perovskite module structures. Source: Modified from Meng et al. [24]. The values of materials cost are assumed by the actual amount of material used in both structures and wholesale price, considering 80% materials usage (Module A: mesoporous structure, B: planar inverted structure). Source: Cai et al. [3]. Licensed under CC BY-4.0. Efficiency: Efficiency progress of PSCs prepared by different coating techniques related to active area (see also Figure 14.1b). Lifetime: Unencapsulated device photostability tests under continuous 1-sun illumination in a home-built chamber filled with N_2. PCE variation of the HTM stacks based on a C_{60}-SAM with a strong binding between the phosphonic acid and the oxide surface as ETM: spiro-MeOTAD/MoO_x/Au (yellow), PDCBT/MoO_x/Au (green), and PDCBT/Ta-WO_x/Au (gray). Source: Hou et al. [25].

vary in detail. Only low to moderate cost materials are acceptable for perovskite solar modules to be cost-competitive with today's silicon technology (Figure 14.4). This applies especially to the hole transport layers (HTL), thus excluding materials such as spiro-OMeTAD, which has been used frequently for high-efficiency lab cells. Even less expensive materials such as P3HT add substantially to the bill of materials (BOM) [3, 30]. Especially for lifetime requirements, there is a strong conflict: the choice of electrode materials for perovskite devices is strongly reduced. Silver and many other metals are prone to corrosion by iodine, whereas relatively stable alternatives such as Au are not viable under cost aspects. This leaves mainly

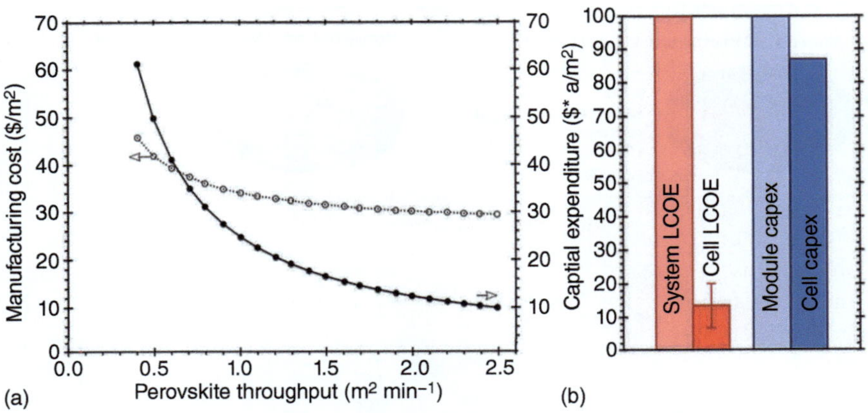

Figure 14.3 (a) Implications of manufacturing throughput on cost and capex of perovskite modules (a) Open circles: manufacturing cost for perovskite modules. Dots: capital expenditure on the perovskite module level. The red dashed line indicates throughput levels attainable by spin coating and conventional (non-rapid) thermal annealing. (b) Breakdown of system-level LCOE and module-level capex into cell-related and other costs (for silicon). Source: Bruening et al. [1].

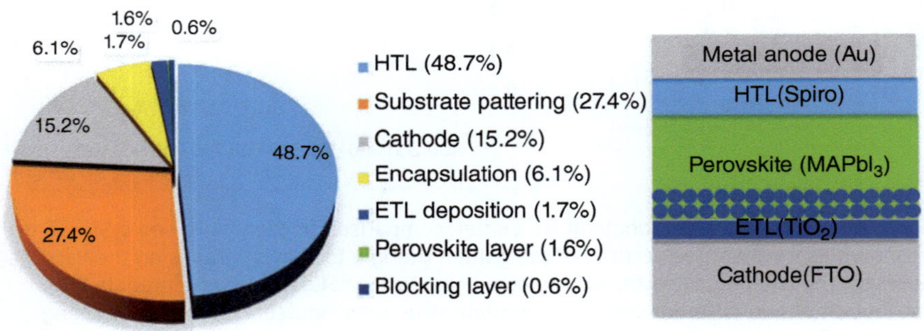

Figure 14.4 Manufacturing costs of perovskite solar modules based on the cell stack shown on the right. Source: Reproduced with permission [4]. Copyright 2019, American Chemical Society.

carbon-based materials as alternatives. Packaging materials also add substantially to the BOM. Glass packaging increases the BOM by about $\$2 \times 10$ per m^2. For flexible packaging, the situation is even more serious, as flexible barrier materials which meet the water vapor transmission rate (WVTR) requirements of maximum $10^{-5}\,g\,m^{-2}\,d^{-1}$ for long-lived modules may add up to $\$2 \times 40$per m^2.

For the second condition, silicon and CdTe are unfavorable technologies because they require excessive capital expenditures (capex) to grow module production at those rates required to fulfill climate targets. While the cell LCOE is low, the system LCOE, module capex, and cell capex are relatively high, determining the selling price of the silicon modules. The main culprit for the high capex of silicon modules is the complex technology required to make the silicon absorber layer (Figure 14.3b) [1]. This is where solution-processed materials, such as perovskites, could drastically lower the capex for cell production [28]. A model proposed by Bruening et al. shows that reducing capex by order of magnitude compared with silicon enables sustainable growth rates of c. 20% annually, which align with climate targets [1]. Moreover, perovskite PVs enable a much shorter energy payback time than Si and CdTe [24].

To exploit this potential, solution deposition and processing methods compatible with high throughput manufacturing are required (Figure 14.3a). While high throughput production may lower specific manufacturing costs by around one-third, capex per unit can be reduced by up to a factor of 10 (Figure 14.3a). However, the production of perovskite solar cells by the most popular technique of spin coating and subsequent thermal annealing is extremely limited in terms of throughput. So far, the scientific community has focused on material development to increase PCE and lifetime. However, almost all reported devices are made in a non-scalable manner, employing techniques such as solvent exchange, spin coating, or prohibitively long annealing steps. The development of scalable fabrication strategies is one of the major hurdles for commercialization. While efficiencies exceeding 20% have become state of the art with small-area lab cells, substantial PCE losses occur upon transfer to high throughput compatible manufacturing methods, large-area modules, and affordable materials (Figure 14.1b).

In this book chapter, after introducing the coating and printing methods available for upscaling, we analyze the current progress of solution-based scalable techniques that allow the fabrication of solar modules. Based on our analysis (Table 14.1), we list the major challenges to overcome to minimize the efficiency gap (i.e. the difference in the performance between small cells and larger module scales) [30]. The first challenge is related to the reproducible preparation of high-quality (i.e. uniformity in coverage and chemical composition, high crystallinity, and low surface roughness over the large area) perovskite films and their compatibility with adjacent functional layers. The second challenge comprises the selection and deposition of charge extraction/transport layers forming the solar module and the perovskite layer. The third challenge is related to optimizing a solar module design capable of attaining high performance.

14.2 Techniques for Upscaling

Due to the chemical nature of organometallic halide perovskites, perovskite films can be deposited from the solution. Solution processing is compatible with roll-to-roll (R2R) manufacturing, which can provide the high throughput at costs required for enabling competitive values of LCOE. At a laboratory scale, the most common solution-based process is spin-coating (Figure 14.5), which has provided the most efficient devices, with a certified PCE of over 25% being the present world record (Figure 14.6). The main merit of the spin-coating technique is that it can easily deposit high quality, well-defined thickness, and different composition perovskite thin films. However, this method wastes more than 90% of the solution and generally shows a process limit at 8 in. of substrate size. Also, the thickness and morphology of perovskite films are determined by the continuous centrifugal force

Table 14.1 Representative reports of large-area PSCs (≥1 cm^2) or Perovskite modules (PSMs) by different coating methods sorted by year.

Device category	Device architecture	Perovskite fabrication method	Strategy to control perovskite film quality	Area (cm^2) (module/active area)	V_{OC} (*V*)	J_{SC} (mA cm^{-2})	I_{sc} (mA cm^{-2})	*FF* (%)	*PCE* (%)	Refs.
Cell	FTO/NiMgLiO/$MAPbI_3$/PCBM/Ti(Nb)Ox/Ag	Spin-coating (one-step)	Antisolvent	—/1.02	1.09	17.35	—	66.8	15	[33]
Cell	FTO/cTiO_2/$MAPbI_3$/spiro-MeTAD/Au	Spin-coating (one-step)	Antisolvent	—/1	1.12	20.5	—	68.0	14.6	[34]
Cell	FTO/cTiO_2/$FAPbI_3$/spiro-MeTAD/Au	Chemical vapor deposition	—	—/2	1.02	19.5	—	53	10.4	[34]
Cell	FTO/NiO/$FA_{0.85}MA_{0.15}Pb(I_{0.85}Br_{0.15})$3-PCBM/PCBM/Ag	Spin-coating (one-step)	Antisolvent + additve	—/1.022	1.08	21.95	—	78.4	18.21	[35]
Cell	ITO/PEI:PCBM/$MAPbI_3$/PTAA/MoO_3/Ag	Spin-coating (one-step)	Antisolvent	—/1	1.10	20.73	—	79.0	18.1	[36]
Cell	ITO/TiO_2Cl/$Cs_{0.05}FA_{0.81}MA_{0.14}PbI_{2.55}Br_{0.45}$/Sprio/Au	Spin-coating (one-step)	Antisolvent	—/1.1	1.20	21.5	—	75.7	19.5	[37]
Cell	FTO/NiO/$MAPbI_3$/PCBM/Ag	Spin-coating (one-step)	Additive	—/1.025	1.12	23.17	—	75.7	19.19	[38]
Cell	FTO/b-TiO_2/mp-TiO_2/$(FAPbI_3)_{1-x}(MAPbBr_3)_x$+$I_2$/PTAA/Au	Spin-coating (two-step)	Additive	—/1	1.1	24.7	—	72.3	19.7	[15]
Cell	ITO/SnO_2/$(FAPbI_3)_{1-X}(MAPbBr_3)_X$/sprio-OMeTAD/Au	Spin-coating (two-step)	—	—/1	1.12	23.45	—	76.85	20.12	[39]
Cell	ITO/PEDOT:PSS/$MAPbI_3$/PCBM/Ca/Al	Spin-coating (two-step)	Additive + vapor anneal	—/1.3	0.99	22.51	—	75.0	16.7	[40]
Cell	FTO/TiO_2/$[CsPbI_3]_{0.05}[(FAPbI_3)_{0.85}(MAPbBr_3)_{0.15}]_{0.95}$/spiro/Ag	Spin-coating (one-step)	Antisolvent	—/1.08	1.17	21.28	—	70	17.37	[41]
Cell	FTO/c-TiO_2/$MA_{1-2x}EDA_xPbI_3$/spiro-OMeTAD/Au	Spin-coating (one-step)	Gas flow	1.04	20.44	1.01	—	73.4	15.15	[42]

Cell	FTO/c-TiO_2/$MAPbI_3$/spiro-OMeTAD/Au	Doctor-blading	Additive + antisolvent	1.2	1.11	23.38	—	72.9	17.3	[31]
Cell	ITO/PEDOT:PSS/$FAPbI_3$/$PC_{61}BM$/Ag	Spin-coating (two-step)	—	1.2	1.05	22.35	—	64.9	15.23	[43]
Cell	FTO/c-TiO_2/$MAPbI_3$/spiro/Au	Blade coating (one-step)	High temperature (150 °C)	1.0	1.08	23.17	—	68.49	17.06	[44]
Cell	ITO/PTAA/$FA_{0.85}MA_{0.15}Pb(Br_{0.15}I_{0.85})_3$/$C_{60}$/BCP/Au	Blade coating (one-step)	High temperature (145 °C)	1.0	1.15	22.3	—	71.3	18.3	[45]
Cell		Blade coating (one-step)	Additive	—/1	1.01	21.7	—	67.15	14.72	[46]
Cell	FTO/NiO/$MAPbI_3$/PCBM/BCP/Ag	Soft-cover deposition (one-step)	210	—/1.0	1.62	21.8	—	78.7	17.6	[47]
Cell	FTO/cTiO_2/mTiO_2/$MAPbI_3$/spiro/Au	Soft-cover deposition (one-step)	50	—/1.0	1.12	22.6	—	76.2	19.3	[48]
Cell	FTO/NiO/$MAPbI_3$/PCBM/BCP/Ag	Soft-cover deposition (one-step)	80	—/5.0	1.03	21.5	—	70.1	15.5	[49]
Cell	FTO/TiO_2/$MAPbI_3$/spiro/Ag	Spin-coating (one-step)	Antisolvent	—/1.2	1.09	21.3	—	66.1	15.3	[50]
Cell	FTO/cTiO_2/$MAPbI_3$/spiro/Au	Spray coating (two-step)	60 °C	—/1	1.032	18.9	—	68.0	13.09	[51]
Cell	ITO/PDDOT:PSS/$FAPbI_3$/CTAB-doped PCBM/Ag	Spin-coating (two-step)	—	—/1.2	1.03	20.70	—	72.34	15.42	[52]
Cell	FTO/cTiO_2/mTiO_2/$FA_{0.81}MA_{0.15}PbI_{2.51}Br_{0.45}$/spiro/Au	Spin-coating (one-step)	Vacuum	—/1	1.143	22.06	—	76.0	19.6	[53]
Cell	FTO/TiO_2/$MAPbI_3$/spiro/Au	Blade coating (one-step)	Gas flow	—/1	1.09	17.62	—	61.2	11.7	[54]
Cell	ITO/SnO_2/$MAPbI_3$/spiro/Au	Blade coating (one-step)	Gas flow	—/1	1.08	23.32	—	75.92	19.12	[55]

(Continued)

Table 14.1 (Continued)

Device category	Device architecture	Perovskite fabrication method	Strategy to control perovskite film quality	Area (cm^2) (module/ active area)	V_{OC} (V)	J_{SC} ($mA\,cm^{-2}$)	I_{sc} ($mA\,cm^{-2}$)	*FF* (%)	*PCE* (%)	Refs.
Module	FTO/cTiO_2/mTiO_2/ $MAPbI_{3-x}Cl_x$/Spiro/Au	Spin-coating (one-step)	—	—/16.8	4.31	—	32.9	60.3	5.1	[56]
Module	ITO/PEDOT:PSS/$MAPbI_3$/ PCBM/LiF/Al	Spin-coating (one-step)	Antisolvent	100/60	0.87	20.7	—	78.3	8.7	[57]
Module	FTO/cTiO_2/$MAPbI_{3-x}Cl_x$/P_3HT/ Au	Spin-coating (two-step)	Gas flow	—/10.08	3.36	13.4	33.8	77.8	13.0	[58]
Module	ITO/TiO_2/mTiO_2/$MAPbI_{3-x}$Clx /spiro/Au	Spin-coating (one-step)	—	—/7.92-flexible	3.39	5.2	—	71	3.1	[59]
Module	ITO/PEDOT:PSS/$MAPbI_3$/ PCBM/Au	Spin-coating (one-step)	—	100/40	10.1	—	80.1	63.7	12.9	[60]
Module	FTO/cTiO_2/TiO_2 nanorods/$MAPbI_{3-x}$Clx/spiro/Au	Spin-coating (one-step)	—	—/10.8	3.37	22.2	57.1	56.0	10.5	[61]
Module	ITO/PEDOT:PSS/$CH_3NH_3PbI_{3-x}Cl_x$+DIO/PCBM/TiPD/Al	Spin-coating (one-step)	—	25/12	2.71	6.48	—	63.8	11.2	[62]
Module	FTO/cTiO_2/$MAPbI_3$/spiro-MeTAD/Au	Spin-coating (one-step)	Antisolvent	25/15.4	0.98	19	—	46	8.5	[63]
Module	FTO/cTiO_2/$FAPbI_3$/spiro-MeTAD/Au	Chemical vapor deposition	—	25/15.4	0.77	15.2	—	49	5.8	[34]
Module	ITO/PEDOT:PSS/$MAPbI_{3-x-y}$ BrxCly/PCBM/Ca/Al	Spin-coating (one-step)	Hot solution (60 °C)	—/25.2	9.05	—	—	74.4	14.3	[64]
Module	FTO/cTiO_2/mTiO_2/$(FAPbI_3)_{0.95}$ $(MAPbBr_3)_{0.05}$/P_3HT/Au	Spin-coating (one-step)	Antisolvent	25/24.97	8.78	2.72	—	71.7	17.1	[16]
Module	FTO/cTiO_2+G/mTiO_2+G/ $(CsPbI_3)_x(FAPbI_3)_{0.83}$ $(MAPbBr_3)_{0.17}$/sprio /fMoS_2/Au	Spin-coating (one-step)	Antisolvent	156.25/108 156.25/82	10.46 10.46	—	173.78 180.53	65.08 60.09	13.56 15.27	[65] [65]

Module	FTO/c-TiO_2/$MAPbI_3$/spiro-OMeTAD/Au	Doctor-blading (one-step)	Additive + antisolvent	12.6/11.09	—	—	—	—	14.06	[31]
Module	FTO/cTiO_2/mTiO_2/$MAPbI_3$/spiro/Au	Blade coating (two-step)	Gas flow	—/10.08 —/100	4.11 9.6	17.17 7.48	43.27 83.09	58.14 53.79	10.26 4.30	[66]
Module	FTO/SnO_2/$K_{0.03}Cs_{0.05}(FA_{0.85}MA_{0.15})_{0.92}Pb(I_{0.85}Br_{0.15})_3$/spiro-OMeTAD/Au	Spin coating (one-step)	Antisolvent	25/10	6.496	3.07	—	62	12.40	[67]
Module	ITO/PTAA/$MAPbI_3$/C_{60}/BCP/Au	Doctor-blading (one-step)	High temperature (145 °C) + additive	110.5/57.8	17.24	—	72.5	−68.9	14.9	[68]
Module	ITO/PTAA/$MAPbI_3$/C_{60}/BCP/Au	Doctor-blading (One-step)	Additive + Vacuum	16/10.08	3.87	4.93	—	59	11.25	[69]
Module	FTO/SnO_2/$GA_{0.12}MA_{0.88}PbI_3$/spiro-OMeTAD/Au	Doctor-blading (One-step)	Air knife	25/16	~8.8	~2.2	—	~71.5	13.85	[70]
Module	ITO/PTAA/$MAPbI_3$/C_{60}/BCP/Cu	Doctor-blading (One-step	Air knife+Additive	—/42.9	13.14	19.71	—	73.5	15.86	[71]
Module	ITO/ZnO/$MAPbI_3$/P3HT/Ag	Slot die coating (Two-step)	N_2 flow	—/47.3	4.35	10.2	—	51.1	4.56	[72]
Module	ITO/c-TiO_2/$CH_3NH_3PbI_{3-x}Clx$/Spiro-OMeTAD/Au	Slot die coating (One-step)	—	149.5/142 168.75/151.9	21.2 20.8	17.3 19	116 114	67.9 70.6	11.8 11.1	[73]
Module	ITO/PTAA/$MAPbI_3$/C_{60}/BCP/Au	Doctor-blading (One-step)	Air knife	—/63.7	18.9	—	74.5	76.2	16.9	[74]
Module	MAPbI3	—	—	—/703 —/802	1.073/cell 1.073/cell	14.36 15.83	—	75.8 68.0	11.7 11.6	[75]
Module	ITO/NiOx/$CH_3NH_3PbI_{2.7}Br_{0.3}$/PCBM/BCP/Ag	Meniscus coating (one-step)	Gas flow + antisolvent	26.7/25	4.5	4.4	—	75	15	[76]
Module	FTO/cTiO_2/mTiO_2/mZrO_2/C/5-AVA)$_x$(MA)$_{1-x}$PbI3	Screen printing (one-step)	—	50/31 100/70	3.72 9.63	19.6 17.72	—	57.5 62.9	10.46 10.74	[78] [78]

(Continued)

Table 14.1 (Continued)

Device category	Device architecture	Perovskite fabrication method	Strategy to control perovskite film quality	Area (cm^2) (module/active area)	V_{OC} (*V*)	J_{SC} ($mA\,cm^{-2}$)	I_{sc} ($mA\,cm^{-2}$)	*FF* (%)	*PCE* (%)	Refs.
Module	FTO/cTiO_2/mTiO_2/mZrO_2/C/5-AVA$)_x$(MA$)_{1-x}$$PbI_3$	Screen printing (one-step)	—	–/4.9 100/49 /198	0.913 9.3 18.2	23.02 23.0 0.5	—	67 53 38.9	14.02 10.4 3.2	[79]
Module	FTO/cTiO_2/mTiO_2/mZrO_2/C/(HOOC($CH_2)_4NH_3)_2PbI_4$-$MAPbI_3$	Screen printing (one-step)	—	100/47.6	7.05	2.247	—	70.4	11.2	[80]
Module	FTO/cTiO_2/mTiO_2/$MAPbI_3$/spiro/Au	Soft-cover deposition (one-step)	50 °C	—/36.13	8.36	—	73	71.5	12.1	[48]
Module	ITO/TiO_2/MAPb($IXBr_{1-X})_3$/spiro/Au	Spray coating (one-step)	50 °C	—/3.8	3.31	—	19.4	70.0	11.7	[81]
Module	FTO/cTiO_2/G+mTiO_2/GO-Li/$MAPbI_3$/spiro-OMeTAD/Au	Spin-coating (two-step)	—	100/50.6	8.57	—	114.8	64.6	12.6	[82]
Module	FTO/SnO_2/$K_xCs_{0.05}(FA_{0.85}MA_{0.15})_{0.95}Pb(I_{0.85}Br_{0.15})_3$/spiro/Au	Spin-coating (one-step)	Antisolvent	36/20	6.804	3.52	—	65	15.58	[83]
Module	FTO/cTiO_2/mTiO_2/$(FAPbI_3)_{0.85}(MAPbBr_3)_{0.15}$/spiro/Au	Spin-coating (one-step)	Antisolvent	—/16	1.13	17.3	—	61.9	12.1	[84]
Module	FITO/ZnO/$MAPbI_3$/C	Slot die coating (one-step)	Vacuum	25/17.6	6.14	3.25	—	53	10.6	[85]
Module	FTO/SnO_2/C_{60}/$Cs_{0.1}FA_{0.9}PbI_{2.9}Br_{0.1}$/spiro/Au	Chemical vapor deposition	—	100/91.8	13.55	1.42	—	59.6	10.37	[86]
Module	FTO/SnO_2/$Cs_xFA_{1-x}PbI_{3-y}Bry$/spiro/Au	Chemical vapor deposition	—	64/41.25	9.18	2.25	—	52.8	12.24	[87]
Module	ITO/TiO_2/$Cs_{0.07}FA_{0.93}PbI_3$/spiro/Au	Chemical vapor deposition	—	25/12	5.84	3.67	—	68.1	14.6	[88]
Module	FTO/TiO_2/$FAPb(I_{0.85}Br_{0.15})_3$/spiro/Au	Chemical vapor deposition	—	25/12	6.29	3.55	—	66	14.7	[89]

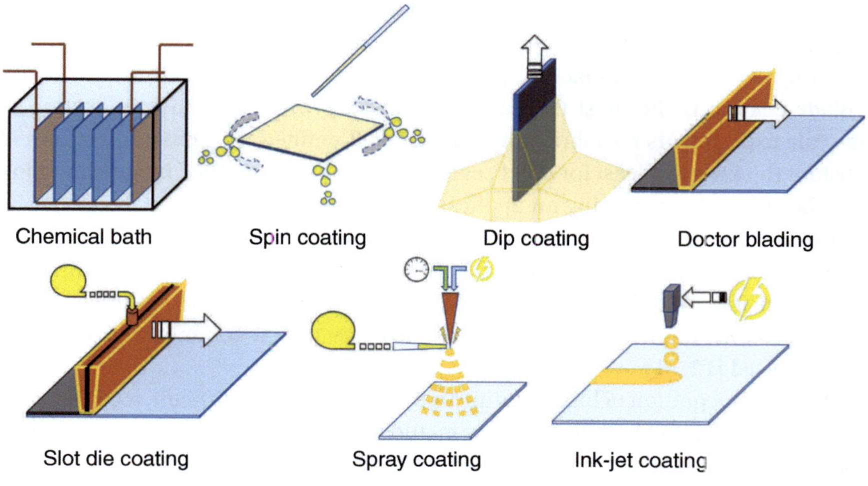

Figure 14.5 Depiction of various solution-based deposition methods.

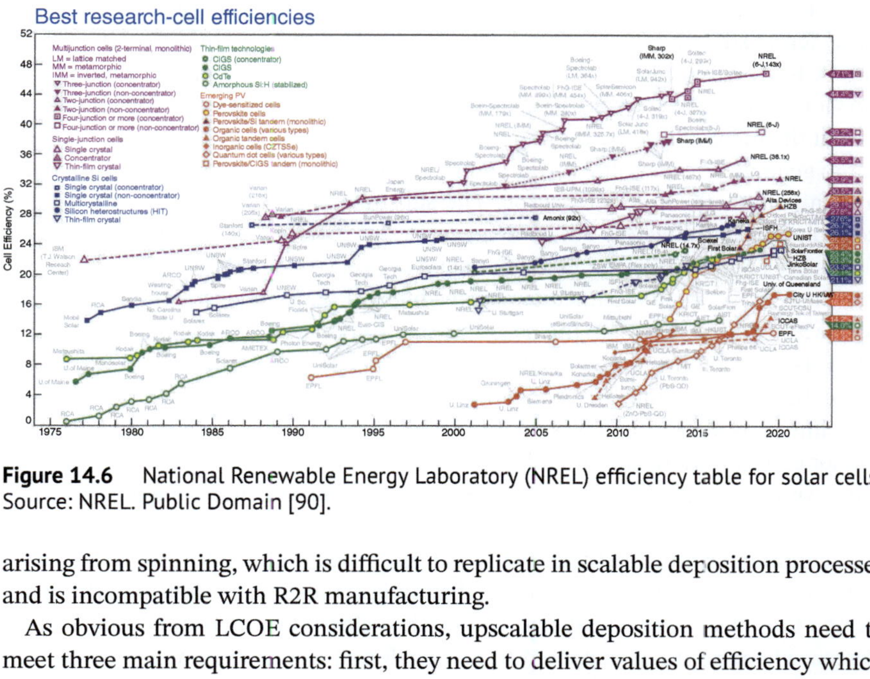

Figure 14.6 National Renewable Energy Laboratory (NREL) efficiency table for solar cells. Source: NREL. Public Domain [90].

arising from spinning, which is difficult to replicate in scalable deposition processes and is incompatible with R2R manufacturing.

As obvious from LCOE considerations, upscalable deposition methods need to meet three main requirements: first, they need to deliver values of efficiency which are at least comparable to those achieved by spin coating, which sets the standards for lab cell efficiencies. Second, they need to maintain a high level of efficiency from small lab cells to large-area modules. Finally, they need to be compatible with high throughput production processes, either sheet-to-sheet (S2S) or, more preferably, R2R.

Therefore, alternative techniques, e.g. doctor-blade coating, slot-die coating, spray coating, and inkjet printing, are more appropriate candidates than spin coating for upscaling perovskite manufacturing (Figure 14.5).

Blade coating is the most frequently employed process next to spin coating in lab-scale experiments for fabricating films of well-defined thickness. A wet film of constant thickness, at least for short coating distances, is deposited by spreading ink over the substrate by moving an applicator (the so-called "knife" or "blade"). The thickness of the resulting film is determined by various factors, such as the distance between the blade and the substrate ("slit"), the applicator speed, concentration, surface tension, and viscosity, as well as the substrate properties. With very little waste of ink during the process, efficiencies above 19% at common cell size have been reported [12, 31].

While the integration of blade coating in R2R equipment is difficult, it can serve as a lab-scale analog to another meniscus method, the slot-die coating process, which works similarly to blade coating. However, a slit in the slot-die head allows continuous ink supply for continuous film deposition. In contrast to blade coating, the thickness of the film can be precisely defined by adjusting the ink-feeding rate, and the speed of the slot-die head with respect to the substrate. Galagan and coworkers have demonstrated large-area modules of $12.5 \times 13.5\,cm^2$ by slot-die coating, showing PCE over 10% with a power output of 1.7 W [73]. The common feature of the previously described coating processes, namely blade coating, and slot-die coating, is that the meniscus of the ink is moved along the substrate by the moving applicator to generate the wet film. A less common process, which may also be considered a meniscus method, is dip coating. The substrate is first dipped into the precursor solution and dragged out at a slower speed, where a film-induced meniscus forms between the solution in the bath and the substrate [32]. Recently, the meniscus method has been modified to achieve highly homogeneous perovskite films of above 51 cm^2 in size [47]. In this so-called soft film cover method, a flexible film covers a specific volume of precursor on the substrate, and the meniscus caused while removing the cover film is used to form a thin layer. In this process, important factors deciding perovskite film thickness are the speed of removal of the cover film, the precursor concentration, and the volume of precursor between the substrate and cover film. Another method for depositing perovskite layers, which may be categorized as a meniscus method, is brush painting [32, 54, 91, 92].

A common printing method, which was also investigated for organic electronics, is screen printing. In screen printing, the ink is applied on the substrate through a screen that has a specific pattern. The mesh is closed by a polymer at areas of the screen where no ink is supposed to be transferred to the substrates. Through the open portions, ink is transferred to the substrate in the desired pattern by squeezing material by passing an applicator ("squeegee"). This can form a very accurate pattern with resolutions in the 20–100 μm range. The thin film made through screen printing is mainly formed around 1–10 μm thickness, depending on the thickness of the screen. It is commonly applied to form electrodes or mesoscopic scaffolds supporting the active layer [93–95]. Screen printing is limited to the rather high viscous ink systems and is therefore dominantly applied for printing electrodes. In

2017, a 2D/3D $(HOOC(CH_2)_4NH_3)_2PbI_4/CH_3NH_3PbI_3$ perovskite junction layer was obtained from screen printing method for the $10 \times 10\,cm^2$ PSC module with PCE of 11.2% by Graetzel and coworkers [80].

In contrast to the methods described above, inkjet printing is a non-contact method for manufacturing (opto-)electronic devices. It uses an array of fine nozzles to deposit small droplets of the precursor solution to form a film with a controlled pattern. Several features, including droplet size, trajectory, and printing speed, influence the final films [96]. The limited printing speed can be compensated by parallelizing of print heads, as in other inkjet applications [96–99]. Recently, Abzieher et al. inkjet printed a $Cs_{0.11}(FA_{0.83}MA_{0.17})_{0.89}$ $Pb(Br_{0.15}I_{0.85})_3$ perovskite layer on top of an undoped electron-beam-evaporated NiO_x hole transport layer. It showed a PCE of 20.7%, proving that inkjet printing perovskite layers are competitive with spin-coating (Figure 14.7), which in this experiment yielded an efficiency of 17.7% [100]. However, there is still a lack of research on suitable perovskite coating conditions or ink compositions. In any printed PV device, the maximum

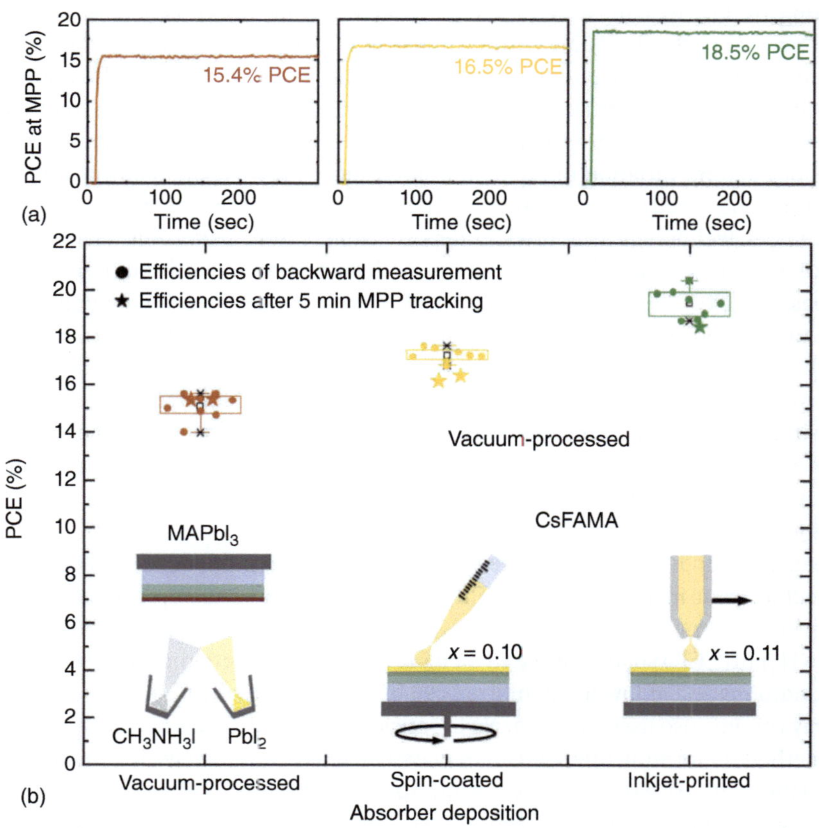

Figure 14.7 Comparison of different approaches for depositing perovskite layers. (a) Stabilized PCEs during MPP tracking for the best device of each approach. (b) Statistical PCE distribution of the solar cells. Source: Abzieher et al. [100].

achievable efficiency is highly dependent on the active layer quality. Most reports on inkjet-printed perovskite solar cells (PSCs) focus on $MAPbI_3$, because it is the best-investigated material in the field and offers variable ways to influence the morphology of the deposited films. In 2014, Wei et al. printed $MAPbI_3$ perovskite layers in a two-step process [99]. In the same year, the first one-step process was introduced by Li et al. who printed precursors prepared with γ-butyrolactone (GBL) as solvent onto mesoscopic TiO_2. They showed a PCE of 12.3% [97]. Later on, Mathies et al. published cells with 11.3% PCE showing improved open-circuit voltage through applying a vacuum [101]. Some of the biggest challenges in inkjet printing of PSCs are solvent evaporation and perovskite crystallization. Inkjet inks are usually slowly drying to achieve good process stability and sufficient nozzle open times. However, using such slowly drying solvents may counteract the development of optimized active layer microstructures. Other processes, such as spin coating, offer centrifugal forces to quickly remove the solvent compared to inkjet printing. In 2018, Liang et al. obtained a pure-phase and high-quality inkjet-printed $MAPbI_3$ perovskite layer. They used a vacuum-assisted process and thermal annealing post-treatment, resulting in devices with a PCE of 17.0% and 13.3% for active area sizes of 0.04 and $4\,cm^2$, respectively [102]. Further progress was reported by inkjet processing of perovskites with mixed compositions. Bag et al. used a desktop multi-channel inkjet printer to deposit different alkyl-ammonium counterions [103]. The solutions were loaded separately into the printer cartridges and deposited onto spin-coated layers of PbI_2. By variation of the composition ratio, they were able to control the structural and electronic properties of their PSCs. In another work by Mathies et al. inkjet technology was used to deposit triple cation perovskite layers with 10% cesium in a mixed formamidinium/methylammonium lead iodide/bromide composite [104]. The good process control offered by inkjet technology allowed them to precisely adjust the layer thickness over a wide range from 175 to 780 nm and the corresponding grain sizes, resulting in PCEs of up to 12.9%. Furthermore, their inkjet-printed solar cells offer improved temperature and moisture stability. In a similar approach, Huckaba et al. inkjet-printed quadruple-cation perovskite layers with additive Cs, formamidinium, methylammonium, and guanidinium ($Cs_{0.10}Gua_{0.05}FA_{0.83}MA_{0.17}PbI_{2.63}Br_{0.37}$) on an mp-$TiO_2$ layer yielding a PCE of 14.1% [105].

Finally, an important work by Gheno et al. should be mentioned [106]. They introduced chlorine to increase the wettability of the perovskite ink to improve its impregnation into porous materials. With this strategy, they achieved for the first time ($CH_3NH_3PbI_{3-x}Cl_x$) solar cells based on a multiple layer inkjet process under ambient conditions and at low temperature (<90 °C) with a PCE of 10.7%.

Spray coating also uses a fine nozzle to spread small ink droplets over a large area. The methods of forming fine sols can be classified into several categories: pneumatic spray (through fast gas flow), ultra-sonic spray (through ultrasonic vibration), or electro-spray (through electrical repulsion). Pneumatic spray and ultra-sonic spray methods have already been used to deposit inorganic oxide layers for supporting the perovskite active layer. In the case of perovskite layers, ultra-sonic spray has been employed [107–111]. For spray coating, the biggest advantage is that it can coat quickly over a very large area. However, the sizes of the ink droplets ejected from the nozzle are very diverse, and the ink droplets can be easily overlapped on the substrate, affecting the final film thickness. New ink droplets deposited on the already formed perovskite will also destroy the existing layer, thereby affecting the crystal quality. Efforts to reduce those effects have also been made mainly by keeping the temperature of the substrate very high to evaporate the solvent rapidly, thus not having the possibility to destroy the layer underneath [110].

14.3 State-of-the-art of Large-Area High-Quality Perovskite Devices

Generally, spin-coating is the most investigated strategy for fabricating cells with uniform and well-controlled crystal growth over the entire device area without pinholes. Currently, the highest efficiency of 25.2% reported for PSCs has been achieved by spin-coating (Figure 14.6). However, the spin-coating method cannot be an effective solution for fabricating large-area perovskite solar modules. To transfer perovskite PV to industrial scale, scalable coating or printing techniques have to be studied. The decision concerning which techniques may ultimately be used in industrial production will be made based on the LCOE considerations described above. However, the PCE of perovskite modules is still significantly lower, as shown in Figure 14.1. High film uniformity of the perovskite layer is one of the main requirements of every scale-up technology. Thus, testing larger PSCs (over 1 cm^2) is more suitable for measuring performance and judging the potential of coating, quenching, and post-treatment methods. In the following, we will give a brief overview of the efficiency development of devices with an area of 1 cm^2 or more. In Table 14.1, the champion performances of large-area perovskite solar cells and solar modules based on different coating techniques are summarized. Figure 14.8 gives a graphical overview of the development, sorting devices according to the techniques for deposition of the perovskite layer. In 2014, the Di Carlo group fabricated a mesoporous structure $MAPbI_{3-x}Cl_x$ perovskite solar module. It showed a PCE of

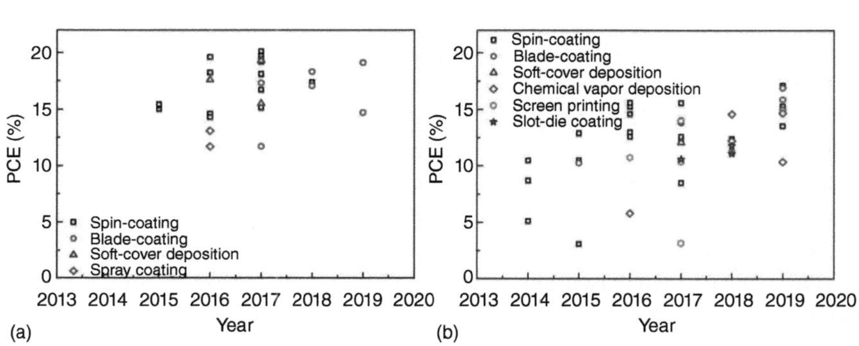

Figure 14.8 Development of efficiencies of large area (active area > 1 cm^2) perovskite solar cells (a) and modules (b) during the period 2013–2020, according to deposition techniques.

5.1% on an active area of 16.8 cm^2 (if not mentioned otherwise, all efficiency values in this section refer to the active area), using both poly(3-hexylthiophene) and spiro-OMeTAD as HTL [56]. Later, the Seok group first reported the fabrication of a planar 10 cm × 10 cm $MAPbI_3$ solar module (active area 60 cm^2), showing 8.7% PCE by using [6,6]-phenyl C61 butyric acid methyl ester (PCBM)/LiF as the electron transport layer [57]. In September of 2014, the Di Carlo group developed innovative and scalable patterning procedures to minimize the series resistance of the interconnections of a $MAPbI_{3-x}Cl_x$ perovskite solar module prepared by two-step deposition to increase its PCE to 13% (on an active area of 10.08 cm^2) [58]. In 2015, PSC modules were fabricated through blade coating with a PCE of 10.4% (active area 10.1 cm^2) and 4.3% (active area 100 cm^2) by Di Carlo and coworkers [66] Also in 2015, through adopting PEDOT:PSS and PCBM as HTL and electron transport layer (ETL) in the p–i–n structure for better charge injection/separation, the Im group prepared a 10 cm × 10 cm $MAPbI_3$ solar module (active area 40 cm^2) which exhibited a PCE of 12.9% [60]. Later, the Han group developed a $MAPbI_3$ perovskite solar cell with a doped inorganic charge extraction layer (LiMgNiO) for rapid carrier extraction. A PCE of over 15% was achieved with a 1.02 cm^2 active area [33]. In 2016, through the addition of chloride to improve the morphology of the film by reducing the crystallization rate of the perovskite and of bromide to increase V_{oc}, Wu fabricated a $MAPbI_{3-x-y}Br_xCl_y$ perovskite solar module by one-step hot solution spin-coating, showing a PCE of up to 14.3% (for an active area of 25.2 cm^2) [64]. In September 2016, the Han group fabricated $FA_{0.85}MA_{0.15}Pb(I_{0.85}Br_{0.15})_3$ perovskite solar cells with a certified PCE 18.21% (active area, 1.022 cm^2) by engineering the perovskite–fullerene heterojunction structure to improve photoelectron collection and reduce recombination losses [35]. In 2017, through combining the polyelectrolyte polyethyleneimine (PEI) with PCBM for a stable and printable ETL with enhanced charge extraction, the Lee group prepared a planar $MAPbI_3$ perovskite solar cell with a PCE of up to 18.1% (active area, 1 cm^2), as described in Figure 14.9 [36]. At the same time, the Sargent group reported a certified PCE of 19.5% for planar $Cs_{0.05}FA_{0.81}MA_{0.14}PbI_{2.55}Br_{0.45}$ perovskite cells with active areas of 1.1 cm^2 by using chlorine-capped TiO_2 colloidal nanocrystal films as an electron selective layer to mitigate interfacial recombination and improve interface binding [37]. Later, by introducing methylammonium acetate and thiosemicarbazone additive into perovskite precursor solution to grow a uniform and large crystal perovskite layer, the Han group achieved a certified PCE of 19.19% for $MAPbI_3$ perovskite solar cells (active area of 1.025 cm^2) (Figure 14.10) [38]. In June 2017, by adding iodide ions into FAI(+MABr) isopropyl alcohol (IPA) solution to modify the two-step spin-coating method through an intramolecular exchange process and decrease the concentration of deep-level defects, the Seok group fabricated $PbI_2/PbBr_2$/FAI/MABr based perovskite solar cells with an active area of 1 cm^2, showing a certified PCE up to 19.7% [15]. By controlling the precise stoichiometry of PbI_2 content in $(FAPbI_3)_{1-x}$ $(MAPbBr_3)_x$ perovskite films prepared by two-step spin-coated and passivated by PbI_2, the You group fabricated 1 cm^2 active area perovskite solar cells with PCE of up to 20.1% [39]. In 2018, the Jung group fabricated

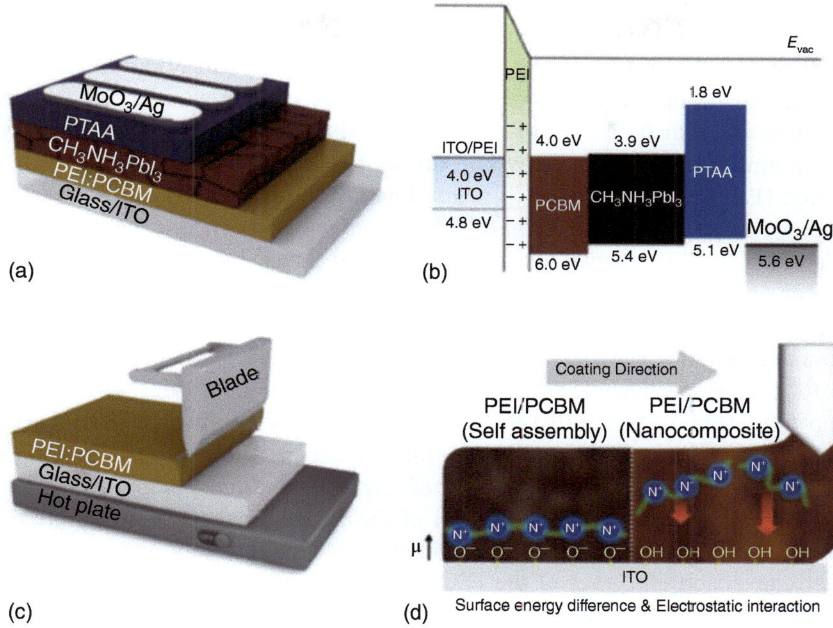

Figure 14.9 (a) Device structure and (b) energy level diagram of the PSC used for PEI:PCBM printing study. (c) Schematic presentation of the coating process. (d) Proposed scheme of the self-assembly of the PEI:PCBM nanocomposite using the doctor-blade printing method. Source: Lee et al. [36].

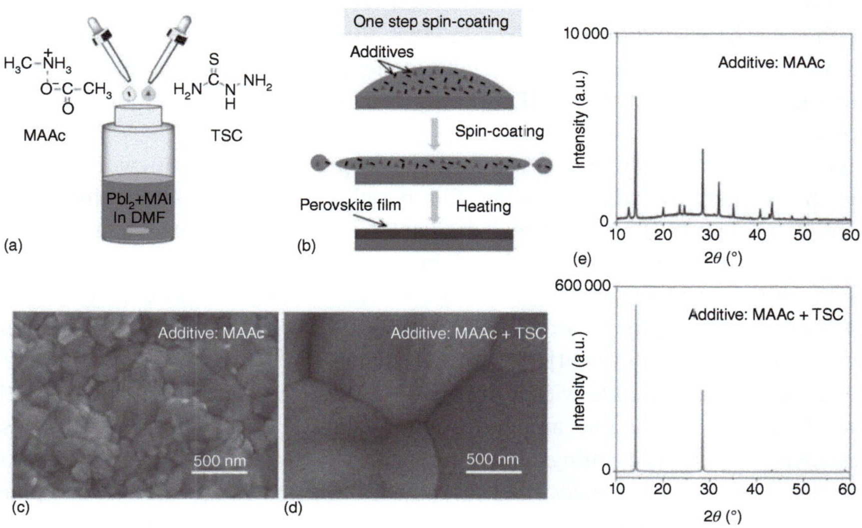

Figure 14.10 (a) Additive engineering by combining MAAc and TSC in the perovskite precursor solution. The chemical structure of MAAc and TSC are shown in this panel. (b) Schematic diagram of the additive-assisted deposition of perovskite layers. (c–e) Top-view scanning electron microscopy (SEM) images of $MAPbI_3$ films fabricated with additives of MAAc (c), MAAc + TSC (d) in the precursor solution, and their X-ray diffraction (XRD) patterns (e). Source: Wu et al. [38].

a large-area high quality $[CsPbI_3]_{0.05}[(FAPbI_3)_{0.85}(MAPbBr_3)_{0.15}]_{0.95}$ perovskite film (14 cm × 14 cm) via a spin-coating method using anisole as antisolvent, with a PCE of 17.37% for 1.08 cm^2 active layer [41]. Compared to other antisolvents (e.g. chlorobenzene), anisole provides an ultrawide process window to fabricate the perovskite thin films. In 2019, through in situ reaction of *n*-hexyl trimethyl ammonium bromide on the perovskite surface to inserting a wide-bandgap halide between $(FAPbI_3)_{0.95}(MAPbBr_3)_{0.05}$ perovskite layer and the P3HT HTL layer for a higher flat-band potential, the Seo group fabricated 5 cm × 5 cm modules with PCEs of up to 17.1% (active area, 24.97 cm^2) [16]. Later, through optimizing the uniform and pinhole-free SnO_2 film by an electrostatic self-assembly method, Jung et al. fabricated 10 cm × 10 cm $[CsPbI_3]_{0.05}[(FAPbI_3)_{0.85}(MAPbBr_3)_{0.15}]_{0.95}$ perovskite modules showing PCEs of 15.3% and 14.03% for the active area 25 and 100 cm^2, respectively [112]. Recently, illustrated in Figure 14.11, the Di Carlo group fabricated 12.5 cm × 12.5 cm perovskite modules by interface engineering mesoporous perovskite solar modules with two-dimensional materials (graphene and functionalized MoS_2) showing PCEs of 15.27% (active area 82 cm^2) and 13.56% (active area 108 cm^2) [65]. The function of the 2D materials is to accelerate the electron and hole extraction, thus improving both the stability and the overall power conversion efficiency of the PSCs compared to standard devices. Through tailoring solvent coordination in the doctor blade-coating method, a large module with an aperture area of 63.7 cm^2 (PCE of 16.4%) has been reported by Deng et al. [74]. In 2017, a 2D/3D $(HOOC(CH_2)_4NH_3)_2PbI_4/CH_3NH_3PbI_3$ perovskite junction layer was obtained by screen printing for the 10 × 10 cm^2 PSC module with PCE of 11.2% by Graetzel and coworkers [80]. In 2018, a large module with an aperture area of 168.75 cm^2 (PCE of 11.1%) had been reported by Galagan et al. prepared by the slot-die coating method [73]. In 2018, through inkjet printing, PbI_2 layers on mesoporous TiO_2 and converting with methylammonium iodide (MAI) powder into the $MAPbI_3$ perovskite, $MAPbI_3$ a PSC with an efficiency of 17.72% on 2.02 cm^2 active area was achieved by Li et al. [98].

From early 2018 to mid-2019, Toshiba Corporation held the world record for the largest $MAPbI_3$ perovskite film-based PV sub-module (Figure 14.12) with 24.15 cm × 29.10 cm (703 cm^2 designated illumination area; definitions of the measurement areas can be found in refs [27, 29]) with a certified PCE of 11.7%, open-circuit voltage (V_{oc}) of 1.073 V, short-circuit current density (J_{sc}) of 14.36 mA cm^{-2}, and fill factor (FF) of 75.8% (all the V_{oc} and J_{sc} values are given per cell in the module) [114]. To further enlarge the area from 703 to 802 cm^2, a perovskite solar module with a certified PCE of 11.6% has been achieved [27]. This large-area submodule and module were fabricated by a two-step process employing meniscus printing technology with PbI_2 and CH_3NH_3I as precursor solutions applied sequentially. It has been mentioned that the goal of the company is to fabricate PV modules with sizes of 900 cm^2, with which it would be possible to achieve a levelized cost of electricity of ~0.06 USD (7 JPY) per kWh by 2030 [114]. In 2019, the record efficiency was increased three times. Currently, Panasonic Corporation holds the world record of 16.1% for a module of >800 cm^2 in size (Figure 14.12).

14.4 Strategies of Upscaling of Perovskite Devices

There are four different general structures of PSCs (Figure 14.13). The mesoscopic n–i–p structure of PSCs (Figure 14.13a) is the indigenous architecture employed because of its higher efficiency [115]. This structure contains a transparent conducting oxide (TCO) (especially fluorine-doped tin oxide (FTO)), a thick layer of electron transport material (ETM, usually TiO_2), having thicknesses ranging from 50 to 70 nm, a mesoporous metal oxide (TiO_2 or Al_2O_3) layer soaked by perovskite material, having a thickness varying from 150 to 300 nm, a general HTL covering perovskite with thickness ranging from 150 to 200 nm, and around 100 nm thick metal electrode (commonly, silver or gold) [115]. Figure 14.13b shows the planar n–i–p structure developed from the n–i–p mesoscopic structure. Since the extraction of holes from the hole transport materials (HTM) surface is more efficient than the movement of the electrons on the ETM interface, mesoporous ETM initially was helpful increasing device efficiency [116]. Nevertheless, current perovskite devices without a mesoporous layer is fabricated under a controlling interface between electron-hole transport layer, perovskite, and electrode. Improved perovskite crystals can help remove the mesoporous layer [116]. In the p–i–n structure, the order in which films are fabricated is inverted, as shown in Figure 14.13c. Normally,

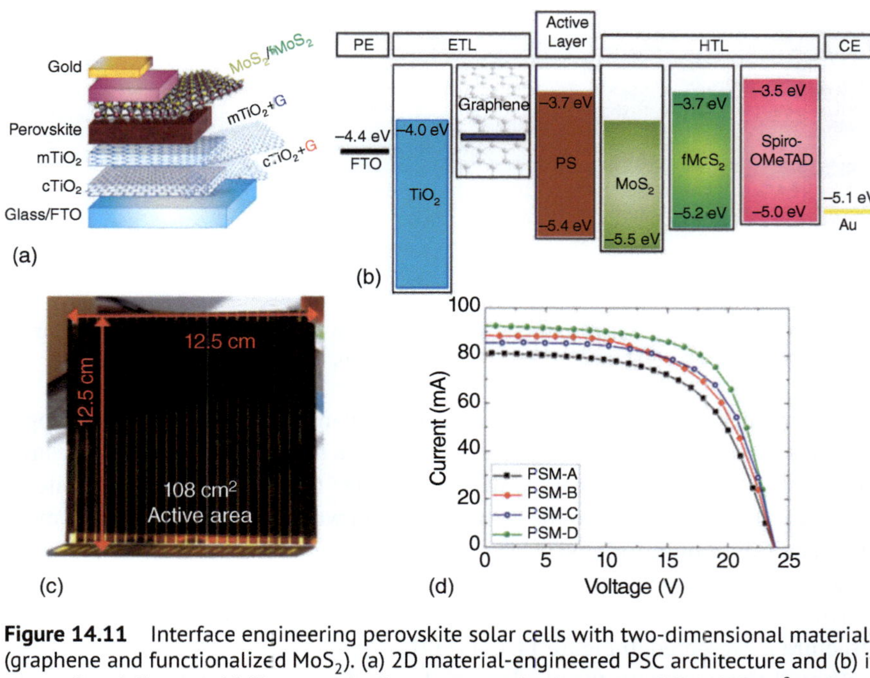

Figure 14.11 Interface engineering perovskite solar cells with two-dimensional materials (graphene and functionalized MoS_2). (a) 2D material-engineered PSC architecture and (b) its energy band diagram. (c) Photograph of a representative large-area PSM (108 cm^2 active area, 156.25 cm^2 substrate area). (d) *J*–*V* characteristics of tested PSMs. Source: Agresti et al. [65].

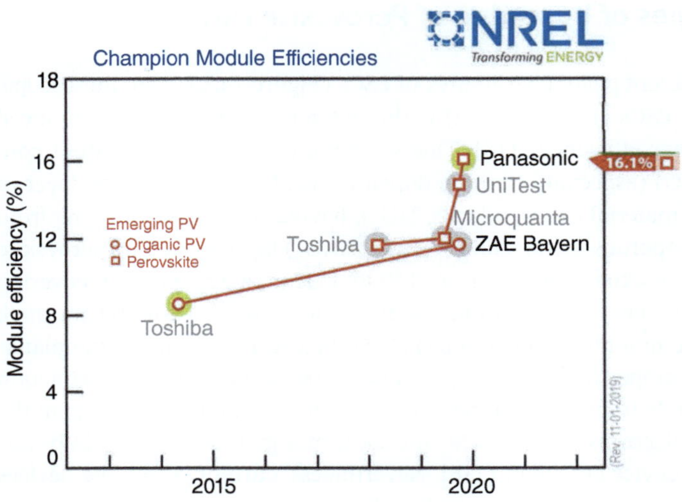

Figure 14.12 NREL module efficiency chart for emerging technologies (modified). Source: Modified from Champion Photovoltaic Module Efficiency Chart [113].

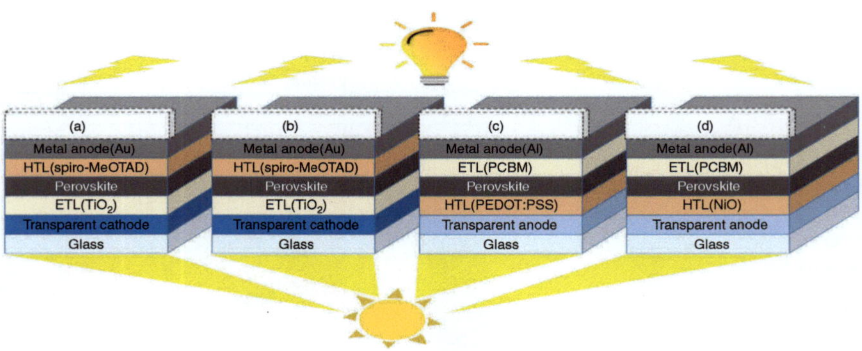

Figure 14.13 Architectures of perovskite solar cells in n–i–p and p–i–n configurations: (a) mesoscopic n–i–p, (b) planar n–i–p, (c) planar p–i–n, and (d) mesoscopic p–i–n.

50–80 nm thickness of PEDOT:PSS conducting polymer is coated on indium-doped tin oxide (ITO) coated substrate as the first layer on top of which a perovskite layer having about 200 nm thickness is deposited. After that, an organic ETL material such as PCBM is coated onto the active layer, followed up by around 100 nm metal electrode (aluminium or gold) [115]. With the recent developments of the p–i–n structure, the organic HTL material has been replaced by inorganic materials such as NiO_x with mesoporous Al_2O_3 or c-NiO_x with mesoporous NiO_x. The detailed structure is presented in Figure 14.13d [117].

We will first deal with the challenges of upscaling perovskite PV, related to the perovskite layer. Subsequently, we will turn to the difficulties encountered when upscaling charge extraction layers and finally to the intricacies of electrodes.

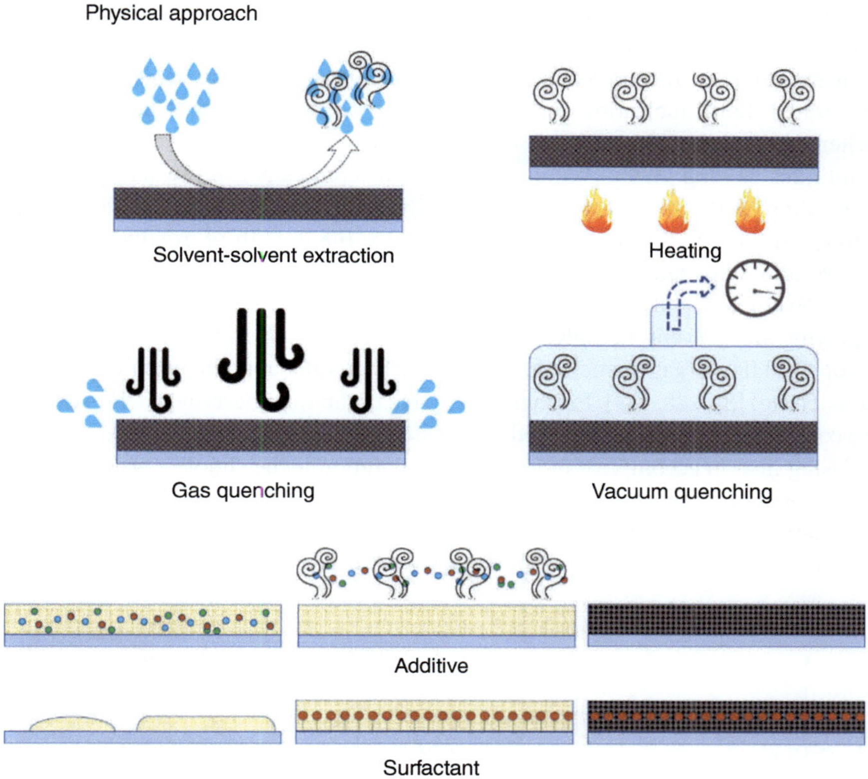

Figure 14.14 Strategies to control nucleation and film formation for scaling up perovskite absorber films.

14.4.1 Strategies for Up-Scaling Perovskite Layers

Even though researchers have invested great efforts into PSCs, there still exists a big efficiency gap between small-area cells on one hand and large area cells or modules on the other hand, as obvious from Figure 14.1. The reason lies in the complex film formation kinetics, which depends sensitively on the rate of solvent evaporation and thus on the area of the substrate [118–121]. The first challenge for upscaling perovskite solar cells is how to prepare a high-quality perovskite film. The typical properties of a perfect perovskite layer include full surface coverage, low roughness, high crystallinity, and low density of grain boundaries. Figure 14.14 shows an overview of the different strategies to control nucleation and film formation for scaling up perovskite absorber layers. Those strategies are mainly divided into two categories, physical methods, and chemical approaches. In the following, we will discuss these strategies in more detail.

14.4.1.1 Physical Methods

Physical methods aim at the fast removal of solvent from the precursor solution to promote supersaturation, thus avoiding dendritic growth and the consequent formation of pinholes. The effect of rapid removal of the solvent is rationalized by the LaMer diagram (Figure 14.14). Evaporation of solvent increases solute concentration to the critical nucleation concentration, above which nuclei are generated. Both nucleation and crystal growth consume solute. Nucleation stops when the concentration of solute drops below supersaturation while crystal growth continues as long as the solute concentration remains above the saturation concentration. If the solvent is removed rapidly, nucleation is favored at the expense of crystal growth. Shorter growth times reduce the growth of existing crystals in the solution, preventing individual nuclei from growing dendritic structures. Such fast solvent removal can be achieved by using high temperature [92, 122, 123], applying anti-solvent, i.e. solvent–solvent extraction (SSE) [11, 124, 125], injecting strong gas flows (gas quenching) [54, 126, 127], and by applying vacuum (vacuum quenching) [53, 128, 129]. Furthermore, some of the ways described before can also be combined to produce improved perovskite layers [130] (Figure 14.15).

Using heat to accomplish fast solvent evaporation is the simplest way to remove the solvent and was very promising because it is relatively easy to apply to all R2R compatible methods such as meniscus coating, inkjet printing, and spray coating [32, 47, 49, 70, 107, 110, 123, 131]. In one example, Nie et al. enhanced performance

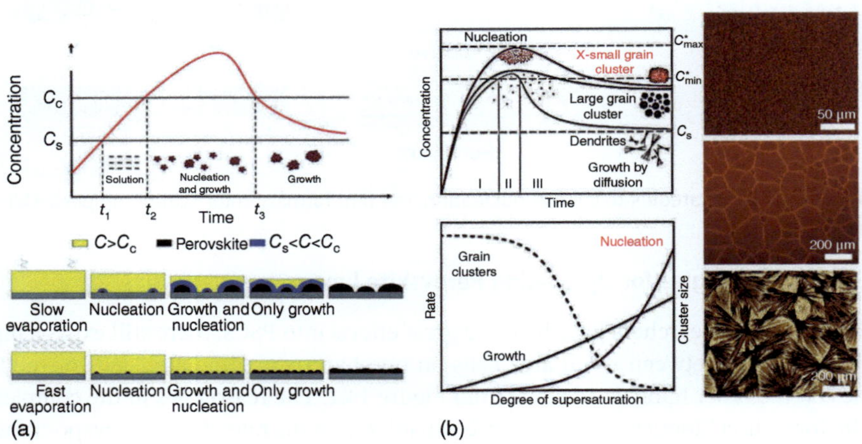

Figure 14.15 (a) Conventional LaMer model: the concentration change of the perovskite precursor solution as a function of time at a constant evaporation rate of the solvent and the resulting nucleation/growth competition model at slow and fast solvent evaporation. (b) The classical LaMer curve divides the crystal growth into three distinct phases. These three phases are presented for films consisting of extra small grain clusters (red trace), large grain clusters (purple trace), and dendrites (grey trace) films. The nucleation (red trace) and growth (blue trace) rates as a function of the supersaturation degree and how this then determines the grain-cluster size (dashed black trace). The resulting films on the right side are categorized as nucleation densities, and material microstructure. The dendrites (top, right), the nucleation density is low and about 100 μm–1 mm of nucleation site forms into needle-like structure. On the other hand, other morphologies such as large (left, middle), medium (right, middle), small (left, bottom), extra small (right bottom) show increased nucleation density with the nucleation site spaced by about 300, 10, 2, and 500 μm, respectively. Source: Ding et al. [130].

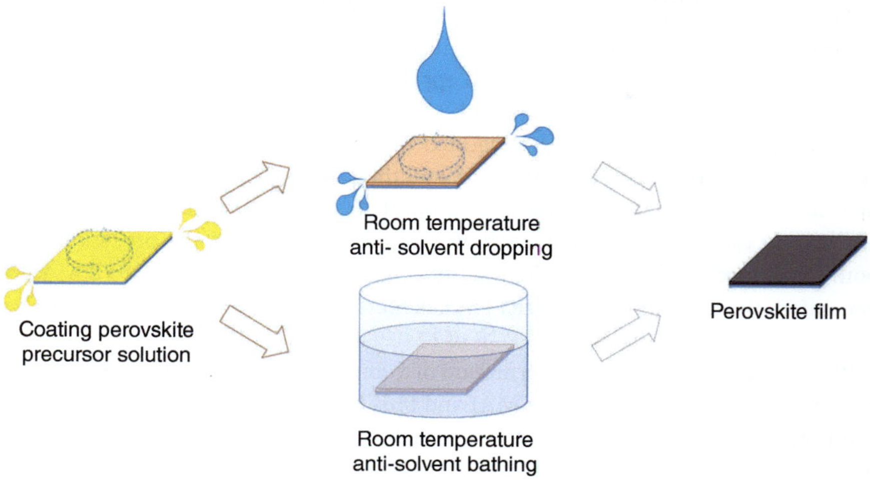

Figure 14.16 Perovskite film formation by solvent–solvent extraction (SSE).

substantially by applying a precursor solution heated up to 70 °C on a substrate kept at 180 °C, which led to fast solvent evaporation from the substrate, achieving a PCE of 12.5% [132]. In another example, Deng et al. deposited a very uniform perovskite film without pinholes by blade coating on a heated substrate close to the boiling temperature of the precursor solution and reported 15.1% of efficiency [123].

Anti-solvent-based methods have been most commonly used to obtain uniform and homogeneous perovskite films (Figure 14.16). The anti-solvent absorbs the solvent in the precursor, whereas the perovskite constituents are insoluble in the anti-solvent. Typically, dimethylformamide (DMF), dimethylsulfoxide (DMSO), *N*-methylpyrrolidone (NMP), and GBL are used as solvents of perovskite precursor, while toluene, chlorobenzene, diethyl ether, ethyl acetate, hexane, and anisole are used as anti-solvents [11, 124, 125, 133–135]. Jeon et al. applied toluene as an antisolvent by dripping it onto the wet precursor layer in the middle of the spin coating process [11]. However, the antisolvent applied by dripping during spin coating caused complicated physical effects with the wet precursor film throughout the process, resulting in low reproducibility with inhomogeneous surface and pinholes. Ye et al. have developed the conventional method of spraying an anti-solvent to produce uniform layers without the physical problems arising from solvent dripping [136]. In particular, Kim et al. achieved 12.1% by spraying anti-solvent on 16 cm^2 large perovskite precursor wet films [84]. However, this method has been tested only for spin coating so far. It has turned out to be difficult to upscale due to inhomogeneous spreading on the substrate, which negatively affects crystallization. To overcome this problem, a method of dipping a substrate covered by a wet precursor film into a container filled with anti-solvent has been developed [125]. By taking the necessary provisions on the equipment side, the dipping process does not have any process limitation with the continuous methods used in R2R manufacturing.

Applying strong gas flow or vacuum has also been used to remove the solvent rapidly. Huang et al. induced rapid crystallization by injecting a strong gas flow onto the substrate with a wet precursor film applied by spin coating [126]. Cotella et al. applied a strong gas flow by an air knife attached to a slot-die coater to fabricate perovskite layers [127]. Li et al. suggested the application of vacuum in order to accelerate solvent evaporation. With this method, they achieved perovskite layers of around 1 cm^2, keeping the efficiency around 20% [53]. By combining both methods, strong vacuum and strong gas flow, Ding et al. developed the process having both advantages. From this method, homogeneous perovskite layers of 1 cm^2 were generated, the champion large-area PSCs achieving a J_{SC} of 22.56 mA cm^{-2}, a V_{OC} of 1115 mV, a FF of 67.80%, and a PCE of 17.05% [130].

Recently, a method using the formation of a complex between methylamine (MA) gas and the perovskite precursors MAI and PbI_2 have been introduced. The viscous solution formed by the reaction of MA with MAI and PbI_2 powders can be diluted in a solvent to adjust the viscosity to the appropriate range for the specific coating device. The adducts of the precursors with MA are soluble in the low boiling point solvent acetonitrile (ACN), in which perovskites and their precursors are insoluble otherwise. Neol et al. synthesized perovskite precursor solutions having fast evaporation rates by the process described above and achieved 18% efficiency without additional processes such as vacuum or gas quenching [137]. The MA process introduced here has been applied to spin coating and doctor blading [138]. It seems plausible that it can also be applied to other solution coating processes for upscaling, such as slot-die coating. It has been reported that exposure of low-quality perovskite crystals to MA gas result in substantial enhancement of crystallinity [139].

14.4.1.2 Chemical Methods

Chemical methods can also be used to improve the quality of perovskite layers. The kinetics of generating nuclei and growing crystals are strongly affected by the chemical composition of the perovskite precursor solution. In particular, the surface morphology of perovskites is affected not only by their composition but also by additives. Snaith et al. reported that adding chloride ions to a perovskite precursor leads to improved smoothness of perovskite surfaces, enhanced grain sizes, and reduced concentration of pinholes [140]. In their experiment, mainly $PbCl_2$ was used as a chloride carrier. Other chloride additives such as MACl or NH_4Cl have also been used and show similar results [141–143]. The chloride anion is generally believed to favor creating perovskite nuclei by retarding lead iodide species. These species are by-products of fabricating perovskite crystals by competing with iodide ions in the complexation of Pb^{2+} ions [144, 145]. In other words, additives containing chloride prevent the crystallization processes forming pin or needle type crystals. The thin film, including the chloride ions, is converted into the perovskite, maintaining the existing homogeneous surface by passing through the intermediate states until reaching the final state [146, 147]. The chloride ions can be removed by annealing as gaseous MACl [147]. Due to these properties, chloride has been used in various R2R compatible coating methods, including blading, slot-die coating, and spray coating [77, 110, 127, 148, 149]. In addition to the chlorides mentioned above, other organic halides or $Pb(Ac)_2$ were reported to form an intermediate state to obtain a homogeneous perovskite film [50, 150]. Moore et al. studied the mechanism of perovskite synthesis in the presence of an additive forming the intermediate state. Later, other groups reported the formation of homogeneous perovskites by using hydrohalic acids as an additive instead of anti-solvent [151–156]. These acids are claimed to have two effects: first, the acids provide anions that form intermediate species with Pb^{2+} to prevent crystallization of lead iodide, and second, they reduce the pH-value of the precursor, thereby influencing the rate of crystal production by destroying small crystals or nuclei of undesired by-products [152, 157].

Many studies have also been done from the point of view of Lewis acid–base interactions. In general, Pb^{2+} as a strong Lewis acid tends to combine with Lewis bases. This interaction improves the solubility of the lead halide in the precursor, thus slowing the rate of nucleation and crystallization. The rate of nucleation and crystallization is determined by the concentration of the Lewis base and the strength of the interaction with Pb^{2+}. To observe this process, Lee et al. improved the quality of perovskite films by adding DMSO and thiourea as the Lewis bases [158]. Also, *N*-cyclyl-2-pyrrolidone (CHP), hexamethylphosphoramide (HMPA), 1,8-diiodooctane (DIO), and thio-semicarbazide (TSC) are Lewis bases. They can be added into perovskite precursor solutions to improve the quality of perovskites, regardless of the coating technique applied [38, 159–161].

Another method suggested for modifying the quality of perovskite films consists of regulating the surface energy of the substrate before coating precursors. This method has been applied mainly to p–i–n perovskite cells, either by coating hydrophobic self-assembled monolayers on the substrate or adding nonpolar modifiers such as perylene into the HTL (PEDOT:PSS) to improve the wettability of the substrate by the precursor [162–164]. Also, the addition of polymers such as poly(vinylpyrrolidone) (PVP) to the precursor solution can improve the coverage of the substrate by perovskite and the smoothness of the final perovskite film [165]. However, there are different opinions about this concept. Bi et al. argued that when the wettability is rather poor, the size of the perovskite crystals will be enhanced, due to the increased space available for individual crystals during crystallization [166].

14.4.1.3 Post-Growth Treatment

It is known that post-growth heat treatment, i.e. annealing, enhances the overall crystallinity of the perovskite films by increasing the size of crystallites and eliminating defects. The heat treatment process is typically performed at temperatures between 150 and 180 °C for around 5–120 minutes. However, longer processing times are difficult to be applied in R2R processes. This is why efforts to reduce the annealing time to less than 10 minutes have been made by controlling the chemical composition of the precursor [31]. Annealing processes in the laboratory are mainly realized by direct contact between substrates and hot plates. However, it is difficult to apply direct contact in R2R processes. Heating the target material by heat radiation is much more convenient for this purpose. By applying near-infrared radiation for 2.5 seconds, Troughton et al. annealed the perovskite active layer effectively and completed solar cells without reducing efficiency [167]. The time was reduced to

1.5 ms by an intensive photonic flash sintering using a Xenon flashbulb. However, the method was somewhat undefined because of poor control of the irradiation intensity that resulted in lower working efficiency [168]. Later, Pool et al. reduced the thermal annealing process time for films of $FAPbI_3$ to 40 seconds by applying halogen lamp radiation [169].

14.4.2 Scalable Charge Extraction Layers

Charge transport materials also play a significant role in the upscaling of perovskite solar cells. They extract the photo-generated charges from the perovskite layer and thus influence the V_{OC} of the cells. The charge transport materials should meet the following requirements for the upscaling of perovskite PV. First, they should have high charge carrier mobility and well-aligned energy levels for extracting the desired type of charge carrier from the perovskite active layer. Second, they should be low cost and easily deposited at low temperature to ensure low-cost, large-scale PSC manufacturing. Third, they should have good stability. In Figure 14.17, some traditional HTL materials are shown. 2,2′,7,7′-tetrakis(*N*,*N*-di-methoxyphenylamine)-9,9′-spirobifluorene (spiro-MeOTAD) material is commonly used as the HTL for giving high efficiency in large-area PSCs [19]. However, only a few works have fabricated spiro-MeOTAD layers with R2R compatible techniques [170]. Also, two drawbacks have restricted the industrialization of spiro-MeOTAD. First, the cost of synthesis and purification of this material is around $500 per g. Second, the hole mobility of neat

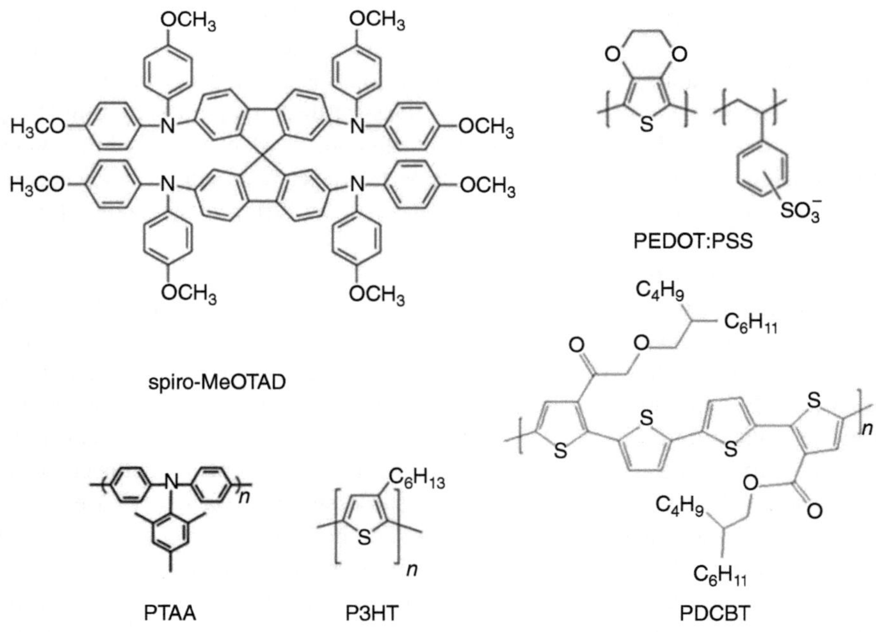

Figure 14.17 Traditional commercial polymeric HTMs for PSCs.

spiro-MeOTAD is only 5.3×10^{-5} cm^2 V^{-1} s^{-1}, requiring dopants (lithium trifluoromethanesulfonimide, 4-*tert*-butylpyridine) which increase the hole mobility but at the same time reduce the PSCs lifetime [171, 172]. Therefore, other inorganic materials and small organic molecular and polymeric materials have been explored as alternatives to spiro-MeOTAD. However, only a few materials are successfully applied in upscaling PSCs owing to their high cost, poor morphology, and low mobility. Using PTAA [173] as HTL also exhibited PCE exceeding 20% but is not applicable for upscaling because of its cost. Until now, for low cost and compatibility with R2R processes, NiO_x [174], CuSCN [175], PEDOT:PSS [176], P3HT [25], and poly[2,2''''-bis[[(2-butyloctyl)oxy]carbonyl][2,2':5',2'':5'',2'''-quaterthiophene]-5,5'''-diyl] (PDCBT) have the most promising perspective for large-scale PSC manufacturing.

Electron transport materials are also quite important for upscaling perovskite solar cells. Low-temperature processable TiO_2 has been developed for flexible perovskite modules [59]. Tin (IV) oxide is an attractive ETL material that provides a higher open-circuit voltage, less hysteresis, and lower photocatalytic effect than TiO_2 [177, 178]. Also, thin SnO_2 films are claimed not to decrease the interconnection quality in a solar module [179]. Zinc(II)oxide also exhibits PCE over 20% in PSCs but has been reported to show detrimental interactions with the perovskite layer, mainly in n–i–p devices. For p–i–n type PSCs, fullerene derivatives (e.g. [6,6]-phenyl-C_{61}-butyric acid methyl ester ($PC_{61}BM$)) are an option for upscaled manufacturing [25].

14.4.3 Scalable Electrodes

14.4.3.1 Bottom Electrode

The bottom electrode is essential to achieve high-performance perovskite photovoltaics. FTO and ITO are commonly recommended for PV because of their unique features, including excellent conductivity, high transmittance, and good compatibility with the perovskite active layer [180]. Most of the devices for which record efficiencies have been reported were fabricated on this type of substrates, such as ITO-coated poly(ethyleneterephthalate) (PET) [181–183].

Especially for R2R compatible perovskite device architectures, the stability of the bottom electrode and the substrate is one of the key factors for a successful process. Generally, ITO-covered polymer substrates are selected as bottom electrodes because of their lower price, flexibility, transparency, and process stability. However, deposition steps that require processing temperatures over 150 °C cause problems when working with polymer substrates. Park and coworkers have coped with this limitation by fabricating high-performance PSCs on CPI (Colorless polyimide)/ITO substrate, which sustains a temperature of 300 °C during the fabrication process yielding a device with a PCE of 15.5% [184].

To avoid the common problem of the temperature sensitivity of polymer substrates, some researchers tried to use metal foils as bottom substrates, which simultaneously take over the role of the bottom electrode. In contrast to polymer substrates, metal substrates have high thermal stability, excellent barrier property, and high conductivity [185]. However, because it cannot be used as a transparent substrate due to the opacity of metal, a transparent top electrode, is essentially required. Troughton et al. addressed perovskite deposited on titanium (Ti)-foil, and as top electrode a (PEDOT:PSS)-covered nickel mesh was laminated onto the HTL. This work achieved 10.3% efficiency with excellent bending stability [186]. Heo et al. suggested graphene layers as a top transparent electrode on perovskite fabricated on flexible Ti substrates, achieving a PCE of 15.0% [187]. Recently, copper foil has also been suggested as an alternative bottom electrode. Nejand et al. fabricated devices with the structure Cu/CuI/$MAPbI_3$/ZnO/AgNWs, showing a PCE of 12.8% [188].

In another direction for fabricating bottom electrodes, carbon-based materials are also applied because of their chemical stability towards perovskite. Qiu et al. deposited a perovskite layer on a multi-walled carbon nanotube electrode which was continuously intertwined with a fiber electrode. The fiber-shaped PSC achieved a PCE of 3.3% with good bending stability [189]. PEDOT:PSS has been deposited on substrates as an electrode for flexible devices. Poorkazem et al. compared PSCs fabricated on two different transparent electrodes, PEDOT:PSS and ITO on PET substrates. It was reported that the PEDOT:PSS electrodes show better mechanical stability than ITO under periodical bending for 2000 cycles without any mechanical defects such as cracks. ITO-based perovskite devices result in extremely high sheet resistance and subsequent device performance degradation [190]. Recently, on ITO coated willow® glass of 100 μm in thickness, gas quenching crystallized perovskite having 42.9 cm^2 aperture area was fabricated by Dai et al. It showed stable flexibility and reasonable working performance (Figure 14.18) [191].

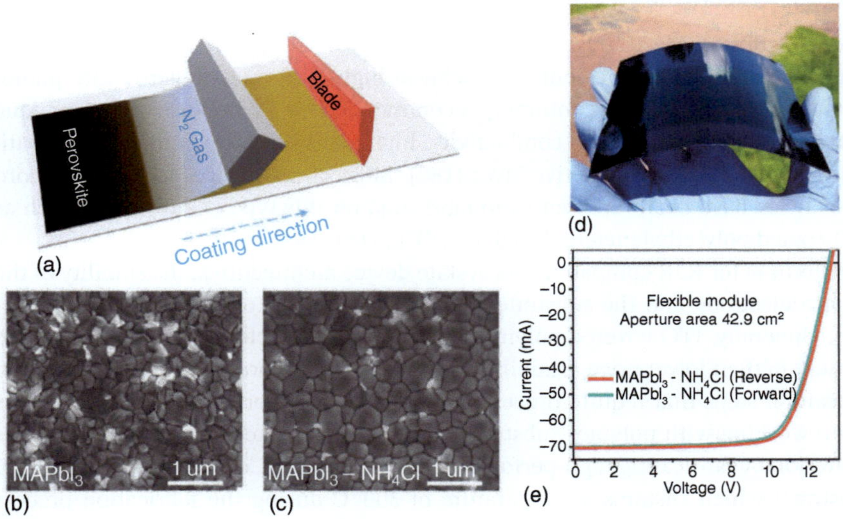

Figure 14.18 Blade coated perovskite films on ITO-coated Willow® Glass with NH_4Cl to improve film morphology and crystallization. (a) Schematic illustration of gas-assisted blade coating of perovskite films. Surface SEM images of the blade-coated films without NH_4Cl (b) and with NH_4Cl (c). (d) Photograph of a flexible perovskite module. (e) J–V curve of the champion flexible perovskite modules. Source: Dai et al. [191].

14.4.3.2 Top Electrode

One of the main obstacles to the commercialization of PV is the high price of the top electrodes made from gold and silver by vacuum evaporation [192]. Because the metal electrode materials such as silver react easily with immigrated halogen atoms from perovskite, it directly deteriorates the photovoltaic performance. As an alternative, due to the low cost, high conductivity, and low processing temperature, carbon-based materials such as graphite/amorphous carbon, graphene, and carbon nanotubes (CNTs) have been addressed in the perovskite field [5, 193, 194]. Another advantage of carbon-based material pertains to the reversible formation of the mono-hydrated phases $CH_3NH_3PbI_3$–H_2O and $(CH_3NH_3)\ _4PbI_6$–$2H_2O$ [195]. In contrast, with other electrode materials, the dihydrate phase is formed, degrading PbI_2 and CH_3NH_3I [196]. Generally, carbon-based materials have a highly hydrophobic nature, leading to the prevention of moisture penetration toward the perovskite layer without any sealing process [197].

In 2018, De Rossi et al. addressed the first A4-size PSC having a carbon electrode with an active area of 198 cm^2 [198]. In Figure 14.19, it is shown that most of the layers were fabricated by spray coating and screen printing. The active layer was fabricated by dripping precursor solution onto the porous carbon film. The perovskite module comprised 22 cells with dimensions of $5 \times 180\ mm^2$ and interconnect spacings of 6 mm. With this large area device, a PCE of 3.2% was reached.

14.5 Module Layout

The third challenge in upscaling PV technologies is how to make an economic module design for practical use. As with other thin-film PV technologies, it is not practical to fabricate a single PSC on a large-area substrate because the power loss due to the resistance of the transparent conducting electrode is substantial over a long transport distance. Resistance losses are reduced by dividing a large cell into smaller sub-cells with series interconnections to form a module. The series interconnections sum the voltages available from the sub-cells of the module. Some authors suggested connecting some cells in parallel, which leads to the summing up of the currents of the individual cells [198]. The interconnections occupy ~1% to 20% of the module area, depending on the geometry of the interconnection layout. The ratio between the photoactive area and the total area of a module is defined as the geometric fill factor (GFF). On one hand, the interconnection area is inactive for power generation and contributes linearly to the power loss in PV modules (Figure 14.20). On the other hand, the narrower the sub-cells, the lower the current, which reduces power losses quadratically with cell length. This leads to an optimum cell length, at which the power losses are minimized. The specific value depends on the current density provided by the respective material and can be calculated, e.g. by finite element modelling (Figure 14.20) [199]. The simulations of Figure 14.20 shows clearly that at current densities of 20–25 mA cm^{-2}, power losses can be reduced to less than 10% of the small cell values, but only if cell lengths are reduced to 4 mm or less. If GFF exceeds 95%, and if the sheet resistance of the transparent front electrode is

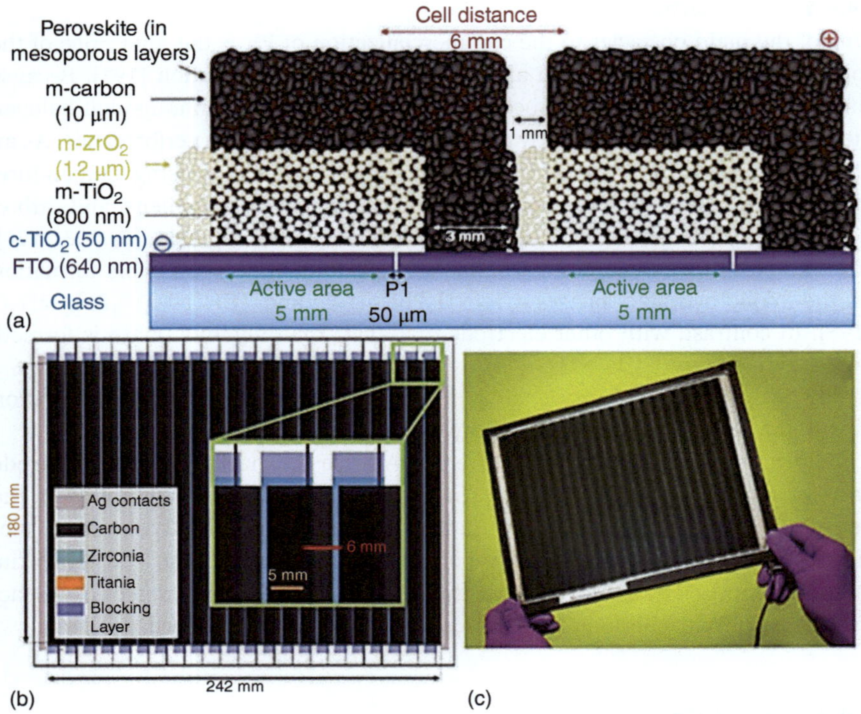

Figure 14.19 (a) Cross-section schematics of adjacent cells in the module produced by De Rossi et al. with a nominal thickness of each layer, highlighting the laser-etched FTO, patterning of TiO_2 blocking layer and the vertical electrical connection, ensured by the carbon back contact; (b) module layout, showing the different overlapping layers, the dimensions of the active area for both the individual single cells and the whole module as well as the distance between adjacent cells (inset); (c) photo of a module; wires have been soldered to the silver-painted busbars to provide more robust electrical contacts. Source: De Rossi et al. [198].

less than 10 Ω/square. The interconnect resistance itself has to be kept well below 0.01 $\Omega\,cm^{-2}$, i.e. care has to be taken to have excellent electrical contact between the bottom and top electrodes in this region.

A possible fabrication scheme for module assembly is to deposit each layer only at desired areas, thus requiring no post-deposition scribing to form the interconnections. Such on-demand deposition has been realized through patterned printing and deposition with masks. However, patterned deposition requires pattern alignment for each layer to be deposited. The spatial resolution is typically low, with GFF not exceeding 80–90%, causing large dead-area power losses (Figure 14.21b) [201].

The most convenient way of fabricating thin-film PV modules (CdTe, copper indium gallium selenide [CIGS], organic photovoltaics [OPV]) is by first depositing the thin-film materials homogeneously on the substrate and then separating individual cells by laser or mechanical scribing [202]. This also applies to PSCs. There are three separation processes involved in module fabrication (Figure 14.20). In the

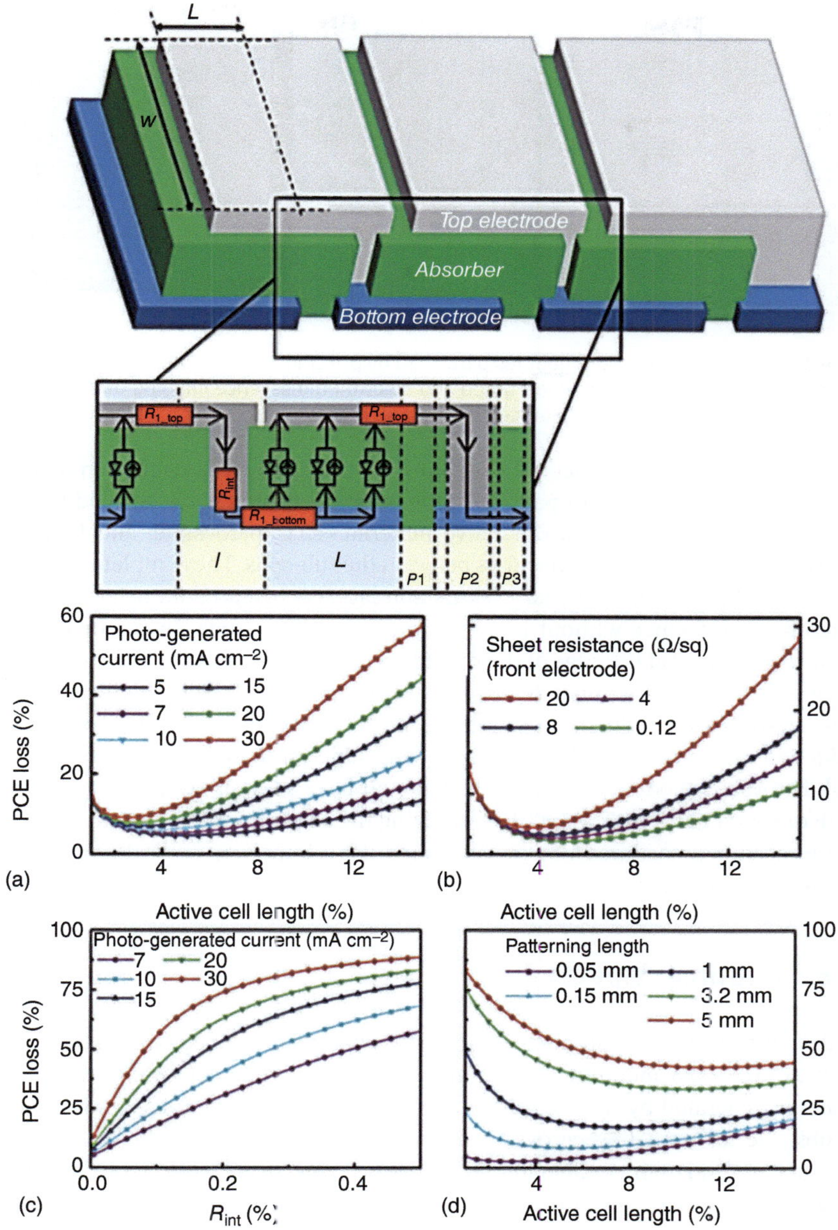

Figure 14.20 Finite element simulation of efficiency losses as a function of module parameters. Top: layout of a module consisting of three cells, with replacement circuit used for simulation. L – active cell length, l – patterning length, $R_{s,top}$ and $R_{s,bottom}$ – series resistance of top and bottom electrodes, respectively R_{int} – interconnect resistance. The photo-generated current is given with respect to the area of a single cell. P1 and P3 – separation of bottom and top electrodes, respectively. P2 – opening of active layer for the formation of contact between the bottom and top electrodes of adjacent cells. Bottom: efficiency losses as a function of (a) active cell length, using J_{sc} as a parameter, (b) active cell length, using sheet resistance of the front electrode as a parameter, (c) interconnect resistance, (d) active cell length, using the patterning length l as a parameter. Source: Lucera et al. [199].

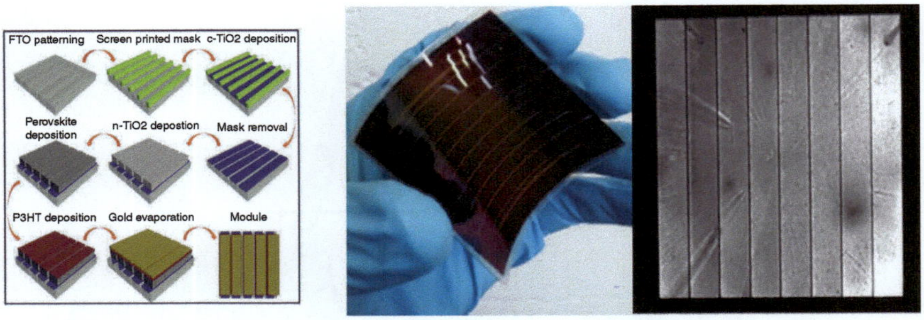

(a) (b)

Figure 14.21 Perovskite solar modules patterned by printing. (a) Fabrication scheme. Source: Matteocci et al. [56]. (b) Photograph and electroluminescence imaging of the flexible perovskite mini-module. Source: Pisoni et al. [200].

first step, the bottom conducting electrode (for example, transparent conductive oxide (TCO)) is divided into parts to form isolated substrates for the individual sub-cells. In the second step, the active material (ETL, perovskite, and HTL) is removed to construct interconnections between the sub-cells. The complete removal of the active layer, including charge extraction layers is essential for achieving low interconnect resistance without removing the bottom electrode. In the final step, the top electrodes and the edge of the module are isolated to separate the individual sub-cells and define the outer boundaries of the module. Scribing with lasers is highly desirable for industrial production, with the advantages of low cost, fast production rate, and high GFF, associated with high precision and fine resolution [184]. Perovskite modules fabricated through all-laser scribing have demonstrated a GFF of 94% (Figure 14.21a) [200]. Optimizing the sub-cell separation and module interconnection procedures for perovskite modules is important for scaling up PSCs and further researched.

14.6 Lifetime Aspects

Besides efficiency and cost, another important parameter to be considered is the lifetime of the solar modules, as the outdoor service lifetime determines the overall energy generated by the module. The instability of perovskite devices is a serious obstacle for upscaling perovskite PV towards the market [203]. The lifetime issues of PSCs and their possible solutions have been nicely presented in Ref [201] so that we will just pick a few aspects to illustrate the challenges in this complex field [204]. As far as known, the longest lifetime of perovskite solar cells is just one year (Table 14.2 [80], which is substantially shorter than the 25 years of commercial Si solar modules. The many factors affecting the lifetime of perovskite solar cells can generally be divided into two categories of external and intrinsic factors. External factors are moisture, oxygen, and ultraviolet (UV) light. When the perovskite materials are exposed to humid air, they decompose into mobile ions, mainly MA^+ and I^-, as well as insoluble PbI_2 [209]. The intrinsic factors are mainly thermal instability and ion migration. Degradation by external factors such as moisture, oxygen,

Table 14.2 Summarizes the state art of power conversion efficiency vs. lifetime.

Device architecture	Area (cm^2)	PCE (%)	Stability	References
FTO/compact TiO_2/mesoporous TiO_2/mesoporous ZrO_2/2D–3D $(HOOC(CH_2)_4NH_3)_2PbI_4$-$CH_3NH_3PbI_3$ perovskite/carbon	47.6 (module)	11.2	>10 000 h zero loss	[80]
FTO/compact TiO_2/mesoporous TiO_2/$(FAPbI_3)_{0.83}(MAPbBr_3)_{0.17}$/CuSCN/rGO/Au	0.16	20.5	Retain >95% of their initial efficiency after aging at a maximum power point for 1 000 h under full solar intensity at 60 °C	[175]
FTO/compact TiO_2/mesoporous TiO_2/RbCsMAFA perovskite/spiro-OMeTAD/Au	0.1225	21.6	Retain 95% of their initial performance at 85 °C for 500 h under full illumination and maximum power point tracking	[205]
FTO/PEDOT:PSS/$((BA)_2(MA)_3Pb_4I_{13}$/PCBM/Al	0.5	12.52	Retain 65% under constant AM1.5G illumination over 2 250 h	[206]
FTO/SnO_2/PCBM/$BA_{0.05}(FA_{0.83}Cs_{0.17})_{0.91}Pb(I_{0.8}Br_{0.2})$ 3/spiro-OMeTAD/Au	0.0919	19.5	Retain 80% under xenon-lamp simulated full-spectrum AM1.5, 76 mW cm^{-2} equivalent irradiance in air (humidity ~45 RH%) without any ultraviolet filter over 1 000 h	[207]
FTO/NiMg(Li)O/$MAPbI_3$/PCBM/Ti(Nb)O_x/Ag	1.02	16.2	With >90% of the initial PCE remaining after 1 000 h of light soaking	[33]
ITO/C_{60}-SAM/$A_{0.83}MA_{0.17}Pb_{1.1}Br_{0.50}I_{2.80}$/PDCBT/Ta-$WO_x$/Au	0.1	21.2	95% of its efficiency after 1 000 h under 1-sun illumination	[25]
FTO/compact TiO_2/mesoporous TiO_2/mesoporous ZrO_2/$(5\text{-AVA})_x(MA)_{1-x}PbI_3$/carbon	0.07	12.8%	Stable for >1 000 h in ambient air under full sunlight	[208]
FTO/TiO_2-Cl/$Cs_{0.05}FA_{0.81}MA_{0.14}PbI_{2.55}Br_{0.4}$/spiro-OMeTAD/Au	1.1	19.5	95% continuous maximum power point (MPP) tracking for 500 h in nitrogen atmosphere under constant simulated solar illumination	[37]

Notes: There are three aspects for improving the lifetime of PSCs, including interface engineering (Interface), multi-dimensional perovskite (MDP), inorganic cations doped in A site for ABX_3 structure (Doping). The hero lifetime of cells or modules is achieved by Nazeeruddin et al. based on the structure of FTO/compact TiO_2/mesoporous TiO_2/mesoporous ZrO_2/2D-3D $(HOOC(CH_2)_4NH_3)_2PbI_4$-$CH_3NH_3PbI_3$ perovskite/carbon showing zero efficiency loss over 10 000 hours.
Source: Modified from Grancini et al. [80].

and UV light can be reduced or eliminated by encapsulation. The issue of thermal instability can be settled by tuning the perovskite composition to increase the decomposition barrier. For example, by increasing the contents of formamidine (FA) and Guanidinium (GA) cations in the perovskite film [210]. The ion migration problem can be solved by modifying the A site component with alkali ions (e.g. K^+, Rb^+) [211], constructing a two-dimensional perovskite to achieve multiple dimensional perovskite films by organic additives [212]. Brabec and coworkers have shown that ion migration can also be eliminated by inserting a SAM of phosphonic acid derivatives of fullerene [213]. The instability of perovskite solar cells may also be caused by the ETL. The commonly used TiO_2 based ETL can induce photocatalytic degradation of the perovskite [214]. Insertion of an interlayer between the TiO_2 layer (e.g. Sb_2S_3 [37], CsBr [215], SAM) or replacing TiO_2 with other UV stable materials (e.g. SnO_2 [216], $BaSnO_3$ [217]) could solve this problem. The most frequently used HTLs spiroOMeTAD or PTAA, also induce instabilities because of the mobility of their dopants (e.g. lithium salt, 4-tert butylpyridine [TBP]) [218]. Inserting buffer layers (e.g. poly(methyl methacrylate) [PMMA]) between perovskite and the spiroOMeTAD layer or replacing spiroOMeTAD with undoped hole transport materials (e.g. P3HT, NiO_x, CuSCN) can enhance device stability [219]. The top electrodes such as Au, Ag, and Al suffer from corrosion by ion migration in the perovskite layer, decreasing the lifetime of perovskite devices [220]. When a perovskite device is exposed to light, migrating ions will gather on the interface, resulting in enhanced charge recombination [220]. Al also suffers from instability due to humidity, as they can easily be oxidized in humid air. Inserting an interlayer between the active layer and metal electrode (e.g. MoOx [175], Ta:WOx [221]) has been reported to hinder the diffusion of metals and ions and thus reduce deterioration. However, the better way is to replace them with stable electrode materials, e.g. by carbon, which is sufficiently inert and abundantly available at a low cost [222].

14.7 Summary and Outlook

The commercial success of perovskite PV as competitive technology to silicon depends mainly on the LCOE, i.e. on how well the parameters of the magic triangle of PV can be optimized. This implies that costs of manufacturing well below 30 ct/Wp must be targeted to be competitive with silicon photovoltaics (SiPV). Direct competition with SiPV may be avoided by developing alternative technologies, such as Si/Perovskite tandem cells, or the development of alternative applications that are not amenable to Si, such as building integration or mobile applications [3, 28].

Although much effort has been invested into upscaling of perovskite PV technology, there is still a large gap between the efficiencies of small lab cells and large area modules (Figure 14.1b). Many of the hero module efficiencies have been achieved with materials that are way too expensive for commercially viable solutions (Figure 14.4). Thus, the successful way to commercialization requires further diligent optimization of both materials and processes.

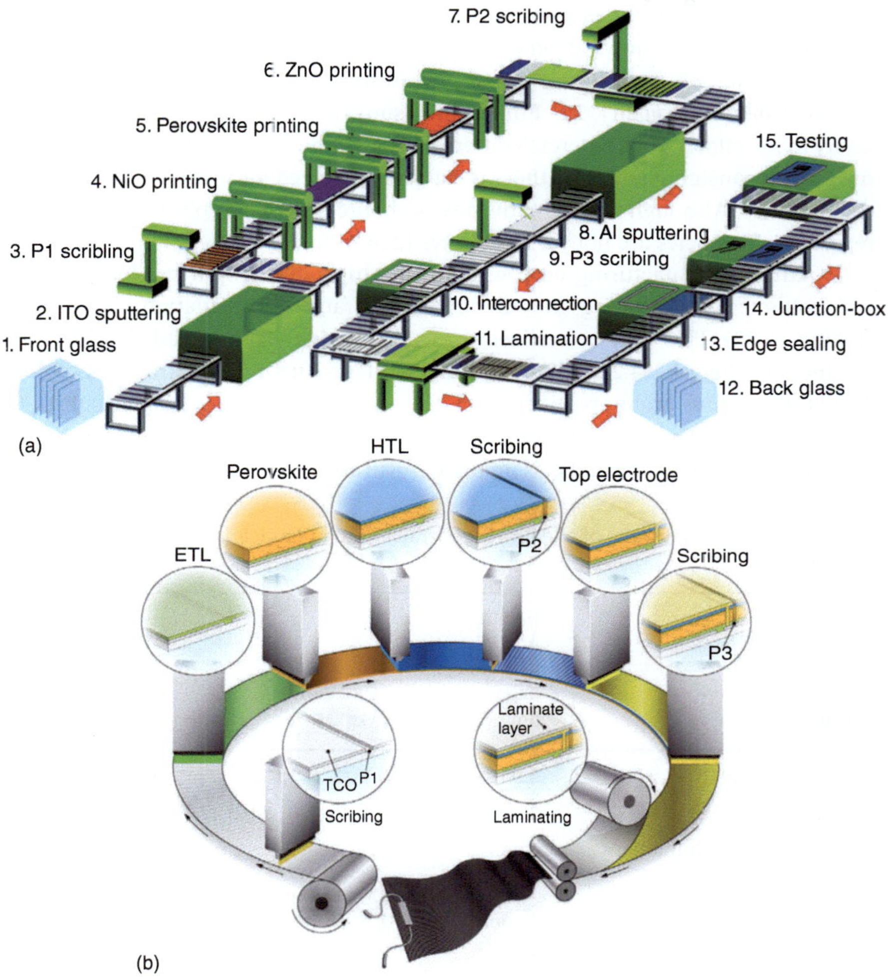

Figure 14.22 Process flow of high throughput S2S (a) and R2R (b) production of perovskite modules. Source: Song et al. [30].

A promising way of accelerating materials development along the lines pointed out above may be the employment of automation and machine learning [223]. Under the aspect of cost reduction, mainly eliminating or replacing of expensive materials for HTLs and partially also for TCOs, without compromising lifetime and efficiency, is the focus of future research. Another cost aspect concerns the development of R2R compatible materials, i.e. materials appropriate for low-temperature processes, including the deposition of transparent electrodes and electron extracting layers. Fabrication on PET substrates excludes sintering temperatures above 140 °C. For polyimide, the acceptable temperature range extends to about 250 °C. An alternative to high-temperature annealing may be the application of metal oxide nanoparticle layers, such as SnO_2 [184]. Another important aspect of R2R processes is developing low-cost, flexible substrates with perovskite compatible transparent electrodes, e.g. introducing Ag diffusion barriers [224]. A further challenge is identifying affordable encapsulation materials or developing in-line deposition of diffusion barriers against water and oxygen [225]. Finally, the issue of lead introduces a factor of uncertainty in the efforts of commercialization. Although the amount of lead in perovskite modules is considered minute, there is certainly the need for further ecotoxicity/life cycle studies. Although there is active research into potential replacements for lead, resulting efficiencies are not yet impressive [226–228].

To reduce manufacturing cost, establishing high throughput production without bottlenecks, including module patterning and encapsulation, is essential (Figure 14.22). This may be realized through the highly automated sheet-to-sheet production (Figure 14.22a) [30] or roll-to-roll production (Figure 14.22b) [26]. The latter ensures higher productivity at less capex but requires flexible substrates and electrodes, narrowing the range of acceptable materials. Avoiding hazardous materials, including solvents, is mandatory (Table 14.3) [26]. Vapor deposition

Table 14.3 Hazardousness of chemicals generally used in fabricating perovskite solar cells.

Reagent	Health Administration permissible exposure limit (PEL)	LD50 dermal ($g\,kg^{-1}$)	Health	Fire	Reactivity
Lead(II) iodide	$0.05\,mg\,m^{-3}$	0.5	2	0	0
Tin(II) iodide	$2\,mg\,m^{-3}$	0.1	2	0	0
1-Methyl-2-pyrrolidinone	1 ppm	8	2	2	0
Methylamine	5 ppm	NA	2	4	0
N,*N*-dimethylformamide	10 ppm	1.5	2	2	0
Chlorobenzene	10 ppm	NA	2	3	0
Acetonitrile	40 ppm	>2	2	3	0
Toluene	200 ppm	12.2	2	3	0
Isopropyl alcohol	400 ppm	12.8	2	3	0
Diethyl ether	400 ppm	14.2	2	4	0
γ-Butyrolactone	None	>5	2	1	0
Methylammonium bromide	None	NA	2	0	0
Methylammonium chloride	None	NA	2	1	0
MAI	None	NA	1	0	0
DMSO	None	>5	0	2	0

Notes: NA, not available; PEL, Occupational Safety and Health Administration (OSHA) permissible exposure limit. LD50 values (the median lethal dose: the amount of an ingested substance that kills 50% of a test sample). Health, fire, and reactivity ratings (0, uncritical; 4,highly dangerous) from the National Fire Protection.
Source: Kim et al. [26].

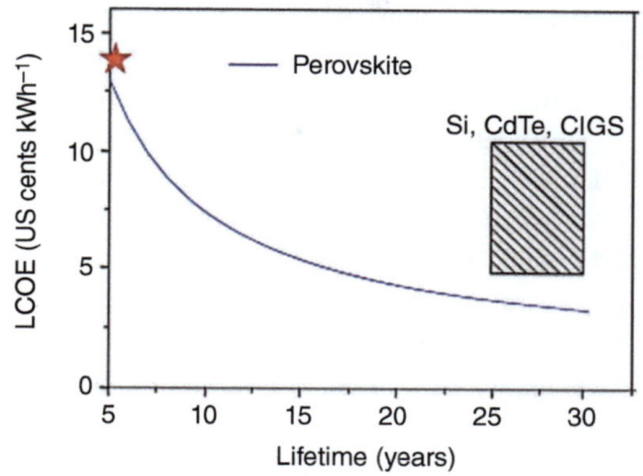

Figure 14.23 Effect of lifetime on LCOE. The star designates the status of PSCs as of 2018. The shaded area gives the situation for SiPV and the most common thin film technologies. Source: Modified from Meng et al. [24].

steps should be avoided because they are costly and reduce throughput. Employing efficient module layouts, optimizing sub-cell separation, and module interconnection procedures for perovskite modules are also required for scaling up perovskite PV. Finally, appropriate in-line characterization has to be developed to establish efficient quality control and maximize the production yield.

Even though hero lifetimes have been reported for small-area cells (Table 14.2), the lifetime of perovskite PV still poses the biggest question mark for the commercialization of this technology. Due to its significant effect on LCOE (Figure 14.23), the lack of a broader data basis on lifetimes under operational conditions, and the uncertainty about how well the devices showing hero lifetimes can be upscaled to large areas [24].

In conclusion, R2R processed modules with efficiencies exceeding 15% and lifetimes of more than 20 years seem to be a realistic option in the near future. However, there is still ample room for developing appropriate material systems, efficient module layouts, and robust processes with wide processing windows.

References

1 Bruening, K., Dou, B., Simonaitis, J. et al. (2018). *Joule* 2: 2464.
2 Li, Z., Klein, T.R., Kim, D.H. et al. (2018). *Nat. Rev. Mater.* 3: 1.
3 Cai, M., Wu, Y., Chen, H. et al. (2017). *Adv. Sci.* 4: 1600269. https://doi.org/10.1002/advs.201600269.
4 Qiu, L., He, S., Ono, L.K. et al. (2019). *ACS Energy Lett.* 4: 2147.
5 Kojima, A., Teshima, K., Shirai, Y., and Miyasaka, T. (2009). *J. Am. Chem. Soc.* 131: 6050.
6 Kim, H.S., Lee, C.R., Im, J.H. et al. (2012). *Sci. Rep.* 2: 1.
7 Lee, M.M., Teuscher, J., Miyasaka, T. et al. (2012). Efficient hybrid solar cells based on meso-superstructured organometal halide perovskites. *Science* 338 (6107): 643–647.
8 Noh, J.H., Im, S.H., Heo, J.H. et al. (2013). *Nano Lett.* 13: 1764.
9 Burschka, J., Pellet, N., Moon, S.J. et al. (2013). *Nature* 499: 316.

10 Liu, M., Johnston, M.B., and Snaith, H.J. (2013). *Nature* 501: 395.

11 Jeon, N.J., Noh, J.H., Kim, Y.C. et al. (2014). Solvent engineering for high-performance inorganic–organic hybrid perovskite solar cells. *Nat. Mater.* 13 (9): 897.

12 Jeon, N.J., Noh, J.H., Yang, W.S. et al. (2015). Compositional engineering of perovskite materials for high-performance solar cells. *Nature* 517 (7535): 476.

13 Yang, W.S., Noh, J.H., Jeon, N.J. et al. (2015). High-performance photovoltaic perovskite layers fabricated through intramolecular exchange. *Science* 348 (6240): 1234–1237.

14 Bi, D., Yi, C., Luo, J. et al. (2016). *Nat. Energy* 1: 1.

15 Yang, W.S., Park, B.W., Jung, E.H. et al. (2017). *Science* 356: 1376.

16 Jung, E.H., Jeon, N.J., Park, E.Y. et al. (2019). Efficient, stable and scalable perovskite solar cells using poly (3-hexylthiophene). *Nature* 567 (7749): 511.

17 Jiang, Q., Zhao, Y., Zhang, X. et al. (2019). *Nat. Photonics* 13: 460.

18 Kim, M., Kim, G.H., Lee, T.K. et al. (2019). *Joule* 3: 2179.

19 Min, H., Kim, M., Lee, S.U. et al. (2019). *Science* 366: 749.

20 Chang, N.L., Yi Ho-Baillie, A.W., Basore, P.A. et al. (2017). *Prog. Photovolt. Res. Appl.* 25: 390.

21 Needleman, D.B., Poindexter, J.R., Kurchin, R.C. et al. (2016). *Energy Environ. Sci.* 9: 2122.

22 Kost, C., Schlegl, T., Thomsen, J. et al. (2018). *Levelized Cost of Electricity Renewable Energy Technologies*. Fraunhofer ISE.

23 Di Giacomo, F., Fakharuddin, A., Jose, R., and Brown, T.M. (2016). *Energy Environ. Sci.* 9: 3007.

24 Meng, L., You, J., and Yang, Y. (2018). *Nat. Commun.* 9: 1.

25 Hou, Y., Du, X., Scheiner, S. et al. (2017). *Science (80-.)* 358: 1192.

26 Kim, D.H., Whitaker, J.B., Li, Z. et al. (2018). *Joule* 2: 1437.

27 Espinosa, N., Serrano-Luján, L., Urbina, A., and Krebs, F.C. (2015). *Sol. Energy Mater. Sol. Cells* 137: 303.

28 Li, Z., Zhao, Y., Wang, X. et al. (2018). *Joule* 2: 1559.

29 Ibn-Mohammed, T., Koh, S.C.L., Reaney, I.M. et al. (2017). *Renew. Sust. Energ. Rev.* 80: 1321.

30 Song, Z., McElvany, C.L., Phillips, A.B. et al. (2017). *Energy Environ. Sci.* 10: 1297.

31 Yang, M., Li, Z., Reese, M.O. et al. (2017). *Nat. Energy* 2: 1.

32 Huang, L., Li, C., Sun, X. et al. (2017). *Org. Electron.* 40: 13.

33 Chen, W., Wu, Y., Yue, Y. et al. (2015). *Science (80-.)* 350: 944.

34 Leyden, M.R., Jiang, Y., and Qi, Y. (2016). *J. Mater. Chem. A* 4: 13125.

35 Wu, Y., Yang, X., Chen, W. et al. (2016). *Nat. Energy* 1: 1.

36 Lee, J., Kim, J., Lee, C.L. et al. (2017). *Adv. Energy Mater.*: 7, doi https://doi.org/10.1002/aenm.201700226.

37 Tan, H., Jain, A., Voznyy, O. et al. (2017). *Science (80-.)* 355: 722.

38 Wu, Y., Xie, F., Chen, H. et al. (2017). *Adv. Mater.* 29: 1.

39 Jiang, Q., Chu, Z., Wang, P. et al. (2017). *Adv. Mater.* 29: 1.

40 Chiang, C.H., Nazeeruddin, M.K., Grätzel, M., and Wu, C.G. (2017). *Energy Environ. Sci.* 10: 808.

41 Zhao, P., Kim, B.J., Ren, X. et al. (2018). *Adv. Mater.* 30: 1.

42 Lu, J., Jiang, L., Li, W. et al. (2017). *Adv. Energy Mater.* 7: 1.

43 Chang, C.Y., Tsai, B.C., Hsiao, Y.C. et al. (2016). *Phys. Chem. Chem. Phys.* 18: 31836.

44 Li, J., Munir, R., Fan, Y. et al. (2018). *Joule* 2: 1313.

45 Zheng, X., Deng, Y., Chen, B. et al. (2018). *Adv. Mater.* 30: 1.

46 Guo, F., Qiu, S., Hu, J. et al. (2019). *Adv. Sci.*: 6, doi https://doi.org/10.1002/advs.201901067.

47 Ye, F., Chen, H., Xie, F. et al. (2016). *Energy Environ. Sci.* 9: 2295.

48 Chen, H., Ye, F., Tang, W. et al. (2017). *Nature* 550: 92.

49 Ye, F., Tang, W., Xie, F. et al. (2017). *Adv. Mater.* 29: 1.

50 Yang, M., Zhou, Y., Zeng, Y. et al. (2015). *Adv. Mater.* 27: 6363.

51 Huang, H., Shi, J., Zhu, L. et al. (2016). *Nano Energy* 27: 352.

52 Chang, C.Y., Huang, W.K., Chang, Y.C. et al. (2015). *J. Mater. Chem. A* 4: 640.

53 Li, X., Bi, D., Yi, C. et al. (2016). *Science (80-.)* 353: 58.

54 Gao, L.L., Li, C.X., Li, C.J., and Yang, G.J. (2017). *J. Mater. Chem. A* 5: 1548.

55 Ding, J., Han, Q., Ge, Q.Q. et al. (2019). *Joule* 3: 402.

56 Matteocci, F., Razza, S., Di Giacomo, F. et al. (2014). *Phys. Chem. Chem. Phys.* 16: 3918.

57 Seo, J., Park, S., Chan Kim, Y. et al. (2014). *Energy Environ. Sci.* 7: 2642.

58 Fabio Matteocci, A.L.P., Cinà, L., Di Giacomo, F. et al. (2016). *Prog. Photovolt. Res. Appl.* 24: 436.

59 Di Giacomo, F., Zardetto, V., D'Epifanio, A. et al. (2015). *Adv. Energy Mater.* 5: 1.

60 Heo, J.H., Han, H.J., Kim, D. et al. (2015). *Energy Environ. Sci.* 8: 1602.

61 Fakharuddin, A., Di Giacomo, F., Palma, A.L. et al. (2015). Vertical TiO_2 nanorods as a medium for stable and high-efficiency perovskite solar modules. *ACS Nano* 9 (8): 8420–8429.

62 Ye, T., Lim, S.L., Li, X. et al. (2018). *Sol. Energy Mater. Sol. Cells* 175: 111.

63 Fu, G., Hou, L., Wang, Y. et al. (2017). *Sol. Energy Mater. Sol. Cells* 165: 36.

64 Chiang, C.H., Lin, J.W., and Wu, C.G. (2016). *J. Mater. Chem. A* 4: 13525.

65 Agresti, A., Pescetelli, S., Palma, A.L. et al. (2019). *ACS Energy Lett.* 4: 1862.

66 Razza, S., Di Giacomo, F., Matteocci, F. et al. (2015). *J. Power Sources* 277: 286.

67 Bu, T., Shi, S., Li, J. et al. (2018). *ACS Appl. Mater. Interfaces* 10: 14922.

68 McGorty, R., Fung, J., Kaz, D., and Manoharan, V.N. (2010). *Mater. Today* 13: 34.

69 Deng, Y., Zheng, X., Bai, Y. et al. (2018). *Nat. Energy* 3: 560.

70 Tang, S., Deng, Y., Zheng, X. et al. (2017). *Adv. Energy Mater.* 7: 1.

71 Lee, D.K., Jeong, D.N., Ahn, T.K., and Park, N.G. (2019). *ACS Energy Lett.* 4: 2393.

72 Vak, D., Hwang, K., Faulks, A. et al. (2015). *Adv. Energy Mater.* 5: 1.

73 Di Giacomo, F., Shanmugam, S., Fledderus, H. et al. (2018). *Sol. Energy Mater. Sol. Cells* 181: 53.

74 Deng, Y., Van Brackle, C.H., Dai, X. et al. (2019). *Sci. Adv.* 5: eaax7537.
75 Green, M.A., Hishikawa, Y., Dunlop, E.D. et al. (2018). *Prog. Photovolt. Res. Appl.* 26: 427.
76 Gotanda, T., Oooka, H., Mori, S. et al. (2019). *J. Power Sources* 430: 145.
77 Barrows, A.T., Pearson, A.J., Kwak, C.K. et al. (2014). *Energy Environ. Sci.* 7: 2944.
78 Priyadarshi, A., Haur, L.J., Murray, P. et al. (2016). *Energy Environ. Sci.* 9: 3687.
79 Hu, Y., Si, S., Mei, A. et al. (2017). *Solar RRL* 1: 1600019.
80 Grancini, G., Roldán-Carmona, C., Zimmermann, I. et al. (2017). *Nat. Commun.* 8: 1.
81 Tait, J.G., Manghooli, S., Qiu, W. et al. (2016). *J. Mater. Chem. A* 4: 3792.
82 Agresti, A., Pescetelli, S., Palma, A.L. et al. (2017). *ACS Energy Lett.* 2: 279.
83 Bu, T., Liu, X., Zhou, Y. et al. (2017). *Energy Environ. Sci.* 10: 2509.
84 Kim, J., Yun, J.S., Cho, Y. et al. (2017). *ACS Energy Lett.* 2: 1978.
85 Cai, L., Liang, L., Wu, J. et al. (2017). *J. Semicond.*: 38. https://doi.org/10.1088/1674-4926/38/1/014006.
86 Qiu, L., He, S., Jiang, Y. et al. (2019). *J. Mater. Chem. A* 7: 6920.
87 Luo, L., Zhang, Y., Chai, N. et al. (2018). *J. Mater. Chem. A* 6: 21143.
88 Jiang, Y., Leyden, M.R., Qiu, L. et al. (2018). *Adv. Funct. Mater.*: 28. https://doi.org/10.1002/adfm.201703835.
89 Jiang, Y., Remeika, M., Hu, Z. et al. (2019). *Adv. Energy Mater.* 9: 1.
90 NREL Best research-cell efficiency chart, https://www.nrel.gov/pv/cell-efficiency.html.
91 Lee, J.W., Na, S.I., and Kim, S.S. (2017). *J. Power Sources* 339: 33.
92 Liao, H.C., Guo, P., Hsu, C.P. et al. (2017). *Adv. Energy Mater.* 7: 1.
93 Bao, Z., Rogers, J.A., and Katz, H.E. (1999). *J. Mater. Chem.* 9: 1895.
94 Liu, S., Cao, K., Li, H. et al. (2017). *Sol. Energy* 144: 158.
95 Liu, L., Mei, A., Liu, T. et al. (2015). *J. Am. Chem. Soc.* 137: 1790.
96 Howard, I.A., Abzieher, T., Hossain, I.M. et al. (2019). *Adv. Mater.*: 31, doi https://doi.org/10.1002/adma.201806702.
97 Li, S.G., Jiang, K.J., Su, M.J. et al. (2015). *J. Mater. Chem. A* 3: 9092.
98 Li, P., Liang, C., Bao, B. et al. (2018). *Nano Energy* 46: 203.
99 Wei, Z., Chen, H., Yan, K., and Yang, S. (2014). *Angew. Chem.* 126: 1.
100 Abzieher, T., Moghadamzadeh, S., Schackmar, F. et al. (2019). *Adv. Energy Mater.* 9: 1.
101 Mathies, F., Abzieher, T., Hochstuhl, A. et al. (2016). *J. Mater. Chem. C* 4: 19207.
102 Liang, C., Li, P., Gu, H. et al. (2018). *Solar RRL* 2: 1700217.
103 Bag, M., Jiang, Z., Renna, L.A. et al. (2016). *Mater. Lett.* 164: 472.
104 Mathies, F., Eggers, H., Richards, B.S. et al. (2018). *ACS Appl. Energy Mater.* 1: 1834.
105 Huckaba, A.J., Lee, Y., Xia, R. et al. (2018). *Energ. Technol.* 6: 1.
106 Gheno, A., Huang, Y., Bouclé, J. et al. (2018). *Solar RRL* 2: 1800191.
107 Chang, W.C., Lan, D.H., Lee, K.M. et al. (2017). *ChemSusChem* 10: 1405.
108 Das, S., Yang, B., Gu, G. et al. (2015). *ACS Photonics* 2: 680.

109 Liang, Z., Zhang, S., Xu, X. et al. (2015). *RSC Adv.* 5: 60562.

110 Heo, J.H., Lee, M.H., Jang, M.H., and Im, S.H. (2016). *J. Mater. Chem. A* 4: 17636.

111 Mohamad, D.K., Griffin, J., Bracher, C. et al. (2016). *Adv. Energy Mater.* 6: 1.

112 Han, G.S., Kim, J., Bae, S. et al. (2019). *ACS Energy Lett.* 4: 1845.

113 NREL Champion photovoltaic module efficiency chart, https://www.nrel.gov/pv/module-efficiency.html.

114 Peters, I.M., Rodriguez Gallegos, C.D., Sofia, S.E., and Buonassisi, T. (2019). *Joule* 3: 2732.

115 Hussain, I., Tran, H.P., Jaksik, J. et al. (2018). *Emergent Mater.* 1: 133.

116 Mali, S.S. and Hong, C.K. (2016). *Nanoscale* 8: 10528.

117 Zheng, S., Liu, B.S., Wang, W.S. et al. (2015). *Adsorpt. Sci. Technol.* 33: 337.

118 Wang, K., Yang, D., Wu, C. et al. (2019). *Prog. Mater. Sci.* 106: 100580.

119 Liu, C., Cheng, Y., and Ge, Z. (2020). *Chem. Soc. Rev.* 49: 8.

120 Hu, H., Ren, Z., Fong, P.W.K. et al. (2019). *Adv. Funct. Mater.* 29: 1.

121 Dai, X., Deng, Y., Van Brackle, C.H., and Huang, J. (2019). *Int. J. Extrem. Manuf.* 1: 022004.

122 Nie, W., Tsai, H., Asadpour, R. et al. (2015). *Science* 347: 522.

123 Deng, Y., Peng, E., Shao, Y. et al. (2015). *Energy Environ. Sci.* 8: 1544.

124 Xiao, M., Huang, F., Huang, W. et al. (2014). *Angew. Chem. Int. Ed.* 53: 9898.

125 Zhou, Y., Yang, M., Wu, W. et al. (2015). *J. Mater. Chem. A* 3: 8178.

126 Huang, F., Dkhissi, Y., Huang, W. et al. (2014). *Nano Energy* 10: 10.

127 Cotella, G., Baker, J., Worsley, D. et al. (2017). *Sol. Energy Mater. Sol. Cells* 159: 362.

128 Ding, B., Gao, L., Liang, L. et al. (2016). *ACS Appl. Mater. Interfaces* 8: 20067.

129 Gao, L.L., Liang, L.S., Song, X.X. et al. (2016). *J. Mater. Chem. A* 4: 3704.

130 Ding, B., Li, Y., Huang, S.Y. et al. (2017). *J. Mater. Chem. A* 5: 6840.

131 Deng, Y., Dong, Q., Bi, C. et al. (2016). *Adv. Energy Mater.* 6: 1.

132 Nie, W., Tsai, H., Asadpour, R. et al. (2015). High-efficiency solution-processed perovskite solar cells with millimeter-scale grains. *Science* 347 (6221): 522–525.

133 Yin, M., Xie, F., Chen, H. et al. (2016). *J. Mater. Chem. A* 4: 8548.

134 Yu, Y., Yang, S., Lei, L. et al. (2017). *ACS Appl. Mater. Interfaces* 9: 3667.

135 Eperon, G.E., Eperon, G.E., Leijtens, T. et al. (2016). Perovskite-perovskite tandem photovoltaics with optimized band gaps. *Science* 354 (6314): 861–865.

136 Ye, J., Zhang, X., Zhu, L. et al. (2017). *Sustain. Energy Fuels* 1: 907.

137 Noel, N.K., Habisreutinger, S.N., Wenger, B. et al. (2019). Interfacial charge-transfer doping of metal halide perovskites for high performance photovoltaics. *Energy Environ. Sci.* 12: 3063–3073.

138 Dou, B., Whitaker, J.B., Bruening, K. et al. (2018). *ACS Energy Lett.* 3: 2558.

139 Zhou, Z., Wang, Z., Zhou, Y. et al. (2015). *Angew. Chem. Int. Ed.* 54: 9705.

140 Podolsky, B., Rosen, N., Clauser, J.F. et al. (2012). Efficient hybrid solar cells based on meso-superstructured organometal halide perovskites. *Science* 338 (6107): 643–647.

141 Zhao, Y. and Zhu, K. (2014). *J. Phys. Chem. C* 118: 9412.

142 Zuo, C. and Ding, L. (2014). *Nanoscale* 6: 9935.

143 Chen, Y., Zhao, Y., and Liang, Z. (2015). *Chem. Mater.* 27: 1448.
144 Krautscheid, H., Vielsack, F., Charactistics, E. et al. (1999). *J. Chem. Soc. Dalton Trans.* 3: 2731.
145 Guo, Y., Shoyama, K., Sato, W. et al. (2015). *J. Am. Chem. Soc.* 137: 15907.
146 Yantara, N., Yanan, F., Shi, C. et al. (2015). *Chem. Mater.* 27: 2309.
147 Yu, H., Wang, F., Xie, F. et al. (2014). *Adv. Funct. Mater.* 24: 7102.
148 Yang, Z., Chueh, C.C., Zuo, F. et al. (2015). *Adv. Energy Mater.* 5: 1.
149 Schmidt, T.M., Larsen-Olsen, T.T., Carlé, J.E. et al. (2015). *Adv. Energy Mater.* 5: 1.
150 Zhang, W., Saliba, M., Moore, D.T. et al. (2015). *Nat. Commun.*: 6, doi https://doi.org/10.1038/ncomms7142.
151 Moore, D.T., Sai, H., Tan, K.W. et al. (2015). *J. Am. Chem. Soc.* 137: 2350.
152 Li, G., Zhang, T., Guo, N. et al. (2016). *Angew. Chem.* 128: 13658.
153 Li, G., Zhang, T., and Zhao, Y. (2015). *J. Mater. Chem. A* 3: 19674.
154 Pan, J., Mu, C., Li, Q. et al. (2016). *Adv. Mater.* 28: 8309.
155 Heo, J.H., Song, D.H., and Im, S.H. (2014). *Adv. Mater.* 26: 8179.
156 Heo, J.H., Song, D.H., Han, H.J. et al. (2015). *Adv. Mater.* 27: 3424.
157 Nayak, P.K., Moore, D.T., Wenger, B. et al. (2016). *Nat. Commun.* 7: 1.
158 Lee, J.W., Kim, H.S., and Park, N.G. (2016). *Acc. Chem. Res.* 49: 311.
159 Liang, P.W., Liao, C.Y., Chueh, C.C. et al. (2014). *Adv. Mater.* 26: 3748.
160 Zhang, Y., Gao, P., Oveisi, E. et al. (2016). *J. Am. Chem. Soc.* 138: 14380.
161 Chen, Q., Zhou, H., Hong, Z. et al. (2014). *J. Am. Chem. Soc.* 136: 622.
162 Wang, Z.K., Gong, X., Li, M. et al. (2016). *ACS Nano* 10: 5479.
163 Gu, Z., Zuo, L., Larsen-Olsen, T.T. et al. (2015). *J. Mater. Chem. A* 3: 24254.
164 Li, B., Chen, Y., Liang, Z. et al. (2015). *RSC Adv.* 5: 94290.
165 Ding, Y., Yao, X., Zhang, X. et al. (2014). *J. Power Sources* 272: 351.
166 Bi, C., Wang, Q., Shao, Y. et al. (2015). *Nat. Commun.* 6: 1.
167 Troughton, J., Charbonneau, C., Carnie, M.J. et al. (2015). *J. Mater. Chem. A* 3: 9123.
168 Troughton, J., Carnie, M.J., Davies, M.L. et al. (2016). *J. Mater. Chem. A* 4: 3471.
169 Pool, V.L., Dou, B., Van Campen, D.G. et al. (2017). *Nat. Commun.* 8: 1.
170 Yang, M., Kim, D.H., Klein, T.R. et al. (2018). *ACS Energy Lett.* 3: 322.
171 Jena, A.K., Numata, Y., Ikegami, M., and Miyasaka, T. (2018). *J. Mater. Chem. A* 6: 2219.
172 Jena, A.K., Ikegami, M., and Miyasaka, T. (2017). *ACS Energy Lett.* 2: 1760.
173 Malinauskas, T., Saliba, M., Matsui, T. et al. (2016). *Energy Environ. Sci.* 9: 1681.
174 Scheideler, W.J., Rolston, N., Zhao, O. et al. (2019). *Adv. Energy Mater.* 9: 1.
175 Arora, N., Dar, M.I., Hinderhofer, A. et al. (2017). *Science (80-.)* 358: 768.
176 Huang, X., Wang, K., Yi, C. et al. (2016). *Adv. Energy Mater.* 6: 1.
177 Jung, J.W., Williams, S.T., and Jen, A.K.Y. (2014). *RSC Adv.* 4: 62971.
178 Liu, X., Tsai, K.W., Zhu, Z. et al. (2016). *Adv. Mater. Interfaces* 3: 1.
179 Qiu, L., Liu, Z., Ono, L.K. et al. (2019). *Adv. Funct. Mater.* 29: 1.
180 Yang, D., Yang, R., Ren, X. et al. (2016). *Adv. Mater.* 28: 5206.
181 Bi, C., Chen, B., Wei, H. et al. (2017). *Adv. Mater.* 29: 1.

182 Yang, D., Yang, R., Zhang, J. et al. (2015). *Energy Environ. Sci.* 8: 3208.

183 Sakamoto, K., Kuwae, H., Kobayashi, N. et al. (2018). *Sci. Rep.* 8: 3.

184 Il Park, J., Heo, J.H., Park, S.H. et al. (2017). *J. Power Sources* 341: 340.

185 Wu, Z., Li, P., Zhang, Y., and Zheng, Z. (2018). *Small Methods* 2: 1800031.

186 Troughton, J., Bryant, D., Wojciechowski, K. et al. (2015). *J. Mater. Chem. A* 3: 9141.

187 Heo, J.H., Shin, D.H., Lee, M.L. et al. (2018). *ACS Appl. Mater. Interfaces* 10: 31413.

188 Abdollahi Nejand, B., Nazari, P., Gharibzadeh, S. et al. (2017). *Chem. Commun.* 53: 747.

189 Qiu, L., Deng, J., Lu, X. et al. (2014). *Angew. Chem. Int. Ed.* 53: 10425.

190 Poorkazem, K., Liu, D., and Kelly, T.L. (2015). *J. Mater. Chem. A* 3: 9241.

191 Dai, X., Deng, Y., Van Brackle, C.H. et al. (2019). *Adv. Energy Mater.* 1903108: 1.

192 Zhang, F., Yang, X., Wang, H. et al. (2014). *ACS Appl. Mater. Interfaces* 6: 16140.

193 Shi, Z. and Jayatissa, A.H. (2018). *Materials (Basel)* 11, doi https://doi.org/10.3390/ma11050729.

194 Meng, X., Cui, X., Rager, M. et al. (2018). *Nano Energy* 52: 123.

195 Christians, J.A., Miranda Herrera, P.A., and Kamat, P.V. (2015). *J. Am. Chem. Soc.* 137: 1530.

196 Leguy, A.M.A., Hu, Y., Campoy-Quiles, M. et al. (2015). *Chem. Mater.* 27: 3397.

197 Zhang, F., Yang, X., Cheng, M. et al. (2015). *J. Mater. Chem. A* 3: 24272.

198 De Rossi, F., Baker, J.A., Beynon, D. et al. (2018). *Adv. Mater. Technol.* 3: 1.

199 Lucera, L., Kubis, P., Fecher, F.W. et al. (2015). *Energ. Technol.* 3: 373.

200 Pisoni, S., Fu, F., Widmer, R. et al. (2018). *Nano Energy* 49: 300.

201 Yang, Z., Zhang, S., Li, L., and Chen, W. (2017). *J. Mater.* 3: 231.

202 Moon, S.J., Yum, J.H., Lofgren, L. et al. (2015). *IEEE J. Photovoltaics* 5: 1087.

203 Ono, L.K., Qi, Y., and (Frank) Liu, S. (2018). *Joule* 2: 1961.

204 Dharmadasa, I.M., Rahaq, Y., and Alam, A.E. (2019). *J. Mater. Sci. Mater. Electron.* 30: 12851.

205 Zhang, M., Yun, J.S., Ma, Q. et al. (2017). *ACS Energy Lett.* 2: 438.

206 Tsai, H., Nie, W., Blancon, J.C. et al. (2016). *Nature* 536: 312.

207 Wang, Z., Lin, Q., Chmiel, F.P. et al. (2017). *Nat. Energy* 2: 1.

208 Mei, A., Li, X., Liu, L. et al. (2014). *Science* 345: 295.

209 Kim, H.S., Seo, J.Y., and Park, N.G. (2016). *ChemSusChem* 9: 2528.

210 Zhang, W., Xiong, J., Li, J., and Daoud, W.A. (2019). *J. Mater. Chem. A* 7: 9486.

211 Li, C., Wang, A., Xie, L. et al. (2019). *J. Mater. Chem. A* 7: 24150.

212 Jena, A.K., Kulkarni, A., and Miyasaka, T. (2019). *Chem. Rev.* 119: 3036.

213 Hou, Y., Scheiner, S., Tang, X. et al. (2017). *Adv. Mater. Interfaces* 4: 1.

214 Wang, S., Liu, B., Zhu, Y. et al. (2018). *Sol. Energy* 169: 335.

215 Li, W., Zhang, W., Van Reenen, S. et al. (2016). *Energy Environ. Sci.* 9: 490.

216 Roose, B., Baena, J.P.C., Gödel, K.C. et al. (2016). *Nano Energy* 30: 517.

217 Shin, S.S., Yeom, E.J., Yang, W.S. et al. (2017). *Science (80-.)* 356: 167.

218 Schloemer, T.H., Christians, J.A., Luther, J.M., and Sellinger, A. (2019). *Chem. Sci.* 10: 1904.

219 Rakstys, K., Igci, C., and Nazeeruddin, M.K. (2019). *Chem. Sci.* 10: 6748.

220 Shao, S. and Loi, M.A. (2020). The role of the interfaces in perovskite solar cells. *Adv. Mater. Interfaces* 7: 1901469.

221 Guerrero, A., Bou, A., Matt, G. et al. (2018). *Adv. Energy Mater.*: 8, doi https://doi.org/10.1002/aenm.201703376.

222 Fagiolari, L. and Bella, F. (2019). Carbon-based materials for stable, cheaper and large-scale processable perovskite solar cells. *Energy Environ. Sci.* 12: 3437–3472.

223 Correa-Baena, J.P., Hippalgaonkar, K., van Duren, J. et al. (2018). *Joule* 2: 1410.

224 Jain, S., Shah, J., Negi, N.S. et al. (2019). *Int. J. Energy Res.* 43: 4743.

225 Channa, I.A., Distler, A., Zaiser, M. et al. (2019). Thin film encapsulation of organic solar cells by direct deposition of polysilazanes from solution. *Adv. Energy Mater.* 9: 1900598.

226 Steirer, K.X., Schulz, P., Teeter, G. et al. (2016). *ACS Energy Lett.* 1: 360.

227 Alsari, M., Bikondoa, O., Bishop, J. et al. (2018). *Energy Environ. Sci.* 11: 383.

228 Davies, P.K., Wu, H., Borisevich, A.Y. et al. (2008). *Annu. Rev. Mater. Res.* 38: 369.

15

Scalable Architectures and Fabrication Processes of Perovskite Solar Cell Technology

Ghufran S. Hashmi

Microelectronics Research Unit, Faculty of Information Technology & Electrical Engineering, University of Oulu, Erkki-Koiso Kanttilankatu 3, Linnanmaa, Finland

15.1 Background

Since the first report [1], perovskite solar cells (PSCs) have received tremendous attention worldwide due to not only their inheritance characteristics such as longer diffusion lengths for the electrons and holes [2], high optical absorption coefficient [3], and possibility of their energy band-gap tuning [4] through the organometallic lead halide light harvesters but also their remarkable enhancements in their solar-to-electrical energy conversions, which has now exceeded over 25% [5].

With these characteristics, numerous device designs of lab-sized PSCs have been demonstrated with several facile fabrication methods. Here, spin coating remained the foremost choice as materials deposition technique for the research labs globally [6–8], which cannot be realized for converting lab-sized PSCs into large-area (>400–800 cm^2) solar modules, thus prompting to develop or adopt alternative and industrially oriented scalable materials deposition methods to achieve large-area perovskite solar module (PSM) production with high reproducibility. Moreover, PSM production with reliable process control has also been considered as one of the key factors for the success of any nanotechnology-based advanced energy and electronic or electrical system [6, 9, 10].

In contrast to earlier developed solar cell technologies such as dye-sensitized solar cells (DSSCs) or organic solar cells (OSCs), scaling up of PSC technology also directly influences solar-to-electrical conversion efficiencies by affecting over individual photovoltaic parameter of large-area PSMs [6]. For instance, achieving high-quality light-absorbing perovskite layer with large grain boundaries over far large area compared with what has been obtained for lab-sized devices has been considered as one of such key factors proving to have a significant influence in achieving higher conversion efficiencies for large-area PSMs [11–13].

Moreover, homogeneous depositions of the electron transporting layer (ETL) have also been thoroughly investigated to ensure the high photovoltaic performance during the production of large-area PSMs [14]. On the other hand, scalable alternatives

Perovskite Solar Cells: Materials, Processes, and Devices, First Edition.
Edited by Shahzada Ahmad, Samrana Kazim, and Michael Grätzel.

to the traditional expensive hole transporting material (HTM) (i.e. spiro-OMeTAD) have also been demonstrated, which showed similar characteristics in achieving impressive solar-to-electrical conversion efficiencies when deployed in large-area PSMs [15].

In addition, other effective solutions such as laser patterning or mechanical scribing have also been developed and demonstrated for PSC technology to minimize dead areas within the active layers for the improvement of the photovoltaic performances of the fabricated modules [16, 17]. On the other hand, careful selection of transparent conducting oxide layers with appropriate sheet resistance (R_{SH}) was also found to keep immense influence on the photovoltaic performance since it directly affects both the light transmittance and the overall cell resistance of the fabricated device geometries [6, 18].

Above all, the final module encapsulation procedures have also been discussed, where numerous encapsulants, laminates, and curable epoxies for preventing the intrusion of moisture and humidity are frequently demonstrated to avoid the possible degradation in the photovoltaic performance [19, 20].

As a result, tremendous progress being made for producing scalable and large-area PSMs both on the rigid glass and flexible polymer substrates. Several research groups and industrial player have now claimed highly stable and long-term photovoltaic performance stability under several simulated and natural climatic conditions, which have been recently standardized for PSC technology through a consensus statement [21].

This chapter systematically covers the scalable device designs, which have been used to produce large-area perovskites based solar modules both on rigid glass-based substrates and flexible polymer-based substrates.

Moreover, the advantages and limitations in adopting possible scalable device designs are also discussed in details, which gives an overview for selecting appropriate device geometries for producing scalable PSCs and modules for specific applications.

Also, long-term device stability reports for these large-area solar modules are gathered, which also provides an overview for the operational limits to opt appropriate device design of a PSC to work under specific conditions.

Finally, the prospects and progress in producing robust PSMs with high durability is discussed in predicting the concrete outcomes regarding utilization of this low-cost-claimed scalable photovoltaic technology.

15.1.1 Configurations and Device Architectures of Perovskite Solar Cells

For lab-sized devices (0.1–1 cm^2), PSCs have been produced by many labs with three fundamental configurations as shown in Figure 15.1 [22]. Among them, the first configuration, also known as n–i–p device structure with a mesoporous layer, remained the best reported device design during the PSC research advancement [23–25].

At the same time, the concept of producing PSC without integrating mesoporous ETL (Figure 15.1b) was also emerged [26], which is now known as planar

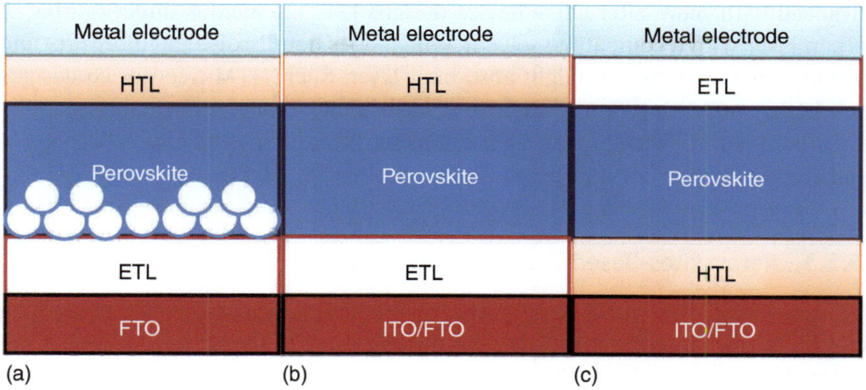

Figure 15.1 Schematic diagrams of different PSC structures. (a) Mesoporous structure, (b) planar n–i–p structure, and (c) planar p–i–n structure. Source: Ali et al. [22].

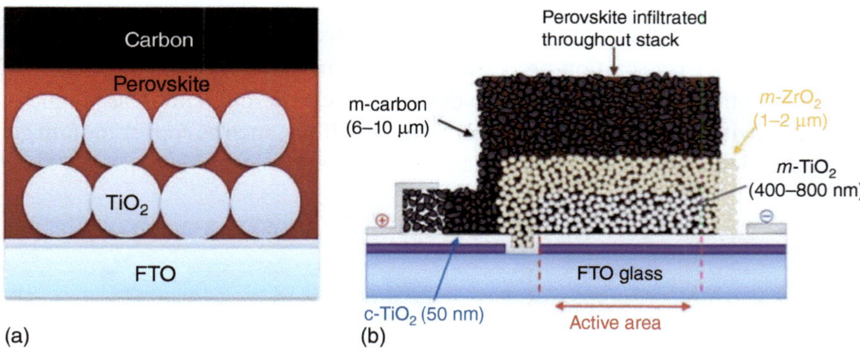

Figure 15.2 Schematic diagrams of HTM-free PSC structures. (a) Carbon layer deposited over perovskite layer. Source: Chen et al. [28]. (b) Triple mesoscopic HTM-free PSCs. Source: Baker et al. [29b].

n–i–p-based PSCs followed by an inverted planar structure, which is also called p–i–n configuration as shown in Figure 15.1 [8, 27, 28].

15.1.2 HTM-Free Device Configurations for Perovskite Solar Cells

On the contrary to the traditional device designs of a PSC, some alternative device designs have also been demonstrated by removing traditional and costly HTMs along with metal electrodes such as HTM-free PSCs shown in Figure 15.2.

Owing to their inherent characteristics such as water resistance and ion migration protection, along with low cost and durability under high temperatures, carbon-based HTM-free PSCs received considerations worldwide for realizing low-cost and stable PSC architectures [28, 29].

In this regard, very earlier device designs of carbon-based PSCs were evolved from monolithic dye-sensitized solar cells (m-DSSCs) [30] and were further advanced with more alternative device designs [31, 32] such as direct coating of carbon layer with a compatible solvent, which does not dissolve the moisture- and solvent-sensitive perovskite light-absorbing layer. Such HTM-free device designs look highly promising due to their long-term photovoltaic performance stability under both natural and simulated environmental conditions and give great hope in realizing robust and low-cost-based PSC technology [8, 29, 31].

15.1.3 Perovskites-Based Tandem Solar Cells

The curiosity of improving solar-to-electrical energy efficiencies of any emerging photovoltaic technology stimulates the development of the structures and novel device designs beyond the state-of-the-art other than single-junction-based photovoltaic devices, which restrain the possibility of achieving high photovoltaic performance with several limiting factors such as limited light absorption, process engineering, and losses due to recombination of carriers at interfaces between active layers [33, 34].

The notable optoelectronic characteristics (as discussed in the earlier sections) combined with easy solution-processable or vacuum deposition methods for PSC fabrication offer striking possibility to develop novel heterojunction-based tandem solar cells (with either existing silicon or other thin films, such as copper indium gallium selenide (CIGS) based photovoltaic solar cell technologies) (Figure 15.3) [35],

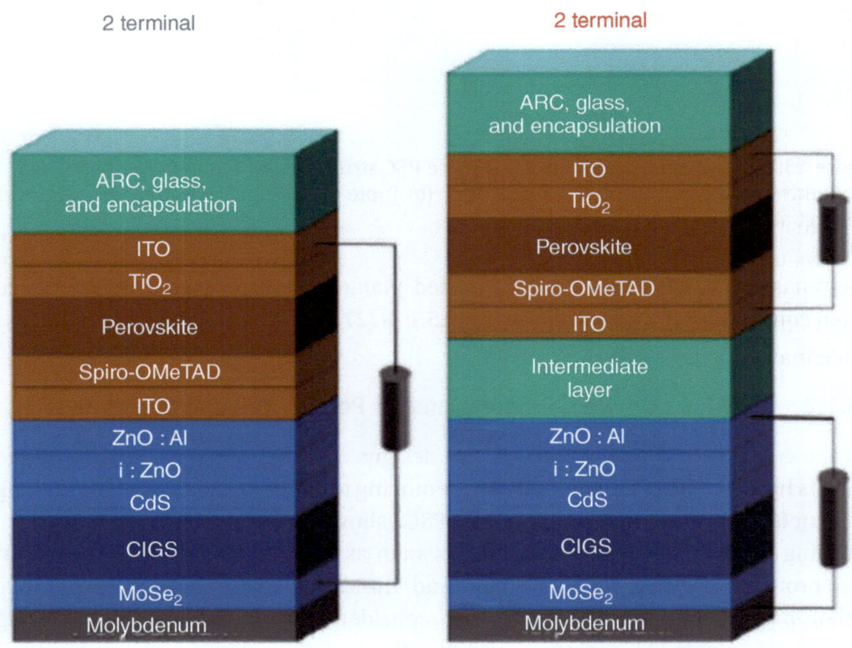

Figure 15.3 An example of perovskite/CIGS tandem configurations. Source: Langenhorst et al. [35].

which keep enormous potential for surpassing solar-to-electrical energy conversion efficiencies beyond 30% [36–38]

These tandem configurations with PSCs have been mainly classified as two-terminal (2T) tandem and four-terminal (4T) tandem solar cells [35, 37], which have also rapidly progressed and are expected to be commercially available in the forthcoming years.

15.2 Scalable Device Designs of Perovskite Solar Cells

The versatility in combination of materials for PSCs has greatly motivated research labs worldwide to authenticate the potential of this emerging solar cell technology via scheming unique device architectures for PSCs, which have been demonstrated both on rigid glass and flexible substrates since the advent of this photovoltaic technology [39, 40].

These efforts have resulted in not only higher solar-to-electrical conversion efficiencies [41] but also impressive long-term photovoltaic performance stability under both natural and simulated climatic conditions [19, 42, 43]. Although showing potential in many research reports for the lab-sized solar cells (~1 cm^2), a few out of many device designs showed potential for upscaling and have been demonstrated by several research labs along with commercial players in recent years. In the following, some of these device designs are briefly reviewed.

15.2.1 Scalable n–i–p Configuration-Based Perovskite Solar Modules

Showing potential for lab-sized PSCs being made with both mesoporous and planar n–i–p device structures, efforts were also given for scaling up these two fundamental device architectures. Figure 15.4 illustrates the scalable device design strategy, which has been opted by many research groups to produce both mesoporous and planar n–i–p device structure-based large-area solar modules. Opting such design allows to produce serially connected individual cell fingers with minimum charge transport distances and minimum parasitic resistance, which restrict to produce high-efficiency-based single solar cell on large substrate [11]. Moreover, several scalable materials deposition processes such as blade coating or slot-die coating can be

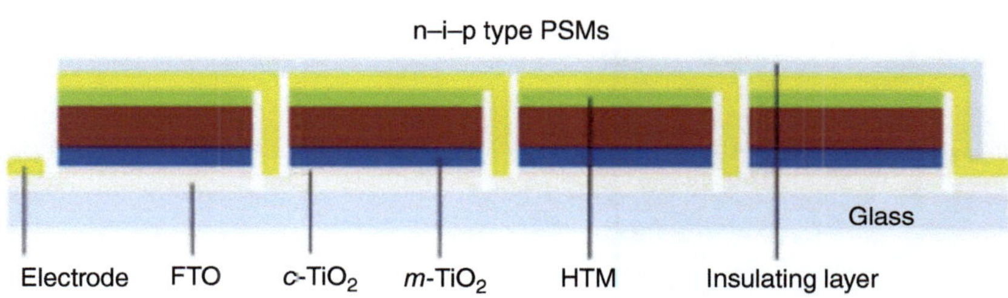

Figure 15.4 Schematic of serially connect n–i–p-type perovskite module. Source: Wang et al. [15].

opted to produce these scalable device configurations. In addition to that, selective laser patterning to produce modules with minimum dead area can also be achieved after coating the active layers on large substrates [44].

Motivated by such characteristics, mesoporous n–i–p device structure-based large-area PSMs on rigid glass substrates have been initially demonstrated [44–46]. Despite opting to traditional blade coating-based depositions for perovskite and HTM along with spray pyrolysis-based scalable methods, some other interesting approaches for achieving pinhole-free and high-quality perovskite layers are also demonstrated such as dropping minute amount of perovskite precursor ink in the center of the active layers followed by pressure application step, which interestingly resulted in defect-free and large crystals grain-based perovskite light-absorbing layer [45]. As a result, the champion mesoporous n–i–p perovskite confirmed very impressive (12.1%) conversion efficiency under full sunlight illumination over an active area of 36.1 cm^2.

On the other hand, mesoporous n–i–p modules were also produced with a two-step process, where the optimization of PbI_2 layer was performed by controlling its crystallization during the blade coating process through an airflow [46]. After that the devices were immersed into methylammonium iodide (MAI) solution to obtain perovskite layer. Conversion efficiencies of 10.1% and 4.3% were achieved for active areas of 10.1 and 100 cm^2, respectively.

Although the integration of mesoporous TiO_2 scaffold layer for the PSC device structure has been considered as one of the key factors for in both achieving controlled perovskite crystal growth and consequently achieving high conversion efficiencies. Nevertheless, reaching mechanically durable and high-performance mesoporous TiO_2 requires a high-temperature sintering method as the diluted paste contains organic binders such as ethyl cellulose and high-boiling solvent such as terpineol, which needs to be removed off through a high-temperature sintering process. This diluted TiO_2 paste does not only hinder the solution processing, which seems possible for other active layers, but also restrict the use of PSC technology into variety of applications and configurations such as flexible PSCs [39] (Figure 15.5).

Figure 15.5 Fabricated planar module with solution-processing schemes. Source: Calabrò [47].

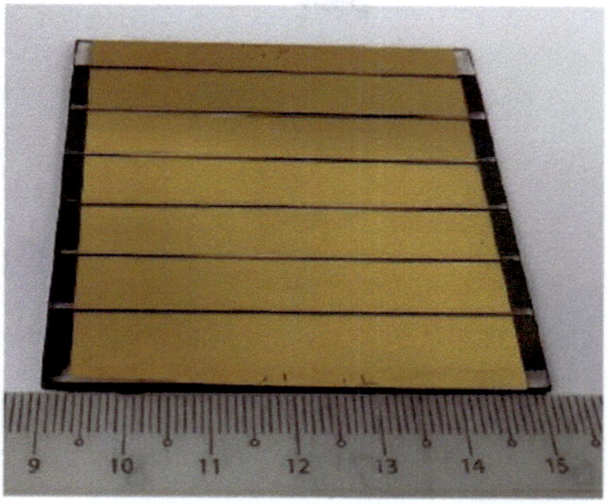

Figure 15.6 Image of an active-area PSM of 21 cm^2 consisting of six series-connected sub-cells. Source: Li et al. [48].

As a result of knowing this limitation, planar n–i–p configuration-based large-area perovskite modules received considerable attention where each active layer can be stacked through solution-processed methods. One of such demonstration has been reported for fabrication of large-area planar n–i–p module by utilizing low-temperature and solution-processable SnO_2 precursor-based ETL as an alternative to TiO_2 ETL followed by all other active layers deposited by spin coating method. After that, the fingers of individual cells were obtained by performing laser patterning on the active layers. A conversion efficiency of 13.1% for an active area of 13.8 cm^2 and a slightly lower conversion efficiency of 11.9% for larger aperture area of 15.2 cm^2 were achieved (Figure 15.5) [47]. More impressively, the fabricated modules maintained the long-term photovoltaic performance and showed high stability when stored under shelf life conditions without any encapsulation (in dark, in air at constant RH of 20%) for 1300 hours (Figure 15.6).

On the contrary to complete solution processing, both solution processing combined with thermal evaporation to produce large- area planar n–i–p modules are is also recently demonstrated (Figure 15.6) where the combination solution-processable SnO_2 and spray-pyrolyzed compact TiO_2 (c-TiO_2) was tested as ETL followed by their further treatment with phenyl-C61-butyric-acid-methyl-ester (PCBM) solution [48].

After that, the desired perovskite layer was deposited via co-evaporating powders of lead iodide (PbI_2) and MAI that was further treated with a solution of potassium acetate (KAc) and MAI to improve the charge transport at the grain boundary. The modules fabricated with such strategy also demonstrated very high >18% solar-to-electrical conversion efficiency over an active area of 21 cm^2. More impressively, these also exhibited very impressive shelf life stability for more than 50 days when stored in a controlled environment with 30% RH and measured in air at 70% RH [48].

Hence, such striking demonstrations proves that upscaling of this emerging technology is possible with very impressive conversion efficiencies and moderate stability, which will be expected to improve further with more stressful conditions in the near future [21].

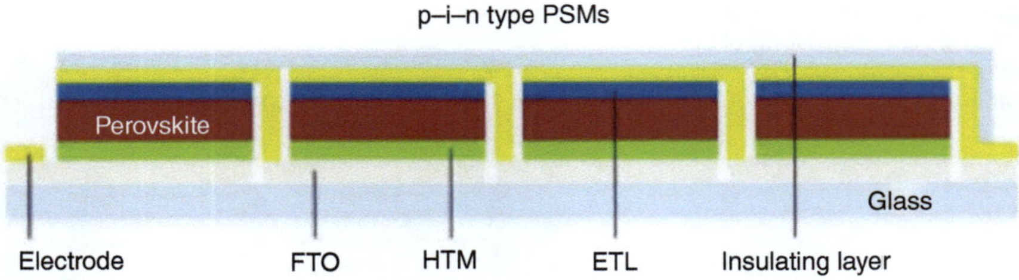

Figure 15.7 Schematic of serially connect n–i–p type Perovskite module. Source: Wang et al. [15].

15.2.2 Scalable p–i–n Configuration-Based Perovskite Solar Modules

Similar to n–i–p device designs, the scaling up of inverted structures called p–i–n structure was also realized for scaling up with the similar serial connections and smart shifting of ETL and HTM as shown in Figure 15.7.

In such device design HTMs like PEDOT:PSS or NiO can be first deposited on transparent FTO–glass substrate [49, 50], which follows some of the same fabrication schemes of deposition of perovskite layers as described in the previous section. After that, ETL and metal electrode can be deposited to complete the structure of the modules (Figure 15.7).

Impressive improvements have also been made toward scaling up such device design [49, 50], where large-area modules employing inverted p–i–n configurations have been fabricated (Figure 15.8) with similar scalable processes as discussed in Section 15.2.1, also exhibiting impressive solar-to-electrical conversion efficiencies along with steady photovoltaic performance stability when subjected to less stressful conditions [50, 52].

15.2.3 Scalable n–i–p and p–i–n Configuration-Based Flexible Perovskite Solar Modules

Although very recently reported, a very impressive demonstration of PSCs on thin flexible glass [53], transferring any photovoltaic technology from rigid to flexible

(a) (b) (c)

Figure 15.8 Highlights of the progress being made for p–i–n configuration-based inverted perovskite solar modules. Source: Reproduced from (a) Li et al. [48], (b) Troughton et al. [49], and (c) Liu et al. [51].

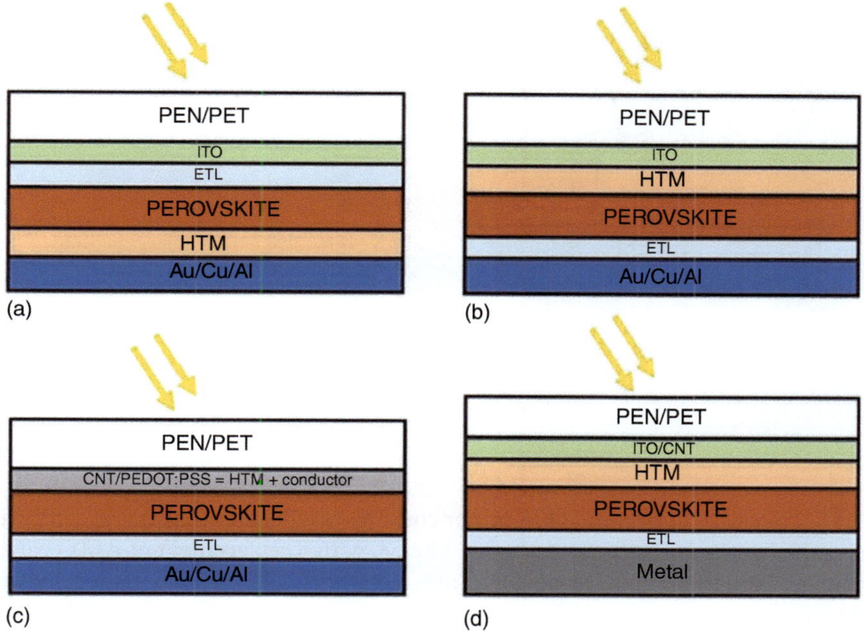

Figure 15.9 Popular device configurations of PSCs on flexible substrates. Source: Lund et al. [7].

substrates, offers many possibilities of their deployment such as curved surfaces, which cannot be realized when fabricated over rigid substrates such as glass. Interestingly, the possibility of low-temperature-based solution processing of almost every active layer of PSCs greatly motivated the researchers worldwide for shifting the technology from the rigid glass to flexible polymer substrates where flexible lab-sized PSCs have also been demonstrated in numerous configurations as shown in Figure 15.9 [7, 39].

Furthermore, flexible conducting polymers, i.e. tin-doped indium oxide on polyethyleneterephtalate (ITO-PET) and tin-doped indium oxide on polyethylenenaphtalate (ITO-PEN), remained the foremost choices for translating the PSC technology on flexible substrates since these conducting polymers have also been successfully used with other type of solar cell technology such as OSCs [54].

One of the initial scaling-up efforts for PSC module on flexible substrates was demonstrated with a mesoporous n–i–p device configuration on ITO-PET polymer substrates [55] where the compact TiO layers were produced with atomic layer deposition (ALD). After that, the mesoporous layer was deposited by diluting a commercial paste in terpineol solvent that was then deposited by screen printing. These screen-printed mesoporous TiO_2 layers were further treated with UV irradiation to avoid incomplete photooxidation [55]. Furthermore, the perovskite deposition was performed via spin coating, and metal contacts were obtained by thermal evaporation. These fabricated flexible module (Figure 15.10) showed 3.1% for an active area of 7.92 cm^2.

Figure 15.10 An image of the four-cell series-connected mesoporous n–i–p module. Source: Di Giacomo [55].

An impressive improvement in power conversion efficiency (8.8%) was achieved after this work over an active area of 12.2 cm^2, with a replacement of ALD-coated compact TiO_2 to solution-processable SnO_2 along with selective laser patterning for obtaining serially connected five cell fingers [17], although solution processing was achieved in this study by spin coating. Nevertheless, laser patterning could be considered here as one of the pioneer works, indicating its potential use in obtaining higher geometrical and active areas and achieving impressive conversion efficiencies.

In addition, planar n–i–p device design-based flexible perovskite modules have also been reported by other novel combination of materials such as fullerene C_{60}, which was proven to work as an efficient ETL [56]. The module fabricated by C_{60} as ETL combined with ultrafast PbI_2 resulted in a conversion efficiency of >8% over an active area of 16 cm^2. Nevertheless, the long-term stability of the module was not mentioned in the presented work.

On the contrary to spin coating or thermal evaporation of ETLs, continuous roll-to-roll coating of ETL with gravure printing has also been demonstrated for n–i–p flexible devices (Figure 15.11) [57]. Although conversion efficiencies of small devices are reported. Nevertheless, these impressive results can be considered as initial steps toward adopting such continuous ETL, which may also be used for producing large-area flexible PSMs.

On the other hand, large-area and flexible PSMs have also been recently demonstrated with inverted p–i–n device structure of PSC where thermal evaporation for the deposition of each active layer was used [58]. To do so, novel hybrid material layer consisting of small organic molecules 2T-NATA and F4-TCNQ was co-evaporated for hole transport. Similarly, perovskite, ETL (C_{60}), and metal electrode were also evaporated to fabricate the devices. The flexible modules (Figure 15.12) exhibited very impressive solar-to-electrical energy conversion of 13.15% for an active area of 16 cm^2 along with remarkable stability upon storing the fabricated module at 25 °C and relative humidity (RH) of 45% for 1000 hours first where the module retained 90% of the initial performance. After that, the module was further exposed to full

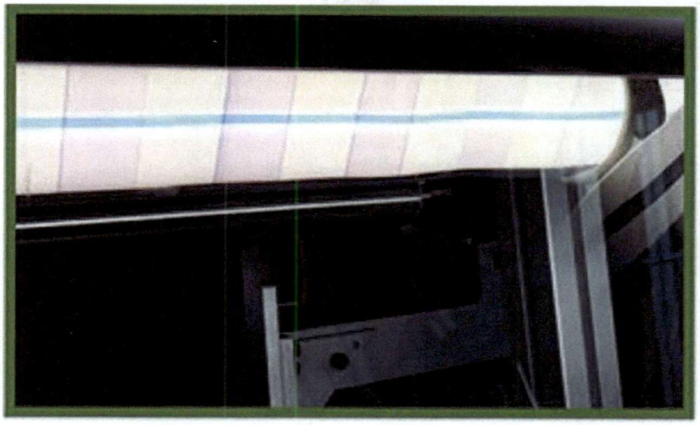

Figure 15.11 Gravure-printed SnO_2 ETL on flexible ITP-PET substrate. (Open access article published by Wiley). Source: Kim et al. [57]

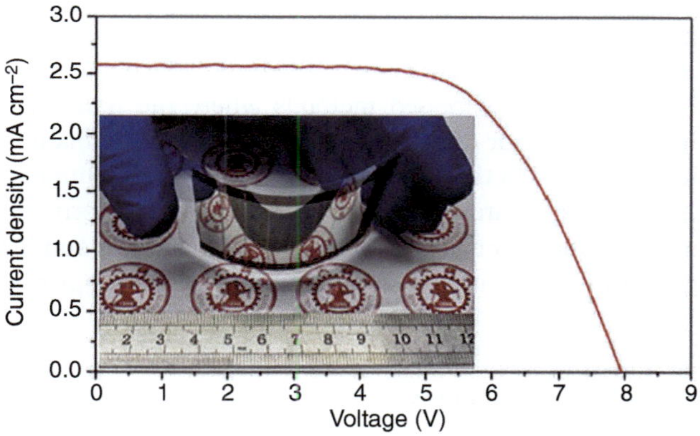

Figure 15.12 Flexible perovskite solar module with inverted-p–i–n configuration. Source: Lei et al. [58].

sunlight illumination in a glove box that again retained 80% of the initial performance compared with the reference modules that showed high degradation rate under similar environmental condition [58].

More strikingly, a remarkable achievement from the perspective of making robust PSMs is recently reported (Figure 15.13) [59] where robust durability inspired by biomimetic mechanisms of the vertebra is replicated by introducing a conductive and glued polymer ((poly(3,4-ethylenedioxythiophene):poly(ethyleneco-vinyl acetate or PEDOT:EVA)) between ITO and perovskite layers, which facilitated impressive crystal growth and grain size of the fabricated perovskite layer over far large area. As a result, the fabricated modules with an effective area of >31 cm^2 gave very high (17.55%) conversion efficiency. More notably, the fabricated devices maintained 85% of original efficiency after 7000 narrow bending cycles with negligible angular dependence.

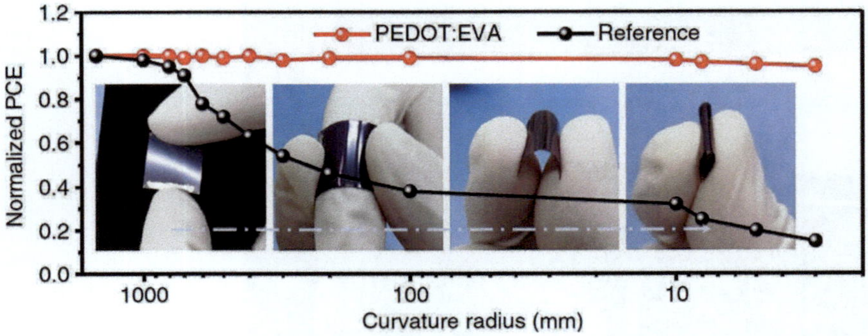

Figure 15.13 Durability of perovskite films tested with 7000 bending cycles. (Open Access article and does not need any permission for re-use). Source: Meng et al. [59].

Hence the aforementioned demonstrations of large-area flexible PSMs indicate steady progress not only for enhancements in solar-to-electrical conversion efficiencies but also for addressing some key challenges such as robust stability of deposited materials against mechanical stresses. Moreover, for the presented module sizes, thermal evaporation of active layers seems more efficient in improving photovoltaic performance compared with solution-processed methods where the removal of solvent after the deposition of precursor of active materials seems problematic in achieving high-quality active layers. On the other hand, in spite of the preliminary evidences, which have been shown regarding long-term photovoltaic performance stability under shelf stored conditions combined with ambient conditions and low humidity, the technology must have to surpass several durability tests for proving itself a commercial hit.

15.2.4 HTM-Free Perovskite Solar Modules

Despite exhibiting impressive conversion efficiencies, the current bottlenecks to make this photovoltaic technology great also include the cost of the active materials, which plays a crucial role in competing with other existing technologies. Most of the progress are being made (with the structures discussed in previous section) by incorporating costly HTMs, i.e. spiro-OMeTAD and metal electrodes (such as gold or silver), which does not only influence on overall production cost but also found unstable in most of the long-term durability tests [8, 60].

Carbon-based HTM-free PSCs can be considered as the most promising candidates to address most of the challenges and bottlenecks of this photovoltaic technology. The pioneer work reported by Mei et al. with regard to the HTM-free and triple mesoscopic carbon-based printable PSCs device [29] (Figure 15.2b) highlighted the promising characteristics such as HTM- and metal electrode-free device designs along with the potential of producing it by established and scalable printing methods such as screen printing and slot-die coating [8, 29, 61]. The high efficiency along with promising stability for lab-sized devices reported in this pioneer work [29] attracted great attention from several research labs and commercial player, and lab-sized solar cells were quickly translated into large-area

Figure 15.14 7 m^2 triple mesoscopic printable perovskite solar panels installed outdoors. Source: Hu et al. [62].

solar modules [62] where demonstrations of outdoor installations (Figure 15.14) [8, 62] and promising photovoltaic performance stability and durability under several simulated and environmental have been tested and reported [8, 19, 43, 62].

Some of the promising efforts being made for the development of these triple mesoscopic carbon-based printable PSC device structures include one of the first demonstrations of inkjet infiltration of perovskite precursor ink [10], promising stability under heavily UV light exposure [63], and discovery regarding perovskite crystal growth upon exposing the triple mesoscopic carbon-based printable PSCs without encapsulation in the thermo-humid environment [63], which currently does not seem possible for any HTM-based PSC device designs.

From the perspective of scaling up, several research groups demonstrated scalable sizes right from $10 \times 10\,cm^2$-based modules [62] to much larger active area-based panels ($198\,cm^2$) [64] with striking solar-to-electrical conversion efficiencies ranging from 6% [64] to 10–12% [62, 65] for different active areas and module sizes. Moreover, compatible encapsulation methods compared with Si-based PV encapsulation schemes have also been developed with novel materials, which have exhibited remarkable stable photovoltaic performance when subjected to long-term performance stability tests [18, 65].

Table 15.1 represents a brief summary for some of the best solar-to-electrical conversion efficiency-based perovskite modules, which have been reported in different configurations and are discussed in the earlier sections of this chapter.

15.3 Critical Overview on Scalable Materials Deposition Methods

It is quite evident that not only every design of PSCs requires several classes of materials but also their deposition methods equally play key roles in achieving satisfactory photovoltaic performance [11, 12]. Scaling-up active materials from their small

Table 15.1 A few best solar-to-electrical conversion efficiency-based perovskite modules reported in recent years during the development of PSC technology.

Device design	Substrate	Active area (cm^2)	Efficiency (%)	Stability	References
n–i–p	FTO-glass	10.1	10.3	Not reported	[50]
n–i–p	FTO-glass	100	4.3	Not reported	[50]
n–i–p	FTO-glass	36.1	12.1	Module ageing at 10 mW cm^{-2} UV filter filtered simulated sunlight at 45 °C in ambient air and maintained at the maximum power point. The module retained 90% of the initial performance after 500 h	[49]
p–i–n	Flexible glass	42.9	15.86	Not reported	[17]
p–i–n	Flexible polymer	16	13.15	Module retained 90% of the initial performance when stored under ambient air (45% average RH) without encapsulation for more than 1000 h	[62]
p–i–n	Flexible polymer	31.20	17.55	Mechanical durability reported with 7000 bending cycles under which, 85% of the initial efficiency remained retain	[10]
n–i–p	FTO-glass	21	18.13	1. Efficiency vs. time under 1 sun continuous illumination measured in air at 70% relative humidity (RH) 2. Efficiency vs. time for over 100 h. PSM were kept at 65 °C in an argon environment and measured in air at 70% RH 3. Efficiency vs. time for over 50 days (shelf-stability). PSM were stored in a controlled environment with 30% RH and measured in air at 70% RH. Devices	[51]
HTM free triple-mesoscopic PSCs	FTO-glass	49	10.4	1. AM1.5 illumination for 1000 h in ambient condition with a temperature of 25 °C and 54% humidity 2. Local environment in Wuhan, China on the roof of our lab, during 1 mo from 22 April to 23 May 2016. The average temperature is around 30 °C and the average humility is around 80% 3. Modules were stored in dark for 1 yr and tested after every 3 mo 4. Devices showed promising stability and did not degrade	[64]
HTM free triple-mesoscopic PSCs	FTO-glass	47.6	11.2	Full sun illumination + 55 °C for >10 000 h. Devices showed no performance degradation	[66]
HTM free triple-mesoscopic PSCs	FTO-glass	60.09	10.09	1. Thermal stability tests at 85 °C for over 325 h 2. Outdoor stability test (location: 39°19′48″N 114°37′26″E), the atmospheric temperature varied from −10 to 35 °C. Devices encapsulated with polyurethane hotmelt did not degrade	[46]

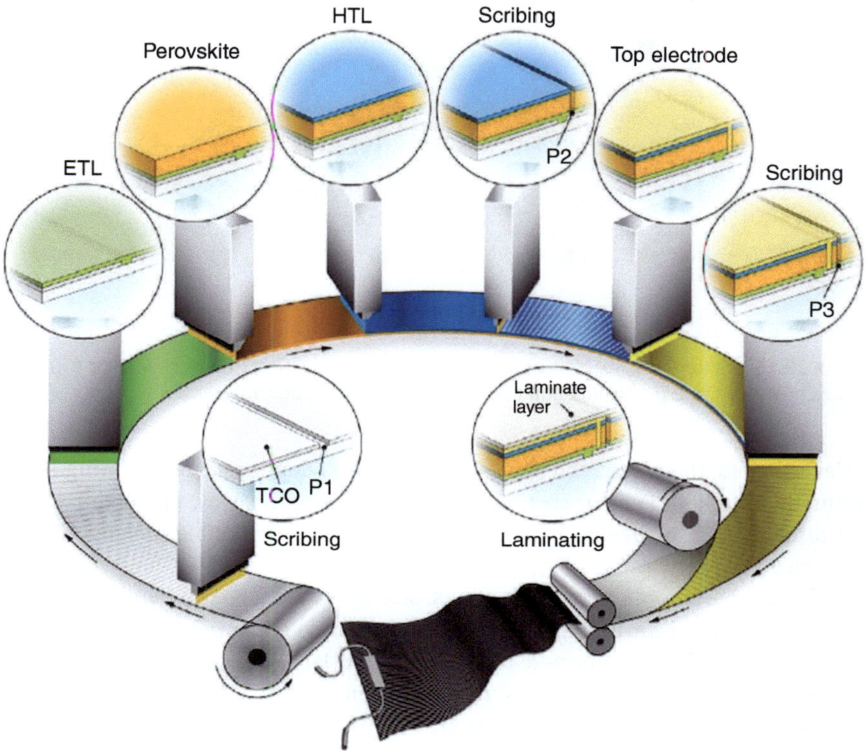

Figure 15.15 Schematic illustration of continuous processing of monolithic-interconnected perovskite solar modules. Source: Kim et al. [67].

sizes and dimensions imposes several challenges, i.e. reproducibility of active layers, solvent quenching for transforming liquid precursors into high-quality crystalline films, and producing stack of active materials with high precision, which cannot be avoided during the fabrication of large-sized modules.

Thanks to the progress being made in developing high-throughput solution-processing methods such as blade coating, slot-die coating, and inkjet printing, these give enormous opportunities to utilize such advancements to produce these PSCs with high reproducibility and low cost. One of such coating schemes to produce PSCs with high throughput has been recently suggested by deploying solution-processed materials (Figure 15.15) [67] and has also been adopted to produce n–i–p-based PSCs on glass substrates [68], which can not only be opted to produce inverted structure but can also be utilized for high-throughput fabrication for flexible substrates as demonstrated before for the roll-to-roll production of OSCs.

On the other hand, batch processing-based fabrication of large-area PSMs in the case of utilization of rigid glass substrates also seems possible where HTM-free triple mesoscopic printable perovskite solar panels have also been impressively scaled up by utilizing established and scalable materials deposition methods such as screen printing as shown in Figure 15.16 [62].

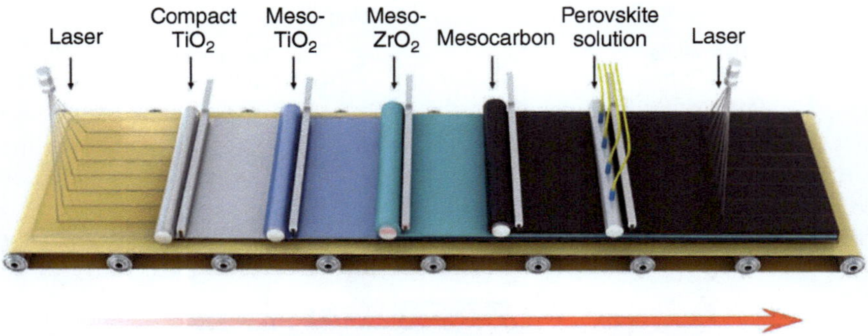

Figure 15.16 Schematic illustration of the proposed production line of HTM-free triple mesoscopic printable perovskite solar panels. Source: Hu et al. [62].

Nevertheless, there has been some limitations for adopting screen printing-based production methods such as cost of screen and its cleaning, slow processing speed, and difficulties in achieving high active area because of possible mismatching in between stacked layers.

With the abovementioned observations, slot-die coating is also demonstrated as fabrication method for producing lab-sized HTM-free triple mesoscopic printable PSCs [69] and is also expected to be opted for large-area module fabrication. Interestingly, slot-die coating has recently been reported for introducing even perovskite precursor solution in large-area HTM-free triple mesoscopic printable PSMs (Figure 15.17), which exhibited >12% solar-to-electrical conversion efficiency for an active area of >60 cm^2 [61]. Hence, these two reports clearly give future directives for realizing high-throughput fabrication of one of the most stable device designs of PSC technology.

Figure 15.17 Demonstration of perovskite precursor casting on a mesoporous scaffold of $TiO_2/ZrO_2/$ carbon. Source: Xu et al. [61].

These advancements toward upscaling PSC technology strongly indicate that the knowledge and understanding for producing large-area PSMs with high-throughput or reproducibility has already been acquired. Moreover, achieving both scaling up and promising stability in some unique device structures seems possible where demonstrations of some of the large deployments of large-area perovskite modules have also surfaced (Figure 15.14) and are getting interested to start exciting commercial activities in the near future.

15.4 Nutshell of Long-Term Device Stability of Perovskite Solar Cells and Modules

Despite all advancements for the development of PSC technology, one of the major bottlenecks regarding its forecasted commercial and mass deployment is unfortunately the long-term operational stability of PSCs due to poor stability of organic–inorganic light-absorbing perovskite semiconductors against moisture and humidity [66, 70]. Being soluble in many solvents including water, it has become inevitable to develop robust encapsulation methods for surpassing stressful and standardized stability testing protocols [21].

Similar to efforts, which have been invested for increasing the photovoltaic performance, research has also been rapidly progressed to improve long-term photovoltaic performance durability through investigating countless novel materials and encapsulants along with novel device encapsulation methods, which have shown capability for retarding degradation in the active area of both lab-sized solar cells and large-area modules [19, 40, 71].

Lab-sized glass-based PSCs have recently surpassed International Electrotechnical Commission (IEC) 61215:2016 Damp Heat and Humidity Freeze tests via incorporating novel polyolefin (PO)-based polymer blanket and polyisobutylene (PIB)-based polymer (PVS 101) blanket encapsulations where the exposed devices did not show any performance degradation during freeze test for continuous 75 cycles [42]. Similarly, encapsulating the exposed PSCs with PIB sealant also showed no photovoltaic performance loss for more than 1800 hours in very high relative humidity (85%) combined with very high temperature (85 °C), thus endorsing the extreme limit of robust durability of this promising solar cell technology.

On the other hand, novel hot-melt polyurethane (PU) was also discovered as a compatible laminate for facilitating the sealing procedures for HTM-free triple mesoscopic printable PSCs [19]. This was found to be more compatible for sealing both lab-sized devices and large-area modules due to the utilization of lower temperature (~80 °C), which tremendously retarded the possibility of water and moisture penetration within the device structure (Figure 15.18). As a result, large-area modules maintain not only the long-term photovoltaic performance stability for 325 hours when exposed to the thermal stability test at 85 °C but also

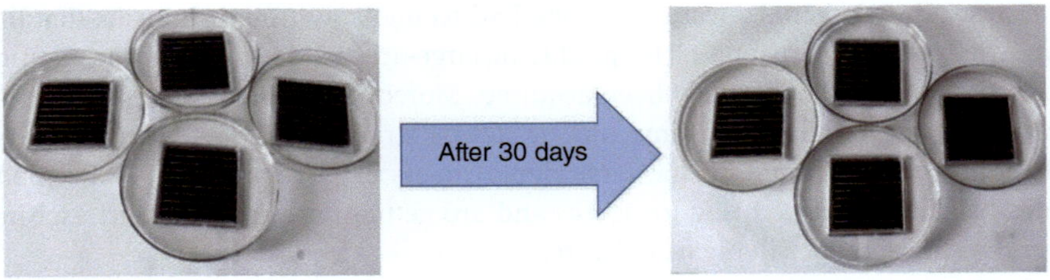

Figure 15.18 The water soaking test of printable mesoscopic PSC modules. Source: Fu et al. [19].

the photovoltaic performance when subjected to an outdoor stability test for a period of more than 2000 hours.

Hence, all these recent advancements clearly indicate that PSC technology can be surely transformed into robust, durable, and reliable photovoltaic technology and keeps immense potential for terawatt-scale-based electricity production [67]. Although strongly criticized for not being commercialized after a decade of research, one after other demonstrations can be counted as growing steps toward futuristic promises for delivering a robust solar cell technology both for bulk electricity production and for building-integrated photovoltaic applications (https://www.oxfordpv.com/; https://sauletech.com/).

15.5 Conclusive Summary and Futuristic Outlook

PSCs have been explored since the past decade. This thoroughly studied photovoltaic solar cell technology has been rapidly surpassing all the existing records of solar-to-electrical conversion efficiencies and is predicted to cross over 30% device efficiency in the near future. Moreover, the technology has been considered to be produced on Mars in the upcoming years, which further indicates its potential crossing boundaries one after another.

These solar cells not only have been visualized for terawatt-scale productions but also show possibilities for indoor applications and IoT nodes, which will also play key roles in the growing data driven-based modern societies. In such cases, the value propositions can be very different from the perspective of robust stability, which has yet to be proven at larger scale for their placements at much larger scale. Nevertheless, for commercial deployment, PSCs must prove its true potential with regard to the efficiency, scalability, and robust stability.

In this regard, decent balance seems logical in between hunting low-cost alternative materials and deep understanding of the underlying mechanisms, which trigger unwanted happenings within the device stack of these solar cells. Nevertheless, abundant availability of the promising alternative materials together with, yet to explore, perovskite semiconductors offers great hope for producing highly efficient and low-cost photovoltaic technology, which could possibly be produced with high throughput by utilizing mass reproducible and scalable fabrication methods. In summary, the future of this promising low-cost solar cell technology seems highly emerging, and most of the bottlenecks are expected to be resolved in the near future, making this technology a commercial hit.

References

1 Kojima, A., Teshima, K., Shirai, Y., and Miyasaka, T. (2009). Organometal halide perovskites as visible-light sensitizers for photovoltaic cells. *J. Am. Chem. Soc.* 131: 6050–6051. https://doi.org/10.1021/ja809598r.

2 Stranks, S.D., Eperon, G.E., Grancini, G. et al. (2013). Electron-hole diffusion lengths exceeding 1 micrometer in an organometal trihalide perovskite absorber. *Science* 342: 341–344. https://doi.org/10.1126/science.1243982.

3 De Wolf, S., Holovsky, J., Moon, S.-J. et al. (2014). Organometallic halide perovskites: sharp optical absorption edge and its relation to photovoltaic performance. *J. Phys. Chem. Lett.* 5: 1035–1039. https://doi.org/10.1021/jz500279b.

4 Prasanna, R., Gold-Parker, A., Leijtens, T. et al. (2017). Band gap tuning via lattice contraction and octahedral tilting in perovskite materials for photovoltaics. *J. Am. Chem. Soc.* 139: 11117–11124. https://doi.org/10.1021/jacs.7b04981.

5 Green, M.A., Dunlop, E.D., Hohl-Ebinger, J. et al. (2020a). Solar cell efficiency tables (version 56). *Prog. Photovolt. Res. Appl.* 28: 629–638. https://doi.org/10.1002/pip.3303.

6 Galagan, Y. (2018). Perovskite solar cells: toward industrial-scale methods. *J. Phys. Chem. Lett.* 9: 4326–4335. https://doi.org/10.1021/acs.jpclett.8b01356.

7 Lund, P.D., Halme, J., Hashmi, G. et al. (2018a). Application of dye-sensitized and perovskite solar cells on flexible substrates. *Flex. Print. Electron.* 3: 013002. https://doi.org/10.1088/2058-8585/aaabc9.

8 Rong, Y., Hu, Y., Mei, A. et al. (2018). Challenges for commercializing perovskite solar cells. *Science* 361: eaat8235. https://doi.org/10.1126/science.aat8235.

9 Asif, A.A., Singh, R., and Alapatt, G.F. (2015). Technical and economic assessment of perovskite solar cells for large scale manufacturing. *J. Renew. Sustain. Energy* 7: 043120. https://doi.org/10.1063/1.4927329.

10 Hashmi, S.G., Martineau, D., Li, X. et al. (2017b). Air processed inkjet infiltrated carbon based printed perovskite solar cells with high stability and reproducibility. *Adv. Mater. Technol.* 2: 1600183. https://doi.org/10.1002/admt.201600183.

11 Li, Z., Klein, T.R., Kim, D.H. et al. (2018). Scalable fabrication of perovskite solar cells. *Nat. Rev. Mater.* 3: 18017. https://doi.org/10.1038/natrevmats.2018.17.

12 Park, N.-G. and Zhu, K. (2020). Scalable fabrication and coating methods for perovskite solar cells and solar modules. *Nat. Rev. Mater.* 5: 333–350. https://doi.org/10.1038/s41578-019-0176-2.

13 Stranks, S.D. and Snaith, H.J. (2015). Metal-halide perovskites for photovoltaic and light-emitting devices. *Nat. Nanotech.* 10: 391–402. https://doi.org/10.1038/nnano.2015.90.

14 Qiu, L., Liu, Z., Ono, L.K. et al. (2019). Scalable fabrication of stable high efficiency perovskite solar cells and modules utilizing room temperature sputtered

SnO_2 electron transport layer. *Adv. Funct. Mater.* 29: 1806779. https://doi.org/10.1002/adfm.201806779.

15 Wang, F., Cao, Y., Chen, C. et al. (2018). Materials toward the upscaling of perovskite solar cells: progress, challenges, and strategies. *Adv. Funct. Mater.* 28: 1803753. https://doi.org/10.1002/adfm.201803753.

16 Palma, A.L., Matteocci, F., Agresti, A. et al. (2017). Laser-patterning engineering for perovskite solar modules with 95% aperture ratio. *IEEE J. Photovoltaics* 7: 1674–1680. https://doi.org/10.1109/JPHOTOV.2017.2732223.

17 Dagar, J., Castro-Hermosa, S., Gasbarri, M. et al. (2018a). Efficient fully laser-patterned flexible perovskite modules and solar cells based on low-temperature solution-processed SnO_2/mesoporous-TiO_2 electron transport layers. *Nano Res.* 11: 2669–2681. https://doi.org/10.1007/s12274-017-1896-5.

18 Galagan, Y., Coenen, E.W.C., Verhees, W.J.H., and Andriessen, R. (2016). Towards the scaling up of perovskite solar cells and modules. *J. Mater. Chem. A* 4: 5700–5705. https://doi.org/10.1039/C6TA01134A.

19 Fu, Z., Xu, M., Sheng, Y. et al. (2019a). Encapsulation of printable mesoscopic perovskite solar cells enables high temperature and long-term outdoor stability. *Adv. Funct. Mater.* 29: 1809129. https://doi.org/10.1002/adfm.201809129.

20 Corsini, F. and Griffini, G. (2020). Recent progress in encapsulation strategies to enhance the stability of organometal halide perovskite solar cells. *J. Phys. Energy* 2: 031002. https://doi.org/10.1088/2515-7655/ab8774.

21 Khenkin, M.V. et al. (2020). Consensus statement for stability assessment and reporting for perovskite photovoltaics based on ISOS procedures. *Nat. Energy* 5: 35–49. https://doi.org/10.1038/s41560-019-0529-5.

22 Ali, J., Li, Y., Gao, P. et al. (2020). Interfacial and structural modifications in perovskite solar cells. *Nanoscale* 12: 5719–5745. https://doi.org/10.1039/C9NR10788F.

23 Kim, H.-S., Lee, C.-R., Im, J.-H. et al. (2012). Lead iodide perovskite sensitized all-solid-state submicron thin film mesoscopic solar cell with efficiency exceeding 9%. *Sci. Rep.* 2: 591. https://doi.org/10.1038/srep00591.

24 Heo, J.H., Im, S.H., Noh, J.H. et al. (2013). Efficient inorganic–organic hybrid heterojunction solar cells containing perovskite compound and polymeric hole conductors. *Nat. Photon.* 7: 486–491. https://doi.org/10.1038/nphoton.2013.80.

25 Burschka, J., Pellet, N., Moon, S.-J. et al. (2013). Sequential deposition as a route to high-performance perovskite-sensitized solar cells. *Nature* 499: 316–319. https://doi.org/10.1038/nature12340.

26 Lee, M.M., Teuscher, J., Miyasaka, T. et al. (2012). Efficient hybrid solar cells based on meso-superstructured organometal halide perovskites. *Science* 338: 643–647. https://doi.org/10.1126/science.1228604.

27 Wang, J., Xu, J., Li, Z. et al. (2020). Front-contact passivation of PIN $MAPbI_3$ solar cells with superior device performances. *ACS Appl. Energy Mater.* 3: 6344–6351. https://doi.org/10.1021/acsaem.0c00561.

28 Chen, H. and Yang, S. (2017). Carbon-based perovskite solar cells without hole transport materials: the front runner to the market? *Adv. Mater.* 29: 1603994. https://doi.org/10.1002/adma.201603994.

29 (a) Mei, A., Li, X., Liu, L. et al. (2014). A hole-conductor-free, fully printable mesoscopic perovskite solar cell with high stability. *Science* 345: 295–298. https://doi.org/10.1126/science.1254763. (b) Baker, J., Hooper, K., Meroni, S. et al. (2017). High throughput fabrication of mesoporous carbon perovskite solar cells. *J. Mater. Chem. A* 5: 18643–18650. https://doi.org/10.1039/C7TA05674E.

30 Ku, Z., Rong, Y., Xu, M. et al. (2013). Full printable processed mesoscopic $CH_3NH_3PbI_3/TiO_2$ heterojunction solar cells with carbon counter electrode. *Sci. Rep.* 3: 3132. https://doi.org/10.1038/srep03132.

31 Zhang, F., Yang, X., Wang, H. et al. (2014). Structure engineering of hole–conductor free perovskite-based solar cells with low-temperature-processed commercial carbon paste As cathode. *ACS Appl. Mater. Interfaces* 6: 16140–16146. https://doi.org/10.1021/am504175x.

32 Wei, Z., Yan, K., Chen, H. et al. (2014). Cost-efficient clamping solar cells using candle soot for hole extraction from ambipolar perovskites. *Energy Environ. Sci.* 7: 3326–3333. https://doi.org/10.1039/C4EE01983K.

33 Chen, J. and Park, N. (2019). Causes and solutions of recombination in perovskite solar cells. *Adv. Mater.* 31: 1803019. https://doi.org/10.1002/adma.201803019.

34 Wolff, C.M., Caprioglio, P., Stolterfoht, M., and Neher, D. (2019). Nonradiative recombination in perovskite solar cells: the role of interfaces. *Adv. Mater.* 31: 1902762. https://doi.org/10.1002/adma.201902762.

35 Langenhorst, M., Sautter, B., Schmager, R. et al. (2019). Energy yield of all thin-film perovskite/CIGS tandem solar modules. *Prog. Photovolt. Res. Appl.* 27: 290–298. https://doi.org/10.1002/pip.3091.

36 https://www.oxfordpv.com/news/oxford-pv-hits-new-world-record-solar-cell (accessed 19 September 2020).

37 Jošt, M., Kegelmann, L., Korte, L., and Albrecht, S. (2020). Monolithic perovskite tandem solar cells: a review of the present status and advanced characterization methods toward 30% efficiency. *Adv. Energy Mater.* 10: 1904102. https://doi.org/10.1002/aenm.201904102.

38 Li, H. and Zhang, W. (2020). Perovskite tandem solar cells: from fundamentals to commercial deployment. *Chem. Rev.* 120: 9835–9950. https://doi.org/10.1021/acs.chemrev.9b00780.

39 Zhang, J., Zhang, W., Cheng, H.-M., and Silva, S.R.P. (2020). Critical review of recent progress of flexible perovskite solar cells. *Mater. Today* S1369702120301747 https://doi.org/10.1016/j.mattod.2020.05.002.

40 Zhao, P., Kim, B.J., and Jung, H.S. (2018). Passivation in perovskite solar cells: a review. *Mater. Today Energy* 7: 267–286. https://doi.org/10.1016/j.mtener.2018.01.004.

41 Green, M., Dunlop, E., Hohl-Ebinger, J. et al. (2021). Solar cell efficiency tables (version 57). *Prog Photovolt Res Appl.* 29: 3–15. https://doi.org/10.1002/pip.3371.

42 Shi, L., Bucknall, M.P., Young, T.L. et al. (2020). Gas chromatography–mass spectrometry analyses of encapsulated stable perovskite solar cells. *Science* 368: eaba2412. https://doi.org/10.1126/science.aba2412.

43 Mei, A., Sheng, Y., Ming, Y. et al. (2020). Stabilizing perovskite solar cells to IEC61215:2016 standards with over 9,000-h operational tracking. *Joule* S2542435120304463 https://doi.org/10.1016/j.joule.2020.09.010.

44 Moon, S.-J., Yum, J.-H., Lofgren, L. et al. (2015). Laser-scribing patterning for the production of organometallic halide perovskite solar modules. *IEEE J. Photovolt.* 5: 1087–1092. https://doi.org/10.1109/JPHOTOV.2015.2416913.

45 Chen, H., Ye, F., Tang, W. et al. (2017). A solvent- and vacuum-free route to large-area perovskite films for efficient solar modules. *Nature* 550: 92–95. https://doi.org/10.1038/nature23877.

46 Razza, S., Giacomo, F.D., Matteocci, F. et al. (2015). Perovskite solar cells and large area modules (100 cm^2) based on an air flow-assisted PbI_2 blade coating deposition process. *J. Power Sources* 6 https://doi.org/10.1016/j.jpowsour.2014.12.008.

47 Calabrò, E. (2018). Low temperature, solution-processed perovskite solar cells and modules with an aperture area efficiency of 11%. *Solar Energy Mater. and Solar Cells* 9 https://doi.org/10.1016/j.solmat.2018.05.001.

48 Li, J., Wang, H., Chin, X.Y. et al. (2020). Highly efficient thermally co-evaporated perovskite solar cells and mini-modules. *Joule* 4: 1035–1053. https://doi.org/10.1016/j.joule.2020.03.005.

49 Troughton, J., Hooper, K., and Watson, T.M. (2017). Humidity resistant fabrication of $CH_3NH_3PbI_3$ perovskite solar cells and modules. *Nano Energy* 39: 60–68. https://doi.org/10.1016/j.nanoen.2017.06.039.

50 Seo, J., Park, S., Chan Kim, Y. et al. (2014). Benefits of very thin PCBM and LiF layers for solution-processed p–i–n perovskite solar cells. *Energy Environ. Sci.* 7: 2642–2646. https://doi.org/10.1039/C4EE01216J.

51 Liu, C., Yang, Y., Zhang, C. et al. (2020). Tailoring C_{60} for efficient inorganic $CsPbI_2$ Br perovskite solar cells and modules. *Adv. Mater.* 32: 1907361. https://doi.org/10.1002/adma.201907361.

52 Hong, S., Lee, J., Kang, H. et al. (2018a). High-efficiency large-area perovskite photovoltaic modules achieved via electrochemically assembled metal-filamentary nanoelectrodes. *Sci. Adv.* 4: eaat3604. https://doi.org/10.1126/sciadv.aat3604.

53 Dai, X., Deng, Y., Van Brackle, C.H. et al. (2020). Scalable fabrication of efficient perovskite solar modules on flexible glass substrates. *Adv. Energy Mater.* 10: 1903108. https://doi.org/10.1002/aenm.201903108.

54 Søndergaard, R., Hösel, M., Angmo, D. et al. (2012). Roll-to-roll fabrication of polymer solar cells. *Mater. Today* 15: 36–49. https://doi.org/10.1016/S1369-7021(12)70019-6.

55 Di Giacomo, F., Zardetto, V., D'Epifanio, A. et al. (2015). Flexible perovskite photovoltaic modules and solar cells based on atomic layer deposited compact layers and UV-irradiated TiO_2 scaffolds on plastic substrates. *Adv. Energy Mater.* 5: 1401808. https://doi.org/10.1002/aenm.201401808.

56 Li, K., Xiao, J., Yu, X. et al. (2018). An efficient, flexible perovskite solar module exceeding 8% prepared with an ultrafast PbI_2 deposition rate. *Sci Rep.* 8: 442. https://doi.org/10.1038/s41598-017-18970-y.

57 Kim, Y.Y., Yang, T., Suhonen, R. et al. (2019). Gravure-printed flexible perovskite solar cells: toward roll-to-roll manufacturing. *Adv. Sci.* 6: 1802094. https://doi.org/10.1002/advs.201802094.

58 Lei, T., Li, F., Zhu, X. et al. (2020). Flexible perovskite solar modules with functional layers fully vacuum deposited. *Sol. RRL* 2000292 https://doi.org/10.1002/solr.202000292.

59 Meng, X., Cai, Z., Zhang, Y. et al. (2020). Bio-inspired vertebral design for scalable and flexible perovskite solar cells. *Nat. Commun.* 11: 3016. https://doi.org/10.1038/s41467-020-16831-3.

60 Han, Y., Meyer, S., Dkhissi, Y. et al. (2015). Degradation observations of encapsulated planar $CH_3NH_3PbI_3$ perovskite solar cells at high temperatures and humidity. *J. Mater. Chem. A* 3: 8139–8147. https://doi.org/10.1039/C5TA00358J.

61 Xu, M., Ji, W., Sheng, Y. et al. (2020). Efficient triple-mesoscopic perovskite solar mini-modules fabricated with slot-die coating. *Nano Energy* 74: 104842. https://doi.org/10.1016/j.nanoen.2020.104842.

62 Hu, Y., Si, S., Mei, A. et al. (2017). Stable large-area (10×10 cm^2) printable mesoscopic perovskite module exceeding 10% efficiency. *Sol. RRL* 1: 1600019. https://doi.org/10.1002/solr.201600019.

63 Hashmi, S.G., Tiihonen, A., Martineau, D. et al. (2017). Long term stability of air processed inkjet infiltrated carbon-based printed perovskite solar cells under intense ultra-violet light soaking. *J. Mater. Chem. A* 5: 4797–4802. https://doi.org/10.1039/C6TA10605F.

64 De Rossi, F., Baker, J.A., Beynon, D. et al. (2018). All printable perovskite solar modules with 198 cm^2 active area and over 6% efficiency. *Adv. Mater. Technol.* 3: 1800156. https://doi.org/10.1002/admt.201800156.

65 Grancini, G., Roldán-Carmona, C., Zimmermann, I. et al. (2017). One-year stable perovskite solar cells by 2D/3D interface engineering. *Nat. Commun.* 8: 15684. https://doi.org/10.1038/ncomms15684.

66 Asghar, M.I., Zhang, J., Wang, H., and Lund, P.D. (2017). Device stability of perovskite solar cells – a review. *Renew. Sustain. Energy Rev.* 77: 131–146. https://doi.org/10.1016/j.rser.2017.04.003.

67 Kim, D.H., Whitaker, J.B., Li, Z. et al. (2018). Outlook and challenges of perovskite solar cells toward terawatt-scale photovoltaic module technology. *Joule* 2: 1437–1451. https://doi.org/10.1016/j.joule.2018.05.011.

68 Yang, M., Kim, D.H., Klein, T.R. et al. (2018). Highly efficient perovskite solar modules by scalable fabrication and interconnection optimization. *ACS Energy Lett.* 3: 322–328. https://doi.org/10.1021/acsenergylett.7b01221.

69 Verma, A., Martineau, D., Hack, E. et al. (2020). Towards industrialization of perovskite solar cells using slot die coating. *J. Mater. Chem. C* 8: 6124–6135. https://doi.org/10.1039/D0TC00327A.

70 Meng, L., You, J., and Yang, Y. (2018). Addressing the stability issue of perovskite solar cells for commercial applications. *Nat. Commun.* 9: 5265. https://doi.org/10.1038/s41467-018-07255-1.

71 Li, N., Niu, X., Chen, Q., and Zhou, H. (2020). Towards commercialization: the operational stability of perovskite solar cells. *Chem. Soc. Rev.* https://doi.org/10.1039/D0CS00573H.

16

Multi-Junction Perovskite Solar Cells

Suhas Mahesh and Bernard Wenger

University of Oxford, Clarendon Laboratory, Parks Road, OX1 3PU, Oxford UK

16.1 Introduction

The barrier of 100 GW growth in the installed capacity of photovoltaic modules was crossed in 2019, and it is forecast that this growth is maintained within two digits over the next decade at least [1]. Monocrystalline and multicrystalline silicon are utilized in a very large majority (≈95%) of these modules [2]. To further facilitate the adoption of solar electricity, the industry and energy policy makers depend on the continuity of the significant decrease in price that has characterized photovoltaics since its inception in the 1950s. However, the analysis of the cost breakdown of a solar module produced in 2019 (Figure 16.1) shows that further reduction of cost will be extremely challenging. The cost of installed solar panels can be divided into the solar module itself, and the so-called balance of systems (BoS) that encompasses all the costs associated to the installation (mounting frame, trackers, landscaping, and so on), and the operation of the solar panels (inverter, cleaning, etc.). On the module itself, the costs are distributed between the barrier materials, interconnections, and finally the solar cell and the base materials. Although the solar industry has championed cost effectiveness at every level of the fabrication chain, there is not plenty of room left for improvement. BoS and module fabrication costs are likely to reach a plateau within the next decade. Consequently, the most promising route for further price reduction and faster adoption of solar electricity is to improve the efficiency of the modules because for a given area (in m^2), the installation costs will be the same but the energy yield (in $kWh\,m^{-2}$) will continue to increase.

According to the International Technology Roadmap for Photovoltaics (ITRPV), monocrystalline silicon solar cells will achieve efficiency approaching 25% within 10 years [1]. However, such efficiencies will approach the practical limits of this technology. Shockley and Queisser proposed in 1961 a theoretical efficiency limit for single junction solar cells [3] in which a maximum efficiency is obtained for a material with a bandgap of $E_g = 1.34\,eV$, where the best compromise is found between thermalization losses of absorbed photons with energy above bandgap and the number of solar photons available at that bandgap [3]. To date, gallium

Perovskite Solar Cells: Materials, Processes, and Devices, First Edition.
Edited by Shahzada Ahmad, Samrana Kazim, and Michael Grätzel.

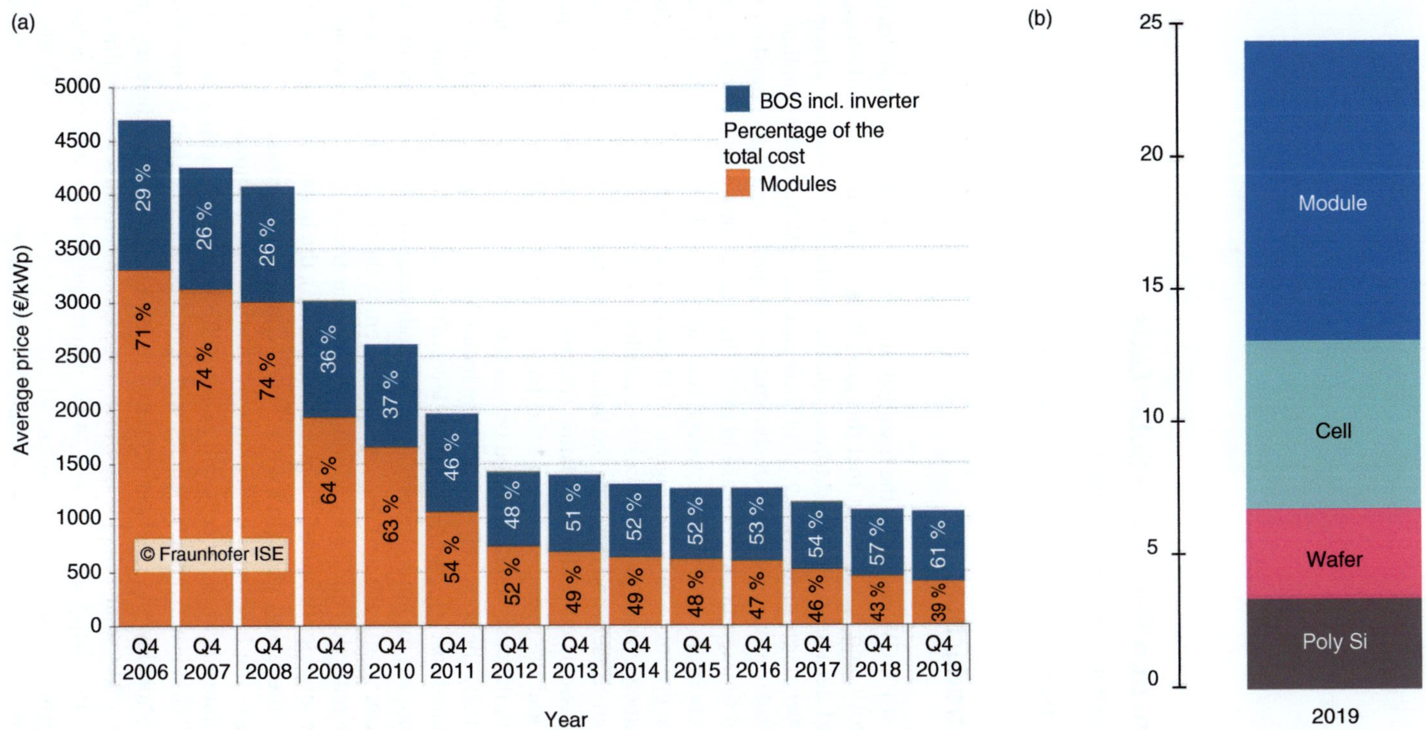

Figure 16.1 (a) Evolution of the average price for a rooftop installation in Germany (10–100 kWp). Since 2013, BoS costs are superior to the module price itself. Source: Philipps and Warmuth [2]. (b) Cost breakdown of PV module in 2019. Source: Adapted from ITRPV [1].

arsenide (GaAs), $CuIn_xGa_{(1-x)}Se_2$ (CIGS: copper indium gallium selenide) and perovskites are the only known semiconductors demonstrated for PV that can be prepared close to this ideal bandgap. Alternatively, several strategies have been proposed to overcome the Shockley–Queisser limit. These include extracting hot carriers to prevent thermalization losses, producing multiple carriers per absorption of a photon (by impact ionization or exciton fission), or multiply the number of bandgaps in the solar absorbing layers. To date, the latter approach is the only practical solution that has been demonstrated and pursued commercially. As we will see below, perovskites, owing to their bandgap tunability and excellent transparency below the band edge, are outstanding candidates for high efficiency multiple-junction (multi-junction) solar cells.

Tandem solar cells are devices comprising a pair of light absorption materials with matched bandgaps stacked on top of each other (see 16.1.2); these have been investigated for many decades. The first demonstration of a tandem solar cell was already achieved by the Research Triangle Institute in 1979 [4]. Compound III–V semiconductors are particularly well suited for such applications due to their bandgap tunability and excellent interfacial properties owing to the epitaxial growth methods developed for these materials. The current world record for the highest solar cell efficiency, an impressive 47.1%, is held by a six-junction device under concentrated illumination (143 suns) [5]. However, due to their prohibitive fabrication cost these, devices are limited to small area applications such as space (where the efficiency-to-weight ratio is key) or for concentrator photovoltaics. Several other absorbing semiconductors have proved to be suitable for use in multi-junction devices. Notably, amorphous silicon could be efficiently combined with microcrystalline silicon in the "Micromorph" concept proposed by Shah et al. in the 1980s [6]. This approach allowed for the scalable fabrication of low-cost, moderately efficient solar cells, which before the advent of multi-crystalline silicon in the late 2000s provided the lowest price-per-watt peak available [7]. Higher efficiencies can be achieved by combining thin film copper chalcopyrite cells such as $CuGaSe_2$ (CGS, E_g = 1.7 eV) with $CuInSe_2$ (CIS, E_g = 1.05 eV) or Cu(In, Ga)Se_2 (CIGS, E_g = 1.15 eV) [8], or as recently demonstrated, III/V semiconductors with c-Si [9].

In this chapter, we discuss the potential of perovskites for use in multi-junction photovoltaic devices in conjunction with various semiconductors. After briefly explaining the theoretical principles, we discuss the different configurations, challenges, and commercial opportunities and review the recent progress in the field.

16.1.1 How Efficient Can Solar Cells Be?

How efficiently can we convert solar energy to work? In the most abstract picture, imagine a solar converter to be a heat engine that circulates a gas of photons between two heat reservoirs, viz. the sun (5800 K) and the Earth (300 K). In this picture, thermodynamics allows us to calculate the limiting power conversion efficiency (PCE) of a solar converter to be 93% ("Landsberg limit" [10]), which

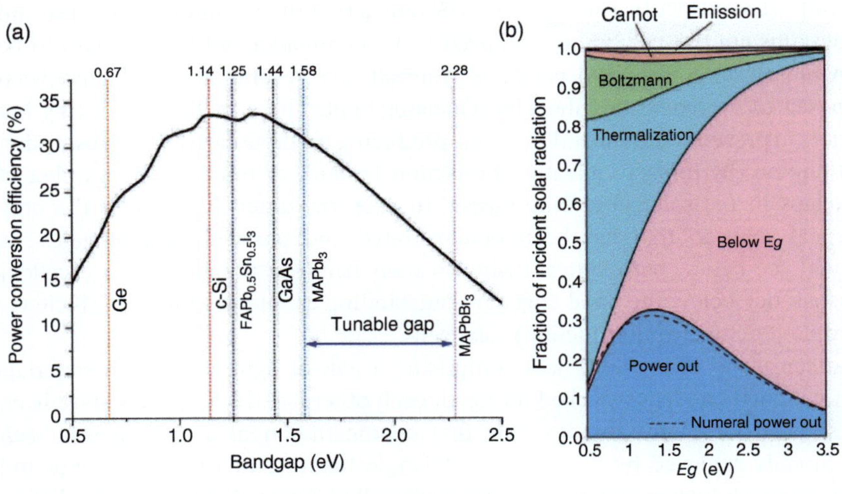

Figure 16.2 (a) The Shockley–Queisser limiting power conversion efficiency for single-junction photovoltaic cells as a function of bandgap. The bandgaps of important solar cell technologies are marked. (b) Intrinsic loss mechanisms in a solar cell as a function of bandgap. Source: Hirst and Ekins-Daukes [11].

is slightly lower than the Carnot limit (95%). Note that this is for a general solar converter, whose mechanism of operation is unspecified. If the solar converter at hand is a photovoltaic (PV) cell involving a semiconductor, additional loss mechanisms begin. Consequently, even the ideal PV cell is much less efficient than 93%. For a single-junction photovoltaic cell, the theoretically maximum PCE under AM1.5G illumination is 33.7% ("Shockley–Queisser Limit" [3]) and is achieved at a bandgap of 1.34 eV (Figure 16.2a). At this ideal bandgap, sub-bandgap photon loss and thermalization loss reach a compromise. Extending the idea slightly to account for non-idealities like non-step function absorptance and auger recombination, the limiting PCE of silicon (29.8%) [12] and ~1.6 eV bandgap perovskite (30.5%) [13] solar cells have also been calculated. The details and assumptions behind the Shockley–Queisser limit are worth examining to understand the loss mechanisms at work. Any strategy to circumvent a loss-mechanism is, in principle, an approach to "break" the Shockley–Queisser limit.

Several intrinsic losses are at work in the ideal bandgap (1.34 eV) single-junction solar cell: Carnot loss (~2%), emission loss (~1%), Boltzmann loss (~10%), thermalisation loss (~30%), and loss of below bandgap photons (~25%) [14] (Figure 16.2b). Of these, the thermalization loss is the most prominent and wastes a third of the incident power as heat. When photons of energy greater than the bandgap are absorbed, they excite carriers into higher energy states. Within a nanosecond, these "hot" carriers cool down ("thermalize") to the band-edges, giving away energy as heat. Thus, if any photon has energy in excess of the bandgap, it is quickly dissipated as heat. If this 30% loss can be circumvented, the limiting efficiency can be improved from 33.7% to over 60%. Calculations show that this is indeed the case. Naturally, efforts to develop more efficient photovoltaic cells must find a way to reclaim this energy. Most proposed "third-generation" technologies [15] attempt to circumvent thermalization in different ways: spectral splitting, multi-junctions, hot-carrier cells, down-conversion, etc. So far, only multi-junction solar cells have been convincingly demonstrated and are industrially scalable.

16.1.2 How Do Multi-Junction Solar Cells Work?

Solar cells suffer the largest loss because they waste (as heat) all photon energy in excess of the bandgap. To put it differently, low-bandgap absorbers are inefficient at utilizing high-energy photons. This inefficiency can be avoided if we use low-bandgap absorbers for low-energy photons and high-bandgap absorbers for high-energy photons (Figure 16.3a). This is the idea behind the multi-junction solar cell. The multi-junction cells consist of several absorbers with different bandgaps, each specializing in photons of a certain energy range. Each absorber acts like an optical filter for the following one. Light is first incident on the widest-gap absorber and is successively filtered down to the lowest-gap absorber. Note that light has to be incident on the widest bandgap first for this concept to work. The sub-cells are all connected electrically in series in a two-terminal multi-junction (see Section 16.1.3 for explanation), resulting in the V_{OC} being the sum of the sub-cell V_{OC}. In essence, this is akin to stacking many solar cells on top of each other and connecting them electrically in series. Though the current is reduced, the increase in V_{OC} more than makes up for it, increasing the net efficiency. This efficiency gain comes from reduced thermalization loss (Figure 16.3b). Thus, a multi-junction of two cells can reach 42%, multi-junction of three cells can reach 49%, and so on, with diminishing returns. In the limiting case of an infinite stack of sub-cells, the thermalization loss is eliminated, and an efficiency of 68% can be calculated. The numbers are higher under concentrated sunlight. Solar cells with up to five sub-cells have been realized so far, yielding a record efficiency of 39% under normal solar irradiance and 46% under concentrated sunlight.

Two-terminal tandems need to be "current matched." This means that bandgaps and layer thicknesses have to be selected such that every sub-cell absorbs the same number of photons and consequently generates the same current. Tandems are at their most efficient when bandgaps are matched.

So far, several kinds of multi-junctions have been demonstrated: III–V, organic–organic, III–V and various multi-junctions involving epitaxially grown III–V semiconductors. These are extremely expensive and have been used primarily for space and concentrator applications.

16.1.3 Multi-Junction: Two-Terminal, Three-Terminal, and Four-Terminal Multi-Junctions

Two kinds of device configurations are popular for distributing photons between several absorbers. In case of tandems, they are termed the two-terminal (2T) and four-terminal (4T) configurations. The names derive from the number of electrical contacts emerging from each cell configuration. We discuss these cases for the case of tandem, but they readily generalize to more junctions.

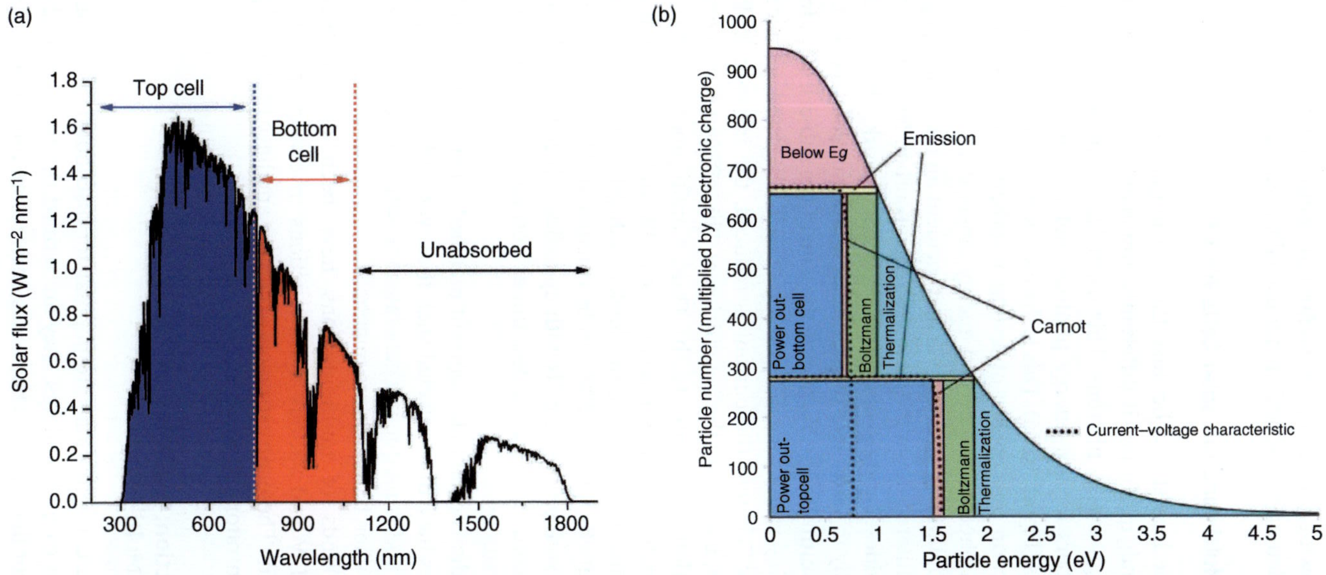

Figure 16.3 (a) Solar irradiance spectrum showing how photons are shared between two absorbers in a perovskite (1.64 eV) on c-Si (1.1 eV) tandem. (b) Intrinsic losses in a tandem solar cell with the ideal bandgap combination, 0.98 and 1.87 eV. The efficiency gain can be seen to come from the reduced thermalization. Source: Hirst and Ekins-Daukes [11].

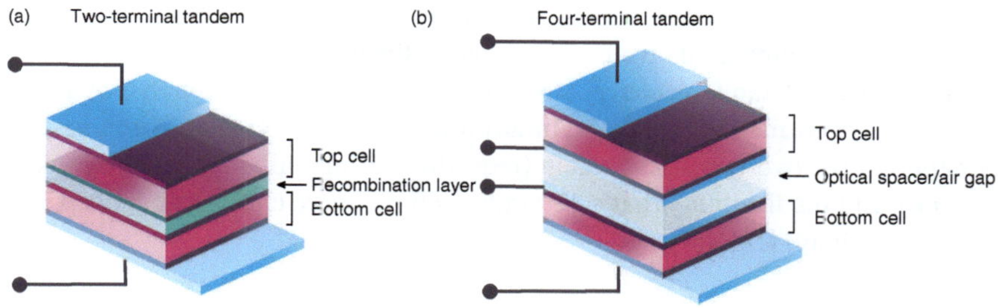

Figure 16.4 (a) Architecture of a two-terminal tandem, showing the two sub-cells separated by a recombination layer. (b) Architecture of a four-terminal tandem.

In the four-terminal tandem (Figure 16.4b), two sub-cells are fabricated on separate substrates and mechanically stacked on top of each other such that the narrow bandgap sub-cell receives light filtered through the wide gap sub-cell. Electrically, both the cells are completely independent, yielding four electrical contacts. The sub-cells are biased independently, with their own maximum power point tracking, inverter, etc. This can cause a substantial increase in cost, since balance of system costs form a large fraction of the total cost of producing solar energy. Since the cells are isolated, each cell is also isolated from any problems or degradation that may occur in the other. Of the four electrical contacts, three must be semi-transparent, leading to some resistive and optical losses in 4T tandems. An advantage of 4T compared to monolithic integration is that the current of the sub-cells does not need to be matched. This results in a small energy yield increase in practical applications, since the efficiency is less affected by changes in incident radiation due to factors such as cloud coverage or the sun's zenith angle [16].

In the two-terminal tandem configuration (Figure 16.4a), the two sub-cells are fabricated on top of each other, such that the two are electrically connected in series. Thus, only two electrical contacts are required. This connection between the sub-cells is usually achieved using a tunnel junction or a low resistance recombination layer. As in the four-terminal tandem, the wide gap cell acts as an optical filter for the narrow gap cell. However, the series connection means that the same current will flow through the two sub-cells (Kirchhoff's [17] law). If the two sub-cells generate different currents, the net current will be determined by the lower of the two. Hence, photocurrents in these cells must be "matched" to ensure the same photocurrent generation at the maximum power point. However, while this requires careful management of layer thickness and bandgaps, there is no current matching requirement in a 4T terminal tandem since the sub-cells are biased independently. Nevertheless, current matching is still desirable for best performance.

The two-terminal multi-junction is more difficult to fabricate than the four-terminal tandem. However, the two-terminal multi-junction is less expensive, both in terms of module cost and balance of system costs. Catastrophic failure is a disadvantage with the two-terminal multi-junction: if any sub-cell incurs a problem, the entire cell stops working. All examples of commercially manufactured tandem solar cells are based on a two-terminal design.

A three-terminal tandem concept has also been proposed, though not much progress has been made [17]. Its main advantage is that it combines the ease of fabrication of the two-terminal tandem (requiring wiring only in front and back of cell stack), and the flexibility that a 4T tandem offers (more choice of bandgap, less spectral sensitivity) [18].

16.1.4 Why Perovskites for Multi-Junctions?

As mentioned earlier, multi-junctions (including tandem) are the optimal strategy to make solar cells with higher efficiencies. While high multi-junction efficiency (>40%) has been achieved, epitaxially grown III–V semiconductor junctions, which can be 2 or 3 orders of magnitude more expensive per watt, are involved. A GaAs-based multi-junction cell costs upto $300 per W [19]. In comparison, utility scale solar is only $1 per W, making epitaxially grown multi-junctions only practical in concentrator PV, or niche applications like space or military. Furthermore, from the established inorganic semiconductors, the choice of bandgaps is limited: Si (1.1 eV), CIGS (1.0–1.7 eV), GaAs (1.42 eV), CdTe (1.5 eV), InGaP (1.4–2.3 eV). Even then, it is not guaranteed that two absorbers with perfectly matched bandgaps can be combined. Furthermore, some of these materials cannot be scaled to terawatt proportions due to elemental scarcity and/or low production capacity.

The bandgap tunability of perovskites provides an answer to this issue. The bandgap tunability of perovskites has opened up the possibility of creating scalable tandems. The bandgap can be tuned across the visible spectrum simply by ionic substitution and mixing. For ABX_3 perovskites, by selecting suitable cations and halides, the bandgap can be tuned between ~1.24 and 3.5 eV. However, to form structurally stable perovskites, the size of the components should be carefully matched. A reasonably good prediction can be obtained by using Goldschmidt's tolerance factor based only on ionic radii. The following ions are the most popular: A = Cs, methylammonium (MA), formamidinium (FA), B = Sn^{2+}, Pb^{2+}, and X = I^-, Br^-, Cl^-. In particular, the mixed-halide I–Br perovskites $APb(I_xBr_{1-x})_3$ have a bandgap in the range 1.48–2.35, making them suitable for making wide gap cells for tandem (Figure 16.5a) and triple junction applications. Perovskite-on-silicon (see 16.2) and CIGS-perovskite cells [25] have been demonstrated using these materials. The $APb_{0.5}Sn_{0.5}I_3$ perovskites (E_g = 1.2 eV) are suitable as a narrow gap absorber, opening up the possibility of designing all-perovskite multi-junction devices (Figure 16.5b). All the elements in perovskites (with the exception of, perhaps, cesium), are available in great abundance and are already produced in copious quantities. The high power/mass ratio also makes it more attractive than Si-based multi-junctions for space applications.

Additionally, high efficiency multi-junctions, which have mostly been epitaxially grown, have been limited to a few niche applications due to the high costs. Concentrator PV has been one such application, where the expensive cell is kept small, and optics are used to gather sunlight over a large area. Perovskite based

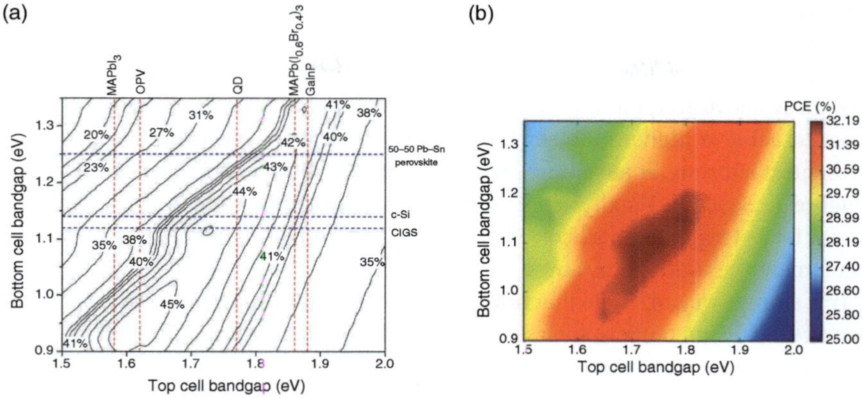

Figure 16.5 (a) Power conversion efficiency of a two-terminal tandem, as a function of the bandgap, calculated in the radiative limit. The representative bandgaps of some PV technologies are indicated: organic PV (OPV) [20], $MAPbI_3$ perovskite, $MAPb(I_{0.6}Br_{0.4})_3$ perovskite, $CsPbI_3$ quantum dots (QD) [21], GaInP [22], $FA_{0.83}Cs_{0.17}Pb_{0.5}Sn_{0.5}I_3$ (50–50 Pb–Sn perovskite), crystalline silicon (c-Si), and copper indium gallium selenide (CIGS) [23]. (b) Modeled power conversion efficiency of perovskite-perovskite tandems, as a function of sub-cell bandgaps. Source: Hoerantner et al. [24].

multi-junction technology, which avoids expensive epitaxy, has the potential to enable concentrators on a large scale. Perovskites cells have been demonstrated to be compatible with concentrated sunlight [26].

The annual energy yields of photovoltaic arrays can be reliably predicted based on optoelectronic models, which take in irradiance data as input [16]. These predictions are quite crucial in an industry that already runs on razor thin margins. For thin-film solar cells, the generalized transfer-matrix method is commonly used to simulate the cell optics. The electronics can be handled either by a simple diode-based, or a drift-diffusion model. Hörantner et al. [16, 24] have calculated that tandems have over 30% higher annual energy yields than single junctions. More precisely, efficiencies of 31.8% (perovskite/Si) and 33.4% (perovskite/perovskite) are possible, within practical limits, if all perovskite sub-cells were state of the art. In the triple junction configuration, even higher efficiencies are possible: 36.6% (perovskite/perovskite/perovskite) and 38.8% (perovskite/perovskite/Si).

16.2 Perovskite-Silicon Tandems

In October 2014, the first report of a perovskite containing multi-junction photovoltaic device, was a thin-film, two-terminal perovskite-kesterite tandem reaching an efficiency of 4.4% [27]. This result was soon overtaken by 4T and 2T perovskite-silicon tandems, with efficiency exceeding 20% by the end of 2015 [28]. Combining perovskites with silicon solar bottom cells is attractive for several reasons. First, silicon PV is the mainstream technology, which has driven the decrease of price over the last decade and accounts for more than 95% of the annual

PV production capacity [2]. Next, the bandgap of the most efficient perovskite absorbers (~1.55 eV) is relatively well suited to the bandgap of crystalline silicon (~1.12 eV). In a two-terminal tandem with such bandgaps, the maximum efficiency is slightly above 30%, mostly limited by the current matching conditions [29]. However, to achieve higher efficiency, tunability of the bandgap toward 1.65–1.70 eV presents a substantial advantage, with maximum theoretical efficiency above 40% (Figure 16.5a) [29]. Finally, since transparent conductive oxide (TCO) are already in use for heterojunction silicon PV, the latter technology is particularly suitable for tandem devices with perovskites.

In a similar way to single junction perovskite photovoltaics, research activities in perovskite/silicon tandem solar cells have been flourishing since the first reports in 2014, with more than 80 reports published by the end of 2019 (see Figure 16.6). This topic has already been reviewed several times, [30–35] and efficiency is increasing steadily (Table 16.1). Therefore, an exhaustive review of the progress will be quickly outdated; we invite the reader to refer to the most recent reviews to learn about the current status of the research, while we concentrate here on the particular challenges needed to be overcome to reach practical power efficiencies above 30%.

16.2.1 Bandgap Engineering

Using the Shockley–Queisser detailed balance model to estimate the maximum theoretical efficiency of a tandem cell based on a silicon bottom cell, the bandgap of the top absorber should be of 1.75 eV. However, when calculating the maximum efficiency using experimental optical constants and reasonable film thicknesses, the optimal bandgap for a two-terminal device is slightly lower, between 1.65 and 1.70 eV, due to the finite slope of the absorption edge [16]. To achieve such bandgaps, several perovskite compositions can be envisaged. For example, cesium lead iodide ($CsPbI_3$) with a bandgap of 1.73 eV would be highly advantageous [50]. Unfortunately, this composition is metastable in its perovskite black phase and is

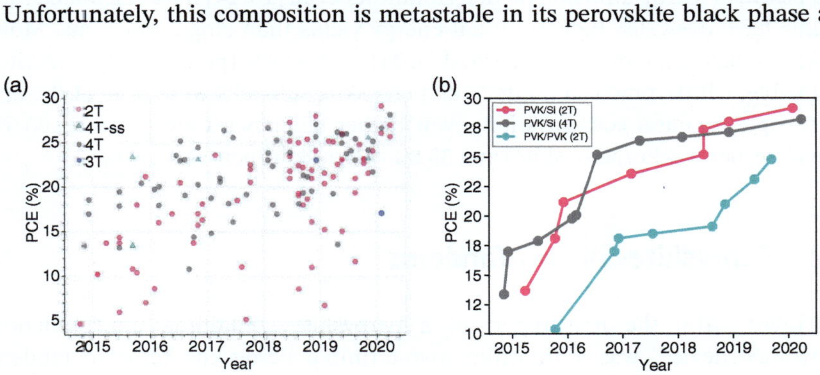

Figure 16.6 (a) Reports of tandem solar cells using at least one perovskite absorber as sub-cell. 2T: two-terminal, 4T-ss: four-terminal with spectral splitting, 4T: mechanically stacked four-terminal, 3T: three-terminal. (b) Evolution of record efficiencies of perovskite/silicon 2T and 4T devices, and all-perovskites two-terminal tandems (including press releases from Oxford PV and Helmholtz-Zentrum Berlin).

Table 16.1 Notable reported two-terminal tandem solar cells that include perovskites in at least one sub-cell.

Top cell	E_g (eV)	Bottom cell	E_g (eV)	Recombination layer	Area (cm^2)	J_{SC} ($mA\ cm^{-2}$)	V_{OC} (V)	*FF* (%)	PCE (%)	Year	References
$MAPbI_3$	1.55	CZTSSe	1.1	ITO	0.45	5.6	1.35	60	4.6	2014	[36]
$MAPbI_3$	1.55	Polymer	≈1.3	PEDOT:PSS	0.1	10.1	1.52	67	10.2	2015	[37]
$MAPbI_3$	1.55	Homojunction	1.1	n^{++}/p^{++} Si	1	11.5	1.58	75	13.7	20115	[38]
$FAMAPbI_{3-x}Br_x$	1.57	SHJ	1.1	ITO	0.16	14.0	1.78	79	18.1	2015	[39]
$MAPbI_3$	1.55	SHJ	1.1	IZO	0.17	15.9	1.69	78	21.2	2015	[28]
$MAPbI_3$	1.55	SHJ	1.1	IZO	1.43	16.4	1.72	73	20.5	2016	[40]
$FA_{0.83}Cs_{0.17}Pb(I_{0.5}Br_{0.5})_3$	1.8	$FA_{0.75}Cs_{0.25}Sn_{0.5}Pb_{0.5}I_3$	1.2	ITO/IZO	0.2	14.5	1.66	70	17.0	2016	[41]
$FA_{0.85}Cs_{0.15}Pb(I_{0.7}Br_{0.3})_3$	≈1.7	$MAPbI_3$	1.55	TaTm:F_6-TCNNQ/ C_{60}:PhIm	0.03	10.1	2.28	76	17.4	2016	[42]
$FA_{0.83}Cs_{0.17}Pb(Br_{0.17}I_{0.83})_3$	1.63	SHJ	1.1	ITO	1	18.1	1.65	79	23.6	2017	[43]
$Cs_xFA_{1-x}Pb(I,Br)_3$	1.6	SHJ	1.1	nc-Si tunnel	1.42	19.5	1.79	73	25.2	2018	[44]
$Cs_{0.09}FA_{0.77}MA_{0.14}Pb(I_{0.86}Br_{0.14})_3$	1.64	CIGS	1.0	BZO/ITO	0.04	17.3	1.77	73	22.4	2018	[45]
$Cs_{0.15}(FA_{0.83}MA_{0.17})_{0.85}Pb(I_{0.8}Br_{0.2})_3$	1.64	SHJ	1.1	ITO	0.42	17.8	1.80	79	25.4	2018	[46]
$Cs_{0.05}(FA_{0.83}MA_{0.17})_{0.95}Pb(I_{0.8}Br_{0.2})_3$	1.63	SHJ	1.1	ITO	1.1	19.0	1.79	74	25.4	2019	[47]
$Cs_{0.05}FA_{0.8}MA_{0.15}PbI_{3-x}Br_x$	1.75	$(FASnI_3)_{0.6}(MAPbI_3)_{0.4}$	1.25	ITO	0.11	15.0	1.94	80	23.1	2019	[48]
$FA_{0.78}Cs_{0.22}Pb(I_{0.82}Br_{0.15}Cl_{0.03})_3$	1.67	SHJ	1.1	ITO	1	19.1	1.89	75	27.0	2020	[49]

promptly converted into a yellow non-perovskite phase upon exposure to ambient air [51]. Nevertheless, several strategies to stabilize $CsPbI_3$ in its perovskite phase have been reported, using layer-by-layer assembly of nanocrystals [52], preferential crystallization in its tetragonal β-$CsPbI_3$ polymorph [53], or addition of small amounts of dimethylammonium together with Cs as an A cation [54, 55]. However, a tandem device with silicon has yet to be demonstrated.

Alternatively, the bandgap can be widened by mixing bromide with iodide in Pb-based perovskite using organic A cations (e.g. MA, FA) or in mixed compositions with inorganic cations (Cs, Rb). This approach provides a large combinatorial parameter space, which has been explored, and will probably be further developed in the future. However, a number of limitations need to be considered when designing an optimal perovskite composition. The thermal stability of perovskites using MA as A cation is generally less than those using FA or Cs, due to its higher volatility and chemical reactivity. On the other side, FA-based perovskite can turn into a yellow non-perovskite phase, in a similar way than the Cs-based analogues. McMeekin et al. discovered that by mixing a small amount of Cs with FA, stable compositions of mixed-halide perovskites could be obtained over a large range of iodide-bromide ratios [56]. Most of the high efficiency perovskite-silicon tandem reported so far (with PCE > 25%; see Figure 16.6) are based on mixed-cation (Cs, MA, FA), mixed-halide (I, Br) perovskites.

Although accurate tuning of the bandgap can be achieved using composition engineering, the efficiency of perovskite solar cells with a significant amount of bromide (typically 40%) is lagging behind neat iodide perovskites [57]. This is particularly expressed by the increasing V_{OC} loss with higher bromide content [57]. Some authors have attributed this loss to the halide segregation effect observed with mixed anion composition. This effect is clearly demonstrated by the reversible shift of the photoluminescence peak toward a lower bandgap under constant illumination. Halide segregation has been imputed to the migration of ions giving rise to iodide-rich regions, and the actual mechanism is currently still under debate despite extensive research effort [58, 59]. However, careful analysis of the effect of halide segregation on the V_{OC} deficit, revealed that the dominant contribution to this loss was rather non-radiative recombination and that segregation played a relatively minor role [60].

16.2.2 Parasitic Absorption

In common with all types of multi-junctions, the optical quality of the perovskite-silicon tandem stack needs to be carefully optimized. In Figure 16.7d, we show a plot of the external quantum efficiency (EQE) illustrating the different loss mechanism limiting the maximum current that can be extracted from a high-efficiency tandem device (Figure 16.7a–c).

First, we discuss the sources of parasitic absorption in a common tandem architecture. In an ideal device, high-energy photons are only and totally absorbed in the top-cell, in this case the perovskite, and the lower-energy photons in the silicon sub-cell, requiring all other layers in the stack to be transparent in a given range of

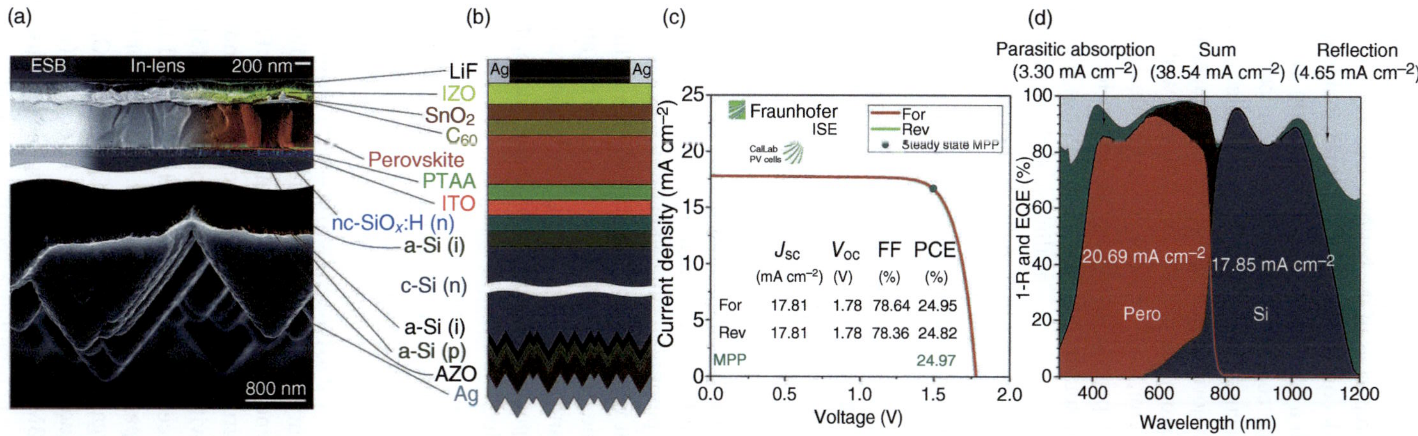

Figure 16.7 Architecture and performance of a two-terminal perovskite on a back-textured silicon heterojunction sub-cell. (a) Colored cross sectional SEM image of the top cell (upper panel) and back side of the bottom cell (lower panel) of a typical monolithic tandem solar cell. The left side of the top cell is recorded with an energy selective backscattered (ESB) detector, the right side with an in-lens detector. (b) Schematic device layout of the tandem architecture. (c) Certified current density–voltage (J–V)-characteristics. (d) Measured external quantum efficiency and reflection spectra with integrated current densities and their sum as indicated. Additionally, the loss in current density due to parasitic absorption (as the difference between sum of the EQE and 1-reflectance) and reflection are shown. Source: Köhnen et al. [61].

the solar spectrum. Materials used in contacts and recombination layers are usually selected to be transparent over the whole spectral range, although this is not strictly required in the blue range for materials in the bottom cell.

Except the interdigitated back contact (IBC) architecture, incident solar light has to be transmitted first through a transparent electrode on top of the photovoltaic device. This is inevitably at the cost of parasitic absorption losses and unwanted reflections, which we will discuss later. The transparent electrode needs to be highly transmissive over the majority of the visible spectral range, as well as in the near-infrared region for the bottom silicon cell. In addition, low sheet resistance is required to maintain the fill factor, most critically when the device area increased. The vast majority of 2T and 4T perovskite/silicon tandem devices, use indium-tin oxide (ITO) as TCO. However, sputtering of ITO can damage the organic charge selective layer, and normally requires an annealing step above 200 °C. Alternative TCOs have been proposed such as IZO [62], AZO, [63] or hydrogenated indium oxide (IO:H) [40]. Another approach is to use transparent metallic electrodes such as silver nanowires [38] or ultrathin evaporated metals such as Au sandwiched between MoO_x [64].

Importantly, the top absorber needs to be highly absorbing above bandgap but completely transparent below bandgap. This is one of the major advantages of perovskites because, as high quality direct gap semiconductors, they have absorption coefficients on the order of $10^4\,cm^{-1}$ near band edge. Moreover, they are highly transparent below bandgap, which is illustrated by a low Urbach energy, the characteristic energy associated with the exponential tail of the absorption spectrum, and the absence of absorbing sub-gap defect states. However, complete absorption of high energy photons in the perovskite sub-cell is limited by the thickness of the film, which must be consistent with the photo-generated charge diffusion lengths. By improving the quality of the perovskites, the thickness of the absorber layer is steadily increasing up to 1 μm and above [62]. We note that while incomplete absorption in the top-cell can be compensated by absorption in the silicon, it still has a cost in terms of maximum power due to the loss in voltage.

All efficient perovskite sub-cells are based on a heterojunction architecture, and thus use p and n selective layers on each side of the absorber. Although, they can be very thin, parasitic absorption in these layers can play a significant role. There is a relatively large choice of organic p-type selective layers that fulfill these conditions, including polymer (e.g. poly[bis(4-phenyl)(2,4,6-trimethylphenyl)amine] [PTAA], poly[*N*,*N*′-bis(4-butylphenyl)-*N*,*N*′-bis(phenyl)-benzidine] [pTPD], poly(3,4-ethylenedioxythiophene) polystyrene sulfonate [PEDOT:PSS]) and small molecules (spiro-MeOTAD, spiro-TTB). There are currently less candidates for inorganic p-type selective layers, with the notable exception of NiO [43]. Interestingly, the current world record device used an ultra-thin self-assembled monolayer as a p-type contact with negligible parasitic absorption [65]. On the other hand, although there is a wider choice of inorganic n-type contacts (e.g. SnO_2, ZnO, TiO_2), highest efficiency devices usually include a fullerene selective layer (C_{60}, (6,6)-phenyl C61 butyric acid methyl ester [PCBM]). Due to a non-negligible parasitic absorption in the fullerenes and evidence for non-radiative recombination at this interface [66], an alternative material would be desirable. In a tandem device using a silicon heterojunction bottom cell, parasitic absorption in the doped hydrogenated amorphous silicon (a-Si:H) carrier selective layers are a significant source of current losses. For a single junction silicon cell, this can correspond to up to 2 mA cm^{-2} [67]. One way to mitigate this parasitic absorption is to replace the p-type amorphous silicon layer with nanocrystalline silicon oxide [47] or transition metal oxides, like MoO_x [67]. In addition, due to effusion of hydrogen from the amorphous silicon layers, the processing temperature of such devices is limited to ≈200 °C. This precludes the use of charge-selective layers that require higher temperature annealing or sintering, like mesoscopic TiO_2 which is usually sintered at 450–500 °C.

Another notable source of parasitic absorption is the recombination layer, which is required to connect electrically the two sub-cells by recombining the hole current from the p-type contact of one cell, to the electron current from the n-type contact of the other cell. Ideally, the recombination layer should be transparent below the bandgap of the top-cell and have a very low electrical resistance. The recombination layer can either be achieved with a tunnel junction consisting of a highly doped p–n junction [38, 68], or by a low-resistance conducting transparent oxide, such as ITO or IZO.

16.2.3 Optical Management

In addition to parasitic absorption losses, reflections play a major role in the efficiency of tandem stacks, in both 2T and 4T configurations. Reflections from the front surface can be reduced relatively easily by deposition of an anti-reflection coating (ARC), such as a thin dielectric layer (e.g. LiF or MgF_2). However, accurate optical management is required in order to optimize the reflections at each interface, so as to design devices with appropriate absorber layer thicknesses. The two main constraints governing the choice of the perovskite layer thickness are (i) the photogenerated carrier diffusion lengths and (ii) the current matching conditions between the sub-cells in a 2T device. To optimize a full device stack, optical modeling is usually performed, using well-established techniques, such as the transfer matrix method.

Although all layers of the stack need to be included to optimize the optical properties of the device, some interfaces play a predominant role. First, when using a TCO as the intermediate recombination layer, strong reflections are generated due to the low refractive index of the oxide (n_{ITO} = 1.4–2) and the strong contrast with the indices of the perovskite (n_{PVK} = 2.6) and silicon (n_{Si} = 3.6). This effect can be reduced with the use of nanocrystalline silicon junctions with higher refractive index (n_{ncSi} = 2.6) [47, 68]. Second, polished silicon surfaces are strongly reflective, and therefore the vast majority of commercial silicon solar cells are textured on both sides via a KOH-etching process. The pyramids generated by this process are typically a few microns high (Figure 16.7a) and therefore set a major challenge for the deposition of the sub-μm continuous layers required for the perovskite sub-cells. Sahli et al. [44] demonstrated an efficient tandem on a fully-textured silicon cell, using a combined vapor and solution processing method. This problem can be mitigated by using back-only textured silicon wafers, and/or with the use of antireflection textured foils deposited on top of the device [49].

16.3 Perovskite–Perovskite Tandems

The idea of an all-perovskite multi-junction solar cell is attractive since these devices would enable over 30% PCE (over 40% in the theoretical limit), while having the advantage of low-temperature processing, high manufacturing throughput, and the ability to use flexible and lightweight substrates (Figure 16.8). The first two-terminal all-perovskite tandem (7% PCE) was realized in 2015, and the current (and rapidly moving) record as of April 2020 is 25% PCE, published by Tan H. and coworkers [70] in October 2019. This cell exhibits a V_{OC} of 1.93 V, J_{SC} of 15.8 mA cm^{-2}, and a fill-factor of 81%. All the perovskite-perovskite cells made so far have followed the template of using a Sn–Pb based perovskite for the low gap cell, and a neat lead mixed-halide perovskite for the wide-gap cell (Table 16.1). The maximum possible short-circuit current for the perovskite–perovskite tandem is set by the low bandgap sub-cell, which is 1.22 eV for the lowest bandgap perovskite ($FAMAPb_{0.5}Sn_{0.5}I_3$). The ideal companion sub cell bandgap for this would be 1.80 eV [24], which is approximately achieved by a perovskite with 40 : 60 Br to I ratio). We note, however, that 1.22 eV is still larger a bandgap than ideal (Figure 16.5b). If a perovskite with a bandgap smaller than 1.22 eV could be found, it would have the dual advantage of moving closer to the thermodynamic optimum and allowing for decreasing the bandgap of the wide gap cell. This is a natural advantage enjoyed by perovskite-on-Si tandems.

The most serious barrier toward 30% PCE is the poor performance of mixed-halide perovskites. Perovskite compositions with higher bromide contents have, so far, failed to achieve the same excellent device efficiencies that distinguish their iodide-rich counterparts. The best iodide-rich perovskites have voltage losses of ~60 mV from the radiative limit [71]. In comparison, bromide rich cells show hundreds of millivolts of voltage loss. The bulk of this loss arises from high trap densities in the absorber and at the perovskite-charge transport layer interfaces [72]. This is further exacerbated by the photo-instability of these phases: they are prone to segregating into iodine-rich and bromide-rich phases upon illumination [58]. This too is a secondary, but significant contributor to the voltage loss.

Electronically, lead-tin perovskites perform extremely well, reaching open-circuit voltages higher than 0.85 V [73–75] from a bandgap ~1.2 eV. The V_{OC} loss from the thermodynamic limit is even smaller than in c-Si cells. However, they present optical challenges: Lead-tin perovskites do not absorb as strongly as the lead perovskites, especially close to the bandgap. The reasons for this are not fully clear. The performance of these cells has been limited by poor absorption in the near-infrared (NIR), which reduces available photo-current by as much as 3 mA cm^{-2}. Tan H. and coworkers have shown that this current can be recovered by fabricating thick films (860 nm) with large diffusion lengths [70]. McGehee and coworkers have passed the IEC61646 standard test (damp heat aging for 1000 hours at 85 °C with 85% relative

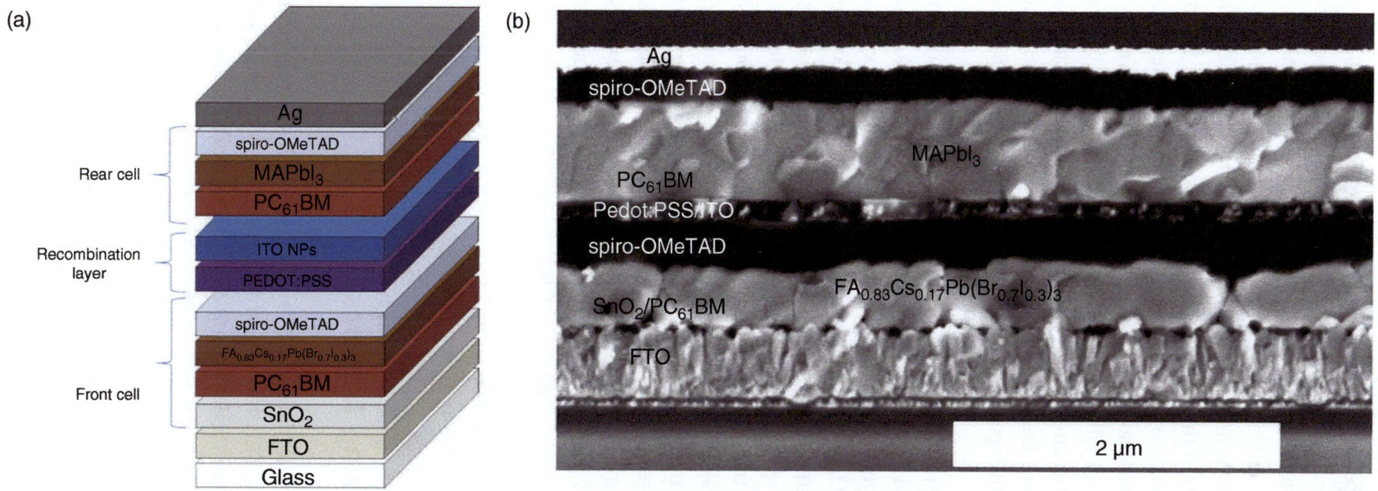

Figure 16.8 Architecture and cross-sectional scanning electron microscopy on a typical two-terminal perovskite–perovskite tandem solar cell. Source: McMeekin et al. [69]. (a) Tandem cell architecture. (b) Cross sectional scanning electron microscopy on the tandem.

humidity) on encapsulated Pb–Sn solar cells that replace MA by Cs and use ITO as the electron blocking layer [76]. These developments should lead to significantly improved tandem cell performances when implemented in cells.

Optical challenges become more severe in perovskite multi-junctions. Layer thicknesses should be selected carefully because of optical interference effects, which become very important in thin-film devices. Reflections at various interfaces in the cell interfere to create dips and valleys in light intensity within the stack. The large numbers of layers (>10) in multi-junctions can make this effect prominent. In ideal circumstances, we would want maximum light intensity with the absorber, and minimum within layers that absorb parasitically. This can be achieved by carefully tuning the layer thicknesses with the aid of optical simulations (typically a Transfer Matrix Model). Layer thicknesses have to also be carefully selected to meet the current matching condition, i.e. both sub-cells must generate the same short-circuit current. The same challenges discussed above for perovskite-Si tandems are also applicable here.

In two-terminal multi-junctions, the sub-cells are electrically connected in series. This is usually achieved through a *recombination layer* that is sandwiched between the charge transport layers of successive sub-cells. For perovskites, this is most commonly chosen to be ITO, for the ease of solution processing succeeding layers. This deposition is made complicated, however, by the possibility of damaging the underlying layers during the sputtering process. Several groups have developed protective buffer layers to protect from sputter damage [28, 39, 41, 43]. MoO_x, NiO, VO_x, AZO, SnO_2 are some materials that have been used. These are typically thin (5–20 nm) to not impede transport. There is also the constraint of low temperature processing (<150 °C) due to the presence of organic selective layers. An organic p–n junction using highly doped n and p organic materials has also been demonstrated in fully evaporated tandems [77]. However, this does not ensure good long-term stability, as the thin organic layers cannot prevent the release and migration of metal and halide ions. In short, the field faces the challenge of developing a stable interlayer with good vertical, and poor horizontal conductivity, which does not damage underlying layers, and is suitable for low temperature deposition.

16.4 Characterizing Tandems

The characterization of tandem solar cells is notably less straightforward than for single junction devices. Therefore, in this section, we introduce some of the concepts and techniques required for accurate and reproducible determination of photovoltaic device performance. For *J-V* curves and steady-state efficiencies, the spectral mismatch between the calibrated reference diode and the device under investigation can be complicated to determine accurately. In addition, the measurement of EQE of the independent sub-cells is challenging, especially when the absorption spectra of the active layers overlap. These issues as well as best measurement practices have been reported previously for III–V multi-junction devices [78–80] and for organic PV cells [81]. In principle, the methods developed for these technologies are applicable and should be accurately followed, considering the particularities of perovskite solar cells when acquiring *J-V* curves for the calculation of the mismatch factors.

Hereafter, we briefly describe the recommended procedure to obtain reliable performance characterization of two-terminal multi-junction solar cells. In order to determine the performance of the device under standard test conditions (i.e. AM1.5G, 100 mW cm^{-2}), the spectrum of the solar simulator has to be adjusted so that each sub-cell generates the same photocurrent as under the AM1.5G reference spectrum. This can be achieved with a dual or multi-zone solar simulator, with each zone chosen to match the spectral response of each sub-cell. Each sub-spectrum can then be adjusted mathematically with a spectral mismatch correction which requires the knowledge of the solar simulator spectrum and the spectral response of each sub-cell. Therefore, the first stage consists in the accurate measurement of the spectral response (or EQE) for each sub-cell. In practice, this can be achieved for a given sub-cell by saturating all other sub-cells with monochromatic bias illumination so that the photocurrent of the full stack is limited by the response of the sub-cell under test. In case of overlap of the sub-cells' spectral range, as commonly observed for organic solar cells, optical modeling is required to determine their relative contributions to the photocurrent. In addition, voltage biasing needs to be applied so that the sub-cell under test runs near short-circuit conditions, and also to prevent artifacts appearing in the case of low shunt resistance or low reverse breakdown voltage [78]. Such artifacts can be identified by performing a dark spectral response (i.e. when no bias light is applied) measurement on the multi-junction device [82]. Once the spectral responses have been obtained, a mismatch factor can be calculated for each sub-cell. Then in order to generate the required photocurrent in each sub-cell, the spectrum of the solar simulator must be adjusted and tested by measuring the J_{SC} of a reference solar cell under the new spectral conditions. New mismatch factors are then calculated. This process must be run iteratively until a consistent photocurrent is obtained for the reference solar cell, and it is only achievable by using a multi-source solar simulator, dual-source for tandem cells, in which the spectra of the light sources are fixed but the intensities can be adjusted independently [79]. If the overall spectrum of the light source is fixed, then although the mismatch factor can be applied to an accurate estimate of the J_{SC}, it is highly likely that the current density will be mismatched between the multi-junctions, and the fill factor (*FF*) of the multi-junction solar cell will be overestimated, in comparison to what would be obtained under true AM1.5G.

16.5 Commercialization

The contribution of the BoS, including inverter, construction, installation and mounting structure, accounts now to more than 50% of the levelized cost of electricity (LCOE) [2]. Therefore, the most attractive way to further decrease the LCOE, is to improve the module PCE since all other costs scale with the installed area. The cost of the module itself will continue to go down, but now mainly due to cost savings as the silicon technology is reaching its practical efficiency limits. Recently, the solar industry adopted widely the PERC (passivated emitter and rear cell) silicon technology, in part due to its relatively low upgrade cost. In 2019, commercial PERC cells reached up to 22% efficiency, with best estimates forecast approaching 24.1% by late 2020 [83]. The world record is currently 26.7%, [84] but seems out of reach for commercial application, and the fundamental thermodynamic limit is 29.4% [85]. Therefore, the potential for significant efficiency increase with silicon technologies is relatively limited. In this context, the prospect of very high efficiencies offered by tandem solar cells including perovskite sub-cells, is highly attractive for the industry. In 2018, Oxford PV demonstrated a perovskite-Si tandem cell with 28.0% efficiency and claim to have a roadmap to achieve a practical limit ~33% [86]. One of the appealing features of the perovskite-Si tandem approach is that, instead of competing with cost-effective, well established technologies, it can be made compatible with existing production lines. Therefore, companies like Oxford PV design their tandem cells to be built on top of commercial $156 \times 156\ mm^2$ solar cells. The success of the manufacturing challenge will depend on the ability to devise processes using available tools, and compatibility with existing module production processes. Therefore, partnerships with established production line equipment manufacturers, as well as joint development agreements with leaders in the solar industry will be important enablers in the commercialization of perovskite/silicon tandem modules.

16.5.1 Reliability

To achieve the reliability required to enter a mature market like silicon PV, perovskite tandem technology is facing a number of challenges, which are not dissimilar to those that this industry had to overcome in the last decades. The stability of silicon modules has progressed to the level that today manufacturers offer typically 25-year warranties, which means the solar modules will still generate about 80–85% of their initial power output after 25 years. To qualify modules, standardized accelerated aging tests have been developed, like the IEC61215 tests which are widely adopted by the industry. Being built on silicon solar cells, current standardized test protocols apply also to perovskite/Si tandem modules, but further research is still required to identify degradation modes that may be specific to perovskites. If these are not revealed by the current accelerated testing standards, additional tests should be included, following the example of the potential induced degradation (PID) probing tests, that were introduced in 2015 in IEC protocols. It is worth noting that silicon manufacturer's often set their internal targets to higher stress levels (typically two to three times) than the standard certifications requirements.

16.5.2 Scalability

The other major challenge the commercialization contenders face is scalability. Even for non-utility applications, solar cells need to be integrated into modules that can take very different shapes depending on the fabrication techniques and end use. For example, flexible and lightweight modules carry potential for new applications in portable devices, transport or building-integrated PV, but come with additional constraints such as the mechanical properties of the substrates and, in most cases, preventing the use of glass, which is an outstanding low-cost, transparent encapsulation material. For perovskite/Si tandems, the most obvious approach, which is adopted by the current commercialization contenders, is to achieve compatibility with the silicon solar cells format (~156 × 156 cm^2, "M2"). This strategy has the advantage of building up on decades of optimization of interconnection and encapsulation processes and materials, and the availability of large-scale production tools. The choice of the right materials and technologies is crucial for a fast market entry. Since 2016, Oxford PV is converting a former thin-film manufacturing facility in Germany into a pilot line for the fabrication of perovskite/silicon modules following this general approach [86].

Technical challenges associated to the increase of the solar cell area are mostly due to the homogeneity of the thin films and are likely to be solved by engineering appropriate deposition processes. Various large-area coating techniques have been demonstrated, including solution phase (slot-die coating, ink-jet printing, etc.) and gas phase deposition methods. The latter are usually favored for perovskite/silicon monolithic tandems, since they are compatible with the fully textured wafers usually required for efficient light absorption in the silicon cell. In addition, to enable integration in a silicon PV production line, takt times need to be compatible with current production rates (~1 wafer per 1.5 seconds), while high yields (~98%) are maintained.

Thin-film multi-junction modules (e.g. PVK/PVK, PVK/CIGS, PVK/PVK/PVK), face scalability challenges from a slightly different perspective. While they carry potential for much higher efficiencies than perovskite/Si tandems, they also offer more flexibility in terms of module design. However, for earliest entry on the market, adoption of manufacturing tools and processes developed for the silicon PV industry will be of high importance. Series resistance can be a major challenge for thin-film modules, while perovskite/Si monolithic tandems benefit from an excellent back contact (the silicon cell) and established TCO materials and deposition processes for the front contact adopted from heterojunction silicon cell technology.

16.5.3 Cost

Perovskite multi-junction technology will only be adopted at the utility scale if it delivers its promise of reduced LCOE via increased energy yields. As of 2020, all companies with declared activities toward commercialization of such technologies are pre-revenue, and therefore need to propose a viable business model to secure capital investment. In a recent review, Ballif and coworkers discussed technoeconomic aspects of the commercialization of monolithic perovskite/silicon tandem cells [31]. They considered the number of process steps required in addition to the fabrication of the bottom silicon cell, which excludes for example metallization and encapsulation steps, assumed to be common for both technologies. Considering investment depreciation, consumables and operational costs, they estimate that, for instance, with a 4% increase in absolute efficiency requiring four additional processing steps, the total processing additional cost is about €10.4 per m^2. In this scenario, the total tandem module production costs are almost on pair with a crystalline silicon module. In addition, the corresponding area-related BoS savings with respect to an 18% efficient silicon module, are of €2.7 ct per W_p and €6.4 ct per W_p for utility scale and roof top installations, respectively. This shows that even if in the short-term it will be difficult to compete with the steadily decreasing production costs for crystalline silicon modules, the additional savings on the area-related BoS costs will allow considerably lower cost of ownership, in particular when BoS costs are high, such as in roof top installations.

Detailed cost analysis have been carried out for mechanically stacked 4T perovskite/silicon modules [87] as well as thin-film perovskite multi-junction modules [88], and all highlight great commercial potential. Therefore, it is not surprising that several companies (e.g. Oxford PV, TandemPV, Swift Solar) and industrial/academic initiatives (e.g. Solliance) are currently actively targeting to bring perovskite tandem modules to the market in the near future.

16.6 Outlook

In this section, we provide some brief suggestions on avenues of priority that will enable the creation of tandems of over 30% efficiency.

Perovskite-on-silicon tandems have already reached 29% PCE, and 30% is not much further away. The tandem current densities predicted by optical simulations for untextured substrates (~19.5 mA cm^{-2}), have already been achieved. Keeping the short-circuit current from a high-performance silicon cell as the benchmark (42.3 mA cm^{-2}) [84], we can estimate 21.1 mA cm^{-2} as a safe upper limit for achievable tandem current. We suggest that the main gains to be made are in the voltage. The 1.65 eV bandgap mixed halide composition are relatively photo-stable (with respect to halide segregation). Unlike higher bromide compositions, they also show low trap densities, as illustrated by high photoluminescence quantum yields. If recombination is further reduced via passivation, or the right choice of transport materials, this represents the most straightforward way to break the 30% barrier.

Perovskite–perovskite tandems are limited by the current poor performance of both wide bandgap (>1.65 eV) and low bandgap (<1.35 eV) absorber materials. To increase the efficiency of all-perovskite tandems, the open-circuit voltage of the wide-gap cell needs to be improved as a matter of urgency. High bromide (>40%) mixed-halide thin-films have high trap densities, which introduces a large bulk recombination loss. The loss is further aggravated by interfacial recombination at the perovskite-transport layer heterojunctions. Improved deposition techniques need to be found, and new charge transport materials with low interfacial recombination and high selectivity, need to be investigated. The goal should be to ensure that the 1.8 eV bandgap cell is electronically at par with the best 1.6 eV bandgap cells (~100 mV loss), yielding a V_{OC} of >1.4 V. Lead-tin cells already perform well electronically ($V_{OC} > 0.85$ V), but waste significant amounts of photons due to the shallow absorption onset of the lead-tin perovskite. Thicker layers (~1000 nm) with suitable diffusion lengths are required to achieve a steep absorption onset. Layer thicknesses also need to be carefully tuned to avoid the omnipresent reflection losses due to the large number of layers. With such improvements, a current of $34\,mA\,cm^{-2}$ is within reach, allowing a matched tandem current of $17\,mA\,cm^{-2}$. A combination of the two improvements – wide-bandgap cell $V_{OC} > 1.4\,V$ and low bandgap cell $J_{SC} > 34\,mA\,cm^{-2}$ – will allow the 30% barrier to be broken for perovskite–perovskite tandems.

References

1 ITRPV (2019). *International Technology Roadmap for Photovoltaic (10th edition)*. Verband Deutscher Maschinen- und Anlagenbau (VDMA) https://pv-manufacturing.org/wp-content/uploads/2019/03/ITRPV-2019.pdf.

2 Philipps, S. and Warmuth, W. (2019). *Photovoltaics Report*. Fraunhofer Institute for Solar Energy Systems ISE, available at https://www.ise.fraunhofer.de/content/dam/ise/de/documents/publications/studies/Photovoltaics-Report.pdf.

3 Shockley, W. and Queisser, H.J. (1961). Detailed balance limit of efficiency of p–n junction solar cells. *J. Appl. Phys.* 32: 510–519.

4 Bedair, S.M., Lamorte, M.F., and Hauser, J.R. (1979). A two-junction cascade solar-cell structure. *Appl. Phys. Lett.* 34: 38–39.

5 Geisz, J.F., France, R.M., Schulte, K.L. et al. (2020). Six-junction III–V solar cells with 47.1% conversion efficiency under 143 Suns concentration. *Nat. Energy* https://doi.org/10.1038/s41560-020-0598-5.

6 Shah, A., Meier, J., Vallat-Sauvain, E. et al. (2003). Material and solar cell research in microcrystalline silicon. *Sol. Energy Mater. Sol. Cells* 78: 469–491.

7 Schmidtke, J. (2010). Commercial status of thin-film photovoltaic devices and materials. *Opt. Express* 18: A477.

8 Nishiwaki, S., Siebentritt, S., Walk, P., and Lux-Steiner, M.C. (2003). A stacked chalcopyrite thin-film tandem solar cell with 1.2 V open-circuit voltage. *Prog. Photovol. Res. Appl.* 11: 243–248.

9 Essig, S., Allebé, C., Remo, T. et al. (2017). Raising the one-sun conversion efficiency of III–V/Si solar cells to 32.8% for two junctions and 35.9% for three junctions. *Nat. Energy* 2 (17144).

10 Markvart, T. (2016). From steam engine to solar cells: can thermodynamics guide the development of future generations of photovoltaics? *WIREs. Energy Environ.* 5: 543–569.

11 Hirst, L.C. and Ekins-Daukes, N.J. (2011). Fundamental losses in solar cells. *Prog. Photovol. Res. Appl.* 19: 286–293.

12 Tiedje, T., Yablonovitch, E., Cody, G.D., and Brooks, B.G. (1984). Limiting efficiency of silicon solar cells. *IEEE Trans. Electron Devices* 31: 711–716.

13 Pazos-Outón, L.M., Xiao, T.P., and Yablonovitch, E. (2018). Fundamental efficiency limit of Lead iodide perovskite solar cells. *J. Phys. Chem. Lett.* 9: 1703–1711.

14 Nayak, P.K., Mahesh, S., Snaith, H.J., and Cahen, D. (2019). Photovoltaic solar cell technologies: analysing the state of the art. *Nat. Rev. Mater.* 4: 269–285.

15 Green, M.A. (2006). *Third Generation Photovoltaics: Advanced Solar Energy Conversion*, Energy Conversion, vol. 10. Springer-Verlag Berlin Heidelberg.

16 Hörantner, M.T. and Snaith, H.J. (2017). Predicting and optimising the energy yield of perovskite-on-silicon tandem solar cells under real world conditions. *Energy Environ. Sci.* 10: 1983–1993.

17 Warren, E.L., Mcmahon, W.E., Rienäcker, M. et al. (2020). A taxonomy for three-terminal tandem solar cells. *ACS Energy Lett.* 5: 1233–1242.

18 Warren, E.L., Deceglie, M.G., Rienäcker, M. et al. (2018). Maximizing tandem solar cell power extraction using a three-terminal design. *Sustain. Energy Fuels* 2: 1141–1147.

19 Meehan, C. (2018). NREL Developing Improved Tech to Lower Costs for Multi-Junction Solar Cells.

20 Green, M.A., Emery, K., Hishikawa, Y. et al. (2015). Solar cell efficiency tables (version 45). *Prog. Photovolt. Res. Appl.* 23: 1–9.

21 Sanehira, E.M., Marshall, A.R., Christians, J.A. et al. (2017). Enhanced mobility $CsPbI_3$ quantum dot arrays for record-efficiency, high-voltage photovoltaic cells. *Sci. Adv.* 3: eaao4204.

22 Green, M.A., Emery, K., Hishikawa, Y. et al. (2016). Solar cell efficiency tables (version 47). *Prog. Photovolt. Res. Appl.* 24: 3–11.

23 Green, M.A., Hishikawa, Y., Dunlop, E.D. et al. (2018). Solar cell efficiency tables (version 52). *Prog. Photovolt. Res. Appl.* 26: 427–436.

24 Hörantner, M.T., Leijtens, T., Ziffer, M.E. et al. (2017). The potential of multi-junction perovskite solar cells. *ACS Energy Lett.* 2: 2506–2513.

25 Guchhait, A., Dewi, H.A., Leow, S.W. et al. (2017). Over 20% efficient CIGS-perovskite tandem solar cells. *ACS Energy Lett.* 2: 807–812.

26 Wang, Z., Lin, Q., Wenger, B. et al. (2018). High irradiance performance of metal halide perovskites for concentrator photovoltaics. *Nat. Energy* https://doi.org/10.1038/s41560-018-0220-2.

27 Todorov, T., Gershon, T., Gunawan, O. et al. (2015). Monolithic perovskite-CIGS tandem solar cells via in situ band gap engineering. *Adv. Energy Mater.* 5: 1500799.

28 Werner, J., Weng, C.H., Walter, A. et al. (2016). Efficient monolithic perovskite/silicon tandem solar cell with cell area >1cm^2. *J. Phys. Chem. Lett.* 7: 161–166.

29 Yu, Z., Leilaeioun, M., and Holman, Z. (2016). Selecting tandem partners for silicon solar cells. *Nat. Energy* 1: 16137.

30 Eperon, G.E., Horantner, M.T., and Snaith, H.J. (2017). Metal halide perovskite tandem and multiple-junction photovoltaics. *Nat. Rev. Chem.* 1.

31 Werner, J., Niesen, B., and Ballif, C. (2018). Perovskite/silicon tandem solar cells: marriage of convenience or true love story? – an overview. *Adv. Mater. Interfaces* 5: 1700731.

32 Wang, Z., Song, Z., Yan, Y. et al. (2019). Perovskite–a perfect top cell for tandem devices to break the S–Q limit. *Adv. Sci.* 6: 1801704.

33 Song, Z., Chen, C., Li, C. et al. (2019). Wide-bandgap, low-bandgap, and tandem perovskite solar cells. *Semicond. Sci. Technol.* 34.

34 Lal, N.N., Chen, C., Li, C. et al. (2017). Perovskite tandem solar cells. *Adv. Energy Mater.* 7: 1602761.

35 Torabi, N., Behjat, A., Zhou, Y. et al. (2019). Progress and challenges in perovskite photovoltaics from single- to multi-junction cells. *Mater. Today Energy* 12: 70–94.

36 Todorov, T., Gershon, T., Gunawan, O. et al. (2014). Perovskite-kesterite monolithic tandem solar cells with high open-circuit voltage. *Appl. Phys. Lett.* 105.

37 Chen, C.-C., Bae, S.H., Chang, W.H. et al. (2015). Perovskite/polymer monolithic hybrid tandem solar cells utilizing a low-temperature, full solution process. *Mater. Horizons* 2: 203–211.

38 Mailoa, J.P., Bailie, C.D., Johlin, E.C. et al. (2015). A 2-terminal perovskite/silicon multijunction solar cell enabled by a silicon tunnel junction. *Appl. Phys. Lett.* 106: 121105.

39 Albrecht, S., Saliba, M., Correa Baena, J.P. et al. (2016). Monolithic perovskite/silicon-heterojunction tandem solar cells processed at low temperature. *Energy Environ. Sci.* 9: 81–88.

40 Werner, J., Barraud, L., Walter, A. et al. (2016). Efficient near-infrared-transparent perovskite solar cells enabling direct comparison of 4-terminal and monolithic perovskite/silicon tandem cells. *ACS Energy Lett.* 1: 474–480.

41 Eperon, G.E., Leijtens, T., Bush, K.A. et al. (2016). Perovskite-perovskite tandem photovoltaics with optimized band gaps. *Science (80-.)* 354: 861–865.

42 Forgács, D., Gil-Escrig, L., Pérez-Del-Rey, D. et al. (2017). Efficient monolithic perovskite/perovskite tandem solar cells. *Adv. Energy Mater.* 7: 1602121.

43 Bush, K.A., Palmstrom, A.F., Yu, Z.J. et al. (2017). 23.6%-efficient monolithic perovskite/silicon tandem solar cells with improved stability. *Nat. Energy* 2: 17009.

44 Sahli, F., Werner, J., Kamino, B.A. et al. (2018). Fully textured monolithic perovskite/silicon tandem solar cells with 25.2% power conversion efficiency. *Nat. Mater.* 17: 820–826.

45 Han, Q., Hsieh, Y.T., Meng, L. et al. (2018). High-performance perovskite/Cu(In,Ga)Se_2 monolithic tandem solar cells. *Science (80-.)* 361: 904–908.

46 Chen, B., Yu, Z., Liu, K. et al. (2019). Grain engineering for perovskite/silicon monolithic tandem solar cells with efficiency of 25.4%. *Joule* 3: 177–190.

47 Mazzarella, L., Lin, Y.H., Kirner, S. et al. (2019). Infrared light management using a nanocrystalline silicon oxide interlayer in monolithic perovskite/silicon heterojunction tandem solar cells with efficiency above 25%. *Adv. Energy Mater.* 9: 1803241.

48 Tong, J., Song, Z., Kim, D.H. et al. (2019). Carrier lifetimes of >1 μs in Sn–Pb perovskites enable efficient all-perovskite tandem solar cells. *Science (80-.)* 364: 475–479.

49 Xu, J., Boyd, C.C., Yu, Z.J. et al. (2020). Triple-halide wide–band gap perovskites with suppressed phase segregation for efficient tandems. *Science (80-.)* 367: 1097–1104.

50 Ahmad, W., Khan, J., Niu, G., and Tang, J. (2017). Inorganic $CsPbI_3$ perovskite-based solar cells: a choice for a tandem device. *Sol. RRL* 1: 1700048.

51 Eperon, G.E., Paternò, G.M., Sutton, R.J. et al. (2015). Inorganic caesium lead iodide perovskite solar cells. *J. Mater. Chem. A* 3: 19688–19695.

52 Swarnkar, A., Marshall, A.R., Sanehira, E.M. et al. (2016). Quantum dot-induced phase stabilization of – $CsPbI_3$ perovskite for high-efficiency photovoltaics. *Science (80-.)* 354: 92–95.

53 Wang, Y., Ibrahim Dar, M., Ono, L.K. et al. (2019). Thermodynamically stabilized β-$CsPbI_3$ – based perovskite solar cells with efficiencies >18%. *Science (80-.)* 365: 591–595.

54 Ke, W., Spanopoulos, I., Stoumpos, C.C., and Kanatzidis, M.G. (2018). Myths and reality of $HPbI_3$ in halide perovskite solar cells. *Nat. Commun.* 9: 4785.

55 Palmstrom, A.F., Eperon, G.E., Leijtens, T. et al. (2019). Enabling flexible all-perovskite tandem solar cells. *Joule* 3: 2193–2204.

56 McMeekin, D.P., Sadoughi, G., Rehman, W. et al. (2016). A mixed-cation lead mixed-halide perovskite absorber for tandem solar cells. *Science (80-.)* 351: 151–155.

57 Yang, T.C.-J., Fiala, P., Jeangros, Q., and Ballif, C. (2018). High-bandgap perovskite materials for multijunction solar cells. *Joule* 2: 1421–1436.

58 Hoke, E.T., Slotcavage, D.J., Dohner, E.R. et al. (2015). Reversible photo-induced trap formation in mixed-halide hybrid perovskites for photovoltaics. *Chem. Sci.* 6: 613–617.

59 Brennan, M.C., Draguta, S., Kamat, P.V., and Kuno, M. (2018). Light-induced anion phase segregation in mixed halide perovskites. *ACS Energy Lett.* 3: 204–213.

60 Mahesh, S., Ball, J.M., Oliver, R.D.J. et al. (2020). Revealing the origin of voltage loss in mixed-halide perovskite solar cells. *Energy Environ. Sci.* 13: 258–267.

61 Köhnen, E., Jošt, M., Morales-Vilches, A.B. et al. (2019). Highly efficient monolithic perovskite silicon tandem solar cells: analyzing the influence of current mismatch on device performance. *Sustain. Energy Fuels* 3: 1995–2005.

62 Hou, Y., Aydin, E., De Bastiani, M. et al. (2020). Efficient tandem solar cells with solution-processed perovskite on textured crystalline silicon. *Science (80-.)* 367: 1135–1140.

63 Fu, F., Feurer, T., Weiss, T.P. et al. (2017). High-efficiency inverted semi-transparent planar perovskite solar cells in substrate configuration. *Nat. Energy* 2: 16190.

64 Yang, Y., Chen, Q., Hsieh, Y.T. et al. (2015). Multilayer transparent top electrode for dolution processed perovskite/Cu(In,Ga) (Se,S)(2) four terminal tandem solar cells. *ACS Nano* 9: 7714–7721.

65 Helmholtz Zentrum Berlin, (2020) World Record: Efficiency of perovskite silicon tandem solar cell jumps to 29.15 per cent. Helmholtz Zentrum Berlin. Available

at: https://www.helmholtz-berlin.de/pubbin/news_seite?nid=21020;sprache=en;seitenid=72384.
66 Stolterfoht, M., Wolff, C.M., Márquez, J.A. et al. (2018). Visualization and suppression of interfacial recombination for high-efficiency large-area pin perovskite solar cells. *Nat. Energy* 3: 847–854.
67 Dréon, J., Jeangros, Q., Cattin, J. et al. (2020). 23.5%-efficient silicon heterojunction silicon solar cell using molybdenum oxide as hole-selective contact. *Nano Energy* 70 (104495).
68 Sahli, F., Kamino, B.A., Werner, J. et al. (2018). Improved optics in monolithic perovskite/silicon tandem solar cells with a nanocrystalline silicon recombination junction. *Adv. Energy Mater.* 8: 1701609.
69 McMeekin, D.P., Mahesh, S., Noel, N.K. et al. (2019). Solution-processed all-perovskite multi-junction solar cells. *Joule* 3: 387–401.
70 Lin, R., Xiao, K., Qin, Z. et al. (2019). Monolithic all-perovskite tandem solar cells with 24.8% efficiency exploiting comproportionation to suppress Sn(II) oxidation in precursor ink. *Nat. Energy* 4: 864–873.
71 Liu, Z., Krückemeier, L., Krogmeier, B. et al. (2019). Open-circuit voltages exceeding 1.26 V in Planar methylammonium lead iodide perovskite solar cells. *ACS Energy Lett.* 4: 110–117.
72 Mahesh, S., Ball, J.M., Oliver, R.D.J. et al. (2019). Revealing the origin of voltage loss in mixed-halide perovskite solar cells. *Energy Environ. Sci.* 50: 675.
73 Li, C., Song, Z., Zhao, D. et al. (2019). Reducing saturation-current density to realize high-efficiency low-bandgap mixed tin-lead halide perovskite solar cells. *Adv. Energy Mater.* 9: 1803135.
74 Zhao, D., Yu, Y., Wang, C. et al. (2017). Low-bandgap mixed tin–lead iodide perovskite absorbers with long carrier lifetimes for all-perovskite tandem solar cells. *Nat. Energy* 2: 17018.
75 Xu, G., Bi, P., Wang, S. et al. (2018). Integrating ultrathin bulk-heterojunction organic semiconductor intermediary for high-performance low-bandgap Pprovskite solar cells with low energy loss. *Adv. Funct. Mater.* 28: 1804427.
76 Prasanna, R., Leijtens, T., Dunfield, S.P. et al. (2019). Design of low bandgap tin–lead halide perovskite solar cells to achieve thermal, atmospheric and operational stability. *Nat. Energy* 4: 939–947.
77 Ávila, J., Leijtens, T., Bush, K.A. et al. (2018). High voltage vacuum-deposited $CH_3NH_3PbI_3$-$CH_3NH_3PbI_3$ tandem solar cells. *Energy Environ. Sci.* 11: 0–8.
78 Meusel, M., Baur, C., Létay, G. et al. (2003). Spectral response measurements of monolithic GaInP/Ga(In)As/Ge triple-junction solar cells: measurement artifacts and their explanation. *Prog. Photovolt. Res. Appl.* 11: 499–514.
79 Meusel, M., Adelhelm, R., Dimroth, F. et al. (2002). Spectral mismatch correction and spectrometric characterization of monolithic III–V multi-junction solar cells. *Prog. Photovolt. Res. Appl.* 10: 243–255.
80 ASTM International, (2019) *Test Methods for Measurement of Electrical Performance and Spectral Response of Nonconcentrator Multijunction Photovoltaic Cells and Modules. ASTM International* (ASTM International). doi:https://doi.org/10.1520/E2236-10R19

81 Timmreck, R., Meyer, T., Gilot, J. et al. (2015). Characterization of tandem organic solar cells. *Nat. Photonics* 9: 478–479.

82 Pravettoni, M. and Mullejans, H. (2012). A method for the detection and quantitative estimation of low shunt resistances via the dark spectral response measurement of multijunction photovoltaic cells: theory and results. In: *37th IEEE Photovoltaic Specialists Conference 001726–001730, 19–24 June 2011, Seattle, WA, USA*. IEEE https://doi.org/10.1109/PVSC.2011.6186287.

83 (2019). LONGi Solar sets new bifacial mono-PERC solar cell world record at 24.06 percent. Available at: https://en.longi-solar.com/home/events/press_detail/id/89.html

84 Yoshikawa, K., Kawasaki, H., Yoshida, W. et al. (2017). Silicon heterojunction solar cell with interdigitated back contacts for a photoconversion efficiency over 26%. *Nat. Energy* 2: 17032.

85 Richter, A., Hermle, M., and Glunz, S.W. (2013). Reassessment of the limiting efficiency for crystalline silicon solar cells. *IEEE J. Photovolt.* 3: 1184–1191.

86 Case, C., Beaumont, N., and Kirk, D. (2019). Industrial insights into perovskite photovoltaics. *ACS Energy Lett.* 4: 2760–2762.

87 Mathews, I., Sofia, S., Ma, E. et al. (2020). Economically sustainable growth of perovskite photovoltaics manufacturing. *Joule* https://doi.org/10.1016/j.joule.2020.01.006.

88 Li, Z., Zhao, Y., Wang, X. et al. (2018). Cost analysis of perovskite tandem photovoltaics. *Joule* 2: 1559–1572.

Index

Perovskite Solar Cells: Materials, Processes, and Devices, First Edition.
Edited by Shahzada Ahmad, Samrana Kazim, and Michael Grätzel.

d

e

f

k

l

m

p

q

r